Safe Science

Workplace Hazardous Materials Information System (WHMIS)

 compressed gas

 dangerously reactive material

 flammable and combustible material

 biohazardous infectious material

 oxidizing material

 poisonous and infectious material causing immediate and serious toxic effects

 corrosive material

 poisonous and infectious material causing other toxic effects

Hazardous h...

Symbol	Danger
	Explosive This container can explode if it is heated or punctured.
	Corrosive This product will burn skin or eyes on contact, or throat and stomach if swallowed.
	Flammable This product, or its fumes, will catch fire easily if exposed to heat, flames, or sparks.
	Poisonous Licking, eating, drinking, or sometimes smelling, this product is likely to cause illness or death.

Practise Safe Science in the Classroom

Be science ready.	Follow instructions.	Act responsibly.
• Come prepared with your textbook, notebook, pencil, and anything else you need. • Tell your teacher about any allergies or medical problems. • Keep yourself and your work area tidy and clean. Keep aisles clear. • Keep your clothing and hair out of the way. Roll up your sleeves, tuck in loose clothing, and tie back loose hair. Remove any loose jewellery. • Wear closed shoes (not sandals). • Do not wear contact lenses while doing investigations. • Read all written instructions carefully before you start an activity or investigation.	• Do not enter a laboratory unless a teacher is present, or you have permission to do so. • Listen to your teacher's directions. Read written instructions. Follow them carefully. • Ask your teacher for directions if you are not sure what to do. • Wear eye protection or other safety equipment when instructed by your teacher. • Never change anything, or start an activity or investigation on your own, without your teacher's approval. • Get your teacher's approval before you start an investigation that you have designed yourself.	• Pay attention to your own safety and the safety of others. • Know the location of MSDS (Material Safety Data Sheet) information, exits, and all safety equipment, such as the first aid kit, fire blanket, fire extinguisher, and eyewash station. • Alert your teacher immediately if you see a safety hazard, such as broken glass, a spill, or unsafe behaviour. • Stand while handling equipment and materials. • Avoid sudden or rapid motion in the laboratory, especially near chemicals or sharp instruments. • Never eat, drink, or chew gum in the laboratory. • Do not taste, touch, or smell any substance in the laboratory unless your teacher asks you to do so. • Clean up and put away any equipment after you are finished. • Wash your hands with soap and water at the end of each activity or investigation.

NELSON
SCIENCE PERSPECTIVES 9

Grade 9 Author Team

Charmain Barker
York Catholic District School Board

Lucille Davies, B.A., M.Sc., B.Ed.
Head of Science, Frontenac Secondary School, Limestone District School Board

Andrew Fazekas
President, Montreal Centre, Royal Astronomical Society of Canada

Douglas Fraser
District School Board Ontario North East

Rob Vucic
Head of Science, T.L. Kennedy Secondary School, Peel District School Board

Senior Program Consultant

Maurice DiGiuseppe, Ph.D.
University of Ontario Institute of Technology (UOIT)
Formerly of Toronto Catholic District School Board

Program Consultants

Douglas Fraser
District School Board Ontario North East

Martin Gabber
Formerly of Durham District School Board

Douglas Hayhoe, Ph.D.
Department of Education, Tyndale University College

Jeffrey Major, M.Ed.
Thames Valley District School Board

NELSON EDUCATION

NELSON EDUCATION

Nelson Science Perspectives 9

Senior Program Consultant
Maurice DiGiuseppe

Program Consultants
Douglas Fraser
Martin Gabber
Douglas Hayhoe
Jeffrey Major

Vice President, Publishing
Janice Schoening

Publisher, Science
John Yip-Chuck

Associate Publisher, Science
David Spiegel

Managing Editor, Development
Susan Ball

National Director of Research and Teacher In-Service
Jennette MacKenzie

General Manager, Marketing, Math, Science, and Technology
Paul Masson

Product Manager
Lorraine Lue

Secondary Sales Specialist
Rhonda Sharp

Program Managers
Christina D'Alimonte
Jennifer Hounsell
Sarah Tanzini

Developmental Editors
Nancy Andraos
Barbara Booth
Lina Mockus
Frances Purslow
Rachelle Redford
Susan Skivington

Authors
Charmain Barker
Lucille Davies
Andrew Fazekas
Douglas Fraser
Rob Vucic

Assistant Editor
Jessica Fung

Editorial Assistants
Vytas Mockus
Amy Rotman
Wally Zeisig

Content Production Manager, Science
Sheila Stephenson

Copy Editor
Holly Dickinson

Proofreader
Judy Sturrup

Senior Production Coordinator
Sharon Latta Paterson

Design Director
Ken Phipps

Interior Design
Bill Smith Studio
Greg Devitt Design

Feature Pages Design
Jarrel Breckon
Courtney Hellam
Julie Pawlowicz, InContext

Cover Design
Eugene Lo

Cover Image
scol22/Shutterstock

Contributing Authors
Sheliza Ibrahim
Barry LeDrew
Don Plumb
Milan Sanader
Michael Stubitsch
Richard Towler
Joe Wilson

Asset Coordinators
Renée Forde
Suzanne Peden

Illustrators
Steve Corrigan
Deborah Crowle
Steven Hall
Stephen Hutching
Sam Laterza
Dave Mazierski
Dave McKay
Allan Moon
Nesbitt Graphics, Inc.
Jan-John Rivera
Theresa Sakno
Ann Sanderson
Bart Vallecoccia
Ralph Voltz

Compositor
Nesbitt Graphics, Inc.

Photo Shoot Coordinator
Lynn McLeod

Studio Photographer
Dave Starrett

Photo/Permissions Researcher
Lynn McLeod
Joanne Tang

Printer
Transcontinental Printing, Ltd.

COPYRIGHT © 2010 by Nelson Education Ltd.

ISBN-13: 978-0-17-635519-7
ISBN-10: 0-17-635519-7

Printed and bound in Canada
1 2 3 12 11 10 09

For more information contact Nelson Education Ltd., 1120 Birchmount Road, Toronto, Ontario M1K 5G4. Or you can visit our Internet site at http://www.nelson.com.

ALL RIGHTS RESERVED. No part of this work covered by the copyright herein, except for any reproducible pages included in this work, may be reproduced, transcribed, or used in any form or by any means—graphic, electronic, or mechanical, including photocopying, recording, taping, Web distribution, or information storage and retrieval systems—without the written permission of the publisher.

For permission to use material from this text or product, submit all requests online at www.cengage.com/permissions. Further questions about permissions can be e-mailed to permissionrequest@cengage.com.

Every effort has been made to trace ownership of all copyrighted material and to secure permission from copyright holders. In the event of any question arising as to the use of any material, we will be pleased to make the necessary corrections in future printings.

REVIEWERS

Accuracy Reviewers

Jean Dupuis, Ph.D.
Program Scientist, Space Astronomy, Canadian Space Agency

Roberta Fulthorpe, Ph.D.
Division of Physical and Environmental Sciences, University of Toronto

Denis Laurin, Ph.D.
Science Manager, Space Astronomy, Canadian Space Agency

John R. Percy, Ph.D.
Department of Astronomy and Astrophysics, University of Toronto

Heather Phillips, Ph.D.
McGill University

Henri M. van Bemmel, B.Sc. (Hons.), B.Ed.
Marc Garneau Collegiate Institute, Toronto DSB

Assessment Consultants

Aaron Barry, M.B.A., B.Sc., B.Ed.
Sudbury Catholic DSB

Damian Cooper
Nelson Education Author

Mike Sipos, B.Ph.Ed., B.Ed.
Sudbury Catholic DSB

Catholicity Reviewer

Ted Laxton
Sacred Heart Catholic School, Wellington Catholic DSB

Environmental Education Consultant

Allan Foster, Ed.D., Ph.D.
Working Group on Environmental Education, Ontario
Former Director of Education, Kortright Centre for Conservation

ESL/Culture Consultant

Vicki Lucier, B.A., B.Ed., Adv. Ed.
ESL/Culture Consultant, Simcoe County DSB

Literacy Consultants

Jill Foster
English/Literacy Facilitator, Durham DSB

Jennette MacKenzie
National Director of Research and Teacher In-Service, Nelson Education Ltd.

Michael Stubitsch
Education Consultant

Numeracy Consultant

Justin DeWeerdt
Curriculum Consultant, Trillium Lakelands DSB

Safety Consultant

Jim Agban
Past Chair, Science Teachers' Association of Ontario (STAO) Safety Committee

John Henry
STAO Safety Committee

STSE Consultant

Joanne Nazir
Ontario Institute for Studies in Education (OISE), University of Toronto

Technology/ICT Consultant

Luciano Lista, B.A., B.Ed., M.A.
Academic Information Communication Technology Consultant
Online Learning Principal, Toronto Catholic DSB

Advisory Panel and Teacher Reviewers

Gabriel Roman Ayyavoo, B.Sc., B.Ed., M.Ed.
Toronto Catholic DSB

Anca Bogorin
Avon Maitland DSB

Christopher Bonner
Ottawa Catholic DSB

Sean Clark
Ottawa Catholic DSB

Charles J. Cohen
Community Hebrew Academy of Toronto

Tim Currie
Bruce Grey Catholic DSB

Greg Dick
Waterloo Region DSB

Matthew DiFiore
Dufferin-Peel Catholic DSB

Ed Donato
Simcoe Muskoka Catholic DSB

Dave Doucette, B.Sc., B.Ed.
York Region DSB

Chantal D'Silva, B.Sc., M.Ed.
Toronto Catholic DSB

Naomi Epstein
Community Hebrew Academy of Toronto

Xavier Fazio
Faculty of Education, Brock University

Daniel Gajewski, Hon. B.Sc., B.Ed.
Ottawa Catholic DSB

Theresa H. George
Dufferin-Peel Catholic DSB

Sharon Gillies
London District Catholic School Board

Stephen Haberer
Kingston Collegiate and Vocational Institute, Limestone DSB
Faculty of Education, Queen's University

Shawna Hopkins, B.Sc., B.Ed., M.Ed.
Niagara DSB

Christopher T. Howes, B.Sc., B.Ed.
Durham DSB

Bryan Hutnick
Formerly of District School Board Ontario North East

Janet Johns
Upper Canada DSB

Michelle Kane
York Region DSB

Roche Kelly, B.Sc., B.Ed.
Durham DSB

Mark Kinoshita
Toronto DSB

Emma Kitchen, B.Sc., B.Ed.
Near North DSB

Erin Lepischak
Hamilton-Wentworth DSB

Stephanie Lobsinger
Lambton College

Alistair MacLeod, B.Sc., P.G.C.E., M.B.A
Limestone DSB

Doug McCallion, B.Sc., B.Ed., M.Sc.
Halton Catholic DSB

Nadine Morrison
Hamilton-Wentworth DSB

Dermot O'Hara, B.Sc., B.Ed., M.Ed.
Toronto Catholic DSB

Mike Pidgeon
Toronto DSB

William J. F. Prest
Rainbow DSB

Zbigniew Peter Reiter
Toronto Catholic DSB

Ron M. Ricci, B.E.Sc., B.Ed.
Greater Essex DSB

Athanasios Seliotis, B.Sc., B.Ed., M.A.
York Region DSB

Charles Stewart, B.Sc., B.Ed.
Peel DSB

Carl Twiddy
Formerly of York Region DSB

Jim Young
Limestone DSB

Contents

Discover Your Textbook xii

UNIT A: INTRODUCTION TO SCIENTIFIC INVESTIGATION SKILLS AND CAREER EXPLORATION 2

Focus on STSE: Science and Your Life 3

CHAPTER 1
Living and Working with Science 4
Key Concepts 5
Engage in Science: The Winds of Change! 6
Focus on Reading: How to Read Non-Fiction Text 7
1.1 Skills of Scientific Investigation 8
 Try This: Questions Leading to Scientific Inquiry 10
 Try This: Analyzing Data 13
 Try This: Getting Your Message Across 15
1.2 Scientific Literacy for Living and Working in Canada 16
Key Concepts Summary 19

UNIT B: SUSTAINABLE ECOSYSTEMS 20

Focus on STSE: Something to Lose? 21
Unit B Looking Ahead 22
 Unit Task Preview: Ontario's Species of Concern 22
 What Do You Already Know? 23

CHAPTER 2
Understanding Ecosystems 24
Key Concepts 25
Engage in Science: Red Crabs and Crazy Ants 26
What Do You Think? 27
Focus on Reading: Making Connections 28
2.1 Life on Planet Earth 29
 Try This: A Scale Model of Planet Earth 31
2.2 Introducing Ecosystems 32
 Try This: Ecosystem ABCs 33
2.3 **PERFORM AN ACTIVITY:**
 Factors Affecting an Aquatic Ecosystem 36
2.4 Energy Flow in Ecosystems 38
 Try This: Products of Cellular Respiration .. 40
2.5 Food Webs and Ecological Pyramids 42
 Try This: Weaving a Food Web 45
2.6 Cycling of Matter in Ecosystems 48
2.7 Biotic and Abiotic Influences on Ecosystems 52
2.8 Major Terrestrial Ecosystems 56
2.9 Major Aquatic Ecosystems 60
Tech Connect: Whales, Darts, and DNA 63
2.10 **PERFORM AN ACTIVITY:**
 Ecosystem Field Study 64
Key Concepts Summary 66
What Do You Think Now? 67
Chapter 2 Review 68
Chapter 2 Self-Quiz 70

CHAPTER 3

Natural Ecosystems and Stewardship . . . 72
Key Concepts 73
Engage in Science: Shark Attack! Overfishing
Large Sharks Impacts Marine Ecosystem 74
What Do You Think? 75
Focus on Writing: Writing to Describe
and Explain Observations 76
3.1 Services from Natural Ecosystems 77
3.2 Equilibrium and Change. 80
3.3 The Importance of Biodiversity 83
   Citizen Action: NatureWatch 86
3.4 Habitat Loss and Fragmentation 87
   Research This: Sweet Grass Gardens 89
3.5 The Introduction of Non-Native Species 91
   Research This: We Do Not Belong! 93
Science Works: Containing the Invasion 95
3.6 Pollution 96
   Research This: Ocean Acidification 98
3.7 Consumption and Resource Management . . . 102
   Research This: Fisheries Mismanagement. . . 105
3.8 PERFORM AN ACTIVITY:
   Comparing Forest Management Practices . . 106
Key Concepts Summary 108
What Do You Think Now? 109
Chapter 3 Review 110
Chapter 3 Self-Quiz 112

CHAPTER 4

Ecosystems by Design 114
Key Concepts 115
Engage in Science: Living the Green Life 116
What Do You Think? 117
Focus on Reading: Asking Questions 118
4.1 Engineered Ecosystems
   and Modern Agriculture 119
   Try This: Watch Where You Step! 119

4.2 Managing the Soil—Controlling
   the Flow of Nutrients and Water 123
   Research This: What Are
   the Chemical Contents? 123
   Try This: Humans Weigh in on
   Soil Compaction 127
Awesome Science: Black Gold 129
4.3 PERFORM AN ACTIVITY:
   Fertile Grounds: The Right Mix 130
4.4 Pests and Poisons 132
4.5 Issues with Pesticides 135
   Try This: Pollutants Follow the Fat 136
4.6 PERFORM AN ACTIVITY:
   Greening the Consumer—Reducing
   Your Ecological Footprint 141
4.7 The Urban Ecosystem 142
   Research This: Conserving the
   Oak Ridges Moraine 143
   Try This: Do You Suffer from NDD? 145
4.8 EXPLORE AN ISSUE CRITICALLY:
   Waste Management or Mismanagement? . . 146
Key Concepts Summary 148
What Do You Think Now? 149
Chapter 4 Review 150
Chapter 4 Self-Quiz 152

Unit B Looking Back 154
   Key Concepts 154
   Make a Summary 155
   Career Links 155

Unit B Task: Ontario's Species
of Concern 156

Unit B Review 158

Unit B Self-Quiz 164

UNIT C: ATOMS, ELEMENTS, AND COMPOUNDS..............166

Focus on STSE: Durable or Degradable?......167

Unit C Looking Ahead..............168

Unit Task Preview: A Greener Shade of Chemistry..............168

What Do You Already Know?..............169

CHAPTER 5
Properties of Matter..............170

Key Concepts..............171
Engage in Science: Denim to Dye For..............172
What Do You Think?..............173
Focus on Writing: Writing a Summary..............174

5.1 From Particles to Solutions..............175
   Try This: How Stretchy Is Your Solder?....177

5.2 Physical Properties..............179
   Try This: Close-Up of a Running Shoe....180

5.3 Chemical Properties..............183
   Try This: Make Common Chemical Changes..............184
   Try This: To Rot or Not To Rot..............185

Science Works: Keeping Baby Dry..............187

5.4 PERFORM AN ACTIVITY:
   Safety in Science..............188

5.5 PERFORM AN ACTIVITY:
   Forensic Chemistry..............190

Awesome Science: Antifreeze Is Not Just For Cars...191

5.6 Characteristic Physical Properties..............192
   Try This: We Scream for Ice Cream......196

5.7 EXPLORE AN ISSUE CRITICALLY:
   Are We Salting or Assaulting our Roads?...199

Key Concepts Summary..............200
What Do You Think Now?..............201
Chapter 5 Review..............202
Chapter 5 Self-Quiz..............204

CHAPTER 6
Elements and the Periodic Table 206
Key Concepts 207
Engage in Science: Mercury and the Mad Hatter 208
What Do You Think? 209
Focus on Reading: Making Inferences 210
- **6.1** A Table of the Elements 211
  - **Try This:** Element Scavenger Hunt 212
  - **Citizen Action:** Recycle Your Cellphone, Save a Gorilla 214
- **6.2** CONDUCT AN INVESTIGATION: Become a Metal Detective 216
- **6.3** PERFORM AN ACTIVITY: Properties of Household Chemicals 218
- **6.4** Patterns in the Periodic Table 220
  - **Research This:** The Periodic Table Is Evolving! 224
  - **Try This:** Solving the Puzzle, Periodically 224
- **6.5** CONDUCT AN INVESTIGATION: Comparing Family Behaviour 226
- **6.6** Theories of the Atom 228
  - **Try This:** Simulating Rutherford's Black Box 231
  - **Try This:** Lines of Light 232
- **6.7** Explaining the Periodic Table 234
  - **Try This:** Family Resemblances in the Periodic Table 237
- **6.8** From Charcoal to Diamonds 241
  - **Try This:** Copper-Plate Your Pencil 242
  - **Research This:** Artificial Diamonds 243

Tech Connect: Stronger Than a Speeding Bullet 245
Key Concepts Summary 246
What Do You Think Now? 247
Chapter 6 Review 248
Chapter 6 Self-Quiz 250

CHAPTER 7
Chemical Compounds 252
Key Concepts 253
Engage in Science: The Attack of Oxygen 254
What Do You Think? 255
Focus on Writing: Writing a Science Report 256
- **7.1** Putting Atoms Together 257
  - **Try This:** Rusting Steel Slowly 259
- **7.2** CONDUCT AN INVESTIGATION: A Race to Rust 262
- **7.3** How Atoms Combine 263
  - **Try This:** When Magnesium Meets Oxygen 264

Science Works: The Tiny World of Nanotechnology 267
- **7.4** PERFORM AN ACTIVITY: Making Molecular Models 268
- **7.5** PERFORM AN ACTIVITY: What Is This Gas? 270
- **7.6** Breaking Molecules Apart: Properties of Hydrogen Peroxide 272
  - **Try This:** When Yeast Meets Bleach 273
- **7.7** PERFORM AN ACTIVITY: What Is Rotten and What Is Not? 274
- **7.8** EXPLORE AN ISSUE CRITICALLY: DDT is Forever 276

Key Concepts Summary 278
What Do You Think Now? 279
Chapter 7 Review 280
Chapter 7 Self-Quiz 282

Unit C Looking Back 284
- Key Concepts 284
- Make a Summary 285
- Career Links 285

Unit C Task: A Greener Shade of Chemistry 286

Unit C Review 288

Unit C Self-Quiz 294

UNIT D: THE STUDY OF THE UNIVERSE 296

Focus on STSE: Encounter with Titan 297
Unit D Looking Ahead 298
Unit Task Preview: Celestial Travel Agency.... 298
What Do You Already Know? 299

CHAPTER 8
Our Place in Space 300

Key Concepts 301
Engage in Science: The Tunguska Event........... 302
What Do You Think? 303
Focus on Reading: Finding the Main Idea 304

- 8.1 Touring the Night Sky.................. 305
  Try This: Create a Horizon Diagram 307
- 8.2 The Sun 309
  Try This: Track Sunspots 310
- 8.3 The Solar System: The Sun and the Planets...................... 313
  Try This: Represent the Sizes of the Planets 314
- 8.4 **PERFORM AN ACTIVITY:** A Scale Model of the Solar System 318
- 8.5 Motions of Earth, the Moon, and Planets...................... 320
  Try This: Modelling the Lunar Phases 325
- 8.6 Patterns in the Night Sky 329
- 8.7 **PERFORM AN ACTIVITY:** Using a Star Map 334
- 8.8 **CONDUCT AN INVESTIGATION:** Modelling Motion in the Night Sky 336
- 8.9 Observing Celestial Objects from Earth ... 338
  Citizen Action: Saving the Night Sky 338
  Try This: Demonstrate Retrograde Motion........................... 340
  Try This: Measuring Altitude and Azimuth........................ 342
- 8.10 **CONDUCT AN INVESTIGATION:** Finding Objects in the Night Sky 344
- 8.11 Satellites............................. 346
  Try This: Orbiting Satellites.............. 348
  Research This: Different Kinds of Satellites...................... 350
  Try This: Satellite Images 351
- 8.12 **EXPLORE AN ISSUE CRITICALLY:** Security Satellites 352

Awesome Science: Mysteries Beyond the Planets 353
Key Concepts Summary 354
What Do You Think Now? 355
Chapter 8 Review............................ 356
Chapter 8 Self-Quiz 358

CHAPTER 9

Beyond the Solar System 360

Key Concepts 361
Engage in Science: Uncovering a Mystery 362
What Do You Think? 363
Focus on Reading: Summarizing 364

9.1 Measuring Distances Beyond the Solar System 365
  Try This: Distances in the Universe 367
9.2 The Characteristics of Stars 370
9.3 CONDUCT AN INVESTIGATION:
  Factors Affecting the Brightness of Stars ... 374
9.4 The Life Cycle of Stars 375
  Try This: Modelling a Supernova Explosion 379
9.5 The Formation of the Solar System 383
9.6 Other Components of the Universe 385
  Try This: Viewing Galaxies 387
  Try This: Classifying Galaxies 387
Tech Connect: Little Satellite, Big Science 392
9.7 The Origin and Evolution of the Universe .. 393
  Try This: The Centre of an Expanding Universe 393
  Try This: Model the Expanding Universe .. 397

Key Concepts Summary 398
What Do You Think Now? 399
Chapter 9 Review 400
Chapter 9 Self-Quiz 402

CHAPTER 10

Space Research and Exploration 404

Key Concepts 405
Engage in Science: The *Apollo 13* Mission 406
What Do You Think? 407
Focus on Reading: Synthesizing 408

10.1 Space Exploration 409
  Research This: See the Space Station 413
10.2 Challenges of Space Travel 419
  Try This: Understanding Free Fall 421
  Research This: Space Travel Safety 425
10.3 Space Technology Spinoffs 426
  Try This: How Does GPS Work? 427
Science Works: Canadian Weather Station on Mars ... 431
10.4 EXPLORE AN ISSUE CRITICALLY:
  Should Canadians Be Paying for Space Research? 432
10.5 The Future of Space Exploration 434

Key Concepts Summary 438
What Do You Think Now? 439
Chapter 10 Review 440
Chapter 10 Self-Quiz 442

Unit D Looking Back 444
  Key Concepts 444
  Make a Summary 445
  Career Links 445

Unit D Task: Celestial Travel Agency 446

Unit D Review 448

Unit D Self-Quiz 454

UNIT E: THE CHARACTERISTICS OF ELECTRICITY 456

Focus on STSE: Powering the World Wide Web 457

Unit E Looking Ahead 458

Unit Task Preview: Communicate with Electricity 458

What Do You Already Know? 459

CHAPTER 11

Static Electricity 460

Key Concepts 461
Engage in Science: Printer Problem 462
What Do You Think? 463
Focus on Writing: Writing a Persuasive Text 464

- 11.1 What Is Static Electricity? 465
  - Try This: Positive and Negative Charges 467
  - Try This: Charging Objects 469
  - Research This: Powder Coating 470
- 11.2 Charging by Contact 472
  - Try This: Charging by Friction 473
  - Research This: Fabric Softener Sheets 476
- 11.3 **CONDUCT AN INVESTIGATION:** Predicting Charges 478
- 11.4 Conductors and Insulators 480

Awesome Science: A Hair-Raising Experience 483

- 11.5 **PERFORM AN ACTIVITY:** Testing for Conductors and Insulators 484
- 11.6 Charging by Induction 486
  - Try This: Bending Water 487
- 11.7 **PERFORM AN ACTIVITY:** Charging Objects by Induction 490
- 11.8 Electric Discharge 492
  - Research This: Electric Discharge 492

Key Concepts Summary 496
What Do You Think Now? 497
Chapter 11 Review 498
Chapter 11 Self-Quiz 500

CHAPTER 12

Electrical Energy Production 502

Key Concepts 503
Engage in Science: Midnight Sun Team Blog 504
What Do You Think? 505
Focus on Reading: Evaluating 506

- 12.1 Introducing Current Electricity 507
  - Try This: Model Electron Flow 508
- 12.2 Electric Circuits 509
  - Try This: Building Circuits 510
- 12.3 Electrical Energy 511
  - Try This: The Pickle Cell 513
- 12.4 Forms of Current Electricity 515
- 12.5 Generating Current Electricity 518
  - Research This: Wind Power on Lake Ontario 526
- 12.6 **EXPLORE AN ISSUE CRITICALLY:** Not in My Backyard! 529

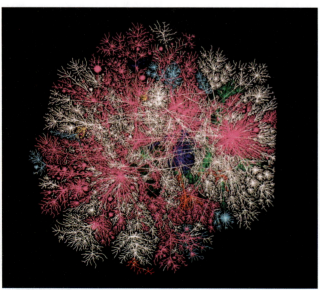

12.7 Electrical Power and Efficiency 530
   Citizen Action: Is Your School
   Conserving Electrical Energy? 535
12.8 PERFORM AN ACTIVITY:
   Examining Electrical Energy Production... 536
Tech Connect: Energy for Spacecraft 538
12.9 PERFORM AN ACTIVITY:
   Performing an Electrical Energy Audit 539
Key Concepts Summary 540
What Do You Think Now? 541
Chapter 12 Review............................ 542
Chapter 12 Self-Quiz 544

CHAPTER 13

Electrical Quantities in Circuits 546
Key Concepts 547
Engage in Science: Suspected Electrical Fire
Destroys Homes 548
What Do You Think? 549
Focus on Writing: Writing a Critical Analysis 550
13.1 Circuits and Circuit Diagrams............ 551
13.2 PERFORM AN ACTIVITY:
   Connecting Multiple Loads 555
13.3 Electric Current 556
13.4 PERFORM AN ACTIVITY:
   Comparing the Conductivity
   of Conductors........................... 558
Tech Connect: Heart Technologies 559
13.5 Potential Difference.................... 560
13.6 PERFORM AN ACTIVITY:
   Measuring Voltage and Current
   in Circuits 562
13.7 Resistance in Circuits 564
   Try This: Measuring Resistance
   with an Ohmmeter 566

13.8 PERFORM AN ACTIVITY:
   Determining the Relationship Between
   Current and Potential Difference 567
13.9 Relating Current, Voltage, and Resistance.. 568
13.10 How Series and Parallel Circuits Differ 571
13.11 CONDUCT AN INVESTIGATION:
   The Effect of Increasing the Number of
   Loads in a Circuit...................... 576
Key Concepts Summary 578
What Do You Think Now? 579
Chapter 13 Review............................ 580
Chapter 13 Self-Quiz 582

Unit E Looking Back 584
   Key Concepts 584
   Make a Summary 585
   Career Links 585

**Unit E Task: Communicate
with Electricity** 586

Unit E Review 588

Unit E Self-Quiz 594

Appendix A: Skills Handbook 596

Appendix B: What Is Science? 646

Numerical and Short Answers 656

Glossary 665

Index 672

Credits 679

Discover Your Textbook

This textbook will be your guide to the exciting world of science. On the following pages is a tour of important features that you will find inside. **GET READY** includes all of the features of the introductory material that come before you begin each unit and chapter. **GET INTO IT** shows you all the features within each chapter. Finally, **WRAP IT UP** shows you the features at the end of each chapter and unit.

Get Ready

Unit Opener
Each of the five units has a letter and a title. Use the photo to help you predict what you might be learning in the unit.

Overall Expectations
The Overall Expectations describe what you should be able to do after completing the unit.

Big Ideas
The Big Ideas summarize the concepts you need to remember after you complete the unit.

Focus on STSE
These articles introduce real-world connections to the science topics you will be learning in the unit.

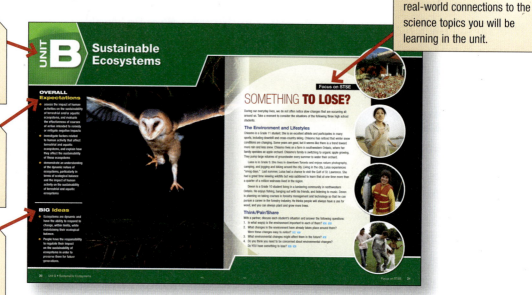

Concept Map
The Concept Map is a description of the topics connected to picture clues to help you predict what you will be learning in the unit.

What Do You Already Know?
This section gives you a list of concepts and skills you developed in previous grades that you will need to be successful as you work through the unit. Use the questions to see what you already know before you start the unit.

Unit Task Preview
Find out about the Unit Task that you will complete at the end of each unit.

Unit Task Bookmark
When you see the Unit Task Bookmark, think about how the section relates to the Unit Task.

Assessment Box
The Assessment Box tells you how you will demonstrate what you have learned by the end of the unit.

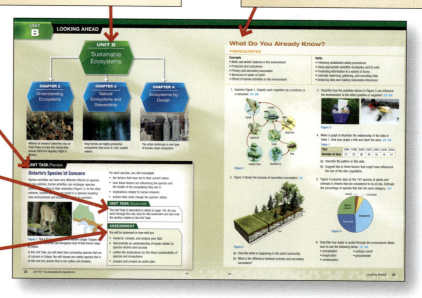

Chapter Opener
Each chapter has a number, a title, and a Key Question which you should be able to answer by the end of the chapter.

Key Concepts
The Key Concepts feature outlines the main ideas and skills you will learn in the chapter.

Engage in Science
These articles connect the topics you will learn in the chapter to interesting real-world developments in science.

What Do You Think?
Using what you already know, form an opinion by agreeing or disagreeing with statements that connect to ideas that will be introduced in the chapter.

Focus on Reading/ Focus on Writing
These reading and writing strategies help you learn science concepts and develop literacy skills in preparation for the OSSLT.

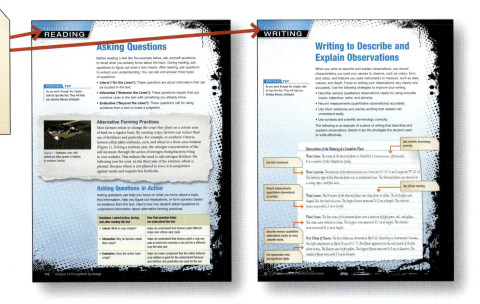

Discover Your Textbook xiii

Get Into It

Weblink
When you see this weblink icon, you can visit the Nelson Science website to learn more about the topic, watch a video, do an online activity, or take a quiz.

Career Link
The Career icon lets you know that you can visit the Nelson Science website to learn about science-related careers.

Try This
These are quick, fun activities designed to help you understand concepts and improve your science skills.

Research This
These research-based activities will help you relate science and technology to the world around you and improve your critical thinking and decision-making skills.

Did You Know?
Read interesting facts about real-world events that relate to the topics you are reading about.

Skills Handbook Icon
This icon directs you to the section of the Skills Handbook that contains helpful information and tips.

Vocabulary
You will learn many new terms as you work through the chapter. These key terms are in bold print. Their definitions can be found in the margins and in the glossary at the back of the book.

Sample Problem
This feature shows you how to solve numerical problems. Make sure to check your learning by completing Practice problems.

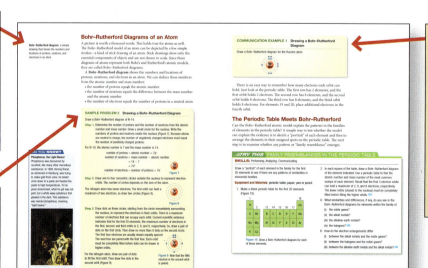

Communication Example
A Communication Example presents a similar problem to the Sample Problem. They are an opportunity for you to follow the steps of the Sample Problem to make sure you understand the process.

xiv   Discover Your Textbook

Citizen Action
These activities encourage you to be a good citizen and a steward of the environment by taking action in the world around you.

Unit Task Bookmark
This icon lets you know that the concepts you learned in the section will help you to complete the Unit Task.

In Summary
At the end of each content section, this quick summary of the main ideas will help you review what you learned.

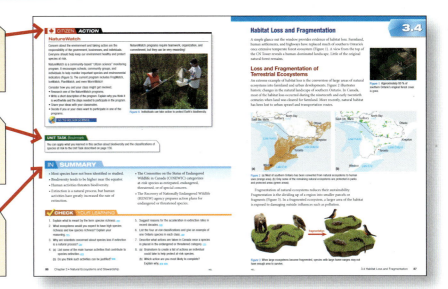

Learning Tip
Learning Tips are useful strategies to help you learn new ideas and make sense of what you are reading.

Reading Tip
Reading Tips suggest reading comprehension strategies to help you understand the science concepts presented in the text.

Math Tip
These math tips are helpful hints that will develop your math and numeracy skills.

Check Your Learning
Complete these questions at the end of each content section to make sure you understand the concepts you have just read.

Magazine Features
Look for these special feature sections in each unit to learn about exciting developments in science, cool new technology, career links, or how science relates to your everyday life.

OSSLT Icon
This icon lets you know that the material will help you develop literacy skills in preparation for the OSSLT.

Discover Your Textbook xv

Perform an Activity
These are hands-on activities that allow you to observe the science that you are learning.

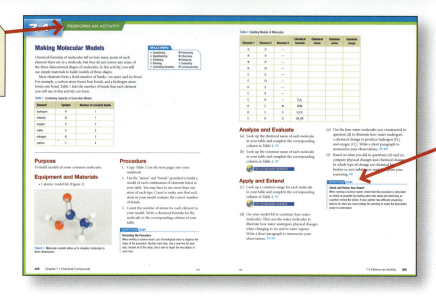

Writing Tip
Writing Tips are useful strategies that help to improve your writing skills.

Explore an Issue Critically
These activities allow you to examine social and environmental issues related to the unit and encourage you to make a difference in your own community by taking action.

Skills Menu
The Skills Menu in each activity lists the skills that you will use to solve the problem or reach a conclusion.

Conduct an Investigation
These experimental investigations are an opportunity for you to develop science process skills.

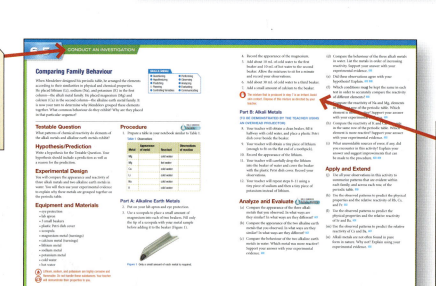

Safety Precautions
Look for these warnings about potential safety hazards in investigations and activities. They will be in red print with a warning icon.

xvi Discover Your Textbook

NEL

Wrap It Up

Key Concepts Summary
The Key Concepts Summary feature outlines the main ideas and skills you learned in the chapter.

What Do You Think Now?
Think about what you learned in the chapter and consider whether you have changed your opinion by agreeing or disagreeing with the statements.

Vocabulary
This feature lists all the key terms you have learned and the page number where the term is defined.

Big Ideas
The checkmark indicates which Big Ideas were developed in the chapter.

Chapter Review
Complete these questions to check your learning and apply your new knowledge from the chapter.

Achievement Chart Icons
All questions are tagged with icons that identify the types of knowledge and skills you must use to answer the question.

Online Quiz Icon
There is an online study tool for each chapter on the Nelson Science website.

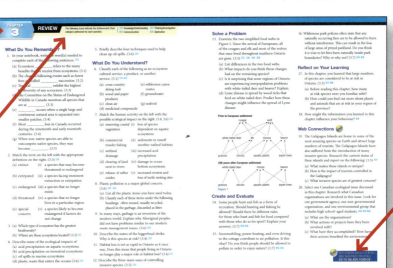

Chapter Self-Quiz
The Chapter Self-Quiz is a helpful tool for you to make sure you understand all the concepts you learned in the chapter.

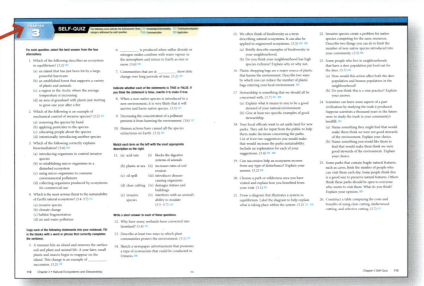

Discover Your Textbook   xvii

Master Concept Map
This feature brings together the Key Concepts from each chapter to summarize all the main ideas in the unit.

Make a Summary
Summarize what you have learned in the unit by completing the Make a Summary activity.

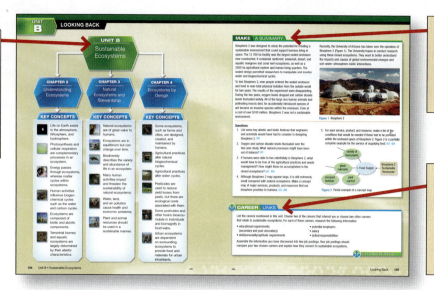

Career Links
Make connections between what you learned in the unit and future careers by completing the Careers Links activity.

Unit Task
Demonstrate the skills and knowledge you developed in the unit by completing the challenge described in the Unit Task.

Skills Menu
The Skills Menu identifies the skills you will use to complete the Unit Task.

Assessment Checklist
This checklist lists the criteria that your teacher will use to evaluate your work on the Unit Task. Read this list carefully before completing the task.

Unit Review
Complete the Unit Review questions to check your learning of all the concepts and skills in the unit.

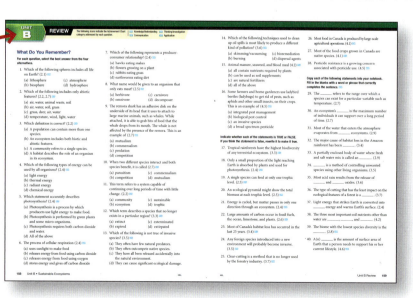

xviii   Discover Your Textbook

Unit Self-Quiz

The Unit Self-Quiz is an opportunity for you to make sure that you understand all the main ideas from the unit.

Glossary

This is a list of all the key terms in the textbook in alphabetical order. Use the Glossary to check your understanding of any key terms you may need to review.

Skills Handbook

The Skills Handbook is your resource for useful science skills and information. It is divided into numbered sections. Whenever you see a Skills Handbook Icon, it will direct you to the relevant section of the Skills Handbook.

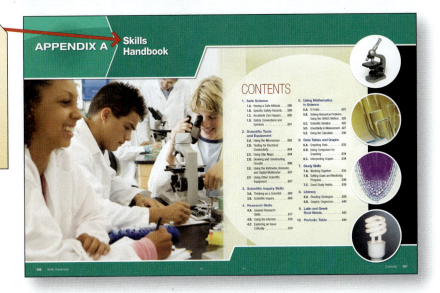

Discover Your Textbook   xix

UNIT A: Introduction to Scientific Investigation Skills and Career Exploration

GOALS of the Science Program

- To relate science to technology, society, and the environment
- To develop the skills, strategies, and habits of mind required for scientific investigation
- To understand the basic concepts of science

Focus on STSE

SCIENCE AND YOUR LIFE

The goals of science education include much more than just the acquisition of scientific facts or knowledge. Your science program and this textbook are designed to help you understand the role of science in your everyday life and the impact of science and technology on society and the environment.

This introductory unit introduces the important scientific investigation skills that you will develop as you study biology, chemistry, Earth and space science, and physics in the following units.

In these four units, you will have many opportunities to learn through scientific inquiry. Through these inquiries you will develop, practise, and refine essential scientific investigation skills. These skills are useful not only in high school science but also in your post-secondary education and in your everyday life. In these units, you will also have opportunities to explore careers that are related to the various science topics.

The main purpose of learning science at this level is to make connections. As you progress through this course, you will develop an understanding of how science, technology, society, and the environment (STSE) are interrelated. You will connect the STSE relationships to your everyday life experiences, and you will develop scientific literacy.

Think/Pair/Share

1. List five ways that science and technology directly or indirectly influence your daily life. A
2. Pair up with a partner and share your ideas. Brainstorm to add further examples. C
3. Join another pair and share your lists. Eliminate duplicate examples and refine your list. Share your list with the whole class. C A

CHAPTER 1

Living and Working with Science

KEY QUESTION: What skills are required to carry out scientific investigations?

Science takes place in all kinds of places—not just in the lab.

UNIT A
Introduction to Scientific Investigation Skills and Career Exploration

CHAPTER 1
Living and Working with Science

KEY CONCEPTS

Science and technology are an important part of our everyday lives.

Scientific inquiry can be conducted in different ways depending on the question to be answered.

All scientific inquiries rely on careful recording of accurate and repeated observations.

Careful analysis and interpretation of observations help to make them meaningful.

Clear communication is important for sharing scientific discoveries and ideas with others.

Scientific literacy is necessary for wise personal decisions, responsible citizenship, and careers.

ENGAGE IN SCIENCE

The Winds of Change!

Wind turbines are becoming a common sight in Ontario, across the country, and around the world. Governments and utility companies are investing in environmentally friendly energy sources and technologies that harness the power of the wind (Figure 1).

Environmental impact studies and the public's reaction to them have helped to convince the Ontario government to make plans to eliminate coal-fired generating stations (Figure 2) by 2014. The plan calls for the use of new renewable energy projects to replace the lost generating capacity. In 2009, approximately 1100 megawatts of electricity will be generated from wind energy in Ontario. This is a significant increase in the amount of energy generated from renewable sources but still represents only a little more than 4 % of peak demand.

A major offshore wind farm project is planned in Lake Ontario approximately 17 to 28 kilometres from the shoreline of Prince Edward County. The farm will contain 142 turbines and produce 710 megawatts of electricity—enough to power about 300 000 homes.

Despite the promise of cost-effective energy with no greenhouse gas emissions and no toxic waste, there is a growing, organized opposition to the use of wind farms to generate electrical energy. Two groups, the local Alliance to Protect Prince Edward County and Wind Concerns Ontario, joined together to oppose the wind projects proposed for Prince Edward County. Their opposition is based on social and environmental concerns such as noise, physical safety, danger to migrating birds, negative effects on tourism and real estate values, and the impact on the aesthetics of the natural landscape.

Who is right in this issue? Is either side right? How does wind compare to fossil fuels or nuclear power as a source of energy? How does it compare to solar energy or hydro-electricity? Do alternative energy sources simply distract us from the importance of conserving energy by changing our lifestyle? These are difficult questions that require research, critical thinking, and common sense. Where do you stand?

Figure 2 A coal-fired generating station

Figure 1 Wind turbines turn generators that convert the energy of the wind into electrical energy.

FOCUS ON READING

How to Read Non-Fiction Text

Non-fiction texts convey ideas and information using words and graphics. Using the following strategies will help you to better understand what you read:

READING TIP

As you work through the chapter, look for tips like this. They will help you develop literacy strategies.

Before Reading

- Think about the topic. Do you have a personal experience that relates to it? Can you use any knowledge you have acquired to understand the topic?
- Preview the text: Scan titles, headings, bolded words, graphics, and captions.
- Identify the organizational pattern of the text and think about how you have approached this type of text pattern before.
- Ask a purpose-for-reading question: Turn a title or heading into a question that you will answer as you read the text.

During Reading

- Find information that answers your purpose-for-reading question.
- Make connections to your knowledge or personal experiences related to the topic.
- Look for relationships among the ideas and information given.
- Ask questions about parts of the text that puzzle you.
- Break the text down into chunks and pause after each one to determine whether or not you understand what you are reading.
- Find the main idea. Sometimes you may have to infer a main idea that is stated indirectly.
- Make jot notes or use sticky notes to keep track of what you understand and what confuses you.

After Reading

- Summarize the main ideas and details that you found important.
- Assess whether you answered your purpose-for-reading question.
- Reflect on what you have learned by forming an opinion, drawing a conclusion, or identifying what you still do not understand.
- Ask yourself how you can apply what you learned from the selection.

1.1 Skills of Scientific Investigation

Scientists assume that the natural world can be explained and understood. Hundreds of years ago, people believed that Earth was the centre of the universe and that all of the heavenly bodies revolved around Earth. Careful scientific observations and technological advances, such as the invention and use of the telescope, have provided evidence that the Sun is the centre of our solar system. Scientists now know that the universe is much different from what they first thought. Because of scientific inquiry, our understanding of the natural world has expanded (Figure 1).

> **READING TIP**
> **Scan for Specialized Vocabulary**
> Before reading, scan the text to find vocabulary highlighted in boldface. Check the margins for definitions of these words. Knowing the meaning of these words ahead of time can help you understand the ideas and information they refer to in the text.

Figure 1 Observations of the night sky have helped us to better understand the universe.

Types of Scientific Inquiry

Scientific inquiry is a set of strategies or methods that scientists use to gather information and explain the natural world. All scientific inquiry uses similar processes to find answers to questions. In most cases, these processes try to find relationships between variables. A **variable** is any condition that could change in an inquiry. A variable that is deliberately changed or selected by the investigator is called the **independent variable**. A variable that changes in response to the independent variable, but is not directly controlled by the investigator, is called the **dependent variable**.

The design of the scientific inquiry depends on the purpose of the inquiry and on the nature of the variables. Two common types of scientific inquiry are the controlled experiment and the observational study.

variable any condition that changes or varies the outcome of a scientific inquiry

independent variable a variable that is changed by the investigator

dependent variable a variable that changes in response to the change in the independent variable

Controlled Experiment

If the purpose of an inquiry is to determine whether one variable causes an effect on another variable, then you can perform a **controlled experiment**. This means that you can set up an experiment in which you can control (change) the independent variable to determine if the change causes a change in the dependent variable. For example, if you want to determine how the diameter of a wire affects how easily electricity flows through the wire, a controlled experiment is the appropriate approach. In this case, the investigator would use wires of different diameters (independent variable) and observe the effect, if any, on the electrical current in the wire (dependent variable).

controlled experiment an experiment in which the independent variable is purposely changed to find out what change, if any, occurs in the dependent variable

Observational Study

Often the purpose of a scientific inquiry is to answer a question by gathering scientific information. **Observational studies** involve observing a subject or phenomenon without interfering with or influencing the subject or phenomenon. Observational studies, like all scientific inquiries, start with observations that lead to a question. Sometimes the researchers start with a specific prediction about the answer to the question. Sometimes they also have an explanation for their prediction—but not always. An explanation might be possible only after evidence is collected.

Sciences such as astronomy, paleontology, and ecology rely heavily on observational studies because it is not possible to conduct controlled experiments. Observational studies are done when an investigator cannot control the variables because of safety, time, distance, cost, or ethical considerations. For example, the question "How long does it take for each of the planets to orbit the Sun?" can only be answered by systematic observation. We now know the answer to this question, but how did we find the answer? Scientists could not travel to any of the planets, so they had to carefully observe the planets from Earth. Scientists determined the orbital period (the time required to complete one orbit of the Sun) of each of the planets by observing them over long periods of time and by carefully recording the time required for a complete revolution around the Sun (Figure 2).

observational study the careful watching and recording of a subject or phenomenon to gather scientific information to answer a question

READING TIP

Scan for Text Features
Text features include headings, sidebars, graphics, and margin definitions. Scan a text before reading to identify the features it uses. This process can help you organize how you read a text and alert you to important information.

Figure 2 Planets farther from the Sun have a longer orbital period than those closer to the Sun. (Art not drawn to scale.)

Scientific Investigation Skills

Regardless of the type of scientific investigation, there are certain skills that are important in the process of conducting the investigation. These skills can be organized into four categories:

1. initiating and planning
2. performing and recording
3. analyzing and evaluating
4. communicating

Figure 3 When scientists first observed the northern lights (aurora borealis), they were probably curious and perhaps proposed a hypothesis to explain what caused them.

hypothesis a possible answer or untested explanation that relates to the initial question in an experiment

prediction a statement that predicts the outcome of a controlled experiment

Initiating and Planning

All scientific investigations begin with a question. The question may have arisen from observations of a natural phenomenon or from an individual's curiosity (Figure 3). Often the question comes from previous experiments or studies.

Some questions cannot be answered through scientific investigation, so it is important to ask the right questions. To lead to a scientific investigation, a question must be testable. Testable questions have certain characteristics:

- They must be about living things, non-living things, or events in the natural world.
- They must be answerable through scientific inquiries such as controlled experiments and observational studies.
- They may be answered by collecting and analyzing data to produce evidence.

If a question suggests that a controlled experiment should be performed, then you may propose a possible answer to the question. The tentative answer, which is based on existing scientific knowledge, is called a **hypothesis**. The hypothesis is directly related to the question.

A hypothesis suggests a relationship between an independent variable and a dependent variable. A hypothesis serves two functions: (1) it proposes a possible explanation along with reasons for this explanation, and (2) it suggests a method of obtaining evidence that will support or reject the proposed explanation.

If you cannot make a hypothesis because you do not have a scientific explanation, then you can make a simple **prediction**. A prediction is a statement that predicts the outcome of a controlled experiment, without an explanation. A prediction is not a guess: it is based on prior knowledge and logical reasoning.

A hypothesis usually includes a prediction. It is often written in the form "If ..., then ..., because ...." The "if ... then" part constitutes the prediction; the "because ..." part is the explanation.

TRY THIS: QUESTIONS LEADING TO SCIENTIFIC INQUIRY

SKILLS: Planning, Communicating

SKILLS HANDBOOK
3.B.4.

The type of question asked in a scientific inquiry determines, to a large degree, the most appropriate method to answer the question. In this activity you will identify the best type of scientific inquiry for answering a series of questions.

Equipment and Materials: notebook or paper; pen or pencil

1. Read the following questions:
   - What effect does pollution have on the plants and animals in the lake outside of town?
   - Does the type of metal a wire is made of affect how well it conducts electricity?
   - What species of birds live in my neighbourhood during the winter?
   - What is the best way to prevent metal from rusting?
   - What are the characteristics of the rings around the planet Saturn?

A. Carefully study each statement and indicate what type of scientific inquiry would be most appropriate.

B. For each question, briefly describe how you would carry out the inquiry.

A hypothesis or a prediction provides the framework for the investigation. It identifies the variables and suggests which is the independent variable and which is the dependent variable. The hypothesis or prediction also suggests an experimental design by which the hypothesis can be tested fairly. The **experimental design** briefly describes the procedure. The value and success of the investigation depend on whether the experiment is fair, so careful planning at this stage is critical.

experimental design a brief description of the procedure by which a hypothesis is tested

Planning the investigation involves the following:

- identifying the independent and dependent variables
- determining how the changes in the variables will be measured
- specifying how to control the variables not being tested
- selecting the appropriate equipment and materials (Figure 4)
- anticipating and addressing safety concerns
- deciding on a format for recording observations

Figure 4 The equipment and materials for scientific inquiry can be very simple or very specialized, such as the equipment used in space exploration.

Performing and Recording

After planning an investigation, it is important to follow the procedures described during the planning stage carefully (Figure 5). That does not mean that the procedures cannot change. If the procedures present problems, they should be modified without changing the overall structure of the investigation. Any modifications to the procedure should be recorded in case you or someone else wants to repeat the investigation. If the problems cannot be overcome, you may have to go back to the planning stage and start again.

When performing the procedure, it is important to be aware of potential safety concerns. Be sure to read the safety guidelines in the Skills Handbook carefully before beginning an investigation, and refer to them frequently if you have any concerns.

While performing an investigation, you will need to make accurate observations at regular intervals and record them carefully and accurately. Observations are any information that is obtained through the senses or by extension of the senses. Observations can be quantitative (numerical) or qualitative (non-numerical).

Figure 5 Two scientists plan their research on crops.

1.1 Skills of Scientific Investigation 11

quantitative observation a numerical observation based on measurements or counting

Quantitative observations are based on measurements or counting (Figure 6). Examples of quantitative measurements include length, mass, temperature, and population counts. Measuring is an important skill in making observations. Not only must the right measuring tool be selected, but it must also be used in such a way as to provide a precise and accurate measurement.

Figure 6 A scientist makes quantitative observations on a polar bear.

qualitative observation a non-numerical observation that describes the qualities of objects or events

Qualitative observations are descriptions of the qualities of objects and events, without any reference to a measurement or a number. Common qualitative observations include the state of matter (solid, liquid, or gas), texture, and odour. These qualities cannot be measured directly or easily.

The method of recording your observations depends on the type of observation. Quantitative observations are often recorded in a data table. Qualitative observations can be written in words or recorded in pictures or sketches (Figure 7). Remember to record your observations clearly and accurately so that you do not have to rely on memory when you report your findings.

LEARNING TIP

Word Origins
The word "quantitative" comes from the word "quantity," which refers to an amount or number. The word "qualitative" comes from the word "quality," which refers to the characteristics of something.

Figure 7 This young scientist records qualitative observations as drawings.

Analyzing and Evaluating

Tables, lists, and drawings of observations are not the final products for the data collected during an investigation. Analyzing, or carefully studying, the observations can provide more information than the raw data itself. In addition, you can plot graphs from the quantitative data to show patterns and trends more clearly.

Analyzing observations also helps to identify any errors in measurement. You should carefully check any measurement that is very different from the others. If the different measurement is caused by an error in measurement, record it but do not include it in your analysis.

TRY THIS: ANALYZING DATA

SKILLS: Analyzing, Communicating

SKILLS HANDBOOK
3.B.7., 6.A.

Data collection is the recording of information in an organized, systematic way. Analyzing data involves studying the data to uncover patterns and trends. In this activity, you will analyze data from an imaginary investigation. The data in Table 1 shows the fuel consumption and the distance travelled during a 10-day period by a courier using a fuel-efficient vehicle.

Equipment and Materials: ruler; graph paper; notebook; pen or pencil

1. Use the data in Table 1 to plot a scatter graph.
2. Starting at the (0,0) point on your graph, use a ruler to draw a straight line through your data so that it is touching (or near to) as many data points as possible. The data points should be evenly scattered around either side of the line you draw.

A. Observe the angle of the line you drew in step 3. What does this line represent?

B. Explain why it is appropriate for the line you drew in step 3 to go through (0,0).

C. Make a general statement about the relationship between the length of the trip and the amount of fuel consumed.

D. Use the graph to estimate how much fuel would be used for a 400 km trip and for a 125 km trip.

Table 1 Fuel Consumption and Distance Travelled per Day

Day	Fuel (L)	Distance (km)
1	24	320
2	13	240
3	29	515
4	19	290
5	21	305
6	35	450
7	10	195
8	40	580
9	44	645
10	18	255

E. On what day(s) did the vehicle have the best fuel efficiency, that is, the most kilometres per litre of fuel? On what day was the vehicle the least fuel efficient?

F. Propose a possible explanation for why all of the points do not fall on a straight line.

A very important skill in a scientific investigation is evaluating the evidence that is obtained through observations. The quality of the evidence depends on the quality of other aspects of the investigation—the plan, the procedures, the equipment and materials, and the skills of the investigator. In order to evaluate the evidence, you need to evaluate all aspects of the investigation.

> **READING TIP**
>
> **Search for the Main Idea**
> During reading, search for the main idea of the text. Check the first, second, and last sentences of a paragraph. Sometimes, you may have to infer a main idea that is not stated directly.

The purpose of analyzing and evaluating observations is to answer the question posed at the beginning of an investigation. You may have evidence that you can use to confidently answer the question. You may conclude, however, that you do not have sufficient evidence to answer the question. If your evidence confirms the prediction, then the hypothesis is supported. The evidence does not, however, *prove* the hypothesis to be true. If your evidence does not confirm the prediction, then the hypothesis may not be an acceptable explanation. Learning that the hypothesis is not supported is just as valuable as learning that it is supported. Rejection of a hypothesis is not a failure in a scientific investigation. It is simply another step along the path of finding an answer to the question.

Work in science seldom ends with a single experiment. Sometimes other investigators repeat the investigation to see if their results are the same. A question in science often sets off a chain reaction that leads to other questions, which then leads to other investigations and other questions. At the end of any investigation, the scientist asks questions such as, What does this mean? Is the information of any practical value? How can the information be used? What other questions need to be answered? What new questions arise as a result of this investigation?

Communicating

One of the important characteristics of scientific investigation is that scientists share their information with the scientific community (Figure 8). Before scientific data can be published in a scientific journal, it must be examined by other scientists in a process called peer review. Other experts check that the data is valid and the science is correct. Clear and accurate communication is essential for sharing information. It is important to share not only the findings, but also the process by which the evidence was obtained. If the investigation is to be repeated by others, sharing the design and procedures is just as important as sharing the findings.

Figure 8 Scientists sharing their information with each other is a key characteristic of scientific investigation.

By sharing their data and the techniques they used to obtain, analyze, and evaluate their data, scientists give others the opportunity to both review the data and use it in future research. The most common method for communicating with others about an investigation is by writing a report or giving a presentation after the investigation is complete.

TRY THIS: GETTING YOUR MESSAGE ACROSS

SKILLS: Evaluating, Communicating

SKILLS HANDBOOK
3.B.8.

Accurate communication is as important in science as it is in everyday life. In this activity, you will work as a class to demonstrate the importance of accurate communication and the need to develop and refine your skills in this area.

Equipment and Materials: paper; pen or pencil

1. Your teacher will create a simple statement, write it on a piece of paper and then whisper it to a student in class.
2. That student will write the message on a piece of paper and then whisper the message to another student.
3. The message will be passed on until every person in the class has received the message. Your teacher will record the sequence of students that the message followed.
4. The last person to receive the message will repeat it out loud to the class.
5. The teacher will then write the original message on the board. All students will then write their version of the message under the previous message.
6. The rules are simple:
   - You must state the message only once.
   - You must write exactly what you hear.
   - You must whisper the message.
   - You cannot show other students what you have written.

A. How closely did the final message resemble the original message? Explain.

B. How does this activity demonstrate the need for clear, accurate communication?

C. Give an example from real life of how the message "received" differs from the message "sent."

D. Based on this activity, suggest a set of guidelines to ensure clear, accurate communication.

IN SUMMARY

- Curiosity about what we see around us often leads to questions that trigger scientific investigations.
- One type of scientific inquiry is the controlled experiment, in which the researcher keeps all but two variables constant, changes one (the independent variable), and observes the other (the dependent variable).
- A second type of inquiry is the observational study, in which the researcher collects data by observing a situation without affecting it.
- Scientific investigation skills include initiating and planning (asking a question and deciding on the best way to find an answer); performing and recording (carrying out the procedure and making observations in an organized way); analyzing and evaluating (searching for patterns in the observations); and communicating (sharing procedures and findings with others).

1.2 Scientific Literacy for Living and Working in Canada

Scientific knowledge and technological innovations play an increasingly important role in everyday life. Science and technology have become so common in our society that we often do not notice them, or we take them for granted. Most new technologies are designed so that the average person can use them without needing to understand how they work. Science, too, is often perceived as too complex to be understood by the average person.

What Is Scientific Literacy?

To make wise personal decisions and to act as a responsible citizen, it is necessary to be scientifically and technologically literate. The Science Teachers' Association of Ontario (STAO) defines a scientifically and technologically literate person as "one who can read and understand common media reports about science and technology, critically evaluate the information presented, and confidently engage in discussions and decision-making activities regarding issues that involve science and technology."

To read famous scientists' statements about the value of scientific literacy,  **GO TO NELSON SCIENCE**

READING TIP

Make Connections
As you are reading, make connections between information in the text and your own experiences, what you have already read, and facts you know about the topic. For example, think about the technology you use and whether you understand how it works. Do you take this technology for granted?

Scientific Literacy for Careers in Science

Look around your classroom at your fellow science students. Some of you may pursue post-secondary education and a career in scientific research. Other classmates will find careers in areas that are related to science. These careers could be in medicine, geology, engineering, or environmental science. They could also be in plumbing, culinary arts, cosmetology, or art (Figure 1). Generally, employers hire individuals with strong critical-thinking and problem-solving skills and the ability to work as part of a team. These skills are emphasized throughout the entire science program.

Figure 1 Scientific literacy is important in many different careers.

Many Canadians are active in a broad range of scientific disciplines. They are world leaders in such areas as astronomy, space exploration, medicine, genetics, environmental science, physics, and information and communication technology. These Canadians are responsible for a long list of scientific discoveries and technological inventions and innovations. The theory of plate tectonics, the discovery of insulin, the invention of the cardiac pacemaker, and the concept of standard time have made a significant contribution to people around the world.

Throughout this book, you will have many opportunities to explore careers that are related to the area of science under study (Figure 2). Wherever you see a Career icon  with the accompanying note in the margin of this book, you will be directed to visit the Nelson website. There you will have an opportunity to research the education and training requirements for these careers and to find out about the roles and responsibilities of these careers.

To learn more about careers in science,
GO TO NELSON SCIENCE

(a)

(b)

(c)

Figure 2 (a) Some scientists conduct their research in laboratories. (b) Others conduct their research in the natural world with (c) living organisms.

Scientific Literacy for Life and Citizenship

Only some people work in jobs that are directly or indirectly related to science. However, every one of us is a citizen. Citizenship comes with certain rights and responsibilities. One of our basic rights is to have access to a full education. Along with that right comes the responsibility to use that education for the benefit of oneself and society. Because science and technology influence our lives, it is important that we recognize and understand this influence. We can then make rational and ethical decisions about issues that affect us as individuals and as a society.

At some point in your life, you will have to make decisions about your own lifestyle. You will need to be aware of the impact your actions have on the world around you (Figure 3). You will likely have to make decisions related to critical issues such as climate change, environmental pollution, depletion of natural resources, protection of species, new medical technologies, space exploration, and world hunger.

Figure 3 Scientific literacy will help you think critically and form opinions about important issues such as the impact of waste and pollution on the environment.

1.2 Scientific Literacy for Living and Working in Canada

> **READING TIP**
>
> **Summarize the Main Idea**
> After reading a selection, reinforce your understanding by summarizing the main idea in your own words and drawing a conclusion about it.

As a society, we need to consider both the positive and the negative impacts of developing and using scientific knowledge and new technology. We cannot foresee all the possible consequences of these achievements (Figure 4). However, we must be aware of the positive and negative implications of new technologies. If we fail to do so, we may experience very unpleasant surprises after the technologies have been adopted.

Figure 4 New computer technology has many positive impacts, but old computers and hard drives often end up in landfills.

It is important to understand the basic concepts of science. However, it is impossible to know everything about science. Likewise, no one will be able to understand every new scientific discovery and technological advance. To achieve scientific literacy, it is just as important to learn about science as it is to learn science. It is important to know what science can do, to know that knowledge produced by science is reliable, and to know that science—despite its limitations—is the best way to learn about the world. It is equally important to be able to find and evaluate information and to use that information to make decisions.

Scientifically literate people understand that the future will be very different from the present. There are always new developments in science and technology. They also understand that society influences science and technology as much as science and technology influence society. Achieving scientific literacy is not the same as preparing to be a scientist. It is important whether you are a small-business person, a lawyer, a construction worker, a mechanic, a doctor, an engineer, or a research scientist. Regardless of your plans and ambitions, achieving scientific literacy is an important goal.

IN SUMMARY

- Canadians have made valuable contributions to the development of science and technology around the world.
- Scientific literacy enables people to understand and evaluate information relating to science and technology in the world around us.
- Being scientifically literate allows us to make informed decisions.
- Specific scientific knowledge and skills are necessary for a wide range of careers.
- A general understanding of science is necessary to be an informed citizen.

LOOKING BACK

CHAPTER 1

KEY CONCEPTS SUMMARY

Science and technology are an important part of our everyday lives.

- All individuals are affected by science and technology every day of their lives.
- Most people do not understand the science and technology that they use in their everyday lives.
- Many social issues, such as alternative energy sources and pollution, have a connection to science and technology.

Scientific inquiry can be conducted in different ways depending on the question to be answered.

- Observations often lead to questions that initiate scientific inquiry.
- The type of scientific inquiry depends on the question.
- Controlled experiments determine how an independent variable affects a dependent variable.
- A hypothesis is a possible answer to, and an explanation of, a scientific question.

All scientific inquiries rely on careful recording of accurate and repeated observations.

- Observations are obtained by using the senses or equipment that extends the senses.
- Observations can be quantitative (involving measuring or counting) or qualitative (involving descriptions).
- All observations should be recorded accurately.
- Repeating observations can eliminate errors and increase the value of the evidence.

Careful analysis and interpretation of observations help to make them meaningful.

- Analysis of observations involves carefully studying data to identify patterns or trends.
- Interpretation of observations involves explaining patterns and trends in data.
- Analysis of observations may lead to conclusions that answer the original scientific question.
- Analyzing and evaluating observations may raise additional questions and lead to further scientific inquiries.

Clear communication is important for sharing scientific discoveries and ideas with others.

- Scientists are expected to share their findings with other scientists and with the public.
- The results of scientific investigations should be reported clearly and honestly.
- Effective communication allows others to repeat scientific inquiries.

Scientific literacy is necessary for wise personal decisions, responsible citizenship, and careers.

- Careers in science and careers related to science require scientific knowledge and skills.
- Scientific knowledge and skills are essential for making logical and reasoned personal decisions.
- Citizens can use their scientific knowledge and skills to make decisions that benefit society.
- A scientifically literate person is better able to understand issues related to science and technology.

UNIT B: Sustainable Ecosystems

OVERALL Expectations

- assess the impact of human activities on the sustainability of terrestrial and/or aquatic ecosystems, and evaluate the effectiveness of courses of action intended to remedy or mitigate negative impacts
- investigate factors related to human activity that affect terrestrial and aquatic ecosystems, and explain how they affect the sustainability of these ecosystems
- demonstrate an understanding of the dynamic nature of ecosystems, particularly in terms of ecological balance and the impact of human activity on the sustainability of terrestrial and aquatic ecosystems

BIG Ideas

- Ecosystems are dynamic and have the ability to respond to change, within limits, while maintaining their ecological balance.
- People have the responsibility to regulate their impact on the sustainability of ecosystems in order to preserve them for future generations.

Focus on STSE

SOMETHING TO LOSE?

During our everyday lives, we do not often notice slow changes that are occurring all around us. Take a moment to consider the situations of the following three high school students.

The Environment and Lifestyles

Chisomo is a Grade 11 student. She is an excellent athlete and participates in many sports, including downhill and cross-country skiing. Chisomo has noticed that winter snow conditions are changing. Some years are good, but it seems like there is a trend toward more rain and less snow. Chisomo lives on a farm in southwestern Ontario, where her family operates an apple orchard. Chisomo's family is switching to organic apple growing. They pump large volumes of groundwater every summer to water their orchard.

Luisa is in Grade 9. She lives in downtown Toronto and enjoys nature photography, camping, and jogging and biking around the city. Living in the city, Luisa experiences "smog days." Last summer, Luisa had a chance to visit the Gulf of St. Lawrence. She had a great time viewing wildlife but was saddened to learn that at one time more than a quarter of a million walruses lived in the region.

Devon is a Grade 10 student living in a lumbering community in northwestern Ontario. He enjoys fishing, hanging out with his friends, and listening to music. Devon is planning on taking courses in forestry management and technology so that he can pursue a career in the forestry industry. He thinks people will always have a use for wood, and you can always plant and grow more trees.

Think/Pair/Share

With a partner, discuss each student's situation and answer the following questions:
1. In what way(s) is the environment important to each of them? C A
2. What changes in the environment have already taken place around them? Were these changes easy to notice? K/U C
3. What environmental changes might affect them in the future? A
4. Do you think you need to be concerned about environmental changes? Do YOU have something to lose? C A

UNIT B

LOOKING AHEAD

UNIT B
Sustainable Ecosystems

CHAPTER 2 — Understanding Ecosystems

Millions of monarch butterflies stop at Point Pelee on Lake Erie during their annual 3000 km migratory flight to Mexico.

CHAPTER 3 — Natural Ecosystems and Stewardship

Kelp forests are highly productive ecosystems that occur in cold, coastal waters.

CHAPTER 4 — Ecosystems by Design

The urban landscape is one type of human-made ecosystem.

UNIT TASK Preview

Ontario's Species of Concern

Human activities can have very different effects on species. At one extreme, human activities can endanger species, potentially resulting in their extinction (Figure 1). At the other extreme, human activities can result in a species invading new environments and increasing rapidly in numbers.

Figure 1 This is the former range of the eastern cougar. Cougars are now absent or extremely rare throughout most of their former range in Ontario.

In this Unit Task, you will select two contrasting species that are of concern in Ontario. You will choose one native species that is at risk and one species that is non-native and invasive.

For each species, you will investigate
- the factors that have led to their current status
- how these factors are influencing the species and the health of the ecosystems they live in
- implications related to human interests
- actions that could change the species' status

UNIT TASK Bookmark

The Unit Task is described in detail on page 156. As you work through the unit, look for this bookmark and see how the section relates to the Unit Task.

ASSESSMENT

You will be assessed on how well you
- research, compile, and analyze your data
- demonstrate an understanding of issues related to species decline and success
- outline the implications for the future sustainability of species and ecosystems
- prepare and present an action plan

What Do You Already Know?

PREREQUISITES

Concepts
- Biotic and abiotic features in the environment
- Producers and consumers
- Primary and secondary succession
- Movement of water on Earth
- Effects of human activities on the environment

Skills
- Following established safety procedures
- Using appropriate scientific vocabulary and SI units
- Presenting information in a variety of forms
- Carefully observing, gathering, and recording data
- Analyzing data and making reasonable inferences

1. Examine Figure 1. Classify each organism as a producer or a consumer. **K/U C**

Figure 1

2. Figure 2 shows the process of secondary succession. **K/U**

Figure 2

 (a) Describe what is happening to the plant community.
 (b) What is the difference between primary and secondary succession?

3. Describe how the activities shown in Figure 3 can influence the environment. Is the effect positive or negative? **A C**

Figure 3

4. Make a graph to illustrate the relationship of the data in Table 1. Give your graph a title and label the axes. **T/I C**

Table 1

Year	1996	1998	2000	2002	2004	2006	2008
Number of deer	21	32	36	44	35	18	12

 (a) Describe the pattern in this data.
 (b) Suggest two or three factors that might have influenced the size of the deer population.

5. Figure 4 presents data on the 197 species of plants and animals in Ontario that are considered to be at risk. Estimate the percentage of species that fall into each category. **T/I**

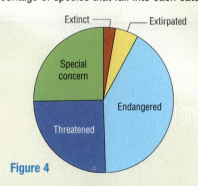

Figure 4

6. Describe how water is cycled through the environment. Make sure to use the following terms. **K/U C**
 - precipitation
 - evaporation
 - condensation
 - surface runoff
 - groundwater

Looking Ahead 23

CHAPTER 2

Understanding Ecosystems

KEY QUESTION: How are organisms influenced by their living and non-living environments?

Millions of monarch butterflies stop at Point Pelee on Lake Erie during their annual 3000 km migratory flight to Mexico.

UNIT B

Sustainable Ecosystems

CHAPTER 2 Understanding Ecosystems

CHAPTER 3 Natural Ecosystems and Stewardship

CHAPTER 4 Ecosystems by Design

KEY CONCEPTS

Life on Earth exists in the atmosphere, lithosphere, and hydrosphere.

Photosynthesis and cellular respiration are complementary processes in an ecosystem.

Energy passes through ecosystems, whereas matter cycles within ecosystems.

Human activities influence biogeochemical cycles such as the water and carbon cycles.

Ecosystems are composed of biotic and abiotic components.

Terrestrial biomes and aquatic ecosystems are largely determined by their abiotic characteristics.

ENGAGE IN SCIENCE

Red Crabs and Crazy Ants

A spectacular event happens on a remote island in the Indian Ocean called Christmas Island. Each year, millions of red land crabs migrate from their burrows in the rainforest floor. They travel over roads, through towns, and down steep cliffs to reach the shoreline. There, they mate and release their eggs into the ocean.

Red crabs are important members of Christmas Island's rainforest ecosystem. A population of 120 million crabs translates to about 13 800 crabs per hectare! The crabs' diet consists of fallen leaves, fruits, flowers, and seedlings. They help maintain the structure of the rainforest ecosystem and regulate nutrient cycles.

Residents of Christmas Island protect the red crabs during their migration. Sections of road are closed during peak migration periods. People build underpasses that allow crabs to cross roads.

Christmas Island is in crisis. The island has been invaded by yellow crazy ants. These ants are native to Africa but were introduced accidentally to Christmas Island from Australia decades ago. The ants attack and feed on the red crabs, spraying them with formic acid and then devouring them. The crazy ants have killed tens of millions of crabs.

Scientists are searching for ways to save Christmas Island from crazy ants. They are testing a poison bait to kill the ants. Ant infestations are under control at three test sites.

The fate of the red crabs and the Christmas Island ecosystem remains unknown. What would happen to Christmas Island if the red crabs disappear? Are the crabs worth saving?

WHAT DO YOU THINK?

Many of the ideas you will explore in this chapter are ideas that you have already encountered. You may have encountered these ideas in school, at home, or in the world around you. Not all of the following statements are true. Consider each statement and decide whether you agree or disagree with it.

1 Oceans make up the majority of Earth's mass.
Agree/disagree?

4 All organisms are helpful in the environment.
Agree/disagree?

2 All of the particles that make up your body are being continuously replaced by new ones.
Agree/disagree?

5 Humans are one of the few species that do not compete with other species.
Agree/disagree?

3 Animals need plants for food, but plants do not need animals.
Agree/disagree?

6 Humans have been successful because they are able to change the natural environment.
Agree/disagree?

FOCUS ON READING

Making Connections

When you make connections with a text, you are really joining what you notice in a new text with what you already know, understand, or have learned. There are three main ways to make connections to help you interpret a text:

- **Text-to-Self Connections:** A text can remind you of something you or someone you know has experienced.
- **Text-to-Text Connections:** A text can remind you of something you understand from reading or viewing another text.
- **Text-to-World Connections:** A text can remind you of something you have learned about science, technology, society, or the environment.

READING TIP

As you work through the chapter, look for tips like this. They will help you develop literacy strategies.

Life on Planet Earth

In our solar system, only a single planet—Earth—is teeming with millions of species. Earth is home to countless organisms and habitat types. A habitat is the place where an organism lives. Habitats may be terrestrial, on the land, or aquatic, in the water. In the oceans, life ranges from colourful fish on coral reefs to strange creatures living in the dark depths. On land, there are cacti and rattlesnakes in the arid deserts. Enormous trees swarm with insects in the tropical rainforests (Figure 1). Low-lying shrubs and herds of caribou abound in the frozen Arctic. What features of Earth permit such a diversity of life to exist?

Figure 1 Earth is home to countless habitats, including tropical rainforests.

Making Connections *in Action*

Making connections can help you understand or respond to a text by visualizing; making predictions, inferences, or judgments; forming opinions; drawing and supporting conclusions; or evaluating the text. Here is how one student made connections with the information about life on planet Earth.

Connections I have made to key ideas or information	How that connection helps me understand or respond to the text
• **Text-to-Self Connection:** between the graphic and an ecotour my family did on Vancouver Island	helps me form an opinion about the importance of protecting old growth forests
• **Text-to-Text Connection:** between description of oceans and movie Finding Nemo	helps me visualize the vast range of plants and creatures in the oceans
• **Text-to-World Connection:** between last sentence and the fact that caribou populations in the Arctic are decreasing	helps me evaluate that the word "abound" makes the reader think that caribou are not an endangered species

Life on Planet Earth

2.1

In our solar system, only a single planet—Earth—is teeming with millions of species. Earth is home to countless organisms and habitat types. A habitat is the place where an organism lives. Habitats may be terrestrial, on the land, or aquatic, in the water. In the oceans, life ranges from colourful fish on coral reefs to strange creatures living in the dark depths. On land, there are cacti and rattlesnakes in the arid deserts. Enormous trees swarm with insects in the tropical rainforests (Figure 1). Low-lying shrubs and herds of caribou abound in the frozen Arctic. What features of Earth permit such a diversity of life to exist?

Figure 1 Earth is home to countless habitats, including (a) deserts and (b) tropical rainforests.

The Spheres of Earth

Earth is a medium-sized planet orbiting our Sun (a star) at a distance of approximately 150 000 000 km. Viewed from space, Earth appears as a pale blue dot (Figure 2). As you near Earth, you will notice it is surrounded by a thin gaseous layer swirling with clouds. Finally, you will be able to distinguish the oceans, land, and ice that cover Earth's surface.

The Atmosphere, Lithosphere, and Hydrosphere

Earth's mass creates a force of gravity strong enough to hold gases near its surface. In contrast, the force of gravity of our Moon is not strong enough to hold gases. Earth's **atmosphere** is the layer of gases extending upward for hundreds of kilometres. It is made up of about 78 % nitrogen gas and 21 % oxygen gas. The remaining <1 % of the atmosphere includes argon, water vapour, carbon dioxide, and a variety of other gases.

Figure 2 Earth looks like a pale blue dot when viewed from space.

atmosphere the layer of gases surrounding Earth

The atmosphere is critical to life on Earth. It acts like a blanket and moderates surface temperatures. The insulation prevents excessive heating during the day and cooling during the night. Without an atmosphere, Earth's surface temperature would drop from the 15 °C average it is now to approximately −18 °C. In addition, Earth's atmosphere blocks some incoming solar radiation, including most ultraviolet light, which is linked to skin cancer. Without the atmosphere, most of Earth's species would be unable to survive.

lithosphere Earth's solid outer layer

hydrosphere all of Earth's water in solid, liquid, and gas form

Scientists use a number of terms to describe Earth's key surface components. The **lithosphere** is the rocky outer shell of Earth. It consists of the rocks and minerals that make up the mountains, ocean floors, and the rest of Earth's solid landscape. The lithosphere ranges from about 50 to 150 km in thickness.

The **hydrosphere** consists of all the water on, above, and below Earth's surface. It includes oceans, lakes, ice, groundwater, and clouds. Nearly all the water on Earth (97 %) is contained in the oceans.

The Biosphere

biosphere the zone around Earth where life can exist

Scientists use the term **biosphere** to describe the locations in which life can exist within the lithosphere, atmosphere, and hydrosphere (Figure 3). Most of the easily observed life forms exist on land and in water, but micro-organisms exist several kilometres beneath Earth's surface.

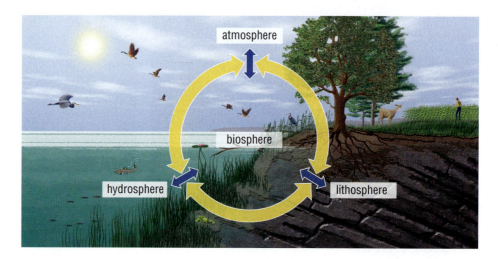

Figure 3 Earth's biosphere is found in regions of the atmosphere, lithosphere, and hydrosphere.

Earth is very large (about 12 700 km in diameter), but the biosphere is very thin by comparison. All conditions required for life must be met and maintained within this thin layer of ground, water, and lower atmosphere.

All living things need space, water, and nutrients to survive. However, the supply of these resources is limited. Ultimately, the availability of resources places a limit on the number of individuals of a species that can survive. All life on Earth is vying for access to these precious resources. The struggle for resources is discussed in Section 2.7.

LEARNING TIP

It's All Greek to Me
Many scientific terms are derived from Greek. *Lithos* means stone, *atmos* means vapour, *hydro* means water, *bio* means life, and *sphere* means ball.

The Gaia Hypothesis

In the 1960s, scientist James Lovelock advanced the Gaia hypothesis. He proposed that Earth, through interactions among the biosphere, lithosphere, atmosphere, and hydrosphere, behaved like a living organism. Lovelock's hypothesis suggested that Earth was capable of responding to changes in its environment (such as incoming sunlight) and maintaining relatively consistent internal conditions over long periods of time—just like a living cell.

The Gaia hypothesis is not widely accepted as a rigorous scientific concept. However, many people feel that thinking of Earth as a living thing may encourage and promote a more caring attitude toward our planet and the life it supports.

READING TIP

Text-to-Text Connections
Brainstorm text-to-text connections by jotting down everything that comes to mind when you think about the selection and texts you have read. You can use a Venn diagram to list similarities and differences between the selection and other texts.

TRY THIS: A SCALE MODEL OF PLANET EARTH

SKILLS: Analyzing, Communicating

SKILLS HANDBOOK
5.A.1.

To appreciate how the lithosphere, atmosphere, and hydrosphere compare to the overall size of Earth, it is useful to consider a scale model of Earth. The diameter of Earth at the equator is approximately 12 700 km. Imagine a model of Earth with a diameter of 1 m, like a very large beach ball. Each distance of 1 mm on the model of Earth would represent a distance of 12.7 km on the true Earth. In this activity, you will create a scale model to compare the size of Earth's components.

1. Draw a table similar to Table 1 and fill in the missing values. For example, the thickest portion of the lithosphere at 150 km would be equivalent to $\frac{150}{12.7}$ or 12.8 mm on our model. T/I C

   MATH TIP
   To keep the model consistent, divide every value by the same number. Therefore, 1 mm always equals 12.7 km.

A. Did the model distances surprise you? Which model distances, if any, were less than you had expected? Which, if any, were greater? A

B. Based on the same scale, the volume of Earth would be 520 L, but the volume of all the world's oceans would be only 640 mL. This is not enough to fill two pop cans. If this is the case, why do you think Earth is often referred to as the watery planet? A

Table 1

Feature measurement	Actual distance	Model distance
thickest portion of lithosphere	150 km	12.8 mm
distance between Toronto and Thunder Bay	1380 km	
average ocean depth	3.7 km	
maximum ocean depth	10.9 km	
height of Mount Everest	8.4 km	
average thickness of Antarctic ice	1.6 km	
thickness of lower atmosphere	20 km	

C. In our Earth model, all life would exist within 1 mm of the surface. Given this information, would it surprise you to learn that many scientists consider the ocean, atmosphere, and biosphere to be very vulnerable to pollution and other forms of damage? A

IN SUMMARY

- Earth's atmosphere is made up of about 78 % nitrogen gas, 21 % oxygen gas, and other gases.
- The atmosphere moderates surface temperatures and blocks some incoming solar radiation.
- The lithosphere is Earth's solid outer shell.
- The hydrosphere is Earth's water in all its forms.
- The biosphere is the area where life can exist within the lithosphere, atmosphere, and hydrosphere.
- The Gaia hypothesis proposes that Earth behaves like a living organism.

CHECK YOUR LEARNING

1. Explain how Earth's mass is related to its ability to have an atmosphere. K/U

2. In what way does the presence of an atmosphere enhance conditions for life on Earth's surface? K/U

3. Define each of the following terms: lithosphere, atmosphere, hydrosphere, and biosphere. K/U

4. Describe the ways in which Earth's "spheres" overlap each other. K/U C

5. If Earth is so large, why do scientists consider the biosphere to be fragile? Explain. K/U A

6. It can be difficult to appreciate the relationships of large objects such as Earth. How does using a scale model make these relationships easier to understand? T/I C

7. What surprised you most about the physical makeup of Earth's spheres and the relationships among them? C

2.2 Introducing Ecosystems

Imagine you are planning a backpacking trip to Algonquin Park in Ontario (Figure 1). You intend to spend four days hiking through the wilderness. To be fully prepared, you must bring proper food, clothing, and equipment (Figure 2). You must also consider the living things you may encounter. Mosquitoes will be abundant, so you pack insect repellent. You will also need a water filter to provide safe drinking water. You know that bears and raccoons are a possibility, so you bring a long rope so that you can hang your food pack in a tree at night. You also pack a stove and matches, and, most importantly, a map and maybe a global positioning system (GPS).

To learn more about hiking in Ontario,
GO TO NELSON SCIENCE

To learn more about becoming a park warden,
GO TO NELSON SCIENCE

Figure 1 Backpacking is a common summer activity.

Figure 2 During a backpacking trip, you must have appropriate equipment and supplies to survive in the environment.

ecosystem all the living organisms and their physical and chemical environment

biotic factors living things, their remains, and features, such as nests, associated with their activities

abiotic factors the non-living physical and chemical components of an ecosystem

Planning for such a trip reminds us of the many factors we encounter in any ecosystem. Scientists define an **ecosystem** as all of the living organisms that share a region and interact with each other and their non-living environment. An ecosystem is composed of both living and non-living components. Some factors, such as the terrain and the weather, are non-living. Other factors, such as insects, bears, and micro-organisms, are alive. The living components, called **biotic factors**, include all organisms, their remains, and their products or wastes. The non-living components, or **abiotic factors**, include physical and chemical components such as temperature, wind, water, minerals, and air.

Some materials, such as a hard coral reef, are not easy to classify. They are built by coral animals and are therefore biotic in origin. Over time, parts of the reef may break down, forming white coral sand, which is usually considered an abiotic factor.

Individual organisms from many species share an ecosystem. Together, all of the individuals of a single species in a particular area make a population. Individuals from all of the populations form the community. An ecosystem is the term given to the community and its interactions with the abiotic environment.

32 Chapter 2 • Understanding Ecosystems

Figure 3 shows how ecosystems are composed of individual organisms, populations, communities, and the physical surroundings in which communities of organisms live.

Figure 3 An ecosystem is composed of populations of plant and animal species and their biotic and abiotic environments.

TRY THIS   ECOSYSTEM ABCs

SKILLS: Analyzing, Communicating

Everything we use and consume in our daily lives comes from the biotic and abiotic parts of our environment. In this activity, you will reflect on our dependence on the environment for the items we use every day.

1. Brainstorm a list of 20 diverse items you have used in the past week. Your list could include foods, fuel, consumer products, packaging materials, or even the sidewalk you walked on.
2. Make a list of separate materials contained in each item. Do not worry if you cannot identify all the materials.
3. Determine whether these materials are a biotic or an abiotic resource.
4. Draw a Venn diagram of two overlapping circles with the headings "abiotic" and "biotic." The overlapping region is for the combined (both abiotic and biotic) resources (Figure 4). Place each of your original items in your Venn diagram. For example, a CD case would be placed in the overlapping portion of the diagram. The paper liner is of biotic origin, while the plastic case is abiotic.
A. Do you depend on both biotic and abiotic resources in your everyday life? Give examples to support your answer.
B. Do you think biotic or abiotic resources are more important for your survival? Explain.

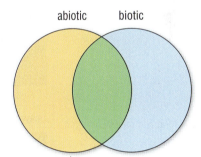

Figure 4 Sample Venn diagram

C. The forestry and mining industries are major employers in Ontario. Identify which of your items were dependent on these two industries for their production.
D. Is it likely that either the forestry or mining industry will become obsolete? Explain.
E. All food items are biotic in nature. Describe how food items themselves are dependent on abiotic resources.
F. Based on this activity and the definition of an ecosystem, do you think the human species is part of Earth's ecosystem? Explain your reasoning.

Describing Ecosystems

Ecosystems are highly variable. They can differ dramatically in size and in their biotic and abiotic features. We generally think of an ecosystem as a fairly large area, such as a forest or a lake. On a much smaller scale, the community of bacteria and fungi living in a rotting log is an ecosystem. In this way, large ecosystems may include many much smaller ecosystems.

> **DID YOU KNOW?**
> **The Heat Is On!**
> Thermophiles live in extremely hot environments near undersea thermal vents. Water is ejected at extremely high pressure, and the temperature can exceed 100 °C. Thermal vents have their own unique ecosystem of bacteria, clams, mussels, and tube worms.

Whatever the size, every ecosystem is characterized by a distinctive set of features. For example, you could describe an ecosystem by its particular organisms, temperature range, or water depth (Table 1).

Table 1 Examples of Large and Small Ecosystems

Ecosystem	Characteristic abiotic features	Characteristic biotic features
coniferous forest	• long, cold winter season • warm summers • moderate rainfall • much of the precipitation falls as snow • snow insulates and protects ground species	• few species • dominated by black spruce forest and bogs • black bear, red squirrels, and moose • many biting flies • short growing season reduces ecosystem productivity
coral reef	• water temperatures from 25–31 °C • water depth from 0–30 m • usually in tropical latitudes • sometimes used as a source of limestone	• wide variety of marine life, including corals, sponges, and fishes • source of many valuable fish species • sensitive to changes in water temperature and chemistry
beaver pond	• shallow water • water warm in summer and ice covered in winter • usually temporary, lasting years to decades • dams may break causing downstream flooding	• variety of aquatic plants, fishes, frogs, turtles, insects, and beavers • surrounding forest used as food supply for beavers • flooding often kills trees that are then used by other species for food or shelter
rotting log	• moist environment • low light • small, temporary ecosystem, lasting years to decades	• variety of decomposing bacteria and fungi • beetles, larvae of various insects • provides cover for small vertebrates such as salamanders

Sustainability of Ecosystems

When we hear the word ecosystem, we tend to think about a natural, pristine environment where it is pleasant to hike, swim, or hang out. Most natural ecosystems are **sustainable**. This means that they maintain a relatively constant set of characteristics over a long period of time.

sustainable ecosystem an ecosystem that is maintained through natural processes

Human activities often change the biotic and abiotic features of an ecosystem. This can render a previously sustainable ecosystem unsustainable. **Sustainability** is the ability to maintain natural ecological conditions or processes without interruption, weakening, or loss of value. We will look at this in detail in Chapter 3.

Other ecosystems are artificially created and maintained by human actions. To create an artificial ecosystem, like an urban park or farm, desired plants and animals are introduced and maintained. Artificial ecosystems are not usually sustainable. They require management to maintain the biotic and abiotic features deemed desirable. We will consider artificial ecosystems in Chapter 4.

Like all species, humans are dramatically influenced by the biotic and abiotic features of the ecosystems that surround them. In northern Ontario, the economy is based on forestry, mining, and tourism. The economy in southern Ontario is based on agriculture and manufacturing. People living in northern environments are more likely to suffer from *seasonal affective disorder* (SAD) associated with low winter light levels. People living in large urban centres, however, are more likely to suffer from breathing problems linked to smog.

sustainability the ability to maintain an ecological balance

READING TIP

Text-to-Self Connections
Brainstorm text-to-self connections by jotting down everything that comes to mind when you think about the selection and your own experiences or those of relatives or friends. You can create a mind map to show the connections.

IN SUMMARY

- Ecosystems are characterized by their biotic and abiotic factors.
- Biotic factors are living components of an ecosystem. Abiotic factors are the non-living physical and chemical components.
- A population is all individuals of the same species living in an ecosystem.
- A community is all organisms living in the same ecosystem.
- Natural ecosystems are generally sustainable, whereas artificially created ecosystems must usually be managed.
- Surrounding ecosystems influence many aspects of our daily lives.

CHECK YOUR LEARNING

1. Classify each of the following as either biotic or abiotic features: temperature, bacteria, wind, sunlight, dead leaves, mosquitoes, sand, milk, hair, ice, plastic, an empty snail shell. K/U

2. Does a community include abiotic features? Explain. K/U

3. Which of the following are considered to be an ecosystem? Explain your reasoning. K/U
    - backyard pond
    - all the cats in your neighbourhood
    - tree
    - schoolyard
    - vase of cut flowers
    - potted plant
    - your digestive system

4. "Human activities change only the biotic features of an ecosystem." Is this statement true or false? Explain why or why not. K/U

5. Would you consider a large city to be a population or a community? Explain your choice. K/U

6. This section describes categorizing all features of the world around us as either biotic or abiotic. Would you find these categories useful personally? Why do you think scientists use these terms? A C

7. In your day-to-day life, do you wonder about whether or not your actions are sustainable? Why is sustainability important in nature? Explain. A C

2.2 Introducing Ecosystems

2.3 PERFORM AN ACTIVITY

Factors Affecting an Aquatic Ecosystem

Aquatic ecosystems, such as ponds, lakes, and rivers, are extremely valuable. Unfortunately, aquatic ecosystems are threatened by human activities. For example, salt spread on highways and fertilizer spread on fields can be washed into nearby waterways, damaging aquatic ecosystems. Acid precipitation impacts aquatic ecosystems because excess acid upsets the chemical balance. In this activity you will consider these effects by building a small model ecosystem and altering the abiotic environment.

SKILLS MENU
- Questioning
- Hypothesizing
- Predicting
- Planning
- Controlling Variables
- Performing
- Observing
- Analyzing
- Evaluating
- Communicating

Purpose

To investigate some effects of altering abiotic features in an aquatic ecosystem.

Equipment and Materials

- eye protection
- lab apron
- large spoon or stirring rod
- 1-L beaker
- funnel
- hand lens
- graduated cylinder
- electronic balance
- eye dropper and Petri dish
- four 2-L plastic pop bottles
- marker
- 4 L of pond water containing assorted aquatic organisms
- liquid plant fertilizer (any general houseplant liquid fertilizer like 10:10:10)
- dropper bottle of dilute sulfuric acid
- weighing papers
- table salt

⚠ Always wear eye protection and a lab apron when working with chemicals.

Procedure

SKILLS HANDBOOK
3.B.

1. You will work as part of a team to examine the influence of three abiotic factors on aquatic organisms. You will build small aquatic ecosystems and alter one abiotic factor in each. You will compare the community of organisms in the natural pond water (the control) with those in which in which acid, salt, or nutrient content has been altered (the treatments).

Your team will need to decide how to alter each variable. You will have to determine how much acid, salt, and fertilizer you will add to each of the three ecosystems. Base your decisions on what you think might occur in a real ecosystem. In your notebook, draw a table similar to Table 1 below and record how much material you will add to each treatment.

Table 1

Treatment	Contents	
	Starting material	Added material
control	1 L of pond water	none
plant fertilizer	1 L of pond water	___ mL liquid plant fertilizer
acid rain	1 L of pond water	___ mL dilute sulfuric acid
salt pollution	1 L of pond water	___ g salt

2. Your team will also need to design a sampling method to estimate abundance of the organisms. How will you compare the variety and numbers of different organisms in your ecosystems? You may want to consider using ranking scales to compare the four ecosystems.
3. Have your teacher approve your design before you proceed.
4. Put on your eye protection and lab apron.
5. Label four 2-L bottles as control, plant fertilizer treatment, acid rain treatment, and salt pollution treatment.
6. Stir the pond water so that organisms are evenly distributed in the water.
7. Using the beaker and funnel, measure 1 L of pond water into each 2-L bottle.
8. Using a hand lens, examine each ecosystem. Make sure that they have approximately equal numbers of organisms. If they seem uneven, add more organisms so that they appear to have approximately equal quantities.
9. In your notebook, describe or draw each type of organism you see and estimate its abundance.
10. Using a graduated cylinder, measure the plant fertilizer and add it to the appropriate ecosystem.
11. Using a graduated cylinder, measure the sulfuric acid and add it to the appropriate ecosystem.
12. Using an electronic balance, measure the salt and add it to the appropriate ecosystem.
13. Design a table to collect your data. Every two or three days, for at least three weeks, record the type of organisms and their abundances in each bottle.
14. Plan how to present your findings. You may want to use summary tables and graphs to present them.
15. You may want to consider using digital images to record and present your findings.

Analyze and Evaluate

(a) Which ecosystem demonstrated the greatest algae growth?
(b) Which ecosystem maintained the greatest number of organisms? Which ecosystems had the most types of organisms?
(c) Which ecosystem had the lowest number of organisms by the end of the experiment? Which had the lowest variety of organisms?
(d) Account for any changes you observed in each ecosystem. Remember to compare your three test treatments with the control.
(e) How did your results compare with those of other teams in the class? Were there any patterns among the quantity of fertilizer, acid, and salt added and their impacts?

Apply and Extend

(f) Assuming there was always a supply of water, which of your ecosystems, if any, do you think could sustain life over long periods of time? Explain your answer.
(g) Why would a scientist use small model ecosystems instead of investigating more complex natural ecosystems?
(h) Based on your observations, what effect do fertilizer, acid rain, and road salt have on the sustainability of aquatic ecosystems?
(i) Human activity can impact large water sources in unhealthy ways. The release of fertilizers or toxins into surface or groundwater can pollute water supplies and harm ecosystems. Road salt and deforestation can also affect water quality. Select a human activity and use the Internet and other sources to answer the following:

   (i) What purpose is served by the activity itself? For example, what is the benefit of using salt on roads?
   (ii) Describe the specific nature of the problem. In what way(s) does this human activity threaten water quality and the sustainability of aquatic ecosystems?
   (iii) What possible alternatives or technological solutions might solve or reduce the problem?
   (iv) Does your research support what you observed in your model ecosystems?

2.4 Energy Flow in Ecosystems

All organisms require energy to stay alive and function. The source of almost all of this energy is **radiant energy**. This is the energy radiated from the Sun. Earth is being continuously bombarded with both invisible radiant energy, such as ultraviolet, and visible radiant energy, or **light energy**, from the Sun. About 70 % of the radiant energy is absorbed by the hydrosphere and lithosphere and converted into **thermal energy**. Thermal energy is what warms the atmosphere, evaporates water, and produces winds. The remaining 30 % of the radiant energy is reflected directly back into space. A small fraction of the radiant energy, a mere 0.023 %, is absorbed directly by living organisms in a process called photosynthesis (Figure 1).

radiant energy energy that travels through empty space

light energy visible forms of radiant energy

thermal energy the form of energy transferred during heating or cooling

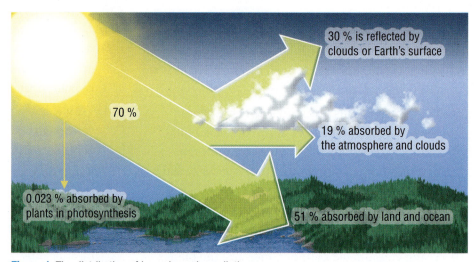

Figure 1 The distribution of incoming solar radiation

Thermal energy keeps Earth's surface warm, but it cannot provide organisms with the energy they need to grow and function. Light energy can be used by some organisms, but it cannot be stored and is not available during the night. In contrast, chemical energy can be stored in cells and then released when needed. Chemical energy is used by all organisms to perform functions, including movement, growth, and reproduction. As chemical energy is used, it must be replaced.

Photosynthesis

Where does the chemical energy used by organisms come from? As you may know, the simple answer is the Sun. Many organisms are able to convert light energy into chemical energy using the process of **photosynthesis**. This conversion of energy is one of the most important chemical processes. Without it, most life on Earth would not exist.

Organisms that photosynthesize make their own energy-rich food compounds using light energy. These organisms are called **producers**. Most organisms that are unable to make their own food through photosynthesis depend on producers for food. You will learn more about this later in the section. On the land, the major producers are green plants. The green colour comes from a chemical called chlorophyll, which captures light energy.

READING TIP

Text-to-World Connections
To make text-to-world connections, sometimes it is necessary to conduct research online or in a library. Use a search engine to gather information about the topic or key ideas. List the three most important points about the topic. Beside each, make a note of the connections you can make.

photosynthesis the process in which the Sun's energy is converted into chemical energy

producer an organism that makes its own energy-rich food compounds using the Sun's energy

In aquatic ecosystems, the main producers are microscopic organisms. These single-celled algae and cyanobacteria also contain chlorophyll (Figure 2). Virtually all of the chemical energy contained in food was once light energy captured in the process of photosynthesis.

Most producers use light energy to convert two low-energy chemical compounds (carbon dioxide and water) into high-energy compounds (sugars). In doing so, they release oxygen gas into the environment as a by-product. The photosynthesis reaction is represented by the following word equation:

$$\text{carbon dioxide} + \text{water} \xrightarrow{\text{light energy}} \text{sugar} + \text{oxygen}$$

The sugar formed in this process contains stored chemical energy. This energy is stored in the roots, stems, leaves, and seeds of plants (Figure 3). Most plants convert the sugar to starch for storage.

Figure 2 Algae are often the dominant producers in aquatic ecosystems.

LEARNING TIP

Reading Word Equations
The arrow in a word equation can be read as "react to form." The word equation for photosynthesis reads "carbon dioxide and water, in the presence of light energy, react to form sugar and oxygen."

Figure 3 Chemical energy is stored in a variety of plant structures.

Not all of the sugar produced through photosynthesis goes toward energy storage. Some sugars are used as building materials. The carbon, hydrogen, and oxygen in the carbon dioxide and water are like building blocks. Using light energy, they are rearranged to form sugars and oxygen gas during photosynthesis. Then, components in the sugars are rearranged to form different combinations. For example, they may form carbohydrates (such as the cellulose used in cell walls) or in combination with nitrogen, proteins (Figure 4).

DID YOU KNOW?

Plants Do Not Always Need the Sun!
Some Ontario plants, including Indian pipe, do not photosynthesize. Instead, this species obtains energy-rich food directly from fungi in the soil. Stranger still, the fungi obtain their food from the roots of neighbouring trees. With no need for green chlorophyll, these white plants have small, functionless leaves.

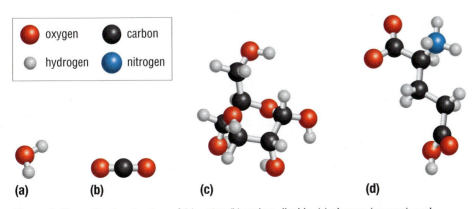

Figure 4 The molecular structure of (a) water, (b) carbon dioxide, (c) glucose (a sugar), and (d) glutamic acid (a building block of protein)

2.4 Energy Flow in Ecosystems

Cellular Respiration

Photosynthesis produces stored energy in the form of sugar. To make stored energy available for use, the plant performs the complementary reaction called cellular respiration. **Cellular respiration** is a chemical process in which energy is released from food. In this process, the sugar and oxygen are rearranged to form carbon dioxide and water. As this reaction takes place, energy is released. The plant is able to use this released energy for any of the activities carried out by its cells. The word equation for cellular respiration is

$$\text{sugar} + \text{oxygen} \longrightarrow \text{carbon dioxide} + \text{water} + \text{energy}$$

The energy originally stored in the sugar is released as a product of cellular respiration. Unlike photosynthesis, cellular respiration occurs continuously. No light energy is required for cellular respiration. Figure 5 illustrates the relationship between the processes of photosynthesis and cellular respiration.

cellular respiration the process by which sugar is converted into carbon dioxide, water, and energy

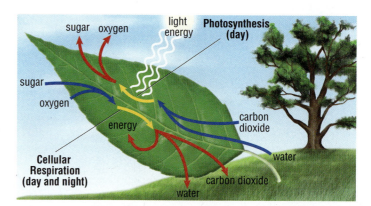

Figure 5 Photosynthesis and cellular respiration are complementary processes.

TRY THIS: PRODUCTS OF CELLULAR RESPIRATION

SKILLS: Questioning, Predicting, Performing, Observing, Analyzing

Cells require a continuous supply of energy to stay alive. Cellular respiration releases this energy from stored food. What observations support these statements? Is there evidence that the cells in your body are continuously performing cellular respiration?

Equipment and Materials: stopwatch or clock with second hand; large test tube or clear plastic cup; scissors; limewater; cardboard or paper disc; drinking straw

1. Review the word equation for cellular respiration noting what is used and what is produced by this chemical process.
2. Try holding your breath for as long as you can. Note the time at which you start to feel a strong urge to breathe. This is an indication of how long it takes for your body to begin running low on the supply of oxygen needed for cellular respiration.
3. Fill a large test tube or cup one-third full of limewater. Limewater turns cloudy when combined with carbon dioxide. Record the appearance of the limewater.
4. Cut an opening in a 5 × 5 cm piece of cardboard or paper just large enough to slip a straw through. This will act as a splash guard. Slide the splash guard halfway along the straw.
5. Place one end of the drinking straw in the limewater and the other end in your mouth. The splash guard should be just above the top of the test tube. Slowly breathe out a steady stream of bubbles through the limewater. You may wish to exhale several breaths through the limewater. Record any changes in the appearance of the limewater.

🚫 Do not suck on the straw.

A. How long do you think human cells can last without oxygen? What evidence supports your answer? T/I
B. By holding your breath, what necessary chemical are you preventing from entering your body? K/U
C. If this chemical is not available, what will happen to the process of cellular respiration? K/U
D. What effect, if any, would this have on your cells' ability to release energy from food? T/I
E. Was there any evidence of limewater combining with carbon dioxide? Explain. T/I C
F. If your body produces carbon dioxide during the process of cellular respiration, what else is most likely being produced? K/U

Many organisms cannot photosynthesize. Therefore, they are not able to make their own energy-rich sugar or building materials. These organisms, called **consumers**, obtain energy and building materials by eating other organisms. To obtain usable energy from food, they undergo cellular respiration. While only producers undergo photosynthesis, both producers and consumers perform cellular respiration. Unless you are a green human with chlorophyll in your skin, you are a consumer (Figure 6). Instead of photosynthesizing, you, like all consumers, obtain energy by eating other organisms or their products.

Perhaps no other set of chemical reactions provides better evidence of our dependence on other living things. Without photosynthesizing producers, we would be without a source of food. A major benefit of plants is that they release oxygen into the environment. However, we would have no use for oxygen if plants did not provide us with the food we need to perform cellular respiration.

Not all organisms require oxygen, and some of these organisms do not rely on photosynthesis either. Many soil micro-organisms, for example, use other chemical pathways to release chemical energy from sugar. This process occurs in the absence of oxygen.

consumer an organism that obtains its energy from consuming other organisms

Figure 6 Humans are consumers, obtaining energy by eating other organisms.

IN SUMMARY

- During photosynthesis, green plants use the Sun's energy to convert carbon dioxide and water into sugar (chemical energy) and oxygen.
- During cellular respiration, sugar and oxygen are converted into carbon dioxide, water, and energy.
- All organisms undergo cellular respiration.
- Producers make their own energy-rich food compounds using the Sun's energy.
- Consumers obtain energy by feeding on other organisms.
- Humans depend on photosynthesizing organisms for food and oxygen.

CHECK YOUR LEARNING

1. How much of the energy reaching Earth is absorbed and converted to chemical energy by the process of photosynthesis? Where does the other energy go? K/U
2. What energy-rich substance is produced by green plants during photosynthesis? K/U
3. Explain how you know plants contain energy-rich substances. T/I
4. Plants do not just use photosynthesis to make sugars for energy storage. Identify other kinds of uses plants have for these substances. K/U
5. How are photosynthesis and cellular respiration related? K/U
6. What chemical process(es) do producers and consumers share? What chemical process(es) do they not share? K/U
7. List five foods that contain the high-energy products of photosynthesis. T/I
8. Prior to reading this section, would you have considered plants as more essential as producers of food or of oxygen? What are your thoughts now? C
9. Were you surprised to learn that plants use oxygen and perform cellular respiration just like animals? Explain. C
10. Animals are unable to make their own energy, yet you obtain energy when you eat animal food products. Explain how this illustrates the flow of energy through an ecosystem. T/I
11. Describe how plants can perform cellular respiration at night when they perform photosynthesis only during the day. K/U
12. It is a major advantage to be able to make your own food using photosynthesis. Can you think of any disadvantages to having this ability? Are there places you could not live or things you could not do? A

2.5 Food Webs and Ecological Pyramids

If you ran blindfolded through a forest in Ontario, you would not likely run into a moose or trip over a raccoon. Instead, you would quickly hit a tree or trip over the plants tangled around your feet (Figure 1).

Figure 1 There are many more large plants than large animals in this ecosystem.

Why are there so many large trees and other plants and so few large animals? Why are you surrounded by hundreds of hungry black flies and mosquitoes but not hundreds of birds or frogs ready to eat them? The answers lie in the relationships between species in ecosystems. Each species is influenced and limited by its surroundings and by the resources it requires. Think about which plants and animals are abundant and which are rare in the ecosystem you live in.

Ecological Niches

Every species interacts with other species and with its environment in a unique way. These interactions define the **ecological niche** of a species. This is the role of a species within its ecosystem. A species' niche includes what it feeds on, what eats it, and how it behaves. No two species occupy identical niches.

The niche of a black bear is as follows: black bears feed on tender plant parts such as nuts and berries (Figure 2). They supplement their diet with insects and other small animals. Bears may carry seeds long distances in their digestive systems before the seeds are expelled and germinate. Bears go into hibernation during the winter. While they have few predators other than human hunters, black bears are themselves fed on by many blood-feeding insects and are hosts to a variety of parasites.

A key feature of any ecosystem is the feeding roles of each species. In Section 2.4, we distinguished between producers and consumers according to how organisms obtain their energy. Consumers can be further subdivided depending on what types of organisms they eat (Table 1).

ecological niche the function a species serves in its ecosystem, including what it eats, what eats it, and how it behaves

Figure 2 Black bears eat berries and nuts.

Table 1 Types of Consumers

Feeding role	Definition
herbivore	animal that eats plants or other producers
carnivore	animal that eats other animals
omnivore	animal that eats both plants and animals
scavenger	animal that feeds on the remains of another organism

Food Chains and Food Webs

The most common interactions between species are through feeding relationships. The easiest way to display these relationships is with **food chains**. Food chains illustrate who eats whom in an ecosystem. In Figure 3, the seeds of a producer (the pine tree) are eaten by a herbivore (the red squirrel), which is in turn eaten by a carnivore (the weasel). The weasel then falls prey to a top carnivore (the goshawk).

food chain a sequence of organisms, each feeding on the next, showing how energy is transferred from one organism to another

pine cone (seeds) → red squirrel → weasel → goshawk

Figure 3 A simple food chain in a forest ecosystem

In this food chain, some of the chemical energy stored in the pine seeds is passed through the red squirrel to the weasel and ends up in the goshawk. In this way, food chains show how energy passes through an ecosystem. Remember that all organisms continually use and release energy to their environment. This means that energy is continuously lost from all levels of the food chain.

Ecologists refer to the **trophic level**, or feeding level, to describe the position of an organism along a food chain. Producers occupy the lowest, or first, trophic level. Herbivores occupy the second trophic level, and carnivores occupy the third and fourth trophic levels (Figure 4).

LEARNING TIP

Reading Food Chains
In a food chain diagram, the arrows show the direction of food and energy flow. Arrows are read as "is eaten by." For example, the red squirrel is eaten by the weasel.

trophic level the level of an organism in an ecosystem depending on its feeding position along a food chain

fourth trophic level
tertiary consumers

third trophic level
secondary consumers

second trophic level
primary consumers

first trophic level
producers

Figure 4 Species can be divided into trophic levels depending on how they obtain their energy.

Food chains do not exist in nature. They are used to show simple feeding relationships. Food chains are part of more complex sets of relationships that exist among species. Many herbivores eat pine seeds. Red squirrels eat a wide range of foods and are themselves food for a variety of predators. A more accurate, but still incomplete, way to illustrate interactions is with a **food web**. This shows a series of interconnecting food chains.

food web a representation of the feeding relationships within a community

2.5 Food Webs and Ecological Pyramids

DID YOU KNOW?

Bizarre Niches
Pearlfish are marine fish with slender bodies. Most species of pearlfish hide inside the digestive system of sea cucumbers (simple marine animals). They enter and exit through the anus! Some species of pearlfish have evolved to live inside the sea cucumber, feeding on its internal organs.

Food webs are highly complex, with consumers feeding on many species (Figure 5). The large number of interactions tends to reduce the vulnerability of any one species to the loss or decline of another species. For this reason, complex food webs are thought to be more stable than simple food webs.

Figure 5 A partial food web of the boreal forest. A complete model of the interactions in this ecosystem food web would show thousands of species including scavengers and decomposers which feed on dead organisms and wastes. Note that the food chain from Figure 3 is embedded in this food web. What would happen if the lynx was removed from the food web?

Food webs are useful tools to figure out what may happen when a species is removed from or added to an ecosystem. For example, if a species is removed from a food web, the species it feeds on may increase dramatically in numbers. Conversely, the population of a newly introduced species may disrupt the entire food chain (Figure 6).

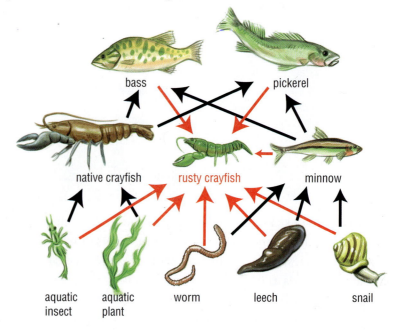

Figure 6 The invasive rusty crayfish (labelled in red) competes with native crayfish for many of the same foods. It also feeds on the eggs of bass and pickerel. Large fish feed on native crayfish but usually avoid eating the rusty crayfish.

44   Chapter 2 • Understanding Ecosystems

TRY THIS | WEAVING A FOOD WEB

SKILLS: Predicting, Analyzing, Communicating

In this activity, you will examine a rainforest food web.

Equipment and Materials: set of rainforest species cards; large piece of paper; glue stick; coloured pencils or markers

1. Obtain a set of "rainforest species cards." Each card will have the name of a species and a brief description of its living habits. For simplicity, decomposers have not been included.
2. On a large piece of paper, spread out the species cards. Place the producers near the bottom, the carnivores near the top, and the herbivores in the middle. Glue each card in position.
3. Use light pencil lines to connect species that feed on other species. Include arrows to indicate the direction of energy flow. The arrows should point from the food supplier to the food consumer.
4. Colour in the pencil arrows using the following colour code: A green line joins any plant to any of its consumers. A blue line joins any herbivore to any of its consumers. A red line joins any carnivore to any of its consumers.
5. Brainstorm three or four other species that live in tropical rainforests. Add them to your food web.

A. How many connections did you include in your initial food web? K/U
B. Which species do you think would be most affected if anteaters were removed from this ecosystem? T/I
C. How would the removal of fig wasps alter this ecosystem? What would happen to the population of fig trees over a long period of time? Explain. T/I C
D. Which species had the most direct connections to other species within the ecosystem? Was this surprising? Explain. T/I C
E. Unlike tropical rainforests, the tundra has relatively few species. How might the loss of a species in the tundra compare to the loss of a species in a tropical rainforest? T/I
F. Describe how food webs show the interdependence of one species on all other species. T/I

Ecological Pyramids

Another way ecologists illustrate how ecosystems function is through ecological pyramids. **Ecological pyramids** display relationships between trophic levels in ecosystems. The three types of ecological pyramids are energy, numbers, and biomass. An energy pyramid illustrates energy loss and transfer between trophic levels (Figure 7).

ecological pyramid a representation of energy, numbers, or biomass relationships in ecosystems

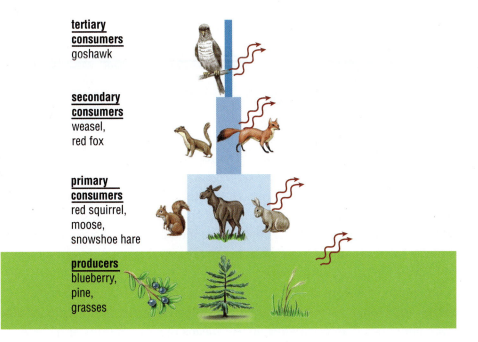

Figure 7 Only a small proportion of the total energy at any given trophic level is passed on to the next level. Energy is used in biological processes such as growth and reproduction and is lost as thermal energy (red arrows).

READING TIP

Making Connections

When trying to make connections with a text, use prompts or questions such as
- I already know about...
- It makes me wonder why...
- Does this text remind me of something I have experienced?
- Does the information in this selection match or differ from information I have read in other texts?

The size of each layer in the energy pyramid represents the amount of energy available at that trophic level. A continuous supply of energy is essential for all living things. By examining how energy flow is depicted in these diagrams, you will gain a better understanding of the relationships between species, including why some species are much more abundant than others.

As one organism consumes another, it obtains both the physical matter (nutrients) and the chemical food energy it needs to survive, grow, and reproduce. However, each time energy is used, some of it is released to the environment as thermal energy. You can feel an example of lost thermal energy by placing your hand on your forehead. Organisms such as plants also release small quantities of thermal energy to the environment.

Plants use the energy they obtain from the Sun for growth, reproduction, and cellular activities. Some of this energy is lost to the environment. As a result, less energy is available for the herbivores that eat the plants. Herbivores use most of the energy they obtain from plants for their own life processes. Only about 10 % of the energy taken in by the individuals at one trophic level is passed on to individuals at the next level. For example, a moose that is eight years old does not possess the food energy of all the plants it has eaten since it was born. A wolf that eats the moose would obtain only a small portion of the lifetime energy consumption of the moose.

Species in the highest trophic levels have less energy available to them than species near the bottom. This often results in their populations being much smaller than species lower in the food chain. This is why an ecosystem will have fewer predators, such as hawks, than herbivores, such as mice.

Populations that occupy different trophic levels vary in their numbers and their **biomass**, which is the total mass of all individuals combined. A pyramid of numbers shows the number of individuals of all populations in each trophic level, whereas a pyramid of biomass shows the total mass of organisms in each trophic level (Figure 8).

biomass the mass of living organisms in a given area

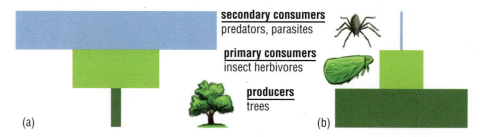

Figure 8 (a) A simplified pyramid of numbers and (b) a pyramid of biomass for a deciduous forest ecosystem

An energy pyramid will always decrease in size from lower to higher trophic levels (remember Figure 7 on page 45). This is not always the case with pyramids of numbers or biomass. In a forest ecosystem, the tiny plant-feeding insects in the second trophic level outnumber the trees in the first trophic level (Figure 9, next page). The biomass of all the trees, however, is much greater than the biomass of herbivores, so this pyramid of biomass will have a typical pyramid shape.

Organisms that feed at lower trophic levels generally have much more energy and biomass available to them. For this reason, herbivores are usually more numerous than carnivores. It also means that disruptions or changes at lower trophic levels can have profound impacts on the entire ecosystem.

Figure 9 There will always be more aphids than plants in a forest, but the biomass of plants will be greater.

As omnivores, humans feed at most trophic levels (although some people choose to be herbivores). As you will learn, the choices we make about our feeding patterns have significant implications for food production and ecosystem sustainability.

UNIT TASK Bookmark

You can apply what you learned about the interactions between species in food webs in this section to the Unit Task described on page 156.

IN SUMMARY

- Every species occupies a unique ecological niche.
- Feeding relationships between organisms can be represented by food chains, food webs, and trophic levels.
- Energy is continuously being lost to the environment by all living organisms.
- Higher trophic levels always have less energy available to them.
- Ecological pyramids can be used to display energy, number, and biomass relationships.

CHECK YOUR LEARNING

1. What is meant by the term *ecological niche*? Describe the ecological niche of humans and of three other species. K/U
2. Explain the differences between food chains and food webs. K/U
3. Describe some possible impacts of adding or removing a species from an ecosystem. K/U
4. As you go up from one trophic level to the next, the amount of available energy decreases. Explain where the energy has gone. K/U
5. What trophic level contains the greatest biomass in most ecosystems? Explain why this occurs. K/U
6. What trophic levels are occupied by carnivores? Provide examples. K/U
7. Bison, zebra, and kangaroos are three large mammals. Explain why they have similar ecological niches but cannot be shown in the same food web. K/U A

8. (a) How would the food web in Figure 10 change if the red fox were killed off by rabies?
   (b) What species would benefit?
   (c) What species might decline? K/U A

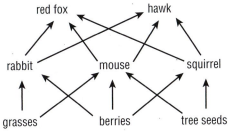

Figure 10

2.5 Food Webs and Ecological Pyramids 47

2.6 Cycling of Matter in Ecosystems

Like humans, all life on Earth requires water and nutrients. Water provides the liquid component that makes up cells. Nutrients are a source of the building materials and chemical energy. Water and nutrients are composed of physical matter. You obtain matter from the food you eat, the water you drink, and the air you breathe (Figure 1). However, the particles of matter do not stay in your body forever. Every part of every cell in your body is replaced over time. Approximately 2 million of your red blood cells are replaced every second of every day. Scientists estimate that, on average, every particle in a human body is replaced at least once every 7 years! You and everyone you know are living recycling machines (Figure 2).

Figure 1 Nutrients from food are used to repair and gradually renew every part of our body.

Figure 2 This woman appears much older than the child, yet almost every particle in her body has been replaced within the last seven years.

Biogeochemical Cycles

The particles that make up matter cannot be created or destroyed. This fact is significant for all life. It means that all water and nutrients must be produced or obtained from chemicals that already exist in the environment. This happens in a series of cycles in which chemicals are continuously consumed, rearranged, stored, and used. Because these cycles involve living (bio) organisms and occur as Earth (geo) processes, they are called **biogeochemical cycles**. Every particle in an organism is part of a biogeochemical cycle.

biogeochemical cycle the movement of matter through the biotic and abiotic environment

The Water Cycle

water cycle the series of processes that cycle water through the environment

The most obvious of the biogeochemical cycles is the **water cycle** (Figure 3, next page). Liquid water evaporates, forming water vapour that moves through the atmosphere. The vapour eventually condenses, forming liquid water or ice crystals, and returns to Earth as rain, hail, or snow. Water falling on land may enter the soil and groundwater or move across the surface entering lakes, rivers, and oceans. Water that is taken in by plant roots may be released from leaves in a process called transpiration. Most of the water that is present in the water cycle is in the abiotic environment.

Hydrologists are interested in the movement of water through the environment. To learn more about becoming a hydrologist,

48 Chapter 2 • Understanding Ecosystems

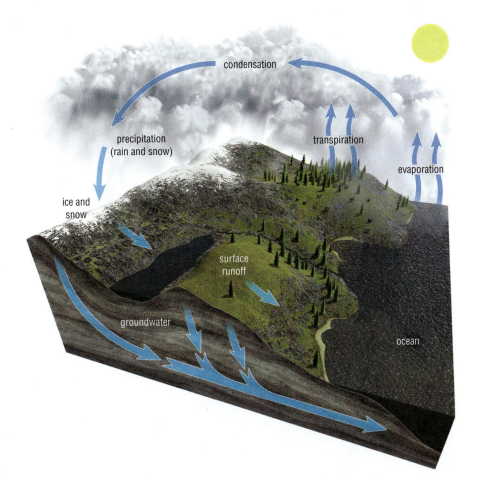

Figure 3 The water cycle

> **DID YOU KNOW?**
> **You Are Who You Eat!**
> Over time, the water in our bodies moves through the biosphere. For this reason, scientists believe that we all contain at least one water particle that was once in the body of Julius Caesar and another from every *Tyrannosaurus rex* that ever lived!

The Carbon Cycle

Carbon moves between the abiotic and biotic parts of an ecosystem in the **carbon cycle**. Most of this exchange occurs between carbon dioxide (either in the atmosphere or dissolved in water) and photosynthesizing plants and micro-organisms.

carbon cycle the biogeochemical cycle in which carbon is cycled through the lithosphere, atmosphere, hydrosphere, and biosphere

CARBON DEPOSITS

While large quantities of carbon cycle through photosynthesis and cellular respiration, most of Earth's carbon is not cycled. Instead, it is stored in carbon-rich deposits. Fossil fuels, such as coal, oil, and natural gas, are the most valuable carbon deposits. They form when decomposed organisms are compressed over millions of years. Carbon is also stored for millions of years as limestone formed from dead marine organisms.

Large quantities of carbon are also contained in plant tissue and as dissolved carbon dioxide in the world's oceans. These locations are referred to as carbon sinks because carbon can enter or leave them over relatively short times.

HUMAN ACTIVITIES CHANGE THE CARBON CYCLE

Human activities have dramatic effects on the carbon cycle (Figure 4). By burning fossil fuels, humans release the stored carbon into the atmosphere. The concentration of carbon dioxide in the atmosphere is now higher than it has been in at least the past 800 000 years. This change is causing global climatic change. Climate change has the potential to alter the most critical abiotic factors in the ecosystems: temperature and water availability. The increase in the average temperature of our atmosphere is melting ice caps and glaciers, causing sea levels to rise, and disrupting ecosystems.

Figure 4 Human activities influence the carbon cycle.

Deforestation also increases the concentration of carbon dioxide in the atmosphere (Figure 5). Large-scale reforestation and a dramatic reduction in the use of fossil fuels are needed to slow the process of climate change. As you study climate change in Grade 10, remember what you have learned about the importance of the carbon cycle and sustainable ecosystems.

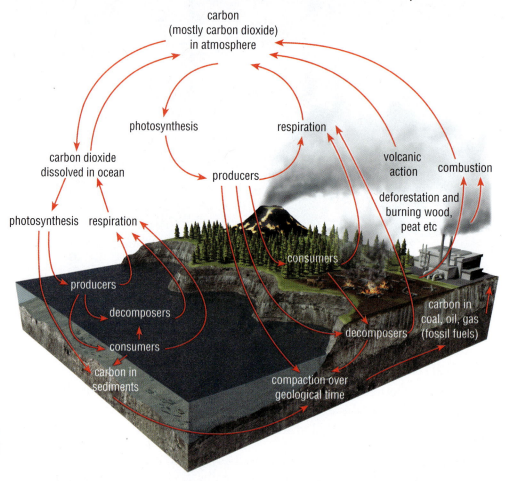

Figure 5 The carbon cycle results in the long-term and short-term storage of carbon.

nitrogen cycle the series of processes in which nitrogen compounds are moved through the biotic and abiotic environment

The Nitrogen Cycle

Nitrogen is extremely abundant in the atmosphere. However, it is not easy to acquire directly from the abiotic environment. Nitrogen enters and leaves the atmosphere through a complex biochemical pathway in the **nitrogen cycle** (Figure 6, next page).

Most of the nitrogen used by living things is taken from the atmosphere by certain bacteria in a process called nitrogen fixation. These micro-organisms convert nitrogen gas into a variety of nitrogen-containing compounds including nitrates, nitrites, and ammonia. Lightning and ultraviolet light also fix small amounts of nitrogen. In addition, humans add nitrogen to the soil as fertilizer.

Once in the soil ecosystem, the nitrogen-rich compounds are available to producers. After the nitrogen is absorbed, it is passed from producer to consumer and on up the food chain. Many animals consume more nitrogen than they can use and excrete the excess in the form of urea or ammonia. A dead organism's nitrogen-rich compounds are taken in by decomposers or are released back into the environment. These compounds are either recycled again by soil micro-organisms or they are converted by denitrifying bacteria back into nitrogen gas which then re-enters the atmosphere.

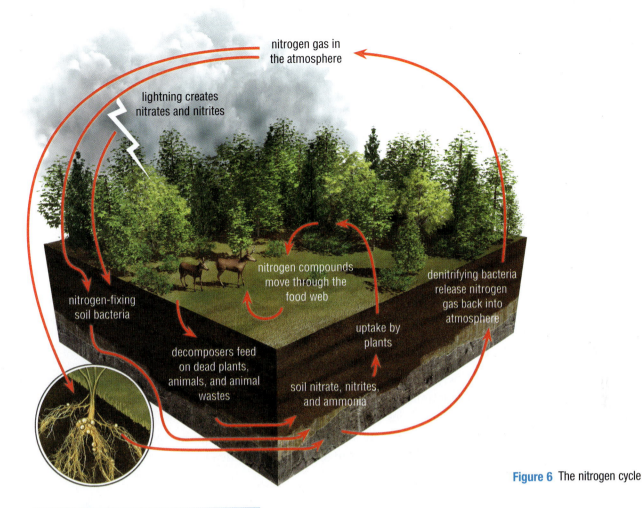

Figure 6 The nitrogen cycle

IN SUMMARY

- Matter is cycled through ecosystems via biogeochemical cycles.
- Water can occur in all states (solid, liquid, and gas) as it moves through the water cycle.
- Carbon moves between the abiotic and biotic components of the ecosystem via photosynthesis and cellular respiration.
- Nitrogen is removed from the atmosphere by soil micro-organisms undergoing nitrogen fixation and returned to the atmosphere by denitrifying bacteria.
- Carbon, oxygen, and hydrogen are readily available to living organisms, but nitrogen is more difficult to obtain.
- Human activities disrupt biogeochemical cycles.

✓ CHECK YOUR LEARNING

1. Describe the main pathways of the water cycle, including how water enters and leaves the atmosphere. K/U
2. Why do you think biogeochemical cycles are considered sustainable? T/I
3. List the main sources of carbon entering the atmosphere. K/U
4. Explain how the carbon cycle is related to energy flow in ecosystems. K/U
5. In what ways do human activities influence the K/U A
   (a) water cycle?
   (b) carbon cycle?
   (c) nitrogen cycle?
6. In this section, you learned that all body matter is eventually replaced. How does this influence your understanding of how your own body is involved in biogeochemical cycles? A
7. Name two human actions needed to slow climate change. K/U
8. Describe some ways that climate change might influence the water cycle. A
9. Plants need nitrogen to produce proteins and other important chemicals. Describe how nitrogen in the atmosphere makes its way into plants. K/U
10. How is the nitrogen in dead organisms released back into the soil? K/U

2.7 Biotic and Abiotic Influences on Ecosystems

What determines the size of a population and where a particular species can and does live? The answer to this is important if we are to live sustainably in our environment. Ideal biotic and abiotic conditions allow a species to flourish. Other conditions may lead to a species' decline or even extinction. Both abiotic and biotic factors determine where a species can live. A **limiting factor** is any factor that places an upper limit on the size of a population. Limiting factors may be biotic, such as the availability of food, or abiotic, such as access to water. Human influences often act as limiting factors.

limiting factor any factor that restricts the size of a population

Influence of Abiotic Factors

Abiotic factors such as temperature, light, and soil can influence a species' ability to survive. Every species is able to survive within a range of each of these factors. This range is called the species' **tolerance range** (Figure 1). Near the upper and lower limits of the tolerance range, individuals experience stress. This will reduce their health and their rate of growth and reproduction. Within a species' tolerance range is an optimal range, within which the species is best adapted. The largest and healthiest populations of a species will occur when conditons are within the optimal range. Each species has a tolerance range for every abiotic factor.

tolerance range the abiotic conditions within which a species can survive

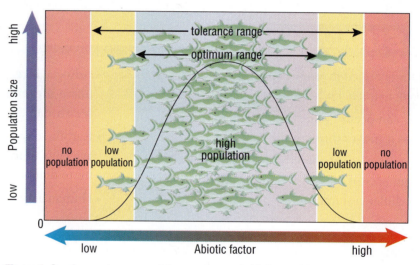

Figure 1 Species can be successful over a range of abiotic conditions. However, they will become stressed and will die out if conditions exceed their tolerance limits.

Some species have wide tolerance ranges, while others have much narrower ranges. Species with broad tolerance ranges will tend to be widely distributed and may easily invade other ecosystems. For example, buckthorn, a small tree native to Europe, has become widespread over much of southern and central Ontario due to its broad tolerance range. Conversely, the showy lady's-slipper orchid has a narrow tolerance range. It is found only in specific types of wetlands.

Chapter 2 • Understanding Ecosystems

The distribution of most terrestrial plant species is largely limited by a combination of temperature, precipitation, and light (Figure 2). The distribution of black spruce in North America, for example, is limited to regions with long, cold winters and moderate precipitation (Figure 3). Other abiotic factors, such as soil type, often have less critical roles.

Figure 2 (a) Cacti can withstand long periods of drought. If overwatered, they may die because their roots cannot survive consistently damp conditions. (b) Aquatic plants, such as water lilies, will perish quickly if the water level drops and the roots are exposed to air. (c) While both cacti and water lilies prefer exposure to full sun, bunchberries are adapted to shade.

The key abiotic factors in aquatic ecosystems are salt concentration and the availability of sunlight, oxygen, and nutrients. Light is abundant in shallow clear water but rapidly decreases with increasing depth. Oxygen concentration is greatest near the water's surface because this is where oxygen enters from the air and where most photosynthesis takes place. Remember that oxygen is released during photosynthesis.

Nutrient availability also varies in aquatic ecosystems. Plants growing in shallow water obtain nutrients directly from the bottom soil. In deeper water, however, the only available nutrients are those dissolved in the water. Table 1 summarizes some of the key abiotic factors and the human activities that can disrupt or alter them.

Figure 3 The distribution of black spruce

Table 1 Key Abiotic Factors of Terrestrial and Aquatic Ecosystems and the Effects Human Activities Can Have on Them

Ecosystem	Key abiotic factors	Human action and result
terrestrial ecosystems	light availability	Clear-cutting and fire remove shade and expose the remaining organisms to much more light.
	water availability	Damming rivers and draining swamps and marshes change water availability. Irrigation increases water availability.
	nutrient availability	Farming practices may increase or decrease nutrient levels in the soil.
	temperature	Global warming is decreasing suitable habitat for many cool-adapted species.
aquatic ecosystems	light availability	Activities that increase erosion or stir up the bottom cloud the water and reduce light penetration.
	nutrient availability	Nutrient runoff from agriculture and urban environments increases the nutrient content of surface water and groundwater, causing algal blooms.
	acidity	Acidic air pollution results in acid precipitation. Carbon dioxide emissions produced by the burning of fossil fuels are increasing the acidity of the oceans.
	temperature	Industries and power plants release heated waste water into lakes and rivers, killing fish and other organisms.
	salinity	Salting highways and long-term irrigation practices can cause salt to accumulate.

Influence of Biotic Factors

While abiotic factors determine where a particular species is able to live, biotic factors often determine the species' success. For example, while deer are able to survive the abiotic conditions in dense forests, they are more abundant in open woodlands. This is where they obtain preferred food species and can watch for predators.

Many key biotic factors involve interactions between individuals. Individuals are often in competition with members of their own species and with other species. They compete for limited resources, such as food, light, space, and mates. For example, a maple tree and a birch tree may compete for sunlight and soil nutrients. Red squirrels compete with each other for pine cones and mates.

Interactions between individuals are not limited to competition. Predation occurs when an individual (the predator) kills and eats another individual (the prey), such as when wolves kill and eat a caribou. Mutualism occurs when two organisms interact, with both benefiting. Lichen, for example, is a mutualistic relationship between algae and fungi (Figures 4(a) and (b)). Parasitism occurs when one organism (the parasite) lives on or in a host and feeds on it. Commensalism occurs when one organism benefits and the other neither benefits nor is harmed (Figure 4(c)). Table 2 lists and describes some common interspecies interactions.

> **DID YOU KNOW?**
>
> **Whose Egg Am I?**
> Cowbirds are brood parasites. They lay their eggs in the nests of other birds. These host birds, such as the eastern phoebe, incubate the eggs and raise the cowbird young.

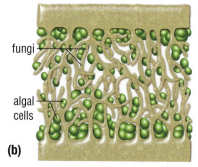

Figure 4 Lichen such as this (a) reindeer lichen is a mutualistic relationship between (b) fungi and algae. Fungi form the structure and absorb nutrients. The algae photosynthesize, providing sugars to the fungi. (c) Barnacles, which are the brown patches on this humpback whale, benefit themselves but neither benefit nor hurt the whale in this commensal relationship.

Table 2 Key Types and Examples of Species Interactions

Relationships	Definitions	Examples
competition	two individuals vie for the same resource	• Foxes and coyotes both feed on common prey such as mice and rabbits. • Humans and insects compete for the same crop plants.
predation	one individual feeds on another	• Lynx prey on snowshoe hares. • Leeches and black flies are "micro-predators" that feed on humans.
mutualism	two individuals benefiting each other	• Nitrogen-fixing bacteria live in the roots of certain plants. The plants provide sugars to the bacteria. The bacteria provide nitrogen to the plant.
parasitism	one individual lives on or in and feeds on a host organism	• Tapeworms are parasites of cats and dogs. • Microbes that cause malaria live within human blood cells.
commensalism	one individual benefits and the other neither benefits nor is harmed	• Many birds nest in particular kinds of trees or in abandoned burrows. • Spanish moss lives on certain tree species.

Carrying Capacity

As a population's size increases, the demand for resources, such as food, water, shelter, and space also increases. Eventually, there will not be enough resources for each individual. Furthermore, as individuals become more crowded, they become more susceptible to predators and diseases. Eventually, these and other factors will result in the population reaching the upper sustainable limit that the ecosystem can support, called the **carrying capacity**.

The carrying capacity can be altered through natural or human activity when resources are removed from or added to the ecosystem. Irrigation can change a desert into a lush oasis because it increases the carrying capacity of the desert (Figure 5). The loss or introduction of a species can change the carrying capacity of the ecosystem for other species in that ecosystem. For example, the removal of wolves by human hunters will increase the carrying capacity of the ecosystem for moose. When new species are introduced, they interact with the original species in many ways. As you will learn, such introductions are among the most serious threats to natural ecosystems.

carrying capacity the maximum population size of a particular species that a given ecosystem can sustain

Figure 5 These green circles, visible from the International Space Station, are formed by an irrigation project in Libya. Irrigation increases the carrying capacity of the desert ecosystem.

UNIT TASK Bookmark

You can apply what you learned in this section about how biotic and abiotic factors limit population size to the Unit Task described on page 156.

IN SUMMARY

- Many factors place limits on the sizes of populations in an ecosystem.
- Tolerance ranges describe the physical conditions under which a species can survive.
- The type of ecosystem that occurs in a particular location is strongly influenced by abiotic factors such as light, water, and temperature.
- Species interactions include competition, predation, mutualism, parasitism, and commensalism.
- Carrying capacity is the maximum population size that an ecosystem can sustain.

CHECK YOUR LEARNING

1. Distinguish between tolerance range and optimal range. K/U
2. List three abiotic factors important to both terrestrial and aquatic ecosystems. Explain your choices. K/U
3. How do human actions increase the carrying capacity of some ecosystems? K/U
4. Give some examples of each of the following: predation, competition, mutualism, parasitism, commensalism. K/U
5. Before reading this section, which types of species–species interactions were you already familiar with? C
6. What species–species relationship or example did you find the most interesting or unusual? C
7. Is it possible to describe abiotic factors as more or less important to an ecosystem than biotic factors? Explain your reasoning. K/U
8. Cedar waxwings are one species of bird that is adapted to withstand our cold winters. Bird watchers in Barrie provide cedar waxwings with seeds at birdfeeders during winter months. K/U T/I
    (a) Would the seeds alter the carrying capacity of the ecosystem? Explain.
    (b) Provide a hypothesis that explains why bird watchers have noted an increase in the falcon population in recent years.

2.7 Biotic and Abiotic Influences on Ecosystems

2.8 Major Terrestrial Ecosystems

Earth's biosphere is home to millions of species. This could lead you to think that there are an unlimited variety of ecosystems composed of different combinations of species. In fact, there are relatively few prominent and easily recognizable types of ecosystems. These prominent types have characteristic features that are observable even without identifying individual species. Deserts, coral reefs, and tropical rainforests have features that are not quickly forgotten.

What is responsible for the occurrence of these characteristic ecosystems and their locations on our planet? Why do we find similar-looking deserts on several continents? Why do coral reefs seem to form in shallow warm waters? Why do tropical rainforests grow in a band around the equator? In this section, you will look at major terrestrial ecosystems. In Section 2.9, you will consider major aquatic ecosystems.

Terrestrial Biomes

The most important factor in determining the location and makeup of a terrestrial ecosystem is climate. On a global scale, the pattern and range of temperature and precipitation cause the establishment of ecologically similar, terrestrial regions called **biomes**. Figure 1 illustrates the relationships between biomes, temperature, and precipitation.

biome a large geographical region defined by climate (precipitation and temperature) with a specific set of biotic and abiotic features

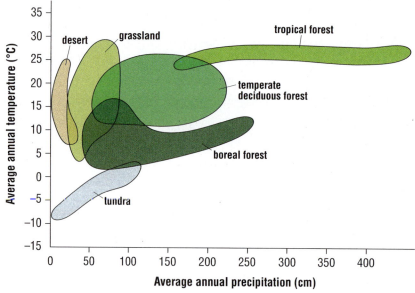

Figure 1 This climatograph shows the influence of precipitation and temperature on biome formation.

The natural landscape in Canada is dominated by four major biomes: tundra, boreal forest, grassland, and temperate deciduous forest. These biomes have characteristic biotic and abiotic features (Figure 2, on the next page). British Columbia contains several smaller biomes, including the mountain forest biome which extends to Alberta, and a narrow strip of temperate rainforest biome along the coast.

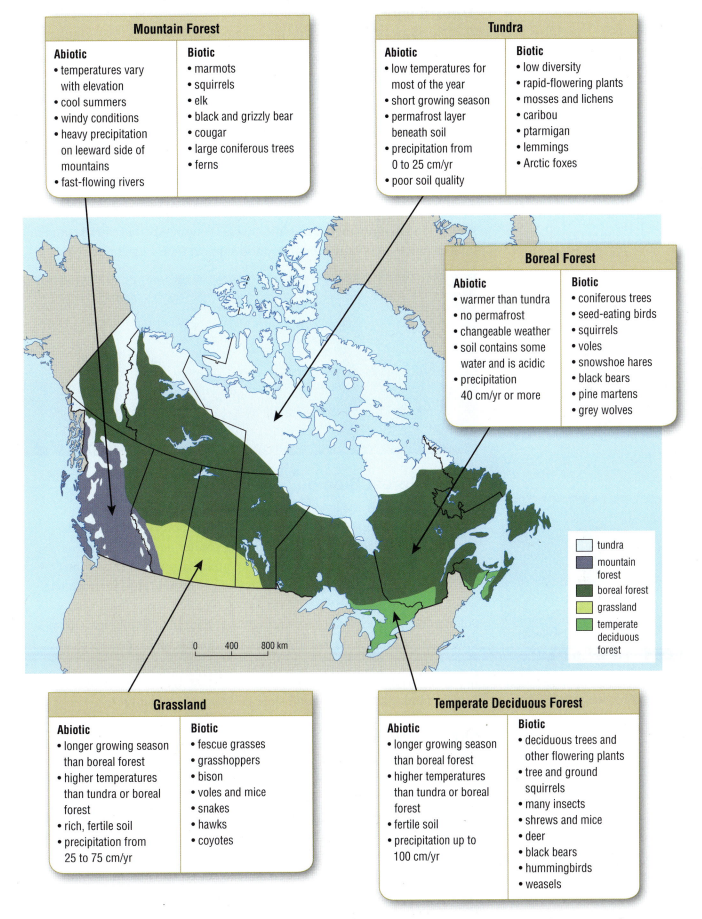

Figure 2 Canada has four major biomes. The mountain forest biome is found in British Columbia and parts of Alberta.

2.8 Major Terrestrial Ecosystems

Tundra Biome

Canada's most northern biome, the tundra, is a cold desert. The extremely short growing season and low temperatures place harsh limits on the kinds of plants that are able to survive here. Because the rate of photosynthesis is reduced, plants in the tundra grow slower than in Canada's other biomes.

Vast regions of tundra have permafrost (permanently frozen ground). During the summer, the soil closest to the surface thaws and forms an active layer in which plant roots can grow. Low soil temperature means that decomposition rates are very slow and nutrients are cycled slowly.

Among the most notable species in the tundra are the barren-ground caribou and the polar bear. The barren-ground caribou travel in large herds feeding on low-growing lichens and mosses (Figure 3). These herds migrate long distances to obtain enough food to support their large body size.

Figure 3 Caribou migrate thousands of kilometres every year in the tundra biome.

Boreal Forest Biome

The boreal forest is the largest biome in Canada. Rainfall and warm summers support the growth of trees. Soil in the boreal forest is acidic because acids are released by decomposing conifer needles. This slows decomposition and limits the variety of plants that grow in this biome.

Conifers are the dominant trees. They can withstand the harsh winters while retaining their needle-shaped leaves. Needles have a thick wax coating that reduces water loss during the winter. In addition, leaves can photosynthesize as soon as the temperature warms up in the spring. The short stature, flexible branches, and conical shape of conifers enable them to both support and shed heavy snow loads (Figure 4). The shaded forest floor is covered in shade-tolerant and slow-growing mosses and ferns.

Figure 4 Conifer branches can support and shed heavy loads of snow.

Grassland Biome

Canada's natural grassland, or prairie, occurs where moderate rainfall supports grasses but cannot support most tree species. The hot, dry summers provide ideal conditions for fires. Fires maintain grassland because they suppress tree growth. The black earth of grassland is among the most fertile of soils in the world. High summer temperatures promote decomposition, which releases nutrients back into the soil.

Canada's grassland biome once extended across much of Manitoba, Saskatchewan, and Alberta. Little of the natural grassland remains (Figure 5, next page). Humans have replaced natural grassland with large fields of crops such as wheat and canola.

The animal communities in grassland have survived in modest numbers. The best known inhabitant of the grassland is the bison, which once numbered in the millions and roamed in vast herds.

DID YOU KNOW?
Driven to the Brink
Bison were almost driven to extinction in the nineteenth century. Massive commercial overhunting decimated this species until only a few small fragmented populations remained. Today, small numbers of bison occur in the wild, while relatively large numbers are raised for meat.

Figure 5 The grassland biome is composed of shrubs, grasses and herbs.

Figure 6 In autumn, the beautiful range of fall colours highlights the variety of this temperate deciduous forest biome.

Temperate Deciduous Forest Biome

The temperate deciduous forest biome is dominated by deciduous trees such as maple, oak, and ash (Figure 6). In this biome, the growing season is longer. Temperatures do not reach the extreme lows found in the boreal region. Decomposition rates are faster.

The deciduous forest plant community is very diverse. It has a layer of canopy trees, understorey trees, shrubs, and non-woody vegetation on the forest floor (Figure 7). The variety of plant life supports a rich variety of animals. Each species is adapted for feeding on or living in a particular portion of the forest.

The climate of this biome has made it attractive to humans. We have replaced large portions of the original temperate forest with farmland, roads, and cities. As you will learn in Section 3.4, these actions have led to the loss of much of the original forest cover and to the division of what remains.

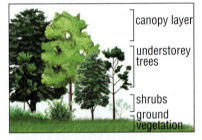

Figure 7 The temperate deciduous forest has distinct layers of tall canopy trees, small understorey trees, shrubs, and ground vegetation.

IN SUMMARY

- Precipitation and temperature are the main abiotic factors influencing biome formation.
- Terrestrial biomes have distinct biotic and abiotic characteristics.
- The five main Canadian biomes are tundra, boreal forest, grassland, temperate deciduous forest, and mountain forest.
- The boreal forest is the largest biome in Canada.

CHECK YOUR LEARNING

1. Order Canada's terrestrial biomes from wettest to driest. Order them from warmest to coldest.
2. Why is fire important to the sustainability of grassland ecosystems?
3. Is the boreal or deciduous forest more diverse? Explain why.
4. Which abiotic factors are most influential in determining what type of biome occurs in a particular region?
5. Describe the terrestrial biome that you live in. Which of the biotic factors listed in Figure 2 have you seen?
6. Explain why conifers are suited to harsh winters in the boreal forest.

2.9 Major Aquatic Ecosystems

Aquatic ecosystems are divided into two broad categories. Freshwater ecosystems have salt concentrations that are typically below 1 %. Ocean, or marine, ecosystems have salt concentrations averaging about 3 %. While the difference in salt concentration may appear small, it has a dramatic influence on the chemical and physical properties of the water. Anyone who has gone swimming in the ocean knows how much more buoyant they are in salt water when compared with fresh water.

Freshwater Ecosystems

Freshwater ecosystems consist of moving bodies of water, such as rivers, and nearly stationary bodies of water, such as lakes. Rivers and streams are unique among ecosystems as they are continuously flushed with a fresh supply of water from upstream. Organisms must either swim continuously against the current or attach themselves to the bottom or some other fixed object (Figure 1).

Figure 1 The water in rivers and streams is constantly moving.

oligotrophic a body of water that is low in nutrients

eutrophic a body of water that is rich in nutrients

Lakes and ponds are classified based on their nutrient levels. **Oligotrophic** bodies of water are low in nutrients (Figure 2). Even with abundant light, photosynthetic organisms have difficulty obtaining enough nutrients to grow. **Eutrophic** bodies of water are high in nutrients. Photosynthetic aquatic plants and algae grow more rapidly and support a large biomass of consumers. Eutrophic bodies of water are often clouded with suspended microscopic plankton.

Figure 2 (a) Oligotrophic lakes are often clear and deep and favoured for swimming and boating. (b) Eutrophic ecosystems have more insects, are murky, have less open water, and may be too shallow for boating.

Wetlands such as bogs and marshes are large areas of shallow water or saturated soils. They are nutrient rich and support large populations of fish, amphibians, insects, and birds. Wetlands act as huge sponges and play a critical role in filtering water in the water cycle.

Watersheds

watershed the land area drained by a particular river; also called a drainage basin

Watersheds are an important characteristic of freshwater ecosystems. A **watershed** is the area of land through which all water drains into a single river or lake. This geographical relationship is important because water always flows downhill. If a pollutant enters a watershed, the areas downstream could become polluted as well. When you manage a water system, you must manage the whole watershed.

Conservation Authorities protect Ontario's watersheds. To learn more about working for a Conservation Authority,

GO TO NELSON SCIENCE

Figure 3 shows the movement of water in a watershed. As you can see, pollutants in the watershed can flow down to other parts of the water system.

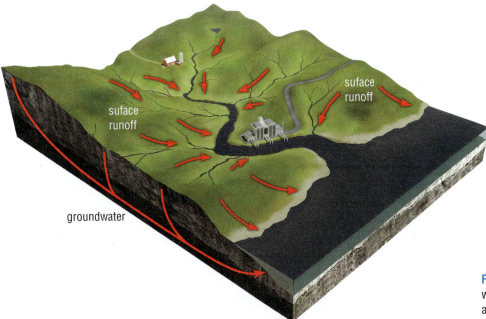

Figure 3 In a watershed, all surface water and groundwater flow downhill and collect into the same river or lake.

Marine Ecosystems

More than 70 % of Earth's surface is covered in oceans. Marine ecosystems are an important part of biogeochemical cycles. Most of the water that evaporates into the air and falls as rain and snow comes from oceans. Marine algae play a critical role in the production of oxygen and the absorption of carbon dioxide gas from the atmosphere.

Much of the ocean supports very little life. The open ocean is nutrient poor and unable to support many photosynthesizing organisms. The deep ocean is a lightless environment, so photosynthesis is impossible. In contrast, the shallow waters near shore are nutrient rich and support abundant life.

Coral reefs develop in warm shallow oceans and support a huge variety of organisms. They are home to commercially valuable species and support a large tourist industry. Coral reefs are extremely sensitive to changes in water temperature, acidity, and pollution.

Estuaries are partially enclosed bodies of water where fresh and salt water mix. They are high in nutrients and often support valuable shellfish, such as clams and scallops. The Gulf of St. Lawrence is the world's largest estuary (Figure 4). Historically, the Gulf supported an important fishery for Aboriginal people. Today, the Gulf supports commercial fishing and is a major shipping route for entry into the Great Lakes basin.

Mangroves are unusual communities that occur along tropical and semitropical sandy shorelines. They contain specialized tree species adapted to live at and beyond the water's edge (Figure 5). The prop roots of the mangroves grow out into the water. This reduces coastline erosion and creates a habitat for several marine organisms. Shoreline developments have destroyed much of the mangrove habitat. This has increased storm damage to shorelines. Replanting efforts in some regions are underway to re-establish this valuable ecosystem.

Figure 4 Pilot whales in the Gulf of St. Lawrence feed primarily on squid and small fishes.

Figure 5 Mangroves protect coastlines from erosion and are refuges for immature individuals of many species.

> **READING TIP**
>
> **Connections and Responses**
> Make a two-column chart labelled "Connections" and "Responses." Use the first column to make text-to-self, text-to-text, and text-to-world connections. In the second column, explain how the connections helped you to better understand the text by visualizing, making predictions, inferences, or judgments, forming opinions, drawing and supporting conclusions, or evaluating it.

The Intertidal Zone

Ocean coastlines are ecosystems that are part-time terrestrial and part-time aquatic. They are home to the unusual communities that occupy the intertidal zone—the area between the low-tide and high-tide lines. Many coastlines exhibit a significant change in water levels approximately four times a day, with two periods of high tide interceded by two periods of low tide. The highest tides in the world occur in the Bay of Fundy, with differences of up to 17 m between low and high tide. Among the most common species that inhabit this unusual environment are seaweeds, barnacles, sea stars, and urchins (Figure 6).

Figure 6 Species that inhabit the intertidal zone like these sea stars and sea anemones must be able to withstand extremely variable conditions, including hot and freezing temperatures and harsh wave action.

Imagine the conditions faced by organisms in the intertidal zone on the coast of Nova Scotia. During the summer they alternate between a life under water and above water where they are exposed to high temperatures, full sun, and strong drying winds. During the winter they alternate between icy ocean water and exposure to bitterly cold air, snow, and ice. Throughout the year, these species must also be able to withstand the daily pounding of wave action. It is not surprising that many intertidal species have protective body coatings and very tough tissues.

IN SUMMARY

- Aquatic ecosystems can be freshwater (for example, lakes and streams) or marine (for example, coral reefs and estuaries).
- The most productive aquatic ecosystems occur in relatively shallow, warm, and nutrient-rich waters.
- The abiotic conditions of intertidal zones continuously alternate between those of a terrestrial and those of an aquatic ecosystem.
- Species living in the intertidal zone are able to survive highly variable abiotic conditions.

CHECK YOUR LEARNING

1. Why do the open oceans not support rich ecosystems and large numbers of fish? K/U
2. What are the most important abiotic features of aquatic ecosystems? K/U
3. Would you rather swim in an oligotrophic lake or a eutrophic lake? Explain why. K/U C
4. What unique challenges might a scientist face when trying to understand coral reefs? A
5. Describe the abiotic conditions of the intertidal zone and how they change. K/U
6. Explain why species living in the intertidal zone have to be unusually tough. K/U
7. The intertidal zone may be sandy, muddy, or rocky. How do you think these conditions might influence what species can live there? T/I

TECH CONNECT — OSSLT

Whales, Darts, and DNA

If you have watched crime shows on television, you will know that criminal cases are now routinely solved using DNA fingerprinting technology. This same technology is now being used to save endangered species of whales.

Scott Baker has a love of science and whales. For years, he studied endangered humpback whales in the Pacific Ocean and documented their population structure and migration patterns (Figure 1). Baker was puzzled because even after a total ban on commercial whaling was announced in 1986, the humpbacks were not recovering. He knew that Japan allowed the sale of whale meat from non-protected species. Baker suspected that endangered whale species were being illegally killed and sold in Japanese fish markets.

Figure 1 At maturity, humpback whales are about 15 m long and weigh about 36 000 kg. They eat krill and small fish.

Collecting Data

Baker needed to gather detailed scientific evidence to document his concerns. He invented a new method of doing that. He designed special darts and a crossbow to remove small samples of flesh from living whales. When fired, the darts hit the whale and bounced off, removing a small tissue sample. He used the samples to produce a DNA fingerprint for each whale.

Over a number of years, Baker gathered DNA samples from whale species and different populations. He compiled a library of their DNA information. Using sophisticated new technology, he was able to identify the species and geographical population from which any sample came.

Fighting Illegal Whaling

In 1993, Scott Baker made a trip to Tokyo. He obtained suspected samples of whale meat being sold at fish markets and then used his hotel room as a lab to determine the species the sample came from (Figure 2). His results confirmed that many of the samples were from highly endangered whale species.

Figure 2 Dr. Scott Baker used his hotel room as a laboratory to identify whale samples.

Baker's efforts led to clamping down on illegal whaling. A new program of random spot-checking of whale meat was started to ensure that protected species are not being harvested and falsely labelled (Figure 3).

Figure 3 Whale meat is still sold in Japanese supermarkets.

2.10 PERFORM AN ACTIVITY

Ecosystem Field Study

Ecosystems are characterized by a particular set of abiotic and biotic factors. In this field study, you will be part of a group that visits a local terrestrial ecosystem and document its abiotic and biotic features.

SKILLS MENU
- Questioning
- Hypothesizing
- Predicting
- Planning
- Controlling Variables
- **Performing**
- **Observing**
- Analyzing
- Evaluating
- **Communicating**

Purpose
To visit a natural or artificial ecosystem and collect data on the biotic and abiotic features.

Equipment and Materials
Note that you may not use all of the equipment listed below. Your teacher will tell you which pieces of equipment you will be using and demonstrate their proper use.
- compass and whistle
- map of field study site
- clipboard
- data recording sheets and field notebook
- anemometer (for measuring wind speed)
- thermometer
- soil thermometer
- sling psychrometer or hygrometer (to measure relative humidity)
- field guides to local flora and fauna
- digital camera
- binoculars
- sweep nets
- forceps
- magnifying glass or bug box
- gloves

Take precautions if you have any particular allergies or reactions, such as to bee stings. Avoid poison ivy (Figure 1).

Take a whistle with you in case you become disoriented.

Figure 1 Poison ivy. Beware!

Procedure

SKILLS HANDBOOK
3.B.6.

Part A: Field Work—Abiotic Features

1. Spend 5 or 10 min walking through your ecosystem to familiarize yourself with its features.

2. If you do not have a map, make a rough outline sketch of your ecosystem indicating major features such as large trees, pathways, roads, and buildings.

3. Choose a location that seems representative of your ecosystem. Mark its location on your map or sketch.

4. Make and record measurements of each of the following abiotic features: T/I
   - wind speed
   - temperature at ground level and at 1.5 m above the ground

- soil temperature
- relative humidity using a sling psychrometer or hygrometer
- percent canopy cover estimate as 0 %, 25 %, 50 %, 75 %, or 100 % (for use in comparing relative light availability)

Part B: Field Work—Biotic Features

5. Spend the remainder of your allotted time observing and documenting what species live in your area. Use field guides to try to identify each species. If you cannot identify some species, just record them as "unknown #1, unknown #2," and so on. Write a brief description of each species in your field notebook. The purpose of this activity is to document and compare the variety of organisms present, not their names or abundance. While you are encouraged to learn the names of these species, their function in the ecosystem does not depend on people knowing their names. If available, use a digital camera to document as many species as you can. **C**

   The following techniques should be used to find and observe life forms: **T/I**

   (i) Use binoculars to observe as many bird and mammal species as you can.

   (ii) Identify and record as many plant species are you can. Be careful not to overlook small plants. Take note of any fungi you see.

   (iii) Closely observe the surfaces of leaves, bark, and the ground for invertebrates. If available, use sweep nets to capture insects on vegetation and in the air. Use forceps and a magnifying glass or a bug box to have a closer view.

   (iv) Carefully look under logs or other objects on the ground in search of salamanders and other organisms. Be sure to return any moved object to where you found it. Handle all animals with gloves. Use great care because they are very easily injured. Return all organisms to where you found them.

   (v) If a small stream or pond is present, look along the water's edge for additional plants and animals.

Part C: Back at School—Ecosystem Show and Tell

6. Working with your group, prepare a "show and tell" audiovisual presentation of your experiences in the ecosystem. Your presentation should portray the diversity of living things. Provide evidence of the abiotic characteristics at that location. Your presentation will be assessed for
   - thoroughness in highlighting both the abiotic and the biotic features of the ecosystem
   - creativity and originality
   - evidence of effective group work and cooperation **T/I** **C**

Analyze and Evaluate

(a) How did the variety and size of producers compare with the variety and size of consumers in your ecosystem? **T/I**

(b) How does your understanding of energy and food pyramids support and explain this observation? **T/I**

(c) What kinds of producers seemed to dominate your ecosystem? **T/I**

(d) Make a food web of the organisms in your ecosystem. **C**

(e) Describe evidence of human influence or intervention in your ecosystem. **A**

(f) If you were able to compare data from two ecosystems, write two summary paragraphs comparing their abiotic and biotic features. **T/I** **C**

Apply and Extend

(g) Is there evidence that this ecosystem has remained unchanged for the last 10 years or 100 years? Explain your answer. **T/I**

(h) Would you expect this ecosystem to look the same in 10 years? Why or why not? What about in 100 years? **T/I**

CHAPTER 2 LOOKING BACK

KEY CONCEPTS SUMMARY

Life on Earth exists in the atmosphere, lithosphere, and hydrosphere.

- Earth's atmosphere is made up of nitrogen gas (78 %), oxygen gas (21 %), and other gases. (2.1)
- The atmosphere acts like a blanket keeping Earth warm and blocking harmful radiation. (2.1)
- The biosphere is the region on Earth where life can exist within the lithosphere, atmosphere, and hydrosphere. (2.1)

Photosynthesis and cellular respiration are complementary processes in an ecosystem.

- During photosynthesis, green plants convert carbon dioxide and water into sugar (with stored chemical energy) and oxygen. (2.4)
- During cellular respiration, sugar and oxygen are converted into carbon dioxide, water, and energy. (2.4)
- Producers and consumers undergo cellular respiration. Only producers photosynthesize. (2.4)

Energy passes through ecosystems, whereas matter cycles within ecosystems.

- Energy passes through ecosystems by means of food webs. (2.5)
- Energy is continuously being lost to the environment resulting in less available energy from one trophic level to the next. (2.5)
- Food webs and ecological pyramids display how energy flows through ecosystems from one trophic level to the next. (2.5)

Human activities influence biogeochemical cycles such as the water and carbon cycles.

- In biogeochemical cycles, matter is continuously consumed, rearranged, stored, and used while moving between the biotic and the abiotic environment. (2.6)
- Carbon is stored for long periods of time in carbon sinks and deposits. (2.6)
- Burning fossil fuels and deforestation release carbon dioxide into the atmosphere. (2.6)
- Humans influence the water cycle when they construct dams and use irrigation systems. (2.7)

Ecosystems are composed of biotic and abiotic components.

- Ecosystems are characterized by their biotic and abiotic features. (2.2, 2.7)
- Natural ecosystems are generally sustainable, maintaining a relatively constant set of biotic and abiotic characteristics. (2.2)
- Artificially created ecosystems are not usually sustainable. (2.2)
- Species interactions include the following: competition, predation, mutualism, parasitism, and commensalism. (2.7)
- The carrying capacity is the maximum number of individuals of a species that an ecosystem can support. (2.7)

Terrestrial biomes and aquatic ecosystems are largely determined by their abiotic characteristics.

- Terrestrial biomes are largely determined by temperature and precipitation patterns. (2.8)
- Canada's major biomes are tundra, boreal forest, grassland, and temperate deciduous forest, and mountain forest. (2.8)
- Aquatic ecosystems are determined by factors such as salt content, light, nutrient and oxygen availability, and the movement of water. (2.9)

WHAT DO YOU THINK NOW?

You thought about the following statements at the beginning of the chapter. You may have encountered these ideas in school, at home, or in the world around you. Consider them again and decide whether you agree or disagree with each one.

1 Oceans make up the majority of Earth's mass.
Agree/disagree?

4 All organisms are helpful in the environment.
Agree/disagree?

2 All the particles that make up your body are being continuously replaced by new ones.
Agree/disagree?

5 Humans are one of the few species that do not compete with other species.
Agree/disagree?

3 Animals need plants for food, but plants do not need animals.
Agree/disagree?

6 Humans have been successful because they are able to change the natural environment.
Agree/disagree?

How have your answers changed since then?
What new understanding do you have?

Vocabulary

atmosphere (p. 29)
lithosphere (p. 30)
hydrosphere (p. 30)
biosphere (p. 30)
ecosystem (p. 32)
biotic factors (p. 32)
abiotic factors (p. 32)
sustainable ecosystem (p. 34)
sustainability (p. 35)
radiant energy (p. 38)
light energy (p. 38)
thermal energy (p. 38)
photosynthesis (p. 38)
producer (p. 38)
cellular respiration (p. 40)
consumer (p. 41)
ecological niche (p. 42)
food chain (p. 43)
trophic level (p. 43)
food web (p. 43)
ecological pyramid (p. 45)
biomass (p. 46)
biogeochemical cycle (p. 48)
water cycle (p. 48)
carbon cycle (p. 49)
nitrogen cycle (p. 50)
limiting factor (p. 52)
tolerance range (p. 52)
carrying capacity (p. 55)
biome (p. 56)
oligotrophic (p. 60)
eutrophic (p. 60)
watershed (p. 60)

BIG Ideas

✓ Ecosystems are dynamic and have the ability to respond to change, within limits, while maintaining their ecological balance.

• People have the responsibility to regulate their impact on the sustainability of ecosystems in order to preserve them for future generations.

CHAPTER 2 REVIEW

The following icons indicate the Achievement Chart category addressed by each question.

K/U Knowledge/Understanding T/I Thinking/Investigation C Communication A Application

What Do You Remember?

1. In your notebook, write the word(s) needed to complete each of the following sentences. K/U

   (a) The solid part of Earth's surface is called the _____. (2.1)
   (b) Oxygen is required by almost all organisms for the process of _____ and is a by-product of _____. (2.4)
   (c) The _____ refers to all water on Earth in the solid, liquid, and gas states. (2.1)
   (d) An ecosystem is _____ if it can continue to function over very long periods of time. (2.2)
   (e) The population that an ecosystem can support continuously is called its _____. (2.7)
   (f) _____ ecosystems are another name for salty ocean ecosystems. (2.9)
   (g) A(n) _____ describes a community of living things and their surrounding physical environment. (2.2)
   (h) Most of the light that reaches Earth's surface is absorbed by the _____ and the _____. This absorbed light energy is then converted into _____ energy and warms Earth's surface. (2.4)
   (i) A very small fraction of the light striking Earth is absorbed by living organisms in a process called _____. (2.4)

2. Why are decomposers important in the carbon and nitrogen cycles? (2.6) K/U

3. Compare photosynthesis and cellular respiration using the following criteria. (2.4) K/U

   (a) What raw materials are needed?
   (b) What are the products?
   (c) Which occurs in plants?
   (d) Which occurs in animals?
   (e) Is light needed?
   (f) Is energy released?
   (g) Is energy needed?
   (h) Is chlorophyll needed?

4. Why is sunlight important for the biosphere? (2.4) K/U

5. In your own words, explain what is meant by the term "trophic level." (2.5) K/U

What Do You Understand?

6. Match the term on the left with the appropriate example on the right. (2.7) K/U

   (a) mutualism   (i) burrowing owls nest in abandoned ground squirrel tunnels
   (b) parasitism   (ii) robins feed heavily on earthworms
   (c) commensalism   (iii) fruit-eating bats feed on figs and spread their seeds
   (d) competition   (iv) the winter tick lives and feeds on moose
   (e) predation   (v) overcrowding can lead to food shortages and starvation

7. Explain the difference between a food chain and a food web. (2.5) K/U

8. In your own words, define oligotrophic and eutrophic. (2.9) K/U

9. List four biotic and four abiotic features you would expect to find in
   (a) a freshwater stream
   (b) an open grassland
   (c) the Arctic tundra (2.8, 2.9) K/U

10. In what ways might individuals of the same species compete with each other if they are (a) herbivores (b) carnivores (c) producers (2.7) K/U

11. If energy is always being lost from organisms in food webs, where does new energy come from? (2.4, 2.5) K/U

12. Use the following data to draw a pyramid of numbers showing each trophic level. (2.5) C

    An ecosystem contains 100 000 grass plants, 30 000 grasshoppers, 5000 snails, 4000 slugs, 80 shrews, 15 moles, and 8 owls. Grasshoppers, snail, and slugs are all herbivores. Shrews and moles are carnivores. Owls are top carnivores.

13. Many animal foods like meat and fat are high in energy content. Where did this energy originate? Explain by referring to food chains. (2.4, 2.5) K/U

Use Figure 1 to answer questions 14 to 18.

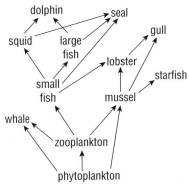

Figure 1 Food web for an estuary ecosystem

14. Classify each organism as a producer, herbivore, carnivore, or omnivore. (2.5) K/U

15. Construct a trophic level pyramid, assigning each organism to a trophic level. (2.5) K/U

16. Which organisms within the food web would be directly affected by the over-harvesting of large fish? (2.5) K/U

17. Predict how the elimination of whales would affect the populations of starfish and squid. (2.5) K/U

18. If the seal population increased significantly, how would other populations be affected? (2.5) K/U

19. Rank the tundra, boreal forest, deciduous forest, and grassland biomes in descending order according to each of the following abiotic factors. Explain your reasoning. (2.8) K/U

    (a) amount of precipitation
    (b) average temperature
    (c) length of growing season
    (d) biodiversity
    (e) total biomass

20. Suggest a reason that carbon cycles more slowly through the tundra biome than through the temperate deciduous forest biome. (2.6, 2.8) T/I

Solve a Problem

21. Use Figure 2 on page 57 of biomes to complete Table 1. (2.8) K/U

    Table 1

Precipitation	Temperature	Possible biomes
low	low	
medium	medium	
high	high	

22. If a place has 35–50 cm of precipitation and an average temperature of 10 °C, what biome might it be part of? Why is it difficult to tell? (2.8) K/U

23. Recall the Engage in Science reading on page 26. How has the introduction of yellow crazy ants upset the ecosystems on Christmas Island? Are only red crabs affected? Explain. T/I

Create and Evaluate

24. As more and more cottages were built around a small lake, people noticed that there was a buildup of green algae in the spring and the water was not as clear as it used to be. They also noticed that fishing was poor. What could account for these environmental changes? (2.7) A

Reflect on Your Learning

25. (a) What information in this chapter did you already know before reading it?
    (b) What information in this chapter was completely new to you?
    (c) How might the new information that you learned affect how you think about Earth? C

26. Think about the ecosystems in your region. What species live there? How do they interact with each other and with you? C

Web Connections

27. Two subspecies of bison live in Canada—the wood bison and the plains bison. Conduct Internet research to determine the historical and current status of these two subspecies. Which has the smallest and most vulnerable population? How are humans responsible for the decline in their numbers? (2.7) K/U T/I

28. Using a map, identify freshwater ecosystems in your area. Conduct Internet research to determine whether or not they are considered to be healthy ecosystems. (2.4–2.7, 2.9) T/I A

CHAPTER 2

SELF-QUIZ

The following icons indicate the Achievement Chart category addressed by each question.

K/U Knowledge/Understanding
T/I Thinking/Investigation
C Communication
A Application

For each question, select the best answer from the four alternatives.

1. Which of these spheres is made up of parts of the other three spheres? (2.1) K/U
   (a) atmosphere
   (b) biosphere
   (c) hydrosphere
   (d) lithosphere

2. Which of Canada's major biomes has the shortest growing season? (2.8) K/U
   (a) tundra
   (b) grassland
   (c) boreal forest
   (d) temperate deciduous forest

3. Which trophic level is made up of herbivores? (2.5) K/U
   (a) producers
   (b) primary consumers
   (c) secondary consumers
   (d) tertiary consumers

4. Which of the following is a key abiotic factor for aquatic ecosystems but not for terrestrial ecosystems? (2.9) K/U
   (a) light
   (b) nutrients
   (c) salinity
   (d) temperature

5. Which term applies to the relationship between two organisms of different species that live together, when both benefit from the relationship? (2.7) K/U
   (a) commensalism
   (b) competition
   (c) mutualism
   (d) predation

6. Which of these factors in a pond ecosystem is a biotic factor? (2.7, 2.9) K/U
   (a) bacteria
   (b) sediment
   (c) temperature
   (d) water

Indicate whether each of the statements is TRUE or FALSE. If you think the statement is false, rewrite it to make it true.

7. Artificial ecosystems, such as city parks and farm fields, are usually sustainable ecosystems. (2.2) K/U

8. Both matter and energy are recycled in an ecosystem. (2.4, 2.6) K/U

9. The Moon has no atmosphere because the force of its gravity is too weak. (2.1) K/U

Copy each of the following statements into your notebook. Fill in the blanks with a word or phrase that correctly completes the sentence.

10. Radiant energy from the Sun enters ecosystems through the process of _____ . (2.4) K/U

11. The word equation for cellular respiration shown here is incomplete.

    sugar + ? → carbon dioxide + water + energy

    The equation will be complete if the question mark is replaced by the word _____ . (2.4) K/U

12. The term population refers to all the individuals of the same _____ living in an ecosystem. (2.2) K/U

Match each term on the left with the most appropriate description on the right.

13. (a) carnivore   (i) eats only plants
    (b) omnivore   (ii) eats only animals
    (c) scavenger   (iii) eats both plants and animals
    (d) herbivore   (iv) eats remains of other organisms (2.5) K/U

Write a short answer to each of these questions.

14. Plants appeared on Earth long before animals. What abiotic factor did plants add to the environment that made animal life possible? (2.4) K/U

15. You are planning an overnight camping trip in the desert. (2.2) T/I
    (a) What are two biotic and two abiotic factors that you may encounter on your trip?
    (b) Describe equipment and supplies that will help you adapt to the factors you named in part (a).

16. Draw a diagram showing the path of energy through an ecosystem from solar radiation to producers to consumers and back to the environment. At each stage, identify the form the energy takes. (2.4, 2.5) C

17. A partial food web is shown below in Figure 1. The mice in this food web are infected by a disease that greatly reduces their population. (2.5) T/I

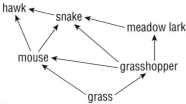

Figure 1

(a) How will the decrease in the mouse population affect the grasshopper population? Explain your answer.
(b) Considering your answer to part (a), how will the decrease in mouse population affect the meadow lark population? Explain your answer.

18. Imagine you have built a dam to create a small pond on your property. You want to create an artificial ecosystem by stocking the pond with two species of fish that occupy different trophic levels. What information would you need to gather about aquatic ecosystems before you buy the fish to put in your pond? (2.9) A

19. You want to study your personal role in the carbon cycle, as an organism that is part of that cycle. Copy and complete Table 1 below to show this information. (2.6) T/I

Table 1

Sources of carbon	Destinations of carbon	Where carbon goes next

20. A friend asks you why commercial fishing boats operate close to shorelines in relatively shallow waters, rather than in deep water or in the middle of the ocean. Explain to your friend why this is so. (2.9) C

21. Carbon, oxygen, and hydrogen are readily available to living organisms, but nitrogen is more difficult to obtain. Explain why this statement is true. (2.6) K/U T/I

22. Corn is a high-protein crop that requires large amounts of nitrogen in the soil to grow well. How might planting a legume, such as soybean, in the same field the previous year improve the growing conditions for corn? (2.6) T/I A

23. You are planning a study of the water cycle in the area where you live. You want to see how water moves from place to place and determine the factors that affect the relative amounts in each place. Describe how you would use data on the following factors, recorded over a ten-year period: (2.6) T/I

   - annual precipitation
   - temperatures
   - levels of lakes, reservoirs, and the water table
   - volume of water taken from various sources for home and commercial use.

24. For the survival of any species, several abiotic factors are required, each of which needs to be within an appropriate range. The optimum range for each factor is the range in which the species thrives best. For the human population on Earth, discuss the range of abiotic factors in which human life can survive. Consider the whole planet as our ecosystem. (2.7) T/I A

   (a) How does the temperature range for human survival compare with the temperature range for survival of most animal species? Justify your answer.
   (b) Select three abiotic factors required by humans, and describe the optimum range of each factor for human survival.
   (c) Give examples of two groups of people who live near the edge of the survival range of two different abiotic factors.

25. You are planning a report on Ontario's two major biomes: boreal forest and temperate deciduous forest. You want to describe each biome in such a way that it cannot be confused with the other. (2.8) K/U A

   (a) Identify two biotic factors that are unique to each biome.
   (b) Identify two abiotic factors that are unique to each biome.

26. (a) Give an example of two organisms whose relationship is one of mutualism. Explain why the relationship is mutual.
    (b) Give an example of two organisms whose relationship is one of commensalism. Explain why the relationship is commensal. (2.7) A

CHAPTER 3

Natural Ecosystems and Stewardship

KEY QUESTION: Why should we value and sustain natural ecosystems?

Kelp forests are highly productive ecosystems that occur in cold, coastal waters.

UNIT B
Sustainable Ecosystems

- **CHAPTER 2** — Understanding Ecosystems
- **CHAPTER 3** — Natural Ecosystems and Stewardship
- **CHAPTER 4** — Ecosystems by Design

KEY CONCEPTS

Natural ecosystems are of great value to humans.

Ecosystems are at equilibrium but can change over time.

Biodiversity describes the variety and abundance of life in an ecosystem.

Many human activities impact and threaten the sustainability of natural ecosystems.

Water, land, and air pollution cause health and economic problems.

Plant and animal resources should be used in a sustainable manner.

ENGAGE IN SCIENCE

SHARK ATTACK!
OVERFISHING LARGE SHARKS IMPACTS MARINE ECOSYSTEM

Tripti and her brother Nico were excited to begin their vacation on a Caribbean island. They knew that a large coral reef existed a short distance from shore. They were anxious to explore the area for sea life. To their disappointment, large patches of the reef were covered with a thick mat of green algae. Many of the corals were dead or unhealthy. The large schools of colourful fish they had expected to see were absent.

Over the next several days, Tripti and Nico discussed their observations with local divers and people selling fish in the local market. They learned that overfishing in the region had reduced the populations of large sharks by more than 95 % (Figure 1). The loss of these top predators had resulted in rapid increases in populations of their favourite prey species, including groupers. The increase in grouper numbers had caused a decline in their primary prey species, parrotfish.

The loss of the parrotfish was particularly devastating to the reef ecosystem. Parrotfish are the main herbivores on the reef, feeding almost exclusively on algae (Figure 2). The removal of most of the parrotfish resulted in the rapid growth of algae mats that smothered the corals and killed them.

Tripti and Nico learned how far-reaching the overexploitation of a living resource can be. By removing the top predators, the entire reef ecosystem had been damaged. In small groups, discuss the long-term impacts this will have on local tourism and the fishing-based economy. Can the reef recover?

Figure 1 Aided by their widely spaced eyes, hammerhead sharks are aggressive predators. They live for 20 to 30 years and are up to 6 m at maturity.

Figure 2 Parrotfish use their rasp-like teeth to extract algae from the coral.

WHAT DO YOU THINK?

Many of the ideas you will explore in this chapter are ideas that you have already encountered. You may have encountered these ideas in school, at home, or in the world around you. Not all of the following statements are true. Consider each statement and decide whether you agree or disagree with it.

1 Forest fires benefit these species.
Agree/disagree?

4 Human demand for safe, clean water is a major concern on a global scale but not an issue in Canada.
Agree/disagree?

2 Pollution is the greatest human-caused threat to natural ecosystems.
Agree/disagree?

5 Ontario has a healthy wildlife population with very few "at-risk species."
Agree/disagree?

3 Lake trout and a panther grouper are able to share the same habitat.
Agree/disagree?

6 Humans enjoy natural ecosystems, but we do not rely on them in our daily lives.
Agree/disagree?

FOCUS ON WRITING

Writing to Describe and Explain Observations

When you write to describe and explain observations, you record characteristics you used your senses to observe, such as colour, form, and odour, and features you used instruments to measure, such as area, volume, and depth. Focus on writing your observations very clearly and accurately. Use the following strategies to improve your writing.

- Describe sensory (qualitative) observations clearly by using concrete nouns, adjectives, verbs, and adverbs.
- Record measurements (quantitative observations) accurately.
- Use short sentences and precise wording that readers can understand easily.
- Use symbols and scientific terminology correctly.

The following is an example of a piece of writing that describes and explains observations. Beside it are the strategies the student used to write effectively.

WRITING TIP

As you work through the chapter, look for tips like this. They will help you develop literacy strategies.

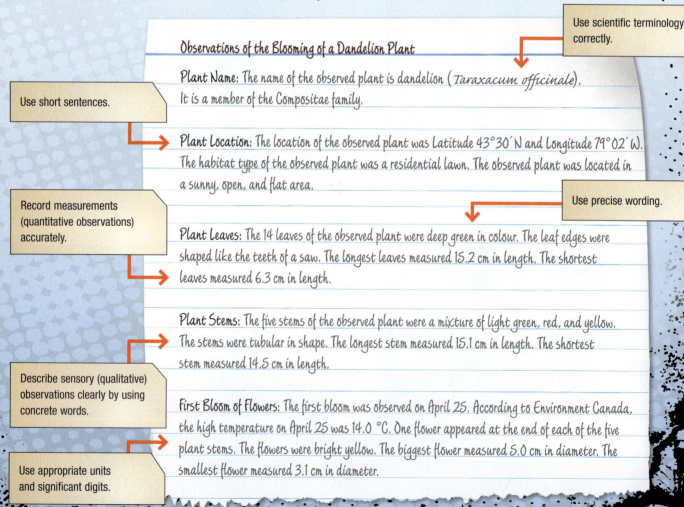

Use short sentences.

Record measurements (quantitative observations) accurately.

Describe sensory (qualitative) observations clearly by using concrete words.

Use appropriate units and significant digits.

Use scientific terminology correctly.

Use precise wording.

Observations of the Blooming of a Dandelion Plant

Plant Name: The name of the observed plant is dandelion (*Taraxacum officinale*). It is a member of the Compositae family.

Plant Location: The location of the observed plant was Latitude 43° 30′ N and Longitude 79° 02′ W. The habitat type of the observed plant was a residential lawn. The observed plant was located in a sunny, open, and flat area.

Plant Leaves: The 14 leaves of the observed plant were deep green in colour. The leaf edges were shaped like the teeth of a saw. The longest leaves measured 15.2 cm in length. The shortest leaves measured 6.3 cm in length.

Plant Stems: The five stems of the observed plant were a mixture of light green, red, and yellow. The stems were tubular in shape. The longest stem measured 15.1 cm in length. The shortest stem measured 14.5 cm in length.

First Bloom of Flowers: The first bloom was observed on April 25. According to Environment Canada, the high temperature on April 25 was 14.0 °C. One flower appeared at the end of each of the five plant stems. The flowers were bright yellow. The biggest flower measured 5.0 cm in diameter. The smallest flower measured 3.1 cm in diameter.

Services from Natural Ecosystems

3.1

On any Friday afternoon, long lines of traffic head out of urban centres (Figure 1). Many people head to the cottage, the beach, or the ski trails to get away from the hustle and bustle of their city. They are heading to peaceful places where they can interact with the natural environment. Humans value natural spaces and often go to great lengths to explore them (Figure 2).

Figure 1 On most weekends, people leave cities to experience the natural environment.

Figure 2 People value the beauty and calm of natural ecosystems.

While Tripti and Nico were on their Caribbean vacation, they benefited from being in a natural ecosystem. What is it about natural spaces and wilderness that draws people with such force? What is the value of an old growth forest, a field of wildflowers, or a coral reef? What is the value of a species that is living in the wild and not used directly by humans? Is there harm in replacing a natural ecosystem with farmland or a housing development?

Ecosystem Services and Products

Natural ecosystems provide services to the biosphere and to humans. These services are the benefits that we receive from ecosystems.

Cultural Services

Cultural services are the benefits relating to our enjoyment of the environment. They include the recreational, aesthetic, and spiritual experiences we receive when we interact with our natural surroundings. Humans get great pleasure from being in natural environments. The environment can be enjoyed by people with different interests. You may prefer wilderness backpacking, while your best friend may prefer a stroll on a beach.

Ecotourism is one example of a cultural service provided by ecosystems. Ecotourists engage in environmentally responsible travel to relatively undisturbed natural areas. An ecotourist tries to leave an area the same as he or she found it, leaving nothing behind and taking nothing away. Canada's wilderness areas are a major destination for ecotourists, so ecotourism benefits our economy with minimal negative impacts on our natural ecosystems.

> **WRITING TIP**
>
> **Use Precise Wording**
> Choose words carefully to describe your observations as accurately as possible. Ask yourself if the words you are using help you to visualize your observation clearly. If your words are vague rather than concrete, your mental picture will be fuzzy rather than clear.

To learn about becoming an ecotour guide,
GO TO NELSON SCIENCE

To learn more about ecotourism,
GO TO NELSON SCIENCE

Ecosystem Products

Humans use products produced by ecosystems. We hunt animals and harvest plants for personal and commercial use (Figure 3). Lakes and oceans supply us with seafood. On a smaller scale, anyone who has gone wild blueberry picking knows the value supplied by these ecosystems. Terrestrial ecosystems are the source of many products, such as medicines, fibres, rubber, and dyes (Table 1).

Figure 3 Humans obtain many food products from the environment, such as (a) fish and (b) wild blueberries. (c) We are also able to obtain materials such as wood.

Table 1 Important Products from Terrestrial Ecosystems

Product	Original source	Use
maple syrup	maple trees	food, sweetener, flavouring
henna, indigo	plant extracts	dyes
latex (rubber) and chicle	assorted tropical trees	hoses, tires, chewing gum, golf balls
acetylsalicylic acid (ASA) (aspirin)	willow tree (original source of salicylic acid)	treats pain, blood thinner
waxes: carnauba, jojoba	carnauba palm leaves, jojoba seeds	commercial wax products, cosmetics, foods, lubricants
vincristine and vinblastine	rosy periwinkle (tropical flowering plant)	treatment of childhood leukemia
digitalis	foxglove (flowering plant)	treatment of heart disorders

Forestry is one of the largest industries and employers in Ontario. It produces over $11 billion worth of products annually. Some trees are grown on tree farms. However, the industry depends on natural ecosystems for most of its wood and wood fibre products.

Other Ecosystem Services

Ecosystems regulate and maintain many important abiotic and biotic features of the environment. As you learned in Chapter 2, ecosystems cycle water, oxygen, and nutrients through the biosphere.

Ecosystems also help protect us from physical threats. Plant communities, for example, protect the soil from wind and water erosion. Ecosystems act as sponges, absorbing water and then slowly releasing it into the groundwater and surface water. This reduces erosion and protects against flooding. It also filters the water in the process. Ecosystems, such as mangroves, also protect land from storms along coasts, where wave damage erodes the shoreline (Figure 5, on page 61).

Monetary Value of Ecosystem Services

It is difficult to put a monetary value on the beauty of wildlife and wilderness experiences (Figure 4). However, it is possible to estimate the value of many of the ecological services provided by natural ecosystems. The dollar value of cleaning the air and water, moderating climate, and providing paper fibre, medicines, and other products is high. On a global scale, the value of natural terrestrial ecosystems ranges into the trillions of dollars per year!

Figure 4 How do you put an economic value on a natural ecosystem?

Ecosystems provide valuable services that are free and renewable. If maintained in proper health, natural ecosystems will continue to provide these services to us indefinitely and at no cost. We are wise to protect our natural ecosystems.

IN SUMMARY

- Cultural services are benefits we obtain related to our enjoyment of the environment.
- Ecosystems provide us with products such as wood fibre, medicines, and food.
- Ecosystems help protect us from some environmental threats.
- Ecosystems have monetary value, although this is difficult to quantify.

CHECK YOUR LEARNING

1. Identify the ecosystem service(s) that each of the following performs: a river through a wilderness park, water in a reservoir behind a dam, wind, wild blueberries, trees planted in a yard. K/U
2. Grass is useful. Describe the ecological services that grass performs in a hay field, on a golf course, or along the shoulder of a highway. K/U
3. Using three examples, explain how your health is influenced by ecosystem services. K/U A
4. List several careers that are closely associated with ecosystem services. A
5. Provide reasons why a renewable service is more valuable than a non-renewable service. K/U
6. Explain what is incorrect about the following statement: "If we did not have the services provided by natural ecosystems, we could just pay the cost and have them done in some other way." A

3.1 Services from Natural Ecosystems   79

3.2 Equilibrium and Change

On a large scale, most natural ecosystems are in a state of **equilibrium**. This means that their biotic and abiotic features remain relatively constant over time. The major biomes, for example, usually maintain a characteristic set of species over hundreds or thousands of years. Changes on a large scale occur slowly and are caused by changes in climatic conditions.

Equilibrium is established when abiotic conditions are stable. Energy flows through the ecosystem. Nutrients are cycled through food webs. In addition, photosynthesis and cellular respiration are balanced. When ecosystems are in equilibrium, populations are healthy and stable.

On the scale of biomes, ecosystems remain relatively unchanged over time. This is not true, however, on a small scale. Smaller ecosystems are in a constant state of change. A forest fire or disease outbreak can cause short-term changes on a local level.

On August 27, 1883, the volcanic island of Krakatoa in Indonesia literally blew up. This explosion produced a sound wave that carried for 4600 km (Figure 1). The island was destroyed, along with every living thing that inhabited its lush tropical forests. The remaining part of the island was buried in more than 40 m of ash and volcanic rock.

Despite this dramatic disturbance, life had returned to the island within 9 months. Seeds were carried from nearby islands by the wind, the sea, and birds. Insects and spiders soon followed. In time, many other organisms returned to the island. Within a hundred years, a lush, rainforest community was re-established. This process of establishing and replacing a community following a disturbance is called ecological **succession** (Figure 2).

equilibrium describes the state of an ecosystem with relatively constant conditions over a period of time

Figure 1 The eruption of Krakatoa was the loudest sound in recorded history.

succession the gradual and usually predictable changes in the composition of a community and the abiotic conditions following a disturbance

Figure 2 After Mount St. Helens, in Washington State, erupted in 1980, lupines were the first plants to appear. Other plant species then arrived, taking advantage of the additional nutrients provided by the lupines. This is an example of ecological succession.

Ecological Succession

Ecological succession is initiated by a disturbance such as a geological event, a fire, or human activity. **Primary succession** occurs on soil or bare rock, where no life previously existed, such as following a volcanic eruption. **Secondary succession** follows a disturbance that disrupts but does not destroy the community. The regrowth of an area following a forest fire is an example of secondary succession. Severe pollution events or industrial

primary succession succession on newly exposed ground, such as following a volcanic eruption

secondary succession succession in a partially disturbed ecosystem, such as following a forest fire

activity, such as surface mining, are human-caused disturbances that initiate secondary succession.

Succession results in gradual changes as plants, animals, fungi, and micro-organisms become established in an area. The typical pattern sees small, hardy plants such as grasses colonizing the open landscape. These plants gradually alter the soil and local abiotic environment and make conditions suitable for shrubs and small trees to grow. These shrubs and trees, in turn, create conditions suitable for large trees that may come to dominate the landscape. Eventually, a relatively stable community may form (Figure 3). While plants are the most visible part of succession, animal species also change.

Aquatic ecosystems also undergo succession. In northern Ontario, bogs form when small lakes of non-flowing water are gradually covered over and filled by vegetation. Bog succession proceeds as sphagnum moss forms a floating mat along the shoreline. Year after year, the floating moss grows further outward from the shore. The dying and decaying moss sinks below the surface, slowly filling in the lake. This living carpet is colonized by other plants. Eventually, the once open body of water becomes completely covered in vegetation.

> **DID YOU KNOW?**
> **Fires Are Important to Succession**
> Fires play an important role in some types of succession. Some natural grassland is maintained by occasional fires. Periodic fires kill off young trees and shrubs that have started to grow. Grasses are able to quickly re-establish themselves.

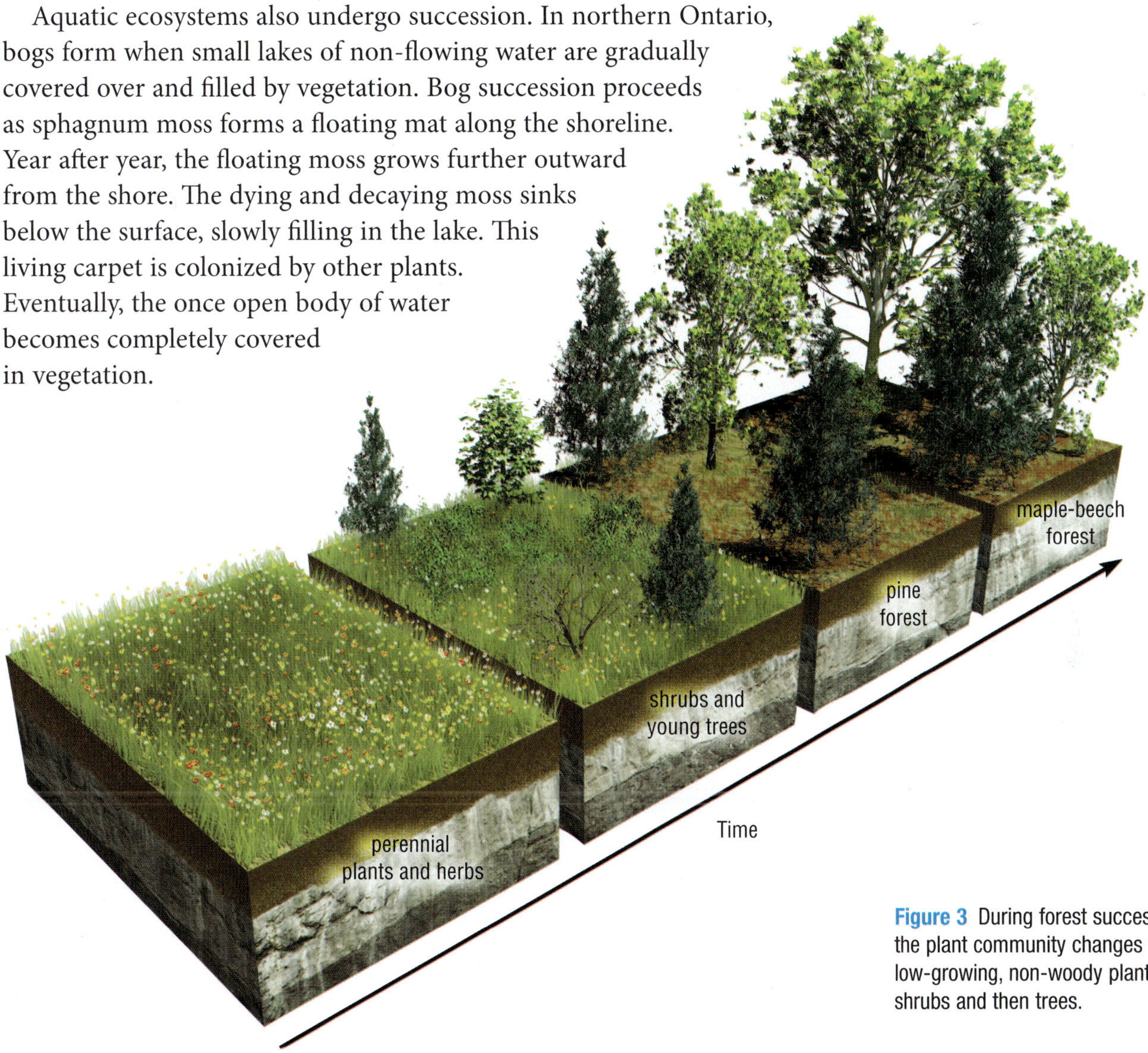

Figure 3 During forest succession, the plant community changes from low-growing, non-woody plants to shrubs and then trees.

Another example of succession occurs along sandy shores of oceans and large lakes, such as along the coast of Lake Huron. Dune succession begins when grasses establish in loose sand. Once the grasses establish, they reduce wind erosion and their roots hold the sand in place. Over time, plant numbers increase and soil characteristics change. Eventually a large sand dune can be transformed into a lush forested ecosystem. Unfortunately, sand dune communities are fragile and easily disturbed by human activity.

Benefits of Succession

Succession provides a mechanism by which ecosystems maintain their long-term sustainability. It also allows ecosystems to recover from natural or human-caused disturbances. Succession offers hope that even severe environmental damage may be reversed (Figure 4). The time needed, however, is very long, and the original cause of the disturbance must be eliminated. Not all human-caused disturbances can be repaired through natural succession. Often disturbances must be repaired through human actions that support the natural processes of succession.

Figure 4 The combination of human actions and succession have helped rehabilitate this old Ontario quarry.

The rest of this chapter looks at several types of human-caused disturbances to natural ecosystems. As you read, think about how communities are disturbed by these activities and how succession can help us to manage ecosystems subjected to disturbance.

IN SUMMARY

- The key biotic and abiotic features of large ecosystems remain relatively constant over time. They are in a state of equilibrium.
- Succession is the gradual changes in the biotic and abiotic features of an ecosystem following a disturbance.
- Succession allows an ecosystem to recover following a natural or human-caused disturbance.

CHECK YOUR LEARNING

1. Explain what it means when we describe an ecosystem as being in equilibrium. K/U
2. Distinguish between primary and secondary succession. K/U
3. Explain which type of succession occurs most often and why. K/U
4. Describe how biotic and abiotic conditions change during secondary succession. K/U
5. From your own experience, list two examples of secondary succession. A
6. Describe how succession can help restore the equilibrium of an ecosystem. K/U
7. Why is succession slower on sand or bare rock than on previously vegetated soil exposed by a fire? K/U
8. The Krakatoa eruption destroyed life and left behind volcanic ash. A large chemical spill can also destroy life. How would succession be very different following these two events? A
9. Why is it reasonable to describe a large ecosystem like a biome as being in equilibrium, but not a very small ecosystem like a rotting log? T/I

3.3 The Importance of Biodiversity

Scientists have only begun to understand the variety of life that exists on Earth. Approximately 1.5 million species have been studied, but this is only a fraction of species that are currently alive. Biologists estimate the total number of living species to be more than 5 million and perhaps as high as 50 million! There is little doubt that most living species have yet to be identified. Canada is home to about 140 000 to 200 000 species of plants and animals. Only 71 000 have been identified.

Biological diversity, or **biodiversity**, is the variety of life found in an area. It is often measured by counting the number of species in a specific habitat or ecosystem. This measurement of species numbers is called **species richness**. A diverse ecosystem will have high species richness.

In general, species richness tends to be higher close to the equator (Figure 1). Tropical rainforests have the highest biodiversity of any ecosystem. In a Peruvian rainforest, scientists identified 283 tree species in a single hectare. A similar-sized deciduous forest in Ontario would have fewer than 15 tree species. The Amazon rainforest is home to more than 200 species of hummingbirds, whereas Ontario has only a single species.

biodiversity the variety of life in a particular ecosystem; also known as biological diversity

species richness the number of species in an area

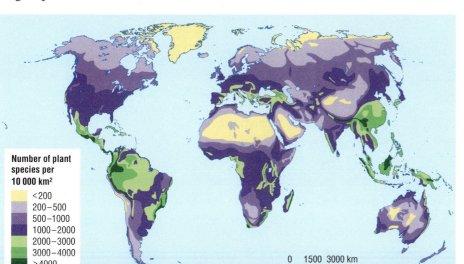

Figure 1 Biodiversity map showing the number of species of plants per 10 000 km²

DID YOU KNOW?
Insects in Extraordinary Numbers
Over half of all known living species are insects. Of Canada's approximately 71 000 identified species, about 30 000 are insects. In Algonquin Park alone, more than 40 species of black flies have been identified, only a few of which feed on humans.

Biodiversity under Attack

Imagine you are watching the burning of the world's greatest library. The building contains a copy of every book ever written. The fire is rapidly destroying the books. Now imagine you are told that most of these books have never been read and that no other copies exist. Such a fire would be a tragic loss of human knowledge.

This scenario can be compared to the current destruction of Earth's biodiversity. Many species are dying out, or going **extinct**. Their habitats are being destroyed through deforestation, urban and agricultural expansion, pollution, and climate change (Figure 2). Like the books in the burning library, most of these species have not been studied. Humans are rapidly destroying Earth's ecosystems without even knowing their biological contents.

extinct refers to a species that has died out and no longer occurs on Earth

Figure 2 Urbanization is a major cause of habitat loss.

WRITING TIP

Using Scientific Terminology
Carefully check your spelling of scientific terms. Since many of these terms are not used frequently in everyday talk or writing, you can benefit from keeping your own glossary of terms such as "biodiversity" and "extinction." This way, you can easily look up the meaning and correct spelling of these scientific terms.

Extinction is a natural process. Over thousands and millions of years, some species become extinct, while new species arise. There have been at least five major extinction events in the past 1 billion years. Extinction events are usually caused by a catastrophic event such as an asteroid impact or massive volcanic eruption. Between such rare events, extinction rates are very low.

Unfortunately, human activity has drastically increased the rate of extinction. When humans first encounter new species, they often overexploit them. Within 200 years of their arrival in New Zealand, humans had caused the extinction of 30 species of large bird, including 26 species of giant moas (Figure 3(a)). About 80 species of mammals and birds went extinct shortly after humans reached North America 10 000 years ago. The extinctions included sabre-toothed tigers, mammoths, camels, and horses. Some ocean species fared no better. The Stellar's sea cow, an unusual marine mammal weighing more than 3 tonnes, was discovered in 1741 and was extinct by 1768 (Figure 3(b))!

Figure 3 The (a) giant moa and (b) Stellar's sea cow became extinct shortly after humans colonized their ecosystems.

In the past 400 years, over 700 species of vertebrates have become extinct. Twelve species have become extinct in Canada in the past 170 years (Figure 4). Unfortunately, the rate of human-caused extinction is increasing.

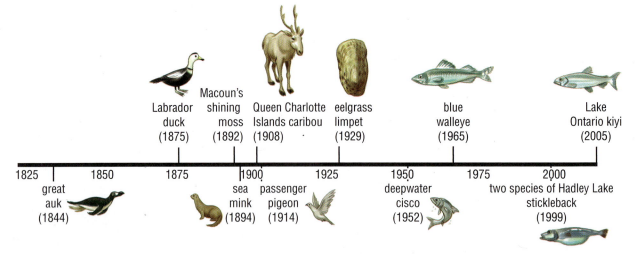

Figure 4 Timeline of recent species extinctions in Canada

84   Chapter 3 • Natural Ecosystems and Stewardship

Species at Risk

Species do not have to be driven to extinction for there to be ecological consequences. When a population's size declines below a critical level, the species will no longer be able to fill its ecological niche. This has consequences for the biotic and abiotic features of the ecosystem. For example, the loss of most, but not all, large sharks from a coral reef ecosystem changes the food web and damages the reef.

In Canada, the status of species is monitored by the Committee on the Status of Endangered Wildlife in Canada (COSEWIC). The committee has members from governments, universities, other agencies, and Aboriginal peoples. Experts on COSEWIC use the data of species at risk to categorize them in one of four categories. Species are classified as **extirpated** when they no longer exist in the wild in a specific area but still live elsewhere. **Endangered** species are in imminent danger of going extinct or becoming extirpated. **Threatened** species are likely to become endangered if current trends and conditions continue. Species of **special concern** may become threatened or endangered because of a combination of factors. Table 1 lists numbers and examples of Canadian species in each COSEWIC classification.

extirpated a species that no longer exists in a specific area

endangered a species facing imminent extirpation or extinction

threatened a species that is likely to become endangered if factors reducing its survival are not changed

special concern a species that may become threatened or endangered because of a combination of factors

Table 1 Examples of Canadian Extinct Species and Species at Risk

Classification	Number of Canadian species (2008)	Examples
extinct	13	• great auk • passenger pigeon • sea mink
extirpated	23	• paddlefish (from all of Canada) • Atlantic walrus (from the Northwest Atlantic)
endangered	238	• barn owl (in some regions) • swift fox • northern cricket frog
threatened	146	• humpback whale • wood bison • Kentucky coffee tree
special concern	157	• polar bear • red-headed woodpecker • Atlantic cod

When a species is placed in the endangered or threatened category, another agency, RENEW (REcovery of Nationally Endangered Wildlife), prepares an action plan to ensure the recovery of the species. As of 2008, 564 plant and animal species are considered at risk in Canada. RENEW has developed recovery plans for more than 300 species.

We must act now to reduce habitat loss, urban expansion, and pollution to protect Earth's biodiversity. In the following sections, you will learn about specific threats to biodiversity. You will also learn what actions are being taken by individuals, organizations, and government agencies to eliminate these threats.

To learn more about Canada's and Ontario's species at risk,
GO TO NELSON SCIENCE

CITIZEN ACTION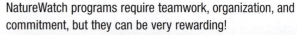

NatureWatch

Concern about the environment and taking action are the responsibility of the government, businesses, and individuals. Everyone should help keep our environment healthy and protect species at risk.

NatureWatch is a community-based "citizen science" monitoring program. It encourages schools, community groups, and individuals to help monitor important species and environmental indicators (Figure 5). The current program includes FrogWatch, IceWatch, PlantWatch, and even WormWatch!

Consider how you and your class might get involved.
- Research one of the NatureWatch programs.
- Write a short description of the program. Explain why you think it is worthwhile and the steps needed to participate in the program.
- Share your ideas with your classmates.
- Decide if you or your class want to participate in one of the programs.

 GO TO NELSON SCIENCE

NatureWatch programs require teamwork, organization, and commitment, but they can be very rewarding!

Figure 5 Individuals can take action to protect Earth's biodiversity.

UNIT TASK Bookmark

You can apply what you learned in this section about biodiversity and the classifications of species at risk to the Unit Task described on page 156.

IN SUMMARY

- Most species have not been identified or studied.
- Biodiversity tends to be higher near the equator.
- Human activities threaten biodiversity.
- Extinction is a natural process, but human activities have greatly increased the rate of extinction.
- The Committee on the Status of Endangered Wildlife in Canada (COSEWIC) categorizes at-risk species as extirpated, endangered, threatened, or of special concern.
- The Recovery of Nationally Endangered Wildlife (RENEW) agency prepares action plans for endangered or threatened species.

CHECK YOUR LEARNING

1. Explain what is meant by the term *species richness*. K/U
2. What ecosystems would you expect to have high species richness and low species richness? Explain your reasoning. K/U
3. Why are scientists concerned about species loss if extinction is a natural process? K/U
4. (a) List some of the main human activities that contribute to species extinction. T/I
   (b) Do you think such activities can be justified? A
5. Suggest reasons for the acceleration in extinction rates in recent decades. T/I
6. List the four *at-risk* classifications and give an example of one Ontario species in each class. K/U
7. Describe what actions are taken in Canada once a species in placed in the endangered or threatened category. K/U
8. (a) Brainstorm to create a list of actions an individual could take to help protect at-risk species.
   (b) Which action are you most likely to complete? Explain why. A C

Habitat Loss and Fragmentation

3.4

A simple glance out the window provides evidence of habitat loss. Farmland, human settlements, and highways have replaced much of southern Ontario's once extensive temperate forest ecosystem (Figure 1). A view from the top of the CN Tower reveals a human-dominated landscape. Little of the original natural forest remains.

Loss and Fragmentation of Terrestrial Ecosystems

An extreme example of habitat loss is the conversion of large areas of natural ecosystems into farmland and urban developments. Figure 2 illustrates historic changes in the natural landscape of southern Ontario. In Canada, most of the habitat loss occurred during the nineteenth and early twentieth centuries when land was cleared for farmland. More recently, natural habitat has been lost to urban sprawl and transportation routes.

Figure 1 Approximately 80 % of southern Ontario's original forest cover is gone.

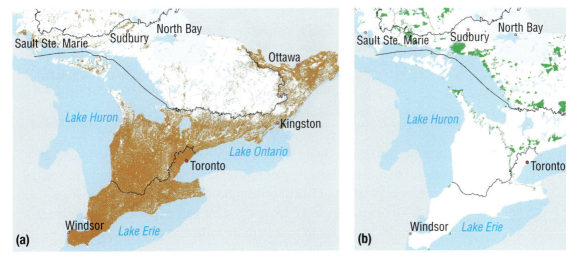

Figure 2 (a) Most of southern Ontario has been converted from natural ecosystems to human uses (orange area). (b) Only some of the remaining natural ecosystems are protected in parks and protected areas (green areas).

Fragmentation of natural ecosystems reduces their sustainability. Fragmentation is the dividing up of a region into smaller parcels or fragments (Figure 3). In a fragmented ecosystem, a larger area of the habitat is exposed to damaging outside influences such as pollution.

Figure 3 When large ecosystems become fragmented, species with large home ranges may not have enough area to survive.

Table 1 describes key factors that enhance ecosystem sustainability. These factors are considered when deciding which areas should be set aside to protect wildlife and ecosystems.

Table 1 Factors that Improve the Sustainability of Habitat Fragments

Factor	Poorer option	Better option	Explanation
size			Large blocks support larger and more stable populations and communities.
number			One large area is better than an equal area composed of many smaller areas because there is less outside influence.
proximity			The closer ecosystem fragments are to each other, the greater the chance populations will be able to use the entire area.
connectedness			Interconnected areas provide wildlife corridors and permit migration between larger blocks.
integrity			Access by roads and trails can increase pollution, hunting, and fishing.

On a global scale, habitat loss and fragmentation are second to climate change as the most serious threats to the sustainability of natural terrestrial ecosystems. Habitat loss is most pronounced in Africa, Latin America, and the Caribbean (Figure 4). Expanding human populations are placing pressure on the land base to supply more food and raw materials.

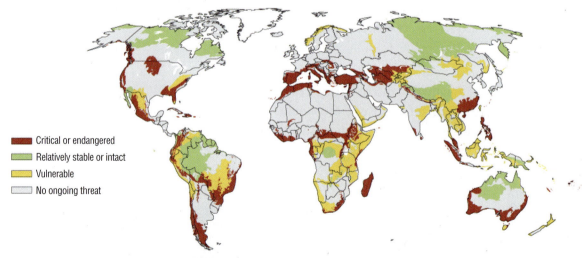

■ Critical or endangered
■ Relatively stable or intact
■ Vulnerable
■ No ongoing threat

Figure 4 Many of Earth's ecosystems are vulnerable or critically endangered.

In the Amazon, which is the world's largest remaining rainforest, clearing and burning are the greatest threats to the ecosystem's sustainability. This is most often done to create pasture for cattle sold to foreign markets.

If we reduce the demand for agricultural products produced in tropical regions, we can reduce rainforest habitat loss (Figure 5).

Ontario is no longer experiencing a rapid loss of native ecosystems, but there is still reason for concern. The remaining threatened areas require wise management. As citizens of a wealthy country, our use of Earth's resources has far-reaching impacts.

The loggerhead shrike is threatened by habitat loss. The range of this small predatory bird once covered much of southern Ontario (Figure 6). Over the past 50 years, changes in agricultural practices have caused the loss of nesting habitat for the shrike. Once a common bird, the population reached a low of 17 breeding pairs in Ontario in 1997. The Loggerhead Recovery Team has developed a habitat conservation plan to promote habitat restoration.

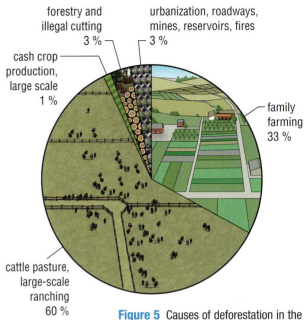

Figure 5 Causes of deforestation in the Amazon, 2000–2005

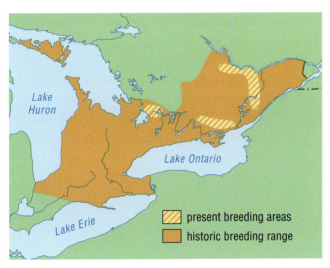

To learn more about the loggerhead shrike and its recovery program,
GO TO NELSON SCIENCE

Figure 6 The former and current distribution of the loggerhead shrike

RESEARCH THIS | SWEET GRASS GARDENS

SKILLS: Researching, Analyzing the Issue, Communicating

SKILLS HANDBOOK
4.B.

The Six Nations people, living in the Grand River region of Ontario, have a close connection with native species and are commited to protecting their local natural ecosystems. This includes the largest remaining stand of Carolinian forest in Canada. In 1996, two members of the Six Nations, Ken and Linda Parker, started Sweet Grass Gardens. This is the first Native-owned and -operated plant nursery in North America.

1. Use the Internet to research Sweet Grass Gardens and Six Nations' initiatives.

 **GO TO NELSON SCIENCE**

A. Sweet Grass Gardens specializes in growing native plant species. In what ways are native plants particularly significant for Six Nations people? How many species does the nursery grow?

B. Describe how this nursery works with the community to enhance education and training related to the environment.

C. Write a summary describing why you think this particular nursery has been so successful.

D. The Six Nations has an eco-centre that runs a forest education program with students. Describe how the program is used to provide plants for restoration projects.

E. List two other ways the eco-centre is helping the environment.

3.4 Habitat Loss and Fragmentation

Loss of Wetlands and Aquatic Ecosystems

Human activities also threaten aquatic ecosystems. In many cases, human activities along shorelines damage neighbouring aquatic ecosystems. Table 2 lists some of the key actions that damage coastlines and aquatic ecosystems.

Table 2 Impact of Human Activities on Wetland and Aquatic Ecosystems

Human activity	Impacts on ecosystem
replacing natural vegetation along coastlines and waterfronts	• habitat destruction; shoreline erosion; loss of some species; loss in breeding areas such as fish spawning beds
dredging to create deeper water for boats	• disruption of bottom-living organisms and spawning beds; habitat destruction
sediment runoff from land-clearing, agricultural, and forestry operations	• sediments may smother natural habitats
commercial fishing	• bottom trawlers and drag lines injure and kill bottom-dwelling organisms; damage to abiotic features
draining wetlands for urban expansion and agriculture	• loss of wetland habitats and associated species

Figure 7 Over 80 % of Ontario's carrots are grown on the Holland Marsh, located 50 km north of Toronto.

Natural wetlands are flat and often have deep, nutrient-rich soil with an abundant water supply. These conditions are ideal for agriculture. The result is that most large wetlands in populated parts of Ontario have been drained and converted to farmland (Figure 7).

There is a push to reverse this trend and re-establish wetlands. Creating new wetlands makes valuable habitats for wildlife and waterfowl. One success story is the 725 hectare Hilliardton Marsh in northeastern Ontario. Constructed on abandoned farmland, the marsh is now a breeding location for two species of at-risk birds, the coot and the black tern (Figure 8).

Figure 8 To date, Bruce Murphy, a local high school teacher, and his students have banded more than 10 000 birds at the Hilliardton Marsh.

UNIT TASK Bookmark

You can apply what you learned in this section on how habitat fragmentation puts species at risk to the Unit Task described on page 156.

IN SUMMARY

- Most of southern Ontario's original natural ecosystems have been replaced by agricultural land and urban centres.
- Habitat loss is one of the most serious threats to Earth's ecosystems.
- Fragmentation reduces ecosystem sustainability.
- The size, proximity, connectedness, and integrity of terrestrial ecosystems will influence their overall sustainability.
- Because of their rich soil, many wetlands have been converted into farmland.

CHECK YOUR LEARNING

1. Explain why one large park is a better refuge than several smaller parks of the same total habitat area. K/U
2. Describe what is being done to reverse the continued loss of wetland habitat in Ontario. K/U
3. Identify the benefits to native species of joining similar habitats with corridors. K/U
4. In what regions of the world is habitat loss most rapid? K/U
5. Describe three human activities that threaten aquatic habitats. K/U
6. Explain why wetland habitats are often preferred for converting to new agricultural land. K/U
7. Do any of your own behaviours or habits contribute to habitat loss or fragmentation? Explain. A

3.5 The Introduction of Non-Native Species

Can you name a species that lives in Ontario but is not native to Ontario? If so, do you think the species has an impact on the natural environment? How and why did the species get here? The introduction of non-native, or exotic, species to an ecosystem by humans is a major cause of species loss.

Introductions of non-native species usually fail because few species can tolerate an entirely new environment. Remember, for a species to survive it must be within its tolerance limits for all abiotic factors (Section 2.7). Even if a species is adapted to the abiotic environment, it may have difficulty finding food or may be unable to compete with native species.

Occasionally, an introduced species is successful in its new environment. The new ecosystem may lack population controls for the new species, such as predators and diseases. Native species might not be able to compete with the introduced species. When a population is unchecked, it gains an advantage over native species. It can increase rapidly and become invasive. **Invasive species** are introduced species with growing populations that spread and have a negative effect on their environment.

There are well over 3000 invasive species in Canada. In the Great Lakes there are over 185 invasive species. Ontario is home to six introduced species of carp (Figure 1(a)). You may also be quite familiar with the European earwig (Figure 1(b)). First documented in Canada about 1830, it is now a widespread pest across southern Canada.

> **DID YOU KNOW?**
> **Killing Oak Trees with Your Feet**
> Sudden oak death is a serious fungal disease of oaks and other tree species. The disease is most common in parks and forests with large numbers of visitors. It is most notable along hiking and biking trails. Humans carry the fungus spores in soil attached to their shoes or vehicles.

invasive species a non-native species whose intentional or accidental introduction negatively impacts the natural environment

Figure 1 (a) Six species of carp and (b) the European earwig are invasive species in Ontario.

A few individuals of an invasive species can have far-reaching impacts. In 1890 and 1891, Eugene Schieffelin released 100 starlings brought from England into Central Park, New York City. He hoped the starlings would control insect pests. He may have chosen starlings because they were mentioned in the works of Shakespeare. Starlings are now one of the most numerous birds in North America. From the original 100 birds, their population has grown to 200 million. Starlings compete with native birds for nesting sites. They have caused a decline in the populations of songbirds, including bluebirds and tree swallows (Figure 2).

Figure 2 The intentionally introduced European starling competes with native songbirds.

Invasive species have significant environmental and economic impacts (Table 1). These species change natural ecosystems and cost Canadians billions of dollars annually just to control their population size. Researchers at the Global Invasive Species Programme estimate the global damage caused by invasive species to be $1.4 trillion annually.

Table 1 Impacts of Invasive Species

Type of impact	Consequences
ecological	• Invasive species compete with or feed on native species, leading to population decline or extinction. • Invasive species change ecosystem dynamics by altering nutrient cycles or energy flow.
economic	• Damage to forests and agricultural crops causes financial losses. • Competition with invasive plants lowers crop yields. • Diseases and pests may destroy livestock and crops, kill trees, and harm important species such as honeybees.
tourism	• Species loss and reduced water quality have negative impacts on wildlife viewing, fishing, and water-based recreation. • Waterways can become choked with invasive aquatic plants, rendering them impassable to boats.
health	• Disease-causing organisms, such as the West Nile virus, are introduced. • Pesticides used to control invasive species cause pollution and are health risks.

Table 2 gives four examples of invasive species from around the world and the impacts they have had on their new environment.

Table 2 Examples of Invasive Species

Invasive species	Location and impact
brown tree snake	• Pacific island of Guam • accidental introduction around 1950 • caused the extinction of 9 of Guam's 12 forest birds and half of the lizard species
kudzu	• eastern United States • intentionally introduced as a forage crop and for erosion control • rapidly spreading vine that kills native trees by shading them
Nile perch	• Lake Victoria, Africa • intentionally introduced to establish a commercial fishery • a large predatory fish that has caused the extinction of more than 200 species of native fishes and decimated the populations of many others
feral goats	• escaped into the wild in many parts of the world • voracious herbivores that threaten many species of native plants with overgrazing • particularly damaging on remote islands where there are no natural predators

DID YOU KNOW?

The Blight of Destruction
The chestnut blight was accidentally introduced when Asian chestnut trees were imported into North America in the early 1900s. The disease, caused by an airborne fungus, spread quickly. In several decades, it had killed billions of American chestnut trees. About 100 American chestnut trees remain.

GO TO NELSON SCIENCE

RESEARCH THIS | WE DO NOT BELONG!

SKILLS: Researching, Communicating

When a non-native species is introduced into an established ecosystem, it may upset food webs or compete with native species for resources. If very successful, these species can become invasive and cause serious damage to an ecosystem. In this activity, you will research an invasive species in Ontario.

1. Use the Internet and other electronic and print resources to compile a list of Ontario's invasive species.

    GO TO NELSON SCIENCE

2. Select two invasive species and record the following information:
   (i) Where did the species originate?
   (ii) How did the species get to Ontario? Were the introductions intentional or accidental? How are they spreading?
   (iii) What is the distribution of the species in Ontario now?
   (iv) What is the main concern with these species? How are they causing harm?

A. Classify Ontario's invasive species using a table similar to Table 3 below. For each species, indicate whether it is a terrestrial (T) or aquatic (A) species.

Table 3

Plants	Fish	Molluscs	Crustaceans	Insects	Other

B. Design an invasive species collector's card. The card should have a picture of the invasive species and an Ontario distribution map on one side and the remaining information on the reverse side.

C. When everyone is done, use the cards to quiz each other or invent a "We Do Not Belong!" card game.

Controlling Introduced Species

It is impossible to predict which introduced species will become invasive. Preventing the accidental introduction of a non-native species is ideal but difficult. Insects can arrive as adults, larvae, or eggs in imported foods and containers from around the world. Small seeds are equally difficult to detect. Species that are introduced intentionally should be studied in advance to determine if they pose a risk.

Once an introduced species is established, it may be difficult to control. Elimination is unlikely. There are three types of control measures.

Chemical Control

Perhaps the most widely used control method is the use of pesticides. Pesticides are used mostly on forest and agricultural pests because trees and crops have significant economic value. Pesticides dramatically reduce crop damage, but there are environmental risks. They may kill non-target native species and pollute the air, water, and soil. You will learn more about pesticides in Chapter 4.

Mechanical Control

Some invasive species can be controlled with physical barriers or removal. Invasive plants can be cut down, burned, or even removed by hand. Invasive animals can be hunted or trapped.

In Hamilton harbour, barriers were constructed to protect Cootes Paradise, a valuable wetland, from carp, an invasive fish species. Smaller fish can swim in and out of the wetland through the barrier. Larger fish, such as carp, swim into a chamber and are sorted so that only native species can enter. In the fall, the barrier is removed so that remaining carp can leave on their migration. The barrier is replaced before they return. (Figure 3).

> **DID YOU KNOW?**
>
> **Underground Invaders**
> Ontario had no earthworms prior to the arrival of European settlers. Europeans introduced a number of species that have spread throughout Canada. While earthworms can benefit the soil in some ways, they have negative impacts on natural ecosystems. They can cause losses of tree seedlings, wildflowers, and ferns by feeding on young plants.
>
> GO TO NELSON SCIENCE

Figure 3 These barriers in Hamilton harbour protect Cootes Paradise from invasive fish species.

3.5 The Introduction of Non-Native Species

WRITING TIP

Use Short Sentences
Break a long sentence with more than one idea into shorter sentences. Instead of "For example, three insect species were released in Ontario to control purple loosestrife, an invasive plant that grows in wetlands," write "Three insect species were released in Ontario to control purple loosestrife. Purple loosestrife is an invasive plant that grows in wetlands."

Biological Control

Biological control is a challenging but effective method of controlling invasive species. Biological control uses intentionally introduced organisms to control the invasive species. For example, three insect species were released in Ontario to control purple loosestrife, an invasive plant that grows in wetlands (Figure 4). Tests conducted before their release indicated that the insects are unlikely to feed on native plants. The results are promising. Although biological control rarely eradicates an invasive species, it may reduce population sizes to ecologically tolerable levels.

Figure 4 Leaf-eating beetles help control purple loosestrife.

UNIT TASK Bookmark

In this section, you looked at invasive species from around the world. In the Unit Task, described on page 156, your class will examine the impacts of invasive species in Ontario.

IN SUMMARY

- Invasive species are non-native species whose introduction negatively impacts ecosystems.
- Invasive species have been introduced intentionally and accidently.
- Invasive species may negatively affect our health and the economy.
- Invasive species can be controlled using mechanical, chemical, or biological methods.

CHECK YOUR LEARNING

1. Are all introduced species invasive? Explain using examples. K/U
2. Explain why most introduced species are not successful in their new environment. K/U
3. How does the success of an invasive species depend on its placement in its new food web? Explain. T/I
4. List and briefly outline the possible ecological consequences of introducing an invasive species. K/U
5. Describe the impacts of invasive species on human society. A
6. Give an example of a domesticated animal that has become an invasive species. K/U
7. Describe three methods used to help control invasive species. K/U
8. How do we sometimes benefit from introduced species that are not invasive? Explain using examples you have witnessed or personally benefited from. K/U A
9. The remote Galápagos islands are famous for their unusual wildlife (Figure 5). In 1900, the Galápagos had 112 invasive species. By 2007, that number had risen to 1321. Why do you think island wildlife might be more threatened by invasive species? T/I

Figure 5 Galápagos tortoise

SCIENCE WORKS OSSLT

Containing the Invasion

The spiny water flea (Figure 1) is a tiny aquatic invertebrate that is causing havoc in Ontario's lakes. It was accidentally introduced in 1982 from Europe in the ballast water of ships. Since then, it has invaded more than 100 lakes in North America. The spiny water flea is less than 1.5 cm long and has a large black eye and long spiny tail.

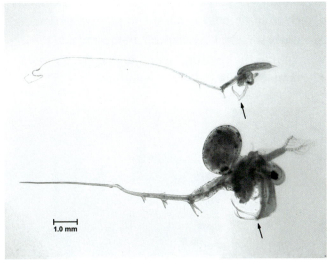

Figure 1 The spiny water flea is a voracious predator, feeding on zooplankton.

The water flea threatens aquatic ecosystems because it competes with native species and upsets the food web. Queen's University biology professor, Dr. Shelley Arnott, is an expert in aquatic biology. She has been studying the ecology of the spiny water flea and methods to control its spread.

Dr. Arnott and graduate student Angela Strecker compared food webs of Ontario lakes invaded by the spiny water flea with the food webs of lakes that had not been invaded. They found that the zooplankton (microscopic primary consumers) population size was 70 % lower in the invaded lakes than in the non-invaded lakes. Arnott suspects that this decline is caused by predation of zooplankton by water fleas.

While the spiny water flea is a tiny organism, it is changing the entire food web. With less zooplankton, there is less food for small fish, such as lake herring. In turn, there is less food for larger fish, such as lake trout that feed on herring. The spiny water flea is not a preferred prey species for small fish. Therefore, the natural food web is changed and energy is diverted away from native species.

Dr. Arnott is concerned that invasive species, such as the water flea, can damage already stressed ecosystems. To understand this problem better, another of Dr. Arnott's graduate students, Leah James, is comparing the growth rate of herring populations in invaded and non-invaded lakes (Figure 2).

Figure 2 Dr. Arnott and graduate student Leah James research fish populations.

Dr. Arnott's research has uncovered some good news. Slowing the spread of the spiny water flea is as easy as letting your boat and equipment dry out before launching in a new lake (Figure 3). When the boat and equipment are thoroughly dry, the adult spiny water fleas and their eggs die.

Figure 3 Let your boat and equipment dry completely before launching in another lake. This simple action will slow the spread of the spiny water flea.

Dr. Arnott explains: "It's such a simple thing for the general public to do, and yet it could make a big difference in the way that our lake ecosystems function."

3.6 Pollution

For thousands of years, Aboriginal peoples lived in relative harmony with their environment. Without modern synthetic materials, they produced little waste compared with modern civilizations. Two factors have changed the relationship humans have with their environment. First, human population size has drastically increased. There are more of us around to create waste. Second, many toxic and persistent chemicals have been invented to support our lifestyle. As a result, each of us has a greater impact than did a single individual of the past.

When released into the environment, toxic materials are called **pollution** (Figure 1). Some polluting materials, such as pesticides and fertilizers, have a specific purpose. Others, such as automobile exhaust and product packaging, are by-products of our consumer society. Even natural, biodegradable products can pollute when put into the environment in high concentrations. In this section, we look at examples of air and water pollution.

pollution harmful contaminants released into the environment

Figure 1 Examples of land, water, and air pollution

Acid Precipitation

Sulfur dioxide and nitrogen oxides are two damaging air pollutants. They are produced in industrial processes and from burning fossil fuels, such as coal and petroleum. Sulfur dioxide and nitrogen oxides combine with water vapour in the atmosphere to form acids. Acids move through the water cycle, sometimes over very long distances, before returning to Earth in the form of **acid precipitation**.

acid precipitation precipitation that has been made more acidic than usual by the combination of certain chemicals in the air with water vapour

The Effects of Acid Precipitation

Acid precipitation, also called acid rain, impacts aquatic and terrestrial ecosystems. As rivers and lakes become more acidic, species decline in numbers and may disappear. For example, microscopic photosynthetic plankton that are at the bottom of the aquatic food web are negatively impacted. Crayfish, clams, and the fish that eat plankton are disappearing as their food supply decreases.

Not all rivers and lakes are equally impacted by acid precipitation. Certain minerals, particularly limestone, are able to **neutralize**, or counteract, acid. Limestone deposits near lakes tend to neutralize acid rain. The Canadian Shield, which is formed of granite, does not neutralize acid. For this reason, acid precipitation causes more damage in northern and southeastern Ontario than in southwestern Ontario.

neutralize counteract the chemical properties of an acid

In terrestrial ecosystems, acid rain chemically changes the soil by depleting the nutrients needed by plants. This damages the vegetation and results in slower growth. Acid rain has affected the soils in parts of Ontario, Quebec, and the Atlantic provinces. Figure 2 shows the effects of acid precipitation on terrestrial and aquatic ecosystems.

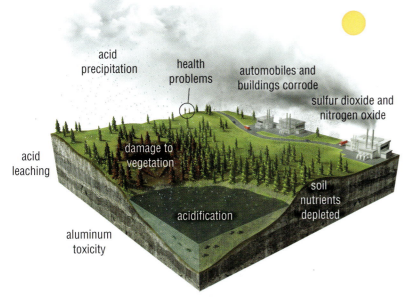

Figure 2 The release of sulfur dioxide and nitrogen oxides into the air causes acid precipitation.

Acid pollution also impacts humans. Sulfur dioxide is toxic and aggravates respiratory problems such as asthma and bronchitis. Acid precipitation speeds up the corrosion of metal, causing automobiles and iron structures to rust. It also dissolves concrete, marble, and limestone, damaging statues and buildings (Figure 3). Acid pollution causes billions of dollars in medical and economic costs each year. Ironically, acid lakes are sometimes considered desirable for recreation because they appear clear and clean (Figure 4).

Figure 3 Acid precipitation dissolves statues.

Figure 4 Acid lakes are clear because few organisms live in them.

Reducing Acid Precipitation

Even if acid emissions were eliminated, it could take hundreds of years for ecosystems to recover. Until then, forests in affected areas must rely on poorer soils. This situation threatens the long-term sustainability of our forests. According to Environment Canada, "If current levels of acid rain continue into the future, the growth and productivity of ~ 50 % of Canada's eastern boreal forests will be negatively affected."

DID YOU KNOW?

Sudbury's Green Solution to Pollution

In Sudbury, Ontario, extensive forestry has removed the area's trees, and emissions from smelting plants have contaminated the soil. In the 1980s, the city undertook a massive re-greening campaign. Lime was applied to neutralize the soil, and trees were planted. The City of Sudbury has won numerous awards for its efforts.

At one time, "the solution to pollution was dilution." Using this reasoning, the concentration of pollutants in the environment was decreased by mixing them with a large volume of air or water. One strategy employed tall superstacks to disperse pollutants over great distances. The result was that rather than the effects being severe and local, the effects were less acute but spread over a greater area. Fortunately, attitudes have changed and progress has been made (Figure 5).

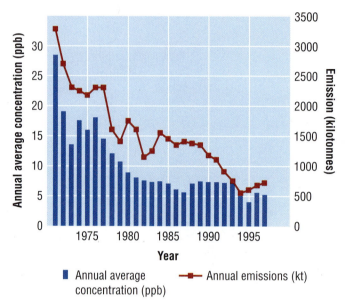

Figure 5 Sulfur dioxide emissions in Ontario have been steadily decreasing since the 1970s.

In eastern Canada, emissions of sulfur dioxide declined by over 60 % between 1980 and 2001. However, even today, the rain in the Muskoka–Haliburton area of Ontario is still 40 times more acidic than natural rain. Furthermore, while Canada has adopted new technologies, much of our acid precipitation comes from the United States, where high-sulfur coal is still widely used.

Now, humans are even changing the acidity of the world's oceans. As you learned in Section 2.6, the burning of fossil fuels is releasing large amounts of carbon dioxide into the atmosphere. Much of this carbon dioxide eventually dissolves in ocean water, where it forms carbonic acid. There is now enough carbonic acid forming to alter the chemistry of the world's largest bodies of water.

To learn more about acid precipitation,
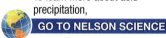

RESEARCH THIS | OCEAN ACIDIFICATION

SKILLS: Researching, Communicating

SKILLS HANDBOOK 4.B.

Human activities have caused acid precipitation damage to thousands of lakes in Ontario.

1. Research the changes that are occurring to the acidity of the ocean. The pH scale is between 0 (highly acidic) and 14 (basic). Acidic water has a pH below 7 (neutral).
2. Research how these changes threaten marine ecosystems.

A. How much has the pH of the ocean changed over the past 100 years?

B. Explain how pH influences marine organisms by considering the following questions:

   (i) What kinds of organisms have shells and external mineral skeletons?

   (ii) How could a more acidic environment affect these organisms?

   (iii) How might the loss of these species impact marine food webs?

98 Chapter 3 • Natural Ecosystems and Stewardship

Laying Waste to Water

Water is our most precious renewable resource. We drink water and use it to grow and cook our food. We use it to generate power and in many industrial and commercial processes. Water is a valuable natural resource, and it is important that we protect it from contamination. Acid precipitation pollutes water from above. Other wastes directly enter and threaten Earth's surface and groundwater.

Oil Spills

The most dramatic water pollution events are large oil spills. Oil tankers navigate the world's oceans. Inevitably, accidents happen. When they do, the results are devastating (Figure 6). Oil is toxic, slow to break down, and difficult to clean up. It forms large slicks that cover the ocean, beach, and seabed.

> **DID YOU KNOW?**
> **Orange Peels Fight Pollution**
> A recent study suggests that orange peel is capable of removing toxic industrial dyes from waste water. It could be a natural alternative to more expensive chemical processing. Orange peel is readily available.

Figure 6 High-pressure steam was used to wash crude oil from the *Exxon Valdez* spill off the rocky coastline of Prince Edward Sound, Alaska, in 1989. More than 40 million litres of crude oil was spilled.

Sea birds and seals are particularly vulnerable to oil spills. When sea birds get oil on their wings, they remove and ingest it while cleaning themselves. The ingested oil damages their digestive tract, liver, and kidneys. When covered with oil, the feathers of birds and fur of seals lose their ability to insulate. The animals lose too much body heat and die of the cold.

Following an accident, a variety of methods can be used to capture, break down, or disperse any oil that cannot be recovered:

- Skimming/vacuuming. Floating oil may be contained and skimmed or vacuumed up into a recovery vessel. This method is used only if sea conditions permit.
- **Bioremediation.** Some micro-organisms are capable of feeding on oil. Scientists are currently studying ways to speed up the rate at which these bacteria are able to break down the oil. The addition of inorganic nutrients is one technique that has shown promise.

bioremediation the use of micro-organisms to consume and break down environmental pollutants

Figure 7 Spilled oil is often burned.

- Burning. Oil floating at the surface is sometimes lit on fire so that it burns away. This prevents oil from sinking or washing up on shore. While this may solve the water pollution problem, it pollutes the air (Figure 7).
- Dispersal agents. Oil may be broken up into small droplets using detergents. This technique allows the droplets to be more easily washed out to sea and dispersed. However, the smaller droplets will spread over a larger area and can be taken in more readily by even more organisms.

Plastics at Sea

On Midway atoll, an island between the United States and Japan, about 500 000 albatross chicks are born each year. About 40 % of them die from dehydration or starvation. Why are so many chicks dying? Researchers have found that the stomachs of chicks that die from these causes are twice as likely to contain plastic as chicks that die from other causes. Where are the albatross getting plastic they feed to their chicks?

Plastic is an extremely useful material. It is inexpensive, lightweight, and strong. The qualities that make the plastic industry so successful also make plastic a harmful part of our waste. Because it does not chemically degrade, plastic that is disposed of today may remain in the environment for hundreds or thousands of years.

Each year, over a billion kilograms (1 million tonnes) of plastics are produced worldwide. Only a tiny fraction is recycled. Most of this plastic ends up in landfills, but some enters waterways and oceans. It may be blown there by the wind, washed into storm sewers, or intentionally dumped from ships. Large plastic commercial fishing nets also break loose and drift through the oceans, unintentionally trapping fish, turtles, birds, and marine mammals like dolphins and whales.

The North Pacific Ocean contains one of several large areas of slowly rotating ocean surface water. Floating debris slowly accumulates in these areas, forming massive mats of plastic trash. The so-called Great Pacific Garbage Patch is twice the size of Texas and is up to 100 m deep (Figure 8).

To learn more about the Great Pacific Garbage Patch,
GO TO NELSON SCIENCE

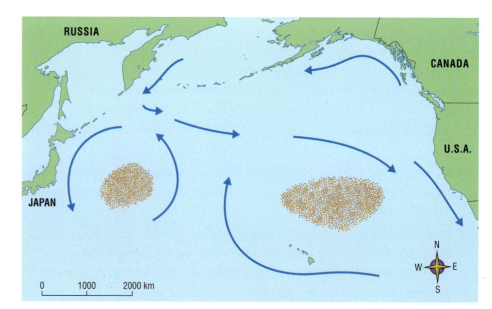

Figure 8 The Western and Eastern Pacific Garbage Patches are sometimes collectively called the Great Pacific Garbage Patch.

Plastics are responsible for significant ecological damage in marine ecosystems. Sunlight and wave action break the plastic into smaller and smaller pieces. These pieces become attractive bite-sized morsels for marine life. Sea turtles may mistake a piece of floating plastic bag for a jellyfish. Fish and sea birds may feed on brightly coloured plastic fragments. The plastic has no nutritional value and may block the digestive systems of the animals. Some animals become entangled in plastic bags or plastic rings (Figure 9).

As plastic gets broken down into smaller fragments, smaller organisms take them in. Scientists have found microscopic particles of plastic in the cells of plankton. Scientists believe that plastic absorbs chemicals from the environment. Proper recycling and disposal of plastics would reduce or eliminate this needless source of ocean pollution.

In addition to the threats they pose to our environment, the manufacturing of plastics creates pollution and results in the consumption of petroleum: a non-renewable resource. One way to avoid some of this waste is simply to use cloth bags when you shop!

Figure 9 Birds and other sea animals easily become entangled in plastic.

IN SUMMARY

- Acid precipitation is precipitation that has been made more acidic than usual by the combination of sulfur dioxide and nitrogen oxides with water vapour.
- Acid rain affects terrestrial and aquatic ecosystems as well as human health and infrastructure such as buildings.
- Acid precipitation can travel a long distance from its source through the atmosphere.
- Oil spills can be cleaned up using skimming, bioremediation, burning, and dispersal agents.
- Plastics cause significant damage to marine ecosystems, especially to aquatic organisms.
- Massive mats of floating plastic trash have formed in the oceans.
- The manufacturing of plastics causes pollution and consumes non-renewable resources.

CHECK YOUR LEARNING

1. Identify two factors that have changed the relationship between people and their environment, resulting in the production of pollution.
2. Name two chemical pollutants that are responsible for acid precipitation. What are their main sources?
3. Describe some of the negative impacts of acid precipitation on ecosystems.
4. What minerals, found in rock formations, are able to neutralize the effects of acid?
5. Explain how oil spills directly harm wildlife.
6. List several methods that can be used to help clean up an oil spill.
7. "The solution to pollution is dilution."
   (a) What does this mean in relation to pollution control?
   (b) Is it a good solution? Explain your answer.
8. Write a short paragraph on the relationship between plastics and ocean pollution.
9. Brainstorm a list of twenty items that you have used recently that contain plastics. Examine your list and consider what items, if any, were wasteful or unnecessary.
10. Do you think humans should attempt to clean up the Great Pacific Garbage Patch? Why do you think this might be difficult?

3.7 Consumption and Resource Management

Natural ecosystems are an important source of valuable resources. Chief among these resources are wood and wood fibre. In Canada, the forestry industry obtains the majority of its wood supply from natural forests. Approximately 60 % of the original forest in Canada has been cut at least once (Figure 1).

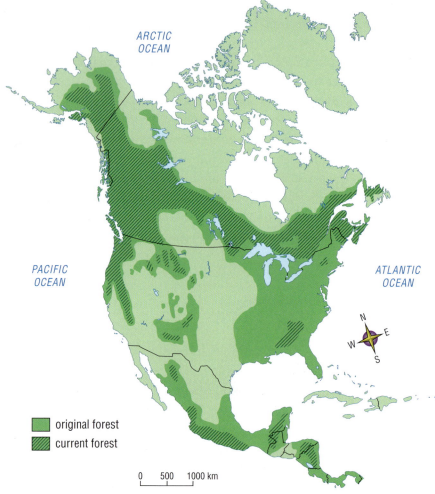

Figure 1 In 1600 CE, there were approximately 6 billion (6 000 000 000) ha of forest on Earth. An estimated 4 billion ha remain.

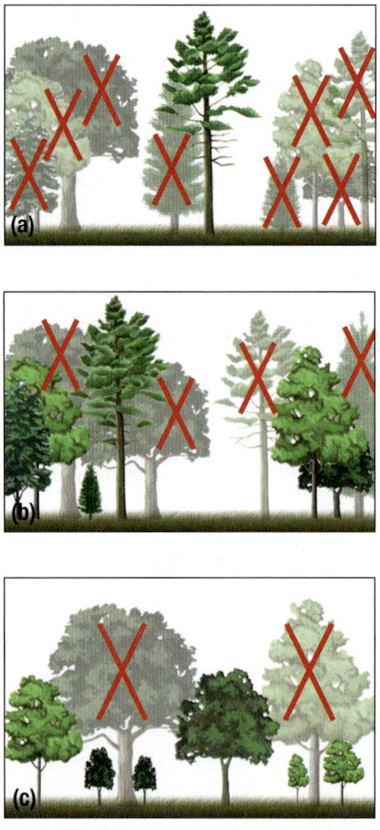

Forestry Practices

Forest harvesting methods fall into three categories: clear-cutting, shelterwood cutting, and selective cutting (Figure 2). Clear-cutting is the removal of all or most of the trees in a given area. This method is economical and efficient. It is intended to recreate the pattern produced by a forest fire. Clear-cuts take the shape of large blocks, strips, or smaller patches. Regeneration can occur naturally or artificially by planting seedlings. Both result in forests of even-aged trees. If natural regeneration is used, some scattered, high-quality "seed trees" may be left standing in the clear-cut. Clear-cutting is the most common method of tree harvesting in Ontario and Canada. It accounts for over 90 % of harvested trees.

In shelterwood cutting, mature trees are harvested in a series of two or more cuts. This permits regeneration under the shelter of remaining trees. Regeneration can be natural or artificial. The shelterwood system can be achieved by making long, narrow, parallel strip cuts.

Figure 2 Forests are harvested using (a) clear-cutting, (b) shelterwood cutting, or (c) selective cutting.

In selective cutting, the forest is managed as an uneven aged system. Foresters periodically come in and harvest selected trees. This is the most costly type of cutting, but it has the least impact on the ecological features of the forest. Selective cutting is often performed on private woodlots. It also occurs where there are high values for individual trees and in parks, where the appearance of the forest is highly valued.

Ecological Issues in Forest Management

Natural forests sustain themselves without being managed. When producing wood fibre, however, management is widely practised. It is difficult to manage forests to meet commercial demands while maintaining ecological values.

Clear-cutting is usually the most profitable harvesting method, but there are drawbacks. Following clear-cutting, nutrients are lost from the soil, and erosion increases (Figure 3). Sediment entering streams harms fish spawning areas, and nutrients increase the growth of algae. Also, only one or two species of tree are planted when artificial regeneration methods are used (Figure 4). This dramatically reduces biodiversity. Shelterwood and selective cutting cause fewer environmental problems but are more expensive. The use of pesticides by the forestry industry is also of concern. This is examined in more detail in Chapter 4.

> **DID YOU KNOW?**
>
> **A Burning Question about the Mountain Pine Beetle**
> The mountain pine beetle has devastated lodgepole pine forests of British Columbia. One cause of this is fire suppression. When we prevent or control natural forest fires, we create forests with old trees that are preferred by the beetles. Allowing natural forest fires to occur helps combat this serious pest.

Figure 3 Exposed soil in clear-cut areas is subject to erosion.

Figure 4 This tree planter is likely planting only one or two species in this area.

Forest practices often include fire suppression because forestry companies do not want to lose valuable timber to wildfires. Firefighting teams attempt to stop the spread of fires as soon as they break out. This practice can have negative ecological impacts because many plant and animal species, such as fireweed and wild turkeys, benefit from fire cycles. The species mix of forests that are not subjected to natural fire cycles can change.

Given the very large areas that are available for cutting, it is vital that sound forestry practices are used so that these ecosystems remain sustainable. Forest certification is a recent positive trend in the industry. In 1993, the International Forest Stewardship Council was founded to set criteria and certify forest management practices that are sustainable. By 2006, over 270 million hectares, or 7 %, of the world's forests were certified, indicating that they are being managed sustainably (Figure 5). Sustainable forest practices balance tree harvests with growth rates, protect fish and wildlife habitat, protect waterways, and maintain biodiversity.

Figure 5 This logo means that the forest products comply with the Programme for the Endorsement of Forest Certification Chain-of-Custody Standard.

To learn about the profession of a forester, **GO TO NELSON SCIENCE**

In our economy, supply and demand influence the prices of goods. When a product is scarce or in high demand, its price tends to rise. As the price rises, there is a strong incentive to produce more of the product. This pattern of supply and demand can be dangerous when applied to natural resources. If the demand for a rare type of tropical wood increases, so will the price (Figure 6). The rise in value will increase the pressure on logging companies to cut more of the rare tree species. Soon the species will be overharvested and may become threatened. Balancing supply and demand is a problem on an international scale because all parties must agree to stop overharvesting the particular species.

Figure 6 High demands for valuable tropical tree species, such as mahogany, increase logging pressure and threaten the sustainability of forest ecosystems.

Wildlife Management

Ontario's wild animals are hunted for food and sport. In southern Ontario, deer and waterfowl are among the most commonly hunted game species. In northern Ontario, moose is a prized species. The impact of hunting depends on a variety of ecological factors. Is the population large and healthy? Is it increasing or decreasing in size? Is the population being properly monitored? Is the hunt uncontrolled, with little understanding of the implications for the ecosystem?

To learn about being a conservation officer, **GO TO NELSON SCIENCE**

Managed hunts are sometimes used to control populations. For example, wolves are now absent from many parts of Ontario. The loss of this top predator has resulted in large deer populations in some regions. As the number of deer increases, they deplete their natural food supply and feed on agricultural crops. The only natural controls on the population are starvation and disease. In such cases, a controlled hunt will bring the population down to a more sustainable size.

Wildlife is a renewable resource when it is consumed sustainably. This means that harvesting must not exceed a population's ability to replace itself. Historically, First Nations and Inuit peoples in Canada harvested wildlife in a sustainable way for many hundreds and thousands of years. Aboriginal peoples had small population sizes and had an intimate knowledge of their natural environment.

DID YOU KNOW?

Changing Fashions Saved the Beaver

Canada's historic fur trade was based on beaver trapping. Beaver pelts were shipped to Europe and made into felt hats. The demand was so great that beavers were driven to near-extinction. A change in European fashion meant that felt hats were replaced by silk hats. This caused the demand for beaver pelts to collapse.

The ethic of **stewardship** and pattern of sustainable harvesting changed dramatically with the arrival of European settlers. Europeans were more interested in resources as a source of revenue. They harvested animals to sell to large European markets rather than to feed and clothe themselves. Without a good knowledge of the land, they sought the expertise of Aboriginal peoples. Aboriginal peoples trapped furs in exchange for European goods such as cotton and rifles. Unfortunately, this resulted in overharvesting of a number of wildlife species.

stewardship taking responsibility for managing and protecting the environment

RESEARCH THIS: FISHERIES MISMANAGEMENT

SKILLS: Researching, Communicating

SKILLS HANDBOOK
4.B.

Fisheries management regulations often include size limits requiring fishers to release individuals that are under a minimum size. The assumption is that harvesting small young fish, lobsters, and crabs is wasteful.

This practice, however, may result in an unsustainable fishery. Instead of helping maintain a strong population, the minimum size limits have actually caused species to become smaller. Scientists now believe the exact opposite approach may be needed—keep the small ones and release the big ones!

1. Use the Internet to learn about fisheries science and how size limit regulations affect fish populations.

   GO TO NELSON SCIENCE

   A. Which individuals are the best reproducing individuals in a population? Describe them in terms of age and size.
   B. How might smaller individuals benefit when only large individuals are caught?
   C. What important Canadian fish species are being affected by current fishing practices? How has the average body size of these species changed over the past few decades?

UNIT TASK Bookmark

You can apply what you learned about how forestry practices and wildlife management influence species distributions to the Unit Task described on page 156.

IN SUMMARY

- Three main forest harvesting methods are clear-cutting, shelterwood cutting, and selective cutting.
- Forests must be managed to reduce ecological impacts such as soil erosion.
- Harvesting of wildlife must be done in a sustainable and ethical manner.
- The impact of hunting on wildlife is influenced by the population's size, health, food supply, and natural predators.
- A supply and demand–based economy increases pressure on rare and valuable natural resources.
- Sustainable forestry practices protect ecosystems and maintain biodiversity.

CHECK YOUR LEARNING

1. Identify the major natural resources derived from natural ecosystems.
2. Describe and compare three common forestry cutting methods.
3. In what ways do humans manage forest ecosystems? Explain.
4. List the benefits of forest product certification.
5. (a) Explain what is meant by "supply and demand."
   (b) How does it influence forestry practices?
6. How did the introduction of livestock change the relationship between people and large predators?
7. Describe how the arrival of Europeans in North America changed the relationships between humans and wildlife.
8. Most Ontario students have grown up in a place that was once a forested region but is no longer. What method of forestry practice produced this change?
9. Modern hunters use all-terrain vehicles, high-powered rifles, and global positioning systems (GPS). How might these technologies influence hunting success?

3.7 Consumption and Resource Management

3.8 PERFORM AN ACTIVITY

Comparing Forest Management Practices

Natural forests are self-sustaining and renewable. Modern forestry practices use natural forests but often manage them to increase the production of wood and wood fibre. These managed forests result in ecological changes that may have long-term consequences.

SKILLS MENU
- Questioning
- Hypothesizing
- Predicting
- Planning
- Controlling Variables
- Performing
- Observing
- Analyzing
- Evaluating
- Communicating

Purpose
To assess some impacts of forestry management practices on natural ecosystems.

Equipment and Materials
- Table 1: Average Nutrient Concentrations in Streams in Undisturbed and Logged Watersheds
- Table 2: Small Mammal Counts in Uncut, Shelterwood, and Clear-Cut Forests in a Maple–Birch Forest in Nova Scotia
- Table 3: Average Number of Breeding Pairs of Birds in Mature and Clear-Cut Forests
- graph paper or graphing-capable software

Procedure

1. Examine Table 1. These values represent the rates at which nutrients were lost by leaching from watersheds in a logged and undisturbed forest in a region northwest of Sudbury.

Table 1 Average Nutrient Concentrations in Streams in Undisturbed and Logged Watersheds

Variable	Undisturbed watershed	Logged watershed
calcium (mg/L)	2.5	2.8
iron (µg/L)	0.07	0.42
potassium (mg/L)	0.24	0.40
magnesium (mg/L)	0.65	0.70
ammonia (µg/L)	5.4	20.4
total nitrogen (mg/L)	170.5	317.1
sodium (mg/L)	0.95	1.03
total phosphorus (µg/L)	6.3	12.5
zinc (µg/L)	6.2	20.5

Source: Freedman, B., Morash, R. and Hanson, A. J. (1981) Biomass and nutrient removals by conventional and whole-tree clear-cutting of a red spruce-balsam fir stand in central Nova Scotia. *Canadian Journal of Forest Research, 11*, 249–257.

2. Plot bar graphs with quantity on the vertical (y) axis and nutrients along the horizontal (x) axis. Place the graph bars of each nutrient from the two watersheds next to each other as shown in the example below (Figure 1). Use two colours, one to represent each watershed. Label the axes appropriately.

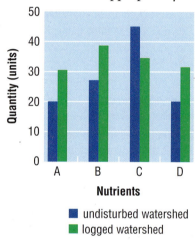

Figure 1 Sample bar graph

3. Examine Table 2. These values represent the average numbers of individuals caught per 100 days of trapping in three different forests in Nova Scotia.

Table 2 Small Mammal Counts in Uncut, Shelterwood, and Clear-Cut Forests in a Maple–Birch Forest in Nova Scotia

Species	Number caught per 100 days of trapping		
	Uncut forest	Shelterwood	Clear-cut
short-tailed shrew	3.9	3.4	3.4
masked shrew	2.2	2.3	4.9
red-backed vole	7.0	4.9	6.0
deer mouse	1.0	3.0	0.7
meadow vole	0.2	0.6	2.8
eastern chipmunk	0.2	0.4	0.2
woodland jumping mouse	2.9	4.7	0.3

Source: Swan, D., Freedman, B. and Dilworth, T. (1984) Effects of various hardwood forest management practices on small mammals in central Nova Scotia. *Canadian Field Naturalist, 98*, 362–364.

4. For each species, calculate the percentage of the total captures it represents in each forest type.

   > **MATH TIP**
   > To calculate percentage, add up the total number of mammals caught for each forest type. Then, divide the number caught for each species by the total for the forest type. Multiply your answer by 100 to get the percentage.

5. For each forest type, create a pie chart using the percentages calculated in step 4 to represent the species share of the pie. Use a different colour to represent each mammal species. (Hint: 10 % represents 36° of the circle.)

6. Examine Table 3. These values represent the number of pairs of each bird species per square kilometre in an uncut and clear-cut forest. The clear-cuts are 3 to 5 years old.

   Table 3 Average Number of Breeding Pairs of Birds in Mature and Clear-Cut Forests

Bird species	Number per square km	
	Mature maple–birch forest	Clear-cut (3–5 years old)
ruby-throated hummingbird	0	23
least flycatcher	137	0
hermit thrush	43	0
red-eyed vireo	53	0
black-throated green warbler	37	0
common yellowthroat	0	152
American redstart	65	0
white-throated sparrow	7	127

   Source: Quinby, P.A., McGuiness, F. and Hall, R. (1995) Forest Landscape Baseline No. 13. Brief Progress and Summary Reports.

7. Count the number of bird pairs and bird species present in each forest type. Record these values.

8. List the species found
   (i) only in the uncut forest
   (ii) only in the clear-cut forest
   (iii) in both the uncut and clear-cut forests

Analyze and Evaluate

(a) How did nutrient losses compare between the natural and clear-cut watersheds?

(b) Compare shrew populations in the three forest types. Were the differences dramatic?

(c) During harvesting, large amounts of nutrients contained in trees are removed from an ecosystem. How then can clear-cut ecosystems release *more* nutrients into surface streams?

(d) Shrews are voracious predators that feed on insects and other invertebrates. How might this influence their success in the different forest types?

(e) Voles, mice, and chipmunks are all rodents. Were they all similarly influenced by the different cutting methods?

(f) Which mammal species benefited most from forestry operations?

(g) How did clear-cutting affect bird species in the study represented by Table 3? Compare the effects on overall numbers with the effects on individual species.

(h) Would you describe most of the bird species as generalists (living in more than one habitat type) or specialists (preferring a single type of habitat)? Explain.

Apply and Extend

(i) Select two species of birds that were found only in the uncut forest and two that were found only in the clear-cut forest. Use the library and Internet resources to find out about their ecological niche. Where do they nest, and what is their main source of food?

(j) How might the long-term loss in soil nutrients influence the selection of forest cutting practices?

(k) Why is it critical to understand the ecological niche of a species in order to predict how it will respond to environmental changes? Give examples to support your answer.

CHAPTER 3 LOOKING BACK

KEY CONCEPTS SUMMARY

Natural ecosystems are of great value to humans.

- Ecosystems provide us with products as well as other ecosystem services. (3.1)
- Ecosystem services are free and renewable. (3.1)
- The total economic value of ecosystems is in the trillions of dollars. (3.1)

Ecosystems are at equilibrium but can change over time.

- Large ecosystems are usually in equilibrium, with biotic and abiotic features remaining relatively constant over time. (3.2)
- Succession is the gradual process of changes in an ecosystem over time. It is initiated by a disturbance. (3.2)

Biodiversity describes the variety and abundance of life in an ecosystem.

- Biodiversity is the variety of life in a particular ecosystem. (3.3)
- Biodiversity of many ecosystems is threatened by human activities. (3.3)
- Many species are going extinct. (3.3)
- At-risk species are categorized as extirpated, endangered, threatened, or of special concern. (3.3)

Many human activities impact and threaten the sustainability of natural ecosystems.

- The major cause of species loss is habitat destruction and fragmentation. (3.4)
- Invasive species can outcompete native species and upset food webs. (3.5)
- Pollution threatens the health of many plant and wildlife populations. (3.6)
- Much of southern Ontario's original forest has been converted to agricultural and urban land uses. (3.4)
- Invasive species have been introduced intentionally and accidentally. (3.5)

Water, land, and air pollution cause health and economic problems.

- The release of harmful materials into the environment is called pollution. (3.6)
- Acid precipitation affects terrestrial and aquatic ecosystems and is detrimental to human health and infrastructure such as buildings. (3.6)
- Oil spills and plastics cause significant damage to marine and freshwater ecosystems. (3.6)
- Massive mats of floating plastic trash have formed in the oceans. (3.6)

Plant and animal resources should be used in a sustainable manner.

- Forests are harvested using clear-cutting, shelterwood cutting, or selective cutting methods. (3.7)
- Harvesting of wildlife must be done in a sustainable and ethical manner. (3.7)
- Sustainable forestry practices sustain ecosystems and maintain biodiversity. (3.7, 3.8)
- A supply and demand based economy can place added pressure on rare and valuable natural resources. (3.7)

WHAT DO YOU THINK NOW?

You thought about the following statements at the beginning of the chapter. You may have encountered these ideas in school, at home, or in the world around you. Consider them again and decide whether you agree or disagree with each one.

Vocabulary

equilibrium (p. 80)
succession (p. 80)
primary succession (p. 80)
secondary succession (p. 80)
biodiversity (p. 83)
species richness (p. 83)
extinct (p. 83)
extirpated (p. 85)
endangered (p. 85)
threatened (p. 85)
special concern (p. 85)
invasive species (p. 91)
pollution (p. 96)
acid precipitation (p. 96)
neutralize (p. 96)
bioremediation (p. 99)
stewardship (p. 105)

1 Forest fires benefit these species.
Agree/disagree?

4 Human demand for safe, clean water is a major concern on a global scale but not an issue in Canada.
Agree/disagree?

2 Pollution is the greatest human-caused threat to natural ecosystems.
Agree/disagree?

5 Ontario has a healthy wildlife population with very few "at-risk species."
Agree/disagree?

3 Lake trout and a panther grouper are able to share the same habitat.
Agree/disagree?

6 Humans enjoy natural ecosystems, but we do not rely on them in our daily lives.
Agree/disagree?

How have your answers changed since then?
What new understanding do you have?

BIG Ideas

✓ Ecosystems are dynamic and have the ability to respond to change, within limits, while maintaining their ecological balance.

✓ People have the responsibility to regulate their impact on the sustainability of ecosystems in order to preserve them for future generations.

CHAPTER 3 REVIEW

The following icons indicate the Achievement Chart category addressed by each question.

K/U Knowledge/Understanding **T/I** Thinking/Investigation **C** Communication **A** Application

What Do You Remember?

1. In your notebook, write the word(s) needed to complete each of the following sentences. **K/U**
    (a) Ecosystem _____ refers to the many benefits that we receive from ecosystems. (3.1)
    (b) The changes following events such as forest fires are called _____ succession. (3.2)
    (c) Tropical _____ exhibit the highest biodiversity of any ecosystem. (3.3)
    (d) The Committee on the Status of Endangered Wildlife in Canada monitors all species that are at _____. (3.3)
    (e) _____ occurs when a single large and continuous natural area is separated into smaller patches. (3.4)
    (f) Most _____ loss in Canada occurred during the nineteenth and early twentieth centuries. (3.4)
    (g) When non-native species are able to outcompete native species, they may become _____. (3.5)

2. Match the term on the left with the appropriate definition on the right. (3.3) **K/U**
    (a) extinct   (i) a species that may become threatened or endangered
    (b) extirpated   (ii) a species facing imminent extinction or extirpation
    (c) endangered   (iii) a species that no longer exists
    (d) threatened   (iv) a species that no longer lives in a particular region
    (e) special concern (v) a species likely to become endangered if factors do not change

3. (a) Which type of ecosystem has the greatest biodiversity?
   (b) Where are these ecosystems located? (3.3) **K/U**

4. Describe some of the ecological impacts of
    (a) acid precipitation on aquatic ecosystems
    (b) acid precipitation on terrestrial ecosystems
    (c) oil spills in marine ecosystems
    (d) plastic waste that enters the oceans (3.6) **K/U**

5. Briefly describe four techniques used to help clean up oil spills. (3.6) **K/U**

What Do You Understand?

6. Classify each of the following as an ecosystem cultural service, a product, or another service. (3.1) **K/U** **T/I**
    (a) cross-country skiing trail
    (b) wood and paper products
    (c) clean air
    (d) medicinal compounds
    (e) wilderness canoe routes
    (f) groundwater
    (g) seafood

7. Match the human activity on the left with the possible ecological impact on the right. (3.4, 3.6) **K/U**
    (a) removing coastal vegetation   (i) loss of species dependent on aquatic ecosystems
    (b) commercial trawler fishing   (ii) sediments in runoff smother natural habitats
    (c) wetland drainage   (iii) increased acid precipitation
    (d) clearing of land next to rivers (iv) damage to ocean-bottom ecosystems
    (e) release of sulfur oxides   (v) increased erosion and loss of turtle nesting sites

8. Plastic pollution is a major global concern. (3.6) **K/U** **T/I**
    (a) List all the plastic items you have used today.
    (b) Classify each of these items under the following headings: often reused, usually recycled, placed in the garbage, discarded as litter.

9. In many ways, garbage is an invention of the modern world. Explain why Aboriginal peoples did not have problems similar to our modern waste management issues. (3.6) **A**

10. Describe the status of the loggerhead shrike. Why is this species at risk? (3.4) **K/U**

11. Habitat loss is not as rapid in Ontario as it once was. Does this mean that people living in Ontario no longer play a major role in habitat loss? (3.4) **K/U**

12. Describe the three main ways of controlling invasive species. (3.5) **K/U**

Solve a Problem

13. Examine the two simplified food webs in Figure 1. Since the arrival of Europeans, all of the cougars and elk and most of the wolves that once lived throughout southern Ontario are gone. (3.5) K/U T/I A C

    (a) List differences in the two food webs.
    (b) What impacts do you think these changes had on the remaining species?
    (c) Is it surprising that some regions of Ontario are experiencing overpopulation problems with white-tailed deer and beaver? Explain.
    (d) Lyme disease is spread by wood ticks that feed on white-tailed deer. Predict how these changes might influence the spread of Lyme disease.

Prior to European settlement

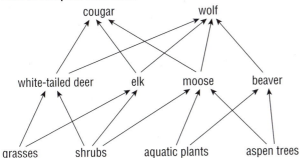

200 years after European settlement

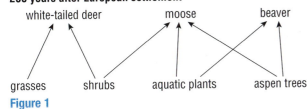

Figure 1

Create and Evaluate

14. Some people hunt and fish as a form of recreation. Should hunting and fishing be allowed? Should there be different rules for those who hunt and fish for food compared with those who do so for sport? Explain your answers. (3.7) A C

15. Snowmobiling, power-boating, and even driving to the cottage contribute to air pollution. Is this okay? Do you think people should be allowed to pollute in order to enjoy nature? (3.7) A C

16. Wilderness park policies often state that any naturally occurring fires are to be allowed to burn without interference. This can result in the loss of large areas of prized parkland. Do you think it is wise to let fires burn naturally inside park boundaries? Why or why not? (3.7) A C

Reflect on Your Learning

17. In this chapter, you learned that large numbers of species are considered to be at risk in Ontario. (3.3) T/I C

    (a) Before reading this chapter, how many at-risk species were you familiar with?
    (b) How could you find out more about plants and animals that are at risk in your region of the province?

18. How might the information you learned in this chapter influence your behaviour? A

Web Connections

19. The Galápagos Islands are home to some of the most amazing species on Earth and attract large numbers of tourists. The Galápagos Islands have also suffered from the introduction of many invasive species. Research the current status of these islands and report on the following: (3.5) T/I

    (a) What makes these islands so unique?
    (b) How is the impact of tourism controlled in the Galápagos?
    (c) What invasive species are of greatest concern?

20. Select one Canadian ecological issue discussed in this chapter. Research what Canadian organizations are involved in this issue. Look for one government agency, one non-governmental organization, and one environmental group that includes high school–aged students. T/I A C

    (a) What are the organizations?
    (b) What actions or projects have they been involved with?
    (c) What have they accomplished? How have their actions benefited the environment?

CHAPTER 3 SELF-QUIZ

The following icons indicate the Achievement Chart category addressed by each question. K/U Knowledge/Understanding, C Communication, T/I Thinking/Investigation, A Application

For each question, select the best answer from the four alternatives.

1. Which of the following describes an ecosystem in equilibrium? (3.2) K/U
   (a) an island that has just been hit by a large, powerful hurricane
   (b) an established forest that supports a variety of plants and animals
   (c) a region in the Arctic where the average temperature is increasing
   (d) an area of grassland with plants just starting to grow one year after a fire

2. Which of the following is an example of mechanical control of invasive species? (3.5) K/U
   (a) removing the species by hand
   (b) applying pesticides to the species
   (c) educating people about the species
   (d) intentionally introducing another species

3. Which of the following correctly explains bioremediation? (3.6) K/U
   (a) introducing organisms to control invasive species
   (b) re-establishing micro-organisms in a disturbed ecosystem
   (c) using micro-organisms to consume environmental pollutants
   (d) collecting organisms produced by ecosystems for commercial use

4. Which is the most serious threat to the sustainability of Earth's natural ecosystems? (3.4–3.7) K/U
   (a) invasive species
   (b) climate change
   (c) habitat fragmentation
   (d) air and water pollution

Copy each of the following statements into your notebook. Fill in the blanks with a word or phrase that correctly completes the sentence.

5. A tsunami hits an island and removes the surface soil and plant and animal life. A year later, small plants and insects begin to reappear on the island. This change is an example of _____ succession. (3.2) K/U

6. _____ is produced when sulfur dioxide or nitrogen oxides combine with water vapour in the atmosphere and return to Earth as rain or snow. (3.6) K/U

7. Communities that are at _____ show little change over long periods of time. (3.2) K/U

Indicate whether each of the statements is TRUE or FALSE. If you think the statement is false, rewrite it to make it true.

8. When a non-native species is introduced to a new environment, it is very likely that it will survive and harm native species. (3.5) K/U

9. Decreasing the concentration of a pollutant prevents it from harming the environment. (3.6) K/U

10. Human actions have caused all the species extinctions on Earth. (3.3) K/U

Match each term on the left with the most appropriate description on the right.

11. (a) acid rain   (i) blocks the digestive systems of animals
    (b) plastic at sea   (ii) increases rates of soil erosion
    (c) oil spill   (iii) introduces disease-causing organisms
    (d) clear-cutting   (iv) damages statues and buildings
    (e) invasive species (v) interferes with an animal's ability to insulate
    (3.5–3.7) K/U

Write a short answer to each of these questions.

12. Why have many wetlands been converted into farmland? (3.4) K/U

13. Describe at least two ways in which plant communities protect the environment. (3.1) K/U

14. Sketch a newspaper advertisement that promotes a type of ecotourism that could be conducted in Ontario. C

15. We often think of biodiversity as a term describing natural ecosystems. It can also be applied to engineered ecosystems. (3.3) T/I A
    (a) Briefly describe examples of biodiversity in your neighbourhood.
    (b) Do you think your neighbourhood has high species richness? Explain why or why not.

16. Plastic shopping bags are a major source of plastic that harms the environment. Describe two ways by which you can reduce the number of plastic bags entering your local environment. A

17. Stewardship is something that we should all be concerned with. (3.7) K/U A
    (a) Explain what it means to you to be a good steward of your natural environment.
    (b) Give at least two specific examples of good stewardship.

18. Your local officials want to set aside land for new parks. They ask for input from the public to help them make decisions concerning the parks. List at least two suggestions you would make that would increase the parks sustainability. Include an explanation for each of your suggestions. (3.4) T/I C

19. Can succession help an ecosystem recover from any type of disturbance? Explain your answer. (3.2) T/I

20. Choose a park or wilderness area you have visited and explain how you benefited from your visit. (3.1) T/I

21. Draw a diagram that illustrates a system in equilibrium. Label the diagram to help explain what is taking place within the system. (3.2) K/U C

22. Invasive species create a problem for native species competing for the same resources. Describe two things you can do to limit the number of non-native species introduced into your community. (3.5) A

23. Some people who live in neighbourhoods that have a deer population put food out for the deer. (3.7) T/I
    (a) How would this action affect both the deer population and human population in the neighbourhood?
    (b) Do you think this is a wise practice? Explain your answer.

24. Scientists can learn some aspects of a past civilization by studying the trash it produced. Suppose scientists a thousand years in the future were to study the trash in your community's landfill. T/I
    (a) Name something they might find that would make them think we were not good stewards of the environment. Explain your choice.
    (b) Name something you would like them to find that would make them think we were good stewards of the environment. Explain your choice.

25. Some parks that contain fragile natural features, such as caves, limit the number of people who can visit them each day. Some people think this is a good way to preserve natural features. Others think these parks should be open to everyone who wants to visit them. What do you think? Explain your opinion. C

26. Construct a table comparing the costs and benefits of using clear-cutting, shelterwood cutting, and selective cutting. (3.7) K/U

CHAPTER 4

Ecosystems by Design

KEY QUESTION: Can Human-Designed Environments Be Sustainable?

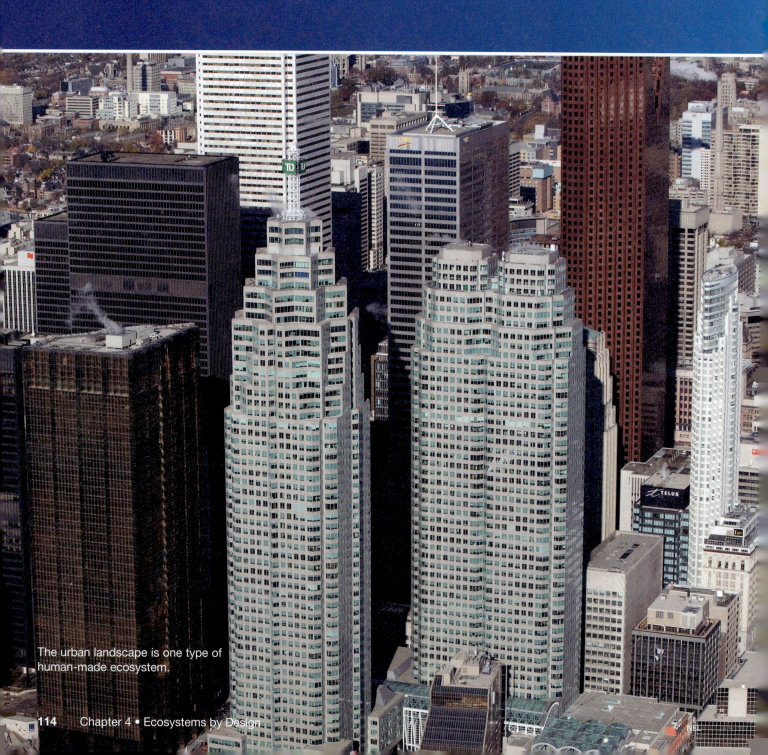

The urban landscape is one type of human-made ecosystem.

UNIT B: Sustainable Ecosystems

- **CHAPTER 2** Understanding Ecosystems
- **CHAPTER 3** Natural Ecosystems and Stewardship
- **CHAPTER 4** Ecosystems by Design

KEY CONCEPTS

Some ecosystems, such as farms and cities, are designed, created, and maintained by humans.

Agricultural practices alter natural biogeochemical cycles.

Agricultural practices alter water cycles.

Pesticides are used to reduce yield losses from pests, but there are ecological costs associated with them.

Some pesticides and other toxins bioaccumulate in individuals and biomagnify in food webs.

Urban ecosystems are dependent on surrounding ecosystems to provide food and materials for urban inhabitants.

ENGAGE IN SCIENCE

Living the Green Life

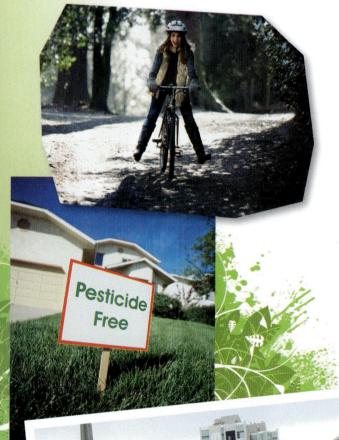

Imagine living in an environmentally friendly city. Consider what you might see or experience in such a community.

As you bicycle through your neighbourhood along a designated bike lane, you notice solar water heaters on every home. Most yards have large trees and pesticide-free gardens. You notice that the commuter traffic is not as congested as in most cities, even at rush hour. Most cars are small, and there are many zero-emission buses.

As you bike through the business section on your way to the beach, you notice that many of the buildings have "green roofs" planted in grasses and wildflowers. The air remains clean and fresh, and there are a number of large wooded parks in the downtown area. They are home to numerous species of wildlife, including songbirds, squirrels, and raccoons.

When you arrive at the beach, you notice that the sand is free of litter and there is a sign stating that the water is safe for swimming. How would you feel about living in such a community? What are some of the benefits of "green" living?

WHAT DO YOU THINK?

Many of the ideas you will explore in this chapter are ideas that you have already encountered. You may have encountered these ideas in school, at home, or in the world around you. Not all of the following statements are true. Consider each statement and decide whether you agree or disagree with it.

1 Agricultural practices closely resemble naturally occurring ecosystem processes.
Agree/disagree?

4 Many of the plants and animals farmed in Ontario are native species.
Agree/disagree?

2 Organic farming has benefits for the environment.
Agree/disagree?

5 The actions of people who live in cities have little impact on wilderness environments.
Agree/disagree?

3 Pesticides are needed to control pests.
Agree/disagree?

6 We can use as many resources as we like because we can recycle or reuse them.
Agree/disagree?

FOCUS ON READING

Asking Questions

Before reading a text like the example below, ask yourself questions to recall what you already know about the topic. During reading, ask questions to figure out what a text means. After reading, ask questions to extend your understanding. You can ask and answer three types of questions:

- **Literal ("On the Lines"):** These questions are about information that can be located in the text.
- **Inferential ("Between the Lines"):** These questions require that you combine clues in the text with something you already know.
- **Evaluative ("Beyond the Lines"):** These questions call for using evidence from a text to make a judgment.

> **READING TIP**
> As you work through the chapter, look for tips like this. They will help you develop literacy strategies.

Alternative Farming Practices

Figure 1 Soybeans, corn, and wheat are often grown in rotation in southern Ontario.

Most farmers rotate or change the crops they plant on a certain area of land on a regular basis. By rotating crops, farmers can reduce their use of fertilizers and pesticides. For example, in southern Ontario, farmers often plant soybeans, corn, and wheat in a three-year rotation (Figure 1). During a soybean year, the nitrogen concentration of the soil increases through the action of nitrogen-fixing bacteria living in root nodules. This reduces the need to add nitrogen fertilizer the following year for corn. In the third year of the rotation, wheat is planted. Because wheat is not planted in rows, it is competitive against weeds and requires less herbicide.

Asking Questions *in Action*

Asking questions can help you focus on what you know about a topic, find information, help you figure out implications, or form opinions based on evidence from the text. Here's how one student asked questions to understand information about alternative farming practices.

Questions I asked before, during, and after reading the text	How that question helps me understand the text
• **Literal:** What is crop rotation?	helps me understand that farmers plant different crops over a three-year cycle
• **Inferential:** Why do farmers rotate their crops?	helps me understand that farmers plant a crop one year to enrich the nutrients in the soil for a different crop the next year
• **Evaluative:** Does the author have a bias?	helps me make a judgment that the author believes crop rotation is good for the environment because less fertilizer and pesticides are used for the soil

4.1 Engineered Ecosystems and Modern Agriculture

As you read this text, you are probably surrounded by a human-designed environment. The room may be lit by fluorescent bulbs instead of the Sun. The floors, walls, and ceiling are not part of a natural environment. They are part of a constructed building. The building has its own energy and water supplies. There is a system for taking away human waste. Whether you are at school or at home, you will find supplies of food in the cafeteria or kitchen.

Unlike most species, humans spend little time in natural ecosystems. Even when you go outside, you are likely to be on a lawn or a sidewalk. Whether you live in a large urban centre or a rural community, the abiotic and biotic features around you have been changed. In some cases, such as a park, the features of an ecosystem are still recognizable. In others, such as a parking lot, most biotic features have been eliminated.

Examine the variety of environments in Figure 1. What do they have in common? What features suggest that they are not natural ecosystems?

In each photo, the work of humans is obvious. The natural vegetation has been removed and replaced with other plants or artificial surfaces. Recreational areas are often covered in gravel, wood chips, or fields of grass. Transportation routes are covered in asphalt. A massive mine has exposed Earth's lithosphere. In all of these examples, the abiotic and biotic features have been changed. Biogeochemical and water cycles have been changed as well.

Engineered ecosystems cover a large portion of Earth's land area. As the human population increases, we replace natural ecosystems with land uses to support our modern lifestyles: farms (or **agroecosystems**), urban centres, and roads.

(a)

(b)

(c)

Figure 1 The biotic and abiotic features of this (a) golf course, (b) open-pit mine, and (c) neighbourhood have been changed.

agroecosystem an agricultural ecosystem

TRY THIS WATCH WHERE YOU STEP!

SKILLS: Observing, Analyzing

In this thinking activity, you will use your knowledge of land uses to obtain a qualitative sense of how Earth's surface has been altered for human use. Think of Earth's surface as a checkerboard or mosaic of different surfaces, some natural and some human-made, but each with distinct characteristics and potential uses.

1. Brainstorm a list of the different surfaces you have used during the last 24 hours, month, and year. Create a table similar to Table 1 at right, and classify the surfaces you have used as either manipulated or natural.

A. During your daily routine, do you make greater use of manipulated or natural surfaces? K/U
B. How many different surfaces have you used in the past year? K/U
C. How many of these surfaces have been altered by humans? K/U
D. Would it have been possible to do these same activities in a completely natural setting? Explain. A

Table 1

	Manipulated surfaces	Natural surfaces
24 hours		
Past month		
Past year		

> **READING TIP**
>
> **Ask Questions in Action**
> If you do not understand something while reading a text, stop reading and ask a question about what you do not understand. Try rereading the previous sentence to see if you missed information that answers your question. Or, you can continue reading to see if the information in the next sentences or paragraphs answers your question.

The ability of humans to manipulate their environment comes from two innovations: tools and concentrated energy sources. Humans have been making tools for thousands of years, but those powered by hand or horse have done less to alter the natural landscape. The recent use of fossil fuels has given us the ability to literally move mountains.

Humans are able to dramatically alter the environment. However, we are not able to change the basic relationships that exist between the environment and ourselves. To live sustainably, we must not disrupt the biotic and abiotic conditions we need to survive. Actions that disrupt water and nutrient cycles, threaten biodiversity, or alter climate patterns jeopardize our ability to survive.

Modern Agriculture—Food for Thought

Eating is one of our fundamental biological activities. Like breathing and drinking, eating food is something we must do. With a population of close to 7 billion people on Earth, producing food is critical to our survival and must continue on a large scale.

Unlike other animals, humans obtain very little of their food from natural ecosystems. Instead, most of our food is produced on large agricultural farms. Most fruits and vegetable crops are grown in large fields. Our meat is provided by a few kinds of domesticated animals that are raised on farms.

Canadian Agriculture Is Based on Imported Species

If you are like most Canadian teenagers, you have probably enjoyed eating pizza. The Hawaiian-style pizza is a popular choice. It consists of tomato sauce, cheese, ham, and pineapple. You probably have not thought about the origin of the ingredients or how their production affects the environment. Even though most of the ingredients were grown in Canada, none of them come from species *native* to Canada (Figure 2).

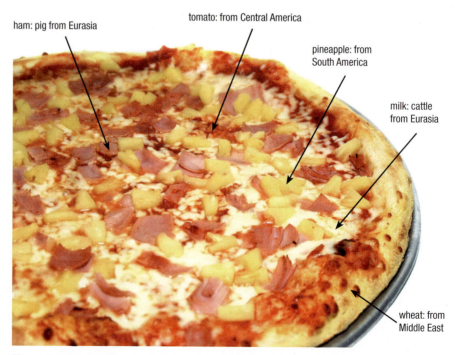

Figure 2 Most of the ingredients in this pizza are derived from species that are not native to Canada.

Like the pizza ingredients, all the major foods we produce in Canada were introduced from other parts of the world. They are all "exotic" species (Table 2). These non-native plants and animals have been domesticated by humans over thousands of years. The results are easy to see. These food species are highly productive, nutritious, and relatively easy to produce in large quantities.

Table 2 Selected Native and Non-Native Crops and Livestock Species Grown and Raised in Canada

	Non-native species	Native species
plant crops	corn, wheat, potatoes, beans, barley, rye, oats, soybeans, canola, tomatoes, apples, squash, beans, peanuts, carrots	strawberries, maple sap, wild rice, blueberries
livestock species	cattle, pigs, chickens, sheep, goats	turkeys

> **DID YOU KNOW?**
> **Grass Seeds Again?**
> About 90 % of our world's food comes from just 15 plant and 8 animal species! The top three food sources are the seeds of rice, wheat, and corn—all of which are grasses.

Agroecosystems Differ from Natural Ecosystems

The presence of non-native rather than native species is just one way in which farmland differs from natural ecosystems. Figure 3 highlights some of the other major differences. In general, when compared to natural ecosystems, agroecosystems have more uniform abiotic features, have lower biodiversity, and are more intensively used by humans. Agroecosystems also require humans to maintain them.

Figure 3 In (a) natural ecosystems, many species interact, participating in and maintaining natural cycles, whereas in (b) agricultural ecosystems, a very limited number of species interact, and most cycles are directly altered by human activity.

On agroecosystems, farmers grow crops in **monocultures** of non-native species where only one species is grown in a large field. In a monoculture, ecological cycles are altered, and biodiversity is reduced. Consumers that feed on crops are considered **pests**. You will learn more about pests and agroecosystems in Section 4.4.

monoculture the cultivation of a single crop in an area

pest any plant, animal, or other organism that causes illness, harm, or annoyance to humans

4.1 Engineered Ecosystems and Modern Agriculture

READING TIP

Ask Evaluative Questions

Sometimes ideas or information in a text makes you ask an evaluative question. For example, you might wonder about converting natural ecosystems to farmland and ask, *Is it worth converting ecosystems to monocultures to grow food? Is a more sustainable solution possible?* You may need to conduct further research on the issue to answer your evaluative questions and make an informed judgment.

The differences outlined in Figure 3 raise concerns about our ability to produce food sustainably. While natural ecosystems sustain themselves over thousands of years, engineered ecosystems such as farmland must be managed. Farmers manage abiotic and biotic conditions to maximize the success of growing monocultures. They attempt to create ideal and uniform growing conditions and to eliminate competitors, diseases, and pests (Figure 4). Ploughing, weeding, fertilizing, irrigating, and the spraying of pesticides are examples of such management techniques.

Figure 4 Most crops in Canada, such as these sunflowers, are grown in monocultures. Farmers must manage monocultures carefully so that only the desired crops will grow.

Remember what you learned about biodiversity. Does the management of a monoculture result in a stable ecosystem? In the next sections, we will consider how these management practices influence our environment.

IN SUMMARY

- Engineered ecosystems, such as farmland and cities, make up a large portion of Earth's land area.
- Most of Canada's crop and livestock species are introduced, non-native species.
- Engineered ecosystems have uniform abiotic features and low biodiversity.
- Most monoculture crops consist of plants of uniform age and size.
- Farmers attempt to create ideal abiotic and biotic growing conditions for crops.

CHECK YOUR LEARNING

1. List six examples where humans have replaced a natural ecosystem with an engineered ecosystem and describe its new purpose. K/U C
2. How do engineered ecosystems differ from natural ecosystems? Refer to both biotic and abiotic ecosystem characteristics. K/U
3. Describe how you use natural ecosystems. A C
4. Describe how you use engineered ecosystems. A C
5. Do Canadians rely on native plant and animal species for all their food? Explain using examples. K/U C
6. How do major human food sources differ from those of most animals? K/U
7. Do you always have to be physically located in an ecosystem to be making use of its services? Explain. K/U
8. Suggest some possible reasons why non-native species are so extensively used in agriculture. K/U A
9. Why do we not milk moose and grow fields of native plants for food? K/U A
10. Construct a table that compares the characteristics of natural ecosystems with the characteristics of agricultural ecosystems. K/U C
11. List some management techniques used by farmers to alter abiotic and biotic growing conditions. K/U

122 Chapter 4 • Ecosystems by Design

4.2 Managing the Soil—Controlling the Flow of Nutrients and Water

One of Earth's most important resources is, quite literally, under your feet. It is the ground you walk on. You may think of soil as the "dirt" you bring into the house on your shoes or something needed for potted plants. For farmers, fertile soil is essential to their livelihood. Plants depend on the soil for their physical support and to provide water, nutrients, and oxygen to their roots. Farmers must consider these functions to manage soil resources sustainably.

Soil is a complex mixture of minerals, water, dissolved nutrients, air spaces, and decomposing organic matter (Figure 1). It is also home to countless organisms, ranging in size from microscopic bacteria to large burrowing mammals. Soils are among the most complex and poorly understood components of Earth's ecosystems, yet they are critical to our survival.

DID YOU KNOW?
The Speed of Soil Formation
The small mineral particles that make up soil are formed from rock. Wind, water, chemical processes, and living organisms gradually cause rock to break up into small particles. It can take 200 years to form a layer of soil just 1 cm thick!

DID YOU KNOW?
The Vital Role of Soil Fungi
Soil fungi can live in a mutualistic relationship with plants. The microscopic filaments of these "mycorrhizal" fungi surround the fine root hairs of their plant partners. The fungi deliver nutrients and water to the plant. In turn, the plant provides energy-rich food molecules to the fungi.

Figure 1 A few grams of healthy soil contain billions of micro-organisms.

RESEARCH THIS: WHAT ARE THE CHEMICAL CONTENTS?

SKILLS: Researching, Communicating

SKILLS HANDBOOK 4.A.

Humans alter the chemical makeup of soils when they add natural or synthetic soil supplements. These may be fertilizers, pesticides, or soil conditioners, such as peat moss. In this activity, you will research the chemical composition of soil supplements.

1. Gather data on 10 commercially available soil supplements. These products are sold at garden and hardware stores. Most of the information you will need is on the ingredient labels. You can also go to online sources.

 GO TO NELSON SCIENCE

A. For each product you researched, list the following information:
   (i) Product name
   (ii) Is the product natural or synthetic?
   (iii) What is the product's intended use? (home and garden, agricultural, and so on)
   (iv) What are the main ingredients in the product?
   (v) What is the intended function of the product? Is it a fertilizer, soil conditioner, pesticide, or combination?

B. Present your findings in table form.

Managing Soil Nutrients

Plants require nutrients to grow. The most important nutrients are nitrogen, phosphorus, and potassium. As you learned in Chapter 2, nutrients are cycled through ecosystems as plants grow and are consumed by animals and micro-organisms as they produce wastes, die, and decompose.

On farmland, or agroecosystems, natural nutrient and water cycles are disrupted. As crops grow, they take up nutrients from the soil and incorporate them into their tissue. However, when farmers harvest crops, the nutrients contained in the crops are removed from the ecosystem (Figure 2). The nutrients enter the human food chain when we eat the crop or the livestock that ate these crops. If human and livestock wastes are not returned to the original farmland, the soil will gradually be depleted of nutrients.

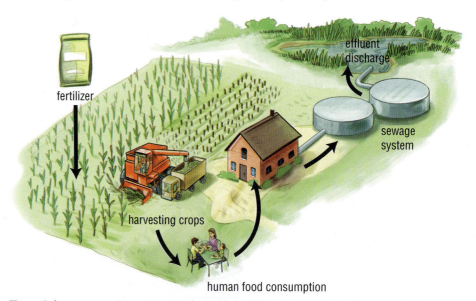

Figure 2 In agroecosystems, many nutrients (shown here as arrows) are added to the soil as fertilizers. Nutrients are taken up by plants and removed from the agroecosystem when we harvest the crop. After we eat the crop, some of the nutrients end up in our waste disposal system.

Natural and Synthetic Fertilizers

How can farmers manage biogeochemical cycles to replace the lost nutrients in an environmentally sustainable way? The most common method of replacing lost nutrients on farmland is to apply fertilizers. All natural ecosystems recycle natural fertilizers in biogeochemical cycles. Most of these ecosystems have remained healthy and sustainable for thousands of years.

For most of human history, farmers have depended on **natural fertilizers** made from materials such as plant and animal wastes. Currently, the most common practice is to apply **synthetic fertilizers**, manufactured by humans. Synthetic fertilizers were largely responsible for the "Green Revolution." In the 1960s, as farmers began using synthetic fertilizers, crop yields dramatically improved. This increased food production and farm profitability.

natural fertilizer plant nutrients that have been obtained from natural sources and have not been chemically altered by humans

synthetic fertilizer fertilizers that are manufactured using chemical processes

Environmental Impacts of Fertilizer Applications

Using fertilizers has drawbacks for both terrestrial and aquatic ecosystems (Figure 3). The nutrients in synthetic fertilizers are highly concentrated. As a result, they may enter the soil rapidly and alter the community of soil organisms. This can lead to soil that has less natural organic matter, as well as stressed soil organisms. Such soils can lose their supply of naturally occurring nutrients. This creates a dependency on synthetic fertilizers. These soils also become susceptible to erosion. Such a situation is not sustainable in the long term.

READING TIP

Read Between the Lines
If you read a word or term that you are unfamiliar with seeing in print, say the word aloud and ask yourself if you have heard the word before or used it when speaking. Also, check the margin and the context of the sentence or paragraph to see if a definition is given.

Figure 3 The use of both (a) natural and (b) synthetic fertilizers increases food production, but has ecological consequences.

Fertilizers, especially concentrated synthetic fertilizers, can have serious impacts on groundwater if they are leached from the soil. **Leaching** occurs when nutrients become dissolved in water and seep out of the soil. Groundwater that has been contaminated with nitrogen compounds is an increasingly serious problem, especially when it is used as drinking water. Drinking water with high levels of nitrogen compounds in it can cause health problems, particularly in infants.

leaching the process by which nutrients are removed from the soil as water passes through it

Aquatic ecosystems are often negatively impacted by agricultural practices. During heavy rain or spring runoff, fertilizers can enter aquatic ecosystems. This effect is more pronounced when synthetic fertilizers are used but also occurs when a large volume of animal manure is applied to a field. Leached nutrients increase the growth of algae, especially in warm, shallow ponds or lakes. The result is an "algal bloom" in which algae grow and then die and decompose (Figure 4). This decomposition is caused by bacteria that use oxygen. These bacteria decrease the level of dissolved oxygen in the water, which kills fish and other aquatic organisms.

Figure 4 Algal blooms decrease the oxygen content of lakes and ponds.

Natural fertilizers tend to be less concentrated than synthetic fertilizers and release nutrients more slowly. Using natural fertilizers is more similar to natural biogeochemical cycles and has less impact on soil quality. However, natural fertilizers can have a negative impact if they are applied inappropriately. Table 1 summarizes some key features of synthetic and natural fertilizers.

Table 1 Advantages and Disadvantages of Using Synthetic and Natural Fertilizers

	Synthetic fertilizers	Natural fertilizers
examples	ammonia, synthetic urea, potash, potassium, commercial chemical fertilizers	animal manure, sludge, plant materials such as seaweed and compost, blood meal, bone meal, wood ashes
advantages	• nutrients are released quickly • amounts of nutrients can be precisely measured • relatively easy to apply	• less danger of overfertilizing • release nutrients slowly • can improve soil structure • benefit soil micro-organisms and nutrient cycling
disadvantages	• production is energy intensive • cause water pollution • nutrients lost from soil through leaching • can cause an imbalance in soil chemistry and upset the balance of soil micro-organisms	• low concentrations of nutrients • release of nutrients may be slower than desired • not easy to measure the quantity of nutrients • may be more difficult to apply

> **DID YOU KNOW?**
> **The Walkerton Tragedy**
> In 2000, the drinking water for the town of Walkerton was contaminated. Heavy rains washed animal waste containing a deadly strain of *Escherichia coli* bacteria into one of the town's wells. At least seven people died, and 2500 others became ill.

Controlling the Flow of Water in Soil

To raise healthy livestock and grow crops, farmers must have a reliable supply of water. To accomplish this, farmers can add water to fields using irrigation or remove water using drainage technologies (Figure 5).

Figure 5 Farmers can (a) add water to agroecosystems using irrigation or (b) remove water using drainage pipes buried below the soil surface.

Farmers obtain water from surface sources such as lakes or rivers or from groundwater. With large-scale irrigation projects, it is possible to even farm a desert. In some cases, irrigation can create ideal growing conditions. In California, for example, by adding water to the otherwise dry environment, farmers have become major producers of rice. This crop requires very wet growing conditions.

To remove water from saturated fields, farmers install drainage tiles. This allows more air to penetrate the soil and gives roots access to oxygen. Drainage is often used to convert natural wetlands into fertile farmland. This has created some of the most productive farmland in Canada. However, this practice has caused a dramatic loss of wetlands and their associated wildlife species.

Irrigation and drainage practices improve our ability to grow food because they allow us to farm land that would otherwise be too dry or too wet. However, altering water levels can have negative consequences. Aquatic ecosystems can be harmed, and water shortages result if too much water is removed from an ecosystem. Water for irrigation may be supplied by reservoirs located behind large dams. Dams and reservoirs alter the aquatic ecosystem that once occupied the natural river. In some places, all of the water that enters a river system is used by humans before reaching the mouth of the river (Figure 6).

Figure 6 At times, none of the water in the Colorado River reaches the ocean. All of it is removed upstream for irrigation, for drinking water supplies, and to meet other human needs.

Soil Air Spaces and Compaction

The air spaces in soil serve two important functions. They allow water and nutrients to pass through the soil to reach roots and they provide oxygen to plant roots and soil organisms. The size of air spaces is influenced by the physical characteristics of soil particles. Soils made up of smaller particles have less air space.

Soil becomes compacted when pressure squeezes soil particles closer together. This reduces air spaces. When compacted, roots may not obtain enough oxygen. Water will have difficulty passing through the soil. Compaction may be caused by heavy equipment or simply by people walking on it. For example, you may have witnessed the impact of compaction along a well-used path.

TRY THIS: HUMANS WEIGH IN ON SOIL COMPACTION

SKILLS: Controlling Variables, Performing, Observing, Analyzing, Evaluating

Human actions often cause soil compaction. This reduces the ability of air and water to move through soil. In this activity, you will examine the impact of soil compaction on water movement through soil.

Equipment and Materials: 2 flower pots with drainage holes; tray; stopwatch; measuring cup or graduated cylinder; potting soil; water

1. Obtain two identical flower pots that have drainage holes at the bottom. Fill both pots three-quarters full of loose potting soil.
2. In the first pot, *gently* push down on the soil surface to remove any large air spaces and even out the soil surface.
3. In the second pot, *firmly* push down to compress the soil into the bottom of the pot. Make sure that the entire soil surface has been compressed.
4. Place the pots on a large tray. Using a stopwatch, begin timing as you slowly pour 150 mL of water onto the soil surface in the first pot. Record the time it takes for all the water to have penetrated below the surface of the soil.
5. Repeat step 4 for the pot containing the compressed soil.

A. Describe the effects of soil compaction on the rate at which water can penetrate the soil surface. T/I A

B. Brainstorm human activities on a farm, around your home, in a park, and in other settings that can cause soil compaction. A

C. How might soil compaction influence root growth? Explain your reasoning. T/I

4.2 Managing the Soil—Controlling the Flow of Nutrients and Water

Figure 7 In no-tillage farming, the stubble of the crop is left on the field after the crop is harvested.

Figure 8 Soybeans, corn, and wheat are often grown in rotation in southern Ontario.

Alternative Farming Practices

Farmers use several methods to reduce the impacts of agricultural practices on biogeochemical and water cycles. Three approaches are no-tillage farming, crop rotation, and crop selection.

In no-tillage agriculture, farmers leave the ground undisturbed after the crop is harvested instead of ploughing the remaining vegetation into the soil (Figure 7). This helps retain soil nutrients, reduces soil compaction and water loss, and improves soil quality. The remaining plant stalks (stubble) protect the soil from wind and water erosion. No-tillage agriculture sometimes requires greater pesticide use because weed populations are no longer controlled by ploughing them under.

Most farmers rotate or change the crops they plant on a certain area of land on a regular basis. By rotating crops, farmers can reduce their use of fertilizers and pesticides. For example, in southern Ontario, farmers often plant soybeans, corn, and wheat in a three-year rotation (Figure 8). During a soybean year, the nitrogen concentration of the soil increases through the action of nitrogen-fixing bacteria living in root nodules. This reduces the need to add nitrogen fertilizer the following year for corn. In the third year of the rotation, wheat is planted. Because wheat is not planted in rows, it is competitive against weeds and requires less herbicide.

Farmers can also choose to grow crops that are better suited to the local growing conditions. For example, growing drought-resistant and heat-tolerant plants in an area with hot, dry summers is more sustainable than growing a crop that needs a lot of water. These crops are dependent on extensive irrigation and may deplete groundwater supplies over time.

IN SUMMARY

- Fertilizers disrupt biogeochemical cycles and can pollute the environment.
- Synthetic fertilizers cause more ecological problems than natural fertilizers.
- Farmers alter the water cycle by irrigating fields or using drainage pipes to remove water.
- Soil compaction limits water flow and harms plants by reducing oxygen and nutrient availability.
- Practices such as no-tillage farming, crop rotation, and crop selection reduce the impacts of agriculture on the soil ecosystem.

CHECK YOUR LEARNING

1. How do biogeochemical cycles in agroecosystems differ from those in natural ecosystems? Explain.
2. Define and give examples of synthetic and natural fertilizers.
3. List the advantages and disadvantages of synthetic versus natural fertilizers.
4. Explain how adding fertilizer to an ecosystem can cause damage to it.
5. Describe how irrigation on farms influences natural sources of water and the ecosystems they support.
6. Identify ways that soil compaction influences nutrient and water cycles.
7. How do humans grow crops on land that would naturally be either too dry or too wet? Explain.
8. Describe three farming methods that can reduce some of the negative impacts of farming on the soil.

AWESOME SCIENCE

Black Gold

In the 1950s a Dutch soil scientist working in the Amazon rainforest discovered unusual pockets of fertile black earth or *Terra preta*. These rich soil deposits were only found where humans had once lived and where the natural surrounding soils were nutrient poor and thin (Figure 1).

Figure 1 (a) Naturally occurring, nutrient-poor soil and (b) dark nutrient-rich *Terra preta*. *Terra preta* means "dark earth" in Portuguese.

More than 1500 years ago, Indigenous peoples living in the rainforest developed a breakthrough in soil conservation and sustainable agricultural practices. The Indigenous peoples used a form of slash-and-burn agriculture in which they cleared the forest by cutting and burning and then planted crops in the forest openings. Normally, the heavy rainfalls would have quickly leached the remaining nutrients from the already nutrient-poor soils. The breakthrough was to add large quantities of charcoal into the soil. This charcoal-rich soil is known today as "biochar."

In short, the Indigenous peoples were growing their food in forest clearings and using the harvested wood for cooking fuel and for charcoal fertilizer production. The result was a human-engineered and extremely fertile black soil. The charcoal itself does not break down readily and may last for hundreds or thousands of years. These Indigenous peoples disappeared 500 years ago, but to this day, the nutrient-rich black soil they created remains.

Adding charcoal has a number of benefits. It increases the water-holding capacity of the soil, it helps hold minerals, and it enhances nutrient uptake by plant roots. It is also beneficial to soil fungi that establish mutualistic relationships with plant roots. The benefits of charcoal supplements in poor soil are so dramatic that some soil scientists refer to it as "black gold" (Figure 2).

Figure 2 Comparison of corn plants grown in soil with (left) and without (right) charcoal supplement.

Today, scientists are excited about the applications and benefits of this technique. At a recent meeting of the American Chemical Society, scientists described biochar as having the ability to improve soil fertility. It may be better than compost, animal manure, and other soil conditioners.

While useful as a soil supplement, biochar's greatest potential might be to capture and store huge amounts of carbon. Charcoal is composed of almost pure carbon that was removed from the atmosphere in the form of carbon dioxide. Carbon dioxide is believed to be largely responsible for recent climate change. Referring to the technique as "black gold agriculture," scientists said this revolutionary technique could provide a cheap method of reducing atmospheric carbon dioxide by simply trapping the carbon in soil (Figure 3).

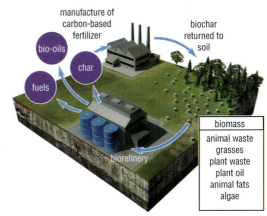

Figure 3 The biochar process

4.3 PERFORM AN ACTIVITY

Fertile Grounds: The Right Mix

Soil is a mixture of minerals, nutrients, water, air, and living and dead organisms. Each of these influences the soil's ability to sustain terrestrial ecosystems. Acid rain also affects soils by altering soil acidity and nutrient content (see Section 3.6).

SKILLS MENU
- Questioning
- Hypothesizing
- Predicting
- **Planning**
- **Controlling Variables**
- **Performing**
- **Observing**
- **Analyzing**
- **Evaluating**
- **Communicating**

Purpose

To determine how soil supplements affect the water retention and holding capacity of soils, and to determine how acid rain influences the acidity and nutrient content of soils.

LEARNING TIP

Measuring Acidity

Scientists measure how acidic a solution is using the pH scale. This scale ranges from 0 to 14. A pH of 0 is the most acidic, and a pH of 14 is the least acidic. A pH of 7 is neutral.

Equipment and Materials

- 6 200 mL planting pots with screens over drainage holes
- electronic balance
- graduated cylinder
- tray
- soil test kit (for testing nitrogen, phosphorus, potassium [NPK] and pH)
- masking tape
- marker
- sandy soil mix
- 2 soil supplements (for example, peat moss, composted manure, crushed charcoal, crushed limestone, slow-release fertilizer)

Note: All supplements must be dry before beginning the experiment.

- plastic spoon
- distilled water
- "acid rain" solution

Procedure

SKILLS HANDBOOK
3.B.4., 3.B.6.

Part A: Water-Holding Capacity of Soils

1. As a group, you will be assigned two soil supplements to test.
2. Obtain three pots and label them "Control" and the names of the two supplements to be tested. Write "Water Test" on each label.
3. Using the information in Table 1, prepare your soil mixtures.

Table 1 Soil Supplement Mixtures

	Control	Peat moss	Manure	Charcoal	Limestone	Fertilizer
supplement	none	50 mL	50 mL	30 mL	30 mL	0.5 mL
sandy soil	150 mL	100 mL	100 mL	120 mL	120 mL	150 mL

4. Thoroughly mix your soil samples using a plastic spoon.
5. Measure and record the total dry mass of each pot and soil mixture. Use a data table similar to Table 2.

Table 2 Water Retention Data Table

Water retention test	Dry mass	Saturated wet mass	Mass after 1 day	Mass after 2 days
control				
supplement 1				
supplement 2				

6. Place the three water test pots in a tray and use distilled water to saturate each soil mixture. This may require you to pour water through the soil mixture several times. When the soil is saturated, discard any extra water from the tray.

7. Determine the mass of each pot and record the saturated wet mass of the water test pots in your table. Over the next two days, you will allow these pots to dry out. You will find and record the mass of each pot each day.

Part B: Effect of Acid Rain on Soil Acidity and Nutrient Content

For this part of the activity, you will design an experimental procedure to compare the effects of neutral rainwater and acidic rainwater on the acidity and nutrient content of soils. You will use a soil test kit to determine the soil pH and to measure the amounts of nitrogen, potassium, and sodium in the soil. You are to conduct this part of the experiment using the same supplements you used in Part A.

8. Within your group, brainstorm how you will design your experiment. Write the steps of your procedure. Be sure your design uses proper controls and includes a description of what information you are going to collect. You may want to consider simulating the effects that several days of rain would have on your soil samples.
9. Add to your procedure any necessary safety precautions.
10. Create a table in which to record your observations.
11. Have your teacher approve your experimental design before proceeding with your experiment.
12. Perform your experiment and record your observations.

Analyze and Evaluate

(a) In Part A, how much water did each sample absorb when saturated?
(b) Calculate the initial water holding capacity of each sample as a percentage of the dry mass using the following formula:

Water holding capacity =
$$\frac{\text{mass of water in saturated sample}}{\text{mass of dry sample}} \times 100\ \%$$

(c) Which type of soil supplement had the greatest water holding capacity?
(d) Compare the ability of supplements to retain water by graphing mass over time. Plot all the data on one graph.
(e) Which supplement had the least water loss? Which had the greatest water loss?
(f) In Part B, which soil sample had the highest initial nutrient levels?
(g) Were any of the nutrient levels influenced by the neutral rain or acid rain? Which soil samples were best able to maintain nutrient levels?
(h) Was the pH of the soil influenced by the simulated neutral rain or acid rain? Explain.
(i) Were any of the supplements better able to neutralize the effects of the simulated acid rain compared with the control? Explain.
(j) After sharing your results with your classmates, rank the soil supplements on the following characteristics:
   (i) ability to absorb and retain soil moisture
   (ii) ability to provide nutrients
   (iii) ability to resist loss of nutrients by neutral rain
   (iv) ability to resist loss of nutrients by acid rain
   (v) ability to resist changes in soil pH

Apply and Extend

(k) What other characteristic(s) do you think growers must consider when choosing supplements to enhance soil quality?
(l) Do you think there might be advantages to combining soil supplements? Which combination do you think might work best? Give reasons for your suggested combination.
(m) Which soil supplements could be obtained from natural sources? How would their use influence sustainability of the agroecosystem?
(n) Although charcoal itself contains no plant nutrients, it is considered to have great potential use as a soil supplement. Research the use of charcoal as a soil supplement and report on your findings.

(o) Do you think that there would be any differences in your observations if plants had been growing in each pot? Suggest some possible effects plants could have caused.

4.4 Pests and Poisons

When walking through a natural ecosystem, have you ever noticed the incredible biodiversity? Your experience would have been much different walking through an agricultural ecosystem (Figure 1). Nothing is more unnatural about agroecosystems than the low biodiversity they support. This is not surprising given that their purpose is large-scale food production. Devoting land to food production has dramatic ecological implications. In addition to habitat loss, it alters food webs as well as water and biogeochemical cycles.

Figure 1 (a) Natural prairie has many types of grasses and other flowering plants. (b) Vegetable crops, such as these potatoes, are usually planted as monocultures.

> **READING TIP**
>
> **Use Sticky Notes**
> Use sticky notes while reading to record your questions beside the part of the text you do not understand. After reading the text, you can check to see if information in the tables, photos, and captions in the rest of the section answered these questions.

How does growing crops alter relationships within food webs? In natural ecosystems, organisms are part of complex food webs where producers and consumers coexist. Producers support the entire community of hundreds or thousands of consumer species. In agroecosystems, these relationships are almost entirely absent. Table 1 compares the food web of a natural ecosystem with that of a typical crop, the potato.

Table 1 Comparison of the Food Web in a Natural Grassland with That of a Potato Field

	Natural grassland	**Potato field**
producers	50 to 100 species of wildflowers and other plants coexist	Monoculture of potato plants. The few other plant species that occur are considered weeds.
herbivores	Numerous species of herbivorous insects, rodents, sparrows, and so on	A few species such as the Colorado potato beetle and the flea beetle. Both are considered pests.
carnivores	Numerous species of carnivorous insects, spiders, snakes, foxes, hawks, and so on	Very few carnivores, if any
food web	Complex web with hundreds of species	Food web effectively eliminated, leaving a single producer with some pest species

Pests

In agroecosystems, thousands or millions of individuals of a single crop species are planted in a monoculture. Then, to maximize growth of the desired crop, we try to eliminate organisms that we consider pests.

Pests are organisms that might compete with or damage crop species. Agricultural pests are plants and animals that reduce crop yields. Weeds are plant pests, mosquitoes are insect pests, and mice are rodent pests. The term pest is used only in reference to human wishes. There are no pests in nature. All organisms are simply producers and consumers within food webs.

By controlling pests, crops grow in the near-absence of their natural consumers and competitors. Farmers are not the only people interested in controlling pests. Insects are considered serious pests in the forestry industry. The spruce budworm, the gypsy moth, and the Asian long-horned beetle are just a few examples of tree-infesting species (Figure 2).

When farmers plant a monoculture, they often create the ideal environment for pests. For example, when a population of Colorado potato beetles finds a field of potatoes, they begin feeding and reproducing rapidly. Uncontrolled, the population could skyrocket and devastate the entire crop.

Figure 2 Caterpillars of the gypsy moth

Pesticides

One of the most common ways to control or eliminate pests is to use poisons that kill pests, or **pesticides** (Figure 3). There are many different kinds of pesticides. Herbicides are used to kill plants. Other pesticides include insecticides, rodenticides, and fungicides. Humans even use molluscicides and piscicides to kill snails and fish, respectively.

pesticide a substance used to kill a pest

Figure 3 Pesticides are toxic, and protective equipment must be worn when they are applied.

DID YOU KNOW?
Livestock and the Drugstore
Although not usually referred to as pesticides, antibiotics are commonly used in agriculture to kill micro-organisms. Antibiotics are also used as a preventive medicine to improve the health of livestock. This large-scale use can lead to micro-organisms becoming resistant to these antibiotics. This is a concern because the same micro-organisms have the potential to infect humans and would then be resistant to the same antibiotics when used for treatment.

Characteristics of Pesticides

Once applied, pesticides vary greatly in how long they persist or remain active in the environment. Some long-lived pesticides persist for many years. Other pesticides are short-lived, lasting only a matter of days before they degrade. In general, pesticides obtained from natural sources are less persistent than synthetic pesticides. However, modern synthetic pesticides are less persistent than those developed 30 or more years ago.

Pesticides vary widely in the number of species they are able to control. **Broad-spectrum pesticides** are toxic to a wide range of species, whereas **narrow-spectrum pesticides** are toxic to a limited number of species. For example, DDT (dichlorodiphenyltrichloroethane), a once widely used insecticide, is toxic to most insect species. Bt, a modern pesticide derived from bacteria (*Bacillus thuringiensis*), is highly toxic only to caterpillars, beetle larvae, and fly larvae. It is not toxic to most beneficial insects.

broad-spectrum pesticide a pesticide that is effective against many types of pest

narrow-spectrum pesticide a pesticide that is effective against only a few types of pest

> **READING TIP**
>
> **Ask Inferential Questions**
>
> Often a text does not supply all the information about a topic. This means you might have to ask an inferential question. You might ask why insects are considered agricultural pests. The text does not say how insects harm plants. Your own experience is that insects eat plants and cause damage. By combining what the text does not say with what you already know, you end up asking an inferential question.

How Do Pesticides Work?

Pesticides work by causing physical or biological harm to the pest organism. Diatomaceous earth, for example, is composed of the fossilized remains of a type of algae called diatoms. This abrasive powder scratches the waxy outer coating of small organisms, such as insects, causing them to dehydrate. Other pesticides may interfere with biological processes, such as photosynthesis, or cause damage to vital organs.

Some pesticides are delivered by contact. In this case, the target pest must be touched by the pesticide. Alternatively, some pesticides are indirectly applied. For example, herbicides can be sprayed on the soil and taken up through the roots of the weed. Or insecticides can be sprayed on a plant and later consumed by an insect.

Table 2 provides an overview of some pesticides. In the next section, you will consider the ecological consequences of pesticide use.

Table 2 Characteristics of Some Pesticides

Pesticide	Origin/source	Use	Important characteristics
DDT	synthetic	broad-spectrum insecticide	• one of the first widely used synthetic pesticides • highly persistent in food chains causing die-off of many predatory birds • widely banned in 1970s but still used in some countries to kill mosquitoes that transmit malaria
rotenone	natural toxin extracted from plant roots	insecticide and piscicide	• highly toxic to many aquatic organisms, including fish • approved for use by some organic farmers
glyphosate	synthetic	broad-spectrum herbicide	• widely used herbicide • very low persistence
Bt	a protein obtained from the bacteria *Bacillus thuringiensis*	narrow-spectrum insecticide	• highly toxic to moth and butterfly caterpillars • safer than most pesticides

IN SUMMARY

- Agroecosystems are often based on a single plant species and the elimination of natural food webs.
- Synthetic pesticides tend to persist longer than natural pesticides.
- Monocultures create ideal conditions for certain pests, which are often controlled using pesticides.
- Pesticides may have a physical effect or disrupt biological processes.
- Pesticides are used to control weeds, insects, rodents, fungi, and other types of pest organisms.

CHECK YOUR LEARNING

1. What is the most unusual characteristic of food webs in agroecosystems? **K/U**
2. Define the term *pest*.
3. List five different types of pesticides. **K/U**
4. Explain what is meant by "broad-spectrum" and "narrow-spectrum" pesticides. **K/U**
5. Do you think pesticides that come from natural sources are better than synthetic pesticides? Explain your reasoning. **C**
6. Explain why the term *pest* is not used in reference to natural ecosystems. **K/U**
7. Why are populations of pests likely to be much larger in agroecosystems than they would be in natural ecosystems? Explain your reasoning. **K/U**
8. Insecticides and herbicides are the most commonly used pesticides. Brainstorm some of the characteristics that make insects and weeds such serious competitors. **T/I**

4.5

Issues with Pesticides

Pesticides have helped farmers reduce crop damage from pests and increase food production. Pesticides have also helped control populations of biting insects, such as mosquitoes, that spread diseases (Figure 1). While such benefits can result in more food and better health for some, pesticide use has a number of environmental costs.

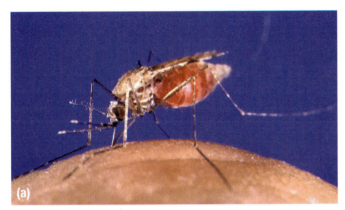
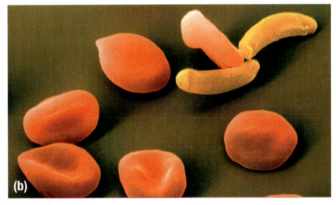

Figure 1 Pesticides have helped control the spread of disease-carrying insects such as (a) this mosquito, which transmits (b) the protozoan parasite that causes malaria.

Pesticides are often applied through aerosols or sprays onto fields, forests, and gardens. A serious drawback of this mode of delivery is that some of the pesticide never reaches the target species because it is carried away by the air or lands on the soil. These pesticides then become potential sources of soil, air, and water pollution. They can also harm other non-target species.

Non-Target Species

Pesticides often kill species they were not intended to kill. Because broad-spectrum pesticides control many different pests, they may kill non-damaging and potentially beneficial organisms. For example, a broad-spectrum insecticide may kill species of predatory insects that might normally feed on pests.

Killing beneficial organisms creates a situation in which farmers become more dependent on pesticides. When natural pest controls, such as predatory insects, are killed, farmers must replace them by using more pesticides.

Improper use of pesticides can also kill non-target species. For example, spraying an insecticide at the wrong time of year may kill honeybees, which are essential for pollinating fruit crops. As a result, less fruit will be produced.

The consequences of non-target killing can be surprising and serious. Consider the dramatic set of events that took place on the island of Borneo. In 1955, the World Health Organization began a DDT spraying program to control mosquitoes that were responsible for spreading malaria. The spraying initially reduced the spread of malaria, but it also caused an unexpected chain reaction on the island. In addition to killing mosquitoes, DDT killed wasps that preyed on thatch-eating caterpillars. Without the wasps, the caterpillars ravaged the thatched homes of the villagers (Figure 2).

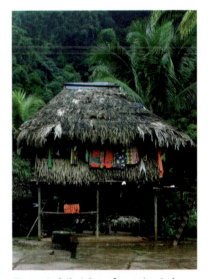

Figure 2 A thatch roof, constructed of palm leaves, can be destroyed by caterpillars. These caterpillars are normally preyed on and controlled by wasps.

DDT also killed cockroaches that were then consumed by lizards. The DDT in the cockroaches damaged the nervous systems of the lizards, making them easy prey for cats. Many cats died from consuming the poisoned lizards.

In a final twist, the villagers were threatened by a new disease. When the cats disappeared, the rat population in the villages increased dramatically. The fleas on the rats carried the plague—a potentially devastating disease. To prevent an epidemic, large numbers of healthy replacement cats had to be brought to Borneo to control the rats.

Bioamplification

One of the most serious side effects of pesticide use is their tendency to accumulate in individual organisms. This happens because some pesticides are not broken down or eliminated with other body wastes. If an individual continues to eat food contaminated with the pesticide, it will accumulate in the body. If the pesticide is long-lived, then the concentration of pesticide in the individuals will increase to levels much higher than in the environment. Pesticides that bioaccumulate do so because they cannot easily be excreted from the body. This is because they are not soluble in water but are soluble in fats and oils. This process is called **bioaccumulation.**

bioaccumulation the concentration of a substance, such as a pesticide, in the body of an organism

TRY THIS: POLLUTANTS FOLLOW THE FAT

SKILLS: Controlling Variables, Performing, Observing, Analyzing, Evaluating

SKILLS HANDBOOK
3.B.

In this activity, you will model the bioaccumulation of toxins in the body. Iodine will represent a persistent pesticide. Water (in the iodine solution) will represent body fluids. Mineral oil will represent body fat.

Equipment and Materials: 3 15-mL screw-top vials; pipette; pipette suction bulb; 6 mL mineral oil; 18 mL 0.1 % Lugol's iodine solution

🛑 Never suck on the pipette with your mouth. Iodine solution is toxic and stains skin and clothing.

1. Record the appearance of the iodine solution and the mineral oil.
2. Using the pipette bulb, pipette 6 mL of iodine solution into each of the three vials.
3. Add 6 mL of mineral oil to a vial. Seal the vial. Shake it vigorously for 1 min while pressing on the vial cover. Let the contents settle for 2 min. Record your observations.
4. Open the first vial and, using a pipette, transfer 4 mL of the oil to the second vial.
5. Repeat step 3 using the second vial.
6. Use a pipette to transfer approximately 2 mL of the oil from the second vial to the third vial. Repeat step 3.

A. What property of iodine allowed you to detect its presence in the water? T/I
B. Was there evidence that iodine moved from the water to the oil? T/I
C. Explain how this process models the accumulation of pesticides in an organism. T/I
D. Describe what happened to the mineral oil as it was transferred to vials 2 and 3 and exposed to more iodine. T/I
E. Was there evidence that the iodine was bioaccumulating in the mineral oil? Explain. T/I
F. Many of your body's waste products are eliminated through your urine. Urine contains no fat or oils. What do your observations suggest about your body's ability to eliminate pesticides that have poor solubility in water? A
G. What does this activity suggest about the risks of eating fatty foods that were produced in contaminated environments? A

bioamplification the increase in concentration of a substance, such as a pesticide, as it moves higher up the food web

All individuals are part of a food chain. As a result, toxins stored in the fats and oils of organisms at one trophic level are passed on to the organisms at the next trophic level. The higher up the food chain, the more concentrated the pesticides become. This process is called **bioamplification.**

If a pesticide bioamplifies in a food chain, it may reach toxic concentrations. As illustrated in Figure 3, the concentration of DDT has increased more than 600 times. At these high levels, the organisms are likely to suffer toxic effects.

MATH TIP

Understanding ppm and ppb
Small concentrations are often measured using units of ppm (parts per million) or ppb (parts per billion). A concentration of 10 ppm is approximately equal to 10 mL of a substance dissolved in 1000 L of water.

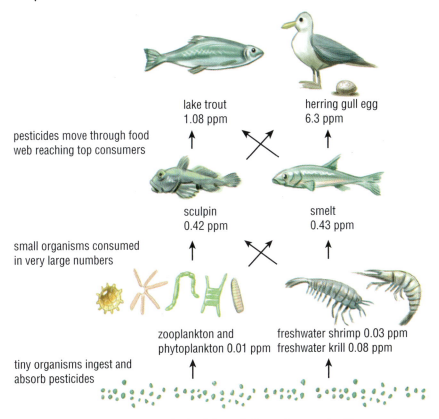

Figure 3 DDT bioaccumulates up the food chain. Gulls require more food because they are warm-blooded and must maintain their body temperature; therefore, they accumulate more DDT.

Many species of predatory birds, including osprey, bald eagles, and peregrine falcons, declined in numbers because DDT bioaccumulated in their bodies (Figure 4). DDT interfered with calcium metabolism and the female bird's ability to produce strong egg shells.

To learn more about bioaccumulation of DDT in herring gulls,

DID YOU KNOW?

Consuming Oils Can Be a Fishy Dilemma
Fish oils, especially from salmon and herring, are high in omega-3 fatty acids, which are considered to be beneficial in the diet. Unfortunately, salmon are near the top of food chains. Therefore, they are more likely to accumulate toxins within the very oils that are beneficial.

Figure 4 Eggshells of many species of birds, such as this peregrine falcon, were softened by DDT.

4.5 Issues with Pesticides 137

Other fat-soluble toxins, such as mercury and polychlorinated biphenyl (PCB), also bioamplify in the food web (Figure 5).

Figure 5 Populations of large predatory birds suffered due to the bioamplification of mercury in aquatic ecosystems.

Arctic ecosystems are particularly vulnerable to biomagnification of toxic substances. Many long-lived top consumers such as whales, polar bears, walrus, and fish live in the Arctic (Figure 6). In addition, Inuit that live in these environments rely on these same species as their traditional food supply—resulting in a potentially dangerous exposure of humans to toxins in their food. This is one danger associated with consuming food from the top of the food web.

Figure 6 Pesticides and mercury have biomagnified in the fatty tissues of the walrus.

Pesticide Resistance

When pesticides are used for long periods of time, some pest species may become resistant to the pesticide. This means that the pesticide is no longer able to control the pest. Individuals that exhibit the greatest resistance are more likely to survive an application of pesticide than those with little or no resistance. The individuals that survive will reproduce and pass on their resistance to their offspring. After many generations, the population can become highly resistant to a particular pesticide.

Weeds and insect pests are likely to develop resistance because they reproduce frequently and produce many seeds or offspring (Figure 7). When resistance develops in a pest, the farmer needs to apply a greater concentration of pesticide to have the same effect or switch to a different pesticide.

On a global scale, pesticide resistance is a serious concern. Figure 8 illustrates the numbers of species that have become resistant to various types of pesticides over a 50-year period.

DID YOU KNOW?
Herbicides Clear the Way
Herbicides are used to maintain right-of-ways along the edges of highways and under power transmission lines. Using herbicides saves on labour costs but may also leave toxic residues. Before going berry picking along the side of a road, you might want to consider whether pesticides have been used.

Figure 7 Populations of green pigweed are resistant to some herbicides.

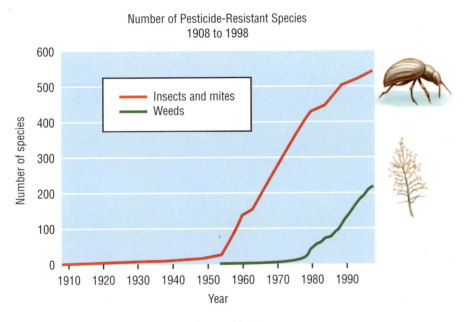

Figure 8 Pesticide resistance is increasing worldwide.

Reducing Our Dependence on Pesticides

There is little doubt that pesticides have dramatically increased global food production. By reducing competition and other pests, crops grow faster and have higher yields. The benefits of using pesticides must, however, be weighed against the risks of pollution, harm to non-target species, bioamplification, and pesticide-resistant species.

One alternative type of agriculture, **organic farming**, uses no synthetic pesticides or fertilizers. Organic farmers sometimes have to accept crop losses to naturally occurring pests. These losses, however, may be offset by the higher price growers get for their organic products, as well as savings from not purchasing synthetic chemicals.

organic farming the system of agriculture that relies on non-synthetic pesticides and fertilizers

To learn more about being an organic farmer,

To learn more about organic farming in Canada,

> **DID YOU KNOW?**
>
> **Pest Baiting**
> The female apple codling moth releases a pheromone, or sex attractant, which attracts mates. When growers place pheromone baits in an orchard, the males become confused and have difficulty finding females. This reduces the population size of the next generation, which means fewer larvae will infest the crop.
>
>

integrated pest management a strategy to control pests that uses a combination of physical, chemical, and biological controls

Organic farmers rely on a range of ecologically sustainable techniques. These techniques are summarized in Table 1.

Table 1 Techniques Used by Organic Farmers

Method	Description
biological control	• Predatory insects, mites, and disease-causing micro-organisms prey on and infect prey species. • Examples include parasitic wasps and ladybird beetles (ladybugs).
altered timing	• Better timing of planting and harvesting can avoid peak pest populations.
crop rotation and mixed planting	• When farmers do not grow monocultures in the same location year after year, pest populations do not have the same opportunities to establish and prosper.
baiting pest	• Pheromone baits can be used to confuse some mating insects.

While organic farming minimizes environmental impacts, it could result in lower crop yields. This may not be acceptable. In such cases, an intermediate approach called **integrated pest management** (IPM) is often employed. IPM takes advantage of all types of management methods. The goal is to maximize efficiency, keep costs low, and reduce harm to the environment. IPM farmers use many of the techniques employed by organic farmers but use synthetic pesticides and fertilizers when necessary.

IN SUMMARY

- Non-target organisms may be harmed when pesticides are released into the air or water.
- Some pesticides and toxins bioaccumulate in the bodies of organisms and bioamplify up the food web.
- Many plant and insect pests are becoming increasingly resistant to pesticides.
- We can reduce our dependence on pesticides by using organic farming or integrated pest management (IPM) methods.

CHECK YOUR LEARNING

1. Outline three key benefits of using pesticides.
2. What is a non-target organism?
3. Describe how the loss of non-target organisms leads to even greater pesticide use.
4. Explain why some pesticides bioaccumulate whereas others do not.
5. Use the example of DDT to explain how a pesticide can bioamplify and damage entire ecosystems.
6. Define pesticide resistance.
7. Explain why pesticide resistance is a major concern.
8. Describe three alternatives to the widespread use of pesticides.
9. Briefly explain the concept of "integrated pest management."
10. Describe how organic farming methods are similar to natural ecosystem processes.
11. Use a graphic organizer to explain what happened on the island of Borneo when DDT was used to control the spread of malaria.

PERFORM AN ACTIVITY 4.6

Greening the Consumer—Reducing Your Ecological Footprint

SKILLS MENU
- Questioning
- Hypothesizing
- Predicting
- Planning
- Controlling Variables
- Performing
- Observing
- Analyzing
- Evaluating
- Communicating

Ecologically, humans are consumers. We depend on the environment for air, water, food, and shelter. We also use resources to improve our quality of life through health care, recreation, and education. These demands affect the environment. Ecologists use the term "ecological footprint" to describe the sum of all the impacts each person has on the environment. It is usually stated as the equivalent surface area of Earth that a person needs to support his or her current lifestyle. In this activity, you will research strategies that can be followed to reduce the ecological footprint of people living in urban settings.

Purpose
To research and select ecological strategies for consumers who live in an urban setting.

In particular, you are to find ways to
- reduce energy consumption in the home
- reduce the environmental impacts of personal transportation
- reduce the ecological impacts associated with food consumption

Equipment and Materials
- computer with Internet connection

Procedure

SKILLS HANDBOOK 4.A., 4.B.

1. Use Internet and library resources to investigate ways to reduce the ecological footprint of individuals. Table 1 provides examples of strategies that you could investigate and consider. Make sure you collect quantitative data on the benefits.

Table 1

energy consumption in the home	• purchase energy-efficient appliances • re-insulate your home • switch to compact fluorescent lighting • conserve water • switch to renewable energy sources
transportation	• use non-motorized transportation • car pool or use public transit • purchase energy-efficient or alternative-fuel vehicles
food consumption	• purchase locally grown foods • purchase organic produce • eat a more vegetarian diet • compost

2. Prepare a report, poster, presentation, or skit to summarize your results. Remember to include quantitative data. Explain how taking "action A" would make your lifestyle more sustainable or how much money it would cost (or save) you to "go green." T/I C A

Analyze and Evaluate

(a) Was it easy to find strategies for consumers to become more "green"? T/I

(b) Are the costs of implementing these strategies high? A

(c) For each of the three main categories listed in Table 1, which single strategy would you recommend? Explain why you are recommending this strategy. T/I

(d) Why do you think more people are not doing many of these actions already? A

(e) Which of your recommended strategies are you doing now or planning on doing? A

Apply and Extend

(f) Write a paragraph summarizing your recommendations. C

(g) How do you think municipal, provincial, and federal governments could encourage the implementation of your recommendations? A

4.7 The Urban Ecosystem

Cities are major sources of air and water pollution. This is not surprising—the population of some cities reaches into the millions (Figure 1). Air pollution is caused by car exhaust, burning of fuels, and gases from industry. Water pollution comes from biological, commercial, and industrial wastes. The ecological footprint of cities reaches well beyond their boundaries. Highways radiate outward from cities like a web. People who live in cities rely on forestry and mining for a variety of materials, on agroecosystems for food, and on natural ecosystems for recreation.

Figure 1 Close to 85 % of Ontarians live in urban settings such as Toronto, Ottawa, and Hamilton. This situation is not unique to Ontario. More than 50 % of the world's population now lives in cities.

An aerial view of an urban ecosystem is very different from a natural ecosystem (Figure 2). In a large city, natural vegetation is replaced by human structures such as buildings, roads, and sidewalks. The remaining vegetation is often lawns that are monocultures of non-native grasses. Table 1 (next page) shows the key differences between natural ecosystems and urban environments.

Figure 2 (a) A natural ecosystem and (b) an urban ecosystem are in stark contrast to one another. The former is dominated by a living carpet of trees and the latter by a layer of concrete and asphalt.

Table 1 Comparison of Urban and Natural Ecosystems

Feature	Urban ecosystem	Natural terrestrial ecosystems
surface features	• dominated by buildings, paved roadways, and parking areas	• dominated by plants
plant species	• low density of plants • mostly non-native species such as grasses, ornamental trees, shrubs, and small flowering plants	• high density and diversity of native species
common non-human animal species	• low diversity and abundance • common species include dogs, cats, house mice, Norway rats, pigeons, starlings, and house sparrows—all are non-native	• high density and diversity of native species
water cycle	• surface water rapidly enters sewers and drainage systems • most of the surface impenetrable to water • human population consumes large quantity of water obtained from surface or groundwater supplies • waste water returned to surface water	• surface water in ponds, streams, rivers, and lakes • most rainfall absorbed at surface • water moves through soil and enters groundwater
flow of materials	• massive influx of materials to support human demand for consumer goods • human body wastes collected, treated, and disposed of in surrounding environment • household and commercial wastes disposed of in surrounding environments • usually minimal recycling and composting	• natural nutrient cycles occur throughout ecosystem
food webs	• dependent on importing food from outside agroecosystems	• complex natural food webs
sustainability	• cannot be sustained without inputs from and outputs to other ecosystems	• independently sustainable

After studying Table 1, you can see that humans cannot survive in an isolated urban environment. Unlike organisms living in a natural ecosystem, humans in an urban setting rely on outside ecosystems.

RESEARCH THIS CONSERVING THE OAK RIDGES MORAINE

SKILLS: Researching, Analyzing the Issue, Defending a Decision

SKILLS HANDBOOK
4.B., 4.C.

Located in south central Ontario, the Oak Ridges Moraine (Figure 3) is an important ecological feature. A moraine is a large deposit of soil and rock left behind by a glacier. The Oak Ridges Moraine is a major source of surface water and groundwater and is home to many ecosystems. It is valued for its attractive landscape and contains prime farmland.

The Oak Ridges Moraine is experiencing pressure from new residential, commercial, industrial, and recreational uses that threaten its sustainability. In this activity, you will research the moraine and the conservation plan that is being implemented to protect this valuable region.

1. Use the Internet and other resources to investigate the Oak Ridges Moraine and the Ontario government's conservation plan for the moraine.

    GO TO NELSON SCIENCE

A. The moraine has been called a "water barrel" for much of southern Ontario. How does the moraine act as a source of clean and abundant water?

B. Describe present-day human and natural communities that occupy the moraine.

Figure 3 The Oak Ridges Moraine

C. Why does the moraine need protection? What activities threaten the sustainability of the region?

D. What are the objectives of the Ontario government's conservation plan?

E. What are "land use designations," and what purpose do they serve?

F. (i) Do you think a conservation plan for the Oak Ridges Moraine is a good idea? Explain your reasoning.

   (ii) Propose changes or suggestions to improve on the plan.

4.7 The Urban Ecosystem

Greener Cities

Humans are becoming more concerned about how we impact the environment. City planners and politicians have developed strategies to make cities more sustainable. They must consider the impact of human actions in cities, as well as in the surrounding ecosystems. This makes planning challenging because it crosses political boundaries, but it also encourages innovation.

Cities also differ in their planning requirements. Iqaluit, a city on Baffin Island, Nunavut, is working on a plan to ban plastic bags. The Greater Toronto Area government has passed a bylaw requiring retail outlets and coffee shops to give customers a discount if they bring their own bags or coffee mugs. Table 2 shows some ways that cities are becoming more sustainable.

Urban planners must understand how to design sustainable cities. To learn more about becoming an urban planner,
GO TO NELSON SCIENCE

DID YOU KNOW?
Green Buildings Are LEEDing the Way
The LEED (Leadership in Energy and Environmental Design) standards have been developed to promote the construction of energy-efficient buildings using environmentally friendly building materials. The city of Kingston is using the LEED standards to construct a four-rink arena and a 5000-seat sports and entertainment complex.

GO TO NELSON SCIENCE

Table 2 Examples of How Some Cities Are Working to Reduce Their Ecological Footprint

Community action	Benefits
use community composting programs to collect food and yard wastes	• reduces need for landfills • produces valuable fertilizer
use light-emitting diodes (LEDs) in traffic lights	• reduces energy consumption • saves on bulb replacement costs
promote green roofs on flat-topped commercial buildings	• improves air quality • reduces building heating and cooling costs
enhance and promote public transit	• reduces air pollution and energy costs
ban the cosmetic use of pesticides	• reduces air and water pollution • reduces health risks associated with exposure to pesticides
enhance green spaces within cities	• improves air quality • encourages participation in healthy outdoor activities
promote shop locally campaigns	• reduces transportation costs

On a larger scale, the United Kingdom plans to give every home in the country a "green makeover" by 2030. The program will include energy-saving measures such as adding insulation, switching to more efficient appliances, and using alternative energy technologies. Close to 30 million homes will receive the makeover. These changes will lower their energy costs and reduce their ecological footprint.

Urban settings also offer environmental benefits. By living close together, we can save energy, resources, and space. Large apartment buildings occupy less space. They also use fewer raw materials and less energy per person than single-family homes. Transportation is more energy efficient when people travel shorter distances and when they have access to large public transportation systems. New green housing developments can benefit from new heating and cooling systems that reduce energy demand.

One promising option is to make public transit free. If successful, such an approach would reduce traffic congestion, pollution, and accidents, lower road maintenance costs, and free up valuable space being used for parking lots.

TRY THIS: DO YOU SUFFER FROM NDD?

SKILLS: Planning, Analyzing, Evaluating

NDD or *nature deficit disorder* is not a recognized medical condition, but people use the term to describe a negative trend in the well-being of children. NDD is the social cost of not spending enough time in nature. Children who do not experience nature are more prone to behaviour and social problems. Think about your own lifestyle and how you interact with natural ecosystems (Figure 4).

Figure 4 How much time do you spend in natural environments?

1. Prepare a survey to determine how familiar your fellow students are with the natural world. Your survey should include 10 questions. They can be yes-or-no, multiple-choice, or numerical-answer questions.

A few sample questions are below:
- When was the last time you went for a walk through a natural habitat?
- Do you know where the closest natural setting is to your home or school?
- How many kinds of amphibians (or reptiles, fish, or birds) can you name that are native or even non-native to Ontario?

2. Give the survey to at least five students. Compare their answers with your own.

A. Did most students have about the same familiarity with nature as you do?

B. Many people believe that we are spending less and less time in natural settings. Do your results support this belief?

C. Make a list of some of the indoor activities that students do in their leisure time. Estimate the time spent on these activities in comparison with time spent outside. Compare inside activities with outside activities.

D. How might enhancing people's experiences with nature influence how they value natural ecosystems?

IN SUMMARY

- Cities are major sources of air and water pollution.
- Urban ecosystems are different from natural ecosystems.
- People living in urban settings rely on outside ecosystems for food, resources, and waste disposal.
- Urban ecosystems impact the surrounding ecosystems that supply them with services.
- Living in urban ecosystems can benefit the environment because high-density housing reduces our consumption of energy, resources, and space.
- There are many ways to enhance urban living spaces and reduce their negative impacts on the environment.

CHECK YOUR LEARNING

1. Ontario is largely an urban population. Explain what this means.
2. Contrast how urban settings differ from natural ecosystems in terms of
    (a) land surface types
    (b) the diversity and origin (native versus non-native) of species
    (c) the water cycle
    (d) cycling of nutrients
3. Describe how a large population living in a large city can actually benefit the environment.
4. How might living in a large city enable a person to be more aware of some environmental issues such as air pollution? Explain.
5. In what ways might living in a large city make it more difficult for a person to be aware of other environmental issues such as habitat loss? Explain.
6. Human populations living in cities would never be sustainable if they were isolated from their surrounding ecosystems. Explain this statement.
7. List five actions cities can take to reduce their ecological footprint and become more sustainable.

4.8 EXPLORE AN ISSUE CRITICALLY

Waste Management or Mismanagement?

Garbage disposal is one of the most visually striking examples of the impact of humans on the natural environment. As you have learned, nothing is wasted in natural ecosystems. The products of one organism are resources for another. Materials are decomposed and reused by living things or undergo processes and cycling through abiotic pathways.

If this is the case for natural ecosystems, what is it about human activities that cause us to produce so much waste? Could we produce and consume products in a way that mimics the biogeochemical cycles of nature?

The amount of waste we produce reflects our consumer-based lifestyles. Waste is produced when we make and use products. More waste is produced when the product is no longer valued and cannot be reused or recycled. Although waste may be produced it is produced in greater amounts when we *mismanage* our use of natural resources.

For thousands of years, humans have disposed of garbage by burning or burying it. Today, we still bury and burn waste, but modern facilities are equipped with technology to reclaim pollutants or produce energy (Figure 1). These modern advances are needed because we produce more garbage than ever before in our history. In addition, our garbage contains toxic substances that we do not want to release into the environment.

> **SKILLS MENU**
> - Defining the Issue
> - Researching
> - Identifying the Alternatives
> - Analyzing the Issue
> - Defending a Decision
> - Communicating
> - Evaluating

The Issue

Why do we produce so much waste or "garbage"? How is it disposed of? Are there ways that we can reduce our impact on the environment by producing significantly less waste?

In this exercise, you will examine the sources of our waste and how we can reuse, recycle, and dispose of it. As part of a class "town hall" meeting, you will formulate and present ways to reduce the amount of waste produced. You will also recommend ways to deal with the waste we do produce. You should consider the following factors:

- Many consumer products come in excess packaging (Figure 2).

Figure 2 Many consumer products come wrapped in unnecessary packaging.

- Many products have a short life expectancy and are designed so that they cannot be repaired.
- "Cradle to cradle" is a zero-waste approach to sustainable manufacturing that models natural processes to reduce waste production.
- Almost all plastic, glass, metal, wood, and paper products can be recycled or reused, but very little of it is (Figure 3).

Figure 1 Modern energy-from-waste facilities use garbage to generate thermal and electrical energy.

Figure 3 These deck chairs are made of recycled plastics.

- Virtually all food scraps and yard wastes could be composted, but few communities have large-scale composting programs.
- We package and market products, such as bottled water, that are almost completely unnecessary.
- Waste incinerators can generate energy but they are very costly and depend on a long-term supply of combustible waste.
- Landfill sites negatively affect adjacent ecosystems, attracting pests, leaching toxic chemicals, and producing greenhouse gases (Figure 4).

Figure 4 The Hagersville tire dump contained 14 million tires when it caught on fire in 1990. It burned for almost two weeks, polluting the air, water, and soil.

Goals

- To understand the negative impacts that resource consumption and waste production have on the sustainability of ecosystems.
- To identify alternatives and strategies to reduce negative impacts.

Gather Information

For this activity, you will represent a particular interest group, such as a consumer, a manufacturer, or a retail store owner. Do research to learn about the major sources of waste generated by your interest group and options for their disposal. Consider the following questions as you gather information:

- Where can you find more information?
- Will you make a visit to a local shopping mall to see what sorts of waste are associated with consumer goods?
- Will you contact your municipality to learn how your waste is disposed of and what sorts of composting and recycling programs are available?
- If you are doing an Internet search, what key words can you use?

Identify Solutions

You may wish to consider the following factors to help identify options for waste reduction:

- How effective would recycling and composting programs be in reducing waste?
- What actions should local, provincial, and federal governments take to reduce the amount of waste going to landfills and incinerators?
- How can I reduce the amount of garbage I produce?

Make a Decision

Make a final list of waste reduction options that you feel will minimize waste produced by your interest group.

Communicate

Present your findings at a model Town Hall meeting held to gather information and recommendations from community members. At the meeting, you will present your ideas and concerns regarding waste reduction.

CHAPTER 4 LOOKING BACK

KEY CONCEPTS SUMMARY

Some ecosystems, such as farms and cities, are designed, created, and maintained by humans.

- Human-engineered landscapes are replacing many of Earth's natural ecosystems. (4.1)
- Engineered ecosystems have lower biodiversity and have more uniform abiotic features than natural ecosystems. (4.1)
- Most of Canada's crop and livestock species are non-native species. (4.1)

Agricultural practices alter natural biogeochemical cycles.

- Soil ecosystems have highly complex and diverse food webs. (4.2)
- Natural fertilizers cause fewer ecological problems than synthetic fertilizers. (4.2)
- Soil compaction limits water flow and harms plants by reducing oxygen and nutrient availability. (4.2)

Agricultural practices alter water cycles.

- Water may be added to soil via irrigation or removed from soil through drainage systems. (4.2)
- Irrigation and drainage alter the water cycle. (4.2)
- No-tillage farming, crop rotation, and crop selection reduce the impacts of agriculture on soil ecosystems. (4.2)

Pesticides are used to reduce yield losses from pests, but there are ecological costs associated with them.

- Pesticides may kill non-target species and may cause air and water pollution. (4.5)
- Pesticide resistance is increasing. (4.5)
- Farming practices such as organic farming and integrated pest management reduce pesticide use. (4.5)

Some pesticides and other toxins bioaccumulate in individuals and biomagnify in food webs.

- Bioaccumulation is the concentration of a pesticide or toxin in the body of an organism. (4.5)
- Bioamplification is the increase in concentration of a pesticide or toxin as it moves higher up the food chain. (4.5)
- Most natural and recently developed synthetic pesticides do not accumulate in food chains. (4.5)

Urban ecosystems are dependent on surrounding ecosystems to provide food and materials for urban inhabitants.

- The ecological footprint of a city reaches well beyond its boundaries. (4.6, 4.7)
- Urban ecosystems are dependent on surrounding ecosystems for food, water, and other materials. (4.7)
- Urban ecosystems can provide environmental benefits by reducing space, energy, and resource consumption. (4.7)
- There are many ways of enhancing urban living spaces and reducing their negative impacts on the environment. (4.7)

WHAT DO YOU THINK NOW?

You thought about the following statements at the beginning of the chapter. You may have encountered these ideas in school, at home, or in the world around you. Not all of the following statements are true. Consider them again and decide whether you agree or disagree with each one.

Vocabulary

agroecosystem (p. 119)
monoculture (p. 121)
pest (p. 121)
natural fertilizer (p. 124)
synthetic fertilizer (p. 124)
leaching (p. 125)
pesticide (p. 133)
broad-spectrum pesticide (p. 133)
narrow-spectrum pesticide (p. 133)
bioaccumulation (p. 136)
bioamplification (p. 136)
organic farming (p. 139)
integrated pest management (IPM) (p. 140)

1 Agricultural practices closely resemble naturally occurring ecosystem processes.
Agree/disagree?

4 Many of the plants and animals farmed in Ontario are native species.
Agree/disagree?

2 Organic farming has benefits for the environment.
Agree/disagree?

5 The actions of people who live in cities have little impact on wilderness environments.
Agree/disagree?

3 Pesticides are needed to control pests.
Agree/disagree?

6 We can use as many resources as we like because we can recycle or reuse them.
Agree/disagree?

How have your answers changed since then?
What new understanding do you have?

BIG Ideas

- ✓ Ecosystems are dynamic and have the ability to respond to change, within limits, while maintaining their ecological balance.
- ✓ People have the responsibility to regulate their impact on the sustainability of ecosystems in order to preserve them for future generations.

CHAPTER 4 REVIEW

The following icons indicate the Achievement Chart category addressed by each question.
K/U Knowledge/Understanding T/I Thinking/Investigation C Communication A Application

What Do You Remember?

1. In your notebook, write the word(s) needed to complete each of the following sentences. K/U

   (a) Any unwanted organisms can be considered a(n) _____. (4.4)
   (b) Manure and ground-up bone meal would be considered _____ fertilizers. (4.2)
   (c) Toxins that kill plants, insects, and rodents are all examples of _____. (4.4)
   (d) Some pesticides and other environmental toxins build up or _____ in living organisms. (4.5)
   (e) _____ pesticides kill only a limited variety of living things. (4.4)
   (f) _____ uses a number of different techniques to control pests. (4.5)
   (g) A(n) _____ is a large concentration of a single species growing in one area. (4.1)
   (h) A(n) _____ ecosystem is one that is designed by humans. (4.7)

2. Identify whether each of the following conditions best represents a monoculture or a natural ecosystem. (4.1) K/U

   (a) numerous species of herbivores present
   (b) food web is almost eliminated
   (c) sustainability requires frequent human intervention
   (d) only one plant species is not considered a weed species
   (e) a small number of herbivores are abundant
   (f) few carnivores are present

3. Examine Figure 1 and list the distinguishing features of an urban environment. (4.7) K/U T/I

Figure 1

4. Match the term on the left with the appropriate definition on the right. (4.1, 4.4) K/U

   (a) DDT   (i) pesticide that remains in the environment for a long time
   (b) narrow spectrum   (ii) pesticide that originated in bacteria
   (c) piscicide   (iii) pesticide that kills fish
   (d) persistent   (iv) pesticide that kills only a few target species
   (e) Bt   (v) pesticide that is banned for use in Canada

5. List ten non-native and five native human food sources produced in Canada. (4.1) K/U

What Do You Understand?

6. (a) List the advantages and disadvantages of broad-spectrum and narrow-spectrum pesticides.
   (b) Identify different cases for which each of these might be preferred. (4.4) K/U A

7. Describe two key innovations that have allowed humans to dramatically alter their natural environments. (4.1) K/U

8. Does most human food come from natural or engineered ecosystems? Suggest reasons for this. (4.1) K/U C

9. Examine Figure 2. List six important differences between these two ecosystems. (4.1) K/U T/I

Figure 2

10. People think of soil as nothing but non-living bits of rock. Provide a more accurate and detailed description of soil composition. (4.2) K/U

11. Explain how leaching can cause problems for both farmers and ecosystems. (4.2) K/U

12. Under what conditions do farmers need to control water availability? Describe two technologies used to accomplish this. (4.2) K/U

13. Figure 3 illustrates the most common way of applying pesticides to fields and forests. List the environmental concerns associated with this application method. (4.4) K/U A

Figure 3

14. What are the advantages of not using pesticides to control pests? (4.5) K/U

15. Why have some farmers decided to use no-tillage techniques? Explain. (4.2) K/U

16. Draw a diagram to illustrate the process of bioamplification. (4.5) K/U C

17. Explain why monocultures are not sustainable on their own, while complex natural ecosystems are sustainable without any human intervention. (4.1) K/U C

Solve a Problem

18. Construct a Venn diagram (such as Figure 4) to compare the features of natural and engineered ecosystems. (4.1) K/U C

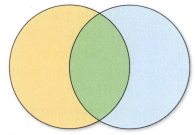

Figure 4

19. Brainstorm a list of five situations in which you would consider an organism to be a pest. For example, a mouse in your cottage could be a pest. For each pest, suggest a way of controlling the pest. (4.1) A C

20. There seems to be "dirt" everywhere you go outdoors. Does this mean that soil is an abundant resource? Explain your reasoning. (4.2) A C

21. London, Ontario, recently banned the sale of water in plastic bottles in city facilities. This sets an example but represents only a small share of the bottles consumed in the city. Should plastic bottles be banned everywhere? Explain your answer. (4.6, 4.7) A C

Create and Evaluate

22. Should golf courses be exempt from recent Ontario legislation regarding cosmetic pesticide use? Explain your answer. (4.5) A C

23. If organic products are grown without the addition of expensive pesticides, hormones, and fertilizers, why do these products cost more? (4.5) A C

24. Urban centres cause harm to the environment, but they also have benefits. Brainstorm some environmental problems that might arise if more people lived in rural settings instead of cities. (4.7) T/I A

Reflect on Your Learning

25. What aspects of human-engineered ecosystems do you think are the least sustainable? Explain. (4.1–4.8) K/U C

26. What choices do you think are most important in making communities more sustainable and environmentally friendly? Explain your reasoning. (4.7) A

27. Has what you have learned influenced your thinking on the use of pesticides or fertilizers? If so, describe how. (4.4, 4.5) C

Web Connections

28. Some wild animals benefit from living in an urban environment. These animals also have impacts as their populations increase. Research the impacts Canada geese or raccoons have in an urban environment. (4.7) T/I C A

To do an online self-quiz or for all other Nelson Web Connections, **GO TO NELSON SCIENCE**

CHAPTER 4 SELF-QUIZ

The following icons indicate the Achievement Chart category addressed by each question.
K/U Knowledge/Understanding **T/I** Thinking/Investigation **C** Communication **A** Application

For each question, select the best answer from the four alternatives.

1. Which statement accurately describes bioaccumulation? (4.5) **K/U**
   (a) Species build up a resistance to certain pesticides over time.
   (b) Pesticides in the soil, air, and water become sources of pollution.
   (c) A substance, such as a pesticide, collects in the body of an organism.
   (d) The concentration of a substance increases as it moves up the food chain.

2. Which of the following is an example of a synthetic fertilizer? (4.2) **K/U**
   (a) ammonia
   (b) manure
   (c) compost
   (d) rotenone

3. Which of the following is true of no-tillage agriculture? (4.2) **K/U**
   (a) Farmers use fewer pesticides to control weed populations.
   (b) Farmers leave the ground undisturbed after harvesting crops.
   (c) Farmers avoid planting crops in the same area every year.
   (d) Farmers plant crops that are suited to local growing conditions.

4. Which of the following is a characteristic of natural fertilizers as compared to synthetic fertilizers? (4.2) **K/U**
   (a) Natural fertilizers release nutrients relatively quickly.
   (b) Natural fertilizers have high concentrations of nutrients.
   (c) Natural fertilizers are a recent agricultural development.
   (d) Natural fertilizers can improve the structure of soil.

Indicate whether each of the statements is TRUE or FALSE. If you think the statement is false, rewrite it to make it true.

5. In general, humans use natural ecosystems more intensively than engineered ecosystems. (4.1–4.8) **K/U**

6. Integrated pest management is a system that uses organic farming methods as well as synthetic pesticides and fertilizers. (4.5) **K/U**

Copy each of the following statements into your notebook. Fill in the blanks with a word or phrase that correctly completes the sentence.

7. Farms, golf courses, and urban centres are all examples of _____ ecosystems. (4.1) **K/U**

8. Some farms use _____, the system of growing only one crop in a certain area every year. (4.2) **K/U**

9. Fertilizers can contaminate groundwater through the process of _____. (4.2) **K/U**

10. _____ is a form of agriculture that avoids the use of synthetic pesticides and fertilizers. (4.5) **K/U**

Match each term on the left with the appropriate definition on the right.

11. (a) altered timing (i) using predatory species to get rid of pests
    (b) biological control (ii) changing the location of monocultures each year
    (c) crop rotation (iii) scheduling planting and harvesting to avoid pests
    (d) baiting pest (iv) confusing mating insects with pheromones (4.5) **K/U**

Write a short answer to each of these questions.

12. Name two ways that communities can reduce their collective ecological footprint. (4.7, 4.8) K/U

13. Why is soil compaction a problem for plant growth? (4.2) K/U

14. How does the food web in a monoculture compare with the food web in a natural ecosystem? (4.1) K/U

15. Farmers use either natural or synthetic fertilizers to improve their crop yields. (4.2) T/I C
    (a) Do you think that natural fertilizers are better than synthetic fertilizers? Write a short argument explaining your opinion.
    (b) How might food production be affected if farmers stopped using fertilizers?

16. Create a chart showing the advantages and disadvantages of using pesticides. (4.4, 4.5) K/U

17. Consider the ecosystems in and around your community. (4.7) A
    (a) Name three natural ecosystems in your area. What are some native species that live in these ecosystems?
    (b) Name three engineered ecosystems in your area. How do these ecosystems show signs of human influence?

18. Which ecosystem is more sustainable: a natural forest or a farm with corn crops? Explain your answer. (4.1) T/I

19. The trout population in a local pond has decreased over the past year. At the same time, the growth of algae in the pond has increased. Explain what may be causing these changes in the pond ecosystem. (4.2) T/I A

20. (a) What are the benefits of spending time in natural ecosystems?
    (b) How could you encourage friends and family members to spend more time in natural ecosystems? (4.7) T/I C

21. How could the agricultural method of crop selection benefit farmers as well as the environment? (4.2) T/I

22. Name and describe some steps your community has taken to promote sustainability. (4.7, 4.8) A

23. (a) Explain why the term *pest* does not apply to natural roles in ecosystems.
    (b) Do you think any species would consider humans to be pests? Explain your answer. (4.1) T/I

24. Your local supermarket does not yet offer organic products. Write a letter to the manager of the supermarket explaining why he should consider selling organic products. (4.5) C

25. Use a Venn diagram or a table to compare the distinctive and shared features of natural and engineered ecosystems. (4.1) K/U

26. Why would a farmer choose to use synthetic rather than natural fertilizers? (4.2) K/U A

27. What are the three main chemical soil nutrients? (4.2) K/U

28. Explain the difference between bioaccumulation and bioamplification. Provide an example of each. (4.5) K/U

29. Many cities and towns in Canada and around the world are working to reduce their ecological footprints. (4.7, 4.8) K/U T/I
    (a) Name three actions that a community could take to reduce its ecological footprint.
    (b) Explain how each action can help to make the community more sustainable.

30. (a) Explain what function air spaces serve in soil.
    (b) Describe how compaction negatively affects the ability of soil to support life. (4.2) K/U

UNIT B

LOOKING BACK

UNIT B
Sustainable Ecosystems

CHAPTER 2
Understanding Ecosystems

CHAPTER 3
Natural Ecosystems and Stewardship

CHAPTER 4
Ecosystems by Design

KEY CONCEPTS

 Life on Earth exists in the atmosphere, lithosphere, and hydrosphere.

 Photosynthesis and cellular respiration are complementary processes in an ecosystem.

 Energy passes through ecosystems, whereas matter cycles within ecosystems.

 Human activities influence biogeo-chemical cycles such as the water and carbon cycles.

 Ecosystems are composed of biotic and abiotic components.

 Terrestrial biomes and aquatic ecosystems are largely determined by their abiotic characteristics.

KEY CONCEPTS

 Natural ecosystems are of great value to humans.

 Ecosystems are in equilibrium but can change over time.

 Biodiversity describes the variety and abundance of life in an ecosystem.

 Many human activities impact and threaten the sustainability of natural ecosystems.

 Water, land, and air pollution cause health and economic problems.

 Plant and animal resources should be used in a sustainable manner.

KEY CONCEPTS

 Some ecosystems, such as farms and cities, are designed, created, and maintained by humans.

 Agricultural practices alter natural biogeochemical cycles.

 Agricultural practices alter water cycles.

 Pesticides are used to reduce yield losses from pests, but there are ecological costs associated with them.

 Some pesticides and other toxins bioaccumulate in individuals and biomagnify in food webs.

 Urban ecosystems are dependent on surrounding ecosystems to provide food and materials for urban inhabitants.

MAKE A SUMMARY

Biosphere 2 was designed to study the potential for creating a sustainable environment that could support humans living in space. The 12 700 ha facility was the largest sealed enclosure ever constructed. It contained rainforest, savannah, desert, and aquatic mangrove and coral reef ecosystems, as well as a 2500 ha agricultural system and human living quarters. The sealed design permitted researchers to manipulate and monitor water and biogeochemical cycles.

To test Biosphere 2, nine people entered the sealed enclosure and lived in near-total physical isolation from the outside world for two years. The results of the experiment were disappointing. During the two years, oxygen levels dropped and carbon dioxide levels fluctuated widely. All of the large non-human animals and pollinating insects died. An accidentally introduced species of ant became an invasive species within the enclosure. Even at a cost of over $100 million, Biosphere 2 was not a sustainable environment.

Recently, the University of Arizona has taken over the operation of Biosphere 2 (Figure 1). The University hopes to conduct research using these closed ecosystems. They want to better understand the impacts and causes of global environmental changes and soil–water–atmosphere–biotic interactions.

Figure 1 Biosphere 2

Questions

1. List some key abiotic and biotic features that engineers and scientists would have had to consider in designing Biosphere 2. K/U

2. Oxygen and carbon dioxide levels fluctuated over the two-year study. What natural processes might have been out of balance? K/U

3. If humans were able to live indefinitely in Biosphere 2, what would have to be true of the agricultural practices and waste management? How might these be accomplished in the closed ecosystems? K/U A

4. Although Biosphere 2 may appear large, it is still extremely small compared with natural ecosystems. Make a concept map of major services, products, and resources that our biosphere provides to humans. K/U C

5. For each service, product, and resource, make a list of the conditions that would be needed if these had to be provided within the enclosed space of Biosphere 2. Figure 2 is a partially complete example for the service of supplying food. K/U C

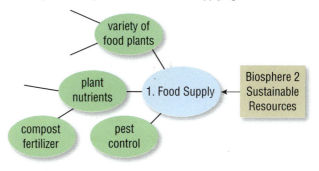

Figure 2 Partial example of a concept map

CAREER LINKS

List the careers mentioned in this unit. Choose two of the careers that interest you or choose two other careers that relate to sustainable ecosystems. For each of these careers, research the following information:

- educational requirements (secondary and post-secondary)
- skill/personality/aptitude requirements
- potential employers
- salary
- duties/responsibilities

Assemble the information you have discovered into two job postings. Your job postings should compare your two chosen careers and explain how they connect to sustainable ecosystems.

UNIT TASK

Ontario's Species of Concern

You are probably familiar with a wide variety of animals, plants, and other organisms that are native to Ontario. Many of these species, such as moose, beaver, and white birch, have large, healthy populations. There are also species that were intentionally introduced into Ontario. Virtually all of our domesticated livestock (cattle, chickens, and sheep), pets (cats, dogs, and guinea pigs), and food crops (wheat, corn, and potatoes) exist in relatively stable and managed populations.

Unfortunately, this situation is not characteristic of all native and non-native species. The distribution of many native species is declining rapidly, placing them at risk. In contrast, the populations of many non-native species are becoming invasive (Figure 1). Collectively, we will refer to them as species of concern.

Figure 1 Sea lampreys are an invasive species in the Great Lakes. These jawless fish are eel-like in appearance and are seen here feeding on a lake trout.

The Issue

Currently, more than 180 species are "at risk" in Ontario. Each of these has populations that are in decline or are experiencing significant environmental threats. The loss of any species is regrettable and can lead to long-term changes in ecosystems that diminish their ability to function.

There are at least as many invasive species in Ontario. They are responsible for damaging ecosystems by infecting, preying on, or competing with native species. Invasive species can destroy valuable resources and upset ecosystems.

SKILLS MENU
- Defining the Issue
- Researching
- Identifying Alternatives
- Analyzing the Issue
- Defending a Decision
- Communicating
- Evaluating

Economic losses from introduced pests can be high. There is often a close association between the introduction of an invasive species and the decline of native species.

Goal

In this Unit Task, you will select two species that are of concern in Ontario: one native species that is at risk and one non-native species that is invasive. For each species, you will investigate the factors that have led to their current status and how these factors influence the species and the sustainability of the ecosystems they live in. You may wish to select species that occur in your own region. This would permit you to conduct field surveys of the species and interview local experts.

Gather Information

Investigate the recent history of each species to learn how its population size and geographical distribution have changed over time. What are the human or natural reasons that one species is declining while the other is increasing in numbers?

Outline the implications for future sustainability of the terrestrial or aquatic ecosystems involved. Consider what other species are being directly and indirectly affected by the loss or addition of these species. Consider implications related to human interests such as economic losses or the degradation of recreational areas.

If possible, conduct field work. You may be able to gather a sample or photographic evidence of an invasive species or an at-risk species in your region (Figure 2, next page). Document threats to the species, such as shoreline developments, habitat loss, or evidence of a disease infecting a tree species (Figure 3, next page). If possible, interview local authorities about your species of concern.

Figure 2 The intentionally introduced dandelion does not pose a threat to any natural ecosystems. It is, however, an agricultural weed, expecially in no-tillage fields.

Figure 3 Human activities such as clearing of land may be one of the threats to your species at risk.

Identify Solutions

- Identify the individuals and organizations most closely associated with these species and the factors influencing their status.
- Find out how the actions of these individuals and organizations could be altered to affect positive change.
- Consider any government, citizen, and First Nations initiatives already in place regarding these species.

Make a Decision

- Choose a few key factors that you believe are the most critical for each species.
- Determine which actions you believe could be used to most effectively address these key factors.
- Develop a Plan of Action that describes how individuals and organization can put these actions into practice.

Communicate

You will prepare and present your Plan of Action to the class. Your plan should explain your strategies and recommended actions for both species. Think about the format you will use to communicate your Plan of Action. A few possibilities include a presentation, a written report, a poster, a documentary, or a news report video. Be sure to include and present your findings in a clear and interesting way. Consider including range maps for each species to illustrate their past and present geographical distribution in Ontario.

ASSESSMENT CHECKLIST

Your completed Performance Task will be evaluated according to the following criteria:

Knowledge/Understanding
- ☑ Research and analyze the factors influencing the population decline or growth of your selected species.
- ☑ Demonstrate an understanding of consequences for ecosystems when a species is at risk.
- ☑ Demonstrate an understanding of the threats posed by invasive species.
- ☑ Research and assess the options available to address problems associated with your species of concern.

Thinking/Investigation
- ☑ Identify the key human factors that are responsible for the decline of at-risk species and the success of invasive species.
- ☑ Devise strategies that will address these factors.
- ☑ Consider and incorporate the initiatives of other individuals and organizations into your Plan of Action.

Communication
- ☑ Prepare and present your Plan of Action in a creative and informative way.
- ☑ Include documentation of the changes in populations over time.
- ☑ Effectively illustrate the relationships between species of concern and Ontario's ecosystems.

Application
- ☑ Demonstrate an understanding of the influence of species of concern on human interests from an ecological, social, and economic viewpoint.
- ☑ Describe how individual and group actions can positively influence the sustainability of ecosystems.

UNIT B REVIEW

The following icons indicate the Achievement Chart category addressed by each question.

- K/U Knowledge/Understanding
- T/I Thinking/Investigation
- C Communication
- A Application

What Do You Remember?

For each question, select the best answer from the four alternatives.

1. Which of the following spheres includes all life on Earth? (2.1) K/U
   (a) lithosphere
   (b) biosphere
   (c) atmosphere
   (d) hydrosphere

2. Which of the following includes only abiotic features? (2.2, 2.7) K/U
   (a) air, water, animal waste, soil
   (b) air, water, soil, grass
   (c) grass, deer, air, water
   (d) temperature, wind, light, water

3. Which definition is correct? (2.2) K/U
   (a) A population can contain more than one species.
   (b) An ecosystem includes both biotic and abiotic features.
   (c) A community refers to a single species.
   (d) A habitat describes the role of an organism in its ecosystem.

4. Which of the following types of energy can be used by all organisms? (2.4) K/U
   (a) light energy
   (b) thermal energy
   (c) radiant energy
   (d) chemical energy

5. Which statement accurately describes photosynthesis? (2.4) K/U
   (a) Photosynthesis is a process by which producers use light energy to make food.
   (b) Photosynthesis is performed by green plants and some micro-organisms.
   (c) Photosynthesis requires both carbon dioxide and water.
   (d) All of the above

6. The process of cellular respiration (2.4) K/U
   (a) uses sunlight to make food
   (b) releases energy from food using carbon dioxide
   (c) releases energy from food using oxygen
   (d) stores energy and gives off carbon dioxide

7. Which of the following represents a producer-consumer relationship? (2.4) K/U
   (a) hawks eating snakes
   (b) flowers growing on a plant
   (c) rabbits eating grass
   (d) earthworms eating dirt

8. What name would be given to an organism that only eats meat? (2.5) K/U
   (a) herbivore
   (b) omnivore
   (c) carnivore
   (d) decomposer

9. The remora shark has an adhesive disk on the underside of its head that it uses to attach to large marine animals, such as whales. While attached, it is able to grab bits of food that the whale drops from its mouth. The whale is not affected by the presence of the remora. This is an example of (2.7) K/U
   (a) mutualism
   (b) commensalism
   (c) predation
   (d) competition

10. When two different species interact and both species benefit, it is called (2.7) K/U
    (a) parasitism
    (b) competition
    (c) commensalism
    (d) mutualism

11. This term refers to a system capable of continuing over long periods of time with little change. (2.2) K/U
    (a) community
    (b) ecosystem
    (c) sustainable
    (d) trophic

12. Which term describes a species that no longer exists in a particular region? (3.3) K/U
    (a) extinct
    (b) expired
    (c) exterminated
    (d) extirpated

13. Which of the following is not true of invasive species? (3.5) K/U
    (a) They often have few natural predators.
    (b) They often outcompete native species.
    (c) They have all been released accidentally into the natural environment.
    (d) They can cause significant ecological damage.

14. Which of the following techniques used to clean up oil spills is most likely to produce a different kind of pollution? (3.6) K/U
    (a) skimming/vacuuming
    (b) burning
    (c) bioremediation
    (d) dispersal agents

15. Animal manure, seaweed, and blood meal (4.2) K/U
    (a) all contain nutrients required by plants
    (b) can be used as soil supplements
    (c) are natural fertilizers
    (d) all of the above

16. Some farmers and home gardeners use ladybird beetles (ladybugs) to get rid of pests, such as aphids and other small insects, on their crops. This is an example of (4.5) K/U
    (a) integrated pest management
    (b) biological pest control
    (c) an invasive species
    (d) a broad spectrum pesticide

Indicate whether each of the statements is TRUE or FALSE. If you think the statement is false, rewrite it to make it true.

17. Tropical rainforests have the highest biodiversity of any terrestrial ecosystem. (3.3) K/U

18. Only a small proportion of the light reaching Earth is absorbed by plants and used for photosynthesis. (2.4) K/U

19. A single species can feed at only one trophic level. (2.5) K/U

20. An ecological pyramid might show the total biomass at each trophic level. (2.5) K/U

21. Energy is cycled, but matter passes in only one direction through an ecosystem. (2.4) K/U

22. Large amounts of carbon occur in fossil fuels, the ocean, limestone, and plants. (2.6) K/U

23. Most of Canada's habitat loss has occurred in the last 25 years. (3.4) K/U

24. Any foreign species introduced into a new environment will probably become invasive. (3.5) K/U

25. Clear-cutting is a method that is no longer used by the forestry industry. (3.7) K/U

26. Most food in Canada is produced by large-scale agricultural operations. (4.1) K/U

27. Most of the food crops grown in Canada are native species. (4.1) K/U

28. Pesticide resistance is a growing concern associated with pesticide use. (4.5) K/U

Copy each of the following statements into your notebook. Fill in the blanks with a word or phrase that correctly completes the sentence. K/U

29. The _____ refers to the range over which a species can exist for a particular variable such as temperature. (2.7)

30. An ecosystem's _____ is the maximum number of individuals it can support over a long period of time. (2.7)

31. Most of the water that enters the atmosphere evaporates from _____ ecosystems. (2.9)

32. The major cause of habitat loss in the Amazon rainforest has been _____. (3.4)

33. A partially enclosed body of water where fresh and salt water mix is called an _____. (2.9)

34. _____ is a method of controlling unwanted species using other living organisms. (3.5)

35. Most acid rain results from the release of _____ and _____ oxides. (3.6)

36. The type of cutting that has the least impact on the ecological features of a forest is a _____. (3.7)

37. Light energy that strikes Earth is converted into _____ energy and warms Earth's surface. (2.4)

38. The three most important soil nutrients other than water are _____, _____, and _____. (4.2)

39. The biome with the lowest species diversity is the _____. (2.8) K/U

40. A(n) _____ is the amount of surface area of Earth that a person needs to support his or her current lifestyle. (4.6) K/U

Match each term on the left with the most appropriate description on the right.

41. (a) oligotrophic
    (b) eutrophic
    (c) limiting factor
    (d) tolerance range
    (e) carrying capacity

    (i) a factor that restricts population size
    (ii) a term that describes bodies of water rich in nutrients
    (iii) the maximum population size of a particular species that a given ecosystem can sustain
    (iv) a term that describes bodies of water low in nutrients
    (v) the abiotic conditions within which a species can survive (2.7, 2.8) K/U

What Do You Understand?

Write a short answer to each of these questions.

42. What is the original source of energy in a food chain? (2.4, 2.7) K/U

43. List 3 biotic and 3 abiotic factors you would find in a forest ecosystem. (2.7, 2.8) K/U

44. Copy Table 1 in your notebook. Complete the table by listing problems that can be solved by the farming methods in the second column. (4.2) K/U

Table 1

Problem	Alternative Farming Practice Solution
	No-tillage agriculture
	Crop rotation
	Crop selection suited to local conditions

45. How do the effects of DDT on an ecosystem illustrate the interdependence of life on Earth? (4.5) T/I

46. What human activities can alter the water, carbon, and nitrogen cycles? (2.6) K/U

47. Why would it be unlikely to find a food chain with 15 trophic levels? (2.5) T/I

48. How do matter and energy differ within a community? Use the words recycle, trophic level, and flow in your answer. (2.4–2.6) C

49. More and more people are living in urban ecosystems. (4.7) K/U
    (a) How do urban ecosystems harm the environment?
    (b) In what ways are urban ecosystems helpful for the environment?

50. Give two examples of each of the following ecological roles: (2.7) K/U
    (a) carnivore
    (b) omnivore
    (c) decomposer
    (d) producer

51. Copy and complete Table 2 in your notebook. (4.2) K/U

Table 2

	Natural fertilizers	Synthetic fertilizers
examples		
advantages		
disadvantages		

52. We intentionally plant grass in our lawns but consider it a weed in a vegetable garden. Explain this distinction in relation to competition. (4.1) A C

53. Distinguish between each of the following pairs of terms: (2.5, 2.9, 4.1, 4.5) K/U
    (a) marine ecosystem and freshwater ecosystem
    (b) energy pyramid and pyramid of biomass
    (c) organic methods and integrated pest management
    (d) natural ecosystem and engineered ecosystem

54. Briefly describe the ways in which human activities affect soils in the urban environment. (4.2, 4.7) A

Solve a Problem

55. In the fall of 2008, astronauts aboard the International Space Station had to repair a new urine recycling device that extracts and purifies water from human urine.
    (a) Why would such a device be vital to the crew?
    (b) In what ways does this example model biogeochemical cycles on Earth? (2.6) A C

56. Calcium levels in lakes in south central Ontario are declining. This is causing a decline in the population size of key microscopic herbivores. (2.5, 2.7, 3.6) A T/I
    (a) How has calcium become a new limiting factor in these lakes?
    (b) How might these changes affect the carrying capacity of the lakes for consumers that feed higher up in the food chain?

57. Lion fish (Figure 1) are a popular marine aquarium fish species. They are native to the tropical waters of the western Pacific Ocean. Unfortunately, they have been released into the water off the southeastern United States and are rapidly spreading through the Caribbean Sea. They are voracious eaters of small fish and have few natural enemies. (2.5, 3.5) T/I A

Figure 1

  (a) What would you expect to happen to the population of this species over the next few years?
  (b) How is this species likely to influence other species and the natural ecosystem?
  (c) Lion fish have poisonous spines. How might this affect their success?
  (d) What can and should be done about this introduction? Consider this question from an educational and a scientific perspective.

58. The human population is growing rapidly. (3.1–3.8, 4.7, 4.8) A C
    (a) Brainstorm and list some of the major impacts this will have on natural ecosystems.
    (b) How might the impacts be different if most of the population lived in cities as opposed to rural settings? Explain.
    (c) As the population grows, which environmental concerns do you think should take priority? Pollution, habitat loss, endangered species, other? Explain your choice.

59. When humans and other animals use fats stored in their bodies, they often release toxins that have accumulated in these tissues. In humans, nursing mothers use their own body fats to produce breast milk. In birds, large amounts of fats are used during egg production. (4.5) A
    (a) Why should pregnant or nursing mothers be concerned about pesticide use?
    (b) How might pesticide use threaten the health of bird populations?
    (c) How can organic foods help address both of these concerns?

60. Your neighbour is a farmer who grows many acres of corn and wheat. He uses synthetic fertilizers and pesticides on his crops. He has noticed that over the past few years, the pond on his property has become overrun with algae and there are very few fish living it anymore. He asks you to help him figure out why his pond has deteriorated. What would you tell him? (4.4, 4.5) A

61. Many environmental organizations are encouraging people to buy food products that are grown locally. Given what you know about crop production and transport, what are two arguments these organizations might use to promote local food products? (4.1–4.6) T/I

62. In April 2009, the Ontario government banned the use of pesticides for cosmetic use. Under this legislation, homeowners can no longer use herbicides to control lawn weeds such as dandelions and crab grass. (4.5) T/I A C
    (a) Create a table outlining the costs and benefits of this legislation.
    (b) Write a letter to your local newspaper supporting or opposing this legislation.

Create and Evaluate

63. Examine the biomass pyramids of a bog and a marine ecosystem (Figure 2). (2.5) T/I

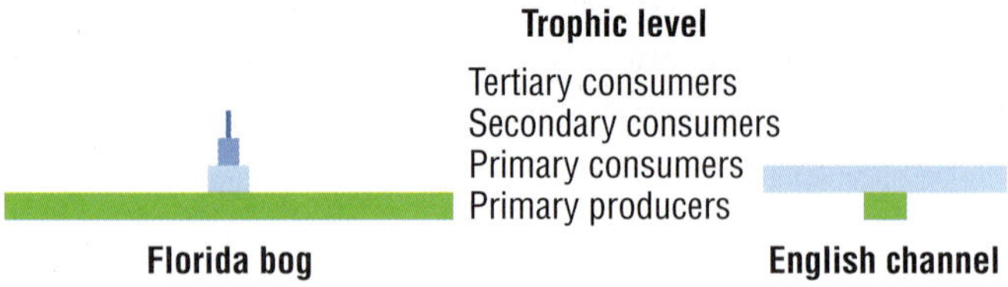

Trophic level
Tertiary consumers
Secondary consumers
Primary consumers
Primary producers

Florida bog   English channel

Figure 2

(a) What is the most striking difference between the two ecosystems?
(b) An ecologist commented that "a marine ecosystem operates like having a small land plant that can produce a large amount of new fruit every single day—enough to feed a large herbivore." How is this statement supported by the biomass pyramid?

64. Thousands of walruses once lived along the coast of eastern Canada. Today, this species would be a major tourist attraction for people visiting the Gulf of St. Lawrence and Maritime provinces. (2.5, 3.3) T/I A C

(a) Why do you think these walruses became extirpated? Explain.
(b) Do you think the loss of the walruses would have influenced any other species? If so, how?
(c) Do you think walruses should be reintroduced into this region? Why or why not?
(d) Do you think people have the right to eliminate an entire population for their own purposes? Explain.

65. Some pesticides come from natural sources. For example, rotenone comes from the roots of certain tropical plants. Some organic food growers use these natural toxins on their crops, but, clearly, applying rotenone is not "natural." Do you think these kinds of pesticides should be allowed to be used to produce organic foods? Explain your answer in a letter to your local member of parliament. (4.4, 4.5) A C

66. Each year, a population of whale sharks, the world's largest fish, migrates to waters off the coast of Christmas Island (Figure 3). Until recently, scientists were puzzled as to the purpose of this migration. After collecting and examining whale shark feces, they were surprised to discover that these massive fish were feeding on millions of tiny, newly hatched red crabs! K/U A C

Figure 3

(a) At what time of year do you think the whale sharks arrive off the coast of Christmas Island? (Hint: Refer back to the Chapter 2 opening.)
(b) How might the invasion of the yellow crazy ants on Christmas Island impact the whale shark population? Explain.
(c) Draw a simple food web that includes the following groups: tree seedlings, yellow crazy ants, whale sharks, red crabs.
(d) How does this example illustrate the link between the sustainability of terrestrial and aquatic ecosystems? Explain.

67. Figure 4 illustrates changes in population sizes of lemmings and snowy owls in the Canadian tundra over a period of many years. (2.7) T/I A C
    (a) What relationship exists between these populations?
    (b) Does the owl population tend to increase before or after there is a increase in the lemming population?
    (c) Is the population size of owls a limiting factor for lemmings, or is the population size of lemmings a limiting factor for owls? Explain your reasoning.

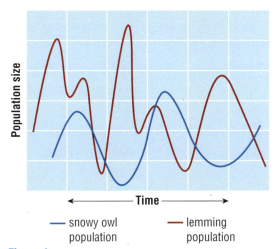

Figure 4

68. Read the following passage describing the feeding relationships in an ecosystem: The primary consumers in this ecosystem are rabbits, deer, and grasshoppers that feed on the grasses. Mice eat the grasshoppers; snakes feed upon the mice and rabbits. Hawks eat the snakes, rabbits, and mice. (2.5, 2.7) K/U A
    (a) Draw a food web for this ecosystem.
    (b) What are the producers in this ecosystem? Why are they so important?
    (c) Name one carnivore and one herbivore in this ecosystem.
    (d) Give an example of competition that might occur in this ecosystem.
    (e) Draw an energy pyramid of the food web.

Reflect on Your Learning

69. Explain how one terrestrial ecosystem and one aquatic ecosystem you learned about in this unit is crucial to life on Earth. K/U A C

70. Upon completing this unit, what sustainable measures will you now employ to minimize your ecological footprint? (4.6) A C

71. Do you think ecosystems, such as farmland and urban environments can be made sustainable? Do you think humans should try and live without engineered ecosystems? Justify your answer. (4.1–4.8) A C

72. What are three changes you can make to live more sustainably? (4.6) A

73. (a) Which concept in this unit did you find difficult to understand? Why?
    (b) What further research can you do to help you better understand this concept?

Web Connections

74. Use Internet and local resources to find out what "green" initiatives are occurring in your community. Find out the status of your community's waste management and recycling systems. Does your community have any plans for environmental actions that will improve the sustainability of your region? Write a brief report summarizing your findings. (4.6–4.8) C

75. Research an endangered species that lives in your area. (3.3) A
    (a) Explain why the species has become endangered.
    (b) What steps can be taken to protect this species?

76. Research the ecology of the pitcher plant (*Sarracenia purpurea*) which grows in northern Ontario. Would you consider the liquid-filled cavity of a pitcher plant to be an ecosystem? Explain your answer in terms of the following: (2.2, 2.5–2.7) T/I A
    (a) biotic and abiotic characteristics
    (b) food webs
    (c) biogeochemical cycles

UNIT B SELF-QUIZ

The following icons indicate the Achievement Chart category addressed by each question.
K/U Knowledge/Understanding **T/I** Thinking/Investigation **C** Communication **A** Application

For each question, select the best answer from the four alternatives.

1. What are producers able to do that consumers cannot? (2.4) **K/U**
   (a) cycle energy in ecosystems
   (b) make their own food
   (c) release energy from food
   (d) generate thermal energy

2. The unique way a species interacts with other species and with its environment is called (2.5) **K/U**
   (a) a trophic level
   (b) a feeding role
   (c) an ecological niche
   (d) a food chain

3. Which of the following is an example of a biotic factor? (2.2, 2.7) **K/U**
   (a) sunlight
   (b) grass
   (c) soil
   (d) water

4. Which of the following is true about pesticides? (4.4, 4.5) **K/U**
   (a) Pesticides are toxic only to the target species.
   (b) Pesticides tend to reduce overall food production.
   (c) Pesticides are used to replace nutrients in soil.
   (d) Pesticides can pass from an organism to its consumer.

Indicate whether each of the statements is TRUE or FALSE. If you think the statement is false, rewrite it to make it true.

5. Primary succession occurs in an ecosystem that has been partially disturbed by an event such as a forest fire. (3.2) **K/U**

6. Fragmentation of an ecosystem occurs when the ecosystem is divided into smaller parcels, making it more sustainable. (3.4) **K/U**

7. A threatened species is a species that may become endangered if current trends continue. (3.3) **K/U**

Copy each of the following statements into your notebook. Fill in the blanks with a word or phrase that correctly completes the sentence.

8. Groundwater can become contaminated with nitrogen compounds from fertilizers through a process called _____. (4.2) **K/U**

9. An environment and all its organisms form a(n) _____. (2.2) **K/U**

Match each term on the left with the most appropriate description on the right.

10. (a) producer
    (b) consumer
    (c) decomposer
    (d) food chain
    (e) food web
    (f) trophic level

    (i) the feeding position of an organism along a food chain
    (ii) an organism that obtains energy by eating other organisms
    (iii) a network of feeding relationships within a community
    (iv) an organism that obtains energy by eating dead organic matter
    (v) an organism that obtains energy by making its own food
    (vi) a sequence of organisms, each feeding on the next (2.4, 2.5) **K/U**

Write a short answer to each of these questions.

11. List three examples of products we obtain from terrestrial ecosystems. (3.1) **K/U**

12. Explain the difference between a population and a community. (2.2) **K/U**

13. Predation and parasitism are two examples of species interactions. (2.7) **K/U**
    (a) How are predation and parasitism alike?
    (b) How are they different?

14. Acid precipitation removes calcium, a critical nutrient for sugar maples, from the soil. However, there is a fertilizer that claims to add calcium back into the soil. Describe a controlled experiment that would determine if this fertilizer counteracts the calcium-depleting effects of acid rain on sugar maples. (3.6) T/I

15. Many people believe that eating lower on the food chain is a more energy-efficient practice. Write an editorial for your school newspaper that supports this position. (2.5, 4.6) C

16. You have recently been elected to the county land planning committee. A woman comes before the committee because she would like to develop a farm. She owns many acres of land that are currently a natural forest ecosystem. She would like to turn this into farmland. What are three disadvantages to this plan from an environmental perspective? (2.8, 3.1, 4.1–4.5) K/U A

17. Hiking is a popular outdoor activity that benefits us. However, it can have negative effects on the environment. A
    (a) When you walk on an established trail in the woods, you do not have to worry about plants or trees growing in the path. Explain why.
    (b) Why is it important to stay on the trail when hiking in the woods?

18. Consider Canada's terrestrial biomes. (2.8) T/I
    (a) In which biome would farming be most successful? Explain your answer.
    (b) In which biome would farming be least successful? Explain your answer.

19. Matter is recycled within the biosphere. (2.6) T/I
    (a) Explain how materials in the biosphere are reused.
    (b) Why is this recycling of matter important?

20. How might massive deforestation alter the following biogeochemical cycles? (2.6) T/I
    (a) water cycle
    (b) carbon cycle
    (c) nitrogen cycle

21. Imagine a hurricane hits an island and kills half of the herbivore species. Predict how this event would affect
    (a) producers on the island
    (b) secondary consumers on the island (2.4) T/I

22. Imagine you are a reporter for an environmental magazine. You have an assignment to interview some farmers who have clear-cut tropical rainforests so they can farm the land. What are three questions you will ask them? (3.1, 3.3) C

23. In the fall and winter, atmospheric levels of carbon dioxide rise. In the spring and summer, these levels are much lower. Use your knowledge of the carbon cycle to propose an explanation for these changes. (2.6) K/U T/I

24. Reptiles have lower metabolic rates than mammals, which means they require less food to sustain them. Imagine the top predators in two different sustainable food chains: a lion and an alligator. How would the ratios of predator to prey compare in these two food chains? (2.5) A

25. Read the following description of a coastal arctic food web. Phytoplankton are the primary producers. Phytoplankton are eaten by zooplankton. Arctic cod, which eat zooplankton, are eaten by beluga whales, ringed seals, and walrus. Polar bears eat ringed seals. (2.5) K/U A
    (a) Draw the food web for this ecosystem.
    (b) Categorize every organism into their trophic level.
    (c) Explain why there are so few polar bears in arctic ecosystems.

UNIT C: Atoms, Elements, and Compounds

OVERALL Expectations

- assess social, environmental, and economic impacts of the use of common elements and compounds, with reference to their physical and chemical properties
- investigate, using inquiry, the physical and chemical properties of common elements and compounds
- demonstrate an understanding of the properties of common elements and compounds, and of the organization of elements in the periodic table

BIG Ideas

- Elements and compounds have specific physical and chemical properties that determine their practical uses.
- The use of elements and compounds has both positive and negative effects on society and the environment.

Focus on STSE

DURABLE OR DEGRADABLE?

Plastic is versatile, flexible, and durable—perfect for making toys, water bottles, and shopping bags. However, because it is so durable, it is not easily degraded. The plastic water bottle that contained the thirst-quenching drink at the sports game will last for generations in a landfill site. Plastics are not the only chemicals with both valuable and worrisome characteristics. Fertilizers and pesticides are used to enhance plant growth and increase food production while posing health risks for humans and the environment.

1. Rate the following choices from 1 to 3 using the following scale:
   1 = strongly agree; 2 = undecided; 3 = strongly disagree
   (a) Plastic goods are essential in our modern world.
   (b) Plastics do not harm the environment if we recycle them.
   (c) We need to invent more biodegradable plastics.
   (d) Plastic products are better than products made of glass, metal, or paper because plastics cost less.

2. Rate the following choices from 1 to 3 using the following scale:
   1 = strongly agree; 2 = undecided; 3 = strongly disagree
   (a) Perfect green lawns increase the economic value of the homes in a neighbourhood.
   (b) Chemical fertilizers and pesticides endanger young children and animals.
   (c) Individual homeowners have the right to decide what chemical products to use on their lawns.
   (d) Chemicals do not harm the soil even if they are harmful to pests.

Think/Pair/Share

3. Discuss the statements above in small groups. Consider the following questions:
   (a) Science and technology have produced many of the chemical compounds found in fertilizers, pesticides, and plastics. These compounds are now posing problems for our environment. Can science help to solve these problems?
   (b) Suggest ways other than science that can help to reverse these problems.

UNIT C

LOOKING AHEAD

UNIT C
Atoms, Elements, and Compounds

CHAPTER 5 — Properties of Matter

Different types of matter have different uses because of their different properties.

CHAPTER 6 — Elements and the Periodic Table

Elements in the same group on the periodic table have similar properties.

CHAPTER 7 — Chemical Compounds

Some compounds break down easily into simpler substances whereas others do not.

UNIT TASK Preview

A Greener Shade of Chemistry

Consumers are becoming increasingly aware that the products they purchase have an impact on the environment. Manufacturers have begun to tap into the growing market of consumers who are environmentally conscious and who want to buy products that are sustainable as well as effective. In this Unit Task, you will have the opportunity to evaluate a traditional consumer product, such as disposable diapers (Figure 1), in terms of its sustainability, or "greenness." You will use what you have learned about the properties of matter and chemical compounds.

You will then research a "green" alternative to this product and evaluate how effectively it can replace the traditional product. A more complete Unit Task outline is found near the end of this unit.

UNIT TASK Bookmark

The Unit Task is described in detail on page 286. As you work through the unit, look for this bookmark and see how the section relates to the Unit Task.

ASSESSMENT

You will be assessed on how well you

- research and classify the properties of a traditional consumer product and a "green" alternative product
- compare the "greenness" of both products
- advertise the "green" product
- understand the threats posed by the "green" product to our health and to the environment

Figure 1 Are "green" disposable diapers better for the environment?

What Do You Already Know?

PREREQUISITES

Concepts
- environmental impacts of disposal of substances
- pure substances versus mixtures
- the particle theory of matter
- viscosity
- density
- states of matter
- compressibility of gases and liquids
- buoyancy
- pressure, volume, and temperature of fluids

Skills
- follow safety procedures
- calculate density
- compare density of different liquids

1. Indicate whether each of the following statements is consistent with the particle theory of matter. Give reasons for your answer. K/U
   (a) All particles have an attraction to each other.
   (b) Particles are in constant motion.
   (c) The attraction between particles is stronger in liquids than in solids.
   (d) Particles in a solid are closer to each other than particles in a gas.

2. Classify each of the following as a pure substance or a mixture. K/U
   (a) distilled water
   (b) air
   (c) gold and copper in a ring
   (d) sugar dissolved in water
   (e) pizza (Figure 1)

Figure 1 Is a pizza a pure substance or a mixture?

3. Answer each of the following questions. K/U
   (a) Which is more viscous, water or ketchup?
   (b) Which is denser, wood or steel?
   (c) Which has more mass, a kilogram of potatoes or two kilograms of wool?
   (d) Which has greater volume, a volleyball or a basketball?

4. Give reasons for each of the following safety procedures. K/U
   (a) Wash hands after handling chemicals.
   (b) Close containers of unused chemicals promptly after use.
   (c) Read safety labels of all chemicals you use.
   (d) Check with your teacher before you dispose of any chemicals in the sink.
   (e) Never pour unused chemicals back into the stock bottle.

5. A block of wood has a density of 0.95 g/cm^3. Will it float or sink when placed on a liquid with a density of 0.88 g/mL? Explain your answer. K/U

6. You are heating a solution and recording its temperature every 2 min during an experiment. Which is your independent variable and which is your dependent variable? K/U

7. Metals such as copper, lead, and nickel are key components of computers and other electronic devices. These metals can leak into our water system and damage our environment when they are disposed of in landfill sites. Discarded computers are still usable, even though they are outdated (Figure 2). Electronic devices sold in Ontario now include additional provincial recycling fees that will pay for diverting electronics from landfills. A
   (a) Identify the main issue in the paragraph above.
   (b) Identify key stakeholders (people who have an interest in the outcome of the issue).
   (c) Choose a specific stakeholder. Identify possible positive and negative impacts of the recycling fee on this stakeholder.

Figure 2 How can electronic devices be diverted from landfills?

CHAPTER 5

Properties of Matter

KEY QUESTION: What properties of substances can you observe, and how do these properties determine the usefulness or hazards of a substance?

In addition to colour, paints have other important physical and chemical properties. For example, they are protective because they do not react with oxygen in the air. However, many paints are toxic.

UNIT C
Atoms, Elements, and Compounds

- **CHAPTER 5** Properties of Matter
- **CHAPTER 6** Elements and the Periodic Table
- **CHAPTER 7** Chemical Compounds

KEY CONCEPTS

Physical properties are characteristics that can be determined without changing the composition of the substance.

Chemical properties describe the ability of a substance to change its composition to form new substances.

Pure substances have characteristic physical properties.

Water has unusual characteristic physical properties.

Physical and chemical properties can be used to identify different substances.

Some common useful substances have negative impacts on the environment.

ENGAGE IN SCIENCE

DENIM TO DYE FOR

Have you spent hours in a fitting room, trying on all the latest styles, in search of the perfect pair of denim jeans? You are not alone. Blue denim jeans have been in style for over 150 years!

Denim fabric is made from fibres produced by the cotton plant. Cotton is strong, durable, and absorbent. Denim fabric has a long history. Its name comes from "de Nîmes," meaning that it is from the town of Nîmes, France. Levi Strauss was the first person to make trousers out of this tough cotton fabric for miners during the California gold rush in the 1850s. Denim jeans caught on as a fashion trend in the 1930s, when Western cowboy movies became popular around the world. The cut and shape of the jeans continued to change over the years, becoming the symbol of the teenage rebel in the 1950s and the hippie in the 1960s (Figure 1).

The traditional deep indigo blue of denim is produced by a natural process that has been used for centuries. Indigo is a natural pigment produced by a plant in the pea family. First, this plant is soaked with wheat bran to ferment the pigment. Next, the fermented plants need to sit for about a week in a large pot of urine. In case this does not appeal to you, a solution of washing soda (containing sodium carbonate) may be used as a substitute. The dye now changes to a new substance with a yellow colour and is ready to use. When the denim fibres are soaked in a vat of this yellow dye, the pigment particles become tightly bound to the fibres, making the dye colourfast. When the cloth is lifted out of the vat and hung to dry in fresh air, something remarkable happens. The yellow dye reacts with oxygen in the air and the colour of the denim cloth slowly changes into a beautiful indigo (Figure 2).

Although it is likely that your jeans were dyed using a less natural process, do you think it is worth knowing why your denims are "to dye for"? As you read the rest of this chapter, think about how the indigo dyeing process relates to the properties of matter.

Figure 1 Today, denim jeans come in many styles.

Figure 2 When using the traditional dyeing method, fibres must be exposed to the air before they turn indigo.

WHAT DO YOU THINK?

Many of the ideas you will explore in this chapter are ideas that you have already encountered. You may have encountered these ideas in school, at home, or in the world around you. Not all of the following statements are true. Consider each statement and decide whether you agree or disagree with it.

1 Some chemicals can be both useful and harmful to us.
Agree/disagree?

4 Hair dyes permanently change chemicals inside each strand of hair.
Agree/disagree?

2 The particles in a solid are closer together than they are in a liquid.
Agree/disagree?

5 In a chemical change, the original substance disappears.
Agree/disagree?

3 A pure substance, such as gold, melts at different temperatures depending on its size.
Agree/disagree?

6 An edible substance, such as table salt, does not pose any harm to you or the environment.
Agree/disagree?

FOCUS ON WRITING

Writing a Summary

When you write a summary, you condense a text by restating only the main idea and key points in your own words. The following is an example of a summary written about the Awesome Science feature *Antifreeze Is Not Just For Cars* near the end of this chapter. Beside it are the strategies the student used to write the summary effectively.

WRITING TIP

As you work through the chapter, look for tips like this. They will help you develop literacy strategies.

Antifreeze Is Not Just For Cars

The proteins fish and insects produce naturally to protect them from the cold are also used by humans. The purpose of this selection is to explain how these proteins work in both nature and human applications. Fish and insects survive cold weather because they produce a specially shaped protein that prevents ice crystals from forming in their tissues. A medical application of this natural kind of antifreeze is to use it to prevent ice crystals from harming human organs while they are being transported to a hospital for a transplant operation. Similarly, the beauty and ice cream industries use these proteins to make their products feel smoother and taste creamier. Scientists can develop beneficial medical and commercial applications from their observations of the natural world.

- Topic sentence paraphrases the author's main idea or message.
- The general word "organs" replaces a list of specific items in the original text.
- Signal word shows the relationship between ideas.
- Closing sentence explains how the ideas are connected together and reinforces the main idea.
- Second sentence identifies the author's purpose for writing the selection.
- Each key point used to support the main idea is summarized in a clear sentence.
- Ideas and information are given in the same order as in the original text.

5.1 From Particles to Solutions

Matter is anything that has mass and takes up space. Chemistry is the study of matter and the changes it undergoes. In this section, you will briefly review some of the important concepts you learned in earlier grades about what matter is made of and how it is classified.

The Particle Theory of Matter

All matter is made of tiny particles. Different kinds of matter are made of different kinds of particles. For example, the particles that make up water are different from the particles that make up the glass containing it (Figure 1). The **particle theory of matter** summarizes what scientists have learned about the particles that make up matter. The main ideas of the particle theory are

1. All matter is made up of tiny particles that have empty spaces between them.
2. Different substances are made up of different kinds of particles.
3. Particles are in constant random motion.
4. The particles of a substance move faster as its temperature increases.
5. Particles attract each other.

According to the particle theory, particles are attracted to each other and are always moving. Particles of a substance form a solid when these forces of attraction are strong enough to hold the particles close together in a rigid shape. When heated, particles gain energy and begin moving faster. When they have enough thermal energy, the particles start sliding past each other because the attraction between particles can no longer hold them together. This is the liquid state. The particles are still very close together, but they are able to flow past one another. If heating continues, the particles gain so much energy that they literally fly apart. The substance is now in the gaseous state, and the particles are so far apart that their forces of attraction have little effect on their behaviour (Figure 2).

particle theory of matter a theory that describes the composition and behaviour of matter

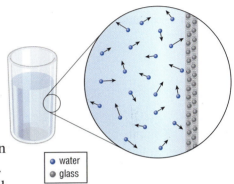

Figure 1 All matter is made up of particles.

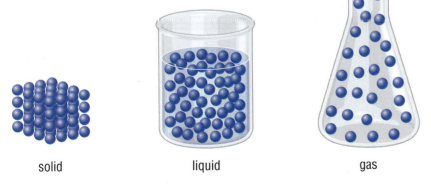

Figure 2 The particle theory describes the different behaviours of solids, liquids, and gases.

Pure Substances

Matter can be made up of many different types of particles. Some types of matter, however, are made up of only one type of particle. For example, pure or distilled water contains only water particles. Distilled water is an example of a **pure substance**—a type of matter that consists of only one type of particle. Water from your tap is not a pure substance because it contains water particles as well as other types of particles, such as dissolved gases.

pure substance a substance that is made up of only one type of particle

Figure 3 Granola bars are mixtures because they are made up of more than one type of particle.

mixture a substance that is made up of at least two different types of particles

mechanical mixture a mixture in which you can distinguish between different types of matter

solution a uniform mixture of two or more substances

Mixtures

When you add a pinch of salt to a glass of distilled water, the salt dissolves and the water tastes salty. The water in the glass is no longer a pure substance but a mixture because it contains salt particles and water particles. A **mixture** contains more than one type of particle. Mixtures can be solids, liquids, or gases. Solid mixtures include cellphones and granola bars (Figure 3). Examples of liquid mixtures are tea and juice. Air is a mixture of different types of gases.

Mechanical Mixtures and Solutions

There are two different kinds of mixtures: mechanical mixtures and solutions. A **mechanical mixture** is a mixture in which the substances in it are distinguishable from each other, either with the unaided eye or with a microscope. Breakfast cereal is an example of a mechanical mixture (Figure 4). A **solution** looks like a pure substance but it contains more than one type of particle. You cannot visually distinguish between the different types of particles in a solution. Clear apple juice is an example of a solution (Figure 5).

Figure 4 When you eat cereal for breakfast, you are eating a mechanical mixture.

Figure 5 Clear apple juice is a solution because you cannot visually distinguish between the different types of particles in it.

There is an easy way to tell whether a liquid or gas mixture is a solution or a mechanical mixture: all liquid and gas solutions are clear! If a liquid or gas mixture appears murky or opaque, it is a mechanical mixture. Fog, milk, and orange juice are examples of mechanical mixtures.

ALLOYS

Tin and lead are pure metals. Each metal by itself is a pure substance because it is made of only one type of particle. When two or more metals are mixed together, the resulting metal is called an **alloy**. An alloy is an example of a solution.

alloy a solid solution of two or more metals

Tin and lead are combined to make a metal alloy commonly called solder (pronounced "sodder"). Solder is used to join together metal components, such as wires in electrical circuits and copper pipes in plumbing. Like glue, solder needs to be fluid so that it can be applied to a joint and fill the spaces in it. It also needs to quickly solidify to firmly hold the parts together. Lead is ideal for this purpose—it is fluid at high temperatures but solidifies quickly.

There is increasing awareness, however, that lead poisoning causes irreversible damage to the brain, kidneys, heart, and reproductive organs, especially in growing children. Many uses of lead have been reduced or eliminated because of this increased awareness. Leaded gasoline and lead bullets used for hunting, as well as lead pipes and leaded solder, have been replaced by less hazardous materials (Figure 6).

DID YOU KNOW?

Lead Poisoning
Before the toxic effects of lead were understood, the seams of metal cans for preserving food were sealed using lead solder. When the cans were heated, a high level of lead leached into the food, particularly if the contents were acidic, such as tomatoes or citrus fruits. It is likely that sailors suffered from lead poisoning on long trips. Fresh meat and vegetables were not available, so sailors ate mostly canned foods. Even today, you should never drink hot water directly from the tap, in case there is lead solder in the plumbing that may be absorbed into the hot water.

Figure 6 A blow torch is used to melt solder. Lead-free solder consists of non-toxic metals such as tin, copper, and silver.

TRY THIS: HOW STRETCHY IS YOUR SOLDER?

SKILLS: Predicting, Controlling Variables, Performing, Observing, Analyzing, Evaluating, Communicating

SKILLS HANDBOOK
3.B.2., 3.B.8.

Have you heard the story about two brothers fighting over a copper penny and neither of them would let go? They ended up holding on to the ends of a copper wire. Metal can be stretched. This property of copper and other metals, called ductility, can be demonstrated using a piece of solder (Figure 7).

Equipment and Materials: marker; ceiling hook or ladder; heavy mass (e.g., bag of books or rocks); tape measure; 1 m long piece of solder

1. Obtain a piece of solder approximately 1 m long.
2. Tie one end of the solder to a ceiling hook or to the top rung of a ladder.
3. Tie the other end around a heavy mass, such as a bag of books or rocks.
4. Use a marker to draw a line near each end of the solder. Measure the distance between the lines. Allow the solder to remain suspended overnight.
5. Measure the length of the solder between the two marked lines again, and compare with the previous length.

Figure 7 Solder is used to form solid connections between wires and copper pipes. You will be measuring the ductility of solder.

A. Did the solder stretch while suspended overnight? How much?
B. Predict whether the solder would stretch more or less if the temperature were increased.
C. Design a way to test your prediction. If possible, test your prediction.
D. What avoidable or unavoidable problems did you encounter in this activity? What improvements could you make to your procedure?

The tree diagram in Figure 8 summarizes the classification of matter you studied in this section.

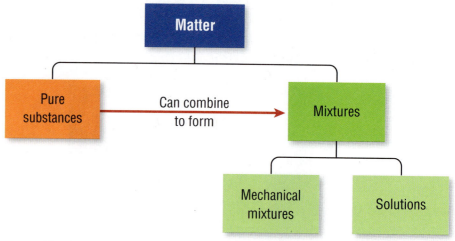

Figure 8 The classification of matter

IN SUMMARY

- The particle theory of matter describes the composition and behaviour of matter.
- A pure substance is made up of only one type of particle.
- A mixture is made up of at least two different types of particles.
- A mechanical mixture contains more than one type of particle, and the different types of particles are visible.
- A solution contains more than one type of particle but the different types of particles cannot be distinguished visually.
- An alloy is a solution composed of two or more metals.

CHECK YOUR LEARNING

1. List the five main ideas of the particle theory. K/U
2. Use the particle theory to explain why water changes from a solid to a liquid when it is heated. K/U
3. Give three examples of a pure substance. K/U
4. Give three examples of a mixture. K/U
5. Describe a mechanical mixture. K/U
6. Describe a solution. K/U
7. Use a Venn diagram to compare mechanical mixtures and solutions (Figure 9). C

8. Identify each of the following as a mechanical mixture or a solution: K/U
   (a) a pane of clear glass
   (b) chocolate chip ice cream
   (c) clear apple juice
   (d) a pizza
   (e) garbage in a garbage can
9. What kind of alloy makes an effective solder material? Explain. K/U
10. Lead is not often used in solder anymore. Explain why not. K/U

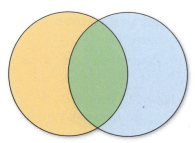

Figure 9 Sample Venn diagram

178 Chapter 5 • Properties of Matter

5.2

Physical Properties

If you have ever lost a piece of luggage at the airport, you know how important it is to provide a good description of its appearance. You need to report its colour, shape, size, and any other identifiable features, such as the colourful name tag that you could use to distinguish your luggage from hundreds of other items (Figure 1). When you describe your luggage in this way, you are reporting the physical properties of your lost item.

Figure 1 Colour, shape, and size are physical properties that help identify objects and substances. Some of these pieces of luggage are easy to identify because of their physical properties.

WRITING TIP

Condensing the Original Text
When writing a summary, find ways to condense the original text. Sometimes several specific words can be replaced by a general word. For example, "colour, shape, size, and any other identifiable features" can be shortened to "physical properties."

Physical properties give us information about what the substance is like. You can determine a physical property by simply observing the substance using your five senses and measuring instruments. Determining physical properties does not involve changing the composition of the substance.

We make direct observations when we are asked to determine the physical properties of a substance. For example, you might describe the substance in Figure 2 as white, odourless, and powdery. These descriptions tell us something about the appearance of the substance—in this case, how it looks, smells, and feels. We make these observations using our five senses. Any property that does not provide numerical information about the substance is called a **qualitative property**. Further, we may take some measurements and note that the substance has a mass of 10.0 g and is at a temperature of 25 °C. These measured physical properties give us numerical information about the substance. These types of information are **quantitative properties** of the substance (Figure 3).

physical property a characteristic of a substance that can be determined without changing the composition of that substance

qualitative property a property of a substance that is not measured and does not have a numerical value, such as colour, odour, and texture

quantitative property a property of a substance that is measured and has a numerical value, such as temperature, height, and mass

Figure 2 Physical properties include qualitative observations such as colour, odour, and texture.

Figure 3 Temperature is a quantitative property that tells us about the energy of particles in a substance.

Figure 4 A bicycle has both quantitative and qualitative physical properties.

Observations that a bicycle has a mass of 10 kg and is 2.0 m long are quantitative physical properties of the bicycle because they include a measurement. Qualitative physical properties of the bicycle are not measured and include that it is red, shiny, and rigid (Figure 4).

So far, we do not have any information about what will happen if we leave this bike out in the rain or whether a cola drink would effectively remove rust from the bike. These properties are chemical properties of the bike, which involve changing the bike's composition. Chemical properties are discussed in Section 5.3.

TRY THIS: CLOSE-UP OF A RUNNING SHOE

SKILLS: Observing, Analyzing, Evaluating, Communicating

SKILLS HANDBOOK 3.B.6.

Do you spend time and effort shopping for the ultimate running shoe (Figure 5)? What physical factors influence your decision? Perhaps comfort and support take priority over breathability and weight. Of course, style and colour are important as well.

In this activity, you will closely examine a running shoe. You will note how the physical properties of the different materials in it determine their specific function in the shoe.

Equipment and Materials: running shoe

Figure 5 A running shoe has important physical properties that enhance its function.

1. Make a table with three columns and record the following information:
   (a) In the first column, list all the different materials used to make each part of the shoe; for example, rubber soles.
   (b) In the second column, record the physical properties of each material listed; for example, waterproof, flexible.
   (c) In the third column, describe the function of each physical property you listed in part (b); for example, keeps the shoe dry, allows the foot to bend.
2. List at least five different quantitative properties of your shoe.

A. Of the physical properties that you listed, which are the most useful in the running shoe?

B. From what you know about the different materials that make up a running shoe, which ones might present a problem to the environment when the shoes are eventually discarded? Explain your answer.

C. There are different brands of running shoes, and they vary in price. What factors might influence the various prices? Compare the physical properties of the materials of several different brands of running shoes. Do they vary significantly? What other factors determine the retail price of running shoes? How important should these other factors be when choosing a pair of running shoes? Make a priority list of factors you will take into consideration when choosing your next pair of running shoes.

WRITING TIP

Restating the Main Idea

When you restate the author's main idea in the topic sentence of your summary, you say it in words that you understand. For example, if the author says that "some physical properties are particularly useful in describing and categorizing substances," you might restate this main idea by saying "physical properties are used to describe substances."

Some physical properties are particularly useful in describing and categorizing substances. Common qualitative physical properties include colour, odour, taste, and texture. Some of the other physical properties of matter are

- **lustre**—shininess or dullness; many silver objects have a high lustre (Figure 6), whereas a rusty nail has low lustre
- **optical clarity**—the ability to allow light through (Figure 7); thin blue glass is clear and transparent, frosted glass is translucent, and a brick wall is opaque
- **brittleness**—breakability or flexibility; glass is brittle (Figure 8) whereas modelling clay is flexible
- **viscosity**—the ability of a substance to flow or pour readily; molasses is viscous (Figure 9) whereas water is less viscous

viscosity the degree to which a fluid resists flow

- hardness—the relative ability to scratch or be scratched by another substance (Figure 10); wax is low on the hardness scale, whereas diamonds are high on the scale because they scratch nearly all other substances
- malleability—the ability of a substance to be hammered into a thinner sheet or molded (Figure 11); silver is malleable whereas glass breaks easily
- ductility—the ability of a substance to be drawn (pulled) into a finer strand; pieces of copper can be drawn into thin wires and are considered ductile (Figure 12)
- electrical conductivity—the ability of a substance to allow an electric current to pass through it; copper wires have high conductivity, whereas plastics do not (Figure 13)

Figure 6 This shiny kettle has high lustre.

Figure 7 The optical clarity of this window allows a lot of sunlight into the room.

Figure 8 Glass is brittle and cracks easily.

Figure 9 Molasses is viscous.

Figure 10 The dark crystal is hard.

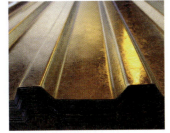

Figure 11 This metal is malleable because it can be made into sheets.

Figure 12 Copper wire is considered ductile.

Figure 13 The copper wires have high conductivity but the plastic switch does not.

As you will learn in Section 5.6, several quantitative physical properties are easy to measure and provide a useful method for identifying a substance. These properties include melting point and boiling point, as well as density.

Physical Changes

If you take a piece of paper and fold it into a paper crane, does the paper undergo a chemical or a physical change? It is true that the paper crane appears to be a new object, but the composition of the paper is not changed. It is still paper, although with a different shape and size (Figure 14). This change is a **physical change**. In a physical change, the composition of the substance remains exactly the same. No new substances are made.

Consider a change of state. If you heat an ice cube until it melts, does it undergo a chemical or a physical change? The ice cube was a cold solid, and now it has changed into a cold liquid. Was a new substance produced?

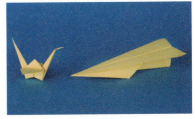

Figure 14 It's a bird! It's a plane! Yes, but it's still paper.

physical change a change in which the composition of the substance remains unaltered and no new substances are produced

5.2 Physical Properties

Figure 15 Wax is wax, whether it is solid or melted.

Recall from the particle theory that all matter is made up of particles that are in constant motion. In the case of melting ice, a new substance was not produced because the particles that make up water did not change. Only the arrangement of the water particles changed. If you were to put the water in a freezer rather than heat it, the water would turn back into ice. A change of state is a physical change—a change in how closely the particles are packed together (Figure 15). Many substances that undergo a physical change can be returned to their original state.

What happens when something dissolves? If you add a teaspoon of sugar to a large pot of hot water and stir, does the sugar undergo a chemical change or a physical change? The sugar seems to have disappeared into the water, which may lead you to think that its composition has changed. All the particles of sugar are close together in the solid state. When they are dissolved, the sugar particles become separated, spread out among the water particles, and are no longer visible in solution. If you taste the water, you can tell that the sugar is still there because the solution tastes sweet. If you allow all the water to evaporate, the sugar reappears at the bottom of the pot. Therefore, dissolving is a physical change.

UNIT TASK Bookmark

You can apply what you learned in this section about physical properties to the Unit Task described on page 286.

IN SUMMARY

- A physical property is a characteristic of a substance that can be determined without changing the composition of that substance. It may be qualitative or quantitative.
- Qualitative physical properties are not measured and include hardness, malleability, and electrical conductivity.
- Quantitative physical properties are measured and include temperature, height, and mass.
- A physical change is a change in which the composition of the substance remains unaltered and no new substances are produced. Examples of physical change are a change of size or shape, a change of state, and dissolving.

CHECK YOUR LEARNING

1. Explain the difference between a qualitative property and a quantitative property. K/U

2. A student recorded the following observations about a T-shirt. Classify each observation as a qualitative property or a quantitative property, and give reasons for your answers. K/U
   (a) It is red and grey in colour.
   (b) It is 60 cm long.
   (c) It is soft and stretchable.
   (d) It will shrink in 70 °C water.

3. What physical properties are important for the materials used to make mountain bikes? A

4. List four physical properties of each of the following: K/U
   (a) a piece of copper wire
   (b) 500 g of butter
   (c) a glass of milk
   (d) a candle
   (e) a piece of aluminum foil
   (f) a spoonful of sugar
   (g) toothpaste

5. In each of the situations below, it seems that a new substance may have been produced. Explain why each situation represents a physical change. K/U
   (a) A tailor makes a new suit out of a piece of fabric.
   (b) A chef makes a salad out of lettuce, tomatoes, and cucumbers.
   (c) A mechanic builds a boat engine out of a lawnmower.
   (d) A chemist boils salt water until only salt crystals are left.
   (e) A child makes juice by adding water to juice concentrate.

5.3 Chemical Properties

A fireworks display is the perfect ending to a Canada Day celebration (Figure 1). A great deal of artistry and planning goes into each show. When the skies are dark, the pyrotechnician ignites the first explosive mixture. The sky lights up with showers of colour and brilliance, accompanied by noise, smoke, and gasps of appreciation from the crowd.

To learn more about becoming a pyrotechnician,
GO TO NELSON SCIENCE

Figure 1 Fireworks traditionally mark Canada Day celebrations.

In scientific terminology, we are seeing the chemical properties of the fireworks. A **chemical property** is a property of a substance that describes its ability to undergo changes to its composition to produce one or more new substances. Fireworks contain ingredients such as metal flakes, fuel, and a bursting charge (Figure 2). These substances react together to produce new substances, some of which are visible in the smoke. The entire reaction releases a great deal of energy, which appears in the form of light, sound, thermal energy, and high-speed motion high into the sky.

All substances have chemical properties. Denim, for example, is resistant to cleaning solutions such as paint removers; that is, the composition of the cotton fabric is unchanged by these chemicals. If you decide to burn your old pair of jeans in a bonfire, you may find that the composition of the cotton changes. Flames engulf the fabric and entirely new substances are produced, called ashes. The resistance to paint removers and the ability to burn are chemical properties of cotton denim.

We take advantage of the chemical properties of substances in our daily lives. We mix different substances together to create products that we want. Baking soda causes a cake to rise, and bacterial cultures turn milk into cheese. We use other chemicals to clean our silver jewellery and clogged shower heads. New chemical products are continually produced to suit our changing lifestyles. Such a wide variety of products is available to us that it is helpful to have a basic understanding of chemical properties.

chemical property a characteristic of a substance that is determined when the composition of the substance is changed and one or more new substances are produced

Figure 2 Inside a fireworks tube

Chemical Changes

chemical change a change in the starting substance or substances and the production of one or more new substances

How can you tell if a chemical change has taken place? A **chemical change** is always accompanied by a change in the starting substance or substances and the production of one or more new substances. The original substances do not disappear. Instead, the components of the original substances are rearranged in the process of forming a new substance or substances.

TRY THIS: MAKE COMMON CHEMICAL CHANGES

SKILLS: Performing, Observing, Analyzing, Evaluating, Communicating

Chemical changes occur everywhere and all the time—at home, at school, and in your own body. In this activity, you will identify any evidence that a substance has changed its composition and something new has been produced.

Equipment and Materials: 2 teaspoons; 3 clear drinking glasses; vinegar; baking soda; lemon juice; strong tea; milk

1. Gather the following items from your kitchen: vinegar, baking soda, lemon juice, strong tea, milk (Figure 3). Combine the items as directed. Record your observations after each step.

Figure 3 Do this activity at home using equipment and materials from your kitchen.

(i) Place a teaspoon of baking soda into a clear drinking glass and pour a teaspoon of vinegar into the glass.

(ii) Fill a second drinking glass about half-full of strong dark tea. Add a teaspoon of lemon juice.

(iii) Fill a third drinking glass about one-quarter full of milk. Add an equal amount of vinegar. Mix gently and allow the mixture to sit for a little while.

A. In each of the activities above, identify any evidence that a new substance was produced. T/I

B. In each of these activities, do you think the process can be reversed so that the end product is changed back to its original form? Explain your answer. T/I

C. Compare the reversibility of these changes to a change of state, such as ice melting. What type of change occurs in a change of state? K/U

D. Compare the reversibility of these changes to a substance dissolving, such as a teaspoon of salt dissolving into a glass of water. What type of change occurs when a substance dissolves? K/U

E. List three differences between physical changes and chemical changes. K/U

WRITING TIP

Identify the Purpose
When writing the second sentence of a summary, consider three common purposes of science texts: to inform, to describe, to explain. For example, a text might describe the process of producing the indigo colour of blue jeans, or it might inform readers of the main ideas of the particle theory.

You can look at the scene of the chemical event for clues about the type of change that has occurred. In the case of the dye that colours cotton denim fibres, the yellow dye in the soaking vat visibly changes when it is exposed to oxygen in the air. A new compound with a new indigo colour is formed right within the fibres of the denim. This colour change is evidence of a chemical change.

Some examples of evidence of chemical change are listed below:
- a change of colour—a new substance has formed that has a different colour than the original substance
- a change of odour—a new substance has formed that has a detectable odour (in scientific language, all smells are called odours, whether they are pleasant or unpleasant)
- bubbles are visible that are not caused by heating—a new substance is produced in the form of a gas
- a new solid is seen—a new substance that is produced does not dissolve in the mixture and shows up as a solid; the solids that are formed in this way are often powdery and are called **precipitates**
- a change in temperature or light—energy is released or absorbed during the chemical change, and is detected as a change in temperature or light

precipitate a solid that separates from a solution

Many chemical changes are easy to observe and occur all around you. Figure 4 shows three examples of common chemical changes.

(a)

(b)

(c)

Figure 4 Evidence of chemical changes: (a) Colourless egg whites turn white during the chemical change called cooking. (b) The new products formed from chemically changed garbage are not always fragrant. (c) Bubbles of carbon dioxide gas are produced when baking soda reacts with vinegar.

TRY THIS TO ROT OR NOT TO ROT

SKILLS: Predicting, Performing, Analyzing, Communicating

SKILLS HANDBOOK
3.B.3, 3.B.7.

Rotting is the smelly way by which substances are naturally reused and recycled so that they can be made into new things. When rotting is contained and controlled, we call it composting (Figure 5). Of course, when we declare that we compost, we are claiming credit that rightfully belongs to the millions of microscopic bacteria and fungi that are doing the actual dirty work. These organisms are called decomposers. The decomposers' role is to break down complex compounds into small reusable components.

Figure 5 Composting is a great way of reducing waste that goes to landfills and at the same time fertilizing your garden.

In this activity, you will choose a variety of materials and investigate whether they can be decomposed in soil. Some items will be natural (from a plant or animal). Other items will be synthetic (made by chemical means). After your experiment is set up, it will be left to decompose. You will retrieve it in Section 7.7 for examination. At that time, you will see what has rotted and what has not rotted.

 Wash your hands thoroughly with soap and water after handling the soil and its contents.

Equipment and Materials: large container (for example, bucket, plastic tub, large pop bottle, small aquarium); garden soil or potting soil with compost starter (from garden store); water; a few small twigs or pebbles; selection of fruit or vegetable peels and scraps (do not include meats); selection of leaves, flowers, and stems; selection of synthetic plastics and polystyrene, cut into small pieces; small pieces of paper towel and newspaper; old spoon

1. Predict which types of items will decompose in the soil and which will not. Give reasons for your prediction.
2. Record the physical properties of each item you have selected to bury in the soil.
3. Add water to the soil. Mix until the soil has the wetness of a wrung-out sponge.
4. Place the twigs or pebbles at the bottom of the container to allow drainage.
5. Pour a layer of soil into the container followed by a layer of selected items. Ensure that each layer contains items of natural and synthetic materials.
6. Repeat step 5 until the container is filled.
7. Cover the container loosely with a lid and allow it to sit in a warm location. Check the moisture level every few days and add water if required.

A. Suggest reasons why the ability to decompose is an important characteristic of consumer goods.
B. Suggest situations where the inability to decompose is an important characteristic of a product.
C. List some examples of objects or appliances at home that are used to slow the decomposition of substances such as food or wood.

UNIT TASK Bookmark

You can apply what you learned in this section about chemical properties to the Unit Task described on page 286.

IN SUMMARY

- A chemical property is a property of a substance that describes its ability to undergo changes to its composition to produce one or more new substances.
- A chemical change is a change in the starting substance or substances and the production of one or more new substances
- Evidence of chemical change includes
  - colour change
  - odour change
  - gas production
  - precipitate production
  - energy change

CHECK YOUR LEARNING

1. Describe the difference between a physical change and a chemical change.
2. Explain why water freezing is not a chemical change.
3. Classify each of the following as a physical or a chemical property. Give reasons for your answer.
   (a) metallic lustre
   (b) boiling point
   (c) explodes when ignited
   (d) changes colour when mixed with water
4. Classify each of the following as a physical change or a chemical change. For each chemical change, explain how you can tell that a new substance has been formed.
   (a) Water boils and turns into steam.
   (b) Wood is sawed and made into a toy box.
   (c) Firewood burns and ashes remain.
   (d) Orange drink crystals are stirred into a pitcher of water.
   (e) Sugar, eggs, and flour are mixed and baked into cookies.
5. What evidence is there that a glowstick works as a result of a chemical change (Figure 6)?

6. When a candle is lit and allowed to burn for 15 minutes, some wax drips and collects at the base of the candle, and the candle becomes shorter.
   (a) Did you observe any physical changes? Explain.
   (b) Why did the candle become shorter? What happened to the missing section of the candle?
   (c) Did you observe any evidence of chemical change? Explain.
7. Think about each of the following situations and describe one chemical change that is occurring. Provide evidence of the chemical change.
   (a) A driver starts the car in the driveway.
   (b) A bathroom cleaning product removes a stain in the sink.
   (c) Bubbles form when baking soda is mixed with lemon juice.
   (d) Cookies baking in the oven give off a delicious aroma.
   (e) A match is struck and ignites.
   (f) Bleach turns a red towel white.
   (g) A banana tastes sweeter as it ripens.

Figure 6 What kind of change causes glowsticks to glow?

SCIENCE WORKS ✓ OSSLT

Keeping Baby Dry

"Necessity is the mother of invention." Where there is a demand, there is probably a product invented to meet that demand and fill that market. The disposable diaper is an invention that was created to meet a demand.

Disposable diapers replace the natural plant fibres of cotton diapers. The typical disposable diaper is made of synthetic materials, mostly plastics made from fossil fuels. Each component of the diaper is designed with properties specific for its particular function (Figure 1).

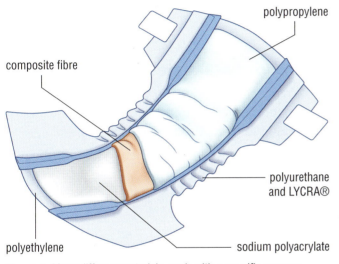

Figure 1 Many different materials, each with a specific purpose, make up a disposable diaper.

- The outer surface is a thin polyethylene film that is waterproof, perfect for preventing accidental leakages onto unsuspecting admirers.
- The inner surface, next to baby's skin, is soft and porous polypropylene, allowing baby's urine to soak into the middle filling of the diaper.
- The leg cuffs and waistband are made of polyurethane and LYCRA®. These materials can expand and contract, letting the baby stretch its little legs with comfort.
- Different types of glues hold the diaper together. The adhesives on the tabs are less sticky, allowing the diaper to be opened and re-opened as needed.
- The hardest working part of the diaper is the inner stuffing, which can absorb many times its own weight of water.

The inner stuffing contains the most important part of the disposable diaper. Nestled inside the stuffing are a few spoonfuls of a white crystalline powder, called sodium polyacrylate. These crystals can absorb up to 400 times their own mass in water. During this process, they grow larger and larger, all the while staying dry to the touch. See this for yourself if you have access to a new disposable diaper. Simply shred the inside filling and shake the crystals into a large bowl. Warning: Do not ingest the crystals or touch your face or eyes if you come into contact with them. The crystals can cause irritation and dehydration. Add tap water and watch the crystals grow. Is this change a chemical or a physical change?

Entrepreneurs have taken advantage of the unusual properties of this material. You can buy toys where a small animal grows to a life-sized pet in a bathtub (Figure 2), and soil mixtures supply an unattended plant with water for months! Sodium polyacrylate is also used to make artificial snow (Figure 3).

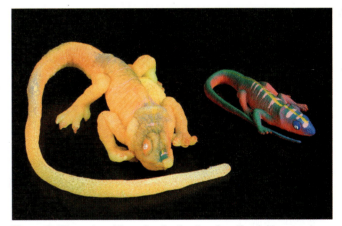

Figure 2 When placed in water, the toy lizard on the right grows to several times its original size (left).

Figure 3 Sodium polyacrylate is used to make artificial snow.

5.4 PERFORM AN ACTIVITY

Safety in Science

Do you know who is responsible for your safety in and out of a science laboratory? You can see that person whenever you stand in front of a mirror. Yes, you are the person who is responsible for your own safety. Everyone who shares the same classroom—your teacher and classmates—shares the responsibility as well as the risks. It is vitally important that you, and the people who are working with you, take safety information seriously and follow safety precautions precisely.

In this activity, you will familiarize yourself with the setup of your classroom and with the safety symbols and general procedures related to lab equipment and techniques.

SKILLS MENU
- Questioning
- Hypothesizing
- Predicting
- Planning
- Controlling Variables
- Performing
- Observing
- Analyzing
- Evaluating
- Communicating

Purpose
To review safety procedures and symbols in the science laboratory.

Equipment and Materials
- notebook
- pen or pencil

Procedure

SKILLS HANDBOOK
1.

Part A: Mapping Your Classroom

1. In your notebook, draw a rough floor plan of your classroom.

2. On your map, label the location of each of the following:
   - emergency exits
   - eye wash station
   - emergency shower
   - first aid kit
   - fire extinguishers
   - fire blanket
   - broken glass container
   - eye protection and lab aprons
   - MSDS binder
   - additional safety equipment

3. Below your map, make notes on the correct use of each piece of safety equipment. The goal is that you will never need to use any of these pieces of equipment because you will practise safety in the lab.

Part B: General Safety Rules

4. Prepare a table with two columns. List laboratory safety rules in the first column and provide the reason for the rule in the second column.

5. Complete the table with safety rules regarding each of the following:
   - eye protection and lab aprons
   - long hair and hats
   - loose clothing
   - personal items
   - food and drink
   - working alone

Part C: Safety Symbols

6. In your notebook, copy each of the WHMIS symbols in the Skills Handbook.

7. Beside each symbol, write the hazard indicated and explain how the symbol represents the hazard.

Part D: Safe Techniques

8. Work with a partner and demonstrate each of the following techniques to each other:
   - smelling a chemical by wafting
   - pouring a liquid
   - lighting a Bunsen burner
   - heating a test tube over a Bunsen burner
   - unplugging an electrical cord
   - removing and disposing of broken glass
   - cleaning up your lab station

Analyze and Evaluate

(a) Why is it important to use standardized safety symbols on all hazardous products? K/U

(b) What does WHMIS stand for? K/U

(c) Describe the WHMIS symbol for each of the following: K/U
   (i) flammable and combustible material
   (ii) poisonous and infectious material causing immediate and serious toxic effects
   (iii) corrosive material
   (iv) biohazardous infectious material

(d) What should you do in each of the following situations? Give reasons for your answers. T/I
   (i) You did not have time to finish your lunch. Should you eat an apple in the science lab before class starts?
   (ii) You accidentally broke a test tube and swept up the broken glass. Should you put the broken glass into the waste container?
   (iii) You are allergic to peanuts. Should you tell the teacher?
   (iv) You are boiling water but not using any hazardous chemicals. Should you wear eye protection and a lab apron?
   (v) Your beaker has a small crack in it. Should you use it anyway?
   (vi) The equipment you used is still hot at the end of class. Should you put it away?
   (vii) You have read the instructions but are still not sure how to do the experiment. What should you do?

(e) Coming to the lab prepared is very important for your safety and the safety of those around you. Explain why this is the case. K/U A

Apply and Extend

(f) List five different occupations in which safety equipment is worn. For each example, explain the safety hazard(s) for which the safety equipment is used. For example, construction workers wear hardhats to protect against head injury from falling or sharp objects. A

(g) Research WHMIS on the Internet or in the library. Write a short paragraph summarizing the purpose of WHMIS and why this system is important in schools and in the workplace. T/I C

GO TO NELSON SCIENCE

(h) WHMIS symbols are applied to dangerous materials used in workplaces. List five examples of workplace products that carry WHMIS labels. Check the containers of these products and record the type of symbol used. A

(i) The following accident report was filed by lab partners Rachelle and Mandeep after a number of students were injured as a result of their actions.

"We were dissolving salt crystals in a beaker of hot water and taking the temperature with a thermometer. We put on aprons but we took off our eye protection because it kept steaming up. Mandeep's sleeve caught the thermometer and knocked the beaker over, splashing the hot water everywhere. The tip of the thermometer broke off, but it was hardly noticeable, so Rachelle put it back in the drawer. We did not have a mop for the floor, but the puddle will dry by itself eventually. Fortunately, we cleaned up the lab bench before the teacher saw us, so we did not get into trouble. Unfortunately, Rachelle had missed breakfast and the cookie she was munching on got totally soaked on the lab bench."

Identify all the errors the students made and the possible consequences of their actions. T/I

5.5 PERFORM AN ACTIVITY

Forensic Chemistry

Somebody eating at a popular restaurant suddenly fell ill. A suspicious white powder was found on the victim. As chief crime scene investigator, your job is to identify the white powder that was collected on the victim. It may be the same substance as one of five different white powders that were stored at the location. This substance may be the cause of the illness, or it may simply be a harmless, edible cooking ingredient. You decide to compare the physical and chemical properties of the unidentified powder to the properties of the five known powders. Once the mystery powder is identified, you will be able to solve the crime—all in the next 60 minutes!

SKILLS MENU
- Questioning
- Hypothesizing
- Predicting
- Planning
- Controlling Variables
- Performing
- Observing
- Analyzing
- Evaluating
- Communicating

To learn more about becoming a forensic chemist,

GO TO NELSON SCIENCE

Purpose
To identify a powder based on its physical and chemical properties.

Equipment and Materials
- eye protection
- lab apron
- well plate
- small toothpick or paper clip for stirring
- 6 white powders: icing sugar, cornstarch, Aspirin powder, baking soda, baking powder, mystery powder
- test solutions in dropper bottles: water, universal indicator, vinegar, iodine solution

⚠ Do not taste any substance in this activity. Iodine solution and some powders are hazardous. Wash your hands after you finish.

Procedure
SKILLS HANDBOOK 3.B.4., 3.B.6.

1. Copy Table 1 into your notebook.

Table 1 Observations

Substance	Physical properties	Chemical properties: reaction with			
		water	universal indicator	vinegar	iodine solution
icing sugar					
cornstarch					
Aspirin powder					
baking soda					
baking powder					
mystery powder					

2. Put on your eye protection and lab apron.
3. Record the physical properties of each powder in your table.
4. Plan a procedure to test the physical and chemical properties of each powder given in Table 1.
5. Perform your procedure once it has been approved by your teacher.

Analyze and Evaluate
SKILLS HANDBOOK 3.B.8.

(a) Identify any evidence of chemical changes in each test. T/I
(b) Use the evidence provided by the tests to identify the mystery powder. T/I
(c) Which properties, physical or chemical, were most useful in identifying the mystery powder? Explain your answer. T/I
(d) How confident do you feel about your identification of the sample? What improvements would you suggest for similar investigations in the future? T/I

Apply and Extend

(e) What other physical properties could have helped to identify the mystery powder? Why were these properties not tested in this activity? T/I
(f) You find a puddle of a clear, colourless liquid on your driveway. It is either water, vinegar, or spilled battery acid from the car. How can you safely test the liquid to determine its identity? Explain how you can interpret your observations. T/I

190 Chapter 5 • Properties of Matter

AWESOME SCIENCE ✓ OSSLT

Antifreeze Is Not Just For Cars

There is no fixed scientific method or scientific way of thinking. We never know what we will find or what we are looking for. When we learn new things about the world, we come up with surprising and exciting ways of putting that new information to good use.

Biochemist Peter Davies from Queen's University wondered why fish in the cold salt waters off Newfoundland continue to swim around all winter without freezing. Unlike warm-blooded animals, fish are at the same temperature as their surroundings. Salt water can reach temperatures lower than 0 °C. Why do the fish not freeze in winter?

Many fish and insects have a defense mechanism against damage from ice crystal formation. When winter arrives, they start making their own antifreeze in their cells. This antifreeze works in the same way as winter windshield fluid for cars—it prevents ice from forming. The cells of these fish and insects produce a protein of a specific shape. This protein attaches to the surfaces of new ice crystals as soon as they begin to form (Figure 1). This protein coating prevents any additional water particles from attaching to the ice crystals, so the crystals cannot grow any bigger. Different types of fish and insects produce different antifreeze proteins, indicating that each species has its own way of coping with harsh surroundings.

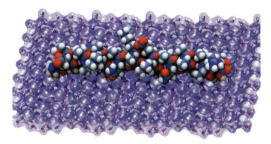

Figure 1 Antifreeze proteins attach to the surface of an ice crystal and stop the crystal from growing larger.

In what ways can this information be useful to us? One important application is the preservation of tissues for organ transplants. Organs, such as the heart, kidneys, and lungs, must be kept cold while in transit. However, there is a danger that water in the cells may freeze and form ice crystals (Figure 2). The sharp edges and pointy shape of ice crystals can break delicate cells and destroy these organs before recipients receive them. Placing the organs in a solution of antifreeze preserves them from damage due to freezing. This technology makes it possible to match donor and recipient tissues that are great geographical distances apart.

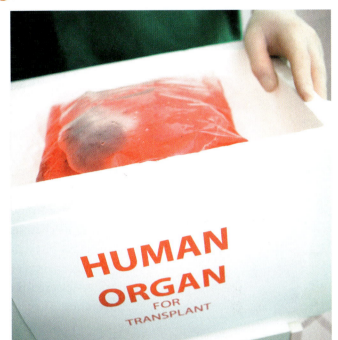

Figure 2 Organs for transplants must be delivered in minimal time to preserve the delicate tissues.

You may have encountered fish antifreeze proteins yourself. Cosmetics companies have added these proteins to their creams and lotions to make them smoother. Ice cream companies add these edible and tasteless proteins to their low-fat ice creams. Lower fat content means higher water content, which can form sharp jagged ice crystals when the ice cream thaws and freezes. The fish proteins prevent the formation of ice crystals and keep the ice cream smooth and creamy. A fish called eelpout is often used as the source of antifreeze, so your delicious cold treat may actually be "vaneela" ice cream (Figure 3)!

Figure 3 Like many cold water fish, the ocean eelpout produces its own antifreeze that stops water particles in its tissues from forming harmful ice crystals.

5.6 Characteristic Physical Properties

If you had to identify a pure liquid, what types of tests could you perform on it? One option would be to test its chemical properties. You could mix it with substances such as vinegar and baking soda and see how it reacts. You would need to perform several tests to narrow down the possible choices. Mixing unidentified substances together can be dangerous. The liquid you are testing may change to form new products with each test. So, the number of tests you can perform is limited by the amount of liquid you have.

You could also examine the physical properties of the pure liquid. However, some physical properties are not useful for identifying a sample. Knowing the volume and temperature of the mystery sample is not a great help because these values are not unique to substances. However, determining properties that are unique, or characteristic, of the pure substance would be much more informative. You could identify the mystery liquid with confidence if you knew its unique physical properties, along with some of its chemical properties (Figure 1).

Figure 1 These liquids look identical but they are not. How can you tell them apart?

Certain physical properties are unique to each pure substance, like fingerprints are unique to each person (Figure 2). These physical properties are unique due to the composition and structure of each substance. These properties are called **characteristic physical properties**, and they can be used with confidence to identify a pure substance. Unlike chemical tests, characteristic physical properties can be determined without changing the composition of the sample, so the test sample is unchanged.

In this section, you will learn about three characteristic physical properties of pure substances: density, freezing/melting point, and boiling point.

Figure 2 Unlike height, fingerprints do not change and are specific to the person. They are a characteristic physical property of a person.

characteristic physical property a physical property that is unique to a substance and that can be used to identify the substance

density a measure of how much mass is contained in a given unit volume of a substance; calculated by dividing the mass of a sample by its volume

Density

As you may recall from previous studies, the **density** of a substance is a ratio of its mass to its volume. The units of density are usually g/cm^3 (for a solid) and g/mL (for a liquid).

Density is the amount of matter per unit volume of that matter. A puffy marshmallow has the same mass as a squished marshmallow. However, a puffy marshmallow has a greater volume, so its mass–volume ratio is different, meaning its density is different. The particles in the puffy marshmallow are farther apart from each other. In other words, the puffy marshmallow is less dense than the squished marshmallow (Figure 3).

DID YOU KNOW?

Mass Spectrometers
Modern labs use advanced equipment, such as mass spectrometers, to identify substances. A very small sample is placed into the machine, which takes only several minutes to output the result. Two of the properties of a substance that a mass spectrometer analyzes are the mass of the particles present in the substance and how much of each particle is present in the substance.

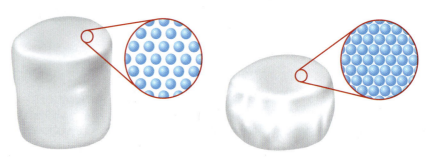

Figure 3 A greater mass of the same substance packed into a smaller volume has a greater density.

SAMPLE PROBLEM 1: Identifying a Metal

Calculate the density of a metal sample that is 18.00 cm long, 9.21 cm wide, and 4.45 cm high and that has a mass of 14.25 kg. What is the identity of the metal?

Given: $l = 18.00$ cm $h = 4.45$ cm
$w = 9.21$ cm $m = 14.25$ kg

Required: density of the metal (d)

Analysis: $\text{density} = \dfrac{\text{mass}}{\text{volume}}$

Solution: Volume of metal sample $= l \times w \times h$
$= 18.00 \text{ cm} \times 9.21 \text{ cm} \times 4.45 \text{ cm}$
$= 738 \text{ cm}^3$

mass of metal sample $= 14.25$ kg $= 14\,250$ g

$\text{density} = \dfrac{\text{mass}}{\text{volume}}$

$= \dfrac{14\,250 \text{ g}}{738 \text{ cm}^3}$

$= 19.3 \text{ g/cm}^3$

To identify the metal, look up the value you obtained for density in Table 1.

Statement: The density of the metal is 19.3 g/cm³. This metal is gold.

MATH TIP

Mass and Volume

Mass is a measure of the amount of matter in a substance. Volume is a measure of the amount of space a substance takes up.

The density of a substance is also related to the mass of the particles of which it is composed. A box filled with billiard balls is denser than an identical box filled with ping pong balls even though the volume of the boxes is the same. This is because the mass of the billiard balls is greater than the mass of the ping pong balls.

The densities of gases are closely related to their temperature and pressure. How closely particles are packed in gases depends on their temperature and pressure. For example, a gas at a higher pressure is denser than the gas at a lower pressure because the particles are packed closer together. Similarly, a gas at a higher temperature is less dense than the gas at the same pressure but at a lower temperature because the particles have more energy to move away from each other.

In liquids and solids, the particles are more closely bound, so changes in density caused by changes in temperature and pressure are minor compared with gases. For this reason, density is a characteristic physical property of pure liquids and solids. Still, temperature affects density, so values of density are often stated for a specific temperature.

Table 1 lists the densities of several common metals. For example, a piece of iron with a volume of 1 cm³, approximately the size of a small sugar cube, has a mass of 7.87 g. A piece of aluminum of the same size has a mass of only 2.70 g—less than one-third the mass of the iron. Imagine how much less an aluminum car would weigh and how much less fuel you would need to drive it. Gold is more than seven times denser than aluminum. It is sometimes called a heavy metal, but it is more correct to call it a very dense metal.

Table 1 Densities of Common Metals (at Room Temperature and Atmospheric Pressure)

Metal	Density (g/cm³)
aluminum	2.70
zinc	7.13
iron	7.87
copper	8.96
silver	10.49
lead	11.36
mercury	13.55
gold	19.32

Freezing Point, Melting Point, and Boiling Point

As Canadians, we know well that 0 °C is a significant temperature. It marks the difference between icy roads and rain puddles, and between skating rinks and fishing ponds (Figure 4). The **freezing point** is the temperature at which a substance turns from a liquid into a solid. It is the same temperature as the melting point of the same substance. **Melting point** is the temperature at which the substance turns from a solid into a liquid. For example, 0 °C is the freezing point of pure water and the melting point of pure ice. Different substances freeze and melt at different temperatures. The temperature at which a substance changes rapidly from a liquid to a gas is its **boiling point** (Figure 5). For example, the boiling point of water is 100 °C.

freezing point the temperature at which a substance changes state from a liquid to a solid; melting point and freezing point are the same temperature for a substance

melting point the temperature at which a substance changes state from a solid to a liquid

boiling point the temperature at which a substance changes state rapidly from a liquid to a gas

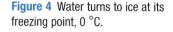

Figure 4 Water turns to ice at its freezing point, 0 °C.

Figure 5 The rapid formation of gas bubbles throughout a liquid occurs only at the boiling point.

The temperature at which a substance changes state depends on the particular composition and structure of the substance, and is thus unique to each substance. So, freezing/melting point and boiling point are characteristic physical properties of a substance. You can use these properties to distinguish between pure substances.

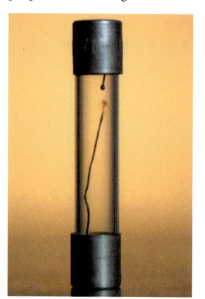

Figure 6 When an electrical circuit is dangerously overloaded, the metal wire in the fuse melts, breaking the circuit and preventing a fire. Sometimes low melting points are a good thing.

Applications of Melting Point

Knowing the melting point of metals is important when selecting a metal for a specific purpose (Figure 6). For example, incandescent light bulbs produce light when an electric current heats a metal wire filament. The hotter the filament is, the more light is given off. Of course, the light bulb will not work if the filament melts. Which metal would you select for the filament? Early inventors of incandescent light bulbs chose platinum for its high melting point. However, this metal is very expensive.

In the late 1870s, Thomas Edison experimented with carbon filaments that glowed well, but they were chemically reactive and produced a black deposit inside the bulb. In 1910, the element tungsten was chosen because of its very high melting point, its low chemical reactivity, and its wide availability (Figure 7). Table 2 lists the melting points of common metals.

Metals with low melting points also have important applications. Mercury is a metal that melts at −39 °C, which means that at normal room temperatures (25 °C), it has already melted. Mercury is the metal of choice in many thermostat designs (Figure 8). Several drops of liquid mercury flow to and from a contact point in the electrical circuit, turning the heater on or off in response to changes in room temperature. This design requires a metal that conducts electricity and is also a liquid at room temperature. Mercury is the only metal that meets both of these criteria. As you will learn in a later section, mercury is hazardous, and its use and disposal must be carefully controlled.

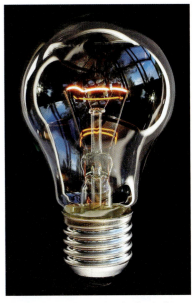

Figure 7 Metals need to be heated to extremely high temperatures before they give off a useful amount of light. Most metals melt before they reach these temperatures. Tungsten is an exception, which is why it is used as the filament in incandescent light bulbs.

Table 2 Melting Points of Common Metals

Metal	Melting point (°C)
mercury	−39
tin	232
lead	328
zinc	420
aluminum	660
silver	961
gold	1065
copper	1085
nickel	1453
iron	1535
platinum	1772
tungsten	3407

Figure 8 When the thermostat cools and tips the vial, mercury completes the electrical circuit and turns on the heat.

In the next section, you will take a closer look at how the characteristic melting point of a pure substance changes when other substances, or impurities, are added to it. This property can also be used to our advantage, especially during Canadian winters.

Salt and Ice

How does salt "melt" snow and ice? As you know, pure water freezes at 0 °C. Any impurity that is added to ice interferes with the freezing process, and water does not become solid ice until the temperature drops even lower. As a result, adding any dissolved impurity such as salt lowers the freezing point of water. Instead of freezing at 0 °C, a mixture of 20 % salt in water stays a liquid until the temperature is as low as −16 °C. This means that, if you add enough salt, the roads will be ice-free even when the temperature dips below 0 °C.

DID YOU KNOW?

Light Bulbs—A Canadian Invention
Canadian Henry Woodward invented an incandescent light bulb in 1874 and sold his U.S. patent to Thomas Edison, an American. This light bulb was very different from the one that was eventually sold commercially in the U.S.

5.6 Characteristic Physical Properties

TRY THIS: WE SCREAM FOR ICE CREAM

SKILLS: Performing, Observing, Analyzing, Evaluating, Communicating

Ice cream has been a favourite treat for hundreds of years. It is simply frozen cream to which fruit or flavourings have been added. Before electrical refrigeration was invented, a mixture of ice and salt was used to achieve temperatures low enough to freeze cream. Just as we sprinkle salt on the roads to melt snow and ice, salt is mixed with ice to lower the melting point of ice. In this activity, you will make a small batch of ice cream.

Equipment and Materials: small sealable plastic bag; large sealable plastic bag; thermometer; one-half cup cream; 2 tbsp sugar; one-quarter tsp vanilla flavouring; 2 cups ice; 2 tbsp salt

1. Place the cream, sugar, and vanilla into the small bag and seal it securely.
2. Place the ice inside the large bag and use the thermometer to measure the temperature of the ice.
3. Add one tablespoon of salt to the ice and mix. Measure the temperature of the ice–salt mixture.
4. Add a second tablespoon of salt to the ice–salt mixture and mix. Measure the temperature again.
5. Place the sealed small bag inside the large bag and seal the large bag securely.
6. Shake the bags gently for about 10 min.
7. When the cream mixture is frozen, open the bags. Be careful to keep the salt and ice out of the ice cream.

A. What happened to the temperature of the ice after salt was added to it? T/I

B. What happened to the state of the ice after salt was added to it? T/I

C. What was the purpose of adding salt to the ice in this recipe? T/I

D. Salt is added to the ice and snow on the roads to keep the roads safe for driving. From what you have learned in this activity, what effect does salt have on snow? Why does spreading salt on roads in the winter improve road safety? A

E. Repeat the experiment, only this time omit the salt. What were the differences in your results? Give reasons for your answer. T/I

WRITING TIP

Finding the Main Idea

Read the first and last sentences of a paragraph to find its main idea. If you cannot find it, read them again and then the middle part of the paragraph. The nucleus of a paragraph is the main idea. You can assume that all paragraphs in a science textbook have a main idea. Your job is to detect it and translate it into your own words.

The Unusual Behaviour of Water

What can be odd about water? It is a most ordinary substance—clear, colourless, odourless, and tasteless. It freezes at 0 °C and boils at 100 °C. The Celsius temperature scale was designed to fit the boiling point and freezing point of water. Water is odd because its solid form floats on its liquid form; that is, ice floats on water (Figure 9)! Very few substances do this. Normally, the particles in a solid state are packed more closely together than they are in a liquid state. Therefore, a solid occupies less space and is denser than its corresponding liquid. A solid usually sinks in a liquid of the same substance.

Water particles are different. Due to their shape and the way they are arranged, water particles occupy more space when they are packed together into solid ice (Figure 10). As an analogy, picture a box filled with L-shaped building blocks. When these blocks are not connected, they can be poured as a fluid. The pieces overlap and each one fits into an empty space.

Figure 9 Water is an unusual substance because its solid form is less dense than its liquid form. This is why ice floats on water.

However, if each L-shaped block were connected end-to-end, in a solid shape, they would not be able to overlap, and would occupy a larger space.

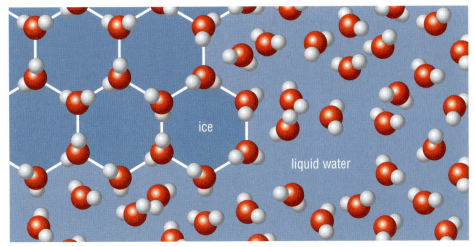

Figure 10 Ice is less dense than water because the particles in ice take up a larger volume than the same number of particles in liquid water. The mass of the particles remains the same.

We should be very grateful for this oddity of water. If water were like other liquids, its solid state would be denser that its liquid state. Water would become denser as it cooled, and sink. A lake would freeze from the bottom up! Luckily for the fish in the lake, this does not happen. Water becomes denser as it cools, but its density is highest at a temperature of 4 °C (Figure 11). Lake water at this temperature sinks but remains a liquid. As the temperature drops below 4 °C, surface water becomes less dense and stays at the surface. If the temperature drops to 0 °C, surface water freezes to form ice.

There are also negative consequences of this property of water. The force of expansion of water as it freezes can cause serious problems (Figure 12). Glass bottles of juice in a freezer can crack, and water pipes at a cottage can burst if the pipes are not emptied before winter.

DID YOU KNOW?

Carbon Freezing
A bottle of carbonated water can be kept at temperatures much lower than 0 °C without freezing. The carbonation comes from forcing carbon dioxide gas to be dissolved in the water under pressure. What happens when the sub-zero-temperature carbonated drink is opened? Try it for yourself and find out. Cool an unopened bottle of carbonated water in a mixture of salt and ice, until it reaches about −8 °C. Then take the bottle out of the ice and open the bottle.

Figure 11 The density of water changes depending on its temperature.

Figure 12 Water that freezes but has no room to expand can cause damage.

5.6 Characteristic Physical Properties

UNIT TASK Bookmark

You can apply what you learned in this section about characteristic physical properties to the Unit Task described on page 286.

IN SUMMARY

- A characteristic physical property is a physical property that is unique to a pure substance and that can be used to identify the substance.
- Density is a measure of how much mass is contained in a given unit volume of a substance. It is calculated by dividing the mass of a sample by its volume.
- Melting point is the temperature at which a substance changes state from a solid to a liquid.
- Freezing point is the temperature at which a substance changes state from a liquid to a solid. Melting point and freezing point are the same temperature for a substance.
- Boiling point is the temperature at which a substance changes state rapidly from a liquid to a gas.
- Unlike other substances, water that is close to its melting point is less dense in the solid state than in the liquid state. Therefore, ice floats on water.

CHECK YOUR LEARNING

1. Explain why the boiling point of water is a characteristic physical property, but the temperature and the volume of a glass of water are not. K/U
2. Which properties of mercury make it a good material to use in a thermostat? K/U
3. If you could place a piece of solid silver into a container of liquid silver, would it float or sink? Explain your answer. K/U
4. A sample of pure iron has a mass of 5.00 g. Calculate its volume. T/I
5. A metal with a mass of 71.68 g occupies a volume of 8.00 cm^3. Calculate the density of the metal. Using Table 1 on page 193, determine the identity of the metal. T/I
6. A sample of pure copper has a volume of 3.75 cm^3. Calculate its mass. T/I
7. A metal with a mass of 1.00 kg occupies a volume of 370 cm^3. Calculate the density of the metal. Using Table 1 on page 193, determine the identity of the metal. T/I
8. A sample of pure zinc has a mass of 4.50 g. Calculate its volume. T/I
9. A metal with a mass of 15.00 g occupies a volume of 1.32 cm^3. Calculate the density of the metal. Using Table 1 on page 193, determine the identity of the metal. T/I
10. Calculate the mass of a gold bar that is 18.00 cm long, 9.21 cm wide, and 4.45 cm high. T/I
11. (a) Explain why water is said to exhibit unusual behaviour.
    (b) Describe some consequences of the behaviour of water. K/U
12. A drinking glass at a crime scene contains a clear, colourless liquid that may be water or alcohol. As the investigator, you know that the densities of alcohol, water, and ice are 0.79 g/mL, 1.0 g/mL, and 0.92 g/mL, respectively. Design a simple method to determine the identity of the mystery liquid. Explain your design. T/I A

EXPLORE AN ISSUE CRITICALLY 5.7

Are We Salting or Assaulting Our Roads?

It is estimated that about 50 % of global salt production is used for de-icing roads in cold climates (Figure 1). Road salt is not friendly to objects that it touches. It speeds up the rusting of doors and frames when it gets splashed onto cars. Salting also increases the number of freezing and thawing cycles of ice in the cracks of roads. The cracks widen from the repeated expansion of water as it freezes. Salt that is washed off the road contaminates nearby soil and damages roadside vegetation. It eventually enters the groundwater, or surface streams, rivers, and lakes, where it threatens aquatic life.

Figure 1 Salt makes winter driving safer, but at a cost.

The Issue

Members of the town council want town crews and homeowners to stop using road salt in winter. At issue is the harmful effect of road salt on vegetation and the environment. Local residents have been invited to express their support or concerns at the next council meeting.

Work in a small group, as assigned by your teacher, to explore the issue. Each member of the group will assume one of the following roles:

- town homeowner
- town parks commissioner
- town roads commissioner
- environmentalist
- lawn care company owner
- school bus driver
- senior citizens' group president
- student

> **SKILLS MENU**
> - Defining the Issue
> - Researching
> - Identifying Alternatives
> - Analyzing the Issue
> - Defending a Decision
> - Communicating
> - Evaluating

The audience will be the mayor and other members of town council who will be voting on the motion.

Goal

To present your viewpoint for or against the use of road salt in winter.

Gather Information

Research the issue in order to make an educated decision about your position in the role that you have chosen. Gather information about

- your town's policies on this issue
- the history of salting roads
- statistics on road accidents or pedestrian injuries related to road conditions
- the financial cost of salting roads
- damage to vegetation caused by road salt
- the effect of spring thaw on salted snow
- the impact on aquatic ecosystems

Identify Solutions

Research and propose alternatives to road salt such as

- using other chemicals or substances to melt ice
- using alternative methods to provide better traction on roads
- increasing snow removal

Make a Decision

Select one of the possible solutions that best serves your role in the community. Prepare reasons for your selection and offer arguments to support your position on the issue. Anticipate the views and arguments of other people.

Communicate

Present your arguments to the mayor and town councillors. Develop your main idea with supporting details, including facts and examples as evidence. Come up with a consensus on what should be done. Write a report on your final decision.

Chapter 5

LOOKING BACK

KEY CONCEPTS SUMMARY

Physical properties are characteristics that can be determined without changing the composition of the substance.

- Physical properties may be qualitative or quantitative. (5.2)
- Examples of qualitative physical properties are lustre, optical clarity, brittleness, viscosity, hardness, malleability, ductility, and electrical conductivity. (5.2)
- Examples of quantitative physical properties are mass, height, and temperature. (5.2)

Chemical properties describe the ability of a substance to change its composition to form new substances.

- A chemical property is a property of a substance that describes its ability to undergo changes to its composition to produce one or more new substances. (5.3)
- Many products are useful to us because of their chemical properties. (5.3)

Pure substances have characteristic physical properties.

- A characteristic physical property is a physical property that is unique to a substance and that can be used to identify the substance. (5.6)
- Characteristic physical properties are density, melting/freezing point, and boiling point. (5.6)

Water has unusual characteristic physical properties.

- Solids are usually denser than liquids, but ice is less dense than water when its temperature is close to its melting point. (5.6)
- This property allows ice to float on water, which allows aquatic life to survive. (5.6)

Physical and chemical properties can be used to identify different substances.

- Substances can be identified by observing the physical and chemical changes that they undergo when they are mixed with other substances. (5.2, 5.3, 5.5)
- Examples of physical change are a change of size or shape, a change of state, and dissolving. (5.2, 5.5)
- Evidence of chemical change includes a colour change, an odour change, gas produced, a precipitate produced, or a temperature change. (5.3, 5.5)

Some common useful substances have negative impacts on the environment.

- Salt lowers the freezing point of water, but it causes corrosion and threatens ecosystems. (5.6, 5.7)

WHAT DO YOU THINK NOW?

You thought about the following statements at the beginning of the chapter. You may have encountered these ideas in school, at home, or in the world around you. Consider them again and decide whether you agree or disagree with each one.

1 Some chemicals can be both useful and harmful to us.
Agree/disagree?

4 Hair dyes permanently change chemicals inside each strand of hair.
Agree/disagree?

2 The particles in a solid are closer together than they are in a liquid.
Agree/disagree?

5 In a chemical change, the original substance disappears.
Agree/disagree?

3 A pure substance, such as gold, melts at different temperatures depending on its size.
Agree/disagree?

6 An edible substance, such as table salt, does not pose any harm to the environment.
Agree/disagree?

How have your answers changed since then? What new understanding do you have?

Vocabulary

particle theory (p. 175)
pure substance (p. 175)
mixture (p. 176)
mechanical mixture (p. 176)
solution (p. 176)
alloy (p. 176)
physical property (p. 179)
qualitative property (p. 179)
quantitative property (p. 179)
viscosity (p. 180)
physical change (p. 181)
chemical property (p. 183)
chemical change (p. 184)
precipitate (p. 184)
characteristic physical property (p. 192)
density (p. 192)
freezing point (p. 194)
melting point (p. 194)
boiling point (p. 194)

BIG Ideas

- ✓ Elements and compounds have specific physical and chemical properties that determine their practical uses.
- ✓ The use of elements and compounds has both positive and negative effects on society and the environment.

Looking Back

CHAPTER 5 REVIEW

The following icons indicate the Achievement Chart category addressed by each question.
K/U Knowledge/Understanding   **T/I** Thinking/Investigation
C Communication   **A** Application

What Do You Remember?

1. Describe several clues that tell you that you are observing a chemical change. Give an example to illustrate each clue you described. (5.3) **K/U**

2. Describe several physical changes that can be made to a silver spoon. Explain why each is a physical change. (5.2) **K/U**

3. List three quantitative characteristic properties of water. Explain why they are considered characteristic. (5.6) **K/U**

4. Explain how the unique properties of water allow it to support life in frozen lakes and ponds during our Canadian winters. (5.6) **K/U** **A**

5. What does WHMIS stand for? (5.4) **K/U**

6. Sketch and describe the WHMIS symbol for each of the following hazards: (5.4) **K/U** **C**
   (a) explosive
   (b) poisonous and infectious, causing immediate and serious toxic effects
   (c) corrosive
   (d) flammable

What Do You Understand?

7. Classify each of the following properties of a cake as qualitative or quantitative. Give reasons for your answers. (5.2) **K/U**
   (a) It is circular in shape.
   (b) Its mass is 1.5 kg.
   (c) It tastes like chocolate.
   (d) It is 30 cm in diameter.
   (e) Its icing is melting.

8. Adding salt to ice melts the ice at a lower temperature than the normal melting point of ice. Explain what changes occur in the process that causes ice to change to water, with and without the presence of salt. (5.6) **K/U**

9. Density is considered a characteristic property of a substance. Explain why its value is stated at a specified temperature. (5.6) **K/U**

10. Classify each of the following changes as physical or chemical. Give reasons for your answers. (5.2, 5.3) **K/U**
    (a) When molasses is warmed, it becomes less viscous.
    (b) When a chair is painted, it has a new colour.
    (c) When sugar is stirred into hot water, it dissolves.
    (d) When egg whites are cooked, they become opaque.
    (e) When wood is sawed, some of it changes to sawdust.
    (f) When wood is burned, ashes remain.
    (g) When vinegar is added to baking soda, bubbles are seen.

Solve a Problem

11. When a sample of blue crystals is heated, a vapour is given off and a white powder remains. Analyze whether a physical or a chemical change occurred. Give reasons for your answer. (5.2, 5.3) **K/U**

12. When an opaque beige solid is heated, a clear colourless liquid is formed. When this liquid is cooled, it returns to being an opaque beige solid. Analyze whether a physical or a chemical change occurred. Give reasons for your answer. (5.2, 5.3) **K/U**

13. Iron pyrite is a lustrous yellow mineral that consists of iron and sulfur (Figure 1). Its common name is fool's gold. Design a procedure to measure one of its characteristic physical properties so that you can distinguish it from real gold. (5.6) **T/I**

Figure 1 Iron pyrite (fool's gold) looks similar to gold.

14. You have learned that adding salt lowers the freezing point of ice. You hypothesize that adding salt will also have an effect on the boiling point of water. To test your hypothesis, you plan to take the temperature of a beaker of salt water every minute as it is heated, until it boils. (5.5, 5.6) T/I

    (a) Identify your independent variable and your dependent variable.
    (b) Explain all the variables that you will control and how you will control them.
    (c) How will you label the axes of the graph you will plot to analyze your results?
    (d) If your results do not support your hypothesis, does that mean your investigation was not successful? Explain.

15. Phone books are updated and changed every year or two. Old phone books pose a massive disposal problem. (5.2, 5.3)

    (a) Research your local recycling program to find out the options available for disposal of your old phone books. T/I
    (b) Brainstorm with a partner to come up with other environmentally acceptable methods of reducing, reusing, or recycling phone books. K/U T/I A

16. Calculate the mass of a liquid with a density of 2.3 g/mL and a volume of 30 mL. (5.6) T/I

17. An irregular object with a mass of 12 kg displaces 1.75 L of water when placed in an overflow container. Calculate the density of the object. (5.6) T/I

18. A piece of wood that measures 3.2 cm by 5.7 cm by 7.3 cm has a mass of 100 g. What is the density of the wood? Would it float on water? (5.6) T/I

19. A plastic ball has a mass of 150 g. If the density of the ball is 0.80 g/cm³, what is its volume? (5.6) T/I

Create and Evaluate

20. Gourmet cooking is the art of blending physical and chemical changes into tasty and nutritious concoctions. As a scientist and chef, your task is to create a scientific recipe that explains the type of change involved in the cooking instructions. Select one of your own recipes that requires baking, or use the following recipe. (5.2, 5.3) T/I A

Quiche Me Quick

- slice 4 cups of mushrooms and dice 2 onions and 2 green peppers
- cook the vegetables in a frying pan with a little oil until just tender
- beat 10 eggs until light and bubbly
- grate 2 cups of cheddar cheese
- cook 8 strips of bacon until crispy, then crumble into small pieces
- mix all ingredients with 1 cup of milk and $\frac{1}{2}$ teaspoon of salt
- bake in a large pie dish at 175 °C until firm to the touch (about 50 min)

(a) Design your scientific recipe so that it is informative, attractive, easy to understand, and clearly explains the type of change (physical, chemical, or both) for each step. T/I
(b) Evaluate and compare the importance of following instructions closely in physical changes and in chemical changes in the recipe. T/I
(c) Create a recipe of your own that involves only physical changes and no chemical changes. Give your recipe a descriptive name. T/I C

Reflect on Your Learning

21. How does the phrase "there are two sides to every coin" apply to the use of common chemicals and their impact on society and the environment? Support your answer with specific examples. A

Web Connections

22. (a) Why is the use of road salt in winter impractical in many northern communities? (5.6, 5.7) A
    (b) What other substance could be used to improve road safety? (5.7) A

23. Research and report on the use of lead and lead poisoning. (5.1) T/I A
    (a) What were common uses of lead in the past?
    (b) What properties of lead made it suitable for each use?
    (c) What are the symptoms of lead poisoning?

CHAPTER 5

SELF-QUIZ

The following icons indicate the Achievement Chart category addressed by each question.

K/U Knowledge/Understanding **T/I** Thinking/Investigation
C Communication **A** Application

For each question, select the best answer from the four alternatives.

1. Which of the following is a solution? (5.1) **K/U**
   (a) sand
   (b) salt water
   (c) orange juice
   (d) granola

2. Which of the following is a chemical property? (5.3) **K/U**
   (a) colour
   (b) density
   (c) boiling point
   (d) ability to burn

3. Which phrase correctly defines density? (5.6) **K/U**
   (a) the ratio of a substance's mass to its volume
   (b) the ratio of a substance's volume to its weight
   (c) the ratio of a substance's length to its volume
   (d) the ratio of a substance's mass to its weight

4. Which of the following is a pure substance? (5.1) **K/U**
   (a) wood
   (b) apple
   (c) gold
   (d) paper

Indicate whether each of the following statements is TRUE or FALSE. If you think the statement is false, rewrite it to make it true.

5. The mass of a substance is an example of a qualitative property. (5.2) **K/U**

6. Adding salt to water causes the water to freeze at a lower temperature. (5.6) **K/U**

Copy each of the following statements into your notebook. Fill in the blanks with a word or phrase that correctly completes the sentence.

7. The particles of a substance move _____ as the temperature of the substance increases. (5.1) **K/U**

8. When a teaspoon of salt is dissolved in a pot of water, the salt undergoes a _____ change. (5.2) **K/U**

Match each term on the left with the most appropriate description on the right.

9. (a) lustre   (i) the ability to flow or pour
   (b) viscosity   (ii) the ability to be pulled into fine strands
   (c) ductility   (iii) the level of breakability or flexibility
   (d) brittleness   (iv) the level of shininess or dullness
   (e) malleability   (v) the ability to be hammered into thin sheets (5.2) **K/U**

Write a short answer to each of these questions.

10. How can you tell that a hamburger is undergoing a chemical change when you cook it? (5.3) **K/U**

11. Name at least two physical properties that are required of a metal used to make kitchen pots and pans. (5.2) **K/U**

12. Which occupies more space: 5 g of solid gold or 5 g of liquid gold? Explain your answer. (5.6) **K/U**

13. Explain why temperature is a physical property but not a characteristic physical property. (5.2, 5.6) **K/U**

14. Give an example of a solid mixture, a liquid mixture, and a gas mixture. (5.1) **K/U**

15. The mass of a sample of aluminium is 12.15 g and its volume is 4.5 cm³. What is the density of aluminium? (5.6) **T/I**

204 Chapter 5 • Properties of Matter

16. The density of sodium is 0.97 g/cm³. If a sample of sodium has a volume of 2.6 cm³, what is the mass of the sample? (5.6) T/I

17. You can separate some substances based on their chemical or physical properties. You are given a mixture of sawdust, small pieces of iron, and small pieces of rock. Devise a procedure to separate the mixture into its three components based on their physical properties. (5.2, 5.3, 5.6) T/I

18. You and a friend watch workmen putting tree limbs into a wood chipper. Large pieces go into the chipper and small pieces come out. Your friend says that because the pieces coming out of the chipper do not look anything like the pieces going in, the tree limbs have undergone a chemical change. Is your friend correct? Explain your answer. (5.2, 5.3) T/I

19. Sea water freezes at lower temperatures than fresh water. Explain why. (5.6) A

20. Imagine that you work for a science magazine. This month, your job is to write a short article that tells readers why water is such an amazing substance. In your article, be sure to mention several properties of water. (5.6) C

21. Name at least three characteristic physical properties that could be used to identify you. Use both qualitative and quantitative properties. (5.6) K/U A

22. It is your turn to make lunch. You decide to heat some canned soup and toast some bread. Identify the chemical change and the physical change that take place as you prepare lunch. Explain your answer. (5.2, 5.3) A

23. Explain the following statement in your own words: "The freezing point of a substance is the same temperature as the melting point of the same substance." (5.6) C

24. The density of liquid mercury is 13.53 g/cm³. The density of solid copper is 8.96 g/cm³. Would you expect a piece of copper to sink or float when placed in a container of liquid mercury? Explain your answer. (5.6) K/U T/I

25. Classify the following properties of hydrogen as physical properties or chemical properties. (5.2, 5.3) K/U
    (a) It is a colourless gas.
    (b) It reacts with oxygen to form water.
    (c) It combines with corn oil to form margarine.
    (d) It floats in air.
    (e) It has a strong odour.

26. Carbon fibre is now replacing aluminum as the material used to make racing bicycle frames. (5.2, 5.3, 5.6) K/U A
    (a) The density of carbon fibre can be less than half the density of aluminum. Why is this property an advantage for the racing bicycles?
    (b) What other physical properties and chemical properties should carbon fibre have if it is to be used to make bicycles?

27. A dented table tennis ball can sometimes be "repaired" by placing it in a cup of hot water. Use the particle theory of matter to explain how this is possible. (5.1) K/U A

28. Backyard barbecues are fuelled by gas. (5.2, 5.3) K/U A
    (a) Identify one physical property that barbecue fuel should have. Explain why this property is important for barbecue fuel.
    (b) Identify one chemical property barbecue fuel should have. Explain why this property is important for barbecue fuel.

CHAPTER 6
Elements and the Periodic Table

KEY QUESTION: Are the elements in the periodic table organized according to their physical and chemical properties, or according to their atomic structure?

The periodic table presents all the known elements. In this chapter, you will learn how and why the elements are organized in this way.

UNIT C
Atoms, Elements, and Compounds

CHAPTER 5 Properties of Matter

CHAPTER 6 Elements and the Periodic Table

CHAPTER 7 Chemical Compounds

KEY CONCEPTS

Elements cannot be broken down into simpler substances.

Metals and non-metals have characteristic physical properties.

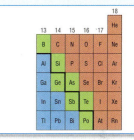

Elements are organized according to their atomic number and electron arrangement on the periodic table.

Atomic models evolved as a result of experimental evidence.

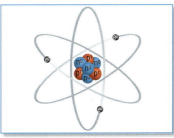

Atoms contain protons and neutrons in a central core surrounded by electrons.

Elements can be both beneficial and harmful to humans and to the environment.

Looking Ahead

ENGAGE IN SCIENCE

MERCURY AND THE MAD HATTER

You may recognize the character named the Mad Hatter from *Alice's Adventures in Wonderland*. What are hatters and what made them mad? A hatter is a person who makes and sells hats. Felt hats were popular in North America and Europe from the 17th to the 19th centuries. Cheap fur underwent a complicated process to be turned into felt. Part of the process involved using a mercury compound.

People of the day noticed that hatters often behaved oddly. "Mad" is a British term that means "behaves oddly, in a crazy, incomprehensible way." They had loss of coordination, slurred speech, memory loss, anxiety, and other personality changes. Today we know that hatters had symptoms of neurological damage caused by mercury poisoning.

Pure mercury is a silvery metal that is a liquid at room temperature. Mercury was commonly used in thermometers, but has been replaced by coloured alcohol. Dental fillings, button cell batteries (used in watches and hearing aids), and fluorescent light bulbs still require mercury or a compound of mercury. About half of our mercury pollution comes from burning coal, which spews the toxic metal into our atmosphere and our waters.

Mercury contamination of our fish and wildlife is a growing environmental and health concern. Government agencies regularly advise consumers on safe quantities of fish to eat.

In 1970, members of the Grassy Narrows First Nation north of Kenora, Ontario, suffered severe mercury poisoning. A chemical plant caused the poisoning when it discharged its contaminated wastewater into the local river system. To protect First Nations communities in the area, the Ontario government ordered them to stop eating fish, their main source of dietary protein. The government closed the commercial fisheries that were run by the First Nations people in the area. This closure escalated the already high unemployment rate and caused economic devastation for the community. In 1985, the Grassy Narrows First Nation received a settlement from the Canadian government, but the mercury has still not been removed from the contaminated river.

What can we do about mercury contamination?

WHAT DO YOU THINK?

Many of the ideas you will explore in this chapter are ideas that you have already encountered. You may have encountered these ideas in school, at home, or in the world around you. Not all of the following statements are true. Consider each statement and decide whether you agree or disagree with it.

1 Water is a common element.
Agree/disagree?

4 It is easy to change one element into another.
Agree/disagree?

2 Gold, silver, and copper are all metals because they are shiny and flexible.
Agree/disagree?

5 "Harmful" is a relative term: a harmful amount of a substance for one person may not be harmful for another person.
Agree/disagree?

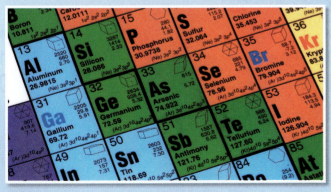

3 Elements are arranged alphabetically on the periodic table.
Agree/disagree?

6 Scientific theories are educated guesses that have been proven to be correct.
Agree/disagree?

FOCUS ON READING

Making Inferences

When you make inferences, you "read between the lines" to figure out what the author is suggesting. Use the following strategies to make inferences:

- Look for context clues such as key words, examples, or explanations.
- Think about what you already know about the topic.
- Combine the clues and your prior knowledge or experience to form an idea of your own.
- Revise your inference if you find new information or clues that alter or disprove it.

READING TIP

As you work through the chapter, look for tips like this. They will help you develop literacy strategies.

Figure 1 Each Canadian diamond is etched with a tiny polar bear as a trademark.

Diamond mining has societal consequences, which are more difficult to measure. Big profits bring business and highly paid jobs in the diamond industry as well as in construction and transportation. The demand for workers in Yellowknife and Iqaluit has left many jobs unfilled in neighbouring towns and cities. The cost of living has skyrocketed, caused in part by the cost of housing. Limited availability of housing has driven rents to unaffordable levels.

Unlike diamonds mined in parts of Africa, Canadian diamonds are much less controversial. They are not "blood diamonds," the name attached to mining operations that finance war and terror. Canadian diamonds are sold with a certificate that guarantees their source. They are promoted as "conflict free" certified. Each stone is etched with a tiny polar bear trademark (Figure 1).

Making Inferences *in Action*

Making inferences can help you respond to a text by making judgments, forming opinions, or drawing conclusions. Here is how one student made inferences as he read about Canadian diamonds.

Clues from Text	+ Prior Knowledge	= Inference
• big profits • societal consequences • cost of living	I read a magazine article about environmental problems in the Alberta tar sands.	It is not right for companies to make profits when they disrupt people's lives.
• not "blood diamonds"	I saw a documentary about how diamonds were used to buy guns for war.	It is acceptable to buy Canadian diamonds.
• certificate • tiny polar bear trademark	I read a news report warning about online sales of fake watches and jewellery.	Certificates could be forged or counterfeit trademarks could be etched on diamonds mined in other countries.

6.1 A Table of the Elements

Perhaps you have already seen the periodic table of elements in the science lab (Figure 1). The periodic table is a work of science—a compilation of years of inquiry and experimentation. It is also a work of beauty—a masterful joining of evidence and theory. As you learn more about the structure of matter, you will discover and appreciate the many layers of information that are stored in the periodic table.

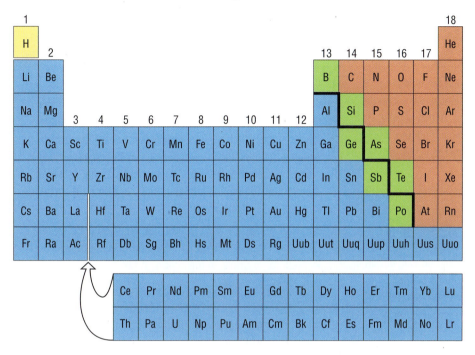

Figure 1 The periodic table of the elements

How does a substance qualify to occupy a spot on the periodic table? Note that the periodic table is a table of elements. An **element** is a pure substance that cannot be broken down into a simpler chemical substance by any physical or chemical means. Consider silver. You cannot carry out any chemical reactions or physical changes that will convert silver into anything chemically simpler. Pure silver is the simplest form of the element silver.

The element symbol Ag is used to represent silver on the periodic table. An **element symbol** is an abbreviation for a chemical element. Ag stands for argentum, which is the Latin word for silver. Some elements on the periodic table have abbreviations that are based on their Latin names.

Now consider water. How can you tell if it is an element? Simple: if it is an element, it is on the periodic table. Conversely, anything that is not on the periodic table is not an element. A quick check of the periodic table reveals that water is not listed, so it is not an element. It must be more complex, made up of two or more different elements. Indeed, if you were to run an electric current through water, you would find that it produces two gases that can be identified as hydrogen and oxygen (Figure 2). Hydrogen and oxygen are both elements on the periodic table. Any pure substance that is composed of two or more different elements that are chemically joined is called a **compound**. Water is a compound of the elements hydrogen and oxygen.

You can already tell from this simple example how useful the periodic table is. It serves as a quick and easy reference to distinguish elements from more complex substances. As Sherlock Holmes would say, "It is elementary, my dear Watson."

element a pure substance that cannot be broken down into a simpler chemical substance by any physical or chemical means

element symbol an abbreviation for a chemical element

compound a pure substance composed of two or more different elements that are chemically joined

Figure 2 When an electric current passes through water, the particles of water are split apart, producing two gases—hydrogen and oxygen.

Elements are the building blocks of all substances. If you think of elements as the letters in an alphabet, then compounds are the words that the letters spell. Think of all the words that can be created from just 26 letters. If there are over one hundred letters in an alphabet, there must be countless words that can be formed. However, just as only certain words exist in a language, only certain combinations of elements are possible. The periodic table can show you the underlying patterns of these combinations. In this chapter, you will evaluate how our theories of the atom can explain these patterns.

TRY THIS: ELEMENT SCAVENGER HUNT

SKILLS: Performing, Observing, Analyzing, Communicating

Elements are the building blocks of everything in the world. However, not many of the everyday pure substances we encounter are elements. Most substances are compounds. In this activity, you will see how many different elements you can find and determine whether any interesting patterns emerge when you arrange them on a periodic table.

Equipment and Materials: Gather as many elements as possible from home: from your kitchen, your tool box, your wallet, or your jewellery box. Some elements that are not available at home may be available at school and provided by your teacher. The following list is a suggested collection. From school, obtain magnesium, sulfur, and silicon. From home, obtain copper, iron, nickel, aluminum, tin, silver, gold, zinc, carbon, and air (containing the elements oxygen, nitrogen, argon) (Figure 3).

Figure 3 Elements are everywhere!

1. If a large wall-size periodic table is available, place it flat on a table or floor. As a class, place small samples of each element in the corresponding location on the periodic table.
2. If a large periodic table is not available, cut out identical squares of paper. Place a small sample of each element on a paper square labelled with the element's name. As a class, arrange the squares on a table or floor to their corresponding location on the periodic table.
3. For each of the following, answer the question and describe the area in the periodic table in which the elements are located; for example, in the centre, in the first column, and so on.

A. Which elements require careful storage and handling? K/U
B. Which elements have been used throughout history to make coins and jewellery? K/U
C. Which elements are metallic in appearance? K/U
D. Which elements are not metallic in appearance? K/U
E. Are there any elements that are difficult to classify as metallic or not metallic in appearance? Which ones? K/U
F. Which elements are gases? Is there a pattern in the way the gaseous elements are arranged on the periodic table? Give reasons for your answer. K/U C

Metals and Non-Metals

When you look at a collection of elements, some elements appear metallic and others do not. For example, copper, silver, and gold have the shiny lustre that we identify as metallic, and so do iron, aluminum, magnesium, nickel, and tin. The elements lithium, sodium, and potassium often have a white coating on the surface. However, when these elements are freshly cut, they reveal their beautiful metallic sheen. In general, **metals** are elements that are located on the left and central parts of the periodic table. They are solids that display a metallic lustre. The metals in Figure 4, from left to right, are beryllium, magnesium, calcium, strontium, and barium.

metal an element that is lustrous, malleable, and ductile, and conducts heat and electricity

Figure 4 Metals are solids that display a metallic lustre.

Non-metals are elements that are not metallic. Non-metals are found in the upper right portion of the periodic table. They are mostly gases and dull powdery solids. The only liquid non-metal is bromine, element 35 (Figure 5). A bold line that resembles a downward staircase, starting at boron, separates the metals and non-metals on the periodic table (Figure 6). Elements located along the staircase line are called **metalloids** because they have properties of both metals and non-metals.

non-metal an element, usually a gas or a dull powdery solid, that does not conduct heat or electricity

metalloid an element that has properties of both metals and non-metals

Figure 5 Non-metals are mostly gases and dull, powdery solids. The non-metals shown here are phosphorus (white and red), oxygen, carbon, sulfur, iodine, and bromine.

Metals exhibit other physical properties besides being shiny. Pots and pans are made of metals because metals are easy to shape and excellent conductors of thermal energy. Copper is used in electrical circuits because it conducts electricity. Copper is also very flexible and ductile, which allows it to be pulled into wires. The lustre and malleability of gold and silver, and their resistance to corrosion, make them ideal for decorative and valuable objects, such as jewellery and coins. Gold is so malleable that it can be pounded into sheets as thin as tissue paper. These sheets of gold foil are used in paintings, sculptures, and decadent desserts (Figure 7).

Non-metals are clearly distinguishable from metals. Many non-metals, such as nitrogen, oxygen, and hydrogen, are gases at room temperature. Non-metals that are solids are not shiny, ductile, or malleable. Consider a charcoal briquette as an example (Figure 8). It is mainly composed of carbon, a non-metal. It is dull and brittle, and shatters easily if pounded or stretched.

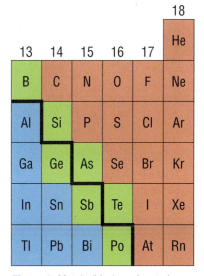

Figure 6 Metals (blue) are located to the left of the staircase line of the periodic table. Non-metals (red) are located on the right. Metalloids (green) are located along the staircase line.

Figure 7 Gold leaf has many interesting applications.

Figure 8 A charcoal briquette is made up of 85 % to 98 % carbon. Carbon, a non-metal, is dull and brittle.

Non-metals are generally poor conductors of thermal energy and electricity. (Carbon is an exception.) We take advantage of this non-conducting property of non-metals to insulate our houses in winter. Inserting a layer of a non-metal, such as argon gas, between the two panes of glass in double-glazed windows greatly reduces the thermal energy loss through glass alone. You apply the same principle when you put on a warm sweater. The layer of air trapped in the loops of yarn, and between the sweater and your body, decreases your body's loss of thermal energy.

CITIZEN ACTION

Recycle Your Cellphone, Save a Gorilla

Do gorillas use cellphones? Not that we know of, but their lives are closely linked to an element called tantalum, a key component of almost all cellphones, pagers, and laptops. Tantalum is a metal that is lightweight and can hold a high electrical charge (Figure 9). These properties make it an ideal material for the miniature circuit boards in electronic devices. Unfortunately, the ore for tantalum is mined mostly in the rainforests of central Africa, where the endangered lowland gorillas live (Figure 10).

Figure 9 Tantalum is used in electronic devices such as cellphones because it is lightweight and can hold a high electrical charge.

Figure 10 The lowland gorillas of central Africa are endangered because of the mining boom for tantalum.

The growing worldwide demand for cellphones, and hence tantalum, has brought thousands of miners into protected parks. Mining destroys habitat and threatens wildlife. Worse, the slow-moving and meaty gorillas are hunted and killed as food for the miners. There has been an estimated 70 % decline in the population of the eastern lowland gorilla since the mining boom began.

Recycling old cellphones allows tantalum to be reclaimed and reused, thus reducing the need for further mining. Cellphones also contain valuable minerals, such as gold, which can be profitably recovered as well. Refurbishing and reusing cellphones also protects landfill sites from hazardous components, including antimony, arsenic, cadmium, copper, lead, and zinc. When these elements are burned or buried in landfills, they pollute our air and leach into our soil and water supplies, eventually entering our food chain.

It is estimated that the average Canadian teenager will buy and discard three cellphones during their years in high school, and there are several million teenagers across Canada. New phone purchases are not made out of necessity, but rather for upgrades as new features and trendy models come on the market.

Organizing a cellphone recycling program can be a simple and effective way to help our environment and reduce mining for tantalum. The gorillas will thank you.

What Can You Do To Help?
- Start a cellphone recycling program at your school.
- Make and display posters educating about cellphone recycling at local businesses in your community.
- Organize a competition to design and build creative collection boxes where used cellphones can be dropped off for recycling. Place these boxes around the school or at local businesses and shopping malls.
- Write an article for your school newspaper or a letter to a local newspaper informing readers about cellphone recycling.
- Encourage family and friends to participate in your school recycling program.
- Most cellphone companies accept used cellphones for recycling. Contact a local store to arrange for pickup of your collected cellphones.
- Contact the Toronto Zoo for information about the EcoCell program, a worldwide cellphone recycling program that helps to restore the habitat of lowland gorillas in Africa.
- Organize a trip to the Toronto Zoo to deliver cellphones collected for recycling and to visit the lowland gorilla exhibit.
- Think of ways to reduce the number of new cellphones you buy.

UNIT TASK Bookmark

You can apply what you learned in this section about elements and compounds to the Unit Task described on page 286.

IN SUMMARY

- An element is a pure substance that cannot be broken down into a simpler chemical substance by any physical or chemical means. Elements are the building blocks of all substances and are arranged on the periodic table.
- A compound is a pure substance composed of two or more different elements.
- A metal is an element that has lustre, is a conductor, and is malleable and ductile.
- A non-metal is an element that is usually a gas or a dull powdery solid. Non-metals are usually poor conductors of heat and electricity.
- A metalloid has both metallic and non-metallic properties.

CHECK YOUR LEARNING

1. Which of the following substances are elements? Explain how you determined your answer.
    (a) bronze
    (b) tin
    (c) chromium
    (d) solder
    (e) propane
    (f) arsenic
    (g) nickel

2. What is the difference between an element and a compound?

3. A white powder, when heated, produces a colourless gas and a black solid. Is the white powder an element? Give reasons for your answer.

4. Explain the significance of the bold staircase line on the periodic table (Figure 11).

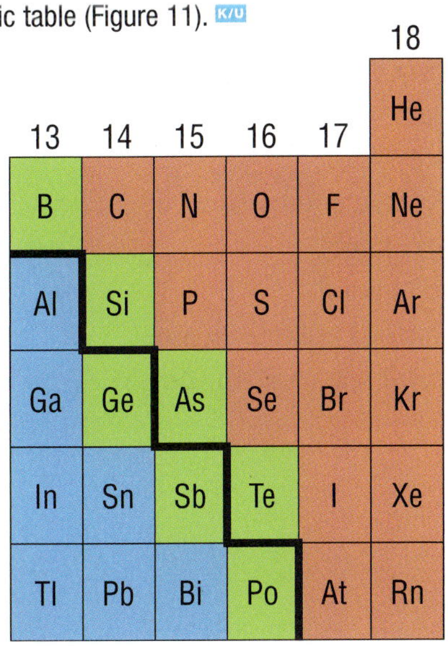

Figure 11

5. Are there more metallic elements or non-metallic elements listed on the periodic table?

6. List three properties of metals.

7. List three properties of non-metals.

8. Create a two-column table in your notebook with the headings "Metals" and "Non-metals." Classify each of the properties below as characteristic of metals or non-metals. Include an example from the periodic table for each property.
    (a) conducts electricity
    (b) is a gas under normal conditions
    (c) can be flattened by hammering
    (d) its symbol is located in the upper-right corner of the periodic table
    (e) shatters when struck
    (f) is a dull yellow powder
    (g) is soft and shiny
    (h) its symbol is located in the first column of the periodic table

9. Identify which properties of each of the following elements make them ideal for their uses.
    (a) copper and aluminum for pots and pans
    (b) silver and gold for jewellery
    (c) argon in double-glazed windows for homes

10. In this section, you were introduced to some of the physical properties of carbon, a non-metal.
    (a) Describe the physical properties of carbon.
    (b) What property of carbon makes it different from other non-metals?

6.1 A Table of the Elements

6.2 CONDUCT AN INVESTIGATION

Become a Metal Detective

In this investigation, you will become a metal detector, or, if you prefer, a metal detective (Figure 1). First, you will perform a series of tests on the collection of chemical elements in your custody. You will make keen observations with your sharp senses. In your detective notebook, you will record and analyze your results for any patterns. Finally, you will confirm the identity of each element as a metal or a non-metal and note its location on the periodic table of elements for future reference.

SKILLS MENU
- Questioning
- Hypothesizing
- Predicting
- Planning
- Controlling Variables
- Performing
- Observing
- Analyzing
- Evaluating
- Communicating

Figure 1 A metal detective

Testable Question
How can you distinguish metals from non-metals?

Hypothesis/Prediction
Predict which of the elements can be classified as metals, and which as non-metals. Give reasons for your prediction.

Experimental Design
In this investigation, you will perform various tests to determine whether selected elements are metals or non-metals. You will note the colour and lustre of each sample. You will also test each sample for malleability, density, magnetism, and electrical conductivity.

Equipment and Materials
- eye protection
- lab apron
- well tray
- magnet
- balance
- low-voltage conductivity apparatus (Figure 2 on the next page)
- fine steel wool
- samples of available elements (for example, Mg, Cr, Fe, Ni, Cu, Ag, Au, Zn, Al, C, Si, Sn, S)

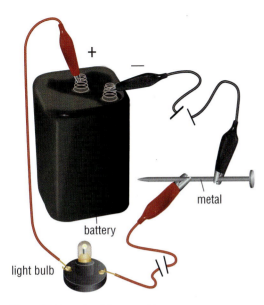

Figure 2 A conductivity apparatus

Procedure

1. Copy Table 1 into your notebook.

Table 1 Observations and Analysis

Element	Properties					
	Colour	Lustre	Malleability	Density	Magnetism	Electrical conductivity
?						
?						

2. Put on your lab apron and eye protection.
3. Obtain a well tray and trace its outline in your notebook.
4. Obtain a sample of each element. Place each element in a separate well in the well tray. Record the name of each element in the corresponding location on the traced outline.
5. Examine each element. Note its colour and lustre. Follow your teacher's instructions for scrubbing a sample with steel wool to expose its surface. Record your observations.
6. Test each element for malleability or brittleness by trying to bend or break it. Record your observations.
7. Estimate the density of each element by comparing its mass with another element sample of similar size. Record your observations.
8. Test each element with a magnet to determine if it is magnetic or non-magnetic. Record your observations.
9. Test each element with the conductivity apparatus to see whether it conducts electricity. If the light bulb lights up, it is a conductor. If not, it is an insulator. Record your observations.
10. When all elements have been tested, follow your teacher's instructions for returning or disposing of all materials and equipment.
11. Wash your hands thoroughly with soap and water.

Analyze and Evaluate

(a) What patterns of behaviour do you observe that allow you to group the elements according to their properties? T/I

(b) Did the evidence you obtained in this investigation support your prediction? Explain why or why not. T/I

(c) Write a report summarizing your findings in this investigation. Follow your teacher's instructions for content and format of your report. C

(d) Which tests, if any, presented problems in making or interpreting the observations? Explain. T/I

(e) Suggest improvements to the procedure that may increase your confidence in the results of the tests. T/I

Apply and Extend

(f) Refer to the elements that you have classified as metals. Use your observations to organize them into subgroups with similar properties. T/I

(g) Explain how the properties of each of the following elements make them suitable for their use in these products: A

  (i) electrical wires made of copper
  (ii) aluminum in drink cans
  (iii) iron needles in compasses
  (iv) silicon in computer chips
  (v) gold foil on paintings and sculptures

6.3 PERFORM AN ACTIVITY

Properties of Household Chemicals

Common household items for sale in a supermarket are usually grouped by their function. Sugar, flour, and salt are found in the baking section, toothpaste can be found in the pharmacy, and candles are found near the florist (Figure 1). You can find sand in the outdoor department in both summer and winter. Like any good chemist, however, you might be more interested in classifying these items by their physical and chemical properties. After all, it is these properties that determine their function!

In this activity, you will plan the steps to determine the properties of a variety of household items. You will use equipment and materials that your teacher makes available to you. After you have received approval from your teacher, you will carry out your planned procedure, following all safety precautions.

SKILLS MENU
- Questioning
- Hypothesizing
- Predicting
- Planning
- Controlling Variables
- Performing
- Observing
- Analyzing
- Evaluating
- Communicating

Figure 1 Supermarkets organize items according to their function. Chemists use physical and chemical properties to classify substances.

Purpose
To determine the physical and chemical properties of some common substances.

Equipment and Materials
- eye protection
- lab apron
- well plate
- balance
- graduated cylinder
- hot plate
- test tubes
- test tube racks
- beakers
- glass stirring rods
- thermometer
- scoopulas
- Bunsen burner
- tongs
- conductivity apparatus
- household substances: table salt, sugar, flour, candle wax, toothpaste, sand, and so on
- water

 Unplug the hot plate by pulling on the plug, not on the cord.

Tie back long hair and loose clothing when using the Bunsen burner.

Procedure

SKILLS HANDBOOK
3.B.4., 3.B.5.

1. Design a procedure to determine the following properties for each of the household substances (Figure 2): colour, hardness, solubility in water, electrical conductivity, melting point, density, combustibility, and reaction with water. Include all safety precautions that are needed.

Figure 2 In this activity, you will determine the properties of common household substances.

2. After your teacher approves your plan, design a table to record your observations.
3. Perform the tests on each of the household items and record your observations.

Analyze and Evaluate

SKILLS HANDBOOK
3.B.7., 3.B.8.

(a) Which of the tests that you conducted determined physical properties? Give reasons for your answer. T/I

(b) Which of the tests determined chemical properties? Give reasons for your answer. T/I

(c) Did you encounter any problems or errors as you completed your procedure? If so, describe and explain whether they were avoidable or unavoidable. T/I

(d) What improvements would you suggest for similar investigations in the future? T/I

(e) Can substances be distinguished by determining several of their physical and chemical properties? Explain your answer based on what you have learned by performing this activity. T/I

Apply and Extend

(f) Were any of the properties that you observed surprising to you? Explain. C

(g) Sugar and salt are similar. They are both used in cooking. If you were standing in a puddle of water in an electrical storm, which would you be safer holding, a handful of sugar or a handful of salt? Explain. (Of course, you should not be standing outside in an electrical storm!) A

(h) What properties of wax make it such a good material for making candles of all different shapes and sizes, even floating ones that light up a bowl of water? K/U

(i) Some foods, such as chestnuts and potatoes, can be cooked in heated sand. Which properties of sand make it suitable for this culinary function? K/U

6.4 Patterns in the Periodic Table

The periodic table is arranged in a particular way. All the elements in the same column have similar physical and chemical properties (Figure 1). If your classroom seating plan were organized in this way, everyone sitting in the same column would look similar and have similar personalities. Each student would still be an individual and different from all the other students. However, if there were an empty seat, you would be able to guess the absent student's appearance and temperament. In this way, we can predict the properties of any element simply by its assigned location in the periodic table. That is the power of the periodic table of the elements.

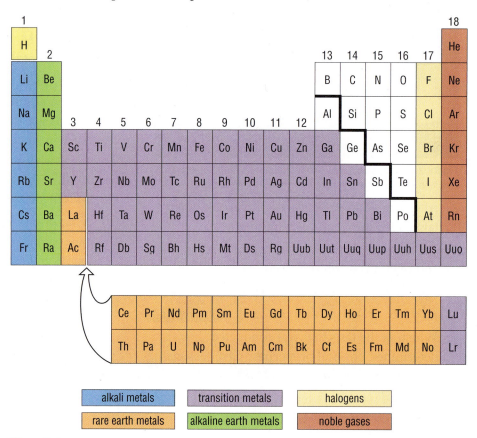

Figure 1 Elements in the same group (column) of the periodic table have similar physical and chemical properties.

Chemical Families

Elements in the same column of the periodic table belong to the same group, or **chemical family**. Families with distinctive properties have been given identifying names. The elements in the first column from the left are called **alkali metals**. They are called Group 1 elements. All columns are numbered from left to right.

chemical family a column of elements with similar properties on the periodic table

alkali metal an element in Group 1 of the periodic table

The alkali metals lithium (Li), sodium (Na), and potassium (K) are shiny, silvery, and soft, but highly reactive (Figure 2). These three elements have relatively low densities and they can float on water. We do not often encounter these elements on their own because they combine readily with other elements and compounds. They are present in many everyday substances such as table salt (NaCl) and baking soda ($NaHCO_3$). Many athletes know that plant foods like oranges and bananas are good sources of potassium. Potassium is important to the function of all living cells.

Elements in the next family, the Group 2 elements, are called **alkaline earth metals**. Beryllium, magnesium, calcium, strontium, barium, and radium are alkaline earth metals. These metals are shiny and silvery but are not as soft or reactive as the alkali metals. As every growing child has been told, calcium (Ca) helps to build strong bones and teeth (Figure 3). Similarly, strontium (Sr) builds a strong shell in coral. Many substances composed of alkaline earth metals burn with bright, colourful flames (Figure 4). As a result of this property, alkaline earth metals such as magnesium are used in fireworks.

Figure 2 The alkali metals are grouped together in Group 1 of the periodic table because they have similar properties.

alkaline earth metal an element in Group 2 of the periodic table

Figure 3 Calcium, in Group 2 of the periodic table, is an important dietary nutrient.

Figure 4 Magnesium, an alkaline earth metal (Group 2), burns with a bright flame.

The elements in the column on the far right of the periodic table are called **noble gases** (Group 18). This name reflects the stable nature or unreactivity of these elements. We have all experienced some of these noble gases. The low density of helium (He) makes it ideal for use in balloons. We use neon (Ne) and other noble gases in neon signs. Although noble gases are colourless, odourless, and tasteless, they glow brightly when an electrical current is passed through them (Figure 5). Argon (Ar) glows blue, krypton (Kr) glows pink, and xenon (Xe) glows purple. Bright red signs, which are the most common, are filled with neon. With the exception of radon, the Group 18 elements are also non-toxic.

noble gas an element in Group 18 of the periodic table

Figure 5 Each noble gas glows with its own distinctive colour when an electrical current is passed through it.

halogen an element in Group 17 of the periodic table

DID YOU KNOW?
Chlorine Warfare
Chlorine was used in chemical warfare for the first time in 1915, during World War I. French soldiers in the trenches in Ypres saw a yellow-green gas cloud approaching from the German side. Thinking it was a smoke screen, they took up positions in the front trenches—with fatal results. The soldiers suffocated as the poisonous gas destroyed their lungs. The Allies provided gas masks that offered limited protection. The soldiers were mistakenly advised that breathing through a urine-soaked cloth would delay the effects of the chlorine. Sadly, the combination of urine and gas produced additional toxic substances. Over 91 000 soldiers were killed by deadly gases during the war.

The **halogens**, also called Group 17 elements, are in the column to the left of the noble gases. The halogens fluorine (F) and chlorine (Cl) are gases at room temperature and atmospheric pressure. Bromine (Br) is a liquid, and iodine (I) and astatine (At) are solids (Figure 6).

All halogens are very reactive. Due to their reactive nature, the halogens are rarely found in elemental form. They often form compounds with alkali metals. Many of the halogens can be poisonous in large amounts. Some, such as chlorine, are poisonous even in small amounts. Chlorine gas was used as a chemical weapon during World War I. At much lower concentrations, chlorine is safely used to kill bacteria in swimming pools.

Iodine, usually dissolved in alcohol, is used to disinfect scrapes and cuts. Other halogens, such as bromine, can be added to light bulbs to increase the brightness and operating life of the bulb. Halogen lamps burn at extremely high temperatures, so care must be taken to keep halogen lamps away from flammable materials.

Figure 6 Samples of the halogens chlorine, bromine, and iodine

Periodic Trends

You have just learned that elements in the same column in the periodic table share some chemical and physical properties. Elements in the same horizontal row in the periodic table show some trends of increasing or decreasing reactivity. For example, Group 1 alkali metals are more reactive than their Group 2 neighbours. Group 17 halogens are more reactive than Group 16 elements in the same row. Therefore, sodium is more reactive than magnesium, and chlorine is more reactive than sulfur. Rows on the periodic table are called **periods**, signifying the recurring nature of these trends in reactivity. The word "period" refers to cycles, such as the classes that recur daily in your school schedule.

period a row on the periodic table

History of the Periodic Table

The current version of the periodic table evolved from the one developed in 1869 by Dmitri Mendeleev (Figure 7), a Russian chemistry professor. He gathered information about the properties of all known elements. At that time there were only 63. The masses of individual elements were known, and Mendeleev arranged the elements by increasing mass. He started with the lightest element, hydrogen, followed by the next lightest, helium, and so forth. He could have chosen to begin a new row after any number of elements. He could have filled rows of ten elements each to produce a perfectly rectangular table of elements, but he did not.

Mendeleev noticed that there were groups of elements with similar physical and chemical properties. In a brilliant move, he decided to arrange elements with similar properties under each other in the same column of the table. To do so, he had to force new rows of different lengths, resulting in the familiar irregular shape of his periodic table.

When he was finished with his table of elements, there were several empty spaces where no known element could fit (Figure 8). Mendeleev suggested that the missing elements did exist but had not yet been discovered. He boldly predicted the physical and chemical properties of the elements that would occupy each vacant spot. Doubters were convinced when new elements, such as gallium (Ga) and germanium (Ge), were discovered several years later with properties almost exactly as Mendeleev had predicted (Table 1).

Figure 7 Dmitri Mendeleev is the father of the periodic table of the elements.

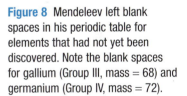

To learn more about becoming a historian of science,
GO TO NELSON SCIENCE

ROW	GROUP I	GROUP II	GROUP III	GROUP IV	GROUP V	GROUP VI	GROUP VII
1	H = 1						
2	Li = 7	Be = 9, 4	B = 11	C = 12	N = 14	O = 16	F = 19
3	Na = 23	Mg = 24	Al = 27, 3	Si = 28	P = 31	S = 32	Cl = 35, 5
4	K = 39	Ca = 40	– = 44	Ti = 48	V = 51	Cr = 52	Mn = 55
5	(Cu = 63)	Zn = 65	– = 68	– = 72	As = 75	Se = 78	Br = 80
6	Rb = 85	Sr = 87	?Yt = 88	Zr = 90	Nb = 94	Mo = 96	– = 100

Figure 8 Mendeleev left blank spaces in his periodic table for elements that had not yet been discovered. Note the blank spaces for gallium (Group III, mass = 68) and germanium (Group IV, mass = 72).

Table 1 Comparison of Mendeleev's Predictions to the Observed Properties of Germanium After Its Discovery

Property	Properties predicted by Mendeleev in 1871	Observed properties of germanium, discovered in 1886
colour	grey	grey
mass of element	72	72.6
density	5.5 g/cm^3	5.4 g/cm^3
melting point	high	947 °C
number of elemental oxygen particles it combines with	2	2
number of elemental chlorine particles it combines with	4	4

Decades later, scientific knowledge and theories about the structure of the elements led to a brilliant explanation of the periodic nature of the elements. You will be exploring these theories later in this chapter.

RESEARCH THIS: THE PERIODIC TABLE IS EVOLVING!

SKILLS: Researching, Communicating, Evaluating

SKILLS HANDBOOK
4.B.

You may have noticed that period 7 of the periodic table is incomplete. The elements have strange, similar names, such as ununbium, and the information about them, such as atomic mass, is missing.

1. Choose one of the elements in period 7.
2. Perform research to find out what is known about this particular element.

A. Write a short paragraph outlining what we know so far about this element. K/U C

B. Explain why the information on the periodic table about this element is incomplete. T/I

C. Who decides which elements get added to the periodic table? T/I

D. Do you think the process by which new elements are added to the periodic table is a good one? Explain why or why not. T/I

TRY THIS: SOLVING THE PUZZLE, PERIODICALLY

SKILLS: Performing, Observing, Analyzing, Communicating

Mendeleev's first attempt at arranging elements into a table was much like trying to put together a puzzle that was missing pieces. In this activity, you will try to assemble the pieces of a puzzle to help you identify the properties of the missing piece.

Equipment and Materials: handouts of puzzles A and B from your teacher

1. Puzzle A shows 19 pieces from an original set of 20 (Figure 9). A single piece is missing. Your mission is to arrange the 19 pieces into rows and columns according to their properties, identifying as many trends as you can.

Figure 9 Puzzle A

2. When all the pieces are in place, the missing piece will become obvious to you, just as the missing elements were clear to Mendeleev. Sketch the missing piece, showing all its properties.

3. Puzzle B has more differentiating properties than does puzzle A, so it is a little more complex (Figure 10). If you enjoy a challenge, you will want to do this one too!

Figure 10 Puzzle B

A. There are many different possible arrangements. When you have finished, share your design with other students so that everyone can admire and learn from each other's thinking processes and ideas. That is how a real scientific community works. T/I

UNIT TASK Bookmark

You can apply what you learned about groups of elements on the periodic table to the Unit Task described on page 286.

IN SUMMARY

- A chemical family is a column of elements on the periodic table.
- Elements in the same chemical family have similar properties.
- The first two families on the periodic table are the alkali metals and the alkaline earth metals.
- The last two families on the periodic table are the halogens and the noble gases.
- A period is a row on the periodic table.
- Mendeleev grouped elements with similar physical and chemical properties in the same column in a periodic table.
- Mendeleev correctly predicted the properties of elements that had not yet been discovered.

CHECK YOUR LEARNING

1. Which of the following statements are correct? Rewrite each false statement to make it true.
   (a) Elements listed in rows on the periodic table are in the same family.
   (b) Elements in the same column of the periodic table exhibit the same physical properties.
   (c) Elements in the same group are in the same family.
   (d) Elements that are side by side on the periodic table belong to the same period.

2. Name the chemical family to which each of the following elements belongs.
   (a) chlorine, Cl
   (b) magnesium, Mg
   (c) potassium, K
   (d) helium, He

3. Read the names of each of the elements in the first and second columns on the periodic table.
   (a) What do the names of these elements have in common?
   (b) Which element is the exception?

4. Sodium is a metal, like copper. Suggest reasons why sodium cannot be used in electrical wires.

5. Identify which properties are common to each of the following chemical families.
   (a) alkali metals
   (b) alkaline earth metals
   (c) halogens
   (d) noble gases

6. Hydrogen is a reactive gas under normal conditions. Based on this property alone, to which group of the periodic table should hydrogen belong? Why? (In Chapter 7, you will learn why hydrogen is placed in Group 1.)

7. Sketch the periodic table and add the following labels: period, family, alkali metals, alkaline earth metals, halogens, noble gases, metals, non-metals, and metalloids.

8. Alkali metals are found in many common substances but are rarely found in pure form. Explain why this is the case.

9. List three useful applications of halogens.

10. List three useful applications of alkaline earth metals.

11. Mendeleev chose to order the known elements by mass. Explain why he did not arrange the elements into rows and columns of equal length.

6.5 CONDUCT AN INVESTIGATION

Comparing Family Behaviour

When Mendeleev designed his periodic table, he arranged the elements according to their similarities in physical and chemical properties. He placed lithium (Li), sodium (Na), and potassium (K) in the first column—the alkali metal family. He placed magnesium (Mg) and calcium (Ca) in the second column—the alkaline earth metal family. It is now your turn to determine why Mendeleev grouped these elements together. What common behaviour do they exhibit? Why are they placed in that particular sequence?

SKILLS MENU
- Questioning
- Hypothesizing
- Predicting
- Planning
- Controlling Variables
- Performing
- Observing
- Analyzing
- Evaluating
- Communicating

Testable Question
What patterns of chemical reactivity do elements of the alkali metals and alkaline earth metals exhibit?

Hypothesis/Prediction
Write a hypothesis for the Testable Question. Your hypothesis should include a prediction as well as a reason for the prediction.

Experimental Design
You will compare the appearance and reactivity of three alkali metals and two alkaline earth metals in water. You will then use your experimental evidence to explain why these metals are grouped together on the periodic table.

Equipment and Materials
- eye protection
- lab apron
- 3 small beakers
- plastic Petri dish cover
- scoopula
- magnesium metal (turnings)
- calcium metal (turnings)
- lithium metal
- sodium metal
- potassium metal
- cold water
- hot water

 Lithium, sodium, and potassium are highly corrosive and flammable. Do not handle these substances. Your teacher will demonstrate their properties to you.

Procedure

SKILLS HANDBOOK
3.B.6.

1. Prepare a table in your notebook similar to Table 1.

Table 1 Observations

Metal	Appearance of metal	Reactant	Observations of reaction
Mg		cold water	
Mg		hot water	
Ca		cold water	
Li		cold water	
Na		cold water	
K		cold water	

Part A: Alkaline Earth Metals

2. Put on your lab apron and eye protection.
3. Use a scoopula to place a small amount of magnesium into each of two beakers. Fill only the tip of a scoopula with your metal sample before adding it to the beaker (Figure 1).

Figure 1 Only a small amount of each metal is required.

4. Record the appearance of the magnesium.
5. Add about 10 mL of cold water to the first beaker and 10 mL of hot water to the second beaker. Allow the mixtures to sit for a minute and record your observations.
6. Add about 30 mL of cold water to a third beaker.
7. Add a small amount of calcium to the beaker.

⚠ The mixture that is produced in step 7 is an irritant. Avoid skin contact. Dispose of this mixture as directed by your teacher.

Part B: Alkali Metals
(TO BE DEMONSTRATED BY THE TEACHER USING AN OVERHEAD PROJECTOR)

8. Your teacher will obtain a clean beaker, fill it halfway with cold water, and place a plastic Petri dish cover beside the beaker.
9. Your teacher will obtain a tiny piece of lithium (enough to fit on the flat end of a toothpick).
10. Record the appearance of the lithium.
11. Your teacher will carefully drop the lithium into the beaker of water and cover the beaker with the plastic Petri dish cover. Record your observations.
12. Your teacher will repeat steps 8–11 using a tiny piece of sodium and then a tiny piece of potassium instead of lithium.

Analyze and Evaluate

(a) Compare the appearance of the three alkali metals that you observed. In what ways are they similar? In what ways are they different? T/I
(b) Compare the appearance of the two alkaline earth metals that you observed. In what ways are they similar? In what ways are they different? T/I
(c) Compare the behaviour of the two alkaline earth metals in water. Which metal was more reactive? Support your answer with your experimental evidence. T/I
(d) Compare the behaviour of the three alkali metals in water. List the metals in order of increasing reactivity. Support your answer with your experimental evidence. T/I
(e) Did these observations agree with your hypothesis? Explain. T/I C
(f) Which conditions must be kept the same in each test in order to accurately compare the reactivity of different elements? T/I
(g) Compare the reactivity of Na and Mg, elements in the same row of the periodic table. Which element is more reactive? Support your answer with your experimental evidence. T/I
(h) Compare the reactivity of K and Ca, elements in the same row of the periodic table. Which element is more reactive? Support your answer with your experimental evidence. T/I
(i) What unavoidable sources of error, if any, did you encounter in this activity? Explain your answer and suggest improvements that can be made to the procedure. T/I C

Apply and Extend

(j) Use all your observations in this activity to summarize patterns that are evident within each family, and across each row, of the periodic table. T/I
(k) Use the observed patterns to predict the physical properties and the relative reactivity of Rb, Cs, and Fr. T/I
(l) Use the observed patterns to predict the physical properties and the relative reactivity of Sr and Ba. T/I
(m) Use the observed patterns to predict the relative reactivity of Cs and Ba. T/I
(n) Alkali metals are not often found in pure form in nature. Why not? Explain using your experimental evidence. T/I

6.6 Theories of the Atom

If you were to take a gold bar and cut it into smaller and smaller pieces, what is the smallest piece of gold that you could get? Would the piece of gold ever become so small that, if you were to cut it further, it would no longer be gold?

These are the questions that occupied Greek philosophers thousands of years ago. Their answers, and those of many curious minds, have shaped our theories about the structure of matter. Let's trace this journey with a brief history of atomic theory.

Theories Evolve

A scientific theory is not a guess. It is not even an educated guess. A scientific theory is an expression of our best understandings of a phenomenon, based on scientific evidence or reasoning. As technologies are improved and new observations are gathered, or when old observations are reinterpreted, old theories may need to be modified. Sometimes old theories need to be discarded, and new theories need to be created.

Theories are constantly evolving as we gather more information and gain a wider perspective. When we begin to "connect the dots" of all the separate pieces of data, we are able to expand and refine our scientific theory. This is certainly true of our model of the structure of the atom.

The Evolution of Atomic Theory

AN INDIVISIBLE PARTICLE—THE ATOM

Around 400 BCE, the Greek philosopher Democritus proposed that all matter can be divided into smaller and smaller pieces until a single indivisible particle is reached. He named this particle the **atom**, which means "cannot be cut" (Figure 1). Without any experimental evidence, he proposed that atoms are

- of different sizes
- in constant motion
- separated by empty spaces (the void)

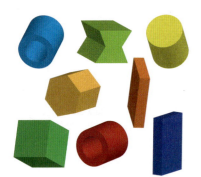

Figure 1 Democritus thought that the atom was indivisible.

atom the smallest unit of an element

EARTH, WATER, AIR, AND FIRE: ARISTOTLE (AROUND 450 BCE)

Another famous Greek philosopher, Aristotle, rejected the idea of the atom. He supported an earlier theory that all matter is made up of four basic substances: earth, water, air, and fire (Figure 2). These substances were thought to have four specific qualities: dry, wet, cold, and hot, respectively. Aristotle's theory of the structure of matter was accepted for almost 2000 years.

Figure 2 According to Aristotle, everything is made of earth, water, air, and fire.

THE BILLIARD BALL MODEL

In 1807, John Dalton, an English scientist and teacher, revived Democritus' theory of the indivisible atom. Dalton proposed that
- all matter is made up of tiny, indivisible particles called atoms
- all atoms of an element are identical
- atoms of different elements are different
- atoms are rearranged to form new substances in chemical reactions, but they are never created or destroyed

Dalton's model is known as the billiard ball model (Figure 3). This model was useful because it could explain many properties of matter. For example, Dalton believed that pure gold samples from different locations had identical properties because the samples contained identical atoms. However, the billiard ball model could not explain why some objects attract each other, while other objects repel each other. For example, rubbing a balloon in your hair causes your hair to be attracted to the balloon (Figure 4). The explanation came almost 100 years later!

Figure 3 In Dalton's model, the atom resembles a billiard ball.

THOMSON'S EXPERIMENTS—THE ELECTRON

In 1897, J.J. Thomson discovered that extremely small negatively charged particles could be emitted by very hot materials. These particles were attracted to the positive end of a circuit (Figure 5). Positive charges and negative charges were known to attract each other, so Thomson concluded that the particles must be negatively charged. These particles were later called **electrons**. Thomson theorized that
- atoms contain negatively charged electrons
- since atoms are neutral, the rest of the atom is a positively charged sphere
- negatively charged electrons are evenly distributed throughout the atom

Figure 4 Dalton's model could not explain attractions between objects.

electron a negatively charged particle in an atom

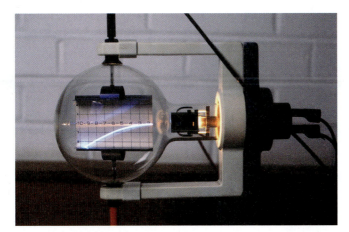

Figure 5 Thomson used a device called a cathode ray tube to conduct his experiments. The particles he detected were attracted to the positive end of the circuit, so they had to be negatively charged.

Thompson's model was called the "plum pudding" model because the electrons embedded in an atom resembled the raisins in a plum pudding (Figure 6). If Thomson had lived in more modern times, he may have called it a "chocolate chip muffin" model.

Figure 6 Thomson's model of the atom resembles a plum pudding.

6.6 Theories of the Atom

READING TIP

Look for Context Clues

Look for context clues such as key words, examples, or explanations to help you make inferences. For example, in the explanation of Rutherford's experiment, you notice the word "reasoned." This tells you that Rutherford did not observe the nucleus of the atom directly, but rather inferred it from the results of the experiment.

THE GOLD FOIL EXPERIMENT—THE NUCLEUS AND THE PROTON

In 1909, Ernest Rutherford supervised an experiment to test Thomson's model of the atom. He predicted that if positive and negative charges were uniformly distributed throughout atoms, then tiny positively charged particles shot at a thin piece of gold foil would pass through the foil. Some of the particles might be slowed down or deflected at very small angles (Figure 7(a)). When the experiment was performed, most of the particles passed through the foil unaffected. Also, a small number of particles were deflected at very large angles, as though something very massive but very small was repelling them (Figure 7(b)). This result was shocking! Rutherford then reasoned that these large angles of deflection were caused by a collision with a small, concentrated, positively charged central mass inside the atom. In Rutherford's revised model,

- the centre of the atom has a positive charge. This centre is called the nucleus. It contains most of the atom's mass but occupies a very small space. The nucleus is what made some particles bounce back during the experiment.
- the nucleus is surrounded by a cloud of negatively charged electrons
- most of the atom is empty space

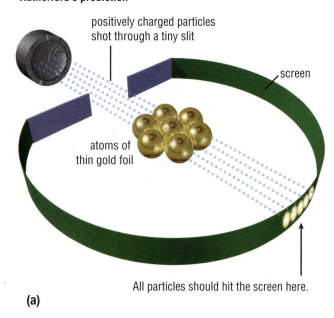

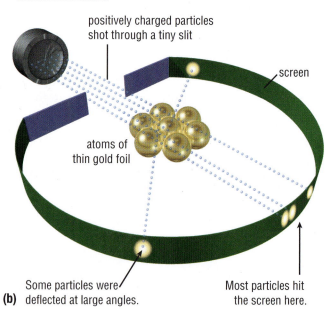

Figure 7 (a) Rutherford's prediction: Particles should pass directly through the gold foil, with very little deflection. (b) Rutherford's results: Most of the particles passed through the gold foil, but a few were deflected at very large angles.

proton a positively charged particle in the atom's nucleus

Rutherford is also credited with discovering the proton in 1920. A **proton** is a positively charged particle that is found in the atom's nucleus. Measurements of atomic mass showed that protons alone could not account for the total mass of a nucleus, given the amount of their charge. Rutherford predicted that there must be a third particle in the nucleus that had about the same mass as the proton but that was neutral in charge.

TRY THIS: SIMULATING RUTHERFORD'S BLACK BOX

SKILLS: Performing, Observing, Analyzing, Evaluating, Communicating

In this simulation of Rutherford's gold foil experiment, you will shoot marbles at unidentified objects that are hidden under a large board. The paths that the marbles take as they travel under the board—straight through or deflected—will allow you to deduce the shape of the contents of this "black box."

Equipment and Materials: large board such as bristol board, foam board, or cardboard presentation board; small objects of similar height such as hockey pucks, small boxes, or Petri dishes; several marbles; large sheet of paper such as newspaper; markers; masking tape

1. Tape one or more objects under a large board. Make sure you perform this step while hidden from view of the participants, all of whom are named "Rutherford."
2. Tape a large sheet of paper to cover the top surface of the board.
3. Bring in the "Rutherfords" but do not allow them to look under the board. Give them marbles and markers. Allow them to roll marbles under the board and trace the paths of the marbles on the paper above the board (Figure 8).
4. After a sufficient number of rolls, ask the "Rutherfords" to draw the shape and size of the objects hidden in the "black box."

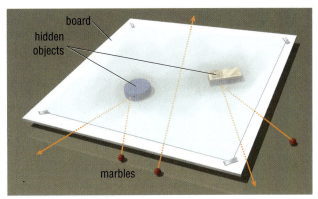

Figure 8 Experimental setup

5. Remove the board and the hidden objects, but do not reveal the objects to the "Rutherfords," just as the actual structure of the atom was never revealed to Ernest Rutherford.

A. In what way is this activity similar to Rutherford's gold foil experiment?
B. In what way is not revealing the contents of the black box similar to scientists trying to discover what is inside the atom?
C. Suggest other items that may be used instead of marbles to produce more informative clues about the hidden objects.

CHADWICK'S EXPERIMENTS—THE NEUTRON

In 1932, James Chadwick, Rutherford's student, found a particle that could penetrate and disintegrate atoms with extraordinary power. Unlike positively charged protons, these particles have zero charge. Therefore, there must be other undetected particles in the atom. These particles must be neutral. Based on this discovery, Chadwick proposed that

- an atom must be an empty sphere with a tiny dense central nucleus
- this nucleus contains positively charged protons and neutral particles called **neutrons**
- the mass of a neutron is about the same as that of a proton (Table 1)
- negatively charged electrons circle rapidly through the empty space around the nucleus (Figure 9)
- a neutral atom has the same number of protons as electrons

This model of the atom is called the planetary model.

neutron a neutral particle in the atom's nucleus

Table 1 Types of Subatomic Particles

	Proton	Neutron	Electron
charge	+	0	−
location	in nucleus	in nucleus	orbiting nucleus
relative mass	1	1	$\frac{1}{2000}$
symbol	p^+	n^0	e^-

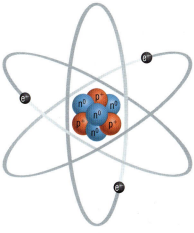

Figure 9 In the planetary model of the atom, electrons orbit the nucleus the way planets orbit the Sun in our solar system.

6.6 Theories of the Atom

ELECTRON ORBITS

Niels Bohr, a Danish scientist, studied the hydrogen atom and the light that it produces when it is excited by thermal energy or electricity. When white light is shone through a prism, a full rainbow of colours is seen (Figure 10(a)). When light produced by hydrogen is examined in the same way, only a few lines of colour are seen. Most colours are missing (Figure 10(b)).

Figure 10 (a) A prism separates white light into a rainbow of colours. (b) Lines of only certain colours of light are emitted by a hydrogen atom. This observation led Bohr to propose a new theory of the atom.

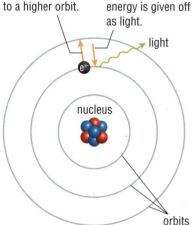

Figure 11 The Bohr–Rutherford model of the atom

In 1913, Bohr used this evidence to propose the following theory:
- Electrons orbit the nucleus of the atom much like the planets orbit the Sun.
- Each electron in an orbit has a definite amount of energy.
- The farther the electron is from the nucleus, the greater its energy.
- Electrons cannot be between orbits, but they can jump to and from different orbits (Figure 11). They release energy as light when they jump from higher to lower orbits, as shown in Figure 11. This energy is the light Bohr observed in his experiments.
- Each orbit can hold a certain maximum number of electrons. The maximum number of electrons in the first, second, and third orbits is 2, 8, and 8, respectively.

This model of the atom (shown in Figure 11) is known as the Bohr–Rutherford model because it is the product of the ideas of these two scientists. This model is useful for explaining the properties of the first 20 elements. However, it does not work as well for explaining the properties of the remaining elements on the periodic table.

TRY THIS: LINES OF LIGHT

SKILLS: Performing, Observing, Analyzing, Communicating

SKILLS HANDBOOK
3.B.6.

In this activity, you will use a device called a spectroscope to analyze the light emitted by two different sources of light. Niels Bohr used evidence from an experiment similar to this one to develop his model of the atom.

Equipment and Materials: fluorescent light source; spectroscope; power supply; hydrogen gas discharge tube; other gas discharge tubes

1. Use the spectroscope to observe the light emitted by the fluorescent lights in the classroom.
2. Your teacher will insert a hydrogen discharge tube into the power supply. Use the spectroscope to observe the light emitted by hydrogen.
3. Use the spectroscope to observe the light emitted by other elements.

A. Compare the appearance of fluorescent light and light from elements as seen through a spectroscope. T/I

B. Why are the lines of light emitted by an element sometimes described as being a "fingerprint" of the element? T/I

The Bohr–Rutherford model completely explained the observations of the light emitted by a hydrogen atom. It is a useful model because of its simplicity. The Bohr–Rutherford model of a central nucleus and orbiting electrons at different electron orbits serves well in explaining and predicting many physical and chemical properties of matter.

IN SUMMARY

- A scientific theory is an expression of our best understandings of a phenomenon, always based on scientific evidence or reasoning.
- John Dalton proposed that matter is made of tiny, indivisible particles called atoms.
- J. J. Thomson proposed that the atom contains negatively charged electrons. If the atom is neutral, the rest of the atom must be a positively charged sphere. The electrons are evenly distributed in the atom, like the raisins in plum pudding.
- Ernest Rutherford's gold foil experiment led him to propose that atoms are mostly empty space. A positively charged centre is surrounded by negatively charged electrons.
- Niels Bohr studied light produced by hydrogen atoms and proposed that electrons occupy fixed orbits around the nucleus.
- The Bohr–Rutherford model of the atom consists of positively charged protons and neutral neutrons in the nucleus of an atom. Negatively charged electrons orbit the nucleus.

CHECK YOUR LEARNING

1. Dalton theorized that matter is made of tiny, indivisible particles called atoms. In what way did the theories of each of these scientists support or differ from Dalton's theory? K/U
   (a) J. J. Thomson
   (b) Ernest Rutherford
   (c) James Chadwick

2. In his experiments, J. J. Thomson discovered a tiny stream of particles. K/U
   (a) Why did he conclude that these particles were negatively charged?
   (b) What were these particles eventually called?
   (c) According to Thomson, where are these particles located in the atom?
   (d) Why did Thomson conclude that atoms also contain a positive charge?
   (e) According to Thomson, where are the positive charges located in the atom?

3. If a neutral atom has three electrons, how many protons does it have? K/U

4. In what ways is Bohr's model of the atom similar to and different from Thomson's model? K/U

5. Rutherford's experiment consisted of beaming positively charged particles at thin gold foil. K/U
   (a) What did he expect to happen to the particles?
   (b) Why did the results surprise him?
   (c) From his results, which particles were proposed as part of an atom?
   (d) According to Rutherford, where in an atom are these particles located?

6. Is it reasonable to expect that these models will change again in time? Explain. K/U

7. Copy and complete Table 2 in your notebook. K/U

Table 2

Scientist	Date	Discovery/idea	Model (diagram)
Democritus			
Aristotle			
Dalton			
Thomson			
Rutherford			
Chadwick			
Bohr			

6.7 Explaining the Periodic Table

In Investigation 6.5, you observed differences in the reactivity of the alkali metals with water (Figure 1). Why do the elements become more reactive as you descend a family in the periodic table? You will learn that the Bohr–Rutherford model of the atom explains this trend, as well as other trends on the periodic table. First, let's take a more detailed look at the contents of the atom.

Figure 1 Lithium (a), sodium (b), and potassium (c) react at different rates with water to produce flammable hydrogen gas. The reactions release so much thermal energy that the hydrogen gas ignites.

Atomic Number

You learned in Section 6.1 that elements are the building blocks of substances. You also learned that pure substances differ because they consist of different elements. You know from Dalton's atomic theory that the atoms of each element are different from the atoms of all other elements.

What makes atoms unique is the number of protons they contain. The number of protons in the nucleus is called the **atomic number**. A hydrogen atom has one proton, so its atomic number is 1. Any atom that has a single proton in its nucleus can only be hydrogen. Any atom that does not have a single proton in its nucleus cannot be hydrogen. The periodic table lists the atomic number for each element in the top left-hand corner of each cell (box) (Figure 2). Chemists have found that when elements are arranged according to increasing atomic number on the periodic table, the elements within each column have similar properties.

The atomic number for gold, Au, is 79. This number tells us that there are 79 protons in every atom of gold. Can we take copper and turn it into gold? The atomic number of copper, Cu, is 29. A copper atom has 29 protons and is 50 protons short of being a gold atom. Where can we find a spare 50 protons? A tin atom, Sn, with an atomic number of 50, contains exactly 50 protons. If we could combine the nucleus of a copper atom with the nucleus of a tin atom, we would get an atom containing exactly 79 protons—a gold atom (Figure 3)!

atomic number the number of protons in an atom's nucleus

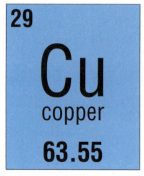

Figure 2 The atomic number is given in the top left-hand corner of each element on the periodic table.

Figure 3 To make a gold atom, we need 79 protons.

234 Chapter 6 • Elements and the Periodic Table

This idea is theoretically brilliant, but practically, it is very difficult to accomplish. Protons are tightly held in the nucleus of an atom. It would take a nuclear reaction (such as that inside an atomic bomb or a nuclear reactor) to combine two nuclei into one. This is not an efficient way to turn copper and tin into gold.

Mass Number and Atomic Mass

In Rutherford's atomic model, the atom is described as mostly empty space. Since electrons have a relatively insignificant mass, the mass of an atom consists of the contents of its nucleus—protons and neutrons. This value is called the **mass number**. Consider the element lithium, Li. The atomic number of lithium is 3, so all lithium atoms contain three protons. Most lithium atoms also contain 4 neutrons. The sum of three and four is seven. Therefore, these lithium atoms have a mass number of 7 (Figure 4).

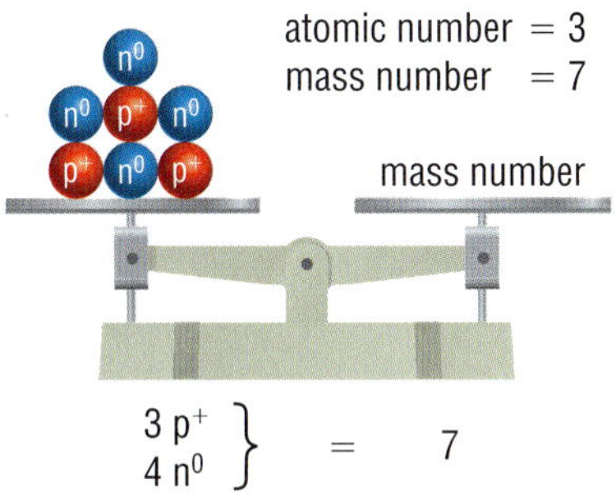

A small number of naturally occurring lithium atoms contain only three neutrons. These lithium atoms have a mass number of 6. Atoms with the same number of protons but different numbers of neutrons are called **isotopes**. Scientists use mass number to distinguish between the isotopes of an element. For example, the lithium isotope with a mass number of 6 is called lithium-6 or Li-6. The lithium isotope that has a mass number of 7 is called lithium-7 or Li-7. Since Li-7 has one more neutron, it is heavier than Li-6 (Figure 5).

The mass of an atom is called the **atomic mass** and is measured in atomic mass units (u). The atomic mass of each element is given below the element symbol on the periodic table. The atomic masses given on the periodic table are not whole numbers. For example, the atomic mass of lithium is 6.94 u (Figure 6). Naturally occurring lithium is a mixture of two isotopes, Li-6 and Li-7. The atomic mass of an element is the weighted average of the masses of its isotopes. Since Li-7 is far more common than Li-6, the average atomic mass for lithium is closer to 7 u than to 6 u. In many cases, you can determine the most common isotope of an element by rounding the atomic mass to the nearest whole number. For example, boron (B) has an atomic mass of 10.81 u. Therefore, the most common isotope of boron is B-11. Once you know the mass number, you can also determine the number of neutrons.

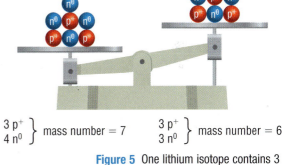

Figure 4 A lithium atom contains 3 protons and 4 neutrons, giving it a mass number of 7.

Figure 5 One lithium isotope contains 3 protons and 4 neutrons, giving it a mass number of 7. The other lithium isotope contains 3 protons and 3 neutrons, giving it a mass number of 6.

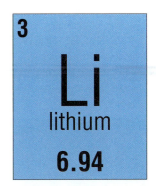

Figure 6 The element lithium has an atomic number of 3 and an atomic mass of 6.94 u.

mass number the number of protons and neutrons in an atom's nucleus

isotope an atom with the same number of protons but a different number of neutrons

atomic mass the mass of an atom in atomic mass units (u)

SAMPLE PROBLEM 1 Finding the Number of Neutrons

Find the number of neutrons in the most common isotope of aluminum.

Given: atomic mass of Al = 26.98 u
atomic number = 13

Required: number of neutrons

Analysis: Round the atomic mass of the element to the nearest whole number to get the mass number of the most common isotope.

mass number of Al = 27 u (rounded up)
mass number − atomic number = number of neutrons

Solution: 27 − 13 = 14

Statement: The most common isotope of aluminum contains 14 neutrons.

Bohr–Rutherford diagram a simple drawing that shows the numbers and locations of protons, neutrons, and electrons in an atom

Bohr–Rutherford Diagrams of an Atom

A picture is worth a thousand words. This holds true for atoms as well. The Bohr–Rutherford model of an atom can be depicted by a few simple strokes—a kind of stick drawing of an atom. Stick drawings show only the essential components of objects and are not drawn to scale. Since these diagrams of atoms represent both Bohr's and Rutherford's atomic models, they are called Bohr–Rutherford diagrams.

A **Bohr–Rutherford diagram** shows the numbers and locations of protons, neutrons, and electrons in an atom. We can deduce these numbers from the atomic number and mass number:

- the number of protons equals the atomic number
- the number of neutrons equals the difference between the mass number and the atomic number
- the number of electrons equals the number of protons in a neutral atom

SAMPLE PROBLEM 2 Drawing a Bohr–Rutherford Diagram

Draw a Bohr–Rutherford diagram of N-14.

Step 1. Determine the number of protons and the number of neutrons from the atomic number and mass number. Draw a small circle for the nucleus. Write the numbers of protons and neutrons inside the nucleus (Figure 7). Because atoms are neutral in charge, the number of negatively charged electrons must equal the number of positively charged protons.

For N-14, the atomic number is 7 and the mass number is 14.

$$\text{number of protons} = \text{atomic number} = 7p^+$$
$$\text{number of neutrons} = \text{mass number} - \text{atomic number}$$
$$= 14 - 7$$
$$= 7n^0$$
$$\text{number of electrons} = \text{number of protons} = 7e^-$$

Figure 7

Step 2. Draw one to four concentric circles outside the nucleus to represent electron orbits. The number of circles depends on the size of the atom.

The nitrogen atom has seven electrons. The first orbit can hold a maximum of two electrons, so draw two circles (Figure 8).

Figure 8

Step 3. Draw dots on these circles, starting from the circle immediately surrounding the nucleus, to represent the electrons in their orbits. There is a maximum number of electrons that can occupy each orbit. Current scientific evidence indicates that for the first 20 elements, the maximum number of electrons in the first, second, and third orbits is 2, 8, and 8, respectively. So, draw a pair of dots on the first circle. Then draw no more than 8 dots on the second circle. The first four electrons are usually drawn equally spaced. The next four are paired with the first four. Each orbit must be completely filled before dots can be drawn in higher orbits.

For the nitrogen atom, draw one pair of dots to fill the first orbit. Then draw five dots in the second orbit (Figure 9).

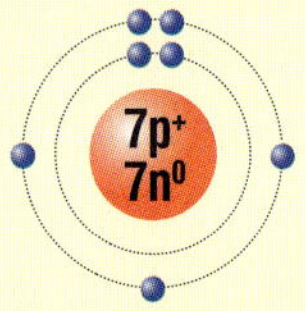

Figure 9 Note that the fifth electron in the second orbit is paired.

DID YOU KNOW?

Phosphorus, the Light Bearer
Phosphorus was discovered by accident, like many other marvellous substances. In 1669, Hennig Brand, an alchemist in Hamburg, was trying to make gold from urine. He boiled urine down to a paste and heated the paste to high temperatures. To his great amazement, what he got was not gold, but a white waxy substance that glowed in the dark. This substance was named phosphorus, meaning "light bearer."

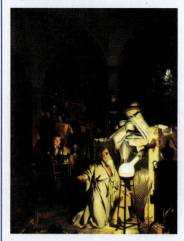

COMMUNICATION EXAMPLE 1   Drawing a Bohr–Rutherford Diagram

Draw a Bohr–Rutherford diagram for the fluorine atom.

There is an easy way to remember how many electrons each orbit can hold. Just look at the periodic table. The first row has 2 elements, and the first orbit holds 2 electrons. The second row has 8 elements, and the second orbit holds 8 electrons. The third row has 8 elements, and the third orbit holds 8 electrons. For elements 19 and 20, place additional electrons in the fourth orbit.

The Periodic Table Meets Bohr–Rutherford

Can the Bohr–Rutherford atomic model explain the patterns in the families of elements in the periodic table? A simple way to test whether the model can explain the evidence is to sketch a "portrait" of each element and then to arrange the elements in their assigned spots on the periodic table. The next step is to examine whether any pattern or "family resemblance" emerges.

TRY THIS   FAMILY RESEMBLANCES IN THE PERIODIC TABLE

SKILLS: Performing, Analyzing, Communicating

Draw a "portrait" of each element in the family for the first 20 elements to see if there are any patterns of similarities in elemental families.

Equipment and Materials: periodic table; paper; pen or pencil

1. Make a blank periodic table for the first 20 elements (Figure 10).

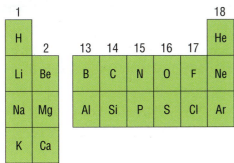

Figure 10 Draw a Bohr–Rutherford diagram for each of these elements.

2. In each square of the table, draw a Bohr–Rutherford diagram of the element indicated. Use a periodic table to find the atomic number and mass number of the most common isotope of each element. Recall that the first 3 electron orbits can hold a maximum of 2, 8, and 8 electrons, respectively. The lower orbits (closest to the nucleus) must be completely filled before filling the higher orbits.

A. What similarities and differences, if any, do you see in the Bohr–Rutherford diagrams for elements within the family of
   (i) the noble gases?
   (ii) the alkali metals?
   (iii) the alkaline earth metals?
   (iv) the halogens?

B. How do the electron arrangements differ
   (i) between the alkali metals and the noble gases?
   (ii) between the halogens and the noble gases?
   (iii) between the alkaline earth metals and the alkali metals?

6.7 Explaining the Periodic Table

DID YOU KNOW?
Poisoning of a Spy
In 2006, 43-year-old Russian spy Alexander Litvinenko was poisoned with a radioactive substance called polonium-210. Extraordinarily high levels of radiation were found in his urine, in his apartment, and at a restaurant where he ate his last lunch. Marie Curie discovered polonium in 1898 and named it after her native land, Poland. Polonium-210 is an element that contains 84 protons and 126 neutrons. Such a high number of protons and neutrons makes its nucleus very crowded and therefore unstable. Litvinenko ate polonium nuclei, which released energy and particles that were absorbed into his tissues. This polonium radiation caused extensive tissue damage that killed Litvinenko.

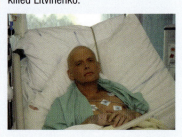

Patterns in the Periodic Table

Did you notice any pattern emerging from the periodic table of Bohr–Rutherford diagrams? Mendeleev would have been fascinated to see such a startling pattern from these "family portraits" of the elements.

- As you go down each family, the number of electron orbits increases—a new orbit is added with each new row. For example, in the alkaline earth metal family, Be has two orbits, Mg has three orbits, and so on.

- Within each family, all atoms have the same number of electrons in their outermost orbits. From the alkali metal family, Li has one outer electron, as do Na, K, and so on. This electron arrangement explains why the reaction of the alkali metals with water becomes more vigorous as you go down the group from lithium to potassium. Evidence suggests that when an alkali metal reacts with water, the alkali metal atoms lose one electron. The most likely electron to be lost is the single electron in the outermost orbit (Figure 11). This electron is farthest from the nucleus, so it has the weakest attraction to the nucleus. Furthermore, since the outermost electron of the sodium atom is farther from the nucleus than the outermost electron of the lithium atom, sodium reacts faster with water than lithium does. So, the reactivity of the alkali metals increases as you go down Group 1.

- The alkali metals undergo similar reactions because they all have one electron in their outermost orbits.

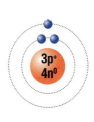 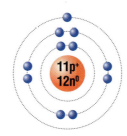 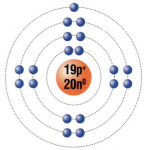

Figure 11 These elements have similar properties because they each have one outer electron.

READING TIP
Use Prior Knowledge
Think about what you already know about an example or description to help you make inferences. For example, you know how bulky a sweatshirt can be, so you can visualize how big a student wearing several sweatshirts would appear. The analogy can help you understand a pattern in the periodic table.

This pattern is similar to a classroom of students who are all wearing sweatshirts with pockets filled with apples. As you go down the column, each student is wearing one additional sweatshirt, so students at the end of the column appear larger than students at the front. Since you can only see their outermost sweatshirts, students in the same column all appear to have the same number of apples in their pockets. The next question is: Does having the same number of electrons in the outermost orbit explain why other groups of elements also have similar properties? You will explore this question in the next chapter.

The Bohr–Rutherford model of the atom is a useful tool to explain the properties of the elements on the periodic table. Table 1 on the next page summarizes some of the evidence you have learned about atomic structure and how the Bohr–Rutherford model explains this evidence.

Table 1 Theory Meets Evidence

Evidence	Theory
Elements have unique properties.	• Each element is made of different atoms.
Some objects can attract each other.	• All atoms contain three subatomic particles: protons, neutrons, and electrons. • Atoms are electrically neutral because they have an equal number of protons and electrons. • Matter is positively charged if it contains more protons than electrons. • Matter is negatively charged if it contains more electrons than protons. • Objects that attract each other are oppositely charged.
In Rutherford's gold foil experiment, positively charged particles were fired at a sample of gold atoms. • Most particles passed through the foil. • Some particles were deflected at large angles.	• Most of the volume of the atom is empty space. • The centre of the atom is a very small, dense, positively charged core called the nucleus.
Elements of the same chemical family have similar properties.	• Electrons exist in specific orbits around the nucleus. • The number of electrons in the outermost orbit determines many elemental properties. • Elements of the same chemical family have the same number of electrons in their outermost orbits.
Each element gives off a unique pattern of coloured lines when it is excited by energy.	• Each electron orbit has a definite amount of energy. • Electrons that absorb energy jump to higher orbits. • Electrons emit energy when they return to lower orbits. • The colour of light observed corresponds to the energy difference between the two electron orbits.

IN SUMMARY

- Atomic number represents the number of protons in an atom. This number is unique to each element.
- In a neutral atom, the number of electrons equals the number of protons.
- Atomic mass is the mass of an atom measured in atomic mass units, u.
- Mass number represents the total number of protons and neutrons in an atom.
- The maximum number of electrons in the first, second, and third orbits is 2, 8, and 8, respectively. Electrons fill lower orbits before filling higher orbits.
- Bohr–Rutherford diagrams of atoms show the number and location of protons, neutrons, and electrons in the atom.
- All atoms within each family of elements have the same number of electrons in their outermost orbits.

CHECK YOUR LEARNING

1. Do all atoms of the same element contain the same number of protons? Explain. K/U
2. Do all atoms of the same element contain the same number of neutrons? Explain. K/U
3. Which of the following statements are correct? Rewrite incorrect statements to correct them. K/U
   (a) The atomic number always equals the number of protons.
   (b) The atomic mass can be smaller than the atomic number.
   (c) The mass number of an atom can be equal to the atomic number.
   (d) The number of protons always equals the number of neutrons in an atom.
   (e) The number of protons always equals the number of electrons in an atom.
   (f) If the number of protons in an atom were changed, it would be a different element.
4. Determine the number of neutrons in the most common isotope of nickel. T/I
5. What is the maximum number of electrons that each of the first three electron orbits can hold, starting from the nucleus? K/U
6. Use a periodic table to complete Table 2 with the correct information about each atom. Round atomic mass to the nearest whole number to obtain the mass number. T/I
7. Draw Bohr–Rutherford diagrams for each of the following atoms. T/I
   (a) sulfur-32
   (b) argon-40
   (c) potassium-39
   (d) oxygen-16
8. Which of the following statements are true? Rewrite false statements to make them true. K/U
   (a) The first four elements in the alkali metal family have the same number of electron orbits.
   (b) The first three elements in the noble gas family have full outermost electron orbits.
   (c) All elements in the second row of the periodic table have the same number of electron orbits.
   (d) The first three elements in the alkaline earth metal family have the same number of electrons in their outer orbits.
9. What evidence is there that chemical properties of elements are related to the number of electrons in the outermost orbit? Choose one chemical family to support your answer. A
10. In Section 6.1, you learned that an element is a pure substance that cannot be broken down into a simpler chemical substance by any physical or chemical means. Write a new definition of an element based on what you have learned in this section. K/U C

Table 2

Element name	Element symbol	Atomic number	Mass number	Number of protons	Number of neutrons	Number of electrons
magnesium						
	Al					
		15				
				50		
						47

6.8 From Charcoal to Diamonds

What do charcoal, a pencil, and a diamond ring have in common (Figure 1)? They are not alike in appearance and certainly quite different in price. Charcoal briquettes are black and opaque, crumble easily, and burn readily. Pencil "lead" is soft enough to leave a trail of black writing when it is gently pressed onto paper. White diamonds are colourless and transparent, sparkle brilliantly, and rank at the top of the hardness scale—and the price scale!

Figure 1 Charcoal, pencil "lead," and diamonds are all made of carbon atoms.

Surprisingly, all three of these substances are mostly made of atoms of exactly the same element: carbon. They differ only in how the carbon atoms are arranged. That difference accounts for their individual properties.

Charcoal

Unlike most other elements, carbon atoms can join with other carbon atoms almost indefinitely, forming unending structures. When this happens, a shapeless, disorganized arrangement of atoms is formed, creating a soft black solid—like the charcoal briquettes that we use in our barbecues. Charcoal consists of up to 98 % carbon mixed with ash and other chemicals.

Graphite

Graphite has an organized structure compared to charcoal. Each carbon atom in graphite joins with three other carbon atoms to form a sheet of interconnected hexagons. A carbon atom is located at each corner of the hexagon. These flat sheets are loosely layered on top of each other (Figure 2). This form of carbon is called graphite. Pencil "lead" is actually graphite. Under slight pressure, the carbon sheets slide across each other, leaving behind the top layer of carbon atoms on the surface of the writing paper. Pencil lead is more accurately named "pencil graphite mixed with clay." The carbon structure of graphite also makes it a good conductor, which is an unusual property for non-metals.

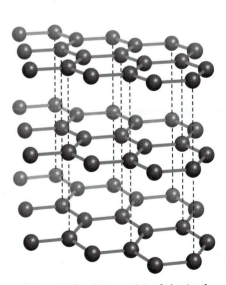

Figure 2 Graphite consists of sheets of carbon atoms.

TRY THIS: COPPER-PLATE YOUR PENCIL

SKILLS: Performing, Observing, Analyzing, Communicating

SKILLS HANDBOOK
1., 3.B.6.

In this activity, you will coat the tip of a pencil with copper. The method you will use is called electroplating—using electrical energy to plate a metal onto a surface. Many beautiful objects, such as silver trays and gold jewellery, are electroplated. An electric current has to run through the object being plated, so the object must be able to conduct electricity.

Equipment and Materials: eye protection; apron; 9 V battery; medium-sized beaker; graduated cylinder; 2 electrical wires; 2 pencils with both ends sharpened (for example, golf pencils); copper(II) sulfate solution

 Copper(II) sulfate solution is a skin and eye irritant. Wear eye protection and a lab apron.

1. Put on your apron and eye protection.
2. Dissolve approximately 5 mL of copper(II) sulfate in 50 mL of water in the beaker.
3. Sharpen both ends of each pencil and connect the electrical wires to one end of each pencil.
4. Connect the electrical wires to the two terminals of the 9 V battery.
5. Complete the circuit by placing the free tips of the pencils into the copper(II) sulfate solution so that the two pencil tips do not touch (Figure 3). Make sure that the wire connectors are above the solution. Secure the wires to the sides of the beaker.
6. After several minutes, disconnect the wires from the battery and remove the pencils from the solution. Rinse the pencils gently in water and observe the pencil tips.
7. Follow your teacher's instructions for the disposal of the materials and equipment.

Figure 3 Copper-plating a pencil

8. Wash your hands thoroughly with soap and water.
A. What did you observe at the pencil tips while there was current in the circuit? T/I
B. Was there any evidence of copper being plated onto a pencil tip? T/I
C. Where did the copper come from in this experiment? T/I
D. What evidence was there to suggest that there was electrical current in the pencil graphite? T/I
E. Is graphite a metal or a non-metal? Explain your answer. T/I
F. Suggest some applications of graphite that make use of its properties. A

Diamond

Under conditions of extremely high temperature and pressure, carbon atoms arrange themselves into regular patterns that are interconnected in three dimensions. These patterns are similar to a playground climbing frame that never ends. This strongly reinforced framework is what gives diamond its remarkable hardness (Figure 4). The closeness of the atoms makes diamond very dense and also allows it to bend light, producing its much-admired sparkle when diamond is cut. This three-dimensional structure does not allow the free flow of electrons, so diamonds do not conduct electricity.

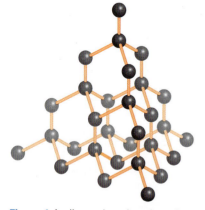

Figure 4 In diamond, each carbon atom is tightly bound to the carbon atoms surrounding it, which accounts for the hardness of diamond.

RESEARCH THIS | ARTIFICIAL DIAMONDS

SKILLS: Researching, Analyzing the Issue, Communicating, Evaluating

SKILLS HANDBOOK
4.A., 4.B.

We usually think of diamonds as jewellery. However, diamonds are also useful in industry because of their hardness. The diamonds used in industry are usually artificial diamonds. In this activity, you will research and report on the industrial use of diamonds. You will also research artificial diamonds and the processes by which they are created.

1. What types of industries use diamond tools?
2. What kinds of processes require the use of diamond tools?
3. How are artificial diamonds manufactured?
4. What are the main applications of artificial diamonds?
5. How are artificial diamonds different from natural diamonds?

GO TO NELSON SCIENCE

A. Artificial diamonds have been available since the 1960s. Should diamond mining be phased out? Give reasons for your answer.
B. List the advantages and disadvantages of artificial diamonds.
C. If you were buying diamond jewellery, would you rather buy real or artificial diamonds? Give reasons for your answer.

Diamond Rush in Canada

In the late 19th century, a gold rush brought tens of thousands of people to the Yukon in search of fortune. Today, mining companies dig for diamonds. Diamonds were first discovered in the Northwest Territories in the 1980s. Since then, Canada has become one of the top three diamond producers in the world (Figure 5). More than ten million carats of Canadian diamonds are produced each year in mines in the Northwest Territories, Nunavut, and Northern Ontario. That is over 15 % of the world's supply!

To learn more about career opportunities in diamond mining in Canada,

GO TO NELSON SCIENCE

Figure 5 The Diavik Diamond Mine on Lac de Gras, Northwest Territories

The diamond rush has benefited the Canadian North in many ways. However, any form of mining disturbs the environment. Layers of topsoil must be removed and carefully replaced to minimize the impact of excavating the land. In the Canadian Arctic, where ecosystems are particularly fragile, strong partnerships between the government and mining companies ensure that strict regulations are closely followed. Compared to gold mining, which uses cyanide, diamond mining has a far smaller environmental imprint. However, it is not without consequences. Diamond mining in Canada's North has resulted in lake drainage and stream destruction. Fish habitat has been lost and water quality has been changed, in some cases irreversibly. Permafrost recovers very slowly, if ever, from damage.

To learn more about becoming an environmental assessment professional,

GO TO NELSON SCIENCE

6.8 From Charcoal to Diamonds

READING TIP

Revise Your Inferences

Sometimes you come across information that conflicts with an inference you have already made. For example, you read that diamond mining creates jobs in Canada and conclude that diamond mining is beneficial. However, after you read about the impact of diamond mining on the cost of living, you may change your opinion.

Diamond mining has societal consequences, which are more difficult to measure. Big profits bring business and highly paid jobs in the diamond industry as well as in construction and transportation. The demand for workers in Yellowknife and Iqaluit has left many jobs unfilled in neighbouring towns and cities. The cost of living has skyrocketed, caused in part by the cost of housing. Limited availability of housing has driven rents to unaffordable levels.

Unlike diamonds mined in parts of Africa, Canadian diamonds are much less controversial. They are not "blood diamonds," the name attached to mining operations that finance war and terror. Canadian diamonds are sold with a certificate that guarantees their source. They are promoted as "conflict free" certified. Each stone is etched with a tiny polar bear trademark (Figure 6).

Figure 6 Each Canadian diamond is etched with a tiny polar bear as a trademark.

IN SUMMARY

- Charcoal, graphite, and diamond are all made of carbon atoms. The carbon atoms are arranged differently in each substance, which accounts for their different properties.

- Diamond mining in Canada's North has brought economic benefits but has raised social and environmental concerns.

CHECK YOUR LEARNING

1. Describe how carbon atoms are joined together in charcoal, graphite, and diamond. K/U

2. Explain how a pencil makes a mark in terms of the arrangement of carbon atoms in graphite. K/U

3. What are some of the economic benefits of diamond mining in Canada's North? A

4. What are some of the environmental drawbacks of diamond mining in Canada's North? A

5. What properties of diamonds make them useful in a variety of applications? K/U

6. Explain which form of carbon is best suited for the following functions: A

   (a) an electrode in a battery

   (b) the tip on a drill bit for drilling through rocks

   (c) fuel in power plants

TECH CONNECT ✓ OSSLT

Stronger Than a Speeding Bullet

It's a vest! It's armour! It's…Kevlar! It is exactly what Superman would wear to protect himself against bullets if he ever lost his super powers. Kevlar® is a fibre that has super properties. It is light as cotton but strong as steel—strong enough to be used in bulletproof vests (Figure 1). It is made of long chains of carbon atoms, with hydrogen, nitrogen, and oxygen atoms attached (Figure 2). Kevlar's extraordinary strength comes from the unique linkages between the chains. It was originally developed by Stephanie Kwolek at DuPont's Pioneering Lab in 1965. Kevlar is now incorporated into numerous everyday products.

Figure 3 A kayak made of Kevlar

Figure 1 A bulletproof vest made of Kevlar

Figure 2 Kevlar is made of long chains of elements spun together into sheets, which are in turn stacked together to form strong fibres. These fibres are woven together to make an extraordinarily tough material—tough enough to stop a speeding bullet!

Hockey sticks made of Kevlar are lightweight and keep their shape better than sticks made of fibreglass or wood. Kevlar sticks break cleanly, without the jagged, sharp edges that can cause injuries. Kevlar is also woven into the nets that protect fans from flying hockey pucks. Sails made of Kelvar are light and strong (Figure 4).

Figure 4 A Kevlar sail for a racing boat

What types of products benefit from Kevlar's strength and lightness? Racing bike helmets are reinforced with Kevlar, allowing for deeper channelled vents, increased ventilation, and minimal weight, with no loss in security. Your feet can feel feather-light in hiking boots that can endure rough use over any terrain. On the ski slopes, you can be fitted out in Kevlar boots, poles, gloves, and even snowboards for excellent speed, stability, and manoeuvrability. Kevlar is added to skateboards to prevent their breakage and prolong their "pop." It is also added to kayaks to reduce weight and increase speed. (Figure 3).

Kevlar saves lives in auto racing. Kevlar supports are used to restrain the driver's head and neck because of their impact-absorbing properties. Kevlar does not shatter, so race car bodies reinforced with it do not leave deadly debris after a crash. This reduces the chance of injury from sharp flying fragments. In Formula 1 cars, the wheels are also strapped to the car with Kevlar to prevent them from bouncing off the track into the spectator stands.

We may not be superheroes, but we can equip ourselves from head to toe with a super fibre made of carbon, hydrogen, nitrogen, and oxygen!

CHAPTER 6 LOOKING BACK

KEY CONCEPTS SUMMARY

Elements cannot be broken down into simpler substances.

- Elements are the building blocks of all substances. (6.1)
- All substances on Earth are made of one or more elements. (6.1, 6.3)
- Elements are arranged on the periodic table. (6.1)
- A compound is a pure substance composed of two or more different elements that are chemically joined. (6.1, 6.3)

Metals and non-metals have characteristic physical properties.

- Metals are elements that have lustre, and are conductors, malleable, and ductile. (6.1, 6.2)
- Non-metals are elements that are mostly gases or dull, powdery solids. (6.1, 6.2)

Elements are organized according to their atomic number and electron arrangement on the periodic table.

- The columns of the periodic table represent groups or families of elements with similar properties. (6.4, 6.5, 6.7)
- Elements in the same family have the same number of electrons in their outermost orbits. (6.7)

Atomic models evolved as a result of experimental evidence.

- John Dalton proposed that matter is made of tiny indivisible particles called atoms. (6.6)
- J. J. Thomson proposed that atoms are positively charged, with negatively charged electrons evenly distributed within them. (6.6)
- Ernest Rutherford proposed that atoms are mostly empty space, with a positively charged centre that is surrounded by negatively charged electrons. (6.6)
- Niels Bohr proposed that electrons occupy fixed orbits around the nucleus. (6.6)

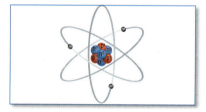

Atoms contain protons and neutrons in a central core surrounded by electrons.

- The Bohr–Rutherford model of the atom consists of positively charged protons and neutral neutrons in the nucleus and negatively charged electrons orbiting the nucleus. (6.6)

Elements can be both beneficial and harmful to humans and to the environment.

- Some elements are harmful to humans in very small quantities. (6.1, 6.4, 6.5, 6.7)
- Some elements are essential nutrients that humans and other living things need in order to survive. (6.1, 6.4)
- Some elements are beneficial to humans, but their extraction can be harmful to the environment. (6.8)

WHAT DO YOU THINK NOW?

You thought about the following statements at the beginning of the chapter. You may have encountered these ideas in school, at home, or in the world around you. Consider them again and decide whether you agree or disagree with each one.

1 Water is a common element.
Agree/disagree?

4 It is easy to change one element into another.
Agree/disagree?

2 Gold, silver, and copper are all metals because they are shiny and flexible.
Agree/disagree?

5 "Harmful" is a relative term: a harmful amount of a substance for one person may not be harmful for another person.
Agree/disagree?

3 Elements are arranged alphabetically on the periodic table.
Agree/disagree?

6 Scientific theories are educated guesses that have been proven to be correct.
Agree/disagree?

How have your answers changed since then? What new understanding do you have?

Vocabulary

element (p. 211)
element symbol (p. 211)
compound (p. 211)
metal (p. 212)
non-metal (p. 213)
metalloid (p. 213)
chemical family (p. 220)
alkali metal (p. 220)
alkaline earth metal (p. 221)
noble gas (p. 221)
halogen (p. 222)
period (p. 222)
atom (p. 228)
electron (p. 229)
proton (p. 230)
neutron (p. 231)
atomic number (p. 234)
mass number (p. 235)
isotope (p. 235)
atomic mass (p. 235)
Bohr–Rutherford diagram (p. 236)

BIG Ideas

✓ Elements and compounds have specific physical and chemical properties that determine their practical uses.

✓ The use of elements and compounds has both positive and negative effects on society and the environment.

Looking Back 247

CHAPTER 6 REVIEW

The following icons indicate the Achievement Chart category addressed by each question.
K/U Knowledge/Understanding
T/I Thinking/Investigation
C Communication
A Application

What Do You Remember?

1. Name the chemical family to which each of the following elements belongs. Describe the location of each named group on the periodic table. (6.1, 6.4) K/U
   (a) iodine
   (b) barium
   (c) francium
   (d) krypton

2. What is the significance of the bold staircase line on the periodic table? (6.1) K/U

3. Who designed the early periodic table on which the modern one is based? (6.4) K/U

4. What is the basis of the way in which the elements are organized on the periodic table? (6.4, 6.7) K/U

5. Explain the difference between the terms in each pair below. (6.1, 6.4, 6.6, 6.7) K/U
   (a) an element and a compound
   (b) a group and a period
   (c) a metal and a non-metal
   (d) a proton and a neutron
   (e) a proton and an electron
   (f) atomic number and atomic mass

6. Explain the similarities and differences between the plum pudding model and the Bohr model of the atom. Draw a sketch to illustrate your answer. (6.6) K/U C

7. What did the surprising results of Rutherford's gold foil experiment lead him to conclude? Explain. (6.6) K/U

What Do You Understand?

8. In Dalton's atomic model, atoms are the smallest possible particles of an element, and they are indivisible. Explain whether our modern atomic model supports or refutes Dalton's model. (6.6) K/U

9. Describe the experimental evidence that led J. J. Thomson to propose that atoms contain particles that are negatively charged. (6.6) K/U

10. In your notebook, copy and complete Table 1. Round atomic masses to whole numbers to obtain the mass number. (6.7) T/I

    Table 1

Element name	Element symbol	Atomic number	Mass number	Number of protons	Number of neutrons	Number of electrons
	Mn					
			73			
mercury						
						36
				15	16	

11. Draw Bohr–Rutherford diagrams for the most common isotope of each of the following in the sequence listed. (6.7) T/I C
    (a) the first three elements in the alkaline earth metal family
    (b) Ne, Ar, Kr
    (c) Li, Be, B, C, N, O, F, Ne
    (d) F, Cl

12. Describe and explain any identifiable patterns that emerge from the series of Bohr–Rutherford diagrams in question 11. (6.7) T/I

13. Complete a summary of similarities and differences in the structure and properties among
    (a) elements in the same family, such as Be, Mg, Ca
    (b) elements in the same period
    (c) elements in the family of noble gases
    (6.2, 6.4, 6.5, 6.7) K/U

14. In the Bohr model of the atom, electrons are found in orbits around the nucleus. (6.6, 6.7) K/U T/I
    (a) What is the maximum number of electrons that can be found in the first three orbits closest to the nucleus?
    (b) Count the number of elements in the first three periods of the periodic table. What is the correlation of this number to the number of electrons in Bohr's orbits?
    (c) Suggest an explanation for the correlation between the rows in the periodic table and the electron orbits.

15. Elements in the same family exhibit similar properties, often to varying degrees. Refer to each of the following families to support the claim that, as far as physical and chemical properties go, "it's all about the electrons." (6.7) K/U T/I
    (a) the alkali metals
    (b) the halogens
    (c) the noble gases

16. Considering our current understanding of atomic structure, why does the atom remain a "black box" to us? (6.6) A

17. As a student, you often work with other students in group activities, sharing ideas and solving problems together. Explain how the history of the development of atomic theory is an example of the importance of collaboration. (6.6) A

18. Theories and models of the structure of matter have changed over time. Most advances are made when scientists use the scientific method. In the investigations that you have designed, in what ways do you incorporate the scientific steps of questioning, predicting, testing, and analyzing? Give an example of each to illustrate your answer. (6.3) T/I A C

Solve a Problem

19. Three different solids, A, B, and C, each with a metallic lustre, were combined individually with water and with an acid. The following observations were made (Table 2):

    Table 2

Solid	Reaction with water	Reaction with acid
A	no change	bubbled slightly
B	no change	no change
C	bubbled slightly	bubbled vigorously

    (a) Which solid was the most reactive? Which was the least reactive?
    (b) If these solids belong to the same chemical family, which solid would you place highest in the column? Which would you place lowest?
    (c) Imagine another solid, X, which is correctly placed between the two top solids in your answer to (b). Predict how solid X would react with water and with acid. Give reasons for your answer. (6.5, 6.7) T/I

Create and Evaluate

20. You have found the following pieces in an old puzzle box. (6.4) T/I A

    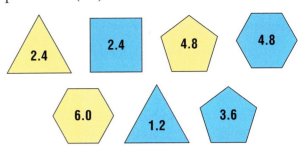

    (a) Consider the different features of the pieces and propose an arrangement that incorporates all these features in a systematic way. Explain why you chose this arrangement. T/I C
    (b) According to your arrangement, is there a possible missing piece? Explain. T/I C
    (c) Predict the shape, colour, and number of one or more additional pieces that would fit your proposed arrangement. The predicted pieces may be filling a gap or extending the pattern you have selected. T/I
    (d) Compare the process you have used in this activity to Mendeleev's arrangement of elements in the periodic table. A

Reflect on Your Learning

21. Does the scientific process require more experimentation or creative thinking? Reflect on what you learned in this chapter and give examples where one or both types of thinking were used.

Web Connections

22. What are the possible disadvantages of Kevlar? Is it biodegradable or environmentally friendly? Is the unregulated addition of Kevlar to consumer goods a wise decision? A

23. Research and report on one discarded (obsolete) scientific theory. Describe the theory and the main reasons it was discarded. Examples include Lamarckism, phlogiston, caloric theory of heat, miasma theory of disease, and the theory of the geocentric universe. T/I A

CHAPTER 6 SELF-QUIZ

The following icons indicate the Achievement Chart category addressed by each question.

K/U Knowledge/Understanding
T/I Thinking/Investigation
C Communication
A Application

For each question, select the best answer from the four alternatives.

1. Which of the following correctly describes the atomic number of an element? (6.6) K/U
   (a) the number of protons in the atom's nucleus
   (b) the number of electrons in the atom's nucleus
   (c) the number of neutrons in the atom's nucleus
   (d) the number of protons, neutrons, and electrons in the atom's nucleus

2. Which phrase best describes a group or family in the periodic table? (6.7) K/U
   (a) elements that have the same melting point
   (b) elements that have the same number of protons in their nucleus
   (c) elements that are all in the same physical state at room temperature
   (d) elements that have the same number of electrons in their outermost orbits

3. From the information in the periodic table, how many neutrons does the most common isotope of carbon have? (6.7) K/U
   (a) 1
   (b) 6
   (c) 12
   (d) 14

4. Which of the following is a compound? (6.1) K/U
   (a) sodium
   (b) chlorine
   (c) water
   (d) oxygen

Indicate whether each of the statements is TRUE or FALSE. If you think the statement is false, rewrite it to make it true.

5. In the periodic table we use today, the elements are arranged by increasing mass. (6.7) K/U

6. Diamonds, the graphite used in pencils, and charcoal are made mostly of carbon atoms. (6.8) K/U

Copy each of the following statements into your notebook. Fill in the blanks with a word or phrase that correctly completes the sentence.

7. Elements classified as metalloids have properties of both _____ and _____. (6.1) K/U

8. The elements in group 17 of the periodic table, such as fluorine and chlorine, are also called _____. (6.4) K/U

9. _____ of an element have the same number of protons but different numbers of neutrons. (6.7) K/U

Match each term on the left with the most appropriate description on the right.

10. (a) atom   (i) a positively charged particle
    (b) proton   (ii) a pure substance listed on the periodic table
    (c) element   (iii) a negatively charged particle
    (d) electron   (iv) a particle that has no charge
    (e) neutron   (v) the smallest unit of an element (6.6) K/U

Write a short answer to each of these questions.

11. Explain the difference between a group and a period on the periodic table. (6.4) K/U

12. (a) A particle contains more electrons than protons. What type of charge does the particle have? (6.7) K/U
    (b) Can the particle described in part (a) be called an atom? Why or why not?

13. Which of the following Group 17 elements would you expect to be the most reactive: chlorine, fluorine, or bromine? Explain your choice based on the arrangement of the element's electrons. (6.7) T/I

14. The mass number of a certain element is 195. The most common isotope of that element has 117 neutrons in each of its atoms. (6.7) T/I
    (a) How many protons does an atom of this element contain? Explain your answer.
    (b) Identify the element.

15. In your own words, explain to a friend the significance of Rutherford's gold foil experiment. (6.6) C

16. How many electron orbits would you expect to find in an atom of aluminum? Explain your reasoning. (6.7) T/I

17. (a) How did Thomson's and Rutherford's experiments provide evidence that part of Dalton's atomic theory was incorrect?
    (b) Was Dalton's atomic theory discarded after Thomson's and Rutherford's experiments? Why or why not? (6.6) K/U T/I

18. Why are electron orbits sometimes referred to as energy levels? (6.6) T/I

19. Many pans used for cooking are made of metal because metals easily conduct thermal energy. (6.1) A
    (a) Why are the handles of these pots usually not made of metal?
    (b) What kind of material would be suitable for the handles of cookware?

20. Dentists sometimes use gold or silver to fill cavities in teeth caused by tooth decay. (6.1) A
    (a) Give at least two reasons why these materials are good choices for tooth fillings.
    (b) Why might these materials not be good choices for tooth fillings?

21. Suppose your science class were to hold an election to decide which element is the most useful chemical element on the periodic table. Choose one element and create an advertisement that illustrates why the chemical element is useful. (6.1, 6.4, 6.8) C

22. How many neutrons would you expect to find in the atom of the most common isotope of hydrogen, H-1? Explain your reasoning. (6.7) K/U T/I

23. In the past, some airships were filled with hydrogen because hydrogen is less dense than air. An airship filled with hydrogen will rise up and easily float through the air. Now airships are filled with helium, which is also less dense than air. Based on their positions on the periodic table, why is helium a better choice than hydrogen to use in airships? (6.7) A

CHAPTER 7

Chemical Compounds

KEY QUESTION: How do we handle environmental problems when things corrode or do not decompose?

Garbage washed up onto Funafuti Atoll, in Tuvalu, Oceania. This tiny island is only 4.5 m above sea level at its highest point. Tuvalu may be the first country to disappear as a result of rising sea levels caused by global warming. During the highest tides, sea water pours through the coral atoll and floods low-lying areas, bringing with it garbage from the Great Pacific Garbage Patch.

UNIT C
Atoms, Elements, and Compounds

CHAPTER 5 — Properties of Matter

CHAPTER 6 — Elements and the Periodic Table

CHAPTER 7 — Chemical Compounds

KEY CONCEPTS

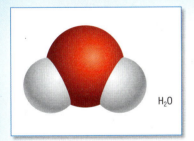

Elements can combine to form compounds.

Metals and non-metals combine to form ionic compounds. Non-metals combine to form molecules.

Some useful compounds have social, environmental, and economic impacts.

Compounds can break apart into simpler substances.

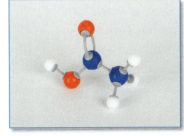

Molecular models are used to represent molecules.

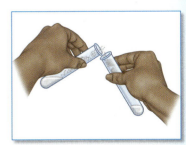

Simple chemical tests can identify common gases.

Looking Ahead

ENGAGE IN SCIENCE

THE ATTACK OF OXYGEN

Countries around the world are all under attack! Nothing is safe against powerful oxygen atoms! When iron or copper atoms meet these highly reactive oxygen atoms, the results are rust or corrosion, respectively. The cost of rusting to our cars and ships, our bridges and monuments is, well... simply monumental.

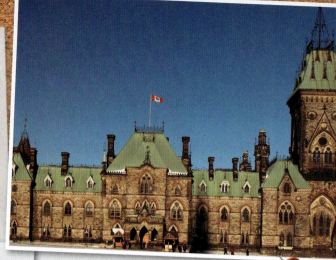

Touching Up the Eiffel Tower

The Eiffel Tower was built for the International Exhibition of Paris in 1889. It is built of cast iron, a metal that readily forms a compound that we call rust. As a result, all 220 000 m² of the iron tower must be painted every seven years! Each painting job requires 60 tonnes of paint, applied by 25 painters working for 15 months, at a cost of five million dollars!

Although oxygen is the main culprit in the rusting process, it needs a partner—water. If the water contains salt, rusting happens even more readily. Here is a tip on how to keep your bicycle rust free: keep it in the moisture-free desert!

Canada's Parliament Buildings

Corrosion is not always a bad thing. Sometimes it can have a protective effect. Consider the Parliament buildings in our nation's capital. The roofs of these historical buildings were originally covered in copper. When this metal is exposed to oxygen in damp air, the copper changes colour from red-orange to brown-black, forming copper oxide. After many years of exposure to air, the copper oxide becomes coated with a green film, copper carbonate. This compound gives the roofs their distinctive green copper film, or patina, also called verdigris. Patina protects the copper from further corrosion.

The Statue of Liberty and Her Ribs of Iron

"Give me your tired, your poor, Your huddled masses yearning to breathe free, The wretched refuse of your teeming shore, Send these, the homeless, tempest-tossed to me, I lift my lamp beside the golden door!" Thus cries the mighty woman with a torch, the symbol of freedom and friendship. The Statue of Liberty was built in Paris in 1884, moved to the United States in 1885 in 214 crates, and dedicated in 1886. The "lady" is constructed of vertical and horizontal "ribs" of iron, covered by a "skin" of copper plates. An inspection in 1981 revealed severe corrosion of the copper skin. Exposure to moisture and to oxygen in the air had caused the iron framework to rust and become weak and brittle. Extensive repairs to the Statue of Liberty began in 1984. The Statue was reopened on July 5, 1986 in a jubilant celebration of the lady's 100th birthday.

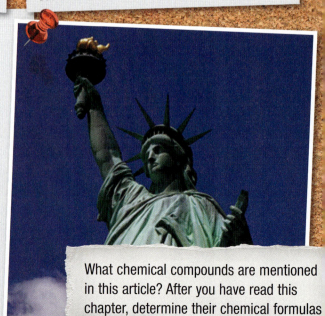

What chemical compounds are mentioned in this article? After you have read this chapter, determine their chemical formulas

WHAT DO YOU THINK?

Many of the ideas you will explore in this chapter are ideas that you have already encountered. You may have encountered these ideas in school, at home, or in the world around you. Not all of the following statements are true. Consider each statement and decide whether you agree or disagree with it.

1 Oxygen can be both beneficial and harmful.
Agree/disagree?

4 All compounds will eventually break down over several years.
Agree/disagree?

2 When oxygen reacts with a substance, flames are always produced.
Agree/disagree?

5 Compounds have different properties than their elements.
Agree/disagree?

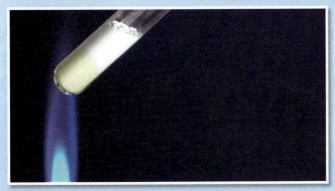

3 Compounds must be heated to break down into simpler substances.
Agree/disagree?

6 The chemical products we use affect only local ecosystems.
Agree/disagree?

FOCUS ON WRITING

Writing a Science Report

When you write a science report you use a specific format with organizational headings to explain to a reader the purpose, procedure, and findings of your investigation. Use the strategies listed next to the report to improve your report-writing skills.

WRITING TIP

As you work through the chapter, look for tips like this. They will help you develop literacy strategies.

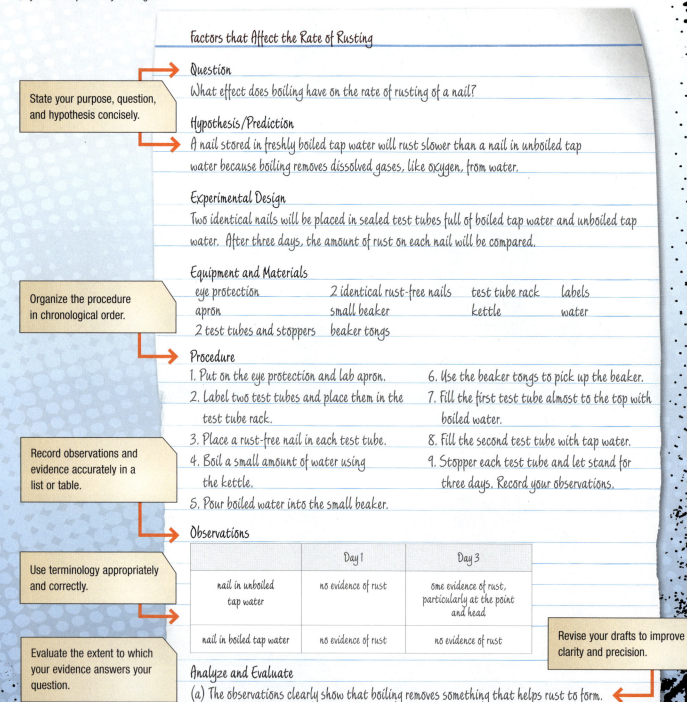

State your purpose, question, and hypothesis concisely.

Organize the procedure in chronological order.

Record observations and evidence accurately in a list or table.

Use terminology appropriately and correctly.

Evaluate the extent to which your evidence answers your question.

Revise your drafts to improve clarity and precision.

Factors that Affect the Rate of Rusting

Question
What effect does boiling have on the rate of rusting of a nail?

Hypothesis/Prediction
A nail stored in freshly boiled tap water will rust slower than a nail in unboiled tap water because boiling removes dissolved gases, like oxygen, from water.

Experimental Design
Two identical nails will be placed in sealed test tubes full of boiled tap water and unboiled tap water. After three days, the amount of rust on each nail will be compared.

Equipment and Materials
- eye protection
- apron
- 2 test tubes and stoppers
- 2 identical rust-free nails
- small beaker
- beaker tongs
- test tube rack
- kettle
- labels
- water

Procedure
1. Put on the eye protection and lab apron.
2. Label two test tubes and place them in the test tube rack.
3. Place a rust-free nail in each test tube.
4. Boil a small amount of water using the kettle.
5. Pour boiled water into the small beaker.
6. Use the beaker tongs to pick up the beaker.
7. Fill the first test tube almost to the top with boiled water.
8. Fill the second test tube with tap water.
9. Stopper each test tube and let stand for three days. Record your observations.

Observations

	Day 1	Day 3
nail in unboiled tap water	no evidence of rust	some evidence of rust, particularly at the point and head
nail in boiled tap water	no evidence of rust	no evidence of rust

Analyze and Evaluate
(a) The observations clearly show that boiling removes something that helps rust to form. Boiling probably does not remove dissolved ions since ionic compounds have large melting points. However, boiling might remove dissolved gases like oxygen and carbon dioxide.

7.1

Putting Atoms Together

Most things are not made of individual atoms. Atoms can chemically join with other atoms to form small units called **molecules**. For example, our air contains many kinds of molecules. It is composed of roughly 80 % nitrogen molecules, 20 % oxygen molecules, and small amounts of water molecules and carbon dioxide molecules.

Nitrogen molecules are composed of two nitrogen atoms joined together (Figure 1(a)). The chemical formula for a single nitrogen molecule is N_2. A **chemical formula** is the notation used to indicate the type and number of atoms in a pure substance. (Recall from Section 5.1 that a pure substance is made up of only one type of particle.) The subscript 2 indicates that there are two atoms of the element preceding it. The chemical formula for an oxygen molecule is O_2, indicating that there are two oxygen atoms joined together in this molecule (Figure 1(b)).

molecule two or more atoms of the same or different elements that are chemically joined together in a unit

chemical formula notation that indicates the type and number of atoms in a pure substance

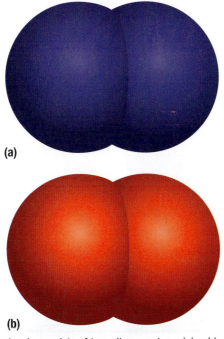

Figure 1 (a) A nitrogen molecule consists of two nitrogen atoms joined together. (b) An oxygen molecule consists of two oxygen atoms joined together.

WRITING TIP

Writing a Science Report
Before you begin writing a science report, say the purpose of the report aloud as clearly as possible in your own words. Turn this statement into a question to use at the beginning of the report.

Molecular Elements

Like a nitrogen atom, a nitrogen molecule, N_2, is an element. There is only one type of atom in this molecule—nitrogen—so N_2 is a **molecular element**, not a compound. (Recall from Section 6.1 that a compound is a pure substance composed of two or more *different* elements that are chemically joined.)

There are seven elements that form molecules consisting of two atoms. These molecular elements are commonly called diatomic molecules, where the prefix *di-* means two. The seven diatomic molecules are H_2, N_2, O_2, F_2, Cl_2, Br_2, I_2.

molecular element a molecule consisting of atoms of the same element

LEARNING TIP

Remembering the Diatomic Molecules
To remember the seven diatomic molecules H_2, O_2, F_2, Br_2, I_2, N_2, Cl_2, use the nonsensical word HOF BrINCl, which rhymes with "Bullwinkle."

7.1 Putting Atoms Together 257

molecular compound a molecule that consists of two or more different elements

To learn more about careers in bioscience,
GO TO NELSON SCIENCE

LEARNING TIP

Element or Compound?
A molecule can be an element or a compound. A molecule is an element when it is composed of only one type of atom. A molecule is a compound when it is composed of two or more different types of atoms.

Molecular Compounds

Most molecules contain more than one type of element. These molecules are called **molecular compounds**. The chemical formula for the molecular compound water (H_2O) tells us that each water molecule contains two hydrogen atoms and one oxygen atom (Figure 2). If a letter in a chemical formula does not have a subscript, it means there is only one atom of that element. Carbon dioxide (CO_2) is a compound that contains one carbon atom and two oxygen atoms in each molecule. The chemical formula for table sugar that you stir into your coffee or tea is $C_{12}H_{22}O_{11}$ (Figure 3). Each molecule of sugar contains 12 carbon atoms, 22 hydrogen atoms, and 11 oxygen atoms. If that sounds like a large molecule, try counting the thousands of atoms in a molecule of protein or DNA!

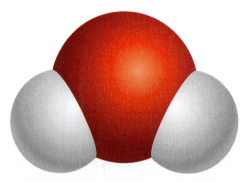

Figure 2 Water, H_2O, is an example of a molecular compound.

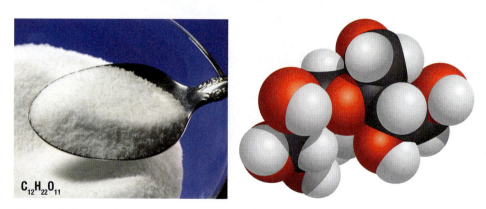

Figure 3 Molecules of sugar consist of atoms of carbon, hydrogen, and oxygen.

Only certain combinations of atoms are found in nature. Atoms of the same element combine in different ratios to form different substances. For example, oxygen atoms can combine in pairs to form oxygen molecules (O_2). They can also combine in triplets to form ozone molecules (O_3). Ozone is an entirely different substance from oxygen. We must breathe in O_2 molecules to keep our cells alive. Ozone is a highly reactive molecule, typically found 10 km to 50 km above the surface of Earth. It filters out harmful ultraviolet rays from the Sun. At ground level, ozone is a harmful air pollutant. It is added to ground-level air by vehicle exhaust and industrial activities.

Atoms of carbon and hydrogen join together in many combinations to form molecules with very different properties. The gas that you use at school in Bunsen burners consists mostly of methane (CH_4). Propane (C_3H_8) is used in barbecue tanks (Figure 4(a)). Octane, C_8H_{18}, is a component of gasoline. Many common substances are composed of three common types of atoms—carbon, hydrogen, and oxygen—combined in different ratios and with different structures. For example, C_2H_6O (alcohol) is found in wine, and $C_2H_4O_2$ (vinegar) can be put on French fries (Figure 4(b)). $C_9H_8O_4$ (Aspirin) can help prevent strokes.

LEARNING TIP

Understanding Molecular Models
In these molecular models, the white spheres represent hydrogen atoms, the black spheres represent carbon atoms, and the red spheres represent oxygen atoms.

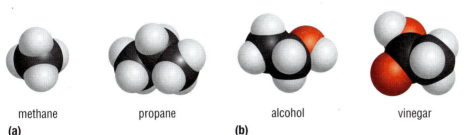

methane   propane   alcohol   vinegar
(a)   (b)

Figure 4 Molecules made from different combinations of the same elements can have widely different properties.

TRY THIS: RUSTING STEEL SLOWLY

SKILLS: Predicting, Controlling Variables, Performing, Observing, Analyzing, Communicating

SKILLS HANDBOOK
3.B.6., 3.B.7.

When iron reacts with oxygen and water, a new compound forms: iron oxide (Fe_2O_3) commonly known as rust. This reaction can happen at different speeds depending on the availability of the reactants. In this activity, you will use a low concentration of oxygen. You will obtain oxygen from the air, which contains about 20 % oxygen.

Equipment and Materials: 2 large test tubes; beaker; super-fine grade iron wool (sold in paint and hardware stores); water

1. Obtain two pieces of iron wool of equal size, each large enough to half-fill a test tube.
2. Keep one piece of iron wool dry. Moisten the other piece thoroughly with water.
3. Pack each piece of iron wool into the bottom of a dry test tube. Label the test tubes "wet iron" and "dry iron" to correspond with the contents.
4. Fill the beaker about half-full with cold water.
5. Invert each test tube into the beaker of water. Make sure that the water in the beaker does not touch the iron wool in either test tube (Figure 5). Allow the test tubes to sit in the beakers overnight.
6. Observe and compare the water levels in the two test tubes and the appearance of the iron wool in each tube.

A. What evidence is there that a new substance has formed? T/I
B. What evidence is there that oxygen is needed to form the new substance? T/I

Figure 5 Make sure that the iron wool does not come into contact with the water.

C. What substances are needed for iron to react with oxygen under these conditions? T/I
D. Predict how the results would be different if an iron nail were used instead of iron wool. Give reasons for your answer. T/I
E. Predict how the results would be different if pure oxygen were used instead of air. Give reasons for your answer. T/I
F. Describe one situation in your everyday life where you can apply what you have learned in this activity. A C

7.1 Putting Atoms Together

ion a particle that has either a positive or a negative charge

cation a positively charged ion

anion a negatively charged ion

Ionic Compounds

Some compounds are not molecules; that is, they are not composed of neutral atoms. Instead, these compounds are made up of charged particles called **ions**. An ion forms when an atom loses or gains one or more electrons without changing its number of protons. When this happens, one of two types of ions results—a positively charged ion, or **cation** (pronounced cat-ion), or a negatively charged ion, or **anion**. If an atom loses an electron, it has one more proton than electrons and therefore has a net positive charge. If an atom gains an electron, it has one more electron than protons and has a net negative charge.

Sodium atoms usually lose one electron when they react with other atoms. The resulting sodium ion contains 11 positive charges (protons) and only 10 negative charges (electrons). Since it has one more positive charge than negative charges, the sodium ion has an ionic charge of +1 (Table 1). Scientists use the symbol Na^{1+} or Na^+ to represent this ion. (Note that the number 1 is usually omitted in chemical symbols.)

When chlorine reacts, it usually gains one electron to form an ion called chloride. Because the chloride ion has one extra negative charge, it has an ionic charge of −1 (Table 1). The chemical symbol of the chloride ion is Cl^-.

Table 1 Sodium and Chloride Ions

	Sodium, Na^+	Chloride, Cl^-
positive charge (protons)	+11	+17
negative charge (electrons)	−10	−18
ionic charge	+1	−1

DID YOU KNOW?
Iodized Salt
Check out your box of salt at home. It is probably iodized, meaning that sodium iodide (NaI) or potassium iodide (KI) has been added to the sodium chloride, NaCl. These iodine-containing compounds are added to supplement our intake of iodine, which our bodies need in very tiny amounts. A deficiency of iodine causes the thyroid gland to swell. This condition is called goitre.

To explain why sodium loses only one electron while chlorine gains one electron, look at the Bohr–Rutherford diagrams of sodium and chlorine (Figure 6). The electron that a sodium atom is most likely to lose is the one farthest from the nucleus: the single electron in its outermost orbit. This electron is the least attracted to the nucleus. A chlorine atom gains only one electron because it can only accommodate one more electron in its outermost orbit, for a maximum of 8 electrons.

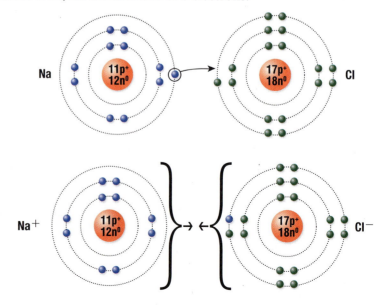

Figure 6 A sodium atom (left) loses the electron in its outermost orbit and becomes a positively charged sodium ion. A chlorine atom (right) gains an electron in its outermost orbit and becomes a negatively charged chloride ion.

Ions of sodium and chloride are oppositely charged, so they attract each other. They combine to form a compound called sodium chloride, commonly known as table salt. Compounds made up of oppositely charged ions are called **ionic compounds**. Aluminum chloride (the white solid in antiperspirants) and iron oxide (in rust) are other common examples of ionic compounds. In Try This: Rusting Steel Slowly, you observed the formation of rust. In Investigation 7.2, you will investigate the factors that determine how fast this reaction occurs.

ionic compound a compound that consists of positively and negatively charged ions

IN SUMMARY

- Atoms can join together with other atoms to form larger units called molecules.
- Molecules may contain atoms of different elements or atoms of the same element.
- Molecules that contain atoms of the same element are called molecular elements.
- Molecules that contain atoms of different elements are molecular compounds.
- A chemical formula indicates the type and number of atoms in a pure substance.
- Some compounds consist of charged atoms called ions. These compounds are called ionic compounds.

CHECK YOUR LEARNING

1. The chemical formula for baking soda is $NaHCO_3$. Answer the following questions about baking soda.
   (a) How many elements does it contain?
   (b) How many atoms of each element are in this formula?
   (c) Does baking soda contain metallic and non-metallic elements? Which ones are which?

2. Consider the following substances: hydrogen gas (H_2), carbon dioxide (CO_2), sulfur (S_8), neon (Ne), and propane (C_3H_8). Which substance(s) are
   (a) elements?
   (b) compounds?
   (c) atoms?
   (d) molecules?

3. What does the term "diatomic molecule" mean?

4. List the seven diatomic molecules.

5. Gasoline contains a mixture of compounds called hydrocarbons. They include molecules with the following formulas: pentane (C_5H_{12}), hexane (C_6H_{14}), heptane (C_7H_{16}), and octane (C_8H_{18}). Explain why "hydrocarbon" is a good name for this group of compounds.

6. (a) Explain the difference between an atom and an ion.
   (b) Give an example of an atom and an ion, and draw a Bohr–Rutherford diagram of each.

7. Write the chemical formula for each of the following compounds. Start the formula with the first element listed in each statement below and follow with the next element. The ratios in which the elements combine are given in each statement.
   (a) Ammonia, commonly found in window cleaning liquids, is formed when one nitrogen atom combines with three hydrogen atoms.
   (b) Carbon dioxide is formed when one carbon atom combines with two oxygen atoms.
   (c) When there is insufficient oxygen, one carbon atom combines with only one oxygen atom, forming the deadly gas carbon monoxide.
   (d) When hydrogen gas is burned, two atoms of hydrogen combine with one atom of oxygen to form a common substance.
   (e) Rubbing alcohol is formed when three atoms of carbon combine with eight atoms of hydrogen and one atom of oxygen.

8. An atom of calcium loses two electrons to become an ion. What kind of ion is a calcium ion?

9. Write the chemical symbol for a calcium ion.

10. Distinguish between molecular elements, molecular compounds, and ionic compounds.

11. State whether each of the following is a molecular element, molecular compound, or ionic compound.
    (a) iodine (I_2)
    (b) sodium chloride
    (c) vinegar
    (d) propane

7.2 CONDUCT AN INVESTIGATION

A Race to Rust

You know that oxygen combines with iron in the presence of water to form iron oxide, or rust: iron + oxygen → iron oxide. What factors affect how readily oxygen combines with iron? In this investigation, you will choose one factor to study.

SKILLS MENU
- Questioning
- Hypothesizing
- Predicting
- Planning
- Controlling Variables
- Performing
- Observing
- Analyzing
- Evaluating
- Communicating

Testable Question

Choose one factor that may affect the rate of rusting. Several suggestions are listed below, but you may choose any other factor of interest to you, with your teacher's approval.

- acidity of the water: for example, the presence of vinegar, carbonated drinks, and household cleaners
- salt in the water: for example, the absence and presence of increasing levels of salt
- temperature of the reaction: for example, warmer or colder than room temperature
- surface area of exposure: for example, fine iron wool, thin iron nails, large iron nails

Write a testable question for your investigation.

Hypothesis/Prediction

Write a hypothesis for your testable question. Include a prediction and a reason for the prediction.

Experimental Design

Create an experimental design for your investigation. Identify the independent and the dependent variables. What variables are controlled? Describe the experimental setup and the overall procedure of your investigation.

Equipment and Materials

You may use any equipment available in the science classroom, as well as common chemicals and materials supplied by your teacher. You may also use common household substances, but you must check with your teacher regarding safety concerns.

- List all the equipment and materials you will need.
- For each material, check and record the safety precautions listed in the Materials Safety Data Sheets.

Procedure

1. Brainstorm how you will investigate your testable question. Present your procedure as a numbered list of steps or as a flow chart.
2. Add any necessary safety precautions to your procedure. Also include any environmental concerns, such as the disposal of used chemicals.
3. List the observations that you will record. Prepare a table in which to record your observations.
4. Ask your teacher to check and approve your experimental design and procedure before you continue. Be sure that you understand and follow your teacher's suggestions. Make any necessary modifications, then carry out your investigation.
5. Record your observations as you carry out your investigation.

Analyze and Evaluate

(a) Analyze your results and discuss the relationship between the factor you studied and the rate of rusting. T/I
(b) To what degree did your results support your hypothesis? If they did not support it, suggest a possible explanation. T/I C
(c) What unavoidable sources of error did you encounter? What improvements to the procedure would you suggest? T/I C

Apply and Extend

(d) In what way(s) might you apply the results of your investigation to a situation in everyday life? A
(e) What careers would make use of the concepts you learned in this activity? A

7.3 How Atoms Combine

Atoms combine to become more stable. The most stable elements in the periodic table are the noble gases (Group 18). They are considered to be the most stable because they have the maximum number of electrons in their outermost orbits: 2 for helium, 8 for the others. Elements that do not have the maximum number of electrons in their outermost orbits combine with other elements to obtain this maximum number of electrons. Atoms become more stable by gaining, losing, or sharing electrons, depending on whether the atoms are metals or non-metals.

Metals and Metals

Metals form mixtures with other metals. These mixtures are called alloys (see Section 5.1). Alloys are created by melting two or more metals and then mixing these hot liquids. After mixing, the alloy is allowed to solidify. Alloys are different from compounds because in compounds, atoms join chemically in specific ratios to form pure substances. Alloys are solutions of metals. The metals do not combine chemically. For example, sterling silver is a solution of 92.5 % silver and 7.5 % copper (Figure 1).

Figure 1 Mixing copper with silver makes the silver harder but also causes it to tarnish.

To learn about becoming a metallurgical engineer,
GO TO NELSON SCIENCE

Metals and Non-Metals

In Section 7.1, you learned about ionic compounds. When metallic atoms such as sodium combine with non-metallic atoms such as chlorine, they usually form compounds made up of charged particles called ions. Metals lose electrons and become cations, whereas non-metals gain electrons and become anions. Cations and anions attract each other because they have opposite charges. When there are millions and billions and trillions of ions, the size of the new compound grows indefinitely. That is how you can grow large crystals of compounds such as sodium chloride (Figure 2).

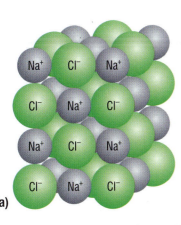

Figure 2 (a) Ions of opposite charges attract each other to form layer after layer of closely packed ions. Eventually, a crystal emerges that has sharp edges, a unique shape, and a beautiful colour. (b) This rock salt, or halite, crystal is made up of sodium chloride.

Like many substances, ionic compounds have both chemical names and common names. To name an ionic compound such as sodium chloride, write the name of the metal first: sodium. Then write the name of the non-metal and change its ending to –ide: chloride. This is just one rule for naming ionic compounds. Other ionic compounds have different rules for naming. Table 1 lists a few ionic compounds, which consist of metals and non-metals. You will learn more about the rules for naming ionic compounds in Grade 10.

WRITING TIP

State the Hypothesis
A hypothesis answers a testable question and gives reasons for the answer. For example, if the question being investigated is "What substance is produced when magnesium reacts with oxygen?" your hypothesis should clearly state what substance would be produced, along with an explanation for why this substance is produced.

Table 1 Ionic Compounds

Chemical formula	Chemical name	Common name	Common use
NaCl	sodium chloride	table salt/road salt	food seasoning, melting road ice
KCl	potassium chloride	potash	fertilizer
CaO	calcium oxide	quicklime	masonry
NaOH	sodium hydroxide	lye	drain cleaner
$CaCO_3$	calcium carbonate	limestone, chalk	building materials
$NaHCO_3$	sodium hydrogen carbonate	baking soda	rising agent in baking
$Mg(OH)_2$	magnesium hydroxide	milk of magnesia	antacid
$CuSO_4$	copper(II) sulfate	bluestone	algicide and fungicide

TRY THIS: WHEN MAGNESIUM MEETS OXYGEN

SKILLS: Predicting, Performing, Observing, Analyzing, Communicating

SKILLS HANDBOOK

You have seen how iron, a metal, and oxygen, a non-metal, combine to form a single compound—rust. In this activity, you will combine another metal and non-metal pair: magnesium and oxygen. This reaction will occur quickly, producing a new substance while giving off a great deal of energy in the form of heat and light.

Equipment and Materials: eye protection; lab apron; beaker; tongs; Bunsen burner; magnesium ribbon

 Do not look directly at the flame while the magnesium is burning. Wear eye protection and a lab apron. Tie back long hair and loose clothing.

1. Put on your eye protection and lab apron.
2. Obtain approximately 2–3 cm of magnesium ribbon. Record its appearance. Predict what will happen when magnesium is heated in the presence of oxygen.
3. Obtain a clean, dry beaker.
4. Light the Bunsen burner.
5. Hold the magnesium ribbon securely with the tongs and ignite it using the flame of the Bunsen burner. **Do not look directly at the bright flame.**
6. Hold the burning magnesium over the beaker so that any product formed falls into the beaker. Again, be careful not to look directly at the flame.
7. Observe and record the properties of the new substance formed.

A. Describe the physical properties of the magnesium ribbon that indicate it is a metal. K/U

B. Describe the physical properties of oxygen that indicate it is a non-metal. K/U

C. Describe the physical properties of the new substance formed.

D. Do the properties indicate that this new substance is a metal, a non-metal, or a compound? Give reasons for your answer. T/I

E. What is the element that usually takes part in all reactions in which burning occurs? K/U

F. Describe some applications of this reaction, which produces extremely bright light. A

Non-Metals and Non-Metals

When non-metallic elements combine with other non-metallic elements, they do not become ions by losing or gaining electrons the way metals and non-metals do. Instead, the nucleus of one atom forms a strong attraction to an electron in the outermost orbit of another atom and vice versa. A "tug of war" for electrons occurs, but neither atom wins. The two atoms share each other's electrons, resulting in a bond that holds the atoms together. A chemical bond that results from atoms sharing electrons is called a **covalent bond**. These bonded atoms form a molecule. Figure 3 shows the covalent bond between hydrogen atoms in a hydrogen molecule. Since the molecule contains two hydrogen atoms, its chemical formula is H_2.

covalent bond a bond formed when two non-metal atoms share electrons

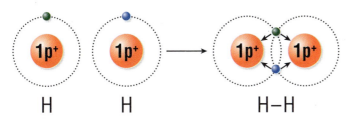

Figure 3 A covalent bond results from the sharing of electrons. In the hydrogen molecule (H_2), the two hydrogen nuclei simultaneously attract both electrons. The notation H–H can also be used to represent a hydrogen molecule. The dash between the bonding elements represents the covalent bond.

Two hydrogen atoms can also form covalent bonds with one oxygen atom to form a molecule of water. The chemical formula of this compound is H_2O. Table 2 shows some common molecules. As you can see, many of these molecules have both chemical names and common names. In the next section, you will build models of some of these molecules.

Table 2 Common Molecules

Chemical formula	Chemical name	Common name	Common use/Source
N_2	nitrogen	nitrogen	• approximately 80 % of air
O_2	oxygen	oxygen	• approximately 20 % of air
O_3	trioxygen	ozone	• in stratosphere • absorbs ultraviolet light
H_2O	dihydrogen oxide	water	• needed in all cells • home for aquatic organisms
CO_2	carbon dioxide	dry ice (solid)	• carbonated beverages • refrigeration
HCl	hydrogen chloride	muriatic acid (solution)	• stomach acid • important industrial chemical
CH_4	methane	natural gas	• fuel
NH_3	nitrogen trihydride	ammonia	• used in fertilizers and household cleaners
C_3H_8	propane	propane	• fuel
$C_2H_4O_2$	acetic acid	vinegar	• used in cooking • preservative
$C_9H_8O_4$	acetylsalicylic acid (ASA)	Aspirin	• blood thinner • for pain

IN SUMMARY

- Metals do not form compounds with other metals. They form solutions called alloys.
- Metals and non-metals combine by forming charged particles called ions. Oppositely charged ions attract and form ionic compounds.
- Table salt, or sodium chloride (NaCl), is an ionic compound.
- Non-metals share electrons with other non-metals to make molecules.
- The atoms in a molecule are held together by covalent bonds.
- An example of a molecule is water (H_2O). Each water molecule consists of two hydrogen atoms and one oxygen atom covalently bonded to each other.
- Many chemicals have both chemical names and names that are used in common language.

CHECK YOUR LEARNING

1. A piece of jewellery that is made of 14 kt gold contains 14 parts gold and 10 parts copper.
   (a) What percentage of the jewellery is gold?
   (b) Do gold and copper form compounds in the jewellery? Explain.
   (c) Explain how the 14 kt gold used to make jewellery is different from the element gold.

2. Metal alloys are sometimes described as solid solutions. Explain how an alloy is similar to a solution such as salt water.

3. Write the common name for each of the following compounds.
   (a) sodium hydroxide
   (b) O_3
   (c) sodium chloride
   (d) CO_2
   (e) sodium hydrogen carbonate
   (f) calcium carbonate

4. Write the chemical name for each of the following compounds.
   (a) muriatic acid
   (b) vinegar
   (c) potash
   (d) quicklime
   (e) milk of magnesia
   (f) natural gas

5. The formula for propane gas, often used as barbecue fuel, is C_3H_8. The formula for butane, the liquid fuel in cigarette lighters, is C_4H_{10}. The wax in a candle is a mixture of molecules, one of which is $C_{30}H_{62}$.
   (a) What is similar and what is different about the chemical compositions of these three substances?
   (b) Consider the physical states of these three substances at room temperature and discuss any relationship you see between the formula and physical state.
   (c) Are these three substances considered molecules? Give reasons for your answer.

6. Draw a Bohr–Rutherford diagram for each of the following molecules.
   (a) fluorine (F_2)
   (b) hydrogen fluoride

7. Explain the difference between a molecule and an ionic compound.

8. Describe how ions are able to form large crystals.

9. In your own words, write a definition for "covalent bond."

10. Explain what types of atoms tend to form covalent bonds.

SCIENCE WORKS ✓ OSSLT

The Tiny World of Nanotechnology

The world of science has entered an exciting new dimension—the tiny world of nanotechnology. Products of this amazing technology can be as small as one fifty-thousandth of the diameter of a human hair.

Strange things happen to particles at the nanometre level. Ordinary physical laws no longer apply. Molecules can be manipulated to have super strength or electrical conductivity. They can be shaped into molecular spheres or tubes (Figure 1).

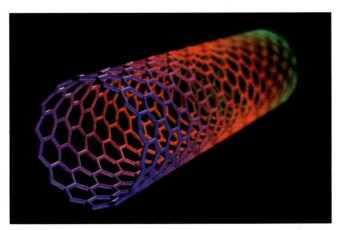

Figure 1 Carbon nanotubes demonstrate extraordinary strength and electrical conductivity. These properties have been put to use in designing combat clothing and solar cells.

Nanotechnology is not science fiction. Canadians are already using hundreds of products that include nanotechnology. We can buy pencils that release nano-scents and toothpaste that foams with nano-whiteners. Nanoparticles are added to fabrics and face creams to resist stains and wrinkles (Figure 2). You have probably used sunscreens that contain invisible nanoparticles of zinc oxide, protecting you from harmful ultraviolet rays.

Figure 2 A water-repellent surface created using inorganic nanoparticles

Nano-sized antibacterial agents are found in baseball caps, sports socks, and athletic supports, taming the familiar sports bag odour. These tiny germ killers are also found in food containers that extend shelf life. Some containers even change colour when the food has gone bad. Invisible antifungal paints are used on both exterior and interior walls of buildings to freshen the air we breathe.

Nanotechnology is opening new frontiers for medical research. At the University of Toronto, biomedical engineer Warren Chan is studying ways of improving the diagnosis of cancer. He is designing tiny nanoparticles called quantum dot probes that can seek out a tumour and accurately signal its location. He takes dots of nano dimensions and attaches them to special proteins that are attracted to tumour cells (Figure 3). These dots light up with different colours depending on where the tumour cells are located. The dots provide valuable information about the stage and spread of the cancer. His team is also researching how nanoparticles might deliver drugs directly to a targeted location without attacking healthy cells surrounding the tumour.

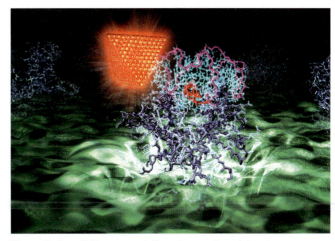

Figure 3 An artist's representation of a quantum dot probe

The future may bring nanorobots, complete with tiny computers and tiny toolboxes, zooming through our bodies. They could scrape away plaque in our arteries and tumours in our organs. However, this magical nano world may bring with it serious concerns. We are inhaling and ingesting these invisible particles. Since they are small enough to enter our cells, they are able to travel anywhere in our bodies and in our environment. Children may be especially vulnerable. Parents may reconsider the benefits of antibacterial baby bottles and soothers, and stain-resistant clothing and toys. As regulation catches up with this technology, consumers will hopefully be better informed and better equipped to make decisions on their nano purchases.

7.4 PERFORM AN ACTIVITY

Making Molecular Models

Chemical formulas of molecules tell us how many atoms of each element there are in a molecule, but they do not convey any sense of the three-dimensional shapes of molecules. In this activity, you will use simple materials to build models of these shapes.

Most elements form a fixed number of bonds—no more and no fewer. For example, a carbon atom forms four bonds, and a hydrogen atom forms one bond. Table 1 lists the number of bonds that each element you will use in this activity can form.

> **SKILLS MENU**
> - Questioning
> - Hypothesizing
> - Predicting
> - Planning
> - Controlling Variables
> - Performing
> - Observing
> - Analyzing
> - Evaluating
> - Communicating

Table 1 Combining Capacity of Some Non-Metals

Element	Symbol	Number of covalent bonds
hydrogen	H	1
chlorine	Cl	1
oxygen	O	2
sulfur	S	2
nitrogen	N	3
carbon	C	4

Purpose
To build models of some common molecules.

Equipment and Materials
- 1 atomic model kit (Figure 1)

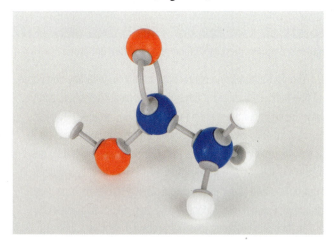

Figure 1 Molecular models allow us to visualize molecules in three dimensions.

Procedure

1. Copy Table 2 (on the next page) into your notebook.

2. Use the "atoms" and "bonds" provided to build a model of each combination of elements listed in your table. You may have to use more than one atom of each type. Count to make sure that each atom in your model contains the correct number of bonds.

3. Count the number of atoms for each element in your model. Write a chemical formula for the molecule in the corresponding column of your table.

> **WRITING TIP**
>
> **Recording the Procedure**
> When writing a science report, use chronological order to organize the steps of the procedure. Number each step. Use a new line for each step. Include all of the steps. Use a verb to begin the description of each step.

268 Chapter 7 • Chemical Compounds

Table 2 Building Models of Molecules

Element 1	Element 2	Element 3	Chemical formula	Chemical name	Common name	Common usage
O	O	—				
H	O	—				
N	H	—				
H	Cl	—				
C	O	—				
C	Cl	—				
H	S	—				
C	H	—				
C	H	—	C_3H_8			
H	C	N	HCN			
H	C	O	H_2CO			
C	H	O	CH_3OH			

Analyze and Evaluate

(a) Look up the chemical name of each molecule in your table and complete the corresponding column in Table 2. T/I

(b) Look up the common name of each molecule in your table and complete the corresponding column in Table 2. T/I

Apply and Extend

(c) Look up a common usage for each molecule in your table and complete the corresponding column in Table 2. T/I

(d) Use your model kit to construct four water molecules. Then use the water molecules to illustrate how water undergoes physical changes when changing to ice and to water vapour. Write a short paragraph to summarize your observations. T/I C

(e) Use the four water molecules you constructed in question (d) to illustrate how water undergoes a chemical change to produce hydrogen (H_2) and oxygen (O_2). Write a short paragraph to summarize your observations. T/I C

(f) Based on what you did in questions (d) and (e), compare physical changes and chemical changes. In which type of change are chemical bonds broken or new substances made? Explain your reasoning. T/I

WRITING TIP

Check and Revise Your Report
When revising a science report, check that the procedure is described as clearly as possible by reading each step aloud and observing as a partner mimes the action. If your partner has difficulty visualizing what to do, then you must change the wording to make the description easier to understand.

7.4 Perform an Activity 269

7.5 PERFORM AN ACTIVITY

What Is This Gas?

Air is a mixture of many gases, including nitrogen, oxygen, and carbon dioxide. Most of the gases in air are colourless and odourless. Although these and other gases are indistinguishable in appearance, they exhibit different chemical properties. These properties can be used to identify the following gases that are often encountered in chemical reactions.

- Oxygen is essential for burning. Things burn more vigorously when they are in pure oxygen than when they are in air, which contains only about 20 % oxygen.
- When hydrogen is ignited in air, the hydrogen atoms combine with oxygen atoms to form molecules of H_2O (water vapour). This rapid reaction is accompanied by a characteristic explosive "pop" sound.
- Carbon dioxide does not burn and does not support combustion. Carbon dioxide will extinguish a flame, making it useful as a fire extinguisher. Carbon dioxide produces a white solid when it reacts with a substance called limewater. A white cloudiness appears in the limewater when this powdery solid is produced. The white solid is calcium carbonate—the main component in chalk and limestone.

SKILLS MENU
- Questioning
- Hypothesizing
- Predicting
- Planning
- Controlling Variables
- Performing
- Observing
- Analyzing
- Evaluating
- Communicating

Purpose
To produce different gases and identify each gas using a chemical test.

Equipment and Materials
- eye protection
- lab apron
- 5 test tubes
- test-tube rack
- metal tongs
- stopper for test tube
- measuring spoon
- 10 mL graduated cylinder
- Bunsen burner
- toothpick
- wooden splints
- candle
- limewater (saturated calcium hydroxide solution)
- manganese dioxide
- hydrogen peroxide (3 % solution)
- hydrochloric acid (5 % solution)
- magnesium ribbon
- sodium hydrogen carbonate (baking soda)

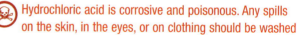 Hydrochloric acid is corrosive and poisonous. Any spills on the skin, in the eyes, or on clothing should be washed 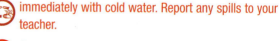 immediately with cold water. Report any spills to your teacher.

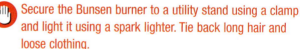 Secure the Bunsen burner to a utility stand using a clamp and light it using a spark lighter. Tie back long hair and loose clothing.

Procedure

1. Copy Table 1 into your notebook.

Table 1 Observations

Reactants	Burning splint test	Glowing splint test	Limewater test	Candle flame test	Identity of the gas
manganese dioxide and hydrogen peroxide					
hydrochloric acid and baking soda					
hydrochloric acid and magnesium ribbon					

2. Put on your eye protection and lab apron.
3. Set up four test tubes in the test-tube rack and label them A, B, C, and D. Fill a fifth test tube about one-third full of limewater.
4. *Manganese dioxide and hydrogen peroxide*: To each of test tubes A, B, C, and D, add a tiny amount of manganese dioxide powder (enough to cover the broad end of a toothpick) and about one-tenth of a test tube of hydrogen peroxide. Allow the reaction to proceed for about 15 s.
5. Ignite a wooden splint with the Bunsen burner and bring the burning splint to the mouth of test tube A. Record your observations in your data table. Safely discard the used splint in this and each of the following steps.
6. Ignite another wooden splint, then blow out its flame. Insert the glowing splint halfway into test tube B. Record your observations in your data table.
7. Carefully pour the gas produced in test tube C into the test tube of limewater (Figure 1). Be careful not to pour any of the hydrogen peroxide and manganese dioxide. Stopper the limewater test tube and shake to mix the gas and liquid. Record your observations in your data table.

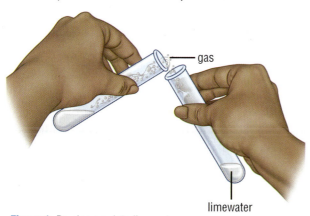

Figure 1 Pouring gas into limewater

8. Light the candle and carefully pour the gas produced in test tube D onto the flame. Record your observations in your data table.
9. Dispose of the contents of your test tubes as directed by your teacher. Clean the test tubes in preparation for the next series of tests.
10. *Hydrochloric acid and baking soda*: Repeat steps 3 to 9, using about 1 mL of baking soda and 4 mL of hydrochloric acid.
11. *Hydrochloric acid and magnesium*: Repeat steps 3 to 9, using 3–5 cm of magnesium ribbon and hydrochloric acid to about one-third of the test tube height.

Analyze and Evaluate

(a) Compare your observations with the description of chemical properties of carbon dioxide, hydrogen, and oxygen given at the beginning of this activity. **T/I**

(b) Complete the last column of Table 1 by identifying each gas produced in the reactions. **T/I**

(c) What gas is produced when manganese dioxide and hydrogen peroxide are mixed? Provide evidence to support your answer. **T/I**

(d) What gas is produced when hydrochloric acid and baking soda are mixed? Provide evidence to support your answer. **T/I**

(e) What gas is produced when hydrochloric acid and magnesium are mixed? Provide evidence to support your answer. **T/I**

(f) In each part of the activity, were the changes physical changes or chemical changes? Give reasons for your answer. **T/I**

(g) What problems, if any, did you encounter in carrying out each of these tests? **T/I**

(h) What improvements, if any, would you make to the procedure in future tests? **T/I**

Apply and Extend

(i) From your observations in this activity, suggest reasons why
  - birthday balloons are filled with helium, not hydrogen
  - during surgery using oxygen cylinders, medical staff wear coverings over their shoes to eliminate sparks produced by static electricity **A**

(j) Design a simple fire extinguisher that produces carbon dioxide gas quickly when needed. Write a paragraph describing your design and how it will work. **A C**

7.6 Breaking Molecules Apart: Properties of Hydrogen Peroxide

Many teeth whiteners contain an ingredient called hydrogen peroxide (Figure 1). The name tells us that this compound is made up of hydrogen atoms and oxygen atoms. The prefix *per–* means *thoroughly*, which in this case refers to the number of oxygen atoms in the molecule. The chemical formula for hydrogen peroxide (H_2O_2) indicates that this molecule has two oxygen atoms compared with the more common H_2O (Figure 2).

To learn more about becoming a dental hygienist or an esthetician,
GO TO NELSON SCIENCE

Figure 1 A brighter smile is the result of a chemical reaction between hydrogen peroxide and stains on your teeth.

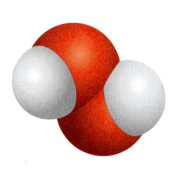

Figure 2 Hydrogen peroxide, H_2O_2

It is this extra oxygen atom that gives hydrogen peroxide its special chemical properties. The molecule easily breaks apart into the more stable H_2O and oxygen (O_2):

$$\text{hydrogen peroxide} \rightarrow \text{water} + \text{oxygen}$$

You have learned that oxygen readily combines with other elements to form new compounds. This process can happen very slowly, such as when iron turns to rust, or very quickly, such as when magnesium is burned to a white powder.

When hydrogen peroxide breaks apart, or decomposes, the oxygen released can bleach a variety of materials. Bleaching occurs when the oxygen reacts with the chemical pigments that give materials their colour. This process can make stained teeth white or turn brown hair blond. Hydrogen peroxide is the key ingredient in many hair bleaches. It is also used in some contact lens soaking solutions. When the lens is placed in the solution, bubbles of oxygen surround the lens and kill bacteria. When the process is complete, it is safe to put that lens back into the eye. Note that this solution is a special mixture. You should never clean your contact lenses with household hydrogen peroxide. It will burn your eyes.

Under ordinary conditions, hydrogen peroxide breaks down into water and oxygen slowly. Nonetheless, precautions must be taken in the storage of this unstable liquid. The bottles are commonly made of plastic to allow for a slight expansion, in case the amount of oxygen gas that is released causes a pressure buildup in the container. Often, the bottles are dark and opaque to reduce the amount of light energy entering the solution, which may speed up its chemical decomposition (Figure 3).

Figure 3 Hydrogen peroxide is kept in special opaque bottles to prevent explosions.

272 Chapter 7 • Chemical Compounds

Large bottles of concentrated hydrogen peroxide are designed with venting caps in case a large volume of oxygen is produced over time. Care must be taken in the use and disposal of hydrogen peroxide because it will react wherever it is spilled or poured.

The decomposition of hydrogen peroxide can proceed much more quickly if you add a catalyst to the liquid. A **catalyst** is any substance that speeds up a reaction without being consumed or chemically altered itself. A catalyst does not actually take part in the reaction, but it provides more favourable conditions for the reaction to happen.

catalyst a substance that speeds up a chemical reaction but is not used up in the reaction

TRY THIS: WHEN YEAST MEETS BLEACH

SKILLS: Predicting, Performing, Observing, Analyzing, Communicating

In this activity, you will use a living organism as the source of a catalyst. Your own body cells contain this catalyst. That is why the cut on your knee bubbles when you put hydrogen peroxide on it, evidence of the oxygen gas being produced. The oxygen bubbles react with any bacteria in the cut and kill them. We will not use your body cells today. We will use yeast instead.

Equipment and Materials: empty plastic or glass bottle, approximately 500 mL (for example, water bottle); small cup; measuring cup or graduated cylinder; spoon; funnel; 125 mL 3 % hydrogen peroxide (for example, from a pharmacy); 50 mL liquid dishwashing detergent; food colouring; 1 package active yeast (approximately 2 tsp); 50 mL warm water (approximately 30 °C)

1. Mix together 125 mL of 3 % hydrogen peroxide, 50 mL of liquid detergent, and a few drops of food colouring in the bottle.
2. Place the bottle and contents in a sink to avoid spilling.
3. Mix the active yeast with 50 mL of warm water in the small cup. Allow to sit for 10 min until the yeast appears to bubble.
4. Use the funnel to pour the yeast mixture into the bottle. Observe.

A. What evidence is there that oxygen was produced in this activity?
B. Describe a test that you can perform to confirm that the gas produced is oxygen.
C. Was oxygen produced before the yeast was added? What test can you perform to find out?
D. What was the role of the yeast in this activity?
E. What was the role of the detergent in this activity?
F. Predict what you would observe if you repeated the activity without the detergent.

IN SUMMARY

- The chemical formula for hydrogen peroxide is H_2O_2.
- Hydrogen peroxide easily decomposes into water (H_2O) and oxygen (O_2).
- The newly released oxygen from the decomposition of hydrogen peroxide reacts with other chemicals, often producing a bleaching effect.
- Some substances are able to speed up chemical reactions. These substances are known as catalysts.

CHECK YOUR LEARNING

1. Explain the hazards of the storage, use, and disposal of hydrogen peroxide.
2. Compare water and hydrogen peroxide in terms of formula, stability and shelf life, uses, and safety precautions in use and disposal.
3. How do the properties of hydrogen peroxide make it suitable for use in hair dye?
4. What properties of hydrogen peroxide solution make it strong enough for killing bacteria on skin cuts?
5. Why is hydrogen peroxide sold in opaque plastic bottles?
6. The contact lens storage cases designed for hydrogen peroxide cleaners contain a platinum catalyst. Explain why platinum is added to the storage container and not to the hydrogen peroxide.
7. Give examples of household substances that are stored in opaque plastic bottles. For each substance, assess its purpose and usefulness as well as the hazards associated with its handling and disposal.

7.7 PERFORM AN ACTIVITY

What Is Rotten and What Is Not?

Think back to the garbage that you buried in soil in Try This: To Rot or Not to Rot in Section 5.3. Now it is time to dig it up and find out what has happened! As you recall, some of the items that you chose to bury were natural: a plant or animal produced them naturally. Other items were synthetic: made using a chemical process, possibly in a factory, from raw materials such as fossil fuels.

SKILLS MENU
- Questioning
- Hypothesizing
- Predicting
- Planning
- Controlling Variables
- Performing
- Observing
- Analyzing
- Evaluating
- Communicating

Purpose
To observe the degree of decomposition of natural and synthetic items.

Equipment and Materials
- lab apron
- rubber gloves
- an old spoon or other digging instrument
- old newspapers
- container of composted material from Try This: To Rot or Not to Rot in Section 5.3

Procedure
1. Put on your lab apron and rubber gloves.
2. Take the container of soil and buried items that you set up in Section 5.3 to a table or an area on the floor.
3. Spread out several layers of old newspaper.
4. Carefully transfer the contents of the container onto the newspaper using a spoon or other instrument.
5. Use the spoon to look through the poured-out contents and sort the buried items into two piles:
   - items that have not decomposed
   - items that show some evidence of decomposition
6. Make a list of the items that did not decompose. Create a table to classify each item as natural or synthetic.
7. Repeat step 6 for the items that showed signs of decomposition.
8. When you have finished, follow your teacher's instructions for the disposal of all materials. You may be able to add some of the contents to a compost container.

Analyze and Evaluate

(a) From your results, describe the degree of decomposition of buried items that are
   - natural
   - synthetic T/I

(b) Research and explain the meaning of the prefix *bio–* as it is used in the words biology, biotechnology, and biodegradable. T/I

(c) Based on what you observed in this activity, which materials, natural or synthetic, are more likely to decompose? T/I

(d) Did any of the synthetic materials show signs of decomposition? If so, explain why you think these particular materials decomposed. T/I

(e) Look back to the physical properties you recorded for the materials in the Try This activity. Does the degree of decomposition depend more on physical properties or on whether the material is synthetic or natural? Explain your reasoning. T/I

(f) Were the predictions you made in the Try This activity accurate? Explain. T/I

WRITING TIP
Analyze and Evaluate Your Results
When you analyze and evaluate your results, explain how they support your prediction. For observations that do not support your prediction, suggest an explanation. Describe any problems you experienced and suggest ways to improve the procedure.

Apply and Extend

(g) Most dog owners take responsibility and "stoop and scoop" after their pet during walks. However, this commendable practice has created a related environmental problem—mountains of non-biodegradable plastic bags containing biodegradable organic waste (Figure 1). Research the biodegradable options that are available.

 GO TO NELSON SCIENCE

Figure 1 How else could we dispose of this biodegradable waste?

(h) Write a headline and a news report for your local daily newspaper based on the photo in Figure 1. Inform your readers about the problems with disposal of plastic bags, particularly those used in dog parks, and the environmentally sound options that are available.

(i) Polyethylene is a versatile, flexible, and durable compound that is made from fossil fuels. Its versatility and flexibility make it ideal for a variety of uses. However, its durability prevents it from being broken down when it has served its function and is discarded. Research the applications of polyethylene and options for its reuse, reduction, recycling, disposal, and substitution.

 GO TO NELSON SCIENCE

(j) Copy Table 1 and record your findings about at least ten items that you or your family might purchase that are made of polyethylene.

Table 1 The Uses and Disposal of Polyethylene

Polyethylene items and their function	Estimated duration of useful life	Options for disposal after useful life	Suggested natural or synthetic substitutes for this function

UNIT TASK Bookmark

You can apply what you learned about the disposal of natural and synthetic materials to the Unit Task on page 286.

7.8 EXPLORE AN ISSUE CRITICALLY

DDT is Forever

It is often thought that in the beautiful Arctic regions of Québec, where there is no heavy industry, Inuit enjoy clean air and water (Figure 1). Dr. Eric Dewailly, a public health official with Québec's Community Health department, began a study to compare the breast milk of women who lived in the unpolluted North to the breast milk of women who lived in the cities of southern Québec.

To learn more about becoming a public health official,
GO TO NELSON SCIENCE

Figure 1 Our polluted Canadian Arctic

SKILLS MENU
- Defining the Issue
- Researching
- Identifying Alternatives
- Analyzing the Issue
- Defending a Decision
- Communicating
- Evaluating

He predicted that the pollution from car exhaust and factory fumes would cause higher levels of toxic chemicals in city dwellers than in Inuit. He was concerned that women in the south would pass these toxins to their breast-fed babies. He expected that Inuit women would be free of the harmful pollutants of the big cities and produce uncontaminated milk.

Much to everyone's surprise, the results of the study showed the exact opposite effect. The concentrations of toxins in Inuit women were five times greater than in women who lived in the polluted south. These chemicals included the extremely poisonous dichlorodiphenyltrichloroethane (DDT). DDT is a powerful insecticide that was widely used from the 1950s to the 1970s. Extensive spraying of DDT reduced malaria infections around the world by killing malaria-carrying mosquitoes. DDT was also used extensively in Canada as an agricultural pesticide.

DDT is a synthetic compound, composed of carbon, hydrogen, and chlorine. DDT does not mix well with water, but dissolves in oil and fats, so it is stored in body fat and breast milk. Studies show that high exposure to DDT is linked to human cancers of the liver, pancreas, and reproductive organs.

The levels of polychlorinated biphenyls (PCBs)—another compound made of carbon, hydrogen, and chlorine—were also much higher in Inuit women than in women in the south. PCBs are found in oils that lubricate electrical equipment, in fire-retardant fabrics, and in certain glues and paints. PCBs impair reproduction and are a probable cause of some cancers. Like DDT, PCBs are fat soluble, so they accumulate in fatty tissue.

Is there an explanation for these unexpected results? The toxins are not in the air or the water. They have been traced to the traditional foods that Inuit women eat—fish, beluga whales, walruses, and seals. These large animals consume countless smaller fish and marine mammals, each of which has eaten smaller contaminated organisms down the food chain. Every molecule of PCB and DDT is stored in the fatty tissues of these sea creatures. The body has no way of breaking down these toxins, so they accumulate, much like discarded plastic water bottles in a landfill site. These stored toxins are passed along from each eaten organism to each eater of the meal. When the top consumer, an Inuit woman, eats whale meat or seal blubber, she is also eating all of the toxins collected in it from polluted waters far from the pristine North. The process that results in a build-up of toxins as you go up the food chain is called bioamplification (Figure 2).

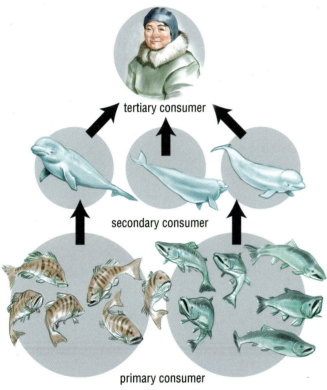

Figure 2 Bioamplification occurs when toxins that are stored in fatty tissue become increasingly concentrated as they move up the food chain.

The Issue

Since Dr. Dewailly's study, Inuit have reduced the amount of high-fat fish and animals in their diet. There is concern that their traditional way of life has been threatened. Most, but not all, developed countries in the world have banned the use of these harmful compounds. The Canadian government banned the use of DDT in the 1980s. Some countries continue to sell DDT to developing nations that have asked for help to prevent the spread of malaria. Is the continued use of DDT worthwhile despite its effects on health and on the lifestyle of Inuit?

Goal

To decide whether the risks of using DDT worldwide ultimately outweigh the benefits.

Gather Information

Choose a stakeholder from the list below, or another stakeholder of your choice. Possible roles include
- a representative of a company that manufactures DDT
- the president of a poor Latin American country with a major malaria problem
- a Canadian health official representing the concerns of Inuit

Research the pros and cons of using DDT from the perspective of your stakeholder.

Identify Solutions

Consider the following questions to help you identify solutions:
- What are the risks of using DDT?
- What are the benefits?
- What are the alternatives to using DDT?

Make a Decision

Make a decision on the issue based on the information you have gathered from the perspective of your stakeholder.

Communicate

Prepare a presentation to the United Nations outlining your arguments for or against the continued use of DDT from the perspective of the stakeholder you have chosen.

UNIT TASK Bookmark

You can apply what you learned about the effects of chemical products to the Unit Task described on page 286.

CHAPTER 7 LOOKING BACK

KEY CONCEPTS SUMMARY

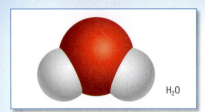

Elements can combine to form compounds.

- Atoms can join together with the same or different atoms to form molecules. (7.1)
- Molecular elements consist of only one type of element. (7.1)
- Molecular compounds consist of two or more different types of elements. (7.1)
- Ionic compounds consist of charged particles called ions. (7.1)
- Rust is an example of an ionic compound. (7.2)

Metals and non-metals combine to form ionic compounds. Non-metals combine to form molecules.

- Metals lose electrons and become cations. (7.3)
- Non-metals gain electrons and become anions. (7.3)
- Cations and anions have opposite charges, so they attract each other and form ionic compounds. (7.1, 7.3)
- Non-metals form covalent bonds with other non-metals by sharing electrons to form molecular compounds. (7.3)

Some useful compounds have social, environmental, and economic impacts.

- DDT and PCBs bioaccumulate and cause health problems for humans and animals. (7.8)
- Compounds such as plastics pose disposal problems because they do not biodegrade. (7.7)

Compounds can break apart into simpler substances.

- Hydrogen peroxide easily decomposes into water (H_2O) and oxygen (O_2). (7.6)
- The oxygen released from the decomposition of hydrogen peroxide reacts with other chemicals, often producing a bleaching effect. (7.6)
- The decomposition of biodegradable materials is an example of compounds breaking down into simpler substances. (7.7)

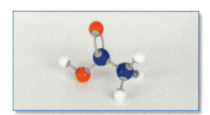

Molecular models are used to represent molecules.

- Models help us visualize the three-dimensional shapes of molecules. (7.4)
- Atoms combine with other atoms to form molecules. (7.3)
- All the atoms of an element form the same number of bonds. (7.4)

Simple chemical tests can identify common gases.

- Oxygen gas can be identified by more vigorous burning. (7.5)
- Hydrogen gas can be identified by a "pop" sound heard when hydrogen atoms combine with oxygen atoms to form water vapour. (7.5)
- Carbon dioxide can be identified by the white solid it produces when it reacts with limewater. (7.5)

WHAT DO YOU THINK NOW?

You thought about the following statements at the beginning of the chapter. You may have encountered these ideas in school, at home, or in the world around you. Consider them again and decide whether you agree or disagree with each one.

Vocabulary

molecule (p. 257)
chemical formula (p. 257)
molecular element (p. 257)
molecular compound (p. 258)
ion (p. 260)
cation (p. 260)
anion (p. 260)
ionic compound (p. 261)
covalent bond (p. 265)
catalyst (p. 273)

1 Oxygen can be both beneficial and harmful.
Agree/disagree?

4 All compounds will eventually break down over several years.
Agree/disagree?

2 When oxygen reacts with a substance, flames are always produced.
Agree/disagree?

5 Compounds have different properties than their elements.
Agree/disagree?

3 Compounds must be heated to break down into simpler substances.
Agree/disagree?

6 The chemical products we use affect only local ecosystems.
Agree/disagree?

How have your answers changed since then? What new understanding do you have?

BIG Ideas

- ✓ Elements and compounds have specific physical and chemical properties that determine their practical uses.
- ✓ The use of elements and compounds has both positive and negative effects on society and the environment.

Looking Back

CHAPTER 7 REVIEW

What Do You Remember?

1. Copy and complete the following table in your notebook. (7.1) K/U

Chemical name	Chemical formula	Atom? Y/N	Molecule? Y/N	Element? Y/N	Compound? Y/N	Total number of atoms
sulfur dioxide	SO_2					
	Cl_2					
	H_2O_2					
	Si					
carbon dioxide						
butane	C_4H_{10}					
cholesterol	$C_{27}H_{46}O$					
	O_3					

2. (a) Explain what is meant by the term "diatomic."
   (b) Write the formulas of seven diatomic elements. (7.1) K/U

3. When each of the following elements combine, what type of compound, ionic or molecular, forms? (7.3) K/U
   (a) potassium and fluorine
   (b) iron and chlorine
   (c) nitrogen and oxygen

4. Identify the colourless gas that produced the following result when tested: (7.5) K/U
   (a) turned the limewater cloudy
   (b) produced a popping sound when a blazing wooden splint was held near it
   (c) reignited a glowing wooden splint

5. Explain the difference between a molecular element and a molecular compound. (7.1) K/U

6. How do ionic compounds form? (7.1) K/U

7. Which types of elements combine to form molecules? (7.3) K/U

8. Which types of elements combine to form ionic compounds? (7.3) K/U

What Do You Understand?

9. When iron combines with oxygen, there are two possible iron oxide compounds that form. One compound contains one atom of iron to each atom of oxygen, and the other compound contains two atoms of iron to every three atoms of oxygen. Write a chemical formula for each compound. (7.1) K/U

10. When a sample of white crystals is heated, a colourless gas is produced that relights a glowing splint. How can you determine from this observation whether the sample of white crystals is an element or a compound? Explain your answer. (7.6) K/U

11. The wildlife in a northern Canadian community has been found to contain high levels of toxic chemicals even though the chemicals are not used in this geographical region. (7.8) A C
    (a) Write to a resident of this community explaining what the source of the chemicals is and why the levels are unacceptably high in the community's food supply.
    (b) What properties of these toxic chemicals make them particularly dangerous and cause them to spread globally and through the food chain?

Solve a Problem

12. Solutions of ionic compounds can conduct electricity, while solutions of molecular compounds often do not. Design a method of identifying a white powder that may be either the simple sugar glucose ($C_6H_{12}O_6$) or the highly poisonous sodium cyanide (NaCN). Include all required safety procedures. (7.1) T/I

13. Use your knowledge about the properties of everyday substances to design a method to safely distinguish among three white powders—chalk dust, icing sugar, and baking soda. Use a variety of physical properties. Include all of the necessary safety precautions. (7.1, 7.3) T/I

14. Carbonated drinks contain dissolved carbon dioxide that is released when the bottle or can is opened. Design a method of collecting and testing the gas to verify its identity. Include all necessary safety precautions. Perform your experiment with your teacher's approval and report your results. (7.5) T/I C

Create and Evaluate

15. The natural gas that your Bunsen burner uses in the laboratory consists mostly of methane. Its chemical formula is CH_4. When this gas burns, it reacts with oxygen (O_2), and produces water (H_2O) and carbon dioxide (CO_2). (7.4) T/I A C

    (a) Using your choice of items to represent atoms (e.g., modelling clay), connect them with toothpicks to make one molecule of CH_4 and two molecules of O_2.
    (b) Break apart the molecules you made in part (a) and rearrange them to make new molecules of H_2O and CO_2. How many of each molecule do you get?
    (c) Consider a situation where the supply of air is limited, such as using a barbecue indoors. Explain why the deadly gas carbon monoxide (CO) is produced instead of carbon dioxide (CO_2).

16. When an electric current is passed through water, two colourless gases are produced. When hydrogen gas makes a popping sound in the presence of a blazing splint, water is produced. Explain how each of these reactions demonstrates that water is a compound, not an element. (7.6) K/U

Reflect on Your Learning

17. Many chemicals, such as plastics and pesticides, impact our lives in beneficial as well as harmful ways. Reflect on the impact of oxygen in our lives. The same chemical reaction that causes oxygen to rust the iron in our cars is also needed for the iron in our blood to carry oxygen to all our cells. Join a small group of students and share your ideas about a planet with an atmospheric oxygen level much lower or much higher than the current 20 %. How would our planet and our lives be different? A C

18. Consider a typical day in your school week. Estimate the number of plastic or other non-biodegradable items that you use. Compare this estimate to the number of biodegradable or reusable items that you use. Why would it be good for the environment to change this ratio? How can you change your daily habits to make these changes? A C

Web Connections

19. Research and report on the history of iodizing table salt. (7.1) T/I C

20. Research the process and risks of body piercing. Develop a list of questions to ask the body piercer before having a piercing done. (7.3) T/I C

21. Research and report on the major concerns about the unregulated use of nanotechnology. T/I A C

CHAPTER 7 SELF-QUIZ

The following icons indicate the Achievement Chart category addressed by each question.
K/U Knowledge/Understanding **T/I** Thinking/Investigation **C** Communication **A** Application

For each question, select the best answer from the four alternatives.

1. Glucose, a type of sugar, has the formula $C_6H_{12}O_6$. What is the total number of atoms in one molecule of glucose? (7.1) **K/U**
   (a) 3
   (b) 6
   (c) 12
   (d) 24

2. Which of these chemical formulas represents an element? (7.1) **K/U**
   (a) O_2
   (b) CO
   (c) CH_4
   (d) CCl_4

3. A sulfide ion has 16 protons and a charge of −2. How many electrons does a sulfide ion have? (7.1) **K/U**
   (a) 8
   (b) 14
   (c) 18
   (d) 34

4. A certain type of brass is 75 % copper and 25 % zinc. Which type of material is brass? (7.3) **K/U**
   (a) an alloy
   (b) an element
   (c) an ionic compound
   (d) a covalent compound

5. What does the formula N_2 represent? (7.1) **K/U**
   (a) an atom and an element
   (b) an element and a molecule
   (c) a molecule and a compound
   (d) a compound and an element

Indicate whether each of the statements is TRUE or FALSE. If you think the statement is false, rewrite it to make it true.

6. Ozone (O_3) is a diatomic molecule. (7.1) **K/U**

7. Non-metals react with other non-metals to form ionic compounds. (7.3) **K/U**

8. Toxic materials such as DDT and PCBs are a hazard for people living in Canada's north. (7.8) **K/U**

Copy each of the following statements into your notebook. Fill in the blanks with a word or phrase that correctly completes the sentence.

9. Propane (C_3H_8), butane (C_4H_{10}), and pentane (C_5H_{12}) all belong to the class of organic compounds called _____. (7.1) **K/U**

10. All the bonds in methanol (CH_3OH) are _____ bonds. (7.3) **K/U**

11. A(n) _____ is formed when an atom loses one or more electrons and retains the same number of protons. (7.1) **K/U**

Match each term on the left with the most appropriate description on the right.

12. (a) rust   (i) sodium chloride
    (b) table salt   (ii) sodium hydrogen carbonate
    (c) baking soda   (iii) sodium hydroxide
    (d) lye   (iv) calcium carbonate
    (e) limestone   (v) iron oxide (7.2, 7.3) **K/U**

Write a short answer to each of these questions.

13. Why are the noble gases the most stable elements? (7.3) **K/U**

14. Explain why all molecules are not considered compounds. (7.1) **K/U**

15. A chemist carries out an experiment in which potassium metal (K) reacts with chlorine gas (Cl_2) to form the salt potassium chloride (KCl), an ionic compound. The salt is then dissolved in water, where it separates into potassium ions (K^+) and chloride ions (Cl^-). (7.1) **T/I**
    (a) Describe what happened in the outermost electron orbits of the potassium and chlorine atoms during the reaction.
    (b) A chloride ion has 17 protons. How many electrons does it have?
    (c) A potassium ion has 19 protons. How many electrons does it have?

16. Just as iron rusts, copper also reacts with oxygen in the air in the process of corrosion. Design an experiment to study how each of the following factors affects the rate at which copper corrodes. (Hint: Pennies are made mostly of copper. Begin with pennies that have been made equally shiny by rubbing with steel wool.) (7.2) T/I
    (a) water
    (b) salt
    (c) acid (in the form of vinegar)

17. Three flasks contain colourless gases. You know that one flask contains pure oxygen, one contains carbon dioxide, and one contains air, but you do not know which is which. (7.5) T/I
    (a) Describe a test you could perform to identify the gases.
    (b) Explain what you would observe when you performed the test on each gas.

18. A student is studying the properties of two metals. Metal A burns brightly when touched to a flame, leaving behind a white powder. Metal B does not burn, but when it is left exposed for several weeks, a dark, scaly coating forms on its surface. (7.2) T/I
    (a) Describe the chemical property that the two metals have in common.
    (b) Describe the chemical composition of the white powder and the dark, scaly coating.

19. Describe three ways you come into contact with alloys in your daily life. (7.3) A

20. Beryllium metal (Be) and fluorine gas (F_2) react to form beryllium fluoride (BeF_2).
    (a) Draw a Bohr–Rutherford diagram to show how electrons are transferred during this reaction. (Hint: draw a beryllium atom, and then draw one fluorine atom on either side of the beryllium atom.)
    (b) Describe what is happening in this reaction in your own words. (7.1) C

21. Oxygen reacts rapidly with some materials and slowly with others. (7.1, 7.2, 7.5) A
    (a) Identify one material with which oxygen reacts rapidly and describe something people do to prevent this reaction.
    (b) Identify one material with which oxygen reacts slowly and describe something people do to prevent this reaction.

22. Consider the type of bonding in Cl_2 (chlorine gas) and in NaCl (table salt). Which substance is a molecule and which substance is an ionic compound? Explain your answer. (7.1, 7.3) T/I K/U

23. The chemical formulas of the four smallest hydrocarbons are shown below:

    $$CH_4 \quad C_2H_6 \quad C_3H_8 \quad C_4H_{10}$$

    The molecules are classified as saturated hydrocarbons because they have the maximum number of hydrogen atoms per carbon atom. Write an explanation or create a formula that can be used to determine the number of hydrogen atoms in a saturated hydrocarbon given the number of carbon atoms. (7.1) T/I C A

24. Draw atomic models of carbon dioxide (CO_2), water (H_2O), ammonia (NH_3), and methanol (CH_3OH) by rearranging the balls shown in Figure 1 below. You will need to use all the balls. (7.4) C

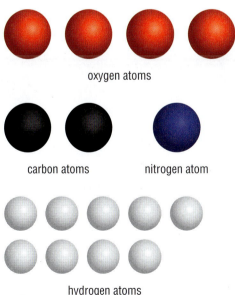

Figure 1

UNIT C

LOOKING BACK

UNIT C — Atoms, Elements, and Compounds

CHAPTER 5 — Properties of Matter

KEY CONCEPTS

 Physical properties are characteristics that can be determined without changing the composition of the substance.

 Chemical properties describe the ability of a substance to change its composition to form new substances.

 Pure substances have characteristic physical properties.

 Water has unusual characteristic physical properties.

 Physical and chemical properties can be used to identify different substances.

 Some common useful substances have negative impacts on the environment.

CHAPTER 6 — Elements and the Periodic Table

KEY CONCEPTS

 Elements cannot be broken down into simpler substances.

 Metals and non-metals have characteristic physical properties.

 Elements are organized according to their atomic number and electron arrangement on the periodic table.

 Atomic models evolved as a result of experimental evidence.

 Atoms contain protons and neutrons in a central core surrounded by electrons.

 Elements can be both beneficial and harmful to humans and to the environment.

CHAPTER 7 — Chemical Compounds

KEY CONCEPTS

 Elements can combine to form compounds.

 Metals and non-metals combine to form ionic compounds. Non-metals combine to form molecules.

 Some useful compounds have social, environmental, and economic impacts.

 Compounds can break apart into simpler substances.

 Molecular models are used to represent molecules.

 Simple chemical tests can identify common gases.

MAKE A SUMMARY

Shuffling the Deck

Prepare 25 cards about the size of playing cards. Write the name of each of the following items on a separate card.

- carbon dioxide
- charcoal
- copper
- DDT
- diamonds
- gold
- graphite
- helium
- hydrogen
- hydrogen peroxide
- iron
- magnesium
- mercury
- oxygen
- plastics
- potassium
- propane
- 10 kt gold
- rust
- salt
- sodium
- sugar
- tantalum
- vinegar
- water

Questions

1. On each card, write the chemical symbol or formula of the item where possible. T/I

2. Sort the cards into the following groups according to their properties. After each sorting, collect and shuffle all the cards and sort again according to the next grouping. In each sorting, explain your reasoning. T/I
    (a) elements, compounds
    (b) atoms, molecules, neither
    (c) metals, non-metals, neither
    (d) chemically reactive, chemically unreactive, unknown

3. Select all the cards of elements. Sort these cards according to chemical families and draw a Bohr–Rutherford diagram of each element that has an atomic number of 20 or lower. T/I

4. Use the Bohr–Rutherford diagrams you drew in question 3 to illustrate the relationship between properties and atomic structure of the element. T/I

5. Select three or more of the cards and explain why the item listed is both useful and harmful to society, the environment, or the economy. Suggest ways in which the harmful impacts may be lessened or reversed. T/I A

CAREER LINKS

List the careers mentioned in this unit. Choose two of the careers that interest you or choose two other careers that relate to Atoms, Elements, and Compounds. For each of these careers, research the following:

- educational requirements (secondary and post-secondary)
- skill/personality/aptitude requirements
- potential employers
- salary
- duties/responsibilities

Assemble the information you have discovered into a poster. Your poster should compare and contrast your two chosen careers and explain how they connect to Atoms, Elements, and Compounds.

GO TO NELSON SCIENCE

UNIT C UNIT TASK

A Greener Shade of Chemistry

Imagine this scenario. Your car is running out of fuel. Instead of stopping at the gas station, you go straight to a restaurant for a fill-up of deep fryer oil and any other greasy mess you can scrounge from the chef. This is exactly what an inventive group of students in the Niagara South School Board have in mind. They plan to convert greasy, energy-rich restaurant waste into a fuel called biodiesel. This fuel could be used to power cars (Figure 1). Local farmers will be able to use this fuel at little or no cost instead of regular diesel to run their machinery. Other than tractors smelling like French fries, the benefits to this plan are clear.

- Biodiesel is inexpensive.
- It is a good use of waste material.
- It reduces the demand for regular diesel fuel.

Figure 1 A driver pours deep fryer oil into his biodiesel-powered car.

This example illustrates a growing trend in our society towards a more environmentally friendly way of making and using consumer products through chemistry. This trend is not limited to Canadian homes and schools. Multinational chemical companies are recognizing that "greening" their products is good for business. What makes a product "green"? Table 1 shows a checklist you can use to determine whether a product is environmentally friendly.

Before making any chemical product, chemists need to have a clear idea of the chemical and physical properties the product should have. They also need to know the properties of the substances used to make the product. However, to be environmentally friendly, or "green," the product should meet as many criteria in Table 1 as possible.

> **SKILLS MENU**
> - Defining the Issue
> - Researching
> - Identifying Alternatives
> - Analyzing the Issue
> - Defending a Decision
> - Communicating
> - Evaluating

Table 1 Environmentally Friendly Product Checklist

Properties of environmentally friendly products	
	• designed to be less hazardous to human health and the environment than regular products
	• can be reused or recycled, or are made from reused or recycled materials
	• made from substances that are less hazardous to humans and the environment
	• manufactured using a process that consumes less energy and material

The Issue

Chemical products are useful, but they can have negative effects on our health and the environment. There are many environmentally friendly alternatives to chemical products. However, environmentally friendly products often cannot be used on a large scale due to practical limitations. For example, it is possible to fuel a few tractors with restaurant grease, but running all the cars in your neighbourhood on it is impractical for many reasons. So how "green" can chemical products become?

Goal

You have been hired as the sustainability expert for a chemical product manufacturer. Your task is to evaluate the "greenness" of a traditional consumer product of your choice using the checklist in Table 1 as a guideline. Sample products include

- computer parts
- diapers
- fuels
- concrete
- gold
- aluminum
- steel
- bleach
- cleaning products
- pesticides
- plastic water bottles
- batteries
- motor oil
- dry cleaning solvents
- paints

286 Unit C • Atoms, Elements, and Compounds

For the product that you have selected, you will need to consider

- the chemical and physical properties of the product
- the chemical and physical properties of the substances used to make the product
- uses of the product
- what happens to the product when it is no longer needed or when it is used up (Figure 2)

Figure 2 Be sure to consider what happens to a product after it is used.

Consider also which aspect of the product makes it problematic to humans or to the environment. Is it the actual product, its container, the production process, the disposal of the product or its container, or another aspect?

You will then consider "greener" alternatives to this product. How environmentally friendly are they? Are there any practical limitations to using these alternatives on a large scale?

Gather Information

Work in pairs or small groups to learn more about

- the chemical and physical properties of a chemical product
- the chemical and physical properties of substances used to make it
- the properties that make the product useful
- the properties of one or more environmentally friendly alternative product(s)

Identify Solutions

Assess the properties and potential uses of "greener" alternatives of the product you chose. What limitations, if any, are there to the widespread use of the more environmentally friendly alternatives?

Make a Decision

Which is the best environmentally friendly alternative product? Why? What next steps do you believe need to be taken for the environmentally friendly alternative to become more popular? Develop a plan of action that describes how individuals and organizations like your company can implement these steps.

Communicate

Prepare an advertisement that your company can use to market the "green" product. The ad should outline the benefits of the environmentally friendly product, both for the consumer and the environment. The ad is intended for the general public and can be in any format, such as a television or radio commercial, a website, a billboard, and so on.

ASSESSMENT CHECKLIST

Your completed Performance Task will be evaluated according to the following criteria:

Knowledge/Understanding
- ☑ Research and classify the properties of the traditional consumer product.
- ☑ Research and classify the properties of important substances used to make the product.

Thinking/Investigation
- ☑ Compare the "greenness" of the traditional product and its more environmentally friendly alternative.
- ☑ Identify any practical limitations to the widespread use of the environmentally friendly alternative.

Communication
- ☑ Prepare and present your advertisement for the environmentally friendly product in a creative and informative way.
- ☑ Include a comparison of both products.

Application
- ☑ Demonstrate an understanding of the threats posed by the product to our health and to the environment.
- ☑ Propose a course of action to promote the use of the environmentally friendly product.

UNIT C REVIEW

The following icons indicate the Achievement Chart category addressed by each question.

K/U Knowledge/Understanding T/I Thinking/Investigation C Communication A Application

What Do You Remember?

For each question, select the best answer from the four alternatives.

1. Which of the following is a list of characteristic physical properties? (5.6) K/U
   (a) acidity, lustre, and viscosity
   (b) colour, boiling point, and reactivity
   (c) density, melting point, and freezing point
   (d) malleability, conductivity, and flammability

2. Which of the following statements is true according to the particle theory of matter? (5.1) K/U
   (a) All particles of matter move.
   (b) Only particles of solids move.
   (c) Particles of matter never move.
   (d) Only particles of liquids and gases move.

3. Which of the following groups of elements is the least reactive? (6.4) K/U
   (a) halogens
   (b) noble gases
   (c) alkali metals
   (d) alkaline earth metals

4. Which of the following is an element as well as a molecule? (7.1) K/U
   (a) CO_2 (c) O_3
   (b) Ca   (d) 14 kt gold

5. Which group of elements belongs in the alkaline earth metal family? (6.4) K/U
   (a) Li, Be, B   (c) Mg, Ca, Sr
   (b) Li, Na, K   (d) O, S, Se

6. In the periodic table, elements are arranged according to
   (a) atomic mass
   (b) number of neutrons
   (c) number of protons
   (d) mass number (6.7) K/U

7. The atomic number tells us the number of
   (a) neutrons
   (b) protons
   (c) protons and neutrons
   (d) electrons in the outermost orbit (6.7) K/U

8. An electron
   (a) carries a negative charge and is located in the nucleus
   (b) carries a negative charge and is very small compared to the proton
   (c) carries a positive charge and is very small compared to the neutron
   (d) carries no charge and orbits the nucleus (6.7) K/U

9. The identity of an element is determined by
   (a) the mass of the atom
   (b) the number of electron orbits in the atom
   (c) the number of neutrons in the atom
   (d) the number of protons in the atom (6.7) K/U

10. The outermost electron orbits of the noble gases all have
    (a) one electron
    (b) two electrons
    (c) seven electrons
    (d) the maximum number of electrons (6.7) K/U

11. From his gold foil experiment, Ernest Rutherford theorized that the atom
    (a) is mostly empty space with positive charges evenly distributed throughout
    (b) is mostly empty space with negative charges evenly distributed throughout
    (c) consists of a dense central nucleus containing positively charged protons
    (d) consists of a dense central nucleus containing negatively charged electrons (6.6) K/U

12. Bohr's atomic model was a result of his experiments, which showed
    (a) negatively charged particles bending in a vacuum tube
    (b) positively charged particles bending in a vacuum tube
    (c) positively charged particles bouncing back from gold atoms
    (d) hydrogen atoms emitting a few lines of colour instead of a complete rainbow (6.6) K/U

Indicate whether each of the statements is TRUE or FALSE. If you think the statement is false, rewrite it to make it true.

13. Mixtures can exist as solids, liquids, or gases. (5.1) K/U

14. In a neutral atom, the number of electrons always equals the number of neutrons. (6.7) K/U

15. The lustre of a substance is its breakability or flexibility. (5.2) K/U

16. In a physical change, a new substance is produced. (5.2) K/U

17. Burning is an example of a chemical change. (5.3) K/U

18. Non-metals are lustrous, ductile, and malleable. (6.1) K/U

19. Compounds are made up of elements. (6.1) K/U

20. Compounds cannot be broken down into simpler substances. (6.1, 7.6) K/U

21. To determine the number of neutrons, subtract the mass number from the atomic number. (6.7) K/U

22. In Bohr's atomic model, the first electron orbit holds a maximum of 8 electrons. (6.6) K/U

23. In the alkali metal family, the elements lower in the column are more reactive. (6.5, 6.7) K/U

24. Examples of non-metals are hydrogen, oxygen, and nitrogen. (6.1) K/U

25. A gas that can re-ignite a glowing splint is oxygen. (7.5) K/U

Copy each of the following statements into your notebook. Fill in the blanks with a word or phrase that correctly completes the sentence.

26. Matter is anything that has _____ and takes up space. (5.1) K/U

27. The _____ of a substance is its mass per unit of volume. (5.6) K/U

28. WHMIS stands for _____. (5.4) K/U

29. The correct technique for smelling a chemical in a lab is called _____. (5.4) K/U

30. A change of state is a _____ change. (5.2) K/U

31. A change of colour is evidence of a _____ change. (5.3) K/U

32. Corrosion is the reaction between a metal and _____. (Ch.7 Engage in Science) K/U

33. The common name for acetic acid is _____. (7.3) K/U

34. The common name for sodium chloride (NaCl) is _____. (7.3) K/U

35. _____ proposed the plum pudding model of the atom. (6.6) K/U

36. The element whose atom has 16 protons is _____. (6.7) K/U

37. The process by which a compound breaks down into simpler substances is called _____. (7.6) K/U

38. The gas that "pops" when ignited with a burning splint is _____. (7.5) K/U

39. Carbon dioxide gas turns _____ cloudy. (7.5) K/U

Match each term on the left with the most appropriate description on the right.

40.
    (a) table salt
    (b) limestone
    (c) vinegar
    (d) baking soda
    (e) stomach acid
    (f) alcohol in wine
    (g) Aspirin
    (h) drain cleaner, lye

    (i) acetic acid
    (ii) hydrochloric acid
    (iii) ethanol
    (iv) sodium chloride
    (v) acetylsalicylic acid (ASA)
    (vi) sodium hydroxide
    (vii) calcium carbonate
    (viii) sodium hydrogen carbonate (7.3) T/I

41.
    (a) element
    (b) solution
    (c) compound
    (d) mechanical mixture

    (i) is composed of multiple pure substances that are chemically joined
    (ii) includes different types of matter that are visibly distinguishable
    (iii) cannot be broken down into a simpler chemical substance
    (iv) contains two or more substances that combine as a uniform mixture (5.1, 6.1) K/U

42. (a) Democritus   (i)   electron orbits
    (b) Dalton   (ii) an indivisible particle called the atom
    (c) Thomson   (iii) the billiard ball atomic model
    (d) Rutherford   (iv) the "plum pudding" atomic model
    (e) Bohr   (v)   the nucleus and the proton (6.6) K/U

Write a short answer to each of these questions.

43. You have mixed eggs, sugar, and flour and baked the batter in the oven. An hour later, delicious cake smells bring you back to the kitchen. List as many pieces of evidence as you can to show that a cake baking is a chemical change. (5.3) T/I

44. Identify each of the following as a physical change or a chemical change, and give reasons for your answer. (5.2, 5.3) K/U
    (a) water droplets forming on the bathroom mirror
    (b) grinding wheat into flour
    (c) toasting marshmallows
    (d) making a peanut butter and jelly sandwich
    (e) lighting a Bunsen burner
    (f) dissolving sugar in tea
    (g) beating a raw egg until it is creamy

45. Complete the following table in your notebook with the words Yes or No. (6.1, 7.1) K/U T/I

Formula	Element	Compound	Atom	Molecule
H_2O_2				
Fe				
N_2				
CH_4				
O_3				

46. Complete the following table with the correct information about each atom. (6.7) K/U T/I

Element name	Element symbol	Atomic number	Mass number	Number of protons	Number of neutrons	Number of electrons
potassium			39			
	Sn		119			
		20	40			
					28	14
				19		9

What Do You Understand?

47. Draw a diagram showing the relationships among the following terms: mixture, matter, solution, pure substance, and mechanical mixture. (5.1) K/U C

48. If metals do not form compounds with other metals, explain the existence of alloys. (7.3) K/U

49. Name two elements, two compounds, and two mechanical mixtures you can find in most kitchens. (5.1, 6.1, 6.3) A

50. Give an example of the following types of solutions. (5.1) K/U
    (a) gas combined with gas
    (b) solid in a liquid
    (c) liquid in a liquid

51. (a) What is the difference between a qualitative property and a quantitative property?
    (b) List two qualitative properties and two quantitative properties.
    (c) How can a property such as optical clarity be either a quantitative or qualitative property? (5.2) K/U T/I

52. Complete the following table. (7.3) K/U

Compound	Types of elements (metal/non-metal)	Type of compound
NaCl		
CO_2		molecular
KF		
SO_2	non-metals	
CaO		

53. Your friend says that matter is "anything you can see and touch." What is wrong with this definition? (6.7) T/I

54. Which number is always equal to or larger than the other, the atomic number or the mass number? Explain your answer. (6.7) K/U

55. Draw a Bohr–Rutherford diagram for each atom below. (6.7) T/I

    (a) Ar-40
    (b) P-31
    (c) H-1

56. A new element has been discovered and evidence shows that it may be an alkali metal. Describe the physical properties and electron arrangements that this element should exhibit for this classification. (6.4, 6.5, 6.7) K/U

57. Use elements in the alkaline earth metal family to explain how electron arrangements within a family are both alike and different. (6.7) K/U T/I

58. Use elements in the second period to explain how electron arrangements across a period are both alike and different. (6.7) K/U T/I

59. Are there more metals or non-metals listed on the periodic table? Explain how you can tell. (6.1) K/U

60. If ice did not float on water, life on our planet would be quite different. (5.6) T/I A

    (a) Explain why ice floating on water is an unusual property for a substance.
    (b) How would the life of a fish in a pond change if ice did not float on water?
    (c) Describe the hazards of water's unusual behaviour to a cottage owner, to the captain of the *Titanic*, and to anyone chilling a bottle of soft drink in the freezer.

61. What is the same about the following processes: bleaching with hydrogen peroxide, burning, and rusting? What is different about them? (7.1, 7.2, 7.5, 7.6) T/I

62. Take a periodic table that is printed on a sheet of paper. Roll the paper like a newspaper so that the alkali metal family is right beside the noble gas family. It is no longer a periodic table, but rather a periodic cylinder. (6.7) T/I

    (a) Use your finger to trace the elements from hydrogen onward in increasing atomic numbers.
    (b) Is this periodic cylinder more logical than a periodic table? Explain.
    (c) Is it less logical? Explain.

63. Draw Bohr–Rutherford diagrams of potassium and chlorine. (6.7) T/I C

    (a) Compare the electron arrangements of potassium and chlorine to that of argon. Suggest how each atom may become chemically stable, as noble gases are.
    (b) Propose an explanation for the stability of the compound potassium chloride, KCl.

Solve a Problem

64. You are given a sample of an unidentified element. (5.6, 6.2) T/I

    (a) Describe a series of tests that would help you determine if the substance is a metal.
    (b) After determining if the substance is metallic, you wish to identify the specific element in the sample. Describe how you could use a characteristic physical property to determine the element's identity.

65. Some transition metals, such as silver and gold, are found in nature in pure form. What can you conclude about the reactivity of these metals? (6.4) T/I

66. (a) You have just chopped several pieces of wood for the fireplace in your home. How does the density of a piece of wood compare before and after it is cut in half? (5.6) A
    (b) Design a test with a sample of wood to validate your answer in part (a). T/I

67. Research and report on safe disposal methods for halogen lamps (Figure 1). (6.4) T/I

Figure 1

68. Research and report on the symptoms and long-term effects of mercury poisoning in living things. What are the main sources of mercury in the environment today? What can be done to reduce or eliminate these sources? (Ch. 6 Engage in Science) T/I A

Create and Evaluate

69. Design and produce a safety poster that can be displayed in the laboratory. (5.4) K/U C
    (a) List the key safety rules of behaviour in the laboratory. Illustrate each rule with a simple diagram.
    (b) List the locations of all emergency exits and safety equipment: fire extinguishers, eye wash, first aid kit, etc.
    (c) Draw the key WHMIS symbols and explain their meaning.

70. Use the evolution of atomic theory to illustrate the notion that advances in technology lead to changes in theory. Is it possible that changes in theory can lead to advances in technology? Explain. (6.6) A

71. Northern Ontario boasts vast regions of forests and lakes, as well as land rich in many precious metals. The Sudbury basin produces 9 % of the world's nickel, as well as considerable amounts of copper, silver, and zinc. In Lac des Iles, Canada's only platinum mine is ranked among the top 10 in the world. Create a table with two columns: one column relating the benefits, and the other column relating the stresses that would result from the presence of a successful mining operation in the region. (6.8) K/U

72. One type of solder is a metal alloy of tin and bismuth. When a length of solder wire is attached to a heavy mass at one end and suspended from the ceiling at the other end, it slowly increases in length. (5.1, 5.2, 7.1, 7.3) K/U T/I A C
    (a) What physical property of solder does this observation demonstrate?
    (b) Design a method to demonstrate the other properties of solder, namely, malleability and electrical conductivity.
    (c) Consider the properties of tin–bismuth solder and of 10 kt gold. Are some properties of the metals in an alloy retained in the alloy itself? Explain.
    (d) Now consider the properties of the compound table salt, sodium chloride (NaCl). Are some properties of the elements in the compound retained in the compound? Explain.

73. Artists rely on compounds and mixtures to create paintings, sculptures, and other works of art. Using supplies available in school or at home, create a work of art based on the content in this unit. Keep a log of the compounds and mixtures you use as you complete the activity. (Hint: You may need to consult the product labels or online resources to determine whether the substances are compounds or mixtures.) When you present your work of art to the class, identify and describe the types of substances you used. (5.1, 6.1) C

74. Two white powdery solids are heated in a burner flame. Solid A quickly turns into a liquid. Solid B releases an invisible gas which pushes the powder out of the test tube.
    (a) Which solid is most likely a compound? Why?
    (b) Which solid is most likely an element? Why?
    (c) Describe a physical test that could be done to help identify the solid in (b). (7.1) K/U T/I

75. Figure 2 on page 211 shows electricity passing through water to produce hydrogen and oxygen.
    (a) Describe a chemical test that could be done to identify which test tube contains oxygen.
    (b) Why is it impossible for one of the gases to be carbon dioxide? (6.1) K/U T/I

76. Eagles are usually the top carnivore in their food chain. A vulture is an example of a bird that feeds on the dead bodies of animals from a variety of levels on a food chain. Which bird has the greatest risk of having the most toxins in its tissues? Why? (7.8) K/U T/I A

Reflect on Your Learning

77. Complete the following statement: "The periodic table is amazing because . . ." A C

78. Complete the following statement: "The periodic table is difficult to use because . . ." A C

79. When people first develop new ideas, they often use old familiar words to verbalize those ideas. Give examples of names used in the atomic model that are borrowed from other areas, such as history, geography, and literature. A

80. If you have a favourite sweater, would you want it to be durable or degradable? Give your reasons and discuss any problems that may arise from your choice. A C

81. Biology is the study of living things, such as plants and animals. Electrostatics is the study of charged particles. Astronomy is the study of stars and planets. Can you think of reasons why chemistry is often called the "central science"? A

82. How does studying matter help you to be a better consumer and a better citizen?

Web Connections

83. A space shuttle gets its propulsion from two tanks of fuel. One tank contains liquid oxygen, and the other tank contains liquid hydrogen. (7.5) T/I A
    (a) From your knowledge of the properties of hydrogen and oxygen, explain why a combination of these two elements makes a good fuel mixture for the space shuttle.
    (b) Research the safety procedures that are needed in the use of hydrogen as a fuel.

84. Although there are many synthetic compounds that are harmful to human health and to the environment, there are many natural compounds with their own hazards. Many plants produce their own toxins as a means of self-defence. A famous example is the hemlock plant, *Conium maculatum*, which contains a deadly poison (Figure 2). Research a natural substance that is harmful to our health or to the environment. Find ways in which the effects of this substance may be controlled. (Unit C Task) A

Figure 2

85. One news report shows wealthy countries banning synthetic pesticides to protect their health and the environment. A second report shows starving children living in a country that cannot grow enough food. Do you think that the use of pesticides is a global problem that requires a global decision? Give reasons to support your answer. (7.8) A C

UNIT C

SELF-QUIZ

The following icons indicate the Achievement Chart category addressed by each question.

K/U Knowledge/Understanding
C Communication
T/I Thinking/Investigation
A Application

For each question, select the best answer from the four alternatives.

1. Which of the following is an example of a solution? (5.1) K/U
   (a) cereal
   (b) carbon
   (c) solder
   (d) water

2. Boiling point is the temperature at which a substance changes rapidly from
   (a) a solid to a liquid
   (b) a liquid to a gas
   (c) a liquid to a solid
   (d) a gas to a solid (5.6) K/U

3. Which statement describes a main idea of the particle theory of matter? (5.1) K/U
   (a) Particles are in constant random motion.
   (b) All matter is made up of the same kind of particle.
   (c) Particles that make up matter naturally repel each other.
   (d) The particles of a substance move faster as its temperature decreases.

4. The property of metals that allows them to be hammered into thin sheets is
   (a) lustre
   (b) malleability
   (c) ductility
   (d) viscosity (5.2) K/U

Indicate whether each of the statements is TRUE or FALSE. If you think the statement is false, rewrite it to make it true.

5. An alloy is a compound formed by chemically joining two or more metals. (7.3) K/U

6. Ionic compounds consist of charged particles called ions. (7.1) K/U

7. Pure water is an example of a mixture. (5.1) K/U

Copy each of the following statements into your notebook. Fill in the blanks with a word or phrase that correctly completes the sentence.

8. Atoms that lose electrons assume a positive charge and become _____. (7.1) K/U

9. Colour and texture are examples of _____ properties, whereas mass and temperature are examples of _____ properties. (5.2) K/U

Match each term on the left with the most appropriate description on the right.

10. (a) copper   (i) solution
    (b) carbon dioxide   (ii) compound
    (c) clear apple juice (iii) element
    (d) trail mix   (iv) mechanical mixture
                          (5.1, 6.1) K/U

Write a short answer to each of these questions.

11. Classify each of the following properties of carbon dioxide as physical or chemical. (5.2, 5.3) K/U
    (a) colourless
    (b) forms carbonic acid in an aqueous solution
    (c) will not burn
    (d) freezes at −78.5 °C
    (e) odourless

12. Name one similarity and one difference between water, water vapour, and ice. (5.1, 5.6) K/U

13. Indicate whether J.J. Thomson would have agreed or disagreed with each of the following statements. Give reasons for your answers. (6.6) K/U
    (a) Atoms of all elements contain electrons.
    (b) Electrons have a negative charge.
    (c) Electrons are confined to the centre of the atom.

14. Explain why the bottom plates on irons are made of metals, but the handles are made of non-metals. (6.1) T/I

15. Two elements are examined. Element A is a colourless gas that causes a red crusty material to form on the surface of steel. Element B is a soft metal that reacts quickly with water to produce a gas. For each element, predict the
    (a) group on the periodic table it belongs to
    (b) number of electrons in its outermost orbit
    (c) charge it acquires when it becomes an ion
    (d) chemical formula of the compound formed when these elements combine (7.2) K/U T/I

16. Write the chemical formulas for compounds with
    (a) one calcium atom and two chlorine atoms
    (b) two nitrogen atoms and four oxygen atoms
    (c) two sodium atoms and two sulfur atoms (7.1) K/U

17. (a) How does the organization of the periodic table allow us to determine the properties of elements?
    (b) Name four properties of calcium you can determine based on its position on the periodic table. (6.1, 6.4, 6.7) T/I

18. How is an element's atomic number like a person's fingerprint? (6.7) K/U

19. Describe how you could use a piece of paper, scissors, and a match to explain and demonstrate physical and chemical changes to someone. (5.2, 5.3) T/I C

20. Dry ice is frozen carbon dioxide. Dry ice changes directly from the solid phase to the gaseous phase in a process called sublimation (Figure 1). Is sublimation a physical change or a chemical change? Explain. (5.2, 5.3) T/I

21. Consider the elements P, Ca, N, K, O, and Si. (6.1, 7.3) K/U T/I
    (a) Classify each as a metal, a non-metal, or a metalloid.
    (b) Choose two elements from this list that would likely form an ionic compound.
    (c) Choose two elements from this list that would likely form a molecular compound.

22. In your own words, explain why the term "periodic" is used to describe the table of elements. (6.4) C

23. Pyrite (fool's gold) has a density of 5.02 g/cm³. Gold has a density of 19.32 g/cm³. Determine whether each of the following samples is pyrite or gold. (5.6) T/I
    (a) A sample has a mass of 2.90 g and a volume of 0.15 cm³.
    (b) A sample has a mass of 55.7 g and a volume of 11.1 cm³.

24. Explain each of the following situations in terms of particle motion. (5.1) A
    (a) Aerosol cans have warnings that state "do not incinerate."
    (b) If a jar lid is stuck, running the lid under hot water allows it to be removed from the jar.
    (c) Expansion joints are needed in buildings and on bridges.

25. (a) Draw the Bohr–Rutherford diagrams of nitrogen and phosphorus.
    (b) How are the diagrams similar and how are they different?
    (c) Predict the charge of the ion that these elements form. (6.7) K/U T/I C

26. Some toxic compounds decompose into safer substances when burned at high temperatures. Why would this method of disposal not work for toxic elements like lead and cadmium? (7.1) K/U A

Figure 1

UNIT D: The Study of the Universe

OVERALL Expectations

- assess some of the costs, hazards, and benefits of space exploration and the contributions of Canadians to space research and technology
- investigate the characteristics and properties of a variety of celestial objects visible from Earth in the night sky
- demonstrate an understanding of the major scientific theories about the structure, formation, and evolution of the Universe and its components and of the evidence that supports these theories

BIG Ideas

- Different types of celestial objects in the Solar System and Universe have distinct properties that can be investigated and quantified.
- People use observational evidence of the properties of the Solar System and the Universe to develop theories to explain their formation and evolution.
- Space exploration has generated valuable knowledge but at enormous cost.

Focus on STSE

ENCOUNTER WITH TITAN

The Cassini Equinox Mission is an 11-year spacecraft mission designed to study Saturn and its moons. In 1997, the *Cassini-Huygens* spacecraft was launched from Earth. In 2004, as it approached Saturn, the spacecraft released *Huygens*, a small probe. The *Huygens* probe was sent to the surface of Titan, Saturn's largest moon, to collect data and take photos, such as the second image shown on the right-hand side of this page.

The mission has revealed that Titan contains two major components of Earth's atmosphere—nitrogen and oxygen. Also, Titan's atmosphere is filled with a brown-orange haze containing organic molecules. In many ways, Titan appears to resemble a primitive Earth.

Although most scientists agree that conditions on Titan are too cold for life to evolve, understanding chemical interactions on the distant moon may help us better understand the chemistry of early Earth and how we came to be. However, the total cost of the mission was more than $3 billion. Many people think that this money would be better spent on education or health care.

Think/Pair/Share

With a partner, discuss each of the following questions:

1. What do you think we might learn about Earth and our own Moon from missions to other planets and moons?
2. Is space exploration worth spending such large sums of money?
3. If you were in charge of spending money on space exploration, how would you ensure that the money was spent in a fair way?
4. Would you support a mission to Titan with people on board? Explain why or why not.
5. If you could go anywhere in the Solar System, where would you want to go and why?

UNIT D — LOOKING AHEAD

UNIT D: The Study of the Universe

CHAPTER 8: Our Place in Space

Observing the motion of celestial objects, such as the Sun, can help us learn about their properties.

CHAPTER 9: Beyond the Solar System

Celestial objects beyond the Solar System have unique properties.

CHAPTER 10: Space Research and Exploration

We must weigh the benefits of space exploration and technologies with the associated costs and hazards.

UNIT TASK Preview

Celestial Travel Agency

In the past few decades, we have greatly increased our knowledge of space and astronomy. While space travel has not yet become commonplace, a few space tourists have paid for the opportunity to view Earth from a different perspective (Figure 1).

(a)   (b)

Figure 1 (a) *SpaceShipOne* completed the world's first commercial (non-governmental) human space flight in October 2004. Breathtaking photos such as the one shown in (b) were taken by passengers.

As you work through this unit, you will develop the skills and knowledge needed to design an advertising campaign for a celestial travel agency. In the Unit Task, you will choose among three space travel options: a short journey travelling into Earth's orbit, a medium journey travelling to another planet, or a long journey into deep space.

Once you have chosen a space travel option to further investigate, you will

- research the destination of the travel option you select
- use a star map to locate the position of your destination
- create an advertising campaign for your space travel option
- present your advertising campaign to the class

UNIT TASK Bookmark

The Unit Task is described in detail on page 446. As you work through the unit, look for this bookmark and see how the section relates to the Unit Task.

ASSESSMENT

You will be assessed on how well you

- research, compile, and analyze your data
- use correct vocabulary and terminology
- use direct observation and star maps to determine the location of your destination
- prepare and present an advertisement that effectively communicates information

What Do You Already Know?

PREREQUISITES

Concepts
- Identify bodies in space that emit light (e.g., stars) and those that reflect light (e.g., moons and planets)
- Physical characteristics of Solar System components
- Effects of the relative position and motions of Earth, the Moon, and the Sun
- How humans meet their basic biological needs in space
- Canadian contributions to space science

Skills
- Identify technological tools and devices used for space exploration
- Investigate scientific and technological advances that allow humans to adapt to life in space
- Assess the impact of space exploration on society and the environment
- Analyze issues related to space exploration
- Use a variety of forms to communicate with different audiences and for a variety of purposes
- Define terms and use them in context

1. Figure 1 shows the Sun and the Moon. Describe the effects of the relative positions and motions of these two celestial objects. K/U

Figure 1 (a) the Sun and (b) the Moon

2. Copy Table 1 into your notebook. Categorize the following objects into the correct column: Mars, the Moon, the Sun, Earth, Jupiter, Mercury, stars. K/U

Table 1

Emits light	Reflects light from the Sun

3. Make a list of all the planets in the Solar System (Figure 2). K/U

Figure 2

4. Explain the astronomical phenomenon shown in Figure 3. K/U

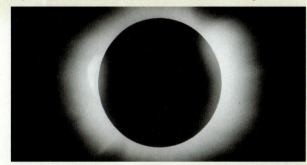

Figure 3

5. Name some of the contributions Canada has made to space exploration (Figure 4). K/U

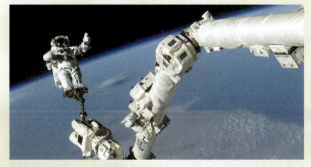

Figure 4

6. Look at Figure 5. Identify the object shown, and its purpose. K/U

Figure 5

CHAPTER 8: Our Place in Space

KEY QUESTION: What are the different celestial objects visible in the sky, and what are their properties?

This multiple exposure image shows the motion of the Sun in the sky as it rises from east to west.

UNIT D
The Study of the Universe

CHAPTER 8 — Our Place in Space

CHAPTER 9 — Beyond the Solar System

CHAPTER 10 — Space Research and Exploration

KEY CONCEPTS

Careful observation of the night sky can offer clues about the motion of celestial objects.

Celestial objects in the Solar System have unique properties.

Some celestial objects can be seen with the unaided eye and can be identified by their motion.

The Sun emits light and other forms of radiant energy that are necessary for life to exist on Earth.

Satellites have useful applications for technologies on Earth.

The study of the night sky has influenced the culture and lifestyles of many civilizations.

ENGAGE IN SCIENCE

The Tunguska Event

This photo shows the Tunguska area after the blast.

The cold, windswept plains of northern Russia are usually empty and quiet. That all changed one June morning in 1908. Indigenous Russians and explorers in the area reported seeing the sky erupt in a huge flash of blue light. The sound of an enormous explosion ripped through the air, and the ground rumbled. A large object from the Solar System had entered Earth's atmosphere and exploded in the air in a giant fireball.

The explosion flattened 12 million trees over a region of 2000 km^2, setting many of them on fire. People in the area were knocked off their feet, and windows in houses were shattered from the force. The explosion could be heard from hundreds of kilometres away. It was equivalent to 10 megatonnes of TNT (an explosive chemical)—about 1000 times as powerful as an atomic bomb. While no people were killed in the blast, many animals lost their lives.

Since that day, scientists have explored the area for clues as to what happened. Based on the patterns of the flattened trees, scientists were able to find the point of explosion. They calculated that the object must have been 5 to 10 km above Earth when it exploded. Tiny chunks of matter from the explosion were found burnt into the trees. The explosion is commonly known as the Tunguska Event.

In 2007, a team of scientists discovered a circular body of water, Lake Cheko, that they thought might be an impact crater caused by the explosion. Scientists have been examining the soil on the lake bottom and in the surrounding region to see if its composition can provide clues as to what exploded in Earth's atmosphere.

Small objects from space enter Earth's atmosphere every day. Most of them burn up in the atmosphere before reaching the ground. However, every hundred years or so, large objects from space hit Earth. In recent years, scientists have started to think about what would happen if such an object hit Earth again, and whether we could do something to prevent it from causing damage.

Can you think of any movies or TV shows that tell fictional stories about the impact of a giant object from space? With a partner, discuss how you might raise awareness that these types of events can happen. Develop an action plan and share it with the class.

The Tunguska blast affected a large portion of Siberia.

Lake Cheko

WHAT DO YOU THINK?

Many of the ideas you will explore in this chapter are ideas that you have already encountered. You may have encountered these ideas in school, at home, or in the world around you. Not all of the following statements are true. Consider each statement and decide whether you agree or disagree with it.

1 Scientists have seen signs of life on other planets.
Agree/disagree?

4 The stars and constellations can be used to map the night sky, as well as for accurate navigation.
Agree/disagree?

2 Many planets in the Solar System have moons.
Agree/disagree?

5 We always see the same side of the Moon.
Agree/disagree?

3 Canada is a world leader in satellite technology.
Agree/disagree?

6 Storms on the surface of the Sun can affect Earth.
Agree/disagree?

FOCUS ON READING

Finding the Main Idea

When you look for the main idea of a text, you identify a general topic and a related key concept. Details are usually given to support the main idea. Use the following strategies to find the main idea

- use clues in the title and the headings
- check the first, second, and last sentences
- look for clues in the figures and captions
- look for signal words (therefore, consequently, for this reason)
- look for repeated words and bolded words

READING TIP

As you work through the chapter, look for tips like this. They will help you develop literacy strategies.

The Force of Gravity

What keeps the planets orbiting the Sun, and the Moon orbiting Earth? There is a force of attraction between all objects in the Universe with mass, known as **gravitational force**, or gravity. The greater the mass of an object, the stronger its gravitational force. The gravitational force of our massive Sun is strong enough to keep Earth in orbit. Without the Sun's gravity, Earth would move away from the Sun into space (Figure 1). Similarly, the strong pull of Earth's gravity keeps the Moon in orbit.

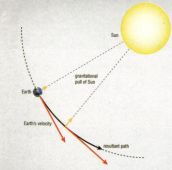

Figure 1 Earth stays in a stable orbit because of the balance between its forward speed and the Sun's gravitational pull.

Finding the Main Idea *in Action*

A main idea narrows a topic to a specific point or opinion. All the details in the body of the text support the main idea with facts, data, or examples. Clues in the body of text can be used to figure out the main idea of a reading. Below is an example of how one student used clues in the above reading to find the main idea.

Clues	Information from the text
Use titles and headings	The heading indicates it is about the force of gravity.
Check the first sentence	It asks a question: "What keeps planets orbiting the Sun, and the Moon orbiting Earth?" The answer will likely be the main idea.
Look for bolded words and repeated words	The term "gravitational pull" is bolded and is repeated in the reading. This must be an important idea.
Look at the photos and captions	Figure 1 is about gravitational pull.
Main Idea: Gravitational force keeps the planets orbiting the Sun, and the Moon orbiting Earth.	

Touring the Night Sky

8.1

People have been fascinated by the wonders of the night sky for thousands of years. **Astronomy** is the branch of science that studies objects beyond Earth, in what we sometimes refer to as "outer space." Any object in space—for example, the Sun or the Moon—is considered to be a **celestial object**. Everything that physically exists is part of what we call the **Universe**. We can see many of the celestial objects in the Universe by simply gazing up at the sky (Figure 1).

astronomy the scientific study of what is beyond Earth

celestial object any object that exists in space

Universe everything that exists, including all energy, matter, and space

Figure 1 You do not need a telescope to begin exploring the night sky.

Stars

Most of the bright points of light that we see in the night sky are stars. A **star** is a massive celestial body composed of hot gases that radiates large amounts of energy. Stars appear tiny in the sky because they are so far away. We are able to see many stars in the night sky because they are **luminous**, which means that they produce and emit light. Astronomers have located billions of stars, many of which we cannot see with the unaided eye.

star a massive collection of gases, held together by its own gravity and emitting huge amounts of energy

luminous producing and giving off light; shining

Our Star, the Sun

The Sun is a star (Figure 2). Compared to other stars, the Sun is average in size. Nevertheless, the Sun has a mass that is almost 340 000 times that of Earth and a volume that is 1 300 000 times the volume of Earth!

These very large numbers (or very small numbers) can also be written in scientific notation—a number between 1 and 9 multiplied by powers of 10. For example, the number 340 000 can be written as 3.4×10^5. To convert numbers written in scientific notation, such as 1.3×10^6, just move the decimal place to the right (for positive exponents) or to the left (for negative exponents) as many places as the exponent indicates. (In this case, move the decimal 6 places to the right: $1.3 \times 10^6 = 1\,300\,000$.)

The Sun appears to be so much bigger and brighter than other stars in the sky because of its proximity to Earth: it is only 1.5×10^8 km away. The next closest star is nearly 300 000 (4.3×10^{13} km) times farther away than the Sun. Since it is relatively close to Earth, we can see details on the surface of the Sun that we cannot see on more distant stars.

MATH TIP

Scientific Notation

To convert numbers written as scientific notation, just move the decimal place to the right (or to the left) as many places as the exponent indicates.

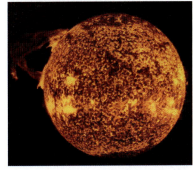

Figure 2 The Sun is a star—an enormous ball of hot, glowing gases.

Life would not be possible on Earth without the energy produced by the Sun. The Sun gives off visible light and other forms of radiant energy, as well as releasing what is called the solar wind—a stream of high-energy particles. Only a small fraction of the Sun's light reaches Earth, but it is enough to keep water in its liquid state and provide life on Earth with the energy needed for survival.

Planets

planet a large, round celestial object that travels around a star

A **planet** is a large celestial object that travels around a star. There are eight planets travelling around the Sun—Mercury, Venus, Earth, Mars, Jupiter, Saturn, Uranus, and Neptune. Each planet differs from the other planets in size, composition, atmosphere, and length of day and year. The four planets closest to the Sun are known as the terrestrial planets. They have a hard and rocky surface similar to Earth's. The next four planets are composed mostly of gases and liquids. They are known as the gas giants.

Planets are non-luminous. Although they do not produce and emit light, we can see planets because they reflect light from luminous objects, such as the Sun and other stars. There are five planets visible with the unaided eye—Mercury, Venus, Mars, Jupiter, and Saturn.

> **DID YOU KNOW?**
> **Only Eight Planets?**
> Until recently, Pluto was considered to be the ninth planet. However, Pluto was reclassified as a "dwarf planet" in 2006.

Our Planet, Earth

Earth is the third planet from the Sun and the fourth largest in the **Solar System**, which consists of the Sun, together with all the planets and celestial objects that travel around it. Earth is a terrestrial planet composed primarily of rock. Like the other planets, Earth is in constant motion. (You will learn more about this in Section 8.5.) Earth differs from the other planets in the Solar System because it has a diversity of life forms and large quantities of water.

Solar System the Sun and all the objects that travel around it

satellite a celestial object that travels around a planet or dwarf planet

orbit the closed path of a celestial object or satellite as it travels around another celestial object

Moons

A moon is a type of **satellite**: a celestial object that travels around a planet or dwarf planet in a closed path. The closed path of a celestial object or satellite as it travels around another (usually larger) celestial object is called an **orbit**. Although some planets have moons in orbit, others do not. For example, Mercury and Venus do not have moons in orbit, whereas Jupiter and Saturn each have 60 or more moons.

Earth's Companion, the Moon

Earth has one natural satellite called the Moon. The Moon is non-luminous. We are able to see it only because sunlight reflects off its surface (Figure 3).

Although the Moon appears to be the biggest and brightest celestial object in the night sky, it is small compared to the planets. It has a diameter four times smaller than that of Earth. It appears large because it is close to Earth. However, it is an average distance of 384 000 km away from Earth—about 55 times the distance between Vancouver, British Columbia, and St. John's, Newfoundland.

Figure 3 The Moon reflects the Sun's light, which makes it appear bright in the night sky.

Galaxies

Within the Universe are huge collections of stars, gas, dust, and planets, which we call **galaxies**. Astronomers looking through telescopes are discovering that there are billions of galaxies scattered throughout the Universe.

Our Galaxy, the Milky Way

Earth is part of the Milky Way galaxy. It contains more than 200 billion stars, including the Sun. It also contains many other celestial objects. From Earth, the Milky Way appears as a hazy band of white light in the night sky.

Astronomers and scientists continue to study stars, planets, galaxies, and other celestial objects in order to learn more about the Universe and our place in space. (Figure 4).

galaxy a huge, rotating collection of gas, dust, and billions of stars, planets, and other celestial objects

READING TIP

Finding the Main Idea
Words that are repeated can provide clues to the main idea. For example, you might notice that in this sub-section the word "galaxies" is repeated in the heading, text, sidebar glossary, illustration, and caption. It is also in bold in the first sentence of the first paragraph. By using these clues, you can focus on determining the main idea about galaxies.

Figure 4 The Universe consists of everything that exists, including galaxies, the Solar System, celestial objects, and even you!

TRY THIS: CREATE A HORIZON DIAGRAM

SKILLS: Observing, Analyzing, Communicating

SKILLS HANDBOOK 3.B.

In this activity, you will create a diagram that shows the positions of celestial objects as seen against the horizon.

Equipment and Materials: flashlight; red cellophane; plain paper and clipboard; pencil; ruler

1. With an adult, go to a dark site to observe the night sky. Before you begin making observations, cover the end of the flashlight with the red cellophane. This will minimize the glare produced by the flashlight.
2. Draw a line at the bottom of your paper to represent the horizon in one direction. Sketch any trees or buildings that you observe along the horizon. You will use these objects as landmarks.
3. Carefully sketch the brightest objects you see in the night sky.
4. Record the date and exact time of your observations, a detailed description of the position from which you made your observations, and the cloud conditions for that evening.

A. See if you can identify any of the celestial objects you have drawn. T/I

B. Is the Moon in your drawing? If so, describe its position and appearance. T/I

C. How do the objects you have sketched differ in their brightness? Why do you think this is so? T/I

8.1 Touring the Night Sky 307

UNIT TASK Bookmark

How can you apply what you learned about horizon diagrams in this section to the Unit Task described on page 446?

IN SUMMARY

- Astronomy is the scientific study of what exists beyond Earth, including stars, planets, and moons.
- Many celestial objects, such as stars, are visible in the night sky. Stars are luminous, whereas planets and the Moon are visible because they reflect light from the Sun.
- The eight planets in the Solar System travel around the Sun.
- Natural satellites, such as the Moon, orbit some planets in the Solar System.
- Within the Universe there are galaxies, such as the Milky Way. Each galaxy contains billions of stars and other celestial objects.

CHECK YOUR LEARNING

1. What do astronomers study? K/U
2. Write a brief paragraph explaining the term "luminous" to a classmate. K/U C
3. If the Moon does not produce its own light, how are we able to see it? K/U
4. Explain the terms "terrestrial planets" and "gas giants," providing examples of each. K/U
5. How are stars different from planets? How are they similar? K/U
6. According to astronomers, what is a satellite? K/U
7. Use an example to define the term "orbit" in your own words. K/U
8. There are more than 100 billion stars, including the Sun, in our solar neighbourhood (Figure 5). What do astronomers call this collection of stars? K/U

Figure 5

9. Arrange the following objects from biggest to smallest: galaxy, moon, star, planet, Universe. K/U

8.2 The Sun

Stars and other celestial objects in the Universe emit energy consisting of electromagnetic waves that travel at the speed of light, known as **electromagnetic (EM) radiation**. Together, these forms of radiation are contained in the **electromagnetic (EM) spectrum**. The EM spectrum consists of radio waves, microwaves, infrared rays, visible light, ultraviolet rays, X-rays, and gamma rays. These waves have energies that become greater as their wavelengths become smaller.

The Sun emits radiation across most of the EM spectrum. Although some of the Sun's energy is absorbed by Earth's atmosphere and some is reflected into space, almost all of the energy that reaches Earth's surface comes from the Sun. EM radiation from the Sun is the driving force behind Earth's weather and climate, and also provides the energy needed for life to exist on Earth (Figure 1).

electromagnetic (EM) radiation energy emitted from matter, consisting of electromagnetic waves that travel at the speed of light

electromagnetic (EM) spectrum the range of wavelengths of electromagnetic radiation, extending from radio waves to gamma rays, and including visible light

Figure 1 Plants and animals could not exist without sunlight.

The Structure of the Sun

The Sun is composed of many layers of gas. Deep inside the Sun's centre is the core, where high temperatures and pressures cause particles to collide with each other at extremely high speeds. This causes the particles to fuse, or join together, in a process called nuclear fusion, which gives off enormous amounts of energy. These high-energy reactions make the core the hottest part of the Sun, reaching a temperature of 15 000 000 °C.

The energy released by nuclear fusion makes its way to the radiative zone, the first layer that surrounds the core. This energy can take up to a million years to reach the next region of the Sun, the convective zone. In the convective zone, hotter substances rise as colder substances fall. Energy continues to move outward until it reaches the photosphere, where light and other types of radiation escape. The photosphere has a temperature of 5500 °C. Above the photosphere lies the Sun's atmosphere. It is divided into two layers: the chromosphere and the corona. The chromosphere makes up the inner atmosphere, and is 60 000 °C hotter than the photosphere. The thin outer layer of the solar atmosphere—the gleaming white, halo-like corona—extends millions of kilometres into space (Figure 2).

To learn more about the Sun's structure and how it works,

GO TO NELSON SCIENCE

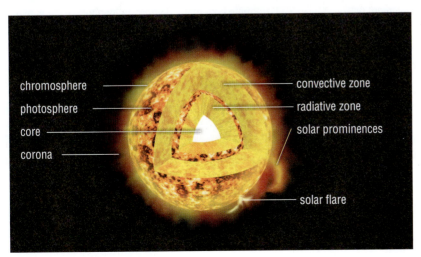

Figure 2 The structure of the Sun

The Sun's Surface

The Sun rotates on its axis, taking approximately 25 days to make one complete rotation. Near its equator, the Sun spins more quickly than at its poles. Early astronomers believed that the Sun's surface was smooth and featureless. Today, with the aid of telescopes, we know that the photosphere has a texture that appears similar to that of a boiling liquid. As heated material rises, it reaches the surface, cools, and sinks back inside. As this happens, convection cells called granules form. This gives the photosphere a grainy appearance. At the centre of each granule, hot solar gases radiate into space. The gases move outward and sink back into the Sun at the darker, cooler boundaries of the granule.

Sunspots, which are darker, cooler areas visible on the Sun's photosphere, appear to move as the Sun rotates. They are caused by disturbances in the Sun's magnetic field. Sunspots vary in size and regularity (Figure 3). Galileo Galilei, an astronomer who lived approximately 400 years ago, was one of the first people to observe and study sunspots in detail. His observations led him to conclude that sunspots must be occurrences taking place on the Sun's surface. To date, the largest sunspot ever recorded covered an area of 1.8×10^{10} km²—this is 36 times the surface area of the entire Earth!

Solar flares are found in active regions near sunspots and release large quantities of gas and charged particles. They are produced by the rapidly changing magnetic fields around sunspots and last only a short time.

Solar prominences are slow, low-energy ejections of gas that travel through the corona. Refer back to Figure 2, which shows a solar flare and a solar prominence on the Sun.

sunspots dark spots appearing on the Sun's surface that are cooler than the area surrounding them

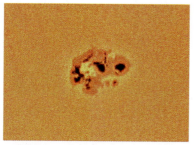

Figure 3 Sunspots consist of a dark central zone, which may be larger than several Earths, and a lighter border.

solar flare gases and charged particles expelled above an active sunspot

solar prominence low-energy gas eruptions from the Sun's surface that extend thousands of kilometres into space

The Sun's Effects on Earth

The Sun experiences powerful activity in its outer atmosphere. Constantly in motion and rotating at different speeds, the gases of the Sun swirl around, causing solar storms. These storms are the cause of solar winds. Solar winds are strongest when there are solar flares and prominences.

TRY THIS: TRACK SUNSPOTS

SKILLS: Observing, Analyzing

SKILLS HANDBOOK 3.B.

In this activity, you will observe and track the movement of sunspots using modern satellite images.

Equipment and Materials: photos of sunspots taken by the SOHO satellite over four consecutive days; overlay of mapping grid; graph paper; pencil

1. Identify the different sunspot groups A, B, and C on each of the photos.
2. Copy Table 1 into your notebook and complete the information for each sunspot group. Use the nearest whole number.

Table 1

Date	Longitude of Sunspot Group A	Longitude of Sunspot Group B	Longitude of Sunspot Group C

3. Determine how many degrees in longitude each sunspot group moved each day. Then calculate the average daily movement of each group.

A. What was the average daily movement of each group? Provide your answer in degrees of longitude. T/I

B. Using the number of degrees that you observed the sunspots to move each day, calculate how long it takes the Sun to make one full rotation on its axis. T/I

C. How many of the sunspots changed in size and shape over the four day period? T/I

D. How might the tracking of sunspots be important? A

THE AURORAS

Earth is surrounded by an atmosphere containing atoms of different gases, such as oxygen, nitrogen, argon, and carbon dioxide. Earth is also surrounded by a magnetic field that is strongest near the North and South Poles (Figure 4). Solar winds travelling toward the Earth become influenced by, and follow, the lines of magnetic force created by Earth's magnetic field. Near the poles, they come in contact with particles in Earth's atmosphere, producing a display of light in the night sky. In the northern hemisphere, we call these colourful displays of light the **aurora borealis** or northern lights (Figure 5). Whenever there is a display of northern lights, there is a simultaneous display near the South Pole called the aurora australis or southern lights.

aurora borealis a display of shifting colours in the northern sky caused by solar particles colliding with matter in Earth's upper atmosphere

To learn more about auroras,

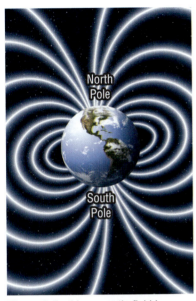

Figure 4 Earth's magnetic field is strongest near the North and South Poles.

Figure 5 The northern lights are featured in many Aboriginal legends. For example, the Cree people call this phenomenon the "dance of the spirits."

COMMUNICATION DISRUPTIONS

Solar activity at the Sun's surface can affect artificial satellites, human-made objects that orbit celestial objects in the Solar System. For example, particles ejected by the Sun can damage the information stored on computer microchips on satellites. This can disrupt cellphone and satellite TV communications (Figure 6).

Figure 6 When directed toward Earth, solar particles can affect communications, navigation systems, and power grids.

8.2 The Sun

DID YOU KNOW?

Observatories in Space
Astronomers use telescopes both on the ground and in space to study the Sun. NASA's recently launched STEREO (Solar TErrestrial RElations Observatory) twin spacecrafts are creating the first ever three-dimensional images of the Sun. The new views will help scientists understand structures in the Sun's atmosphere and improve space weather forecasting.

To learn more about becoming a solar scientist,

The temperature and density of Earth's upper atmosphere, where satellites orbit, is sometimes increased by solar radiation and storms. The friction caused by the denser atmosphere slows down the satellites and can alter their orbital path. Scientists constantly monitor satellites and boost them back into their orbits to prevent them from falling and burning up in the atmosphere.

RADIATION HAZARDS

On July 14, 2000, a powerful solar storm occurred. It is referred to as The Bastille Day Event. Charged solar particles entered Earth's atmosphere and disrupted signals from communications satellites orbiting the planet. Solar scientists warned that people travelling in airplanes could receive a higher than usual dose of radiation because of their high altitude. The radiation in solar storms can also be harmful to astronauts during a space walk or through the walls of their spacecraft while they orbit Earth because they are not protected by Earth's atmosphere.

UNIT TASK Bookmark

How can you apply what you learned about radiation hazards in this section to the Unit Task described on page 446?

IN SUMMARY

- The Sun emits radiation across most of the EM spectrum. Some of the Sun's energy is absorbed by Earth's atmosphere and some is reflected into space. Almost all of the energy that reaches Earth's surface comes from the Sun.
- The Sun is necessary for life on Earth.
- The Sun has various layers with different temperatures and properties.
- The Sun constantly undergoes changes due to storms and a turbulent atmosphere that also affects Earth.
- The aurora borealis (northern lights) and aurora australis (southern lights) are caused by the interaction between solar particles and the Earth's atmosphere near the North and South Poles.
- Solar radiation causes communication disruptions and can be hazardous to human health.

CHECK YOUR LEARNING

1. In a short paragraph, explain the types of radiation given off by the Sun.
2. Why is the Sun so important for life on Earth?
3. Draw and label a diagram of the different layers of the Sun. Briefly describe each layer.
4. What range of temperatures are found on, and within, the Sun?
5. How long does it take for the Sun to rotate on its axis? How do astronomers know this?
6. Who was the first astronomer to observe sunspots in detail? What did he conclude from his observations?
7. Compare and contrast solar prominences, solar flares, and solar winds.
8. What are the auroras and where on Earth are they visible?
9. How are communication satellites affected by activity on the surface of the Sun?

The Solar System: The Sun and the Planets

8.3

The Solar System consists of the Sun, the eight planets and their moons, and billions of other smaller celestial objects. All of these celestial objects orbit the Sun (Figure 1).

Some planets are relatively close to the Sun. Mercury is just 58 million km away. Other objects are much farther away. Neptune is almost 4 billion km from the Sun! It would take a spacecraft travelling at 28 000 km/h almost 50 years to cross the Solar System.

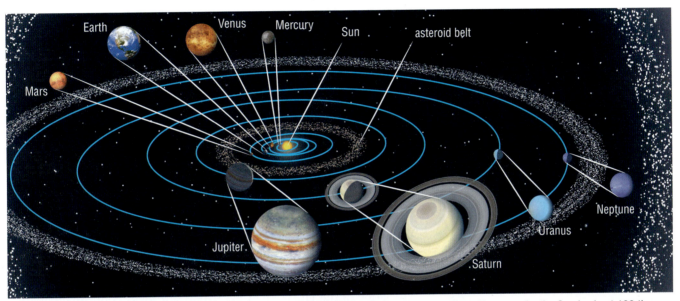

Figure 1 This drawing shows the Solar System, but does not illustrate true relative distances or sizes. For example, the Sun is about 100 times larger than Earth.

Measuring Distances in the Solar System

Distances in the Solar System are so great that astronomers must use a more convenient unit than the kilometre to measure them. The **astronomical unit**, or AU, is the average distance between the Sun and Earth—approximately 150 000 000 km. The AU provides a more manageable way to measure astronomical distances. For example, the planet Jupiter is 780 million km from the Sun. This equals 5.2 AU, which is more than five times farther from the Sun than Earth!

astronomical unit approximately 150 million kilometres; the average distance from Earth to the Sun

Planets Big and Small

The Sun is the largest object in the Solar System. The next largest objects in the Solar System are the planets. On a clear night, you can sometimes see planets such as incredibly hot Venus, desert-like Mars, and monster-sized Jupiter. In the night sky, the planets appear as only bright points of light because they are so far away.

The four planets nearest the Sun are Mercury, Venus, Earth, and Mars. These small, rocky planets are considered part of the inner Solar System. The four planets beyond Mars are Jupiter, Saturn, Uranus, and Neptune. These planets lie in the outer Solar System and are known as the gas giant planets. As their name implies, they are all big. Each of the eight planets has unique properties that make it different from the others.

TRY THIS: REPRESENT THE SIZES OF THE PLANETS

SKILLS: Planning, Performing, Observing

SKILLS HANDBOOK
3.B.

In this activity, you will create scale models of the planets to help you visualize the size differences between them.

Equipment and Materials: metre stick; calculator; ruler; modelling clay; balloons; masking tape

1. Your teacher will draw a circle on the board, measuring 1.0 m in diameter. This represents a scale model of the Sun, with a diameter 100 times that of Earth.
2. Look at the equatorial diameter values in Table 1. These values represent the diameter of each planet in relation to Earth's diameter. For example, Mars has a value of 0.53, which means that its diameter is about half that of Earth.
3. Use modelling clay to create a model of Earth with a diameter of 1 cm. Compare the size of your model Earth with the Sun on the board and note your observations in your notebook.
4. Use the equatorial diameter data in Table 1 to create scale models of the other planets. Choose modelling clay or balloons as appropriate.
5. Use masking tape to label each planet.

A. Compare the size of Jupiter with the size of the Sun. Estimate how many Jupiters would fit across the diameter of the Sun. Write the answer in your notebook. Then measure to see whether your estimate was correct. T/I

B. The largest known star is VY Canis Majoris, which is 2100 times the size of the Sun. How big would its diameter be at the scale of this activity? T/I

C. Using the equatorial diameter data from Table 1, create a bar graph with the planet names on the x-axis and diameters on the y-axis. C

Table 1 Properties of Planets in the Solar System

Properties	Mercury	Venus	Earth	Mars	Jupiter	Saturn	Uranus	Neptune
equatorial diameter (Earth diameters)	0.382	0.949	1.0	0.53	11.21	9.44	4.01	3.88

Dwarf Planets

As our knowledge of the Universe grows, so do our ideas about the planets, stars, and other celestial objects. At the end of the 19th century, the word "planet" only applied to celestial objects in the Solar System. Over time, however, astronomers began to discover other celestial objects. These discoveries led to a change in the definition of "planet" in 2006. To be considered a "planet" a celestial object must

- be in orbit around a star (such as the Sun)
- have enough mass to be pulled into a stable sphere shape by gravity
- dominate its orbit (i.e., its mass must be greater than anything else that crosses its orbit)

dwarf planet a celestial object that orbits the Sun and has a spherical shape but does not dominate its orbit

From its discovery in 1930 until 2006, Pluto was considered the ninth planet in the Solar System. However, the new definition of planet excluded Pluto and added it to the category of "dwarf planet." **Dwarf planets** orbit the Sun and have a spherical shape. However, they do not dominate their orbits (Figure 2).

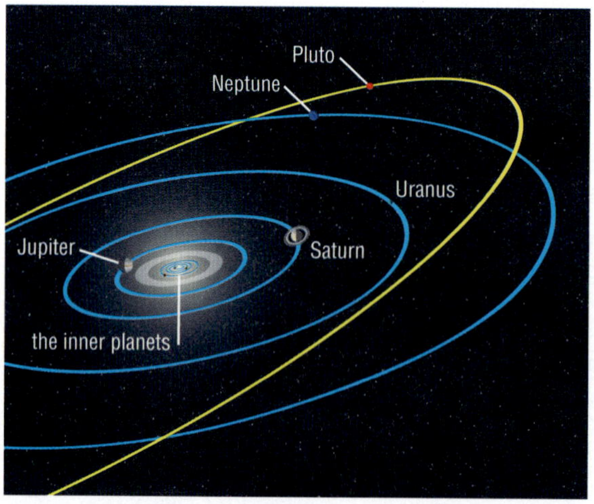

Currently, there are five recognized dwarf planets—Ceres, Pluto, Haumea, Makemake, and Eris. However, only two of these—Ceres and Pluto—have been observed in enough detail to demonstrate that they fit the definition. Most of the dwarf planets discovered to date lie beyond the orbit of Neptune. Astronomers suspect that up to 2000 dwarf planets exist, with as many as 200 of these located within a region of the outer Solar System known as the Kuiper belt.

Figure 2 Pluto does not meet the criteria for a planet because its tilted orbit crosses Neptune's orbit.

Smaller Members of the Solar System

In addition to the Sun and planets, the Solar System contains billions of smaller celestial objects. Some are made of rock and metal; others are composed of ice.

Asteroid Belt

Asteroids are small celestial objects in the Solar System composed of rock and metal (Figure 3). Although they also orbit the Sun, they are too small to be considered planets. The vast majority of asteroids lie in an area known as the asteroid belt, located between the orbits of Mars and Jupiter. Asteroids vary in size and can have a diameter of up to 950 km. The largest asteroids are round, but most are irregularly shaped.

Figure 3 At least 40 000 asteroids, approximately 800 m across or larger, make up the asteroid belt.

Meteoroids

A meteoroid is a piece of metal or rock in the Solar System that is smaller than an asteroid. Most meteoroids are the size of dust particles, but some can be as large as a car or building. Meteoroids sometimes get pulled in by Earth's gravity. As they are pulled down into Earth's atmosphere, friction causes them to burn up, creating a bright streak of light across the sky, known as a meteor. This phenomena is commonly referred to as a "shooting star." Meteors enter Earth's atmosphere at speeds of more than 1.5×10^5 km/h. On rare occasions, larger meteors do not burn up completely in the atmosphere and their remains, which we call meteorites, crash to the ground (Figure 4).

Several large meteorites have been known to create craters on impact. Lake Cheko, for example, is believed to have been formed by an impact crater during the Tunguska Event mentioned at the beginning of this chapter.

Canada is home to more than two dozen identified impact craters. The Manicouagan Crater, in northern Quebec, was formed from the impact of a meteorite with a 5 km diameter approximately 212 million years ago. The resulting crater is 100 km wide (Figure 5). The Sudbury Basin, located in northern Ontario, was formed from a 10 km meteorite impact that occurred 1.85 billion years ago. It is the second largest impact crater in the world, measuring 62 km long by 30 km wide. Much of the nickel mined in the Sudbury area originated from this meteorite.

DID YOU KNOW?
Meteoroids
About 100 tonnes of meteoroids as small as dust particles fall to the surface of Earth every day.

DID YOU KNOW?
Impact Event
A leading theory on the extinction of dinosaurs suggests that a 10 km meteorite impact is what caused the extinction of these animals 65 million years ago.

Figure 4 On November 20, 2008, a meteor flashed across the Prairie provinces. A few days later, this meteorite was found on a frozen pond at Lone Rock, Saskatchewan.

Figure 5 This satellite image shows the Manicouagan Crater, in northern Quebec. The crater is now filled with water, forming the ring-shaped Lake Manicouagan.

8.3 The Solar System: The Sun and the Planets

METEOR SHOWERS

On a clear night there is, on average, one meteor flash in the sky every 15 minutes. However, on certain dates during the year, a number of meteors can be seen radiating from one point in the sky. We call this a meteor shower. For example, the Leonid meteor shower occurs every year around November 17th. Meteor showers mainly occur when clouds of particles left behind by comets enter the atmosphere of a planet, such as Earth (Figure 6).

Figure 6 To observe a meteor shower, you need to watch from a dark location, outside city limits.

comet a chunk of ice and dust that travels in a very long orbit around the Sun

> **DID YOU KNOW?**
> **Naming Comets**
> Most comets are named after their discoverers. For example, Comet Hale-Bopp was named after Alan Hale and Thomas Bopp, both amateur astronomers in the United States who discovered the comet in 1995.

Comets

Comets are large chunks of ice, dust, and rock that orbit the Sun. They range in size from less than 100 m to more than 40 km across. Some comets take a few years to travel around the Sun, whereas others take hundreds of thousands of years.

Comets are classified as either short- or long-period comets. Short-period comets originate from a region just beyond the orbit of Neptune and travel around the Sun in less than 200 years. Halley's Comet is the most famous example of a short-period comet, taking 75 to 76 years to make one trip around the Sun. The last visit of Halley's Comet was in 1986; it will return in 2061. Long-period comets originate from a spherical cloud of debris much farther away than Pluto and take more than 200 years to orbit the Sun. Comet Hale-Bopp is one of the most recent long-period comets to be observed from Earth. It takes about 2380 years for Hale-Bopp to make one trip around the Sun.

When a comet gets close enough to the Sun, its outer surface begins to sublimate—changing state from a solid to a gas—and its icy nucleus heats up. As this occurs, gases and dust escape. These gases and dust form a gaseous cloud around the nucleus called a **coma**, which can be thousands of kilometres wide (Figure 7).

As a comet approaches the Sun, radiation and solar wind from the Sun exert a force on the coma, which causes a gaseous tail to form, pointing directly away from the Sun. In addition, a dust tail forms in the direction from which the comet originated (Figure 8).

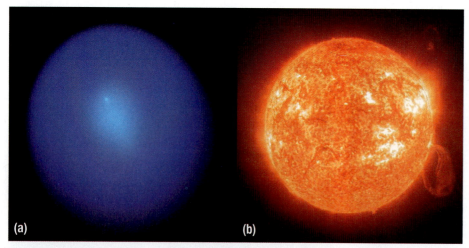

Figure 7 In 2007, Comet Holmes released large mounts of gas and dust, forming a coma (a) that expanded the diameter of the comet to be greater than that of the Sun (b).

Figure 8 Most comets have two tails—a gaseous tail and a dust tail.

If a comet ventures too close to a planet, it can be pulled in by the planet's gravity. This is what happened to Comet Shoemaker-Levy 9 when it crashed into Jupiter in July 1994 (Figure 9). Satellites that survey the Sun have even photographed some comets falling into the Sun.

UNIT TASK Bookmark

How can you apply what you learned about the planets and other members of the Solar System in this section to the Unit Task described on page 446?

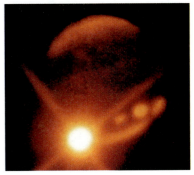

Figure 9 The fireball created when Comet Shoemaker-Levy 9 hit Jupiter on July 18, 1994 grew to over 20 000 km in diameter.

IN SUMMARY

- The average distance between the Sun and Earth is defined as one astronomical unit (AU).
- The four planets closest to the Sun are the terrestrial planets. These small, rocky planets are considered part of the inner solar system. The four planets farthest from the Sun are the gas giants. These large, gaseous planets are part of the outer Solar System.
- Pluto has been reclassified as a dwarf planet, based on its physical properties and motion.
- Besides the planets, there are smaller objects in the Solar System that orbit the Sun, such as asteroids, comets, and meteoroids.

CHECK YOUR LEARNING

1. Draw a picture of the Solar System from the "top" looking down onto the North Pole of Earth. Include and label the eight planets, plus Pluto, the asteroid belt, a comet, and a meteoroid.

2. Copy Table 2 into your notebook. Complete the table, naming the planets that are terrestrial and those that are gas giants. Then compare the two types under the remaining headings.

   Table 2

Names of planets	Large or small	Rock and metal or gas	Inner or outer Solar System

3. (a) Why do astronomers use astronomical units to measure distances in the Solar System?
   (b) How many kilometres make up 1 AU?

4. Why is Pluto no longer considered a planet in the Solar System? What kind of object is it considered to be now?

5. Identify the errors in the following quotations by referring to the definitions of "meteoroid," "meteorite," and "meteor." Then rewrite them with the correct vocabulary.
   (a) "I saw a really bright meteorite flash across the sky!"
   (b) "The meteoroid made a huge hole in the ground when it crashed into Earth."
   (c) "Meteors sometimes hit spacecraft when they are travelling through deep space."

6. What evidence exists in the Canadian landscape to suggest that large meteorites have crashed into Earth?

7. Describe what you think meteor showers look like and explain why they occur. You may use a diagram in your explanation.

8. Halley's Comet is one of the most famous objects in the Solar System, even though we have not seen it since 1986.
   (a) In what year will we next be able to see Halley's Comet?
   (b) In 1997, people could see Comet Hale-Bopp. In what year will this comet return?

9. Why do comets appear to have large tails flowing away from the Sun?

8.3 The Solar System: The Sun and the Planets

8.4 PERFORM AN ACTIVITY

A Scale Model of the Solar System

When scientists are dealing with very large objects, they often create scale models to help them view a system as a whole. Unfortunately, the enormous distances in space make constructing a scale model of the Solar System a challenging task. To put this into perspective, the distance between Earth and the Sun is almost 12 000 times the diameter of Earth!

In this activity, you will attempt to build a scale model of the Solar System to represent the distances between the Sun and the planets.

SKILLS MENU
- Questioning
- Hypothesizing
- Predicting
- ● Planning
- Controlling Variables
- ● Performing
- ● Observing
- ● Analyzing
- Evaluating
- ● Communicating

Purpose
To build a scale model of the Solar System using data and calculations.

Equipment and Materials
- calculator
- 100 m tape measure or trundle wheel
- pictures of objects in the Solar System

Procedure

SKILLS HANDBOOK

1. Work in small groups. Your teacher will provide each group with a picture of a celestial object in the Solar System.

2. The radius of the Solar System is roughly 50 AU, which is the distance to the farthest dwarf planets that have been observed. In the scale model you will produce, 2 m will be equal to 1 AU. This means that if the Sun is placed at one edge of the model, then the farthest dwarf planet will be 100 m away (Figure 1).

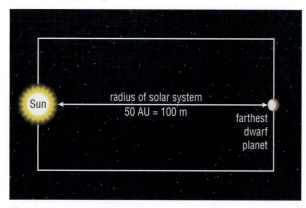

Figure 1

MATH TIP
To convert distances using a scale, remember that

$$\frac{2 \text{ m}}{1 \text{ AU}} = \frac{X \text{ m}}{\text{actual distance}}$$

Simply cross multiply to obtain the value for X in metres.

3. Copy Table 1 into your notebook. (Include only those objects that your class is working with.) Using the information given in step 2 and the values given in Table 1, convert the actual distance to your object into a scale distance. Record this value in the third column of Table 1.

Table 1 Planetary Distances from the Sun

Object	Actual distance (AU)	Scale distance (m)
Sun	0	
Mercury	0.39	
Venus	0.72	
Earth	1.0	
Mars	1.5	
asteroids	2.5	
Ceres	2.8	
Jupiter	5.2	
Saturn	9.5	
Uranus	19	
Neptune	30	
Pluto	40	

4. Share the values for your celestial objects with those of other groups in the class. Add the celestial objects and values found by other groups to Table 1 in your notebook.

5. Take the picture of your object to the location your teacher has designated as the model (your school football field, for example). Use the tape measure or trundle wheel to measure out the scale distance from the picture of the Sun to your object. Place your picture on the ground at that distance.

6. Once all of the groups in the class have placed their objects on the field, walk through the scale model of the Solar System as a class, identifying the objects as you go.

Analyze and Evaluate

(a) From your model, what do you notice about the distance between the terrestrial planets compared with the spacing between the gas giants? T/I

(b) Consider all the information presented in Table 2.
   (i) Which planet is most similar to Earth? Which is most different? Explain your reasoning. T/I
   (ii) Which planets have densities much lower than that of Earth? What can you conclude about these planets? T/I
   (iii) What are some ways in which you could classify the planets, using the data presented in Table 2? T/I
   (iv) Which planet is most different from the other planets? Explain how it is different, using examples from Table 2. T/I

Apply and Extend

(c) All of the planets except Venus and Mercury have moons. Select one planet, and research how far one of its moon is. Then calculate how far its moon would be at this scale. T/I

(d) For the planet you chose in (c), determine whether it has a solid surface or a gaseous surface. Research three other physical properties of the planet, such as colour, the elements that make up its atmosphere, its temperature, and its density. Present your information to the class. T/I C

(e) The closest star to our Sun is Proxima Centauri. It is approximately 268 000 AU away from the Sun. How far away would this be from the Sun in your scale model? Convert your answer to kilometers. T/I A

(f) In your scale model, you likely placed the celestial objects in the Solar System in a straight line. Conduct research to find out if the objects actually line up like this in reality. What would be a more accurate way of depicting the objects in the Solar System? You may use a diagram to aid in your answer. A C

UNIT TASK Bookmark

How can you apply what you learned about distances in the Solar System in this section to the Unit Task described on page 446?

Table 2 Properties of the Planets in the Solar System

Properties	Mercury	Venus	Earth	Mars	Jupiter	Saturn	Uranus	Neptune
Average distance from the Sun ($\times 10^6$ km)	57.9	108	150	228	778	1427	2870	4497
Time for one rotation	59 d	243 d	24 h	24 h 39 min	9 h 50 min	10 h 39 min	17 h 18 min	15 h 40 min
Average diameter (km)	4880	12 100	12 750	6790	142 980	120 540	51 120	49 530
Mass (kg)	3.3×10^{23}	4.9×10^{24}	5.9×10^{24}	6.4×10^{23}	1.9×10^{27}	5.7×10^{26}	8.7×10^{25}	1.0×10^{26}
Mass (Earth = 1)	0.06	0.8	1.0	0.1	318	95	15	17
Density	5.44	5.25	5.52	3.95	1.31	0.70	1.27	1.64
Surface gravity (Earth = 1)	0.39	0.90	1.0	0.38	2.53	1.06	0.90	1.14
Range of surface temperature (°C)	−173 to 427	462	−88 to 58	−90 to −5	−148	−178	−216	−214
Main substances in the atmosphere	none	carbon dioxide, nitrogen	nitrogen, oxygen	carbon dioxide, nitrogen	hydrogen, helium, methane	hydrogen, helium, methane	hydrogen, helium, methane	hydrogen, helium, methane
Number of moons (as of 2009)	0	0	1	2	63	60	27	13

The symbol "d" stands for one Earth day.

8.5 Motions of Earth, the Moon, and Planets

If you get up early in the morning and look for the Sun, you will see it low in the eastern sky. At sunset, the Sun is once again low in the sky, but in the west. This is why people often say that the Sun "rises in the east and sets in the west." Of course, the Sun does not really move across the sky, it only appears to do so.

Earth's Rotation

The apparent motion of the Sun in the sky is caused by the rotation of Earth on its axis. This motion is similar to the spinning of a top (Figure 1). Earth makes one complete rotation, in a west-to-east direction, once each day. During its rotation, the portion of Earth that faces the Sun experiences daylight. The portion facing away from the Sun experiences darkness.

Earth's Revolution and Planetary Orbits

While Earth spins on its axis, it also revolves, or travels around the Sun. Earth's orbit, like that of the other planets, is elliptical.

The distance of each planet from the Sun changes as it completes its orbit around the Sun. This is because the Sun is located closer to one end of the elliptical path. Earth, for example, is farthest from the Sun around July 4, and closest to the Sun around January 3. Astronomers calculate the average distance from the Sun for each planet and call this value the planet's **orbital radius**.

The shape and size of a planet's orbit affects the time that it takes it to complete a revolution around the Sun, a value which we call an orbital period. The farther a planet is from the Sun, the slower it moves along its orbit and the longer it takes to orbit the Sun (because its orbit is larger). Mars, for example, takes 687 days to orbit the Sun whereas Mercury orbits the Sun in only 88 days. Earth completes one revolution around the Sun approximately every 365.25 days (Figure 2). If you could look down on the Solar System (such that the North Pole of Earth is pointing toward you), you would see that the planets revolve in a counterclockwise direction.

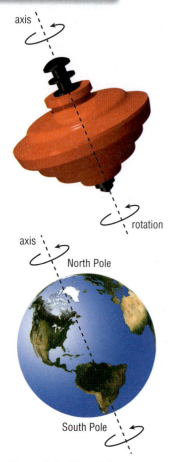

Figure 1 Earth's axis is an imaginary line that goes through the planet from pole-to-pole.

orbital radius the average distance between an object in the Solar System and the Sun

DID YOU KNOW?
Leap Years
Most of our calendar years are 365 days. Since Earth completes one revolution around the Sun in 365.25 days, we fall 1/4 of a day behind each year. The extra quarter-days are added up every four years and added on as an extra day (February 29) in leap years. By inserting a 366th day, our calendar year closely matches Earth's revolution around the Sun.

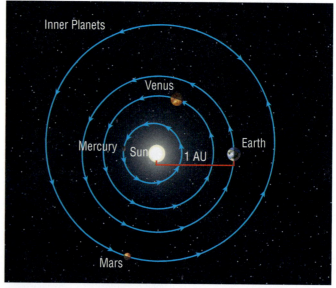

Figure 2 Earth has an orbital radius of 1 AU and an orbital period of 365.25 days.

320 Chapter 8 • Our Place in Space

Effects of Earth's Revolution

Earth's revolution affects our view of celestial objects in the sky. You will learn more about this in Section 8.9.

Motions of the Moon

The Moon also rotates on its axis. As it rotates, the Moon also revolves around Earth. The Moon completes one rotation on its axis in about the same time it takes to complete one revolution around Earth. The result of this is that the same side of the Moon faces Earth at all times. Together, the Earth-Moon pair revolve around the Sun.

The Force of Gravity

What keeps the planets orbiting the Sun, and the Moon orbiting Earth? There is a force of attraction between all objects in the Universe with mass, known as **gravitational force**, or gravity. The greater the mass of an object, the stronger its gravitational force. The gravitational force of our massive Sun is strong enough to keep Earth in orbit. Without the Sun's gravity, Earth would move away from the Sun into space (Figure 3). Similarly, the strong pull of Earth's gravity keeps the Moon in orbit.

> **DID YOU KNOW?**
> **Earth's Speed**
> Earth spins on its axis at a speed of about 1600 km/h at the equator. Meanwhile, Earth revolves around the Sun at a speed of 108 000 km/h.

gravitational force the force of attraction between all masses in the Universe

> **DID YOU KNOW?**
> **Universal Law of Gravity**
> Isaac Newton (1643–1727) realized that the gravity that kept the Moon in orbit was the same force that made objects on Earth fall to the ground. He developed the Universal Law of Gravitation, which states that every object in the Universe attracts every other object. The more massive the objects are and the closer they are to each other, the greater the attraction.

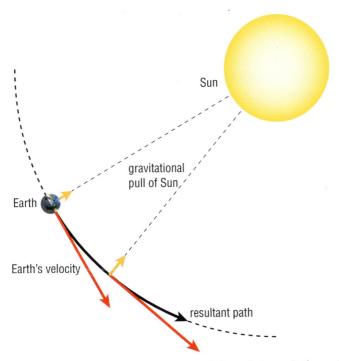

Figure 3 Earth stays in a stable orbit because of the balance between its forward speed and the Sun's gravitational pull.

Explaining Motion in the Solar System

Although today we understand that the planets, including Earth, revolve around the Sun, this was not always understood. About 2000 years ago, the Greek astronomer Claudius Ptolemy (87–150 CE) believed that the Sun and the planets revolved around Earth.

This type of Solar System model is called a geocentric model of the Solar System (Figure 4). Ptolemy observed the apparent motion of the Sun and planets in the sky and assumed that these objects were revolving around Earth. This idea was widely accepted until the Middle Ages, when Nicholas Copernicus (1473–1543) proposed a model placing the Sun at the centre of the Solar System, called the heliocentric model of the Solar System (Figure 5). This model was much better at explaining the motion of the planets than the geocentric model.

LEARNING TIP

Word Origins

Geo means "Earth" and *centric* means "centre," so the term *geocentric* means "Earth centred." Likewise, *helios* means "Sun," so *heliocentric* means "Sun centred."

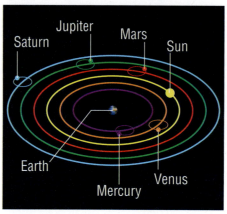

Figure 4 A geocentric model of the Solar System.

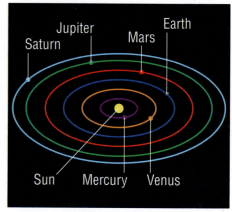

Figure 5 A heliocentric model of the Solar System.

In the early 1600s, Galileo was an outspoken supporter of the heliocentric model. In 1610 he used the telescope he invented to observe four moons orbiting Jupiter. A planet with other celestial objects orbiting it did not conform to the geocentric model, so many astronomers of the time did not believe his discoveries. Galileo's further observations of other planets, especially his observations of Venus, eventually led to the conversion of the scientific community to the heliocentric model. The heliocentric model is the basis of our modern understanding of how the Solar System and other planetary systems in the Universe work.

Earth's Tilt

Earth's rotational axis is tilted 23.5° from the vertical when compared to the plane of Earth's orbit around the Sun (Figure 6). This tilt affects the average daytime temperature experienced by Earth's hemispheres.

The Reason for Seasons

As Earth revolves around the Sun, the northern and southern hemispheres experience the seasons—spring, summer, autumn, and winter. Many people mistakenly believe that the seasons are caused by Earth's distance from the Sun. Although it is true that Earth's distance from the Sun changes over the course of the year, the changes in Earth-Sun distance are not enough to cause the changes of the seasons. Seasons change primarily because of Earth's tilt.

When Earth is farthest from the Sun, the northern hemisphere is tilted toward the Sun and sunlight spreads over a relatively small area of Earth's surface (Figure 7(a)). This causes intense heating of Earth's surface and atmosphere. During this time, the Sun appears to travel its highest path in the sky, and there are more hours of daylight.

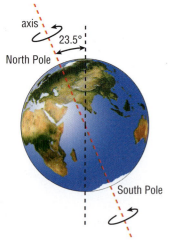

Figure 6 Earth's axis is tilted 23.5° from the vertical.

When Earth is closest to the Sun, the northern hemisphere is tilted away from the Sun and sunlight spreads over a larger area of Earth's surface (Figure 7(b)). This causes less heating of Earth's surface and atmosphere. During this time, the Sun appears to travel a lower path in the sky and there are fewer daylight hours.

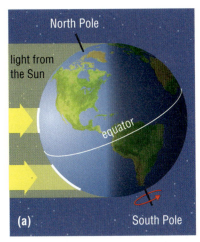 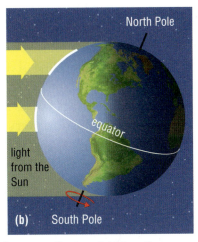

Figure 7 (a) The northern hemisphere receives more direct sunlight than the southern hemisphere when Earth is tilted toward the Sun. (b) The reverse effect occurs when Earth is tilted away from the Sun.

When Earth's axis is most inclined toward or away from the Sun, we call it a **solstice**. Solstices occur twice each year. Around June 21, Earth's northern hemisphere is tilted toward the Sun as much as possible. This is the longest day of the year, and is considered the first day of summer in the northern hemisphere. Around December 21, the northern hemisphere is tilted away from the Sun as much as possible. This is the longest night of the year, and is considered the first day of winter in the northern hemisphere.

When the northern hemisphere is tilted toward the Sun, the southern hemisphere is tilted away from the Sun, and vice versa. This means that when the northern hemisphere experiences summer, the southern hemisphere experiences winter, and vice versa. Between the solstices are two days of equal daytime hours and nighttime hours called the **equinoxes**. In the northern hemisphere, the vernal equinox occurs on about March 21 and the autumnal equinox occurs on about September 21 (Figure 8).

solstice an astronomical event that occurs two times each year, when the tilt of Earth's axis is most inclined toward or away from the Sun, causing the position of the Sun in the sky to appear to reach its northernmost or southernmost extreme

equinox the time of year when the hours of daylight equal the hours of darkness

Figure 8 The seasons in the northern hemisphere.

LEARNING TIP

Night and Day
The word "equinox" comes from the Latin words *equi* meaning "equal" and *nox* meaning "night." The equinox is a period of "equal night," in which there are 12 hours of daylight and 12 hours of darkness.

Figure 9 This time lapse photo shows stars revolving around Polaris (indicated by arrow).

precession the changing direction of Earth's axis

Precession—Earth's Wobble

If you extend the line of Earth's axis from the North Pole into space, it would pass very close to the star Polaris. This is why Polaris is often referred to as the North Star, or Pole Star. Astronomers call this point in the sky the North Celestial Pole. For anyone living in the northern hemisphere, the entire sky appears to rotate around this one star (Figure 9).

Polaris has not always been the pole star. As Earth rotates, its axis slowly turns, pointing in different directions over time. This change in direction of Earth's rotational axis is called **precession**, and is a motion similar to the wobble of a spinning top. If you watch a spinning top closely, you will see its upper and lower tips trace circles as they wobble (Figure 10). Earth's axis also wobbles and traces a circle every 26 000 years. Precession will result in the North Celestial Pole pointing near the star Vega in about 12 000 years and back to Polaris in another 26 000 years.

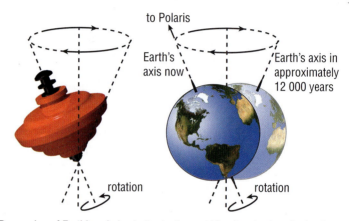

Figure 10 Precession of Earth's axis is similar to the wobble of a slowly spinning top.

Phases of the Moon

The Moon, like all celestial objects in the Solar System, is illuminated by the Sun. However, the illuminated side does not always face Earth, which means that we see different amounts of the lit side as the Moon orbits Earth. Over a period of about 4 weeks, the amount of the illuminated surface of the Moon we see (called phases) follows a predictable pattern. The eight phases of the Moon that we see over this period of time make up the **lunar cycle** (Figure 11).

The lunar cycle begins with the new moon, when the Moon is not visible from Earth. We do not see the Moon during this phase because the side that is illuminated faces away from Earth. After this phase, the illuminated portion of the Moon visible from Earth waxes (increases in size). This occurs during the first half of the lunar cycle. First, we see the waxing crescent—a curved sliver of light. About a week after the new moon, the Moon reaches the first quarter phase. This appears as a half moon in the sky because half of the illuminated portion of the Moon is facing Earth. When the waxing gibbous phase arrives, we see more than half of the Moon's illuminated surface. The full moon phase appears as a lit circle in the sky, with the Moon's illuminated side facing Earth.

During the second half of the lunar cycle, the illuminated portion of the Moon visible from Earth wanes (decreases in size). The lunar cycle progresses to waning crescent moon and then begins the lunar cycle again.

READING TIP

Finding the Main Idea
The first sentence of a paragraph is called the topic sentence and this is where the main idea usually is stated. For example, *the lunar cycle begins with the new moon.* Sometimes the first sentence gives an introduction to the topic, so check the second sentence to see if it states the main idea.

lunar cycle all of the phases of the Moon

DID YOU KNOW?

Misconception
Some people mistakenly believe that the phases of the Moon are caused by Earth's shadow. This is incorrect. The phases of the Moon are caused by how much of the Moon's illuminated surface we can see.

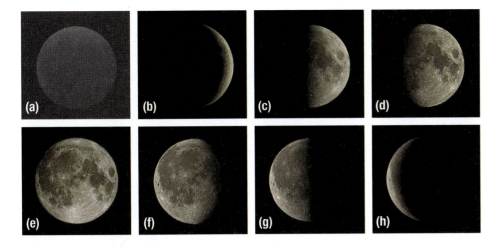

Figure 11 The phases of the Moon as seen from Earth.
(a) New moon (darkened image to represent the new moon, which we cannot see)
(b) Waxing crescent
(c) First quarter
(d) Waxing gibbous
(e) Full moon
(f) Waning gibbous
(g) Third quarter
(h) Waning crescent

TRY THIS: MODELLING THE LUNAR PHASES

SKILLS: Observing, Communicating, Analyzing

SKILLS HANDBOOK 3.B.

You will be creating a model of the Earth–Moon system in order to view the Moon cycle through its phases (Figure 12).

Equipment and Materials: a lamp with a 40- to 60-watt light bulb (lamp shade removed); polystyrene ball; pencil

1. Choose one person in your group to represent Earth.
2. Have another member of your group stick the pencil into the polystyrene ball. This is the Moon.
3. Turn out the classroom lights and turn on the lamp, which represents the Sun.
4. Ask Earth to stand several metres away from the Sun.
5. Have the Moon revolve around Earth in a counterclockwise direction, while the Earth-bound observer watches the lighted part of the Moon and determines the phases visible as the Moon continues in its orbit.
6. Have each member of your group take a turn being Earth and observing the Moon go through its phases.

A. Starting with the new moon phase, use the model to describe the main phase of the Moon during one lunar cycle. T/I
B. How could the activity be changed to make the phases of the Moon appear more distinct? T/I

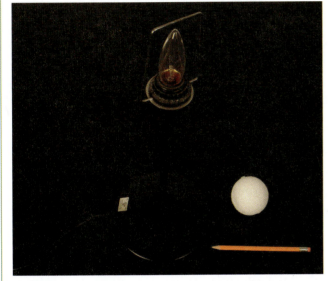

Figure 12

Eclipses—In the Shadow

Eclipses are spectacular astronomical events that occur when the position of one celestial object blocks, or darkens, the view of another celestial object from Earth. Throughout history, these events have been interpreted in different ways by various cultures. Solar eclipses were often associated with bad omens, such as disease epidemics, deaths of kings, and wars. Lunar eclipses were believed by some to be caused by a mythological creature swallowing the Moon.

eclipse a darkening of a celestial object due to the position of another celestial object

Solar Eclipse

The Sun has a diameter 400 times greater than the Moon. It is also 400 times farther from Earth than the Moon is. As a result, the Moon and the Sun appear approximately the same size in the sky.

> **DID YOU KNOW?**
> **Moon Drift**
> The Moon is slowly drifting away from Earth. As a result, in about 3000 years, a total eclipse will no longer be possible.

When the Moon is aligned between Earth and the Sun, it blocks the Sun from being observed from Earth–an event that we call a solar eclipse (Figure 13(a)). A solar eclipse is only possible during a new moon phase and is quite rare. During a total solar eclipse, only the outer atmosphere of the Sun, the corona, remains visible. This is because of the relative sizes of the Sun and the Moon in the sky. Total solar eclipses allow scientists to study the Sun's corona (Figure 13(b)). To view a total eclipse from Earth, observers must be located along the lunar shadow path, which is only a few dozen kilometres wide.

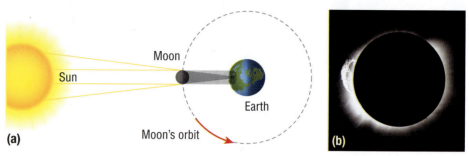

Figure 13 (a) This view of the Sun and Earth, looking down from the North Pole, shows a total eclipse. A total eclipse occurs somewhere on Earth once every two years. (b) During a total eclipse, astronomers can safely study solar flares from the Sun without damaging their eyes because the Sun is covered by the Moon.

Partial solar eclipses occur when the Moon does not cover the entire Sun (Figure 14). Solar eclipses should be viewed through a suitable solar filter or by projection, and not with the naked eye. The Sun's powerful rays can damage your eyes when it is not fully hidden behind the Moon.

Lunar Eclipse

A lunar eclipse occurs when Earth is positioned between the Sun and the Moon, casting a shadow on the Moon (Figure 15(a)). During a total lunar eclipse, the entire Moon passes through Earth's shadow. If only part of the Moon passes through Earth's shadow, a partial lunar eclipse occurs. A lunar eclipse can be seen anywhere on Earth where the Moon is visible above the local horizon.

A lunar eclipse can appear orange or red (Figure 15(b)). This colouration is caused by the refraction of sunlight as it passes through Earth's atmosphere. This is also why sunsets appear orange-red.

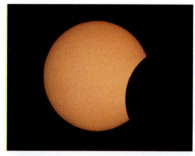

Figure 14 A partial solar eclipse does not fully block out the Sun and should not be observed without special eye protection.

> **DID YOU KNOW?**
> **Volcanic Eruption**
> In June 1991, Mount Pinatubo, a volcano in the Philippines, erupted in the second largest volcanic explosion of the twentieth century. The eruption ejected billions of tonnes of ash and dust high into Earth's atmosphere. This debris circled Earth for more than a year. The lunar eclipses that followed in 1992 were such a deep red that some people saw the eclipsed Moon completely disappear in the sky.

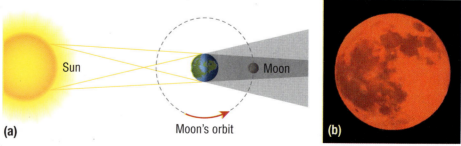

Figure 15 (a) This view of the Sun and Earth, looking down from the North Pole, shows a total lunar eclipse. A total lunar eclipse can last up to an hour. (b) The colour of the Moon during a total eclipse depends on the amount of dust and clouds in Earth's atmosphere.

Tides—The Pull of the Moon

If you have spent much time at an ocean beach, you may have noticed that the beach gets wider or narrower at certain times of the day. This change is due to the rise and fall of the water level, or tides. **Tides** are the rising and falling of the surface of oceans, and are caused by the gravitational pull of the Moon and, to a lesser extent, the Sun. The Moon's gravitational force pulls Earth and its oceans toward it. This causes a bulge of water to form on the side of Earth facing the Moon. As Earth is pulled toward the Moon, a bulge of water also forms on the opposite side of Earth, where the Moon's gravitational force is weakest. (Figure 16). This results in two high tides and two low tides on Earth each day. The time between low and high tide is approximately six hours.

tide the alternate rising and falling of the surface of large bodies of water; caused by the interaction between Earth, the Moon, and the Sun

To find out more about tides,

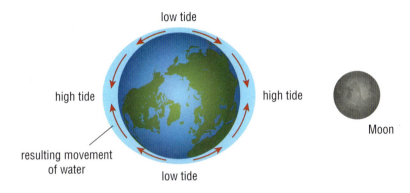

Figure 16 When the Moon is directly overhead, coastal areas experience high tide. At the same time, coastal areas on the opposite side of Earth also experience high tide. The regions on Earth between the two tidal bulges experience low tide.

The Sun also has an effect on tides, but its tidal effect is much smaller because it is so much farther away than the Moon. During the new and full moon phases of the lunar cycle, Earth, the Moon, and the Sun are aligned. The combined gravitational force of the Moon and the Sun causes very high tides, called spring tides, to form (Figure 17(a)). When the Moon and the Sun are perpendicular to each other with respect to Earth (during the quarter phases) the gravitational pull of the Sun somewhat counteracts the gravitational pull of the Moon. This forms weaker tides, called neap tides (Figure 17(b)).

DID YOU **KNOW?**
Bay of Fundy
Canada is home to the world's highest tides. The Bay of Fundy, a narrow finger of salt water between New Brunswick and Nova Scotia, experiences tides that can rise and fall the height of a four-storey building. Twice a day, 100 billion tonnes of water flow into and out of the bay. Researchers hope to one day use this massive daily flow of water as a renewable energy source.

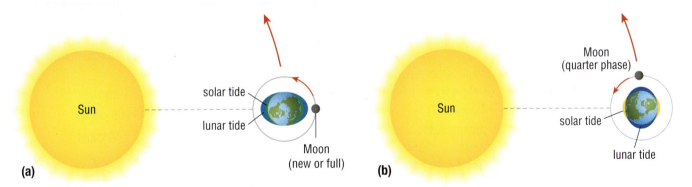

Figure 17 (a) When the Moon, the Sun, and Earth are aligned, spring tides occur. (b) When the Moon and the Sun are on perpendicular sides of Earth, weaker neap tides occur.

UNIT TASK Bookmark

How can you apply what you learned about gravity and the motion of celestial objects in this section to the Unit Task described on page 446?

IN SUMMARY

- Earth rotates around its own axis once a day and revolves around the Sun once a year.
- Precession is the gradual change in direction of Earth's rotational axis.
- Gravity causes all of the planets in the Solar System to orbit the Sun at an average distance known as their orbital radius.
- Seasons are caused by the tilt of Earth's axis.
- Equinoxes and solstices are astronomical events that that can be tracked based on the changing length of daylight.
- As the Moon orbits Earth, we see it change phases, based on the relative positions of the Sun, the Moon, and Earth.
- The alignment of the Sun, the Moon, and Earth also results in solar and lunar eclipses.
- A combination of the Sun's gravity and the Moon's gravity causes tides on Earth.

CHECK YOUR LEARNING

1. Which planets have larger orbital radii, terrestrial planets or gas giants? K/U
2. Imagine that you are explaining to Grade 3 students why it gets dark at night and bright during the day. What would you tell them? K/U C
3. Using Earth as an example, explain the difference between the terms "rotation" and "revolution." C
4. Explain Earth's motion, using the terms "precession" and "rotation" in your answer. C
5. (a) What type of force is responsible for keeping the Moon in orbit around Earth? K/U
   (b) What evidence is there that this force also pulls on objects closer to the surface of Earth? K/U
   (c) Who was the scientist who explained this force? K/U
6. Where is the North Star located, and what is its other name? K/U
7. Create two drawings that show the difference between the geocentric model of the Solar System and the heliocentric model of the Solar System. K/U C
8. Many people think that because summer is the warmest season, it must be when Earth is closest to the Sun. Explain why this is incorrect and give the proper explanation for the seasons on Earth. K/U
9. (a) When do the solstices occur? K/U
   (b) Compare the length of daylight during the winter and summer solstices. K/U

10. Figure 18 shows Earth in its orbit around the Sun. In your notebook, write down which season is represented at each of the four positions.

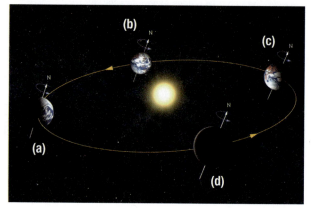

Figure 18

11. Match each of the following terms with the correct length of time: K/U

    (a) revolution of Earth   (i) 27.3 days
    (b) rotation of Earth   (ii) 26 000 years
    (c) revolution of Moon   (iii) 23 h 54 min 4.1 s
    (d) precession of Earth   (iv) 365.25 days

12. How many phases are there in the lunar cycle? K/U
13. What are the two kinds of eclipses we can see from Earth? Explain how they occur, using diagrams to show the positions of the Sun, the Moon, and Earth. K/U
14. What is the cause of the spring tides? Do these tides occur only in the spring season? K/U

328 Chapter 8 • Our Place in Space

8.6 Patterns in the Night Sky

For thousands of years, people the world over have been fascinated by the night sky. Many civilizations, both in the past and present, believe that celestial objects are connected to events that occur on Earth. For example, many First Nations groups noted that the appearance of certain patterns of stars marked the changing seasons. This information helped them determine when to plant and when to harvest crops.

Stars were also used for navigation. People living in areas surrounded by water had no land to use as a reference point, and often used stars for navigation. Polynesians, for example, navigated among the Pacific Ocean islands by observing and recording patterns of stars.

Constellations

Over the centuries, many cultures have noticed that some stars in the night sky appear to form patterns. They began naming these star patterns after their heroes, mythical monsters, and animals, such as Leo the lion (Figure 1). In everyday language, we often call these star patterns constellations. However, in astronomical terms, constellations are regions of the sky. It is the stars found within a **constellation** that can appear as a pattern in the sky. Ursa Major, for example, is the constellation where the Big Dipper can be found. There are 88 constellations recognized by the International Astronomical Union (IAU).

Historically, people formed these star patterns using their imagination to link up the stars in the sky like connect-the-dots puzzles. Different cultures linked up the stars in different ways. For instance, ancient Greek astronomers thought that the stars that make up the constellation Orion depicted a great hunter (Figure 2(a)), whereas some First Nations peoples of North America thought that the same stars looked like a canoe floating down a river (Figure 2(b)). Many of the official constellation names that exist today come from Greek or Arabic mythology.

> **DID YOU KNOW?**
> **88 Constellations**
> There are as many constellations in the sky as there are keys on a piano—88.

Figure 1 Ancient Egyptians worshipped the stars of Leo because when the Sun was among its stars, the Nile River rose.

constellation a grouping of stars, as observed from Earth

To learn more about the history and legends of constellations,

GO TO NELSON SCIENCE

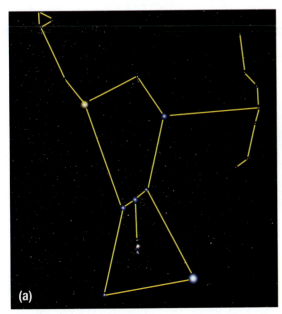

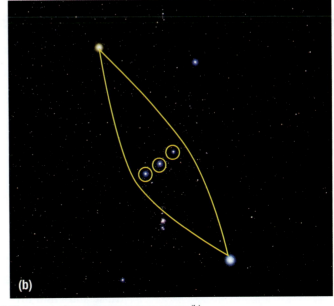

Figure 2 The constellation Orion can appear in the pattern of a hunter (a). The same stars can appear as a canoe (b).

Although we can mark out the same constellations our ancient ancestors saw thousands of years ago, their component stars are not in exactly the same location as they were then. Precise observations of stars reveal that they move relative to each other in space, but these changes in position occur slowly, over many years. Since most stars are so far away from Earth, when we observe them in the night sky, they do not appear to be moving relative to one another. These changes can be seen only in photographs taken through high-powered telescopes over many years. Resulting shifts in star patterns such as the Big Dipper will become noticeable only in tens of thousands of years (Figure 3).

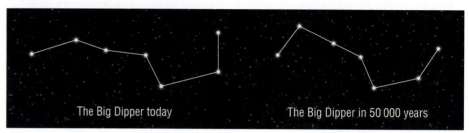

Figure 3 The stars of the Big Dipper are moving, but so slowly that the change is only noticeable over many years.

Astronomers also noticed that celestial objects that are closer to the Sun appear to move faster relative to more distant ones. This visual effect is similar to the effect seen while driving in a car along a country road. Nearby trees appear to move more rapidly by your window, whereas more distant trees appear to move by much more slowly.

To learn more about the movement of the Sun and stars,

MAPPING THE STARS

A star map is a map of the night sky that shows the relative positions of the stars in a particular part of the sky (Figure 4). Some star maps show only those objects that can be seen with the unaided eye, while others show objects that can only be viewed using a telescope or other instrument. A star map can be used to recognize celestial objects in the sky, and to observe the motions of these objects. By making use of star maps, people are able to determine both their location on the Earth's surface and the direction in which they are going. For this reason, star maps have been used for navigation for centuries. To learn more about star maps, go to the Skills Handbook, section 2.C.

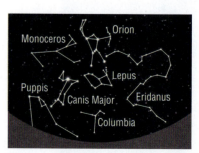

Figure 4 Star maps show the stars in a certain region of the sky.

A planisphere is a very useful type of star map that is used to display only those stars that are visible at a given date and time. A planisphere can be used for locating stars, constellations, and galaxies.

The Celestial Sphere

From Earth, we observe the sky as it appears, not as it is. In the Middle Ages, people believed the sky to be a solid sphere with celestial objects in fixed positions. This imaginary sphere onto which all celestial objects are projected is called the **celestial sphere**. Today, we know that it is Earth that rotates, and not the sky. The celestial sphere extends around Earth. However, an observer on Earth can only see half of the sphere, the same way you can only see what is in front of you, and not behind (Figure 5).

celestial sphere the imaginary sphere that rotates around Earth, onto which all celestial objects are projected

DID YOU KNOW?
See the Night Sky Inside
A planetarium offers people the opportunity to view the stars and constellations projected onto a domed ceiling, which mimics the night sky. A projector can show the night sky seen from any part of the world, at any time of night, on any day, including the past, present, and future.

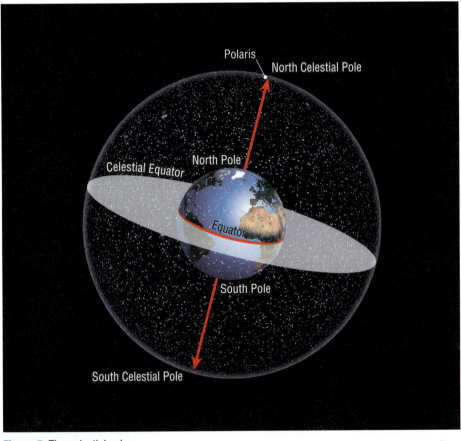

Figure 5 The celestial sphere

Early astronomers realized that the celestial sphere is a useful tool for determining location using the stars, known as **celestial navigation** (Figure 6).

The celestial sphere is divided into northern and southern hemispheres by the celestial equator, which is an imaginary extension of Earth's equator, projected into space. Astronomers mark the position of stars and other objects in the sky of each hemisphere using a grid system similar to latitude and longitude used on Earth. The North and South Celestial Poles are the imaginary points where Earth's axis of rotation extends out onto the celestial sphere above the North and South Poles on Earth. For observers in the northern hemisphere, the North Celestial Pole points very close to Polaris, the North Star. Even today, celestial navigation is widely used by sailors.

celestial navigation the use of positions of stars to determine location and direction when travelling

Figure 6 An astrolabe measures angles to determine location.

Calendars Based on the Sky

Scientific historians recognize that many cultures have developed calendars by observing the sky. Historically, keeping accurate calendars was important for timing crucial agricultural and religious events. Ancient Egyptian farmers noticed that the annual flooding of the Nile River, used for crop irrigation, would occur once every 365 days when the Sun was passing through the constellation Leo. Other ancient peoples marked the beginning of the summer season using stones. By lining up the stones with the Sun's path in the sky, they were able to track the start and end of each season.

To find out how to be a scientific historian,

 GO TO NELSON SCIENCE

Cultural Significance of Solstices and Equinoxes

For many cultures, solstices and equinoxes have historically been astronomical events that held significant spiritual meaning. On the Yucatán Peninsula of Mexico, the ancient Mayans built a giant pyramid, El Castillo, that is aligned to the Sun's movement in the sky. At sunset on the spring and fall equinoxes, a corner of the structure casts a shadow that resembles a plumed snake slithering down the pyramid steps (Figure 7).

Another famous ancient monument marking the solstice is the prehistoric astronomical observatory known as Stonehenge, in England (Figure 8). Stonehenge is a circular arrangement of giant stones and boulders that may have been used as an astronomical calendar to predict the movement of the Sun, the Moon, and their eclipses. Who built Stonehenge, and why, remains a mystery, but it is known that the alignment of the monument is so precise that at sunrise on the summer solstice every year, the Sun's first rays strike a particular stone.

READING TIP

Finding the Main Idea
Use the title and headings to get a sense of what a passage is about. For example, "Calendars Based on the Sky" suggests the 365-day orbit of Earth around the Sun. Similarly, "Cultural Significance of Solstices and Equinoxes" helps you distinguish the main idea in the first two sentences from the examples in the remaining sentences of the paragraphs.

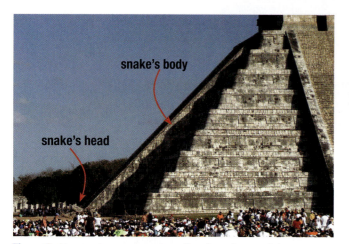

Figure 7 At sunset on each equinox, the plumed serpent, known as "Kukulcan," forms along the west side of the north staircase at El Castillo, Chichen Itza.

Figure 8 An early observatory called Stonehenge located near Salisbury, England

Aboriginal Views of the Sky

Cultures around the world have passed stories from generation to generation to explain what they see in the sky. Some legends are about astronomical phenomena, such as eclipses of the Moon or the Sun.

Other legends offer various explanations for the formation of celestial objects, such as the Moon and stars. A West Coast Tsimshian legend tells of the stars being formed when the sleeping Sun's mouth spewed sparks through the smoke hole of its house. As the Sun slept, its brother, the Moon, rose in the east. In other First Nations legends, lunar eclipses are the result of monsters in the sky swallowing the Moon.

Some Aboriginal legends in Canada view the stars as living beings who wander the skies forever. According to an Iroquois legend, the stars that make up the bowl of the Big Dipper are, in fact, a giant bear. The stars of the handle are three warriors hunting the bear. As the constellation sets close to the horizon in autumn, the hunters injure the bear. The blood of the injured bear turns the leaves of the trees on Earth red.

UNIT TASK Bookmark

How can you apply what you have learned about star maps to the Unit Task described on page 446?

IN SUMMARY

- The sky is divided into 88 regions associated with patterns of stars, or constellations.
- Star maps show the arrangement of celestial objects in the night sky.
- The celestial sphere is a convenient way to describe the positions of stars and their apparent movement in the sky.
- Ancient peoples often built structures to acknowledge and utilize their understanding of the movements of celestial objects.
- Aboriginal peoples in Canada, as well as other cultures, have observed the patterns in the night sky for cultural, agricultural, and navigational reasons.

CHECK YOUR LEARNING

1. Give three reasons why ancient peoples carefully observed the patterns in the night sky. K/U
2. Do the stars that make up constellations change their position over time? How might this change the shape of the constellation? K/U
3. Using a globe as a guide, describe where you might find the celestial sphere, celestial equator, and celestial poles. C
4. The North Celestial Pole is very close to the star shown in Figure 9. What is its name? K/U
5. Explain why it might be beneficial to use the geocentric model when describing the positions and motions of the stars in the night sky. T/I
6. Name two ways that ancient cultures recognized the importance of solstices and equinoxes. K/U
7. Identify two structures built by ancient peoples to reflect their beliefs about the motions of objects in the sky. K/U
8. In your own words, describe two First Nations legends about astronomical phenomena as if you were explaining to a friend. K/U

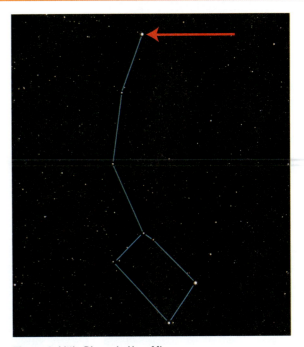

Figure 9 Little Dipper in Ursa Minor

8.6 Patterns in the Night Sky

8.7 PERFORM AN ACTIVITY

Using a Star Map

In this activity, you will learn how to use a star map. Star maps show the constellations and bright stars in the sky that are visible from either the northern or southern hemispheres at a certain time of the year. Although some star maps are specific to each month, the star map you will use can be used throughout the year and is specific to the northern hemisphere. See "Using Star Maps" on page 605 in the Skills Handbook.

SKILLS MENU
- Questioning
- Hypothesizing
- Predicting
- Planning
- Controlling Variables
- Performing
- Observing
- Analyzing
- Evaluating
- Communicating

Purpose

To determine how the stars of the Big Dipper can be used to locate other constellations in the northern hemisphere.

Equipment and Materials

- star map (to be provided by your teacher)
- ruler

Procedure

Part A: Circumpolar Constellations

1. Look at the Big Dipper in Figure 1. Using a ruler, draw lines on your star map to join the stars that create the image of the Big Dipper.

2. Look at the Little Dipper in Figure 1. On your star map, draw in the lines for the Little Dipper. Label the star at the tip of the handle "Polaris." Polaris is also called the North Star because you need to face toward the North Pole in order to see it.

3. On your star map, locate the two stars of the Big Dipper that are farthest from the handle. These two stars are called pointer stars because they can be used to point toward other constellations. Using the pointer stars, draw a straight dashed line to Polaris.

4. Continue your dashed line until you reach Cassiopeia, a constellation that resembles a stretched out M or W (Figure 1(c)). Join the stars of Cassiopeia.

5. Label all three constellations on your star map. Check with your teacher to make sure that you have labeled your star map correctly so far.

6. There are several other constellations that are visible in the Canadian night sky during different seasons (Figure 2). On your star map, locate the three stars in the constellation Orion that make up the imaginary hunter's belt. Draw in this constellation. Orion can be seen most easily during the month of December. Label the stars Rigel and Betelgeuse within Orion too.

Figure 1 (a) The Big Dipper (b) The Little Dipper (c) Cassiopeia

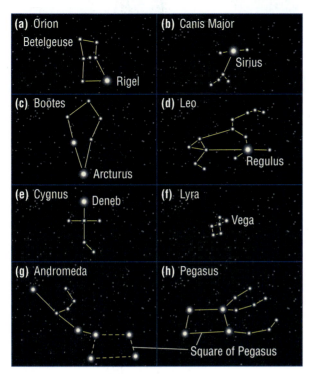

Figure 2

7. Draw a dashed line to show how you would use Orion's belt to locate the brightest star in the night sky, Sirius. Draw Canis Major, the constellation that contains Sirius.

8. Locate and draw the constellation Boötes on your star map. To find Boötes, first look for the star Arcturus, which is part of this constellation. Draw a dashed line to show how you would use the handle of the Big Dipper to find Arcturus.

9. Draw the constellation Leo on your star map. Leo contains the star Regulus. Draw dashed lines to show how you would use two stars of the Big Dipper as pointer stars to Regulus.

10. The three bright stars Deneb, Vega, and Altair form the Summer Triangle. Join them with dashed lines. Deneb and Vega are in the constellations Cygnus and Lyra. Draw these constellations on your star map.

11. Draw the lines for the remaining constellations in Table 2 on your star map. Follow the order of the constellations as they appear in Table 2. For each constellation, use a dashed line from another constellation. When you have finished locating and drawing all the constellations, show your teacher your map and explain how you used pointer stars. You will share your results with the class.

Part B: Locating the Constellations of the Zodiac

12. You will notice there are still many unlabelled constellations on your star map. These are the zodiac constellations. They are visible during specific months, and the planets can be seen passing through them. You have already labeled one of the zodiac constellations, Leo. There are 11 more for you to find. The months in which they are best viewed are shown in the margin of the star map. Starting with Leo, move in a counterclockwise direction and label the rest of the zodiac constellations as follows: Virgo, Libra, Scorpius, Sagittarius, Capricornus, Aquarius, Pisces, Aries, Taurus, Gemini, and Cancer.

Analyze and Evaluate

(a) Which constellations are visible from Canada in all seasons?

(b) Using your star map, identify two constellations that you can see from Canada during each of the four seasons.

(c) Describe how you used pointer stars in the Big Dipper to find three other constellations.

Apply and Extend

(d) Look at your star map again. Can you find any other uses for pointer stars? Draw these uses with dashed lines, and describe them.

(e) To locate the constellations at night, you stand facing north and hold the star map with the current month at the top. Why do you hold the star map this way?

UNIT TASK Bookmark

How can you apply what you learned about using a star map in this section to the Unit Task described on page 446?

8.8 CONDUCT AN INVESTIGATION

Modelling Motion in the Night Sky

The motion of the stars in the night sky is caused by Earth's rotation around its axis and its revolution around the Sun. These are predictable movements. With careful observation, scientists can create models of the motion of celestial objects and predict where these objects will be at a later date. In this investigation, you will construct a planisphere and use it to predict the positions of constellations in the night sky.

SKILLS MENU
- Questioning
- Hypothesizing
- Predicting
- Planning
- Controlling Variables
- Performing
- Observing
- Analyzing
- Evaluating
- Communicating

Testable Question
Can a star map be used to predict which constellations are visible in the night sky at a certain location and time?

Hypothesis
If the motions of the stars in the night sky are consistent night after night, then we can predict the positions of the constellations in the night sky using an adjustable star map.

Equipment and Materials
- glue stick
- scissors
- copy of a star map
- copy of a star map frame
- two file folders
- split pin

Experimental Design
SKILLS HANDBOOK 3.B.4.

Using the materials listed, you will create a planisphere and use it to locate constellations in the night sky. You should be able to notice the patterns the stars make in the night sky, how they change over the course of a night, and how they change over the course of a year.

Procedure
1. Your teacher will distribute copies of a star map and a star map frame.
2. Glue the star map to one side of a file folder and cut it out.
3. Glue the star map frame to the second file folder so that the spine of the file folder is at the bottom of the frame.
4. Cut around the edge of the star map frame through both layers of the file folder (Figure 1).

Figure 1 step 4

5. Cut out the inside of the star map frame through only the upper side of the file folder, so the backing is still uncut and shows through the inside of the frame (Figure 2).

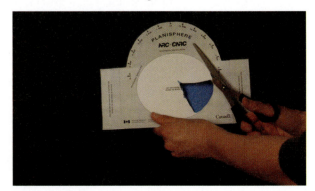
Figure 2 step 5

6. Wrap the tabs around the back of the frame and glue them in place to the backing (Figure 3).

Figure 3 step 6

7. Slide the star map into the frame so that the constellations are visible through the frame (Figure 4).

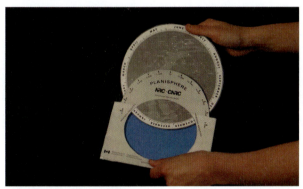

Figure 4 step 7

8. Make a small hole through the star map and the backing at the location of the star Polaris.

9. Insert the split pin through the hole and fasten it so that the map can spin around this point (Figure 5).

Figure 5 step 9

Analyze and Evaluate

(a) Review the Testable Question and your hypothesis. Explain whether your results support the hypothesis. Remember to state your evidence.

(b) Why do the constellations appear to rotate around Polaris? K/U

(c) Rotate the star map so that April 15 lines up with 8 p.m. Which constellations are visible on the southern horizon? Which ones are directly overhead? T/I

(d) Rotate the star map so that October 15 lines up with 8 p.m. How has the sky changed from your observations of April 15? T/I

(e) Rotate the star map so that today's date lines up with 8 p.m. Slowly rotate the wheel through 9 p.m., 10 p.m., and so on until morning. Describe the change in the position of the constellations throughout the night. C T/I

Apply and Extend

(f) Based on what you learned in Section 8.5, adjust your star map so that it matches the date of the winter solstice. Which constellations are visible on the southern horizon? Do the same for the summer solstice. T/I

(g) Based on what you learned in Section 8.5, adjust your star map so that it matches the date of the vernal equinox. Which constellations are visible on the southern horizon? Do the same for the autumnal equinox. T/I

(h) Rotate the star map so that is matches today's date. Which constellations will be visible tonight in the southern sky at 10:00 p.m.? T/I A

(i) Why is the Sun not on the star map? T/I

(j) Why are the planets not on the star map? T/I

> **UNIT TASK Bookmark**
>
> How can you apply what you learned about using a star map in this section to the Unit Task described on page 446?

8.9 Observing Celestial Objects from Earth

Celestial objects are visible from Earth both by day and by night. In the daytime you can see the Sun and, sometimes, the Moon. Looking up at the night sky on a cloudless night far away from bright cities, you can see as many as 2000 stars. Even in the city, where light pollution makes it difficult to see stars, you can still observe dozens of celestial objects.

Most of the pinpoints of light in the sky are stars—similar to our Sun, but much farther away. Some of the brighter pinpoints of light, such as Venus and Jupiter, are planets (Figure 1).

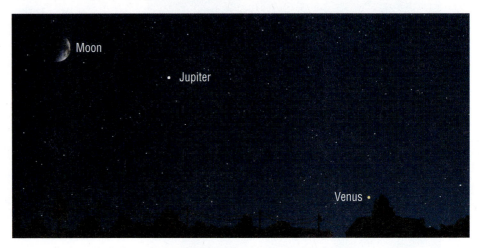

Figure 1 Venus and Jupiter are two of the brightest objects in the night sky.

CITIZEN ACTION

Saving the Night Sky

SKILLS HANDBOOK
4.A.7., 4.C.1., 4.C.3.

How many stars you can see on a clear night depends on where you are on Earth (Figure 2). From a sports park filled with floodlights, you may see only a few stars. From a dark, remote countryside location in northern Ontario, this number jumps to more than 2000 stars.

Figure 2

Unfortunately, many major observatories are fighting light pollution from nearby growing towns and cities. Some have lost the battle and are closing. The David Dunlap observatory just north of Toronto is one such victim of light pollution. Others are adapting by using new technologies and observing only certain celestial objects that are not greatly affected by light pollution. Light pollution is also a significant waste of energy because the light is released up into the sky, where it is of no use.

Use the Internet and other electronic and print resources to investigate the causes and effects of light pollution, and ways that individuals and municipalities in Canada are reducing it.

 **GO TO NELSON SCIENCE**

Then choose one of the following:

A. Compile a list of different ways that you personally can help cut down on light pollution in your neighbourhood.

B. Write an article for your school newspaper suggesting how the school can reduce light pollution.

C. Write a letter to your town or city council suggesting measures they could take to reduce light pollution.

The Ecliptic

Over the course of a year, the Sun appears to move against the constellations of the celestial sphere. The path taken by the Sun (as it appears to us on Earth) across the celestial sphere is called the **ecliptic**. This apparent motion of the Sun is caused by Earth's revolution around the Sun. The Moon, planets, and some constellations also appear to move from east to west along the ecliptic. The ecliptic is best observed when looking up at the sky toward the southern horizon (Figure 3).

ecliptic the path across the sky that the Sun, the Moon, the planets, and the zodiac constellations appear to follow over the course of a year

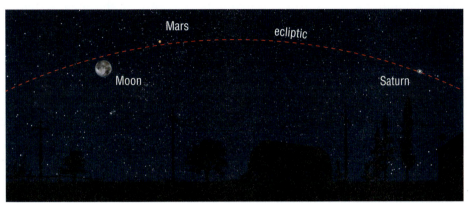

Figure 3 The ecliptic is an important reference pathway that helps us locate the positions of the Sun, the Moon, the planets, and the constellations.

DID YOU KNOW?

Ecliptic
The word "ecliptic" is derived from the word "eclipse" because eclipses of the Sun and the Moon always occur along the ecliptic—the imaginary line in the sky along which the Sun and the Moon appear to move.

Changing Views of the Night Sky

Earlier, you learned that Earth's rotation from west to east causes celestial objects to appear to move across the sky from east to west. This explains the apparent motion of the Sun, the Moon, and other celestial objects during a 24 h period (which is the time it takes for Earth to complete one full rotation).

The distance of celestial objects can also play a role in how they are observed from Earth. For example, because the planets are much closer to Earth than the stars, their path along the ecliptic over time appears to change with respect to the constellations. A planet can be distinguished from a star in the night sky by observing its motion over weeks or months.

In addition, our view of the night sky changes with each passing season due to Earth's revolution around the Sun. For example, if we observe the northern hemisphere's night sky in winter we see stars that are opposite the Sun. The stars that are in the same sky as the Sun will only be in the northern hemisphere's sky during the day and will not be visible (Figure 4). Therefore, the constellations we see in the night sky are not the same at all times of the year.

READING TIP

Finding the Main Idea
Normally, the last sentence concludes a paragraph by reminding the reader of the main idea. Sometimes words or phrases such as "therefore" or "as a result" are used to signal a conclusion. The last sentence of a paragraph is a good place to check for a rewording of the main idea.

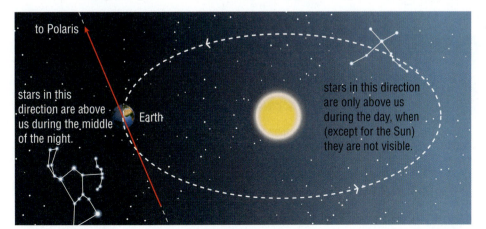

Figure 4 Orion is visible in the northern winter sky because it is opposite the Sun at that time. Cygnus is in the northern winter sky during the day, however, and cannot be seen. Stars that are close to Polaris are visible in all seasons.

8.9 Observing Celestial Objects from Earth 339

retrograde motion the apparent motion of an object in the sky, usually a planet, from east to west, rather than in its normal motion from west to east

Some stars appear to never set, due to their proximity to one of the celestial poles. For example, Polaris is near the North Celestial Pole. From North America, Polaris is visible for the entire night on every night of the year. The same is true for constellations near the celestial poles. In the northern hemisphere, these include Cassiopeia, Cepheus, Draco, Ursa Minor, and Ursa Major.

Retrograde Motion

Ancient astronomers observed an usual planetary motion that was not explained until the Sun was discovered to be the centre of the Solar System. Certain planets appeared to change their path in the sky relative to the background stars, slowing to a stop, reversing direction, and then looping across the sky—a phenomenon known as **retrograde motion**. This apparent motion occurs because Earth travels around the Sun faster than the outer planets. For example, as Earth passes Mars in its orbit, Mars appears to at first stop and then travel backward in the sky. Earth continues past Mars, and forward motion appears again in the sky (Figure 5). We observe retrograde motion of only those planets that are farther from the Sun than Earth.

We can observe the retrograde motion of Mars, Jupiter, and Saturn with the unaided eye. However, we need to use a telescope to observe the retrograde motion of Uranus and Neptune. Each of the outer planets retrogrades for different lengths of time. Jupiter spends 4 of every 13 months in retrograde motion, for example, whereas Neptune retrogrades for 5 months of every year.

Figure 5 A time lapse photo of the retrograde motion of Mars

DID YOU KNOW?

Planet Wanderers
The word "planet" comes from the Greek word *planetes*, which means "wanderer."

To learn more about retrograde motion, **GO TO NELSON SCIENCE**

TRY THIS: DEMONSTRATE RETROGRADE MOTION

SKILLS: Observing, Communicating, Analyzing

SKILLS HANDBOOK 3.B.

Occasionally, a planet undergoes retrograde motion. When this happens, its path in the sky appears as a series of loops. In this activity, you will model this phenomenon.

Equipment and Materials: paper; markers

1. Find a partner. One of you will represent Earth and one will represent Mars. Use the paper and markers to create labels. Each person will hold the label indicating what planet they represent.
2. Stand beside your partner. Mars walks slowly forward from the starting point. Earth stays at the starting point, pointing at Mars as Mars moves forward (Figure 6).

Figure 6 step 2

3. Earth leaves the starting point, moving quickly to overtake the slower-moving Mars.

4. As Earth overtakes Mars, Earth begins to point backwards (Figure 7). This illusion of backwards motion is retrograde motion.

Figure 7 step 4

5. Repeat the experiment, switching roles with your partner.

A. Mars undergoes retrograde motion only when it is very close to Earth. Do you think that Mars appears at its brightest or dimmest when it retrogrades? Explain.

GO TO NELSON SCIENCE

B. The farther a planet is from the Sun, the slower it moves. Why is retrograde motion visible only with planets that are farther from the Sun than Earth is? T/I

C. While in retrograde motion, in which direction does Mars appear to move with respect to the background stars? T/I

Navigating the Night Sky

Finding your way around the night sky can seem confusing at first. The first step is to learn how to describe the location of celestial objects that you can see with the unaided eye.

In geography, latitude and longitude are used to pinpoint a place or object on Earth. In astronomy, celestial coordinates called azimuth and altitude can be used to describe the position of a celestial object in the sky relative to an observer on the ground for a particular time and place.

Azimuth and Altitude

Azimuth is the distance measured from north along the horizon to a point directly below the celestial object. North has an azimuth of 0°, east has an azimuth of 90°, south has an azimuth of 180°, and west has an azimuth of 270° (Figure 8). **Altitude** is the angular height of a celestial object, measured from the horizon. Figure 8 shows how the degrees of azimuth and altitude are measured.

azimuth the horizontal angular distance from north measured eastward along the horizon to a point directly below a celestial body

altitude the angular height a celestial object appears to be above the horizon; measured vertically from the horizon

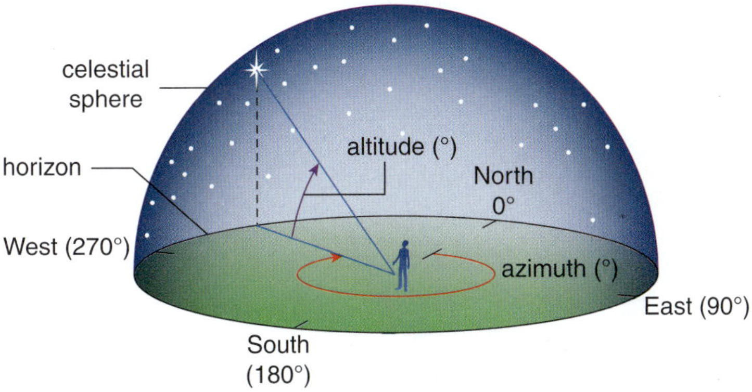

Figure 8 Altitude and azimuth are measured with reference to the horizon.

You can use your hands to measure angles for determining the positions of celestial objects. For instance, when you hold your hand out at arm's length, the width of your index finger is approximately equal to 1° (Figure 9). The width of a fist at arm's length is approximately equal to 10° (Figure 10). If you extend the fingers of your hand, the width from the tip of your pinky finger to the tip of your thumb is approximately equal to 20° (Figure 11). In this way, you can estimate the altitude and azimuth coordinates of any celestial object in the sky. Data obtained in this manner are relative to your position on Earth.

Figure 9 The width of a finger measures approximately one degree.

Figure 10 The width of a closed fist measures approximately ten degrees.

Figure 11 The width of an outstretched hand measures approximately twenty degrees.

TRY THIS | MEASURING ALTITUDE AND AZIMUTH

SKILLS: Performing, Observing, Analyzing, Communicating

SKILLS HANDBOOK
3.B.

In this activity, you will determine the altitude and azimuth of the Moon using your hands.

Equipment and Materials: compass; pencil; notebook

1. On the date and time determined by your teacher, go outside to an open area such as a park or parking lot.
2. Using the compass, determine the direction north.
3. Begin by facing north with your fists outstretched at eye level.
4. Keeping your arms straight out in front of you at eye level, use your fists as measuring tools while you slowly turn in a clockwise direction toward the east (Figure 12). Continue measuring until you reach the spot on the horizon directly below the Moon. Remember that each fist represents 10°.

Figure 12

5. You can also use your fingers to measure smaller increments as you get closer to the Moon. Each finger width is 1°.
6. When you reach the spot directly below the Moon, stop counting. The value you have counted is the azimuth of the Moon from your location on Earth.

7. With your arms still outstretched at eye level, begin counting degrees with your fist in an upward direction until the top of your fist reaches the Moon (Figure 13). Again, use your fingers to measure smaller increments as you get closer to the Moon. When your fist reaches the Moon, you have determined the altitude of the Moon from your position on Earth.

Figure 13

8. Record your results.

A. What was the altitude and azimuth of the Moon from your position on Earth? T/I
B. The next day, compare your answers with those of your classmates. Why do you think they are not exactly the same? T/I
C. Would the altitude and azimuth of the Moon be the same for people in different parts of the country? Why or why not? T/I

GO TO NELSON SCIENCE

UNIT TASK Bookmark

You can apply what you learned about measuring azimuth and altitude in this section to the Unit Task described on page 446.

IN SUMMARY

- The Sun appears to follow an annual path in the sky called the ecliptic.
- Some planets exhibit retrograde motion—a reversal of direction in their apparent path.
- Astronomers can describe the positions of celestial objects with altitude and azimuth.
- Stargazers can use their hands to approximate the altitude and azimuth of celestial objects.
- Light pollution can affect our view of the night sky in cities and other areas with a lot of light.

CHECK YOUR LEARNING

1. Name three different objects you can see in the night sky even if you are in a city.
2. What is meant by the term "light pollution," and how does it affect our view of the night sky (Figure 14)?

Figure 14

3. What is the name of the apparent annual path the Sun, the Moon, and the planets follow across the sky?
4. How does Earth's rotation affect your view of stars and constellations in the night sky?
5. How does Earth's revolution affect your view of stars and constellations in the night sky?
6. In a brief paragraph, explain the astronomical phenomenon shown in Figure 15.

Figure 15

7. Explain why planets farther from the Sun than Earth exhibit retrograde motion. Use a diagram to aid in your explanation.
8. Write an explanation of the terms "altitude" and "azimuth."
9. If you are facing north and the Moon is directly behind you over the southern horizon, what is its azimuth?
10. If you are observing a star that is directly overhead, what is its altitude?
11. Describe two different ways of using your hands to determine the position of an object with an azimuth of 42°.
12. An astronomer sees an interesting star in the night sky and wants to know its azimuth. She holds her hand outstretched at arm's length and notices that the width of her finger fits four times between the point on the horizon below the star and the direction north. What is the star's azimuth?
13. A stargazer uses his fist to determine the altitude of the Moon. He notes in his journal that the altitude is approximately 30°. How many times does his fist fit between the Moon and the horizon?

8.10 CONDUCT AN INVESTIGATION

Finding Objects in the Night Sky

In Section 8.1, you sketched the positions of objects in the night sky using a horizon diagram. In Section 8.8, you used a star map to predict the motions of the stars throughout the night and as the year progresses. In Section 8.9, you practiced estimating the positions of objects in the night sky using altitude and azimuth. In this activity, you will combine these skills to sketch, describe, and identify objects in the night sky from your location in Ontario.

SKILLS MENU
- Questioning
- Hypothesizing
- Predicting
- Planning
- Controlling Variables
- Performing
- Observing
- Analyzing
- Evaluating
- Communicating

Purpose
To sketch objects in the night sky, describe their position using altitude and azimuth, identify them, and predict their motions using a star map.

Testable Question
How can the positions of celestial objects be predicted?

Hypothesis/Prediction
In this investigation, you will make your predictions in Part B.

Equipment and Materials
- compass
- flashlight
- paper
- pencil
- ruler
- star map

Procedure
SKILLS HANDBOOK 3.B.5, 3.B.6

Part A: Sketch and Describe
1. Your teacher will choose a date and time for you and your classmates to go outside and observe the night sky. Find an open area where you can see as much of the horizon as possible.
2. Using your compass, find north.
3. Turn around 180°, so that you are facing south.
4. On a piece of paper, sketch trees or buildings that are along the southern horizon. You will use these as landmarks to draw the objects in the night sky.
5. Draw the brightest objects you see in the night sky as they appear against the horizon. Make your drawing as accurate as possible.
6. Determine the altitude and azimuth of the celestial objects. Write down the altitude and azimuth measurements next to the objects on your diagram.
7. Be sure to identify the location, date, and time on your horizon diagram.

Part B: Identify and Predict
SKILLS HANDBOOK 3.B.3

8. Set the star map you created in Section 8.8 to the date and time of your observation.
9. Using the star map, identify some of the stars or constellations you drew on your horizon diagram.
10. Using the star map, predict how your sketch would change if you observed from the same location two hours later.
11. Go to the same location two hours later than you did in Part A and repeat your observations. Did your observations support your predictions?
12. Continue observing from the same spot for a few nights in a row *at the same time*. Sketch and identify common celestial objects, such as the Moon; the planets Jupiter, Mars, and Venus; and the stars Polaris, Sirius, and Vega if these are visible. Sketch a new horizon diagram each night.

13. In your notebook, create a table similar to Table 1 and record your data.

Table 1 Altitude and Azimuth of Night Sky Objects

Date	Celestial object	Altitude	Azimuth
	Moon		
	Venus		
	Sirius		

14. Compare your horizon diagrams with those of other members of your class.

Analyze and Evaluate

(a) What was the brightest object you could identify? T/I
(b) What was the dimmest object you could identify? T/I
(c) How does the azimuth of the Moon change from night to night? T/I
(d) How does the azimuth of the planets change from night to night? T/I
(e) How does the azimuth of the stars change from night to night? T/I
(f) How did the positions of the celestial objects in your horizon diagrams compare with those of other members of your class? T/I
(g) Answer the Testable Question. T/I

Apply and Extend

(h) The horizon diagram in Figure 1 shows the Big Dipper and the Little Dipper as seen in the northern part of the sky in late autumn. Draw another horizon diagram showing both constellations as they would appear several hours later. T/I

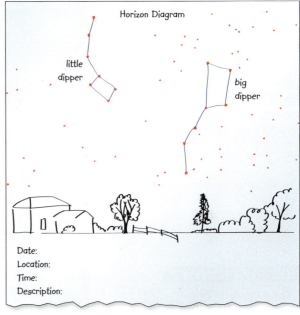

Figure 1 Ursa Major and Ursa Minor are commonly known as the "Big Dipper" and "Little Dipper," respectively.

(i) Identify a planet in the night sky and draw a horizon diagram for that planet. Make sure to note the date and time of your observation on your horizon diagram. Observe the same planet on the same (or close to the same) day of the month for 4 to 5 months, drawing a labelled diagram each time. After the 4 to 5 month period, compare your horizon diagrams with each other. How has the location of the planet changed over time? How does this change compare to the changes in the position of the celestial objects in (d)? T/I

UNIT TASK Bookmark

How can you apply what you learned about horizon diagrams and altitude and azimuth in this section to the Unit Task described on page 446?

8.11 Satellites

As you know, Earth has one natural satellite orbiting it—the Moon. Earth also has thousands of other satellites circling it at different altitudes and orbits, but these are all made by humans (Figure 1).

Artificial satellites help forecast weather, monitor agriculture, aid in telecommunication or navigation, assist military activities, and explore the Universe (Figure 2).

Figure 1 Artificial satellites provide valuable monitoring and communication services for humankind.

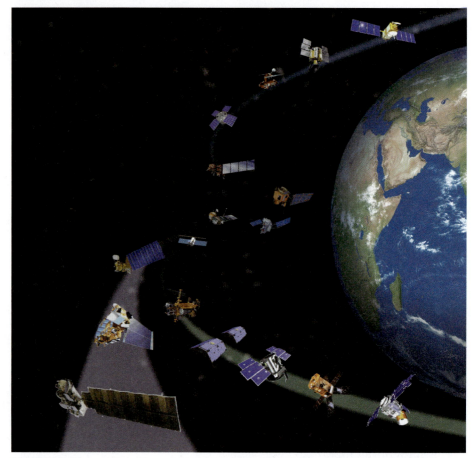

Figure 2 Today, more than 40 countries have put over 3000 functioning satellites into orbit around Earth.

Human-occupied spacecraft, such as the Space Shuttle, and space facilities, such as the International Space Station, also function as artificial satellites.

In 1957, the first artificial satellite (*Sputnik 1*) was sent into space by the Soviet Union. Its mission was to orbit Earth. Five years later, Canada's first satellite (*Alouette 1*) was launched.

Canada has been a world leader in developing satellite technology over the last 50 years. We have become renowned for building some of the most powerful telecommunications and Earth observation satellites. Canadian engineers and scientists employ this expertise to design and build satellites used in astronomy and space exploration—for everything from measuring characteristics of planets around distant stars to searching for potentially dangerous asteroids.

DID YOU KNOW?

Space Debris

In February 2009, two communications satellites (one American and one Russian) collided over Siberia. The resulting debris added to the 300 000 pieces of space junk already orbiting Earth. What safety hazards does this space debris pose? Will more satellites in orbit contribute to an increased incidence of these collisions?

To learn more about the satellites in Figure 2,
GO TO NELSON SCIENCE

READING TIP

Finding the Main Idea

The main idea gives the author's thoughts about a topic or key concept. Start by identifying the topic or key concept of the text (*satellites*) and whether the text breaks it down into subtopics (*natural and artificial satellites*). Then check if the author gives a perpective on the topic (*Canada is a world leader in developing artificial satellites*).

Staying in Orbit

How do all of these satellites keep orbiting Earth without plunging back to the ground? The force of Earth's gravity continuously pulls the satellite toward Earth. However, the forward motion of the satellite and the curvature of Earth prevent the satellite from getting any closer to the surface.

Imagine firing a powerful cannon as pictured in Figure 3(a). The cannonball would fly through the air until it curved toward the ground, pulled down by gravity. If the cannonball was fired with more velocity, it would travel farther before coming to rest on the ground. It would still curve toward Earth, but Earth also starts to curve beneath it because Earth is shaped like a sphere (Figure 3(b)). If we could fire the cannonball with enough velocity, it would fall toward Earth but never actually hit it because its flight would extend around the curve of Earth (Figure 3(c)). When engineers launch a satellite into Earth orbit, they launch it much like the powerful cannon. The satellite needs to attain sufficient velocity to make sure that when Earth's gravity pulls it down, it continuously falls around Earth's curvature (Figure 3(d)). It is called a continuous Earth orbit.

LEARNING TIP

Diagrams
Diagrams often help clarify what is explained in the text. How did the diagrams in Figure 3 help you understand about satellites staying in orbit?

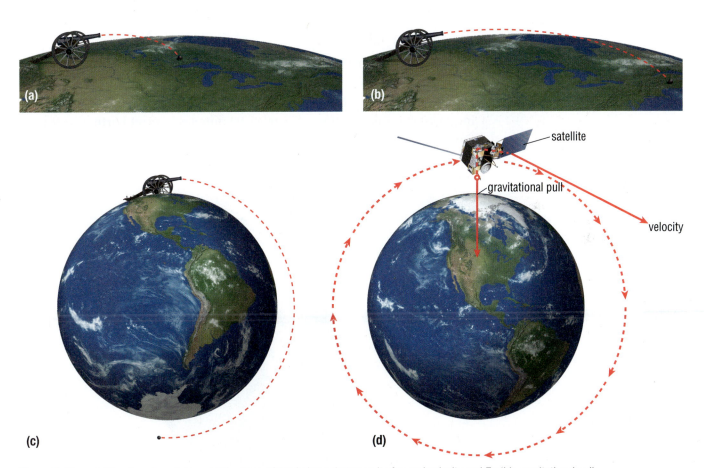

Figure 3 The satellite stays in a stable orbit because of the balance between its forward velocity and Earth's gravitational pull.

TRY THIS: ORBITING SATELLITES

SKILLS: Controlling Variables, Observing, Communicating

SKILLS HANDBOOK
3.B.2.

In this activity, you will model and observe how satellites stay in Earth orbit.

Equipment and Materials: computer with Internet access

1. Research some websites where you can simulate the launch of a satellite from Earth.

    GO TO NELSON SCIENCE

2. Select a website applet (short for application) that has various settings to adjust, such as launch speed, launch height, or gravity. There might be other variables to adjust. Play with the variables and see what happens when you launch satellites at different speeds.

3. Adjust the variables until the satellite is successfully in orbit around Earth.

4. Try and find other stable orbits at different altitudes.

A. How do you know if the launch speed of the satellite is too high? T/I

B. How do you know of the launch speed of the satellite is too low? T/I

C. In which direction does gravity pull the satellite? T/I

D. In which direction does the satellite's velocity point? T/I

E. To send a satellite into a high-altitude orbit, how would you change the location or speed of your launch? T/I

F. If you wanted to send a spacecraft to the Moon or to another planet, how would you change the location or speed of your launch?

Types of Orbits

Artificial satellites orbit outside Earth's atmosphere at altitudes of 200 km to more than 35 000 km. The higher the satellite is, the longer the orbital period—the time it takes to circle Earth. At an altitude of about 350 km, the International Space Station (ISS) takes 90 minutes to orbit Earth, whereas Canada's MOST space telescope satellite completes one orbit 820 km above Earth's surface in 101 minutes (Figure 4).

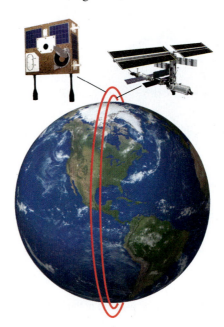

Figure 4 Canada's MOST space telescope was designed to probe stars outside the Solar System by measuring changes in the light they emit.

Low Earth Orbit Satellites

Satellites are placed into different orbits around Earth, depending on their function. For instance, most human-occupied spacecrafts and those conducting Earth observations are set at low-altitude orbits. These low Earth orbit satellites revolve around our planet at altitudes up to 2000 km.

Satellites that are required to see every part of the planet as they orbit it are placed in a special type of low Earth orbit called a polar orbit (Figure 5). A satellite in polar orbit generally travels at an altitude of 200 to 900 km in a path that takes it over both the North Pole and the South Pole. This type of orbit allows the satellite to view all parts of Earth. As the planet rotates beneath the satellite, a new pole-to-pole slice of Earth is monitored with each orbit of the satellite. These satellites have many uses, including military and Earth observation. Canada's RADARSAT satellites are in polar orbits to keep an eye on a variety of natural and human-made events. They chart icebergs in Canada's far Arctic oceans, monitor shifting patterns in agriculture in Africa, and play an important role in natural disaster response in Asia.

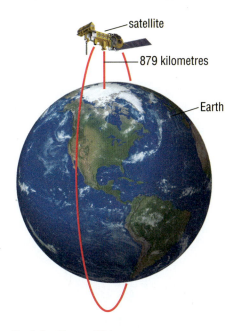

Figure 5 A polar orbit provides the best global coverage.

Medium Earth Orbit Satellites

Medium Earth orbit satellites travel at altitudes up to 35 000 km. Two dozen of these satellites are part of the **global positioning system (GPS)**. GPS satellites travel in medium Earth orbits at about 11 000 km (Figure 6). They aid in navigation by transmitting signals down to GPS receivers on the ground, providing them with precise geographical coordinates of their location. The orbital height of GPS satellites is governed by the fact that at least three satellites must be visible at any one time by a GPS receiver anywhere on the ground.

global positioning system (GPS) a group of satellites that work together to determine the positions of given objects on the surface of Earth

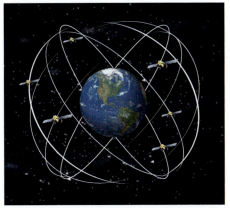

Figure 6 Six orbital planes host 24 GPS satellites.

8.11 Satellites 349

geostationary orbit an orbital path directly over Earth's equator with a period equal to the period of Earth's rotation

> **DID YOU KNOW?**
>
> **Satellite Television**
> The advantage of satellite TV is that it provides a wide coverage area, with the signal reaching millions of homes across large regions of the world. Some of the latest satellites can transmit more than 200 high-definition channels simultaneously. Launched in September 2008, Canada's communications satellite Nimiq 4 sends TV broadcast signals across most of North America.

Geostationary Orbit Satellites

One type of satellite orbits Earth at a distance of 35 790 km, which is about a tenth of the way to the Moon. This altitude is significant because it produces an orbital period equal to the period of the rotation of Earth. When a satellite is orbiting at this height directly above the equator, it is said to be in **geostationary orbit**. Satellites in geostationary orbit appear motionless in the sky, which makes them useful for communications and other commercial industries because they can be linked to antennas on Earth. Weather satellites, for example, track weather in this manner (Figure 7).

Communication industries use geostationary satellites for satellite broadcast television and radio. Television satellite dishes attached to the sides of homes are able to receive the satellite signal by aiming at a fixed spot in the sky (Figure 8).

Figure 7 This satellite image shows Hurricane Katrina off the coast of Louisiana in August 2005.

Figure 8 The satellite dishes on these apartments are aimed at a satellite in geostationary orbit.

RESEARCH THIS: DIFFERENT KINDS OF SATELLITES

SKILLS: Researching, Evaluating

SKILLS HANDBOOK 4.A.5., 4.A.6.

The many different satellites orbiting Earth perform various functions. In this activity, you will research a satellite and describe its function and what effect it has on the lives of people.

1. Research a satellite that interests you and conduct research to answer the following questions.

   A. What function does your satellite perform?

   B. In your opinion, does your satellite perform a valuable function?

   C. What kind of orbit is your satellite in: low, medium, or high (geostationary)?

   D. RADARSAT-1 is Canada's first commercial Earth observation satellite. Research its function, orbit, and scientific and commercial applications. Summarize your findings and present them to the class in a pamphlet, short story, or visual presentation.

GO TO NELSON SCIENCE

TRY THIS: SATELLITE IMAGES

SKILLS: Performing, Observing, Analyzing

SKILLS HANDBOOK 3.B.

The altitude of a satellite determines the speed at which it travels and is also an important factor in determining how much of Earth's surface it can see at any given time. Low orbit satellites, such as GPS satellites and weather satellites, are much closer to Earth's surface and cannot see as much of its surface as higher altitude satellites in geostationary orbit.

In this activity you will use a globe to understand how much of Earth's surface different satellites can see. Then you will classify pictures of Earth taken by various satellites.

Equipment and materials: standard-sized (30 cm diameter) globe; metre stick; pictures of Earth

1. To get a sense of the height of the different orbits, use the metre stick to find the following distances above the surface of the globe. They represent the altitudes of the different orbits at the same scale of the globe.

    low Earth orbit: 4 cm
    medium Earth orbit: 20 cm
    geostationary orbit: 84 cm

2. Place your eyes at the given distances and look at the surface of the globe. Move your eyes around at each altitude to determine how much of Earth's surface you can see at each orbital height.

3. Look at the pictures your teacher has provided.

A. Classify the pictures in one of the three categories: low Earth orbit, medium Earth orbit, or geostationary orbit. T/I

B. If you were developing a satellite to observe the entire continent of South America, which orbit would you pick? Why? A

C. If you were developing a satellite to observe weather patterns over the city of Hamilton, which orbit would you pick? Why? A

UNIT TASK Bookmark

How can you apply what you learned about travelling in Earth orbit in this section to the Unit Task described on page 446?

IN SUMMARY

- Many countries, including Canada, have launched artificial satellites into Earth orbit to study Earth and objects beyond Earth.
- Objects need to be sent into orbit with enough velocity to avoid being pulled back to Earth's surface by gravity.
- Different types of orbit (low, medium, and geostationary), are categorized based on the satellite's altitude.
- Satellites have many different applications for technologies on Earth, such as satellite television and global positioning systems.

CHECK YOUR LEARNING

1. What is the difference between an artificial satellite and a natural satellite? Give one example of each. K/U

2. Describe three ways in which artificial satellites benefit society. A

3. (a) When was the first artificial satellite launched into Earth orbit? What country was responsible for this?
   (b) When was Canada's first artificial satellite launched into Earth orbit? What was its name? K/U

4. Identify three artificial satellites that Canada has launched into Earth orbit, and describe their function. K/U

5. How does Earth's shape enable engineers to put satellites into stable orbit around our planet?

6. Identify four different reasons for launching a satellite into orbit around Earth. K/U

7. Compare and contrast the three types of Earth orbit regarding their altitude and what they are used for. K/U

8. Sketch Earth with the three different types of Earth orbit. Label the altitudes and a typical satellite found at each level. K/U C

9. What does the acronym GPS stand for, and what is it used for? K/U

8.11 Satellites 351

8.12 EXPLORE AN ISSUE CRITICALLY

Security Satellites

The first artificial satellite, *Sputnik*, was sent into Earth orbit in 1957 by the Soviet Union. The United States soon followed with the launch of its own satellite, *Explorer 1* in 1958. Canada was the third country in the world to design a satellite for Earth orbit in 1962, although it was launched by a U.S. rocket.

Today, there are thousands of satellites orbiting Earth. They are used for many purposes, such as monitoring weather patterns on Earth and observing distant stars. Some of the satellites are communications satellites used for relaying cell phone signals, and some are military satellites used for security purposes. Some of these satellites orbit Earth at the same rate of speed that Earth rotates, so they can continuously observe the same place on Earth.

Canada has many satellites in orbit, including RADARSAT-1 and RADARSAT-2, which observe Earth. RADARSAT-2 observes the Arctic in order to protect Canada's claim on northern lands and the north Pole. Some countries, including Russia, have claimed ownership over the North Pole, so Canada is monitoring the waterways and landforms of the north to look for activity.

The Issue

Canada is concerned that without a strong presence in the Arctic it might lose its claims of sovereignty over the north. Recently, Canada has been using satellite technology such as RADARSAT-2 to monitor the Arctic and the Northwest Passage waterway. Many people think Canada needs to launch another satellite, so the government can observe the north around the clock. However, satellite technology is expensive. RADARSAT-2 cost a total of $524 million, with $60 million spent by the Department of Defense to use it in the Arctic.

Goal

To recommend to the Canadian government whether Canada should spend the money to develop and launch another satellite into orbit similar to RADARSAT-2 in order to monitor the waterways and land of the Arctic.

SKILLS MENU
- Questioning
- Hypothesizing
- Predicting
- Planning
- Controlling Variables
- Performing
- Observing
- Analyzing
- Evaluating
- Communicating

Gather Information

Work with a partner and research more about satellite research-and-development costs. Find more information about RADARSAT-2 and how the Canadian Defense Department has used the satellite to monitor the Arctic. Investigate why it is important for the Canadian government to have our claims for the Arctic respected. What would the Arctic or the Northwest Passage be worth to the country who claimed it? Which other countries are claiming the Arctic as theirs?

Identify Solutions

To make your decision, you need to balance the factors of cost, security, and the contribution to the scientific community. You need to weigh whether a second satellite in polar orbit is a worthy reason to spend such a lot of money. Consider the advantages and disadvantages to the Canadian people.

Make a Decision

Decide on a final recommendation for the government on their decision to launch another security satellite to observe the Arctic. Which criteria did you use to decide?

Communicate

You need to decide on a method for delivering your recommendation. You may write a report, make an oral presentation, or include pictures in a PowerPoint presentation to make your point. Present your findings to the class. Do your classmates agree with your conclusion? Can you reach a compromise?

AWESOME SCIENCE ✓ OSSLT

Mysteries Beyond the Planets

The farthest reaches of the Solar System, past the orbit of Neptune, are a cold, dark place. Billions of kilometres away from the warmth and light of our Sun lies Pluto, in the most unexplored region of our solar neighbourhood. In just the past decade, however, astronomers have found strange new celestial objects in this dark and cold region. Scientists think these mysterious objects may unlock clues about how planets were formed.

Once thought to be a lonely outpost at the edge of the Solar System, this region is not so empty after all. Astronomers probing this icy realm believe that Pluto and several other dwarf planets are members of a far-off collection of celestial objects that belong to the Kuiper Belt (Figure 1).

Figure 1 Astronomers have begun finding Kuiper Belt objects only in the last decade.

The Kuiper belt is located 4.5 billion to 7 billion kilometres from the Sun. It is a doughnut-shaped ring of debris made up of icy celestial objects, some of which can be up to 50 km in diameter.

Short-period comets, such as Halley's Comet, originated in the Kuiper Belt before they were knocked out of their stable orbits and pulled in by the Sun's gravity. Many of these comets are the same age as Earth. Astronomers believe that this comet reservoir may be a frozen time capsule of what the Solar System was like in its infancy. Kuiper Belt objects (KBOs) are far from the Sun and do not change much over time. The Kuiper Belt may therefore hold the key to understanding how planets were formed. NASA's *New Horizons* robotic probe will be swinging by Pluto and the other Kuiper Belt objects in 2015 (Figure 2).

KBOs are so faint that astronomers were unable to detect them until 1992, when they began using digital camera technology attached to a 2.2 m telescope on top of Hawaii's Mauna Kea. New telescope technology enables farther and smaller objects to be

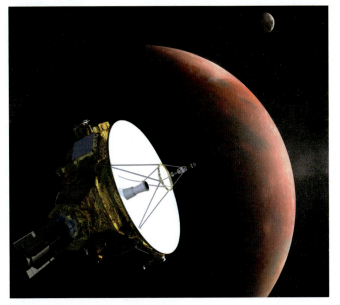

Figure 2 *New Horizons* robotic probe

detected. So far, more than 1200 KBOs have been catalogued, including the biggest dwarf planet, Eris, which is larger than Pluto!

Far beyond the Kuiper Belt lies an even more mysterious region of the Solar System called the Oort Cloud. Because it is so distant, no one has ever seen this region, but astronomers believe it is where long-period comets such as Hale-Bopp (1997) and Comet McNaught (2007) originated (Figure 3).

Orbiting the outer limits of the Solar System at about 5000 to 100 000 AU, the Oort Cloud is thought to be a giant, spherical shell of frozen chunks of ice and dust. The Oort Cloud is a frozen storage area for trillions of comets. Like fossils, they may hold clues to the early history of the Solar System.

Some experts say that while striking the early Earth billions of years ago, these long-period comets may have delivered frozen water and organic molecules to our planet, triggering the process of the origins of life.

Figure 3 Comet McNaught in 2007

CHAPTER 8

LOOKING BACK

KEY CONCEPTS SUMMARY

Careful observation of the night sky can offer clues about the motion of celestial objects.

- The stars appear to move across the night sky from east to west, except for the North Star (Polaris), which appears stationary. (8.5–8.7)
- The Moon rises in the east and sets in the west. Sometimes the Moon is visible in the daytime. (8.9)
- Different constellations are visible at different times of year. (8.7, 8.8)
- Lunar and solar eclipses sometimes occur. They are due to the alignment of Earth and the Moon with respect to the Sun. (8.5)

Celestial objects in the Solar System have unique properties.

- There are eight planets in the Solar System. The physical properties of the four terrestrial planets are distinct from the physical properties of the four gas giant planets. (8.1)
- The planets all orbit the Sun in ellipses, in the same direction but at different distances. (8.3–8.5)
- Asteroids, meteoroids, comets, and dwarf planets all orbit the Sun but are smaller and have properties that distinguish them from the planets. (8.3)

Some celestial objects can be seen with the unaided eye and can be identified by their motion.

- Mars, Venus, Saturn, and Jupiter can be easily seen in the night sky without a telescope. They move eastward from night to night. (8.1, 8.3)
- The positions of the objects in the night sky can be specified by their altitude and azimuth. (8.8, 8.9)
- The planets that are farther from the Sun than Earth sometimes exhibit retrograde motion. (8.9)
- Artificial satellites orbiting Earth have different uses, such as communications, weather monitoring, and military applications. (8.11)

The Sun emits light and other forms of radiant energy that are necessary for life on Earth to exist.

- Solar weather interacts with Earth's magnetic field, causing auroras around the North and South Poles. (8.2)
- Satellites in Earth orbit can be damaged by solar wind. (8.2)
- Various types of radiation from the Sun bring light and warmth to the surface of Earth, providing support for life on Earth. (8.1, 8.2)

Satellites have useful applications for technologies on Earth.

- Galileo Galilei first observed the Moon and the planets in the night sky with a telescope, obtaining evidence to support the heliocentric model of the Solar System. (8.5)
- RADARSAT-1 and RADARSAT-2 are successful Canadian satellites that observe the surface of Earth. (8.11)
- Astronomers gather information on the Sun with special observatories like SOHO and STEREO. (8.2)

The study of the night sky has influenced the culture and lifestyles of many civilizations.

- Patterns of stars can be used for navigation. (8.6)
- Aboriginal peoples in Canada have a long history of observing patterns in the night sky for navigational and cultural purposes. (8.6)
- Cultures across the globe have legends and stories specific to their culture based on patterns of the stars in the night sky. Sometimes they built structures that aligned with the positions of certain stars. (8.6)

WHAT DO YOU THINK NOW?

You thought about the following statements at the beginning of the chapter. You may have encountered these ideas in school, at home, or in the world around you. Consider them again and decide whether you agree or disagree with each one.

1 Scientists have seen signs of life on other planets.
Agree/Disagree?

4 The stars and constellations can be used to map the night sky, as well as for accurate navigation.
Agree/Disagree?

2 Many planets in the Solar System have moons.
Agree/Disagree?

5 We always see the same side of the Moon.
Agree/Disagree?

3 Canada is a world leader in satellite technology.
Agree/Disagree?

6 Storms on the surface of the Sun can affect Earth.
Agree/Disagree?

How have your answers changed since then?
What new understanding do you have?

Vocabulary

astronomy (p. 305)
celestial object (p. 305)
Universe (p. 305)
star (p. 305)
luminous (p. 305)
planet (p. 306)
Solar System (p. 306)
satellite (p. 306)
orbit (p. 306)
galaxy (p. 307)
electromagnetic (EM) radiation (p. 309)
electromagnetic (EM) spectrum (p. 309)
sunspots (p. 310)
solar flare (p. 310)
solar prominence (p. 310)
aurora borealis (p. 311)
astronomical unit (p. 313)
dwarf planet (p. 314)
comet (p. 316)
orbital radius (p. 320)
gravitational force (p. 321)
solstice (p. 323)
equinox (p. 323)
precession (p. 324)
lunar cycle (p. 324)
eclipse (p. 325)
tide (p. 327)
constellation (p. 329)
celestial sphere (p. 331)
celestial navigation (p. 331)
ecliptic (p. 339)
retrograde motion (p. 340)
azimuth (p. 341)
altitude (p. 341)
global positioning system (GPS) (p. 349)
geostationary orbit (p. 350)

BIG Ideas

✓ Different types of celestial objects in the Solar System and Universe have distinct properties that can be investigated and quantified.

CHAPTER 8 REVIEW

The following icons indicate the Achievement Chart category addressed by each question.

K/U Knowledge/Understanding   T/I Thinking/Investigation
C Communication   A Application

What Do You Remember?

1. Using your own words, define the terms "Universe," "solar system," and "astronomy." (8.1) K/U

2. Identify the parts of the Sun (Figure 1). (8.2) K/U

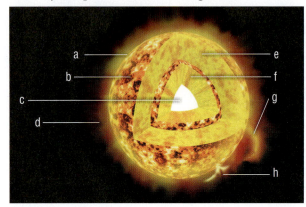

Figure 1

3. Describe the way the stars, the Moon, and the planets move across the sky throughout the night. (8.8) K/U

4. Draw a diagram illustrating how the phases of the Moon change throughout a month. (8.5) K/U C

5. Write a paragraph retelling a First Nations legend about a constellation visible in the northern hemisphere. (8.6) K/U C

6. Make a table comparing the terms "meteoroid," "meteor," and "meteorite." (8.3) K/U

7. Describe two reasons humans have sent artificial satellites into orbit around Earth. (8.11) A

What Do You Understand?

8. What is an asteroid? Where in the Solar System would you find asteroids? (8.3) K/U

9. What is a comet? Where in the Solar System would you find comets? (8.3) K/U

10. What is retrograde motion and how does it occur? (8.9) K/U

11. Describe how the Moon and the Sun cause tides on Earth. (8.5) K/U

12. Explain how the particles released by the Sun cause the aurora borealis and the aurora australis. (8.2) K/U C

13. What is the difference between a lunar eclipse and a solar eclipse? (8.5) K/U

14. Explain how to determine the altitude and the azimuth of an object in the night sky. (8.9) K/U T/I

15. Describe two ways in which radiation from the Sun supports life on Earth and two ways in which the Sun causes harm to people and technologies on Earth. (8.2) C A

16. What are RADARSAT-1 and -2, and what do they do? (8.11) K/U

Solve a Problem

17. How do the rotation and the revolution of Earth affect the apparent motion of celestial objects in the night sky? (8.5) T/I

18. Compare the terrestrial planets with the gas giant planets. List two similarities and two differences. (8.3) K/U

19. (a) Compare the model that people used to explain the Solar System 1500 years ago to the one we use today.
    (b) Do you think the geocentric model was flawed? Explain your reasoning. (8.5) K/U C

20. Your friend tells you that Earth is hot in summer and cold in winter because Earth is closer to the Sun in summer and farther in winter. Write a letter to your friend explaining the true causes of the seasons. (8.5) K/U C

21. How is Mars similar to Earth? How is it different? (8.3) K/U

22. Even without written records, scientists know that ancient peoples observed the motions of celestial objects in the night sky. How do they know this? (8.6) T/I

23. A company wants to launch a series of communication satellites so that they continue to orbit above the same location on Earth at all times. At approximately what altitude would these satellites need to orbit Earth? Explain. (8.11) T/I

24. You observe a bright object in the night sky and want to research what it is. Which two measurements would you need to make to describe its position to an astronomer? What else would you need to know aside from the object's position? (8.8) T/I

25. Imagine you are lost in the woods of northern Ontario. What star could you use as a navigational guide? How could you find that star among the thousands of stars that are visible on a clear night? (8.8) T/I

Create and Evaluate

26. A satellite with booster rockets is slowing down while in Earth orbit. How would this affect the satellite's orbit? How could astronomers fix this problem? (8.11) T/I

27. Imagine you are the head of the Canadian Space Agency. In your notebook, list the following research projects in order of priority. Explain your reasoning. (8.11) C A
    - a spacecraft designed to analyze atmospheric conditions on Mars
    - a space station for astronauts to perform experiments in Earth orbit
    - a spacecraft designed for passage to the Moon to analyze the Moon's surface
    - a spacecraft designed to collect data about the atmosphere of Venus

28. More than 2200 years ago, Aristarchus of Samos, an ancient Greek astronomer, suggested that the Sun was the centre of the Solar System. It took more than 1800 years for this theory to be revived by Copernicus. Why do you think it took so long for people to accept that the Sun was at the centre of the Solar System? (8.6) T/I

29. A new object is discovered in the Solar System and needs to be classified as a comet, an asteroid, a planet, or a dwarf planet. What information about the object would you want to collect to properly classify it? (8.3) T/I

30. If you were to build a machine that was solar powered, where on Earth would you build it? Why? (8.5) T/I

Reflect on Your Learning

31. In what ways could you use your new knowledge of the positions of the objects in the night sky?

32. How do you think you would feel if you discovered a comet or asteroid? What name would you give it and why?

33. Identify one topic covered in this chapter that you found difficult to understand.
    (a) Why was this topic difficult for you to understand?
    (b) What did you do to help overcome your difficulty?

Web Connections

34. Research the constellations that are visible in the night sky in a country in the southern hemisphere, such as Australia (Figure 2). T/I

Figure 2

35. Research some of the star stories that North American Aboriginal peoples have been telling for hundreds of years. How do they compare with the stories from Ancient Greece? Can you see any similarities in the stories? T/I

36. Research and write several paragraphs about how tides affect the lives of people who live and work on or near the sea. T/I C A

To do an online self-quiz or for all other Nelson Web Connections, **GO TO NELSON SCIENCE**

CHAPTER 8 SELF-QUIZ

The following icons indicate the Achievement Chart category addressed by each question.

K/U Knowledge/Understanding T/I Thinking/Investigation C Communication A Application

For each question, select the best answer from the four alternatives.

1. In which region of the Sun do nuclear fusion reactions take place? (8.2) K/U
   (a) radiative zone
   (b) photosphere
   (c) corona
   (d) core

2. Which statement best explains why people on Earth always see the same side of the Moon? (8.5) K/U
   (a) A day is the same length on Earth as it is on the Moon.
   (b) It takes 365 days for the Moon to complete one revolution around Earth.
   (c) Earth revolves around the Sun more slowly than the Moon revolves around Earth.
   (d) The Moon rotates on its axis and revolves around Earth in the same amount of time.

3. Under what circumstances can we observe a solar eclipse? (8.5) K/U
   (a) when the Moon is aligned between Earth and the Sun
   (b) when the Sun is aligned between Earth and the Moon
   (c) when Earth is aligned between the Sun and the Moon
   (d) when the Moon, Earth, and the Sun move out of alignment

4. Which statement best explains why Pluto is no longer considered a planet? (8.3) K/U
   (a) Pluto is not in orbit around a star.
   (b) Pluto does not dominate its own orbit.
   (c) Pluto is not close enough to the Sun.
   (d) Pluto does not have a spherical shape.

Indicate whether each of the statements is TRUE or FALSE. If you think the statement is false, rewrite it to make it true.

5. During the new moon phase, the surface of the Moon is not visible from Earth. (8.5) K/U

6. In North America, the North Star is visible only at night during spring and summer. (8.6) K/U

Copy each of the following statements into your notebook. Fill in the blanks with a word or phrase that correctly completes the sentence.

7. The _____ is at the centre of the Solar System. (8.2) K/U

8. During our _____ season, Earth's northern hemisphere is tilted toward the Sun to a greater degree than it is at other times of the year. (8.5) K/U

9. Stars are _____ objects, which means that they produce and emit light. (8.1) K/U

Match each term on the left with the most appropriate description on the right.

10. (a) star
    (b) asteroid
    (c) comet
    (d) meteorite
    (e) satellite

    (i) a small celestial object made of rock and metal that orbits the Sun between Mars and Jupiter
    (ii) a chunk of ice, rock, and dust that travels in a very long orbit around the Sun
    (iii) a celestial object, sometimes human-made, that travels around a larger celestial object
    (iv) a lump of metal or rock that has hit Earth's surface
    (v) a celestial body composed of hot gases that radiates large amounts of energy (8.1, 8.3) K/U

Write a short answer to each of these questions.

11. Why does a planet's orbital radius represent that planet's average distance from the Sun and not its exact distance? (8.5) K/U

12. (a) Neptune's average distance from the Sun is 4 497 000 000 km. What is Neptune's average distance from the Sun in astronomical units (AU)? (8.3, 8.5) T/I

    (b) The average distance between Venus and the Sun is 0.72 AU. What is this distance in kilometres? (8.3) T/I

13. At its closest point, Mercury is approximately 0.31 AU from the Sun. Use this information to describe how Mercury's orbit compares to Earth's orbit. (8.3) T/I

14. Imagine that you see the aurora borealis for the first time while on vacation. Write a postcard to a friend describing the event and explaining the natural conditions that caused it to happen. (8.2) K/U C

15. Solar scientists study space weather, which includes natural events such as solar storms. Based on what you have learned in this chapter, do you think it is important to forecast space weather? Explain your answer in a brief paragraph. (8.2) C A

16. Canada has several time zones. When it is noon in Ottawa, it is 9 a.m. in Vancouver. If noon is the time when the Sun is highest in the sky, explain why Canada needs more than one time zone. (8.5) A

17. You are a member of a committee that will recommend a location for a new observatory. There are two choices for the location. One choice is a site in a large city where many people would have easy access to the observatory. During construction, workers and materials could be transported to the site easily. The other choice is a site on a mountain range far from any major city. Workers, materials, and visitors would have to travel long distances to reach the observatory site. Which site would you recommend? Give at least two reasons for your choice. (8.9) C A

18. A friend tells you that because a geostationary satellite always remains over a specific point on Earth's surface, the satellite does not move in the sky. Explain to your friend why this explanation is incorrect. (8.11) K/U C

19. Is it possible to have a solar eclipse and a spring tide at the same time? Draw a diagram to explain your answer. (8.5) K/U C

20. Although solar eclipses occur about as often as lunar eclipses, more people can observe a lunar eclipse than a solar eclipse. Explain why this is true. (8.5) K/U C

21. Imagine that you are stranded on an island in the ocean without any timekeeping devices. Describe how you could keep track of days and months using the movements of Earth and the Moon. (8.5) C A

22. If people in your community were to name constellations today, do you think they would name them after wild animals, mythical figures, and common tools such as a dipper? Explain your answer. (8.6) C A

23. Will a GPS receiver function anywhere on Earth? Explain your answer. (8.11) K/U A

24. Why must a planet's moon contain less mass than the planet itself? (8.5) K/U

CHAPTER 9

Beyond the Solar System

KEY QUESTION: What are the properties of the celestial objects beyond the Solar System?

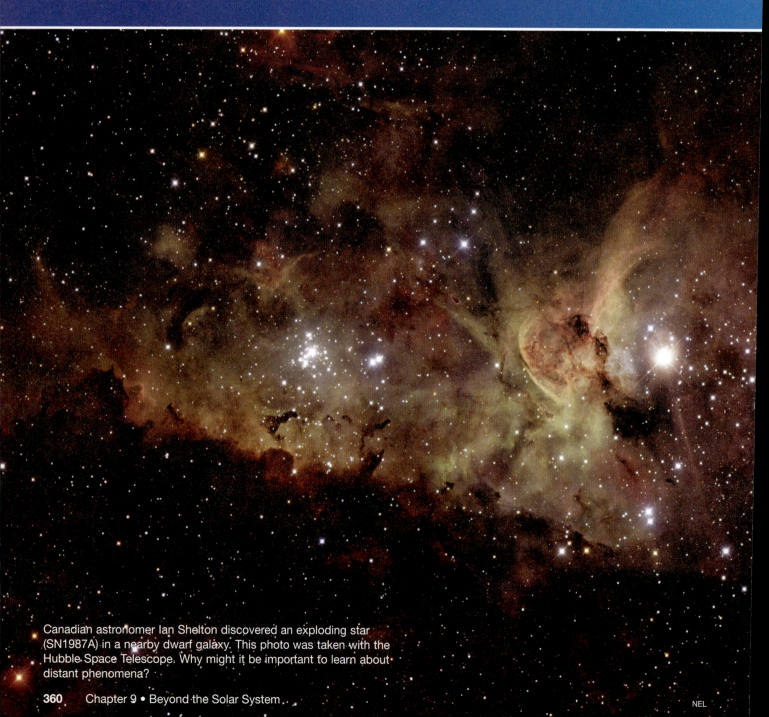

Canadian astronomer Ian Shelton discovered an exploding star (SN1987A) in a nearby dwarf galaxy. This photo was taken with the Hubble Space Telescope. Why might it be important to learn about distant phenomena?

UNIT D
The Study of the Universe

CHAPTER 8 — Our Place in Space

CHAPTER 9 — Beyond the Solar System

CHAPTER 10 — Space Research and Exploration

KEY CONCEPTS

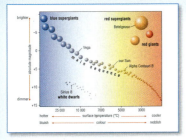

Stars differ in colour, size, and temperature.

Stars change their physical properties over time.

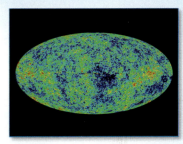

There is strong evidence that the Universe had its origin around 13.7 billion years ago.

Stars are clustered into galaxies, which we can observe using telescopes.

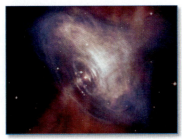

Black holes, neutron stars, and supernovas are part of the life cycle of massive stars.

Evidence collected by satellites and Earth-based observations support theories of the origin, evolution, and large-scale structure of the Universe.

ENGAGE IN SCIENCE

UNCOVERING A MYSTERY

It was 1971, and 28-year-old astronomer Charles Thomas Bolton was working at the David Dunlap Observatory, in Richmond Hill, just north of Toronto. In recent years, astronomers had observed star systems consisting of pairs of stars orbiting around their common centre of mass, much like a couple dancing. Bolton, too, was interested in this type of star system.

On a hot, dark, summer night that year, Dr. Bolton swung the giant telescope—almost two metres in diameter—toward a star in the constellation Cygnus. As the young astronomer observed the stars of this constellation, he noticed that a star in one of the pairs seemed to be moving much too quickly around its companion. What could be causing it to move so fast? Bolton reasoned that there must be an object with a very large mass nearby whose strong gravitational pull was affecting the star's movement.

Over the next few months, the data Bolton gathered on the moving star showed that the nearby object was very strange indeed: not only did it have an extremely strong gravitational pull, but it was also invisible! What was this astronomical phenomenon with such unusual properties?

Bolton had discovered a region in space with a gravitational pull so strong that even light cannot escape. This discovery—the first of its kind—has since become known as Cygnus X-1, a black hole. Since 1971, several other black holes have been discovered. Scientists around the world continue to study these strange and fascinating celestial objects (which you will learn more about later in this chapter).

How do you think Bolton felt when he discovered the first black hole? If astronomers cannot see black holes directly, how do you think they detect their presence? Can you think of any references to black holes in popular movies, television shows, or video games? Do you think they are accurate in their portrayal of black holes?

An artist's rendering of the black hole Cygnus X-1. The black hole is to the right and is pulling material off a nearby star because of its massive gravity. You can see matter swirling into the black hole like water down a drain.

WHAT DO YOU THINK?

Many of the ideas you will explore in this chapter are ideas that you have already encountered. You may have encountered these ideas in school, at home, or in the world around you. Not all of the following statements are true. Consider each statement and decide whether you agree or disagree with it.

1 Stars differ only in their size and how far they are from Earth.
Agree/disagree?

4 Scientists can detect everything that there is in the Universe.
Agree/disagree?

2 The Big Bang theory explains how the Universe began.
Agree/disagree?

5 All galaxies in the Universe are like the Milky Way.
Agree/disagree?

3 Spacecraft orbiting Earth can observe objects that are billions of kilometres away.
Agree/disagree?

6 Telescopes allow astronomers to see all of the celestial objects in the Universe.
Agree/disagree?

FOCUS ON READING

Summarizing

When you summarize a text, you shorten it by restating only the main idea and key points in your own words. Specific facts, examples, and questions are not included. Use the following strategies when summarizing a text:

- determine the main idea and key points using text features such as the title, headings, topic sentences, and signal words such as "thus," "therefore," and "in other words"
- omit details that do not expand your understanding of the main idea
- use the same organizational pattern as the text uses (cause/effect, concept/definition, descriptive, compare/contrast)
- replace several specific words with one general word

READING TIP
As you work through the chapter, look for tips like this. They will help you develop literacy strategies.

Galaxy Clusters

Figure 1 The Local Group, made up of many small irregular and elliptical galaxies, is dominated by two large spiral galaxies—the Milky Way and Andromeda.

The Milky Way is part of a group of more than 35 galaxies called the Local Group (Figure 1). This collection of galaxies is about 10 million ly across. Our largest and nearest galactic neighbour is the Andromeda galaxy—a slightly larger version of the Milky Way. The Andromeda galaxy is a spiral galaxy that contains at least 300 billion stars. Lying about 2.6 million ly away, it is the farthest object the unaided human eye can see. It is amazing to think that the light we see from this galaxy travels for 2.6 million ly before reaching Earth. To put this into perspective, this means that the light we see today left the Andromeda galaxy when our early ancestors were just beginning to use stone tools. Since then, we have refined our tools and developed telescopes powerful enough to view this distant galaxy!

Summarizing *in Action*

The key to summarizing is paraphrasing, which means using your own words to restate the main idea. Rewording the original text is crucial. Identify the main idea and briefly highlight the key points that support the main idea. Omit information that is not necessary to understand the main idea. Here is how one student used the strategies to summarize the selection about galaxy clusters.

Clues	Unnecessary Information	Text Pattern
The title suggests that galaxies are in a group	number of galaxies in cluster number of stars in Andromeda	Descriptive (organizes facts about the defining features or the parts that make up a whole)
The Milky Way and Andromeda galaxies are part of the Local Group.	distance from Andromeda to Milky Way tools and technology used on Earth	

Summary: The Milky Way and Andromeda galaxies are part of a cluster of galaxies called the Local Group.

Measuring Distances Beyond the Solar System

9.1

The Sun, our nearest star, appears as a large, bright disc in the sky because it is relatively close to Earth—approximately 150 million (1.5×10^8) km away. The next nearest star to the Sun is Proxima Centauri, which appears as a twinkling dot of light. Proxima Centauri is approximately 40 trillion (4.01×10^{13}) km from Earth, which is about 267 000 times farther away from Earth than the Sun. Most stars are more than 100 trillion (1.0×10^{14}) km from Earth.

Distances beyond the Solar System are so vast that scientists cannot measure them in the same way they measure distances on Earth. For example, if you were to measure the distance from the Sun to Proxima Centauri using kilometres it would be like measuring the distance from Vancouver Island to Prince Edward Island in centimetres!

In Chapter 8 you learned about the astronomical unit (AU). The AU can be used to measure distances within the Solar System, but it is too small for measuring the enormous distances beyond the Solar System. Instead, scientists use more convenient units of distance to measure the distances between stars.

The Light Year

Astronomers use **light years** to measure the distance to stars or other celestial objects outside of the Solar System. Although it sounds like a unit of time, the light year is, in fact, a unit of distance. A light year is the distance that light travels in a vacuum (empty space) in one year. In space, light travels at a constant speed of approximately 300 000 km/s. This means that 1 light year (ly) is approximately equal to 10 trillion (9.46×10^{12}) km. Table 1 shows the approximate distances of some celestial objects beyond the Solar System.

We can use the following equation to determine the distance between two celestial objects in light years:

$$1 \text{ ly} = 9.46 \times 10^{12} \text{ km}$$

> **LEARNING TIP**
>
> **Large Numbers**
> A trillion is a million million. That is 1 000 000 000 000 (1.0×10^{12}).

> **DID YOU KNOW?**
>
> **Proxima Centauri**
> Proxima Centauri is the nearest star to the Sun and is only visible in the southern hemisphere. If you could travel to Proxima Centauri in a vehicle going 1000 km/h, it would take almost 46 million years to get there!

light year the distance that light travels in a vacuum in 1 year (365.25 days); equal to 9.46×10^{12} km

> **MATH TIP**
>
> **SI Units**
> 1 km = 1×10^3 m
> 1 AU = 1.5×10^{11} m
> 1 ly = 9.46×10^{15} m

SAMPLE PROBLEM 1 Calculating the Distance to a Star in Light Years

If Proxima Centauri is 4.01×10^{13} km from Earth, what is its distance from Earth in light years?

Given: distance to Proxima Centauri = 4.01×10^{13} km
1 ly = 9.46×10^{12} km

Required: distance to Proxima Centauri in light years (ly)

Analysis: distance in light years = 4.01×10^{13} km $\times \dfrac{1 \text{ ly}}{9.46 \times 10^{12} \text{ km}}$

Solution: distance to Proxima Centauri = 4.24 ly

Statement: The distance from Earth to Proxima Centauri is 4.24 ly.

Practice

Polaris is 400 ly from Earth (Table 1). Calculate the distance from Earth to Polaris in kilometers.

Table 1 Distance of Some Celestial Objects

Star or galaxy	Approximate distance from Earth (ly)
Alpha Centauri	4.3
Vega	25
Polaris	400
Deneb	1400
Andromeda Galaxy	2 600 000

Figure 1 Vega

The farther a star is from Earth, the longer it takes for the light from that star to reach Earth. For example, the star Vega is approximately 25 ly from Earth (Figure 1). The light from Vega travels at the speed of light, which means that it takes the light from Vega approximately 25 years to reach Earth. When you observe Vega in the night sky, you see it as it was 25 years ago, not as it is today! Some celestial objects are thousands or even millions of light years from us, so when astronomers look at them through telescopes, they are actually looking back in time thousands or millions of years.

While the astronomical unit is useful for measuring distances within the Solar System, the light year is typically used to measure distances to stars and other celestial objects beyond the Solar System. Figure 2 shows which units are used for various distances.

DID YOU KNOW?
Distance to the Sun
The Sun is approximately 0.000 016 ly away from Earth. Light from the Sun takes approximately 0.000 016 years (or 8 minutes) to reach us.

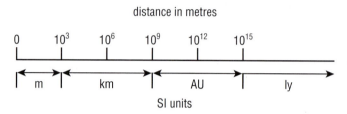

Figure 2 This chart (which is not to scale) shows which units are appropriate for different distances. The top line shows distances, written in scientific notation and measured in metres. The bottom line shows the SI units that are used for different distances.

The light year can be modified to produce values that are easier to work with. For example, remembering that 1 ly is equal to 9.46×10^{12} km and that there are 60 minutes in an hour and 60 seconds in a minute, we can determine the distance between two celestial objects in light minutes (lm) or in light seconds (ls).

DID YOU KNOW?
Light Nanosecond
In one billionth of a second, light travels approximately 30 cm. This is called a light nanosecond. Computer memory speed is measured in nanoseconds, as are many other modern technologies.

Radar uses the light nanosecond to measure the distance to objects in the air. A radar antenna sends out radio waves (which travel at the speed of light) and then counts the number of nanoseconds it takes for the waves to bounce back from the object in the air. The distance to the object can then be determined by dividing the number of nanoseconds by two.

SAMPLE PROBLEM 2 Converting Light Years to Light Seconds

If 9.46×10^{12} km = 1 ly, how many kilometres are there in 1 ls?

Given: 1 ly = 9.46×10^{12} km

Required: the number of kilometres that make up 1 ls

Analysis: 1 ly = $(60 \times 60 \times 24 \times 365)$ ls
1 ly = 3.15×10^7 ls
$$1 \text{ ls} = \frac{9.46 \times 10^{12} \text{ km}}{3.15 \times 10^7 \text{ ls}}$$

Solution: 1 ls = 3.0×10^5 km

Statement: There are 3.0×10^5 km in 1 ls.

Practice
Betelgeuse is 5.7×10^{15} km from Earth. How many light seconds away is Betelgeuse?

TRY THIS: DISTANCES IN THE UNIVERSE

SKILLS: Analyzing, Evaluating

SKILLS HANDBOOK

In this activity, you will determine the distances of some objects in the Solar System using a variety of units based on the speed of light. You will learn more about probes in Chapter 10.

Equipment and Materials: computer; calculator

1. Copy Table 2 into your notebook.

   Table 2 Distances of Robotic Probes

Probe	Distance (km)
New Horizons	
Stardust	
Voyager 2	
Dawn	
Cassini	

2. Using the Internet, research the current distances (in kilometres) to the robotic probes listed in Table 2. Record the data in Table 2.

   GO TO NELSON SCIENCE

A. Calculate the distance, in light years, from Earth to each of the five robotic probes. T/I

B. Convert the distances you calculated in A into light minutes. T/I

C. Which unit would be best for describing the distances to probes travelling in the Solar System? Why? T/I

D. The Sun is 1.0 AU from Earth. Convert this distance into light seconds. T/I

Stellar Distances

Early astronomers discovered that over a period of weeks and months, the planets appeared to move against a background of stars. This led them to believe that stars are much farther away than the planets. However, early astronomers did not have a useful way of measuring the distance between celestial objects or the distance between these objects and Earth.

Scientists have since developed various methods of measuring interstellar distances—the distances between stars in the Universe. One method, parallax, relies on geometry. **Parallax** is the apparent change in position of an object against a fixed background when viewed from two different lines of sight (or locations). You can observe parallax when you hold your thumb upright at arm's length and view it first with one eye and then with the other (Figure 3).

parallax the apparent change in position of an object as viewed from two different locations that are not on a line with the object

Figure 3 (a) A student lines up his thumb with an island in the distance. (b) When the student views the same island through only the right eye, the island appears to move to the left. His thumb is no longer lined up with the island.

9.1 Measuring Distances Beyond the Solar System

READING TIP

Restating an Idea

Paraphrasing involves restating an idea in your own words. Read a text silently. Ask yourself what the main idea and supporting details are. Look away from the text. Say aloud what you think the main idea is. For example, "Astronomers observe parallax and then use triangulation to determine how far away a star is from Earth." Compare your version with what is in the text to make sure you have captured the main idea.

Each eye views your thumb (the object) from a slightly different direction. This results in an apparent change in the position of your thumb when you close one eye. When you have both eyes open, your brain uses parallax to get a sense of the distance between your eyes and your thumb. If you move your thumb halfway to your nose, you can see that parallax increases as distance decreases. Parallax is measured in angles (Figure 4).

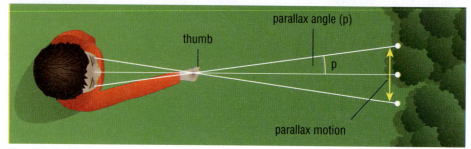

Figure 4 Parallax is the angle measured between the two lines of sight, divided by two.

Parallax can be used to calculate the distance to stars using a method known as triangulation. Triangulation uses the principle that if you know the length of one side of a triangle and the angles of two corners, you can calculate all the dimensions of the triangle.

Figure 5 shows how astronomers use triangulation to calculate the distance from Earth to a star. The diameter of Earth's orbit is the "known" side of the triangle. The star is then observed from two different positions in Earth's orbit around the Sun. To do this, astronomers observe the star and then wait six months to observe it again while Earth is at the opposite end of its orbit. The "triangle" formed by these observations can be used to calculate the distance to the star.

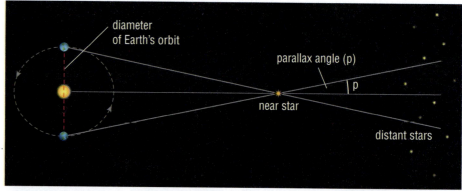

Figure 5 Because stars are distant, astronomers need to measure parallax from positions that are very far apart.

DID YOU KNOW?

HIPPARCOS Satellite

Launched in 1989 by the European Space Agency, the HIgh Precision PARallax COllecting Satellite (HIPPARCOS) was the first satellite dedicated to accurately measuring star positions. After the mission ended in 1993, astronomers published the Hipparcos catalogue, which lists the most detailed position measurements ever made of 100 000 stars. The satellite's name honours a Greek astronomer named Hipparchus who used triangulation in 150 BCE to determine the Moon's distance from Earth.

With the aid of extremely large telescopes, astronomers are able to use parallax to estimate the distances to stars as far as 100 ly away. For distances beyond this, using parallax becomes too imprecise. In the early 1990s, a unique satellite called the HIPPARCOS satellite was sent into space on a star-mapping mission. It was able to accurately measure star positions, parallaxes, and motions of stars up to 1000 ly away. The HIPPARCOS satellite was able to make more accurate measurements than observatories on Earth. In part, this was because it did not have to contend with the blurring effect of Earth's atmosphere.

UNIT TASK Bookmark

You can apply what you have learned about distances in space to the Unit Task described on page 446.

IN SUMMARY

- The distances to stars and galaxies can be measured in light years.
- Astronomers use parallax to determine the distances to nearby stars, based on the apparent change in position of stars as seen from different locations along Earth's orbit.

CHECK YOUR LEARNING

1. Why do astronomers use special units such as light years to measure distances in space? K/U
2. Betelgeuse is 600 ly from Earth. Calculate the distance in kilometres. T/I
3. Polaris is 4.07×10^{15} km from Earth. Calculate the distance in light years. T/I
4. If you hold your finger out at arm's length and view it only with your left eye and then only with your right eye, it appears to jump sideways. Imagine that you have been asked to write a paragraph for a children's science magazine explaining this phenomenon. Include a diagram if you wish. K/U C
5. The HIPPARCOS satellite orbited Earth for four years observing stars. What properties did the satellite measure? K/U
6. The most distant objects in the Universe known to astronomers are 14 billion ly away. Convert this value to metres. T/I
7. The largest known star, VY Canis Majoris, is an incredible 9.18×10^{12} m in diameter. What is VY Canis Majoris's diameter in AU? T/I
8. Why would it not be practical to use light years to describe the distance from your school to your home? T/I
9. Many people think that a light year is a unit of time. Write a short letter to a family member explaining the importance of the light year as a unit of distance for astronomers studying distant objects. Include an example in your explanation. K/U C
10. How far does light travel in one nanosecond? K/U
11. Which unit would you use to measure the distance to the following celestial objects? T/I
    (a) an asteroid found orbiting the Sun 4.7×10^{11} m away
    (b) a meteor burning up in Earth's atmosphere 7.8×10^3 m above us
    (c) a dwarf planet found in the Kuiper Belt, 6.6×10^{12} m from the Sun
    (d) the next nearest star to our Sun, Proxima Centauri, which is 4.0×10^{16} m away
    (e) a spy satellite orbiting Earth at an altitude of 3.5×10^7 m

9.2 The Characteristics of Stars

From Earth, most stars appear as twinkling dots of light. Most stars appear similar, but you may have noticed that some appear to be brighter than others. What causes this difference in brightness? Stars also differ in their colour. Some stars, like our Sun, appear yellowish-white, whereas other stars may appear reddish or bluish. In this section, you will learn about the characteristics of stars, including brightness, colour, temperature, composition, and mass.

Star Brightness

In Chapter 8, you learned that stars are giant balls of hot gases that produce and emit energy—much of it in the form of visible light. The total amount of energy produced by a star each second is referred to as its **luminosity**. The luminosity of a star is measured by comparing it with the luminosity of the Sun, which is assigned a luminosity of 1. Sirius, the brightest star in the night sky, has a luminosity of 22. This means that Sirius gives off 22 times more energy each second than the Sun. So why does the Sun appear so much brighter than Sirius in the night sky?

The Sun appears brighter because it is much closer to Earth than Sirius. Sirius is approximately 9 ly from Earth, whereas the Sun is only 0.000 016 ly away. You may have noticed that luminous objects, such as flashlights and headlights, appear much brighter when they are closer to you than when they are farther away. This is because when the light source is farther away, the light spreads out over a larger area and becomes more diffuse. When the light source is closer, the light is concentrated in a smaller area and appears brighter.

Imagine that you are looking at two light sources—a small flashlight located close to you and a car headlight located one or two kilometres away. The car headlight has a higher luminosity than the flashlight, but both lights appear to have the same brightness. The same is true for stars. To properly compare the brightness of the two lights, you would need to place them at equal distances from the observer (Figure 1).

luminosity the total amount of energy produced by a star per second

DID YOU KNOW?
Sirius
Sirius is the brightest star in the constellation Canis Major and is often referred to as the "Dog Star." Located only 8.6 light years away, it is the brightest star visible in the northern night sky.

Ancient Egyptians worshipped Sirius as the Nile Star. Its appearance in the dawn skies of summer coincided with the annual flooding of the Nile River. Later, the Romans thought that because Sirius was seen as a bright star in summer, its light added to the heat of summer. This led to the saying "dog days of summer."

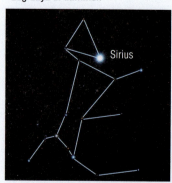

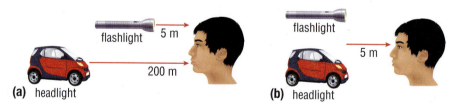

Figure 1 (a) When located at different distances from the observer, two lights with different luminosities can appear to have the same brightness. (b) When placed at equal distances from the observer, the light with a higher luminosity will appear brighter.

As illustrated in Figure 1, a star with a lower luminosity can appear brighter than a star with a higher luminosity if the more luminous star is much farther away. Therefore, the brightness of a star depends on both its luminosity and its distance from the observer.

Apparent Magnitude

The brightness of a star, as seen by an observer on Earth, is called the star's **apparent magnitude**. About 2100 years ago, Greek astronomer Hipparchus created an apparent magnitude scale that could be used to measure the apparent magnitude of celestial objects (Figure 2). Hipparchus classified the brightest stars in the night sky as "magnitude 1" stars, or first magnitude stars, and the faintest stars as "magnitude 6" stars, or sixth magnitude stars.

According to this scale, brighter stars have a lower apparent magnitude value and dimmer stars have a higher value. The apparent magnitude scale is still used today but has been extended to include fainter celestial objects that are visible to us only through telescopes, as well as brighter celestial objects such as the Sun. The Sun has an apparent magnitude of about −27. The unaided eye can see stars as dim as magnitude 6.

Absolute Magnitude

Stars are scattered across the Universe at different distances from Earth. The varying distances make it difficult to visually compare stars to determine which are emitting more light and which are emitting less. Although apparent magnitude values help us classify stars according to their observed brightness, they do not help us distinguish between stars that actually give off more energy and those that give off less energy. To overcome this problem, astronomers have developed an absolute magnitude scale. The **absolute magnitude** of a celestial object equals the apparent magnitude the object would have if it were 33 ly from the observer.

If the Sun were 33 ly from Earth, it would have an apparent magnitude of 4.8. Therefore, the Sun's absolute magnitude is 4.8. Polaris is approximately 430 ly from Earth and has an apparent magnitude of 2. Astronomers have calculated that if Polaris were 33 ly from Earth, then its apparent magnitude would be about −3.63. This means that Polaris's absolute magnitude is −3.63.

If the Sun and Polaris were side by side, Polaris would appear to be much brighter than the Sun, since its absolute magnitude is −3.63 and the Sun's absolute magnitude is +4.8. (Remember: the smaller the magnitude number, the brighter the celestial object appears to the observer.) As you can see, absolute magnitude gives you a much better idea of the actual luminosity of a celestial object than does the apparent magnitude and allows you to compare stars as if they were the same distance away.

Table 1 shows the apparent and absolute magnitudes of several stars. Which star gives off the most light?

Figure 2 Hipparchus (190 BCE–120 BCE)

apparent magnitude the brightness of stars in the night sky as they appear from Earth; for example, Polaris's apparent magnitude is 1.97

absolute magnitude the brightness of stars as if they were all located 33 ly from Earth; for example, Polaris's absolute magnitude is −3.63

DID YOU KNOW?

Apparent Magnitudes
Looking through a small backyard telescope, we can see stars as faint as magnitude 12. Meanwhile, the largest observatory telescopes can record objects as faint as magnitude 30. This is about 2 billion times dimmer than the faintest object visible to the unaided eye.

Table 1 Apparent and Absolute Magnitudes of Some Stars

Star	Apparent magnitude	Absolute magnitude
Sun	−26.8	4.83
Sirius	−1.45	1.5
Vega	0.04	0.5
Betelgeuse	0.41	−5.6
Deneb	1.99	−7.5

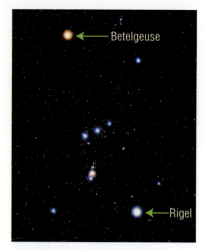

Figure 3 Betelgeuse, with a surface temperature of 3000 °C, appears reddish-white (upper left). Rigel, with a much higher surface temperature of almost 11 000 °C, appears bluish-white (bottom right).

Star Colour and Temperature

The colour of a star gives scientists an indication of the star's surface temperature. A relatively hot star appears bluish, whereas a relatively cool star appears reddish. Betelgeuse and Rigel, for example, are two stars in the Orion constellation that differ in their apparent colour. The star with the higher surface temperature appears bluish-white, whereas the star with the lower surface temperature appears reddish-white (Figure 3). Our Sun, with a surface temperature of 6000 °C, appears yellowish-white and falls midway between bluish stars and reddish stars.

Table 2 lists the approximate temperature ranges of different colours of stars and gives examples of stars we can see in the night sky.

Table 2 Colour and Temperature Ranges of Some Stars

Colour	Temperature range (°C)	Example(s)
bluish	25 000–50 000	Zeta Orionis
bluish-white	11 000–25 000	Rigel, Spica
whitish	7500–11 000	Vega, Sirius
yellowish-white	6000–7500	Polaris, Procyon
yellowish	5000–6000	Sun, Alpha Centauri
orangish	3500–5000	Arcturus, Aldebaran
reddish	2000–3500	Betelgeuse, Antares

The Composition of Stars

Scientists use special instruments to analyze the light emitted by the Sun and other stars. One such device, called a spectrograph, splits light energy into patterns of colours for observation. Recall from Chapter 6 that when white light is shone through a prism, a full rainbow of colours is seen, whereas the light emitted by hydrogen produces only a few lines of colour (Figure 4). Most colours are missing—this is because each element emits light energy only at certain characteristic frequencies. These frequencies are influenced by the unique electron energy levels within each atom. Therefore, each element has its own unique spectrum.

Figure 4 (a) The spectrum of light from the Sun (b) the spectrum of hydrogen light

To determine the composition of a star, scientists use a spectrograph to analyze its spectrum. They can then compare the star's spectrum with the known spectra of the elements. Much of what we know about stars today has resulted from using the spectrograph. The spectrum of a star can tell us which elements make up the star and can also be an indicator of the star's temperature.

The Mass of a Star

The mass of our Sun is 2×10^{30} kg, which is referred to as one **solar mass**. Scientists use solar mass as the standard to which all other stars are compared. Star masses vary and can range between 0.1 to 120 solar masses. One of the most massive stars, known simply as A1, has a solar mass of almost 118! The mass of a star does not always reflect its size. You will learn more about this in Section 9.4.

solar mass a value used to describe the masses of galaxies and stars other than our Sun; equal to the mass of the Sun (2×10^{30} kg)

UNIT TASK Bookmark

How can you apply what you learned about the properties of stars to the Unit Task on page 446?

IN SUMMARY

- The brightness of stars can be measured in luminosity, absolute magnitude, and apparent magnitude.
- There is a relationship between the colour of a star and its temperature.
- Scientists can use a star's spectrum to infer its composition.
- Solar mass is a value used to compare the masses of stars and galaxies.

CHECK YOUR LEARNING

1. Not all stars in the Universe shine with the same amount of energy and have the same colour. Why do they differ? K/U
2. What do astronomers mean when they refer to a star's luminosity? K/U
3. Why was the star Sirius significant to some ancient cultures? K/U
4. What was the name of the astronomer who first devised the system of magnitudes? How long ago did he live? K/U
5. How can two stars have the same apparent magnitude yet emit different amounts of light? K/U
6. Astronomers use a standard distance to determine a star's absolute magnitude. What is this standard distance? Is it farther than the star Sirius? K/U
7. Figure 5 shows four coloured spheres, representing the colours of four stars. In your notebook, list which colour you think corresponds with the following temperatures: 3100 °C, 4800 °C, 8000 °C, 10 200 °C. T/I

Figure 5

8. Compare the stars Betelgeuse and Rigel in terms of their colour and temperature. K/U
9. Describe our Sun in terms of its colour, temperature, apparent magnitude, and absolute magnitude. K/U
10. Compare the terms "apparent magnitude," "absolute magnitude," "luminosity," and "brightness." K/U
11. Use the terms "absolute magnitude" and "luminosity" in the same meaningful sentence. K/U C

9.3 CONDUCT AN INVESTIGATION

Factors Affecting the Brightness of Stars

You have learned that stars vary in brightness. These differences can be tricky to spot without specialized instruments. However, even with the unaided eye, we notice that stars appear to vary in brightness. Astronomers can tell a lot about a star based on how bright it appears. In this investigation, you will look at the factors that affect the apparent brightness of a light source.

SKILLS MENU
- Questioning
- Hypothesizing
- Predicting
- Planning
- Controlling Variables
- Performing
- Observing
- Analyzing
- Evaluating
- Communicating

Testable Question
How does the distance of a star affect its brightness as seen from Earth?

Hypothesis
Make a hypothesis based on the Testable Question. Decide which factors might affect how bright a light source appears. Your hypothesis should include a prediction and reasons for your prediction.

Equipment and Materials
- metre stick
- two light sources of differing brightness (for example, two flashlights, two LEDs, etc.)
- pencil
- masking tape

Experimental Design
You will be given two light sources that give off different amounts of light. In small groups, you will design a procedure to determine how distance affects the brightness of a light source. Consider how you might arrange the light sources so that they appear to have the same apparent brightness.

Procedure
SKILLS HANDBOOK 3.B.3., 3.B.4.

Design a procedure that allows you to test the light sources at different distances in order to test the factors that affect how bright the light sources appear. Be as detailed as possible in your procedure.

⚠ Light bulbs can become hot very quickly. If you are using light bulbs, allow them to cool before touching them.

Analyze and Evaluate
SKILLS HANDBOOK 3.B.7., 3.B.9.

(a) What were the independent variables in this investigation? What was the dependent variable? T/I

(b) Under what conditions did a light source appear at its brightest? T/I

(c) Under what conditions did a light source appear at its dimmest? T/I

(d) Answer the Testable Question: What are the factors that affect how bright a light source appears? T/I

(e) How do your observations compare to your hypothesis? Do your observations support your original prediction? T/I

(f) Were there any sources of error that might have affected the outcome of your investigation? Describe some ways you could improve your procedure if you were to perform this investigation again. T/I

Apply and Extend

(g) How could you use the information you learned in this investigation to estimate the distance to a car at night based on observations of its headlights? A

(h) Describe how the factors you isolated in the last section affect how astronomers see stars in the night sky. C

(i) How do astronomers rank stars based on how bright they appear? K/U

(j) How do astronomers rank stars based on how bright they would be from the same distance? K/U

9.4 The Life Cycle of Stars

Every star has a life cycle: a beginning, a middle, and an end. The life of a star may last billions of years. Our Sun, for example, has been around for almost five billion years and is not yet near the end of its life cycle. Using modern instruments, scientists have been able to study stars at different stages of their life cycle. Our knowledge of gravitational forces has also greatly contributed to understanding the life cycle of stars. (Recall from Chapter 8 that a star is a massive collection of gases held together by its own gravity.)

Star Beginnings

A star has its beginning deep inside a massive cloud of interstellar gases and dust called a **nebula** (Figure 1). A nebula consists primarily of hydrogen and helium.

nebula a massive cloud of interstellar gas and dust; the beginning of a star

 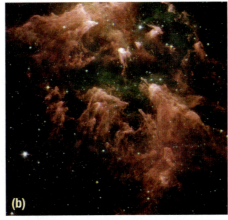

Figure 1 (a) As illustrated by this false-colour Hubble telescope image, the Orion Nebula offers one of the best opportunities to study how stars begin, partly because it is the nearest large star-forming region. (b) The Carina Nebula is another star-forming region. This false-colour telescope image shows young stars (yellow or white) among clouds of thick dust (pink).

LEARNING TIP

Word Meaning
"Proto-" means "first," "original," or "on the way to becoming something."

Stars are formed when parts of nebulas collapse in on themselves. Nebulas extend over vast distances—thousands of light years in space—and the gases within them are unevenly distributed. When a nebula reaches a certain density, gravitational forces begin to pull the gas and dust particles close together, causing clumps to form inside the main cloud of the nebula. As the clumps draw in gas and dust from the cloud, they become more massive and have increasingly stronger gravity. Over time, this gravity causes regions of greater density to form within the nebula. For about a million years, these dense regions continue to pull in gas and dust from the less dense regions of the nebula, forming a **protostar**.

As the mass and gravity of a protostar increase, it becomes a tightly packed sphere of matter, drawing more and more matter into its core. The force of gravity eventually causes the atoms in the core of the protostar to become so tightly packed that the pressure in the core rises and nuclear fusion begins (Figure 2).

protostar a massive concentration of gas and dust thought to eventually develop into a star after the nebula collapses

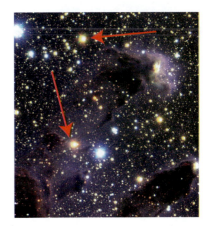

Figure 2 Arrows point to egg-shaped clouds with protostars embedded inside as they bud off from the Eagle Nebula.

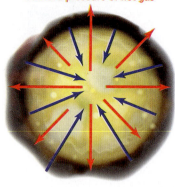

Figure 3 The red arrows represent the outward pressure of hot gas; the blue arrows represent gravity pulling toward the centre of the star.

Nuclear Fusion

For millions of years, the core of a protostar continues to contract due to the pull of gravity. The core temperature rises until it meets a critical temperature of 15 million °C (1.5×10^7 °C). At this temperature, nuclear fusion begins. Hydrogen atoms in the core fuse to form helium atoms, producing an enormous amount of energy. This energy rushes outward from the core of the star, counteracting the gravitational forces that caused the protostar to form (Figure 3).

The new star, buried inside the nebula, emits radiation in the form of heat, light, X-rays, gamma rays, and other energetic particles. Energy generated at the core makes its way to the surface and is radiated away at the photosphere. This radiation causes gases surrounding the star's core to glow, or shine. The star eventually stabilizes at a particular size. Our Sun went through this process, most likely taking up to 30 million years to condense and begin "glowing."

All stars begin in the same way. However, the life of a star is determined by its mass—the more massive the star, the faster its rate of fusion, which results in a shorter life cycle.

The Hertzsprung–Russell Diagram

Uncovering the mystery behind the life cycle of stars became much easier when scientists realized that a star's mass determines its brightness, colour, size, and how long it will "live."

Early in the twentieth century, Danish astronomer Ejnar Hertzsprung and American astronomer Henry Norris Russell organized this information into a diagram called the Hertzpsrung–Russell (H–R) diagram. The H–R diagram plots absolute magnitude against star surface temperature (Figure 4).

> **DID YOU KNOW?**
> **Star Colour**
> Although the H–R diagram plots the temperature of stars from blue to red, our eyes perceive the colour of stars slightly differently. Blue stars appear bluish-white to the eye, and red stars appear pale pink or orange.

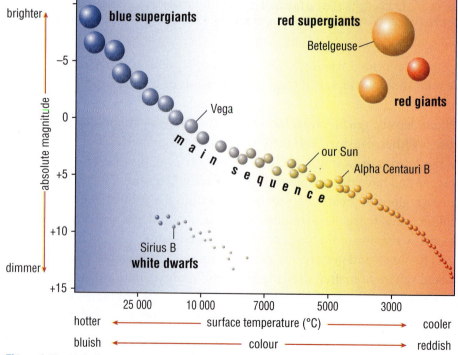

Figure 4 The H–R diagram is a visual representation of stellar evolution.

When astronomers studied the distribution of stars on the H–R diagram, they realized that the diagram offered clues to the evolution of stars. For example, they noticed that as the physical characteristics of a star change over time, the position of the star on the H–R diagram also changes. Hence, the H–R diagram can be used to describe the evolution of stars. In the last hundred years, thousands of stars have been plotted on the H–R diagram.

To explore a career as an astrophysicist,

Position on the H–R Diagram

Astronomers noticed that 90 percent of stars plotted on the H–R diagram fit into a diagonal band which they called the **main sequence**. In the lower right are the cooler, reddish stars that are dim. Moving along the main sequence to the upper left are the very luminous, hot, bluish stars. They also noticed that while some stars are dim and hot, others are luminous and cool. These stars are located on the H–R diagram off the main sequence, in the upper right corner and lower left corner. How could cooler stars, which likely produce less energy per unit area, be more luminous than hotter stars? It was reasoned that these cooler stars have a greater surface area than the hotter, dim stars, resulting in more light being produced. The large, bright, cool stars are called red giants. The small, dim, hot stars are called white dwarfs. The hottest, most luminous stars are very large stars called blue supergiants. You will learn more about these types of stars later in the section. As you can see from the H–R diagram, our Sun is an average star.

main sequence the stars (including the Sun) that form a narrow band across the H–R diagram from the upper left to the lower right

MAIN SEQUENCE STARS

Most stars, including our Sun, can be found along the main sequence band. These main sequence stars vary in surface temperature, as you can see from the shape of the band in the H–R diagram in Figure 4. The hotter these stars are, the more luminous they are. Astronomers have determined that hotter, more luminous main sequence stars are more massive, while cooler, less luminous stars are less massive.

Main sequence stars fuse hydrogen to produce helium in their cores. Stars do not move through the main sequence as they age. Once a star is formed with a particular mass, it stays in one position on the main sequence until all its hydrogen is used up. Most stars spend the bulk of their existence as main sequence stars.

DID YOU KNOW?

Red Supergiant
The largest known star is VY Canis Majoris, a red supergiant. Recent calculations suggest that it is more than two thousand times bigger than the Sun, which makes it about one quarter the size of the Solar System.

The Death of a Star

Billions of years after forming, a star begins to burn out as it nears the end of its life cycle. Depending on what type of star it is, this can happen in two different ways.

Stars Like the Sun: Red Giant to White Dwarf

After spending approximately 10 billion years as a main sequence star, a star's available hydrogen will have been converted to helium by nuclear fusion. This results in the formation of a helium-rich core, surrounded by an outer layer of hydrogen. With less hydrogen to burn, the outward flow of energy slows and the core begins to contract. This contraction heats the core, which heats up the remaining hydrogen enough for fusion to restart in the outer layer. While the core contracts and gets hotter, the outer layers of the star expand and then cool, becoming a **red giant** (Figure 5).

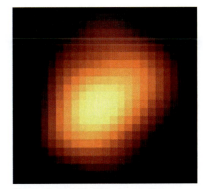

Figure 5 This Hubble Space Telescope image shows a red giant star, Mira (Omicron Ceti). It is approximately 700 times the size of the Sun!

red giant a star near the end of its life cycle with a mass that is equal to or smaller than that of the Sun; becomes larger and redder as it runs out of hydrogen fuel

red supergiant a star near the end of its life cycle with a mass that is 10 times (or more) larger than that of the Sun; becomes larger and redder as it runs out of hydrogen fuel

In about 5 billion years, our Sun will become a red giant. A star with a mass that is equal to or smaller than that of the Sun becomes a red giant, whereas a star with a mass that is 10 times (or more) larger than that of the Sun becomes a **red supergiant**.

As a red giant ages it consumes the remaining supply of hydrogen and the core contracts further. This causes the temperature and pressure in the core to once again rise, and the helium-rich core begins to undergo fusion. The fusion of helium continues the expansion of the red giant and produces heavier elements, such as carbon (Figure 6). The red giant is now fully formed, and its luminosity has increased by several thousand times. Examples of red giants we can see in our night sky include Aldebaran and Betelgeuse.

As a red giant expands, it sends gas and dust into space and begins to lose mass. The increased surface area of the star has caused it to move off the main sequence band in the H–R diagram.

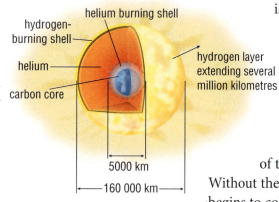

Figure 6 In main sequence stars, the helium burns in a shell around a carbon core, expanding and cooling the star even more as it ages.

white dwarf a small, hot, dim star created by the remaining material that is left when a red giant dies

A star with a mass that is equal to or smaller than that of the Sun is said to "die" when nuclear fusion stops occurring. Without the outward pressure generated by nuclear fusion, the star's core begins to collapse due to its own gravity. The outer layers of the star drift away and, eventually, the hot core is all that remains of the star. The matter that remains is known as a **white dwarf**—a small, dim, hot star. After developing into a red giant, our Sun will eventually become a white dwarf. Astronomers believe that once a white dwarf cools down, all that remains is dark, cold matter, which they refer to as a black dwarf.

The hot white dwarf emits ultraviolet light that collides with the gas and dust shed in the last stages of its life as a red giant. The energy illuminates the clouds of dust and gas and creates a beautiful planetary nebula, which is a nebula that results from the death of certain stars (Figure 7). A white dwarf will continue to radiate its energy into space, becoming cooler and dimmer, until its light goes out.

Figure 7 This Hubble Space Telescope image shows the Cat's Eye nebula, 3000 ly from Earth. The planetary nebula is illuminated by the white dwarf at its centre.

Stars More Massive than the Sun

Not all stars end up as white dwarfs. Some stars can be tens of times more massive than the Sun (but these high-mass stars are rare). Recall that the mass of a star determines how long a star will "live." A star with a high mass consumes hydrogen much faster than a star with low mass, resulting in a shorter life cycle.

When a massive star runs out of hydrogen for fusion, it begins to fuse helium into carbon (like our Sun). The core of a massive star becomes so hot that when helium is no longer available for fusion, carbon undergoes fusion. This produces heavier elements, beginning with oxygen and up to iron. Once iron is produced in the core, fusion can no longer occur. (This is because fusing iron requires more energy than it releases.)

Once fusion stops, the star collapses under its own gravity and the iron core increases in temperature. The inward rush of gas is suddenly halted by the core, and, like a rubber ball bouncing off a brick wall, the gases bounce back outward with great force. The outer layers of the star explode outwards in what is called a **supernova**, sending out a series of shock waves. This creates a rapidly expanding nebula of gas and dust.

The energy released by such massive supernova explosions is capable of causing many fusion reactions. These fusion reactions are responsible for the formation of all the additional elements in the periodic table.

Supernovas are rare astronomical events. One of the few supernovas that have been observed is the supernova that created the Crab Nebula in the year 1054 (Figure 8). Since the invention of the telescope 400 years ago, only one supernova has ever been seen with the unaided eye. This was in 1987, when Canadian astronomer Ian Shelton discovered a supernova now known as Supernova 1987A.

> **DID YOU KNOW?**
> **We Are Made of Star Dust**
> The nickel we use for making coins and the silver and gold we wear as jewellery were all created by fusion reactions inside a supernova explosion long before the formation of our solar system. In fact, with the exception of hydrogen, every atom in your body was originally formed inside a star or an exploding star!

supernova a stellar explosion that occurs at the end of a massive star's life

Figure 8 The supernova that formed the famous Crab Nebula was so bright that it was visible even during daylight hours.

TRY THIS: MODELLING A SUPERNOVA EXPLOSION

SKILLS: Predicting, Observing, Analyzing

When the core of a star collapses, it does so with such enormous force that it rebounds. As the core collapses, all the outer atmospheric layers are also collapsing. The less dense outer layers are still falling in toward the core when the core rebounds. The rebounding core collides with the outer layers with enough energy to blow the atmospheric layers away from the star. This is the supernova explosion. In this activity, the basketball models the core, and the tennis ball models the star's outer atmospheric layers.

Equipment and Materials: basketball; tennis ball

1. Drop the basketball and then the tennis ball. Record your observations.
2. Place the tennis ball on top of the basketball, and hold them out in front of you. Predict how each ball will bounce.
3. Let go of both balls at the same time. Record your observations.

A. How far above the floor did each ball rebound in step 1?
B. What did you observe when both balls hit the floor?
C. What is the source of the extra energy that caused the result you observed?
D. How is this model like a supernova explosion?
E. What parts of this model are not like a supernova explosion?

NEUTRON STARS

When a star with an initial mass between 10 and 30 solar masses explodes into a supernova, the core left behind becomes a **neutron star**—an extremely dense star composed of tightly packed neutrons. Although these stars are tiny (only about 10 km across—the size of a large city on Earth), the neutrons within them are so tightly packed that the mass of a neutron star can be more than 10^{30} kg. A spoonful of a neutron star would weigh as much as Mount Everest! The gravity of a neutron star is 300 000 times that of Earth.

Some neutron stars spin very quickly—hundreds of times per second at first. As they spin, they emit high-frequency radio waves, which we detect intermittently as pulses—like the beam from the rotating light of a lighthouse. For this reason, these neutron stars are also called pulsars, and they can be seen from thousands of light years away in spite of their small size (Figure 9).

BLACK HOLES

When a star with an initial mass larger than 30 solar masses dies, it leaves behind a core so massive that it collapses under its own gravity into a **black hole**—a quantity of matter so dense and with gravity so strong that not even light can escape. The word "hole" is misleading because it sounds like there is nothing there. In reality, there is a huge amount of matter packed into a dense core.

Astronomers cannot see black holes because they do not allow any light to escape. How do they know that black holes exist? They see the gravitational effect that a quantity of matter believed to be a black hole has on the surrounding area. The gravity of a black hole is so strong that it pulls in any nearby matter. The matter forms a disc of gas and dust around the black hole, much like water swirling down a drain (Figure 10). Just before the matter spirals into the black hole and disappears, it heats up and emits powerful X-ray radiation. This radiation can be detected by instruments on a satellite.

Figure 11 shows the different paths that a star's life can take, depending on its mass.

neutron star an extremely dense star made up of tightly packed neutrons; results when a star over 10 solar masses collapses

Figure 9 NASA's Chandra X-ray Observatory and the Hubble Space Telescope captured this image of a pulsar at the centre of the Crab nebula. The bright rings are made up of high-energy particles that are propelled outward by the Crab pulsar.

black hole an extremely dense quantity of matter in space from which no light or matter can escape

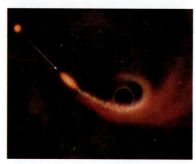

Figure 10 This illustration depicts a star (orange circle) close to a massive black hole. The enormous gravity of the black hole stretched the star until it was torn apart, pulling in some of the star's mass while the rest is flung out into space.

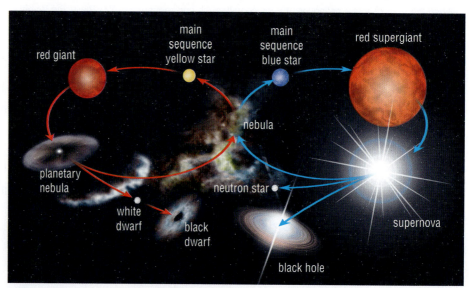

Figure 11 The red arrows show the life cycle of low-mass stars like the Sun. The blue arrows show the life cycle of large-mass stars that are more massive than the Sun.

380 Chapter 9 • Beyond the Solar System

Every star has its own unique life cycle. Although scientists have learned much about the characteristics of stars, they have only recently uncovered the mystery of star evolution. Scientific theories are always changing based on new observations. Table 1 summarizes what scientists have learned about the life cycle of stars, based on mass.

Table 1 The Evolution of Different Types of Stars Based on Mass

	Small to medium star (<5 solar masses)	Large star (10–30 solar masses)	Extremely large star (>30 solar masses)
Birth and early life	Forms from a small- or medium-sized portion of a nebula; gradually turns into a hot, dense clump that begins producing energy.	Forms from a large portion of a nebula, and in a fairly short time turns into a hot, dense clump that produces large amounts of energy.	Forms from an extremely large portion of a nebula, very quickly turning into a hot, dense clump that produces very large amounts of energy.
Main sequence phase	Uses nuclear fusion to produce energy for about 10 billion years if the mass is the same as the Sun's, or 100 billion years or more if the mass is less than the Sun's.	Uses nuclear fusion to produce energy for only a few million years. It is thousands of times brighter than the Sun.	Uses nuclear fusion to produce energy for only a few million years. It is extremely bright.
Old age	Uses up hydrogen and other fuels and swells to become a large, cool red giant.	Uses up hydrogen and other fuels and swells to become a red supergiant.	Uses up hydrogen and other fuels and swells to become a red supergiant.
Death	Outer layers of gas drift away, and the core shrinks to become a small, hot, dense white dwarf star.	Core collapses inward, sending the outer layers exploding as a supernova.	Core collapses, sending the outer layers exploding as a very large supernova.
Remains	White dwarf star eventually cools and fades.	Core material packs together as a neutron star. Gases drift off as a nebula to be recycled.	Core material packs together as a black hole. Gases drift off as a nebula to be recycled.

Note: these drawings are not to scale.

UNIT TASK Bookmark

How could you apply the information you learned about the life cycle of stars as you work on the Unit Task described on page 446?

IN SUMMARY

- Stars are formed inside giant clouds of gas and dust called nebulas.
- Stars shine through a process known as nuclear fusion.
- When stars run out of fuel, they turn into white dwarfs, black holes, or neutron stars.
- Astronomers use the Hertzsprung–Russell diagram to identify where stars are in their life cycle based on their luminosity and temperature.
- The mass of a star determines its life cycle or evolution.

CHECK YOUR LEARNING

1. At what temperature do stars begin the process of nuclear fusion?
2. In what ways do stars change their physical properties over time?
3. Using the H–R diagram as a guide, list three physical properties of the Sun.
4. (a) What is the factor that determines whether a black hole will form at the end of a star's life?
   (b) Will the Sun form a black hole at the end of its life? Explain your reasoning.
5. In a paragraph, describe the importance of nuclear fusion. Use the title "Why Do Stars Shine?"
6. Compare a red giant with a white dwarf by referring to some of the physical properties of the stars covered in this section.
7. (a) What unit do astronomers use to measure the masses of stars and star-like objects in the Universe?
   (b) Why do they use this unit instead of something more traditional, such as kilograms?
8. How can the evolution of a star over its lifetime be tracked using the H–R diagram? Give a specific example of a category of star on the H–R diagram.
9. How can astronomers detect black holes if they do not give off visible light?
10. Re-examine Table 1 and identify three ways in which a high mass star evolves differently than a low mass star.

9.5 The Formation of the Solar System

There have been many attempts over the years to explain how the Solar System formed. One early scientific theory involved collisions between celestial objects that left behind debris that eventually came together to form planets. Another theory suggested that the gravitational force of stars close to our Sun pulled matter away from the Sun, which then formed the planets and moons. Today, the theory that astronomers accept is called the Solar Nebula theory, because there is much observational and theoretical evidence to support it.

The Solar Nebula Theory

According to the Solar Nebula theory, the Solar System was formed approximately 5 billion years ago from a massive cloud of gas and dust (the solar nebula) that began to contract. Some astronomers suggest that a shockwave from a nearby supernova may have triggered the collapse of the nebula. Inside this giant nebula, clumps of gas and dust began to form. One of these dense, massive clumps became a protostar, and much of the remaining gas and dust was pulled in by gravity. As the protostar contracted, the temperature and pressure in its core increased. More particles were attracted to the protostar and it continued to contract under its own gravity. The counterclockwise spin of the protostar became amplified, like a figure skater spinning faster as she makes her body more compact (Figure 1). As the cloud collapsed, it spun faster, further increasing the temperature and pressure at its core. Over time, the cloud became hot enough for nuclear fusion to begin. This caused the contracting material to form a disc of gas and dust around the core of the protostar.

Using large telescopes, astronomers have discovered that some nearby young stars have discs that stretch for billions of kilometres into space (Figure 2). This swirling material may be the birth of new planetary systems. These observations that similar processes are occurring elsewhere provides support for the Solar Nebula theory.

READING TIP

Analogies
Analogies are used to help the reader understand an abstract idea by describing something concrete that the reader may already know and be able to visualize. An example of an analogy is "like a figure skater spinning faster." When summarizing a text, leave out supporting details such as analogies.

(a)

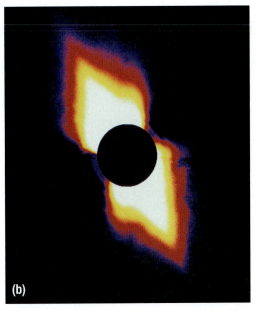
(b)

Figure 1 (a) When an ice skater draws in her arms while spinning, her speed increases, similar to what happens in a spinning gas cloud that contracts. (b) This infrared light image of the newly formed star Beta Pictoris reveals a disc of dust around it.

Figure 2 The planets in the Solar System began to form out of the same disc of gases and dust as the Sun. Early in the evolution of our solar system, there may have been millions of small clumps of material, which slowly collided to form the eight planets we see today.

To see more images and interactive content about the formation of the Solar System,

GO TO NELSON SCIENCE

The Sun formed from the disc in the centre of the whirling mass of material. While the Sun was forming, lighter gases (hydrogen and helium) were pushed away from it by intense solar winds. The lighter gases accumulated in the outer solar system, forming the gas giant planets (Figure 2). Denser clumps of solid matter remained behind in the inner solar system. It is from these clumps of solid matter that the terrestrial inner planets—Mercury, Venus, Earth, and Mars—are thought to have formed. Over millions of years, the clumps of matter collided with each other, accumulating more matter and eventually forming planets.

The time it takes the protostar to collapse from a cool, dark gas cloud to a hot, bright star depends on its mass. As you learned in Section 9.4, the more mass a star has, the greater the gravitational force and the faster it contracts. Astronomers estimate that our Sun took about 30 million years to become a newborn star. In comparison, a 15 solar mass star may take less than 200 000 years to contract, while a 0.2 solar mass star can take up to a billion years.

Evidence for the Solar Nebula Theory

The following observational evidence supports the Solar Nebula theory for the formation of planetary systems:

- The theory explains key aspects of the Solar System, including the observation that the planets orbit the Sun in the same direction; the planets are all in the same plane, with the Sun at the centre; most planets rotate in the same direction on their axes; and the terrestrial planets are closer to the Sun than the gas giants are.
- We observe other planetary systems in various stages of this formation process.
- Computer simulations result in planetary systems that look like the systems we see today.

UNIT TASK Bookmark

You can apply what you have learned about the Solar Nebula theory to the Unit Task described on page 446.

IN SUMMARY

- According to the Solar Nebula theory, the Solar System formed in a giant disc of spinning gas and dust called the solar nebula.
- Evidence from a variety of sources supports the Solar Nebula theory.

CHECK YOUR LEARNING

1. Create a series of drawings that show how the Solar System formed. Include a written explanation for each drawing. K/U C
2. How does observing other planetary systems provide evidence for the Solar Nebula theory? K/U
3. What characteristics of our solar system are explained by the Solar Nebula theory? K/U
4. According to the Solar Nebula theory, how did the terrestrial planets form? K/U
5. What were the main elements that made up the gas giants? K/U
6. What caused the nebula that formed our solar system to rotate? What caused the nebula to spin even faster? K/U

9.6 Other Components of the Universe

The most common type of celestial object astronomers see in space is a star. Most stars appear to be gravitationally bound together into groups, and some groups are more numerous than others. The following categories describe various types of star groupings.

Star Clusters

All galaxies contain **star clusters**. Star clusters are groups of stars that develop together from the same nebula, are gravitationally bound, and travel together. There are two types of star clusters:

- Open star clusters are collections of six to thousands of usually young stars.
- Globular clusters are ball-shaped collections of thousands to millions of very old stars (Figure 1).

star cluster a group of stars held together by gravity

Figure 1 (a) The stars in the open cluster Pleiades can be seen with the unaided eye. (b) Omega Centauri is an example of a globular cluster with a tightly packed group of stars.

Approximately 20 000 open star clusters are found within the main disc of the Milky Way. Globular clusters are scattered like a halo above and below the disc of the Milky Way. Astronomers believe globular clusters date back to the formation of the Milky Way galaxy. The largest globular cluster in the Milky Way is Omega Centauri (Figure 1(b)), which was discovered by Edmund Halley in 1677. This 12 billion-year-old globular cluster orbits the Milky Way at a distance of 18 300 ly from Earth. You will learn more about the Milky Way later in this section.

Types of Galaxies

Galaxies are collections of millions to hundreds of billions of stars, planets, gas, and dust, measuring up to 100 000 ly across. They come in different shapes and sizes and are spread across the Universe. The combined light emitted by the stars in a galaxy defines its size and shape, as observed from Earth.

Figure 2 Edwin Hubble made the key discoveries that the Universe is filled with galaxies and is expanding.

elliptical galaxy a large group of stars that together make an elliptical or oval shape

spiral galaxy a large group of stars that together make a spiral shape, such as the Milky Way

lenticular galaxy a large group of stars that together make a shape that has a central bulge but no spiral arms

irregular galaxy a large group of stars that together make an irregular shape

In the 1920s, astronomer Edwin Hubble (Figure 2) changed the way in which scientists view the Universe. Through the use of the most powerful telescope at the time—the Hooker telescope located at the Mount Wilson Observatory in California—he discovered that galaxies besides the Milky Way existed. By 1936, Hubble had created a classification for these new galaxies based on the photographs of the images he saw. He grouped them by shape into four categories: elliptical, spiral, lenticular, and irregular.

- **Elliptical galaxies** vary in shape from spherical to a flattened oval (Figure 3). We know that they are older galaxies with very little gas, dust, or young stars. Ellipticals account for more than half of all the galaxies we can see.

- **Spiral galaxies** look like spinning pinwheels—flattened discs with a central bulge, and two to four spiral arms (Figure 4). The central core is made of old, red stars. The spiral arms contain clouds of gas and dust along with new and young stars. A subclass of spirals are barred spiral galaxies (Figure 5), which are similar to spiral galaxies but have a central bar pattern running down the middle. Spiral arms trail from the ends of these bars. Our galaxy, the Milky Way, is a barred spiral galaxy.

- **Lenticular galaxies** have a central bulge surrounded by a flattened disc of gas and dust but have no spiral arms (Figure 6). These galaxies are thought to be spiral galaxies that have lost their gas and dust. Most lenticular galaxies are composed of older, red stars.

- **Irregular galaxies** have no definite shape (Figure 7). They contain even more gas and dust than their spiral galaxy cousins. They have no spiral arms or central nucleus and make up at least 10 percent of all galaxies. Irregular galaxies usually contain only 100 million to 10 billion stars.

Figure 3 Elliptical galaxy

Figure 4 Spiral galaxy

Figure 5 Barred spiral galaxy

Figure 6 Lenticular galaxy

Figure 7 Irregular galaxy

To learn more about classifying galaxies,
GO TO NELSON SCIENCE

TRY THIS: VIEWING GALAXIES

SKILLS: Performing, Observing, Analyzing

SKILLS HANDBOOK
3.B.

Although astronomers can classify galaxies into different types based on their shape, it can be difficult to get clear images of their shape. Galaxies can be blocked by gas and dust or other celestial objects. Even when we get a clear view, the galaxy can be difficult to classify because of the angle we are observing it from.

1. Your teacher will give you four blank overhead projector sheets. Draw one type of galaxy in as much detail as you can on each of the sheets.
2. Have a classmate hold the sheets up one at a time on the other side of the room so they are facing you. Identify each type.
3. Have your classmate tilt each of the sheets at various angles while you try to identify the galaxy types.
4. Finally, have your classmate hold each sheet parallel to the floor while you try to identify the galaxy types.

A. Were the galaxies easy to identify on the other side of the room when they were facing you? T/I
B. As your classmate tilted the sheets, did it become easier or harder to identify the types of galaxies? T/I
C. When the sheets were parallel with the floor, could you see any detail that might help you decide what kinds of galaxies they were? T/I
D. Explain how the angle of observation might make it difficult for astronomers to distinguish spiral galaxies from barred spiral galaxies or lenticular galaxies. T/I

TRY THIS: CLASSIFYING GALAXIES

SKILLS: Observing, Evaluating, Communicating

SKILLS HANDBOOK
3.B.3., 3.B.5.

For years, powerful telescopes sitting on mountain tops and spacecraft orbiting Earth have been taking pictures of galaxies found in deep space. To fully understand the structure of these distant galaxies, they first need to be identified and classified based on their appearance.

1. Copy Table 1 into your notebook.

Table 1 Shapes of Galaxies

Galaxy type	Numbers
spiral	
elliptical	
irregular	
lenticular	

A. Look at the shapes of the various galaxies in Figure 8. The illustration is representative of a random sample of galaxies from the Universe. Decide whether each one is a spiral, elliptical, irregular, or lenticular galaxy. Then write its number in the appropriate row. T/I
B. Write one line of description for each type of galaxy, explaining your choice and the features you identified in the picture. T/I
C. How many of each type of galaxy were there? T/I
D. Based on this data, make a prediction about the most common type of galaxy in the Universe. T/I
E. Conduct research to find out if your prediction was correct. T/I

GO TO NELSON SCIENCE

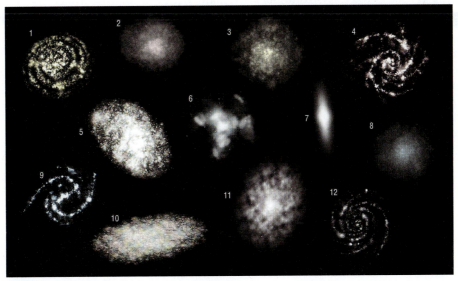

Figure 8

DID YOU KNOW?

Orbiting the Milky Way
Our Sun takes about 200 million years to complete one orbit around the centre of the Milky Way galaxy. Dinosaurs roamed Earth the last time we were in the same position as we are today.

The Milky Way Galaxy

Our understanding of the Milky Way comes from many years of research by astronomers. We now know that the Milky Way consists of more than 200 billion stars and is more than 100 000 ly across. In 1935, Canadian astronomer John S. Plaskett conducted the first detailed study of the structure of the Milky Way using the Dominion Astrophysical Observatory in British Columbia. He discovered that the Sun lies about 30 000 ly from the centre of the galaxy.

You can see parts of our galaxy if you look in the moonless night sky away from the city lights during the winter or summer. It looks like a hazy band of light that arches in the dark sky (Figure 9). Legends and stories of various cultures describe the Milky Way as the road to a heavenly palace, a pool of cow's milk, and a giant spine that keeps the stars in place in the sky above.

Figure 9 Seen from Earth, the Milky Way Galaxy appears as an arch of light in the night sky.

DID YOU KNOW?

The Spitzer Telescope
Launched in 2003 by the United States, this 900 kg satellite is the largest infrared observatory ever put into space. The Spitzer telescope can detect young stars not visible to optical telescopes because they are hidden behind giant dust clouds.

Astronomers originally believed that the Milky Way had four spiral arms. However, in 2005, astronomers using infrared images from the Spitzer Space Telescope discovered that the Milky Way has just two spiral arms. The Spitzer Space Telescope observes the Universe in the infrared part of the spectrum, which cannot be seen from the ground because it is blocked by Earth's atmosphere. (You will see an infrared image of the Milky Way taken by the Spitzer Space Telescope in Section 10.1.)

The Milky Way consists of stars of all ages. The central bulge is a huge collection of older stars (Figure 10(a) on the next page). The disc of much newer stars spins around the bulge. Young and middle-aged stars, along with clumps of gas and dust, form the bulk of what makes up the spiral arms of the disc. Our solar system is located on the inner edge of one of the spiral arms, about 26 000 ly from the central core of the galaxy (Figure 10(b) on the next page).

At the centre of the Milky Way, a supermassive black hole exists. Astronomers know the black hole is there because they see the effects of its gravitational pull on the movement of stars close to it in the Milky Way.

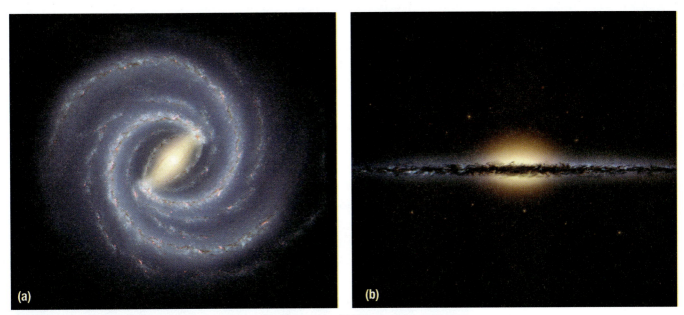

Figure 10 (a) A computer-generated top view of the Milky Way galaxy. (b) This computer-generated image shows that stars in the central bulge of the Milky Way galaxy appear close together because they are so numerous. They are actually very far apart.

Quasars—Powerhouses of Energy

Of all the celestial objects known, **quasars** are considered the most powerful energy producers of all. Their name comes from "**quas**i-stell**ar**" because quasars looked like stars when they were first observed through optical telescopes. In fact, quasars are galaxies with a central region that is more luminous than normal. The large amount of energy emitted by a quasar is in large part caused by a supermassive black hole at its centre. Astronomers observe that matter pulled into these giant black holes by their strong gravity fields is converted into powerful energy waves that are released back out into space (Figure 11). Quasars emit 100 to 1000 times more energy than the entire Milky Way.

quasar a distant, young galaxy that emits large amounts of energy produced by a supermassive black hole at its centre

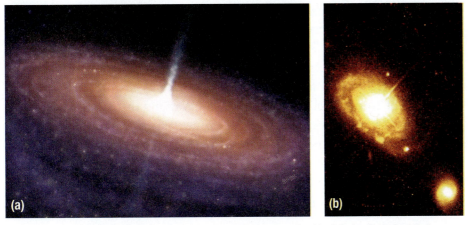

Figure 11 (a) Astronomers think quasars are powered by massive black holes that eject their energy into space from jets located at their top and bottom. (b) A real telescope image of a quasar.

To learn more about quasars,
GO TO NELSON SCIENCE

Astronomers have discovered more than 200 000 quasars. They are all between 2 and 12 billion ly away, so observing them is like looking back in time billions of years. Quasars are thought to be primitive galaxies that formed early in the existence of the Universe.

Galaxy Clusters

The Milky Way is part of a group of more than 35 galaxies called the Local Group (Figure 12). This collection of galaxies is about 10 million ly across. Our largest and nearest galactic neighbour is the Andromeda galaxy—a slightly larger version of the Milky Way. The Andromeda galaxy is a spiral galaxy that contains at least 300 billion stars. Lying about 2.6 million ly away, it is the farthest object the unaided human eye can see. It is amazing to think that the light we see from this galaxy travels for 2.6 million ly before reaching Earth. To put this into perspective, this means that the light we see today left the Andromeda galaxy when our early ancestors were just beginning to use stone tools. Since then, we have refined our tools and developed telescopes powerful enough to view this distant galaxy!

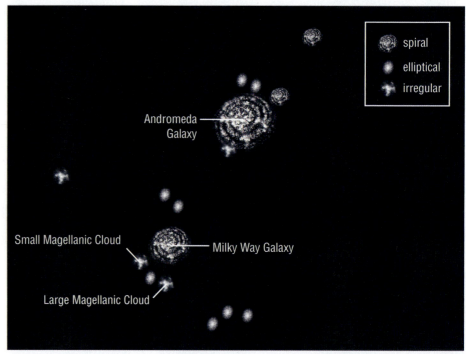

Figure 12 The Local Group, made up of many small irregular and elliptical galaxies, is dominated by two large spiral galaxies—the Milky Way and Andromeda.

Using supercomputer models of galaxy movements, astronomers at the University of Toronto believe that our Milky Way and the Andromeda galaxy are on a collision course with each other. However, this event will not happen for 5 billion years.

Our Local Group belongs to a much larger collection of galaxies called the Virgo Supercluster (Figure 13 on the next page), which is just one of millions of superclusters of galaxies in the Universe.

Figure 13 The Virgo Supercluster is a collection of thousands of galaxies loosely bound together by gravity. It includes our own Local Group. The Virgo Supercluster is just one of millions of superclusters scattered across the Universe.

UNIT TASK Bookmark

How can you apply what you have learned about types of galaxies to the Unit Task described on page 446?

IN SUMMARY

- Stars occur in star clusters and galaxies.
- The Milky Way galaxy contains our solar system and billions of other stars.
- Galaxies can be classified based on their shape as elliptical, spiral, lenticular, or irregular.
- Quasars are extremely high energy galaxies.
- Galaxies occur in clusters and superclusters.
- The Milky Way is part of a galaxy cluster known as the Local Group. The Local Group is part of the Virgo Supercluster.

CHECK YOUR LEARNING

1. Compare and contrast open star clusters and globular star clusters. K/U

2. Create a top view drawing of the Milky Way galaxy. Use the values given in this section for the diameter of the Milky Way and the distance of the Sun from the centre. Label the location of the Sun, old stars, young stars, and supermassive black hole. T/I C

3. What is astronomer Edwin Hubble's contribution to the understanding of galaxies? K/U

4. Astronomers know that the Sun is approximately 30 000 ly from the centre of the Milky Way and that the Milky Way has two spiral arms. K/U
   (a) When were these discoveries made?
   (b) In each case, what technology was used that allowed the discoveries to take place?

5. Create sketches of the following types of galaxies: spiral, barred spiral, elliptical, lenticular, irregular. Be sure to label them. K/U C

6. Where does the term "quasar" come from? Is this an accurate name for these objects? K/U

7. Compare and contrast a quasar with our Milky Way galaxy. T/I

8. What is the name of the galaxy cluster that the Milky Way is part of, and how many galaxies make up this cluster? K/U

9. What is the name of the galaxy supercluster that the Milky Way is part of, and how many galaxies make up this supercluster? K/U

TECH CONNECT ✓ OSSLT

Little Satellite, Big Science

Although it may be tiny and economical as space telescopes go, an innovative Canadian satellite is making big discoveries that are out of reach of even the powerful Hubble. It is also helping to change the future of satellite design. MOST (Microvariability and Oscillation of Stars) is Canada's first space telescope (Figure 1). Launched in 2003, its mission is to uncover the secrets behind the life cycles of Sun-like stars.

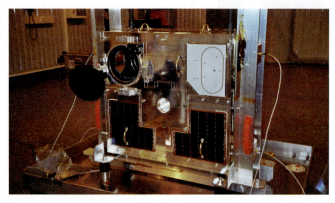

Figure 1 MOST, launched in June 2003, is 65 cm × 65 cm × 30 cm.

With its onboard telescope, MOST measures minute changes in the brightness of stars. Although the telescope is only the size of a dinner plate, it is so sensitive that it can detect a change in the brightness of a star as small as one part per million (Figure 2). That is like looking at a skyscraper lit up at night and measuring the change in overall brightness when someone lowers a shade at one window by 3 cm. No other space telescope can do that.

MOST is able to monitor the same star for seven weeks and then send the data to Canadian-based ground stations. In comparison, the Hubble telescope can keep a star in view for only six days.

MOST has discovered new data that give astronomers clues about the Sun in its youth. For 29 days, the space telescope observed a "teenage" Sun-like star in the constellation Cetus (Whale), as well as a middle-aged star that is similar to the Sun in the constellation Canis Minor (Little Dog). MOST's observations support predictions of what we think the Sun was like when it was young.

Astronomers are now using the innovative telescope to search for planets orbiting stars beyond our solar system. Scientists may even be able to detect clouds and monitor the weather on some of these planets.

Weighing only 56 kg, MOST belongs to a new class of compact satellites called microsatellites that are being placed in Earth orbit. These lightweight satellites cost much less to design, build, and launch into space than their larger counterparts.

MOST, which is about the size of a large suitcase, cost less than $10 million to develop and manufacture, making it the world's smallest and cheapest space-based observatory. In comparison, its larger cousin, the Hubble Space Telescope, is about the size of a school bus and cost more than $2 billion to develop. Astronomers and satellite engineers from the Canadian Space Agency, the University of British Columbia, and the University of Toronto, as well as Mississauga, Ontario–based company Dynacon Enterprises, worked together to design and build MOST.

Now Canada is becoming a global leader in developing new miniaturized satellites. The country's next mission is a miniaturized satellite called NEOSSat (Near Earth Object Surveillance Satellite), involving University of Western Ontario astronomers. Scheduled for launch in 2010, this new microsatellite uses the same miniaturized technology as MOST but will be the world's first space telescope to hunt for potentially dangerous asteroids and comets that may hit Earth in the future.

GO TO NELSON SCIENCE

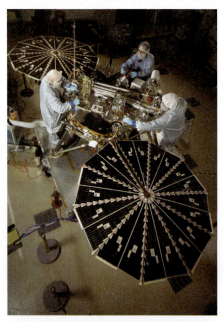

Figure 2 The telescope housed in MOST is 15 cm in diameter.

The Origin and Evolution of the Universe

People have been wondering about the Universe for a long time. They have asked questions such as Where did the Universe come from? How big is it? What will happen to it? In the last 100 years, technologies have been developed that help scientists find answers to these questions. By studying objects and events that occur in the Universe, astronomers continue to develop new theories and ideas about its origin and future.

Our Expanding Universe

Until the 1920s, astronomers thought that the Milky Way was the entire Universe. When they observed other galaxies, they thought that the galaxies were nebulas containing only gas and dust. The acceptance of this concept was so strong that when Albert Einstein's equations predicted that the Universe might be expanding, he changed them to reflect this belief that the Universe was unchanging.

Then Edwin Hubble made one of the most important discoveries in science. He identified individual stars inside the Andromeda galaxy (Figure 1) and realized that it was a separate galaxy far away from our Milky Way. When he determined that the Universe was filled with more galaxies, he noticed something even more surprising: all galaxies appeared to be moving away from each other, and the farther away a galaxy was, the faster it was moving away.

Figure 1 Andromeda is a large spiral galaxy that can be seen with binoculars on a clear autumn night.

TRY THIS: THE CENTRE OF AN EXPANDING UNIVERSE

SKILLS: Observing, Analyzing, Communicating

SKILLS HANDBOOK
3.B.6., 3.B.7.

As the Universe expands, the space between all distant objects increases. This makes it appear as though the Universe is moving away from the observer in all directions. You will examine a pair of images that represent a collection of distant galaxies for evidence of an expanding universe.

Equipment and Materials: two overhead images representing galaxies; overhead projector

1. Your teacher will display an overhead image that contains dots representing distant galaxies. Notice that some galaxies are bigger and some are smaller.
2. Next, your teacher will overlay another image on the previous one, lining up the galaxies at point A. The new one represents the view of distant galaxies taken at a later time. This is how objects in space will appear to have moved relative to an observer at point A.
3. As your teacher slides the second overhead image to line up both at point B and then point C, observe how the patterns on the overhead change.

A. From the point of view of an observer at point A, where is the "centre of the Universe"? K/U

B. From the point of view of an observer at point B and at point C, where is the centre of the Universe? T/I

C. What can you conclude about the location of the centre of the Universe? T/I

D. How far do objects, near an observer at point A, appear to have moved compared with objects that are further away from the observer? How does this compare with Hubble's observations? T/I

Red Shift

Hubble made his revolutionary discovery by studying the patterns of light emitted from galaxies. He observed the following phenomena:

- Each galaxy emits its own distinctive spectrum of light (Figure 2(a)).
- Light spectra shift, depending on whether the light source is moving or stationary.

While observing the light of distant galaxies, Hubble noticed a **red shift**—the light from the galaxies shifted toward the red end of the spectrum. This indicated that the galaxy itself is moving *away* from the Milky Way (Figure 2(b)).

red shift the phenomenon of light from galaxies shifting toward the red end of the visible spectrum, indicating that the galaxies are moving away from Earth

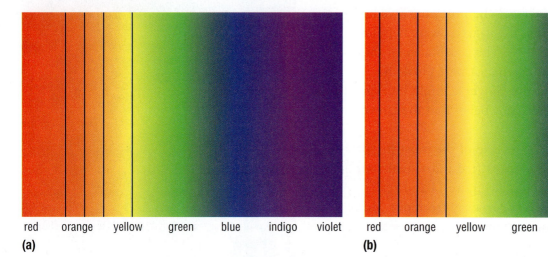

Figure 2 (a) The spectrum of a stationary galaxy (b) The same spectrum with the galaxy moving away from Earth. If a galaxy moves away from us, its spectrum shifts toward the red. The instruments used to detect this effect record the spectrum of light from red to violet. However, all of this light has shifted. We could not detect this shift except for the presence of dark lines in the solar spectrum. The dark lines are caused by gases in the upper atmosphere of stars absorbing some of the light emitted by the stars. This unique pattern of dark lines depends on the types of elements in the stars' atmosphere. Although the spectrum we see still looks largely like a rainbow, these lines have shifted toward the red because the galaxy is moving away from us.

Today, scientists observe the red shift in visible light and other radiation emitted from distant galaxies. The red shift indicates that the galaxies move farther apart from each other and from the Milky Way. This is evidence that the Universe is expanding. Hubble found that the farther away the galaxy, the greater the red shift, and, therefore, the faster it appeared to be moving. This relationship is called Hubble's Law.

Dark Matter

Stars travel at tremendous speeds within the galaxies themselves, and this speed depends on the overall mass of the galaxy to which they belong. However, astronomers were puzzled by the fact that these speeding stars were moving faster than expected, given the mass of the stars and matter that they could see. They concluded that there must be ten times more mass in the galaxy than we can detect with telescopes. It is the gravity of this hidden mass that is holding the galaxies together. Astronomers called the mass "dark matter" because it does not emit or absorb light (Figure 3).

Figure 3 Scientists hope to use specialized devices like this one to detect and study dark matter

This explains why astronomers have not been able to observe this matter with optical or other telescopes: it is not visible. We now understand that dark matter is the most abundant form of matter in the Universe, but little is known about its properties or behaviour.

Dark Energy

Before 1998, astronomers thought that the pull of gravity would cause the expansion of the Universe to slow down and eventually reverse. However, in 1998, astronomers noticed through observation of supernovas in distant galaxies that not only is the Universe expanding, but the rate of expansion is increasing. This discovery suggested that there is some strange form of energy working against the force of gravity. Called "dark energy," this phenomenon is causing the Universe to expand faster and faster. We do not yet understand the nature of dark energy or how it is able to oppose the effects of gravity.

The Big Bang Theory

Cosmologists—people who study the components, evolution, and physics of the entire Universe—now believe that all matter and energy in the Universe expanded from a point that was smaller than the period at the end of this sentence. This theory is called the **Big Bang theory**.

Around the same time that Hubble was studying red shift, Georges Lemaître, a Belgian priest and astrophysicist, suggested that all matter and energy in the Universe expanded from a hot, dense mass with an incredibly small volume. This expansion is known as the Big Bang. Astronomers today estimate the Big Bang occurred between 13.6 and 13.8 billion years ago.

At this first instant of time and space, the Universe was extremely hot, and energy was spreading outward very quickly. As the Universe cooled, energy began turning into matter—mainly hydrogen. Over hundreds of millions of years, this matter formed clumps, which eventually formed the stars and galaxies we see today (remember the Solar Nebula theory from Section 9.5).

The Big Bang theory has become the most widely accepted scientific explanation of the origin of our universe. It makes predictions about the Universe's origin that can be verified by observation. For instance, Canadian astrophysicist Jim Peebles used the Big Bang theory to show that in the first few minutes of the Big Bang, about a quarter of the matter fused into helium, whereas the rest remained as hydrogen. This is confirmed by evidence from the observation of the oldest stars.

Big Bang theory the theory that the Universe began in an incredibly hot, dense expansion approximately 13.7 billion years ago

Evidence for the Big Bang Theory

In 1965, strong evidence for the Big Bang was accidentally detected by Arno Penzias and Robert Wilson. The two physicists were conducting experiments with a supersensitive antenna in Crawford, New Jersey (Figure 4). They thought that the antenna was malfunctioning when they kept detecting radiation from all directions in the Universe.

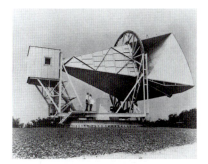

Figure 4 The antenna that provided Penzias and Wilson with evidence to support the Big Bang theory

To learn more about how Penzias and Wilson discovered microwave static radiation,

GO TO NELSON SCIENCE

DID YOU **KNOW?**

Television and Static
Some of the static on your TV set is caused by leftover radiation from the Big Bang.

This radiation "interfered" with their radio experiments. Other scientists determined that the static interference represented the remnants of the energy released by the initial expansion of space that followed the Big Bang.

The quest for more evidence of the Big Bang theory continued. Physicists John Mather and George Smoot tried to determine what happened during the first trillionth of a second after the Big Bang. They researched the initial expansion of the Universe and the time when the first stars began to shine. In 1989, the Cosmic Background Explorer (COBE)—a satellite designed by Mather—precisely measured the temperature of background microwave radiation. The measurements matched the evidence collected by Penzias and Wilson. These temperature variations are similar to an imprint of the beginning of structure in the Universe, much like fossils are an imprint of past life on Earth.

In 2001, NASA launched a new cosmological satellite called the Wilkinson Microwave Anisotropy Probe (WMAP) (Figure 5). It can detect variations in temperatures in space as small as a millionth of a degree. The data collected so far provides more information about the early stages of our universe. WMAP found evidence that the first stars began to shine about 200 to 300 million years after the Big Bang.

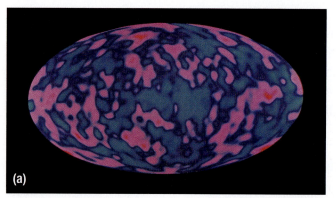
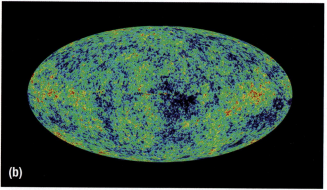

Figure 5 Readings from the (a) COBE and (b) WMAP satellites mapping the radiation in the Universe left over from the Big Bang. Examining the patterns in these images provides scientists with clues to the formation of the very first stars and galaxies, and supporting evidence for the Big Bang theory.

The Future of the Universe

Astronomers believe that over the course of the Universe's lifetime, there has been a constant battle between the gravitational pull of all matter—including dark matter—and the repulsive force of dark energy. Astronomers have discovered that the effect of dark energy was being felt 9 billion years ago and has steadily increased over time. Observations in 2006 confirmed that the expansion rate of the Universe began speeding up about 5 to 6 billion years ago. That is when astronomers believe that dark energy's repulsive force overcame the gravitational pull of matter. If their predictions are correct, over the course of the next billions of years, dark energy will continue to grow stronger, eventually pulling apart galaxies and leaving the Universe filled with even more cold, empty space.

TRY THIS: MODEL THE EXPANDING UNIVERSE

SKILLS: Planning, Performing, Observing, Analyzing, Communicating

SKILLS HANDBOOK
3.B.

The Big Bang theory states that the Universe started as a tiny dot of matter that expanded rapidly and continues to expand to this day. In this activity, you will investigate the expansion of the Universe and its effect on the movement of galaxies.

Equipment and Materials: round balloon; felt tip pen; tape measure

 Be careful when blowing up and writing on balloons. Pressing too hard on the balloon may cause it to burst and possibly cause injuries.

1. Fill your balloon with air until it is the size of a large apple. Then twist the end and hold it closed with one hand. Do not tie it shut.
2. Use the pen to draw four galaxies on the balloon in a line, with 1 cm of space between each one (Figure 6).

Figure 6

3. Label the galaxies A, B, C, and D.
4. Now inflate the balloon to the size of a volleyball and tie it off (Figure 7).

Figure 7

5. Measure the distances between the galaxies again and write down the measurements.

A. What happened to the distances between galaxies as you blew up the balloon? K/U T/I

B. Imagine that you are standing on galaxy "A" while the balloon is expanding. Which galaxy would appear to move away from you more quickly? T/I

C. Which one would appear to move more slowly? T/I

UNIT TASK Bookmark

How can you apply what you have learned about the Big Bang theory to the Unit Task described on page 446?

IN SUMMARY

- The observed red shift of distant galaxies suggests that the Universe is expanding at an increasing rate.
- Dark matter and dark energy are phenomena needed to explain the observed behaviour of distant galaxies.
- The Big Bang theory suggests that the Universe was born in a giant expansion 13.7 billion years ago.

CHECK YOUR LEARNING

1. How does red shift help astronomers determine the movement of distant galaxies? K/U
2. Summarize the conclusion that Edwin Hubble made about the Universe and the evidence he had to support his claim. K/U
3. Compare and contrast dark matter with dark energy. K/U
4. If dark energy has never been directly seen, what evidence do astronomers have that dark energy exists? K/U
5. According to the Big Bang theory, how long ago did the Universe begin? Who was the first person to suggest this theory? K/U
6. After the Big Bang, did stars and galaxies emerge fully formed? Why or why not? K/U
7. You have a friend who doubts that the Universe started billions of years ago during the Big Bang. Write a letter to your friend outlining two pieces of evidence that support the Big Bang theory. K/U C
8. Name two satellites that gathered evidence in support of the Big Bang theory. What kind of data did they capture? K/U
9. Describe the experiments Mather and Smoot performed that provided evidence for the Big Bang theory. K/U
10. What do astronomers predict will happen to our universe billions of years in the future? K/U

CHAPTER 9 LOOKING BACK

KEY CONCEPTS SUMMARY

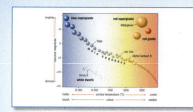

Stars differ in colour, size, and temperature.

- The colour of a star gives astronomers a clue to its surface temperature. (9.2)
- The relationship between the absolute magnitude and temperature of stars can be seen on the Hertzsprung–Russell (H–R) diagram. (9.2, 9.4)
- Stars can be millions of times bigger in volume than our Sun or hundreds of times smaller. (9.4)

Stars change their physical properties over time.

- Stars begin to shine when nuclear fusion occurs. (9.4)
- Stars spend most of their life as "main sequence" stars, as represented on an H–R diagram. (9.4)
- Depending on its mass, a star can end up as a white dwarf, a black hole, or a neutron star. (9.4)

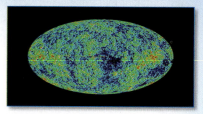

There is strong evidence that the Universe had its origin around 13.7 billion years ago.

- Radiation from the Big Bang can still be detected in the Universe. (9.7)
- Distant galaxies are moving away from us, which is evidence that the Universe is expanding. (9.7)

Stars are clustered into galaxies, which we can observe using telescopes.

- Galaxies are immense and can be made up of hundreds of billions of stars. (9.6)
- Distances to stars and galaxies are measured in light years. (9.1)
- Galaxies can be classified based on their shape as spiral, elliptical, lenticular, or irregular. (9.6)
- Quasars are mysterious galaxies that emit large amounts of energy. (9.6)

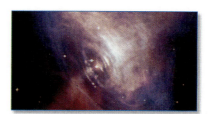

Black holes, neutron stars, and supernovas are part of the life cycle of massive stars.

- Black holes are celestial objects that have so much gravity that even light cannot escape. (9.4)
- Pulsars are spinning neutron stars that emit light that sweeps across Earth at regular intervals. (9.4)

Evidence collected by satellites and Earth-based observations support theories of the origin, evolution, and large-scale structure of the Universe.

- The Hubble Space Telescope has taken remarkable pictures of objects near the edge of the observable Universe. (9.4)
- The Cosmic Background Explorer (COBE) and the Wilkinson Microwave Anisotropy Probe (WMAP) created maps of the distribution of radiation in the Universe from the Big Bang expansion. (9.7)

WHAT DO YOU THINK NOW?

You thought about the following statements at the beginning of the chapter. You may have encountered these ideas in school, at home, or in the world around you. Consider them again and decide whether you agree or disagree with each one.

1 Stars differ only in their size and how far they are from Earth.
Agree/disagree?

4 Scientists can detect everything that there is in the Universe.
Agree/disagree?

2 The Big Bang theory explains how the Universe began.
Agree/disagree?

5 All galaxies in the Universe are like the Milky Way.
Agree/disagree?

3 Spacecraft orbiting Earth can observe objects that are billions of kilometres away.
Agree/disagree?

6 Telescopes allow astronomers to see all of the celestial objects in the Universe.
Agree/disagree?

How have your answers changed since then?
What new understanding do you have?

Vocabulary

light year (p. 365)
parallax (p. 367)
luminosity (p. 370)
apparent magnitude (p. 371)
absolute magnitude (p. 371)
solar mass (p. 373)
nebula (p. 375)
protostar (p. 375)
main sequence (p. 377)
red giant (p. 377)
red supergiant (p. 378)
white dwarf (p. 378)
supernova (p. 379)
neutron star (p. 380)
black hole (p. 380)
star cluster (p. 385)
elliptical galaxy (p. 386)
spiral galaxy (p. 386)
lenticular galaxy (p. 386)
irregular galaxy (p. 386)
quasar (p. 389)
red shift (p. 394)
Big Bang theory (p. 395)

BIG Ideas

✓ Different types of celestial objects in the Solar System and Universe have distinct properties that can be investigated and quantified.

✓ People use observational evidence of the properties of the Solar System and the Universe to develop theories to explain their formation and evolution.

• Space exploration has generated valuable knowledge but at enormous cost.

CHAPTER 9 REVIEW

The following icons indicate the Achievement Chart category addressed by each question.

K/U Knowledge/Understanding T/I Thinking/Investigation
C Communication A Application

What Do You Remember?

1. What is a nebula? (9.4, 9.5) K/U
2. Define the term "red shift." (9.7) K/U
3. What are the four types of galaxies, based on their shape? (9.6) K/U
4. What are the two factors that affect how bright stars look in the night sky (i.e., apparent magnitude)? (9.2, 9.3) K/U
5. Describe the process that allows stars to create their own energy. (9.4) K/U
6. Define the term "dark matter." (9.7) K/U
7. Where in our galaxy might you find a globular cluster? (9.6) K/U
8. What is a quasar, and where does the name come from? (9.6) K/U
9. What is the purpose of the Hertzsprung–Russell diagram? (9.4) K/U
10. What does the term "parallax" mean? (9.1) K/U

What Do You Understand?

11. What is the difference between a globular cluster and an open star cluster? (9.6) K/U
12. Place the following in the order that represents the life cycle of a star: supernova, neutron star, nuclear fusion, nebula, protostar. (9.5) K/U
13. Explain the difference between a galaxy cluster and a galaxy supercluster, with reference to the Milky Way. (9.6) K/U
14. If an astronomer finds a star that is 623 ly away, how far away is it in kilometres? (9.1) T/I
15. Describe how you could use the method of parallax to find the distance to a star. (9.1) T/I
16. Name two satellites orbiting Earth discussed in this chapter that have made important observations of distant stars and galaxies. What contributions have each one made? (9.1, 9.4, 9.6, 9.7) A
17. List as many physical properties as you can that describe a red supergiant star. (9.4) K/U
18. Describe the relationship between star colour and temperature. Use the following words in your answer: blue, red, yellow, orange, hottest, medium, cool, coolest. (9.2, 9.4) K/U C
19. What are two pieces of evidence that support the Big Bang theory, and which technologies have been used to collect this data? (9.7) K/U

Solve a Problem

20. Compare and contrast a neutron star with a black hole. (9.4) K/U
21. Arrange the following stars in order from coolest to hottest: blue supergiant, white dwarf, red giant, yellow main sequence star, orange main sequence star. (9.4) T/I
22. The oldest galaxies ever observed have been found 11.4 billion ly from the Milky Way. How far is this distance in kilometres? (9.1) T/I
23. What was the cause of the static interference that Arno Penzias and Robert Wilson detected in 1965? (9.7) K/U
24. Based on the H–R diagram in Section 9.4, determine the temperature of a main sequence star with absolute magnitude −2.2. (9.4) T/I
25. Based on the H–R diagram in Section 9.4, determine the absolute magnitude of a star with a temperature of 4500 °C. (9.4) T/I
26. Arrange the following objects in order from largest to smallest: white dwarf, nebula, neutron star, spiral galaxy, Virgo supercluster, red supergiant. (9.4, 9.5, 9.6) K/U T/I
27. What type of galaxy is shown in Figure 1? (9.6) K/U

Figure 1

Create and Evaluate

28. Consider the terms astronomers use to describe stars and planets forming in nebulas. Then suggest what term should be used to describe stars coming together to form a new galaxy. Explain your choice. (9.4, 9.5)

29. Imagine that you need to explain the structure of the Milky Way galaxy to a Grade 4 class. Describe how you might construct a three-dimensional model of the Milky Way detailing its structure so it is easy to understand. Be sure to describe the position of our solar system. (9.6)

30. The Hubble Space Telescope has cost $6 billion. Many people think that this money could have been better spent elsewhere. Do you think that the money was well spent? Support your opinion with information from this section and your own personal experience. (9.4, 9.7)

31. Copy Table 1 into your notebook. Then fill in the second column with the correct units used to describe the measurements of celestial objects from this chapter. (9.1)

Table 1

Measurement	Unit
diameter of the Milky Way galaxy	
distance from Earth to the Sun	
distance to the star Vega	
distance to a quasar	
diameter of a neutron star	
distance to the MOST satellite	
distance to the star Betelgeuse	

32. When we observe the light coming from the Andromeda galaxy, we see "blue shift" in that light. Based on what you know about galaxies that show red shift, what can you conclude about the motion of the Andromeda galaxy with respect to the Milky Way? (9.6, 9.7)

33. Imagine that a new star is discovered only 128 ly away. How could you tell that this star has just been created? What kinds of physical properties could you gather about this star? How would you go about collecting your data? Write a report to the Canadian Space Agency explaining the properties of this star and why it is important to perform research on distant stars. (9.1, 9.2, 9.4, 9.5)

Reflect on Your Learning

34. Find a partner. Find a page in your respective notebooks where you both took notes on the same topic on the same day.
    (a) Make a list of the ways in which your notes are similar and different.
    (b) What can you conclude from this about your different learning styles?
    (c) Why is it important to keep good notes?

Web Connections

35. Many of the images in this chapter come from the Hubble Space Telescope. List three of your favourite images from this chapter and write down one sentence describing each.

36. Astronomers need your help! In the last few decades, billions of stars have been found in galaxies all over the Universe. Many of these stars have planets orbiting them, and perhaps some of them could support a form of life. Join the Search for Extra-Terrestrial Intelligence (SETI) project by helping search through data collected from distant stars for evidence of intelligent life. After completing your research, present your findings in a format of your choice.

CHAPTER 9 SELF-QUIZ

The following icons indicate the Achievement Chart category addressed by each question.

K/U Knowledge/Understanding **T/I** Thinking/Investigation **C** Communication **A** Application

For each question, select the best answer from the four alternatives.

1. Which of the following lists units for measuring distances between celestial objects from smallest to largest? (9.1) K/U
   (a) km, ly, AU
   (b) ly, AU, km
   (c) AU, km, ly
   (d) km, AU, ly

2. Sirius, the brightest star in the night sky, has a luminosity of 22. This means that Sirius
   (a) has an apparent magnitude 22 times greater than the Sun's
   (b) gives off 22 times more energy than the Sun
   (c) developed 22 years before the Sun
   (d) is 22 times farther from Earth than the Sun (9.2) K/U

3. A lenticular galaxy
   (a) has a central bulge but no spiral arms
   (b) does not have a regular shape
   (c) varies in shape from spherical to flattened oval
   (d) includes a central bulge and spiral arms (9.6) K/U

4. Which of the following best describes a protostar? (9.4) K/U
   (a) a cloud of gas and dust that forms during the birth and death of a star
   (b) the hottest, most luminous type of star in the Universe
   (c) a stellar explosion that occurs at the end of a star's life
   (d) a dense concentration of gas and dust that eventually forms a star

Indicate whether each of the statements is TRUE or FALSE. If you think the statement is false, rewrite it to make it true.

5. The astronomical unit (AU) is the most convenient unit for expressing distances to celestial objects outside our solar system. (9.1) K/U

6. Apparent magnitude is a measure of the brightness of stars in the night sky as they appear from Earth. (9.2) K/U

7. Dark matter is difficult to observe because it neither emits nor absorbs light. (9.7) K/U

Copy each of the following statements into your notebook. Fill in the blanks with a word or phrase that correctly completes the sentence.

8. Astronomers use _____ to determine the distances to celestial objects based on the apparent change in their positions as seen from two different locations. (9.1) K/U

9. The brightness of a star depends on both its _____ and its distance from the observer. (9.2) K/U

10. Some _____ are called pulsars because they emit high frequency radio waves as they spin. (9.4) K/U

Match each term on the left with the most appropriate description on the right.

11. (a) luminosity (i) the brightness of a star as if it were located 33 ly from Earth
    (b) colour (ii) indicates a star's surface temperature
    (c) absolute magnitude (iii) determines the length and stages of a star's life cycle
    (d) mass (iv) the total amount of energy produced by a star (9.2) K/U

Write a short answer to each of these questions.

12. What is the difference between the Solar Nebula theory and the Big Bang theory? (9.5, 9.7) K/U

13. What was Greek astronomer Hipparchus's contribution to the understanding of the characteristics of celestial objects? (9.2) K/U

14. A misnomer is a misleading or inaccurate term applied to a person or object. Explain why the term "black hole" is a misnomer. (9.4) T/I

15. Your class is discussing the difference between scientific theories and personal beliefs. Write a brief paragraph explaining why the Big Bang concept is considered a theory rather than a belief or an opinion. (9.7) T/I C

16. If Polaris is 400 ly from Earth, what its distance from Earth in kilometers? (9.1) T/I

17. Create a concept map showing the organization of objects of the Universe. Use the following terms in your concept map: star, star cluster, galaxy, and galaxy cluster. (9.4, 9.6) K/U C

18. Describe two celestial bodies that may form in the final stages of the life cycle of a star that is more massive than the Sun. (9.2) K/U

19. Write a paragraph describing the relationships among our solar system, the Milky Way, the Local Group, and the Virgo Supercluster. (9.6) K/U C

20. (a) Quasars are some of the most distant objects in the Universe. In which direction do the light waves from quasars most likely shift?
    (b) A scientist observing a quasar in 2009 is actually seeing the quasar as it existed in the early Universe. Explain why this is so. (9.6, 9.7) A

21. An astronomer observes several stars and records their respective colours, as shown in Table 1 below. Using this data, order the stars by temperature from warmest to coolest. (9.4) A

    Table 1

Star	Colour
10 Lacertra	bluish
Antares	reddish
Aldebaran	orange-ish
Canopus	bluish white

22. (a) Name some words that contain the prefix "pro" or "proto."
    (b) Explain how someone who did not know the meaning of the word "protostar" could derive its meaning by looking at the word itself. (9.4) C

23. Which unit would be most appropriate for measuring each of the following distances? Choose between kilometers, AU, and light years. (9.1) T/I
    (a) the distance from the Milky Way to the Canis Major Dwarf Galaxy
    (b) the distance from the Sun to Charon, one of Pluto's moons
    (c) the distance from Earth's surface to the International Space Station

24. (a) How does the phenomenon of red shift provide evidence of an expanding universe?
    (b) A scientist observes blue shift in the edge of a rotating galaxy. What can you conclude about the galaxy's movement? (9.7) T/I

25. Create a list of characteristics used to describe stars. Beside each characteristic, describe how the Sun compares with other stars. (9.2) K/U T/I

26. Refer to Table 2 on page 372. Describe how you think conditions on Earth would be different from conditions today if the Sun were
    (a) as hot as a bluish star
    (b) as cool as a reddish star (9.2) K/U A

Chapter 9 Self-Quiz 403

CHAPTER 10

Space Research and Exploration

KEY QUESTION: How have humans explored the Universe beyond Earth?

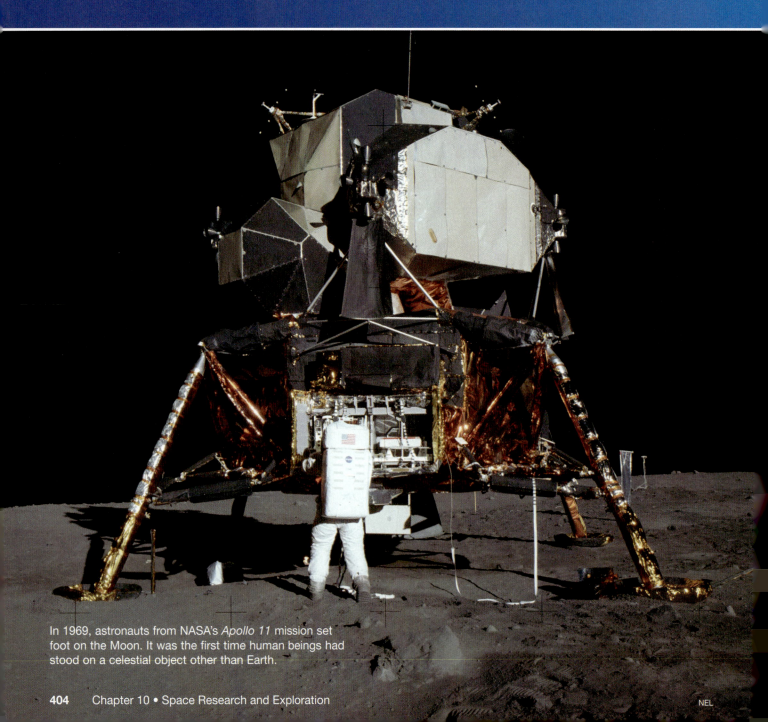

In 1969, astronauts from NASA's *Apollo 11* mission set foot on the Moon. It was the first time human beings had stood on a celestial object other than Earth.

UNIT D: The Study of the Universe

- **CHAPTER 8** — Our Place in Space
- **CHAPTER 9** — Beyond the Solar System
- **CHAPTER 10** — Space Research and Exploration

KEY CONCEPTS

Humans use telescopes that gather energy from the entire EM spectrum to learn about objects in the Universe.

Humans use space probes to learn more about the Solar System.

Canada is a world leader in developing technology for space exploration.

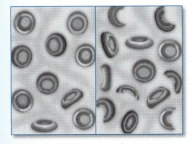

The environment in space is hostile to living things.

New technologies and information from the space program have benefited our everyday lives.

Space exploration is technically challenging and expensive.

ENGAGE IN SCIENCE

The Apollo 13 Mission

All space missions have a mission patch worn by the astronauts. This one, from *Apollo 13*, has a phrase in Latin that says "From the Moon, knowledge."

Imagine yourself strapped onto 2000 tonnes of explosive fuel, waiting for it to ignite and send you into space. On April 11, 1970, three lucky astronauts were in exactly that position, crammed into the tiny command module of NASA's third attempt to put humans on the Moon. If everything went according to plan, two of them would be walking on the Moon in four days.

The giant *Saturn V* rocket lifted the astronauts from the ground in a giant spectacle, sending out clouds of engine exhaust for miles around the launch site in Florida. As their families watched nervously from the VIP stands, the astronauts of *Apollo 13* began their journey to the Moon.

The first part of the flight went exactly as planned. The astronauts checked their equipment and performed experiments. A couple of days into the flight, more than halfway to the Moon, the crew heard a loud bang. One of them radioed back to mission control in Houston, Texas: "Houston, we've had a problem here."

An electrical fault had caused an oxygen tank to explode, and this resulted in a loss of power to the spacecraft. With no power to remove carbon monoxide from the air, oxygen dropped to dangerously low levels. The astronauts realized that they would have to abort their mission; they would not be walking on the moon.

The astronauts soon found themselves floating in space, more than 300 000 km from Earth, with little oxygen and no power. They quickly moved into the lunar module *Aquarius*, which was designed for transporting the astronauts from the spacecraft to the surface of the Moon. Squeezed into the lunar module, the astronauts relied on mission control in Houston to redirect them around the Moon and back to Earth.

As the crew approached Earth, they relied on Earth's gravity to pull them in. Tired, cold, and hungry, the astronauts splashed into the Pacific Ocean six days after they had left. They had not made it to the Moon, but they were alive. Thanks to the dedicated work of engineers on Earth, the astronauts survived a life-threatening accident in space.

Although *Apollo 13* did not put astronauts on the Moon, there were a number of successful Apollo missions between 1968 and 1972. One of the most famous is *Apollo 11*, the first human spaceflight mission to land on the Moon. On July 20, 1969, the astronauts who set foot on the Moon's surface took pictures, conducted experiments, and collected samples of rock from the Moon's surface. They then safely returned to Earth.

What kinds of things do you think the scientists learn from missions to outer space? What does this story reveal about the problems of sending human beings into space for research purposes?

With no power to remove carbon monoxide from the air, the *Apollo 13* astronauts followed the instructions of NASA controllers in order to create a makeshift device.

WHAT DO YOU THINK?

Many of the ideas you will explore in this chapter are ideas that you have already encountered. You may have encountered these ideas in school, at home, or in the world around you. Not all of the following statements are true. Consider each statement and decide whether you agree or disagree with it.

1 Sending humans into space is important for research and discovery.
Agree/disagree?

4 Spending billions of dollars on robots that explore the surface of Mars is worthwhile.
Agree/disagree?

2 All celestial objects can be viewed from Earth using a telescope.
Agree/disagree?

5 Technology developed for space exploration has many practical uses here on Earth.
Agree/disagree?

3 Space exploration encourages international cooperation and understanding.
Agree/disagree?

6 Astronauts "float" in space because there is no gravity.
Agree/disagree?

FOCUS ON READING

Synthesizing

When you synthesize, you combine different sources of information in a way that brings you to a new understanding. Use the following strategies when synthesizing a text:

- make connections between new information in the text and what you already know
- explain relationships among ideas in the text and the ideas in other texts
- monitor how the new ideas and information change your perspectives
- consider how features such as graphs and illustrations combine with the text
- draw conclusions about what you have learned

READING TIP
As you work through the chapter, look for tips like this. They will help you develop literacy strategies.

Mars Exploration

Two of the most recent successful planetary explorers have been the twin Mars rovers, *Spirit* (Figure 1) and *Opportunity*. They were originally designed to last only three months, but more than five years later, they are still working and sending back data from the surface of the Red Planet.

Landing on opposite sides, the twin rovers have each travelled more than 8 km. *Spirit* has been exploring volcanic rocks and climbing hills, whereas *Opportunity* has descended into craters and travelled across deserts. Both have found evidence in Martian rocks that more than a billion years ago, the planet was covered by a vast body of salty water.

Figure 1 This photo was taken through the "eyes" of the *Spirit* rover. *Spirit's* robotic arm has a small drill and microscope used to inspect Martian rock.

Synthesizing *in Action*

Synthesizing helps you combine new information in a text with what you already know about a topic to draw your own conclusions. Here is an example of how one student used the strategies to synthesize the selection about Mars exploration.

Information in the Text	What I Already Know	What I Think Now
Spirit and *Opportunity* are still working after more than five years.	Mars is cold and has huge dust storms that last for weeks.	The rovers have proved they can survive harsh conditions on Mars.
Rovers have explored over 16 km of the surface of Mars.	Rovers have cameras, a robotic arm, and scientific instruments to analyze soil and rock samples.	Robots are a safe way of gathering data on the surface of celestial objects.
Rovers found evidence of salty oceans on Mars in the past.	Water is necessary for life.	Life forms may have existed on Mars in the past.

Space Exploration

10.1

Technological developments over the past century have helped us explore outer space and push the boundaries of the known Universe. In the past five decades, humans have expanded their presence from our planet to the edge of the Solar System and beyond. We have built scientific laboratories in space (Figure 1), and spacecraft have landed on other planets, moons, and asteroids. We place instruments into space that gather information we would not be able to access from our planet's surface. Can you think of any objects or materials in everyday life that may have been developed as a result of space exploration?

Figure 1 The International Space Station (ISS) is a research facility in Low Earth Orbit.

Exploring Space from Earth

The science of astronomy is different from many other sciences. Scientists cannot directly experiment on the objects they study because they are so far away. They must use many different types of instruments and devices to make accurate observations and predictions about celestial objects.

Telescopes are instruments that detect and collect different types of electromagnetic (EM) radiation. Large telescopes are housed in observatories. In the past, all observatories were large buildings on the ground. During the last few decades, a growing number of telescopes and observatories have been launched into orbit around Earth. One example of such a telescope is the Spitzer Space Telescope, which detects the radiation given off by young stars that are difficult to see because they are usually hidden in clouds of dust and gas (Figure 2). Another telescope, the XMM-Newton Observatory (Figure 3), examines the X-ray radiation emitted by objects such as black holes.

DID YOU KNOW?

International Year of Astronomy
The year 2009 was declared the International Year of Astronomy by the United Nations because it marked the 400th anniversary of Galileo's telescope. All over the world, people attended "star parties," where they could look through telescopes and observe real images of distant celestial objects.

GO TO NELSON SCIENCE

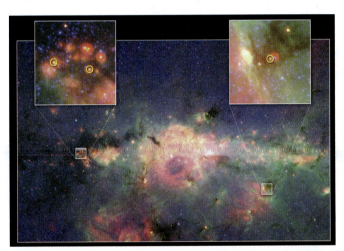

Figure 2 This Spitzer Space Telescope image shows three new stars in the Milky Way galaxy (circled in yellow). These stars were previously undetected because they are obscured by dust, but the Spitzer can detect them by monitoring infrared radiation.

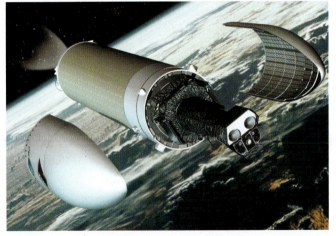

Figure 3 The XMM-Newton, the European Space Agency's largest astronomical satellite, is being released from its *Ariane 5* launch vehicle.

Early Telescopes

The first telescope was invented in 1608 in the Netherlands by glassmaker Hans Lippershey. The following year, astronomer Galileo Galilei refined the design. Merchants and sailors used Galileo's telescopes to look out to sea and spot distant ships coming to land, but Galileo used his own telescope to explore the night sky. Using the telescope's magnification and resolving power, Galileo revolutionized our view of the Universe.

refracting telescope an optical telescope that uses glass lenses to gather and focus light

reflecting telescope an optical telescope that uses mirrors to gather and focus light

The first optical telescopes used curved lenses to collect beams of light. This type of telescope, called a **refracting telescope**, gathers and focuses light from the visible light spectrum to form an image (Figure 4(a)).

In 1668, Sir Isaac Newton built the first astronomical telescope with mirrors. Known as a **reflecting telescope**, it uses a series of mirrors instead of glass lenses to collect and focus light from celestial objects (Figure 4(b)).

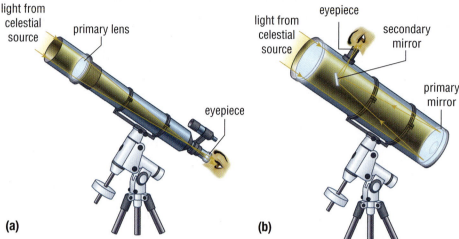

Figure 4 Stargazers still use (a) refracting and (b) reflecting telescopes to get a closer, clearer, and brighter image of stars and planets in the night sky.

Ground and Orbital Observation

Within the last century, astronomers have developed new technologies to build different types of telescopes. These new telescopes specialize in looking at various parts of the EM spectrum outside the range of visible light, such as radio waves, X-rays, and gamma rays.

RADIO TELESCOPES

Radio waves are easy to detect, are unaffected by weather, and can pass through concrete structures. Astronomers use radio telescopes to collect radio waves emitted from distant stars and galaxies. Similar to tuning a radio to different stations, astronomers can tune to different radio frequencies and "listen" to the stars and galaxies. Most radio telescopes look like a large curved satellite dish made of wire and metal (Figure 5). They collect radio waves from space and focus them on the telescope's receiver (Figure 6). The receiver amplifies the signal and transmits it to a computer for analysis.

> **DID YOU KNOW?**
>
> **Square Kilometre Array Telescope**
> Canada is one of 20 countries working together to build the world's largest telescope to be completed by 2012. Called the Square Kilometre Array, this radio telescope will be 100 times larger and 10 000 times more powerful than the largest observatory today. Consisting of thousands of radio dish antennas, it will be able to "listen" to radio waves travelling from the edge of the Universe and the beginning of time.

Figure 5 The Arecibo radio telescope is the largest single radio telescope in the world. It is situated in a natural valley in Puerto Rico.

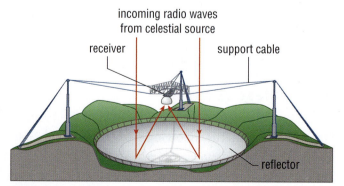

Figure 6 The receiver of the Arecibo radio telescope is suspended above the reflector.

X-RAY TELESCOPES

Some types of EM radiation emitted by distant celestial objects, such as X-rays, are not detectable from the surface of Earth because they are filtered out by our atmosphere. X-rays are emitted from remnants of massive dying stars, such as neutron stars and black holes. To intercept the X-rays, astronomers place telescopes equipped with X-ray detectors into orbit around Earth. One such telescope, the Chandra X-ray telescope, has four long, smooth mirrors made of a hard, silver-like metal called iridium. This telescope can detect extremely hot, invisible objects in space.

GAMMA RAY TELESCOPES

Gamma ray observatories, such as Europe's INTEGRAL space telescope and the United States' Swift space telescope, track intense surges of gamma ray radiation from distant regions in the Universe. Astronomers think that these mysterious bursts of gamma rays may be caused by pairs of black holes or neutron stars colliding billions of light years away.

HUBBLE SPACE TELESCOPE

The Hubble Space Telescope (HST) was named after the famous astronomer Edwin Hubble. The HST orbits Earth (Figure 7), which allows astronomers to view objects at distances far greater than they could from telescopes on the ground. Hubble's spectral range extends from the ultraviolet through the visible, and into the near-infrared. The HST has a variety of telescopes and instruments on board.

DID YOU KNOW?

Looking into Space
Astronomers have learned more about celestial objects by comparing images of them using different telescopes. These images of the Sun were taken using (a) optical and (b) X-ray telescopes.

(a)

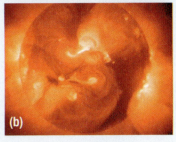
(b)

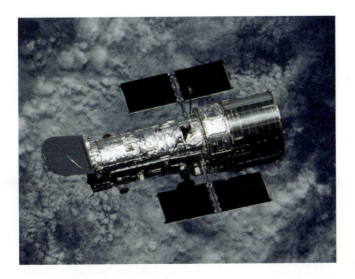

Figure 7 The HST has revolutionized astronomy by providing unprecedented deep and clear views of the Universe. Many of the images in this unit were taken by the HST.

Human Space Exploration

A **spacecraft** is a vehicle designed for travel in space. It can be operated directly by humans, or it can be a robotic vehicle that is propelled into space and operated from an Earth-based control centre. Spacecraft are made up of various parts. One part—the rocket body or launch vehicle—overcomes the force of gravity on Earth and launches the object toward its destination. On most human-occupied missions, astronauts sit in another area where they pilot the spacecraft. Another part of the spacecraft is designed to carry cargo of scientific equipment, supplies, and sometimes even other spacecraft, such as lunar or planetary landers and rovers.

spacecraft a human-occupied or robotic vehicle used to explore space or celestial objects

To learn more about being a design engineer for spacecraft or robotics,

GO TO NELSON SCIENCE

Figure 8 In 2009, Julie Payette, who is an electrical and computer engineer, manoeuvred three robotic arms to help build a "porch" on the ISS.

READING TIP

Changing Your Perspective

When synthesizing a text, look for clues relating to what you already know about the topic. For example, if the text gives information about how long the space shuttles have been in operation and how many accidents there have been, you may draw on your knowledge of the dangers of space flight, such as being hit by space debris, and conclude that the space shuttle has a good track record.

Until the 1980s, spacecraft were designed to be used only once. The *Apollo* missions that sent humans to the Moon used a multistage launch vehicle called *Saturn V*. These giant rockets carried the human crew, the lunar lander, and the command module toward the Moon. After the *Apollo* program ended, the United States began to develop and build a partially reusable spacecraft called a space shuttle. Although the external tank of a space shuttle is expendable, the two solid rocket boosters and the orbiter are reusable. The Space Transportation System (STS), also called the Space Shuttle program, launched with the Space Shuttle *Columbia* (STS-1) in April 1981.

Since Marc Garneau became the first Canadian in space in 1984, Canadian astronauts have flown on more than a dozen space shuttle missions. In 1999, the space shuttle *Discovery* lifted off with Canadian Space Agency (CSA) astronaut and mission specialist Julie Payette (Figure 8). In 2009, Payette once again went into space aboard space shuttle *Endeavour* on a mission to continue building the International Space Station (ISS). This was the 127th space shuttle mission.

Working and Living in Space

The International Space Station (ISS) is the ninth space research facility to be built in orbit (Figure 9). Since its launch in 1998, more than 16 countries have participated in its construction and scientific use. The ISS, weighing 245 735 kg, is the largest human-made object to orbit Earth. It travels at an average speed of 27 700 km/h and completes 15.7 orbits per day at an altitude of 350 km.

Figure 9 The ISS as it will appear in 2010 after construction is complete

Chapter 10 • Space Research and Exploration

Approximately 50 space shuttle missions and more than 150 space walks (when astronauts put on spacesuits and leave the spacecraft) will be needed to complete the construction of the ISS. When completed in 2010, it will be as big as a football field. The station can be seen with the unaided eye. It looks like a bright star gliding across the night sky.

Two astronauts at a time began living on the ISS in the year 2000, staying aboard the station for up to six months. In 2009, six crew members lived on board together for the first time, one of which was Canadian veteran astronaut Robert Thirsk (Figure 10). There have also been visits by six space tourists for week-long stays.

Figure 10 In 2009, Robert Thirsk became the first Canadian to participate in a long-duration stay as part of a six-person crew onboard the ISS.

Astronauts face daily challenges when living in space. Water is a precious resource in space, and there is a limited supply available. Every drop of water—from water vapour that astronauts exhale, to urine they produce—is put through a water-purifying unit to be recycled. This process produces cleaner water than many municipal water treatment plants on Earth. To keep clean, astronauts use special kinds of soap and shampoo that do not need water to rinse.

Crews of astronauts and scientists from various countries are conducting experiments on the ISS and in space to study a variety of medical, chemical, and physical science questions. The ISS provides a unique location for conducting long-term science experiments under effective low-gravity conditions. Canadian experiments on the station include studying the physics and chemistry of how different liquids undergo mixing, which will enable the oil industry to operate more efficiently. Other experiments investigate how blood pressure and fainting can affect both space travellers and people back on Earth.

The 16 countries who contributed $100 billion to build the space station are committed to keeping the station going until 2015 and possibly extending its life to 2020 to conduct more research. This project encourages international cooperation and understanding among countries.

RESEARCH THIS: SEE THE SPACE STATION

SKILLS: Researching, Evaluating

SKILLS HANDBOOK
4.A., 4.D.

As the ISS increases in size with the addition of new solar panels and laboratory modules, it becomes more visible from Earth. When completed, the ISS may become the third-brightest point of light in the sky—brighter than all celestial objects except the Sun and the Moon. The ISS orbits Earth in about 1.5 hours, so it moves fairly quickly and is easily seen without a telescope. In this activity, you will research how, when, and where to observe the International Space Station.

1. Conduct research to discover the exact time of the next overhead pass of the International Space Station for your location.

 GO TO NELSON SCIENCE

2. Note the azimuth and altitude of the ISS at an exact time so you know where to look. You might want to sketch the path you expect the ISS to take on a star map and use it to locate nearby constellations for reference.

3. On a clear night, look for the ISS at the appropriate time.

A. Why does the space station shine so brightly? K/U T/I

B. What is the altitude and azimuth of the ISS at the exact time you observed it? T/I

C. As you observe the space station, why does it disappear from view at the horizon? T/I

D. How long would you have to wait to see the ISS at the same spot again? T/I

LEARNING TIP

DEXTRE
DEXTRE is short for "Special Purpose Dexterous Manipulator." The word "dexterous" means "careful and skillful use of the hands," and is an appropriate nickname for the ISS's new robotic hand.

READING TIP

Making Connections
Ask yourself what you already know about a topic that relates to the information in a text. For example, you probably have some knowledge of the anatomy of a human arm. You can use this prior knowledge to help you understand the components of the Canadarm and how it works. You might draw a conclusion about how robots are designed to mimic functions that humans perform.

DID YOU KNOW?

Boom Sensor System
During its launch in 2003, the space shuttle *Columbia* suffered damage to one of its wings, causing it to burn up during re-entry. The damage had been suspected, but there was no easy way of checking. Since then, all space shuttles are outfitted with a Canadian-made orbital Boom Sensor System. A special scanner attached to its end can be used to visually check the shuttle exterior for damage. This 15-metre boom with cameras at its end attaches to the Canadarm, effectively doubling its length and giving it the necessary reach to view the wings and the underbelly of the shuttle.

Space Tools

For three decades, Canada has played an important role in developing robotic tools for use on space shuttle missions and the ISS. The Canadarm, Canadarm2, and DEXTRE are three such tools. Each of these mechanical devices is used in the construction and maintenance of the ISS. During space shuttle missions that bring up new components for the station, these robotic cranes and hands work together to pass loads to one another.

Canadarm

Since 1981, space shuttles have been equipped with the Canadarm, a 15.2 metre–long robotic arm designed and built by Canadians (Figure 11). In the harsh environment of space, this mechanical arm works with the dexterity of a human arm. Its "skin" is made of three types of material—titanium metal, stainless steel, and graphite—and an insulating blanket that protects the internal parts of the arm from the extreme hot and cold temperatures of space. Its "nerves" are copper wiring. Its "bones" are graphite fibre. Its "muscles" are electric motors, and its "brain" is a computer. Canadarm also has six joints: two "shoulders," one "elbow," and three "wrist" rotating joints that allow it to move loads in space. The robotic arm is hollow and would not be able to support its own weight on Earth, but it works very well in space. Attached to a shuttle, the Canadarm is used like a construction crane to lift parts, capture and deploy satellites, and assist astronauts moving in space.

Figure 11 Weighing 450 kg, the Canadarm can pick up and move objects that weigh as much as 29 tonnes. Astronauts use computers within the space shuttle to control this robotic crane.

The first Canadian to operate the Canadarm in space was astronaut Chris Hadfield in 1995. Canadian astronaut Marc Garneau used the Canadarm to deploy and retrieve a satellite on his second space mission in 1996. Four years later, Garneau used the Canadarm to lift solar panels from the space shuttle *Endeavour* to the ISS.

Next-Generation Arm

Canadarm2—a technologically more sophisticated Canadarm—was installed on the ISS in 2001 with the assistance of CSA astronaut Chris Hadfield. It has one more joint than the Canadarm, and neither end is permanently attached to the outside of the ISS. At 18 m in length, the Canadarm2 can manipulate 115 tonnes of equipment and transfer cargo from space shuttles to the ISS. It can crawl along the ISS, capture satellites, and assemble pieces to build and maintain the ISS (Figure 12). It can also work in tandem with the original Canadarm, which is attached to every space shuttle.

To learn more about Canada's contribution to space exploration and the ISS,

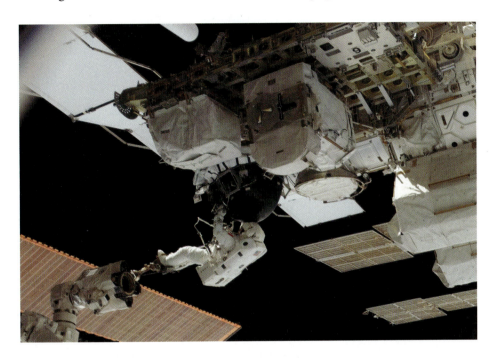

Figure 12 Anchored to the foot restraint on the Canadarm, Canadian astronaut Dave Williams spent 6.5 hours replacing a faulty piece of equipment on the ISS.

Canada Lends a Hand

In March 2008, Canada installed a robotic hand onto the International Space Station's Canadarm2. This special purpose dexterous manipulator, known as DEXTRE, is a two-armed robot that is used to do construction and repair work on the outside of the orbiting space station (Figure 13). It can be positioned along various worksites around the station. DEXTRE is outfitted with movable jaws at the ends of its 3.5-metre, multi-joined arms. It also has video cameras and a toolkit around its waist. DEXTRE's main function is to regularly replace 100 kg batteries on the outside walls of the space station, saving astronauts from engaging in dangerous spacewalks.

Figure 13 DEXTRE is another of Canada's robotic contributions to the ISS.

Robotic Exploration

Scientists have been able to learn a lot about the planets just by observing them with telescopes on Earth. However, **space probes** have revolutionized our understanding of the Solar System in the last 50 years. These robotic spacecraft, partially controlled by scientists and engineers on the ground, are filled with instruments designed to make close-up observations of objects in the Solar System. Since signals take so long to travel to and from the probes, many of the probes' activities are controlled by onboard computers. The first planets to be explored in this manner were the ones closest to Earth—Mars and Venus.

space probe a robotic spacecraft sent into space to explore celestial objects such as planets, moons, asteroids, and comets

In the 1960s and 1970s, the *Mariner* and the *Viking* series space probes explored Mars by orbiting the planet and landing on its surface. Since then, hundreds of probes have been sent into space to collect information on neighbouring worlds as far as the gas giant planets—Jupiter, Saturn, Uranus, and Neptune (Figure 14).

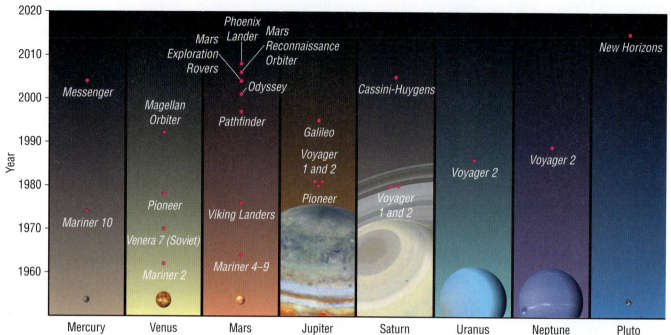

Figure 14 This timeline graph shows when some space probes explored or will explore other planets and the dwarf planet Pluto.

A robotic explorer called *New Horizons* is on its way to the dwarf planet Pluto. Because Pluto is more than 6 billion km away, *New Horizons* will not reach it until 2015. After reaching Pluto, the probe will continue on its journey toward the outer edge of the Solar System.

Mars Exploration

Space agencies in Europe, the United States, Russia, and Japan have been sending spacecraft to fly by, orbit, and land on Mars for five decades. However, not every probe has been successful. Of the 42 missions launched, 22 never sent back any data. In 1997, NASA's *Mars Pathfinder* became the first remote-controlled rover sent to another planet. The tiny robotic car successfully explored the soil and rocks in the landing area for three months.

Two of the most recent successful planetary explorers have been the twin Mars rovers, *Spirit* (Figure 15) and *Opportunity*. They were originally designed to last only three months, but more than five years later, they are still working and sending back data from the surface of the Red Planet.

Landing on opposite sides of Mars, the twin rovers have each travelled more than 8 km. *Spirit* has been exploring volcanic rocks and climbing hills, whereas *Opportunity* has descended into craters and travelled across deserts. Both have found evidence in Martian rocks that more than a billion years ago, the planet was covered by a vast body of salty water.

Figure 15 This photo was taken through the "eyes" of the *Spirit* rover. *Spirit*'s robotic arm has a small drill and microscope used to inspect Martian rock.

Both rovers recharge their batteries through their solar panels. Over the years, dust storms have covered parts of their solar panels with dust, reducing the amount of power available to the rovers. However, scientists have been surprised to find that small Martian tornadoes, or dust devils, occasionally blow some of the dust off the panels (Figure 16).

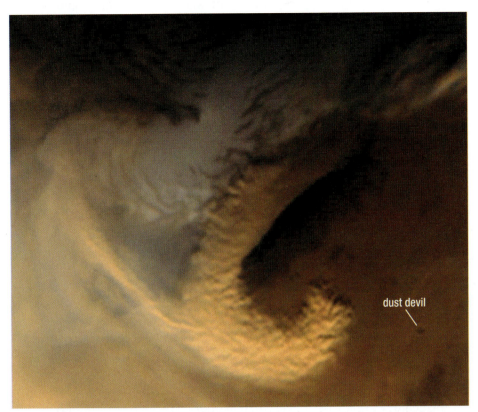

Figure 16 This photo of Mars shows a swirling pink storm at its north pole and dozens of dust devils farther south.

The latest robotic explorer of Mars was the NASA *Phoenix Mars Lander* probe that blasted into space in August 2007. It successfully landed in the Martian Arctic 10 months later, in May 2008. When the probe used its robotic arm equipped with a scoop to examine the soil, it discovered the first visual evidence of water-ice on Mars (Figure 17). This was a significant discovery: the presence of water on Mars suggests that the planet's conditions are (or were) favourable to life. A Canadian-designed meteorological station on board monitored Mars' weather on a daily basis. For the first time in the history of planetary exploration, we had daily weather reports on temperature, cloud cover, humidity, and wind speed on another planet.

Figure 17 The *Phoenix* lander dug a series of trenches with its robotic arm in the Martian northern Arctic region and discovered a layer of water-ice just a few centimetres below the surface.

Saturn Exploration

Astronomers, engineers, and other scientists from around the world continue to share their expertise and resources to form cooperative space missions. For example, 17 nations collaborated to engineer the space probe *Cassini–Huygens* to explore the distant ringed planet Saturn and its moons. Although the radio signals from this probe travel through space at the speed of light, they still take 68 to 84 minutes to reach Earth, depending on our exact distance.

Figure 18 This side-by-side image shows a Cassini radar image (on the left) of what is the largest body of liquid ever found on Titan's north pole, compared to Lake Superior (on the right). This image offers strong evidence for seas (most likely liquid methane and ethane) on Titan.

Just before *Cassini* went into orbit around Saturn in 2004, it released a probe, *Huygens*, to explore Saturn's moon, Titan. Titan's surface has remained a mystery for decades because it is surrounded by a thick, smog-like atmosphere. The largest moon orbiting Saturn and the second largest in the Solar System, Titan is the only moon with an atmosphere.

The *Huygens* probe snapped photos of dark, river-like channels and bright hills as it floated down to Titan's surface using a parachute system (Figure 18). With batteries lasting only 90 minutes, the *Huygens* probe transmitted pictures of a dark plain with scattered, light-coloured water-ice pebbles.

The main *Cassini* spacecraft continues to orbit Saturn, using its cameras and radar system to study Titan and neighbouring moons. It has discovered that Titan may have lakes of liquid methane, whereas the tiny moon Enceladus is covered by water-ice and has ice geysers with plumes projecting 400 km into space.

UNIT TASK Bookmark

You can apply what you have learned about telescopes, Canadian contributions, and the challenges and dangers of space travel as you work on the Unit Task described on page 446.

IN SUMMARY

- Telescopes that detect various types of radiation are key tools for astronomers to research celestial objects.
- Robotic spacecraft have been sent throughout the Solar System to explore different objects.
- There have been several successful human space flight missions, including the Apollo and Space Shuttle missions, and expeditions to the ISS.
- Canada has contributed to space exploration with projects such as the Canadarm.
- Saturn and Mars are two planets in the Solar System that are currently being explored by spacecraft.

CHECK YOUR LEARNING

1. Compare the design of refracting telescopes and reflecting telescopes.
2. Name two kinds of EM radiation other than visible light that are used to observe distant objects in the Universe.
3. Why do X-ray telescopes need to orbit Earth beyond the atmosphere to make observations? Name an example of an X-ray telescope.
4. Name three telescopes that are currently collecting data in orbit around Earth.
5. In a paragraph, describe the contributions of one of the Canadian astronauts who has flown in space with the Space Shuttle program.
6. Compare and contrast the *Saturn V* rocket with a space shuttle.
7. How many countries participated in the construction of the ISS?
8. Describe how the Canadarm works. In which ways is the Canadarm like a human arm?
9. Describe two experiments currently being performed on the ISS.
10. Describe three ways that Canadians have contributed to space exploration.
11. What are two planets that humans have explored with space probes? What do scientists hope to learn from these missions?

418 Chapter 10 • Space Research and Exploration

Challenges of Space Travel

10.2

Engineers and scientists perform the difficult job of designing spacecraft that safely take humans to their celestial destinations and back (Figure 1). What challenges do you think spacecraft designers must overcome?

Figure 1 A wire frame computer graphic image of a spacecraft.

Getting into Space

Spacecraft must be equipped with powerful rockets that can reach a speed that will take the spacecraft beyond Earth's atmosphere and escape Earth's gravitational pull (Figure 2). To maintain its orbit, a spacecraft must travel at precisely the right speed. The acceleration of the rocket booster has to be carefully calculated based on the overall weight of the entire spacecraft. There cannot be too much vibration or gravitational pressure put on the spacecraft and its astronauts. Space shuttles, satellites, and other spacecraft require massive amounts of fuel to propel them into space. They may require additional fuel depending on their destination, trip time, and the mass of their cargo, or payload.

Figure 2 To escape Earth's atmosphere, rockets have to reach a speed of 7.7 km/s.

Traditional Fuel

Traditional launch vehicles are powered by a chemical reaction between fuel and oxygen. The fuel is heavy and most of it is used to overcome the pull of gravity during liftoff, so there is not much left over for the spacecraft to use for the remainder of the journey. Once the vehicle is in space, fuel is burned only in spurts for correcting or altering course. The spacecraft must coast between spurts in the desired direction. New space propulsion technologies using lightweight fuels are being developed to enable spacecraft to travel farther. However, traditional fuel is still the main way to launch spacecraft from Earth into orbit.

Alternative Fuel

Instead of using traditional fuel, a number of small robotic planetary probes now fire atoms of xenon gas to provide the thrust they need to navigate through the Solar System. This new technology allows small spacecraft to travel great distances at relatively little cost.

READING TIP

Drawing Conclusions
After reading a text, you might need to do some research to help you synthesize and draw conclusions. For example, if you wonder how much a space shuttle weighs, you might find the information and calculate the total weight to be 2050 metric tonnes. You might draw a conclusion that future spacecraft carrying humans to Mars will need to be launched from space.

DID YOU KNOW?

Powering Space Probes
Solar energy alone could not be used to provide all the electricity necessary for the instruments onboard *Cassini–Huygens*, a probe sent to Saturn in 2004. Because of the size of the probe and the weakness of solar radiation at Saturn's great distance from the Sun, scientists needed to use electricity generated by nuclear power. They were very concerned about the nuclear fuel at launch. If there had been a failure or malfunction in the launch systems, the radioactive fuel onboard the probe might have polluted the surface of Earth and the atmosphere around the launch site. Scientists conducted an environmental impact study that resulted in a number of recommendations aimed at preventing environmental damage.

Nuclear energy, which is energy released from the nucleus of radioactive atoms, can be used to power the instruments on space probes. The major advantage of this type of fuel is that a small mass produces enormous amounts of energy. By carrying a smaller mass of fuel, the space probe could carry more cargo and travel much faster. Using smaller masses of fuel also lowers the cost of space travel.

The next Mars rover, for NASA's Mars Science Laboratory mission, will use nuclear electrical energy to power scientific instruments. The compact car–sized rover will be able to conduct scientific experiments day and night and is expected to drive hundreds of kilometres on the Martian surface over many years (Figure 3).

Figure 3 Three generations of Mars rovers: the two on the left used solar panels to generate power; the new, larger Mars rover uses a nuclear generator, giving it many times more power.

Space engineers estimate that nuclear propulsion may reduce the travel time to Mars from 8 months to as little as 4 months. New technologies that reduce travel times decrease the risks to astronauts during long-term space travel, such as weakening bone strength and other health issues. Using nuclear power also means that a spacecraft does not have to rely on solar energy. This is particularly important where the Sun's radiation is low, such as in the outer parts of the Solar System—Jupiter and beyond.

Nuclear fuel is not without risks. If an explosion occurred at a launch site, the surrounding environment would be contaminated with nuclear material. Today, jet propulsion technologists and design engineers are developing even safer nuclear power systems for human-occupied spacecraft. Other options that are being explored include solar and electrical propulsion systems.

To learn more about rocket engines and rocket fuel,
GO TO NELSON SCIENCE

To learn more about being a jet propulsion technologist or design engineer,
GO TO NELSON SCIENCE

A Feeling of "Weightlessness"

Another challenge of space travel is the apparent "weightlessness" of objects and people for the duration of a space voyage. Since the pull of gravity inside the orbiting ISS and space shuttle is only slightly less than it is at Earth's surface, why do astronauts and other objects appear to float inside these spacecraft?

When a spacecraft is placed into orbit, it rises to a predetermined height above the atmosphere and then turns and travels in a direction parallel to Earth's surface. If its forward speed is too slow, the spacecraft will collide with the ground. If its forward speed is too fast, the spacecraft may move away from Earth. When the pull of gravity and the forward speed are perfectly balanced, the spacecraft will fall toward Earth at the same time as it is trying to speed away from Earth. (Recall the cannonball diagrams from Chapter 8.)

The continuous falling motion, or free fall, gives the spacecraft and everything in it the feeling of weightlessness—the same feeling of weightlessness you experience in a plunging roller coaster. Although the force of gravity is still relatively strong, the continuous falling motion causes unsecured objects inside the spacecraft to appear to float (Figure 4). Sometimes this is called a **microgravity environment** because in this environment, unsecured objects behave as they would in an environment with very little gravity.

Figure 4 U.S. astronaut Mike Fincke with fresh fruit in microgravity conditions aboard the ISS.

microgravity environment an environment in which objects behave as though there is very little gravity affecting them

TRY THIS UNDERSTANDING FREE FALL

SKILLS: Controlling Variables, Performing, Observing, Communicating

SKILLS HANDBOOK
3.B.6.

When astronauts are in orbit around Earth, they appear to float because they are in free fall. Gravity accelerates all objects toward the centre of Earth at the same rate, which is why astronauts inside the ISS appear to float. In this activity, you will experiment with water inside a falling pop bottle to demonstrate the concept of free fall.

Equipment and Materials: geometry compass; empty 1 L pop bottle; pen; duct tape; water

1. Poke a small hole near the bottom of the pop bottle with the compass tip. Widen the hole with a pen, so the hole is as wide as the pen's body.

   🛑 Compasses are very sharp! When pushing the compass tip into the pop bottle, be careful that it does not slip.

2. Cover the hole with a piece of duct tape and fill the pop bottle with water.
3. Holding it by the neck, position the bottle over a sink or take it outside. Pull off the tape and, without squeezing the bottle, watch the water stream out.
4. When the water has drained out, put tape over the hole and fill the bottle with water again.
5. Now raise the bottle as high as you can. As you peel the tape off the hole, drop the bottle and watch what happens to the water streaming out of the hole.

A. The first time you held the bottle up without squeezing it, the water flowed out. What was the force responsible for pushing the water out of the hole in the bottle? K/U T/I

B. When you dropped the bottle, what happened to the stream of water? T/I

C. Why did the water behave this way after you dropped the bottle? T/I

D. What do you think it would feel like if you were inside a giant container that was falling like this? T/I

E. Write a concluding statement about how this activity demonstrates the concept of free fall. T/I C

Health and Other Risks

The cells, tissues, and organ systems in your body rely on the force of gravity to assist in their function. In conditions of microgravity, astronauts may experience a range of health issues. About 50 % of astronauts experience dizziness, disorientation, and nausea during the first few days of exposure to microgravity. Researchers believe that this is caused by the balance organs in the inner ear being disoriented when there is little sensation of gravity.

Dr. Roberta Bondar was Canada's first woman in space. The experiments she conducted on an eight-day trip in 1992 revealed that exposure to microgravity can lead to some of the same symptoms as brain damage from strokes.

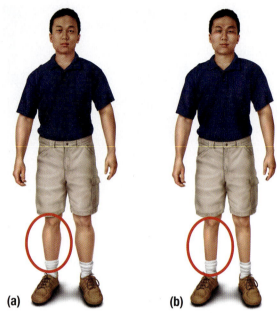

Figure 5 (a) Person on Earth (b) The same person under microgravity conditions. This effect is sometimes called the "puffy-face, bird-leg look."

Microgravity can also affect the appearance of the human body in space. For example, blood vessels in a person's legs have one-way valves to prevent blood from pooling in the feet due to the downward pull of gravity (Figure 5(a)). When astronauts are in space, blood pools in the upper part of their bodies (Figure 5(b)). This affects the body's ability to regulate water and can cause dehydration.

Blood cells have specific shapes required to carry out their functions (Figure 6(a)). In space, their shapes change, which affects their ability to function (Figure 6(b)). Because we are always working against it, the force of gravity helps keep our bones and muscles strong. Astronauts' bones and muscles weaken in space, where little force is needed to breathe, to move blood through the body, or to lift objects. As well, in space, the discs of the spinal column expand, making astronauts taller by about 2 to 4 cm while in orbit. This can lead to lower back pain in some taller astronauts.

To stay healthy and help avoid any permanent damage to their bones and muscles, astronauts spend time in space working their muscles with resistance devices and running on treadmills.

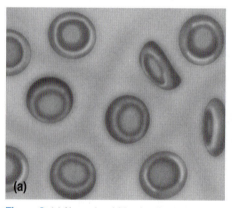 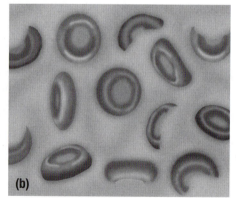

Figure 6 (a) Normal red blood cells (b) The shape of red blood cells in space

Bone Loss

Astronauts lose up to 2 % of their bone mass for each month they spend in microgravity conditions. Thus, a space environment offers the perfect conditions for researchers to study osteoporosis—an Earthly disease that causes a loss of bone mass. One in four women and one in eight men over the age of 50 suffer from osteoporosis. Treatment for osteoporosis costs Canadians more than $600 million annually. By focusing on the accelerated process of bone loss in space, scientists are able to conduct their studies more quickly and cheaply than on Earth.

Canada is a world leader in using space research to understand how bone loss happens in astronauts as well as in people here on Earth. Working with scientists in Ontario, the CSA has developed a unique mini-lab called OSTEO that has flown aboard the Space Shuttle and the ISS (Figure 7).

Figure 7 In 1998, 77-year-old astronaut John Glenn conducted the first OSTEO experiments to determine the rate of bone loss and help understand osteoporosis.

Medical Monitoring

When you are sick, you can go to a doctor and receive medicine to help you get better. Astronauts face routine illnesses and much more serious health issues without direct access to a doctor. For this reason, health care specialists on Earth use electronic communications to monitor an astronaut's heart and brain functions, body temperature, and respiration to anticipate and avoid serious health problems.

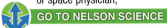

To learn more about being a health care technician or space physician,
GO TO NELSON SCIENCE

Radiation Alert

Earth's atmosphere and magnetic field help protect you from the Sun's radiation. In space, astronauts are not protected by the atmosphere. Protective shields on their space vehicle provide some safeguards from this harmful radiation. Their spacesuits are also specially designed to protect them from some of the incoming harmful radiation during their space walks. Astronauts on missions outside of Earth's magnetic field—such as those to the Moon or farther—have to take extra precautions. They must monitor the Sun's radiation and may have to consider aborting missions when radiation levels are too high.

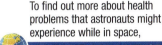

To find out more about health problems that astronauts might experience while in space,
GO TO NELSON SCIENCE

Humans on Mars

Human-occupied spacecraft have landed on the Moon. The next step for future exploration of the Solar System is to safely land humans on the Moon to set up a permanent lunar base and then land on a nearby planet, such as Mars. A mission to Mars will require the cooperation of many governments, businesses, scientists, engineers, and health care professionals. All will work together to ensure the crew's protection from the Sun's radiation and other hazards of space (Figure 8).

Figure 8 This NASA image shows an artist's concept of a possible exploration program on Mars. Astronauts on Mars will probably live in a self-contained structure to protect them from harmful EM radiation and other hazards. Similar structures have already been tested in the Canadian Arctic.

The long trip to Mars will be one of the biggest challenges in space exploration and will be full of potential hazards. The spacecraft must carry enough fuel for the ship plus food and water for the crew during their interplanetary journey. A second robotic craft filled with all the supplies needed for a month-long stay would have to safely land on Mars before the astronauts arrive. The craft design has to be such that it could protect the astronauts from potentially hazardous amounts of solar radiation, especially during solar storms.

Before the astronauts leave Earth, scientists will need to determine how to keep them physically and mentally healthy, in spite of the long-term physical effects of microgravity, the lack of on-site medical care, and the psychological effects of being isolated from Earth while living in a confined area. Despite the many challenges of such a mission, the exploration of Mars will provide us with more clues to our understanding of the Universe, give us opportunities to develop new technologies, and satisfy our desire to explore.

Space Junk

Countries around the world have launched thousands of satellites, space probes, and telescopes into orbit since the 1950s. What happens when they no longer work or a new, improved model is developed? Some of this space technology becomes waste that stays in space.

Some satellites may remain in orbit for hundreds of years or more. As they deteriorate or collide with other objects, they create space debris or **space junk**. Space junk can also be tools or other materials that astronauts accidentally misplace as they work in space (Figure 9). For example, in November 2008, an astronaut working on the ISS let go of a backpack-sized toolbag. The orbiting toolbag can still be seen in the night sky.

space junk debris from artificial objects orbiting Earth

Figure 9 This artwork shows some of the many objects—from functioning satellites to old rocket parts—orbiting Earth. Some space junk or debris that circles Earth can cause re-entry problems for spacecraft.

Space junk travels at speeds up to 40 000 km/h and can damage the hundreds of functioning satellites or the ISS orbiting Earth. Scientists are trying to determine how to get rid of this space junk.

Astronomers use radar to monitor space junk. More than 100 objects have broken up in Earth's orbit, resulting in hundreds of thousands of pieces of debris in orbit, such as the upper stages of launch vehicles, nuclear reactor coolant, metal, nuts, bolts, and paint from deteriorating satellites, telescopes, and probes. Some of this debris lands on Earth. In early 2009, scientists were monitoring a Russian rocket that was entering Earth's atmosphere. Falling debris from the re-entry—10 square metres in size—appeared to be heading for southern Alberta. The city of Calgary was only minutes away from issuing an impact alert. Luckily, the trajectory shifted and the debris broke up over the ocean without incident.

Other space junk becomes dangerous when spacecraft such as the space shuttle are launched, as their trajectory must avoid hitting this floating junk. Astronomers, other scientists, and organizations such as NASA and the CSA are now studying the orbits and effects of space junk in space and on Earth.

DID YOU KNOW?

How Much Junk Is Too Much?
Astronomers are tracking any piece of space junk that is larger than a basketball. Currently, about 18 000 objects are being watched. Astronomers estimate that there may be more than 300 000 objects larger than 1 cm orbiting Earth at approximately 700 to 2000 km above the ground. That is a lot of junk!

To learn more about space debris sightings,

RESEARCH THIS | SPACE TRAVEL SAFETY

SKILLS: Defining the Issue, Researching, Defending a Decision, Communicating

SKILLS HANDBOOK
4.B., 4.C.4.

Nearly 500 humans have been in space, and space traveller numbers are growing rapidly. However, space travel is risky and is still considered experimental. It is not as safe as flying in an airplane. Thirty-two astronauts worldwide have lost their lives in space exploration accidents. Astronauts also experience serious health risks when exploring space.

In this activity, you will research the causes of a space shuttle accident. You will also research the safety requirements and considerations that are now in place for future space travel.

1. Using the Internet, research the missions of the *Challenger* in 1986 or the *Columbia* in 2003. What were the purposes of these missions?
2. Both of these missions ended in disaster. Research the causes of these disasters.
3. Research what safety measures were put in place by space safety engineers after the accident.

GO TO NELSON SCIENCE

A. How could space travel be made safer in the future? T/I
B. What is Canada's contribution to new safety measures for the space shuttle fleet? T/I
C. Could there have been other ways to achieve the goals of the *Challenger* and *Columbia* missions without putting human lives at risk? A
D. Write a paragraph or a list explaining why you think it is (or is not) worth risking people's lives to perform research in outer space. In a second paragraph or list, compare space travel with other explorations in history. C A

UNIT TASK Bookmark

You can apply what you have learned about Canadian space research in microgravity and the physical challenges of space travel when you work on the Unit Task described on page 446.

IN SUMMARY

- It takes an enormous amount of energy to travel in space, and various technologies have been developed to perform this task.
- Astronauts in a microgravity environment orbiting Earth appear to float because they and their spacecraft are in free fall around Earth.
- The microgravity environment causes serious health concerns for astronauts.
- Space debris orbiting Earth can create risks for further exploration.

CHECK YOUR LEARNING

1. Explain how spacecraft are launched into orbit. What are some of the challenges that must be overcome? K/U
2. Identify the type of probe that uses xenon gas for propulsive power. K/U
3. What are the potential dangers of nuclear power for spacecraft? What have scientists done to help reduce the possibility of a nuclear fuel accident? A
4. The next generation of Mars explorers, such as the Mars Science Laboratory, will be able to travel much farther on Mars than the current explorers. Explain why. K/U
5. Describe what scientists mean when they refer to a "microgravity" environment. How might you experience what a microgravity environment feels like? K/U
6. Describe three health problems astronauts experience when they are working aboard the ISS. K/U
7. Write a brief newspaper article (maximum 100 words) about Canada's contribution to osteoporosis research aboard the ISS. K/U C
8. How do astronauts protect themselves against radiation hazards when they are working outside the ISS? K/U
9. What are some of the challenges of sending people to Mars? Is sending humans to Mars necessary? How else could scientists explore the planet without actually going there? A
10. What is space junk, and why do scientists monitor its position in space? In your answer, give four examples of objects considered to be "space junk." K/U

10.3 Space Technology Spinoffs

If countries had never put money into space exploration, would your life be any different? You might be surprised to learn that there are many **spinoffs**, or offshoots, of space technology that you encounter in your everyday life—from the bar codes used to manage foods at your local grocery store to the cordless power tools in your home (Figure 1). In this section, we look at a few of these space technology spinoffs.

spinoff a technology originally designed for a particular purpose, such as space technology, that has made its way into everyday use

Canadarm Robotic Technology

Canada's own Canadarm robotic technology has been used in the design of prosthetic hands and to clean up radioactive waste and other toxic wastes. The space vision technology that controls the robotic movement of the Canadarm has been adapted to minimize the amount of pesticides used when spraying farm crops. Testing is now underway to use the same space robotics to conduct remote robotic surgery (telesurgery) (Figure 2). Recently, researchers at the University of Calgary and at McMaster University conducted experimental projects in which the surgeon directed medical staff from an operating console while conducting robotic surgical operations on patients hundreds of kilometres away in a hospital in North Bay, Ontario.

Figure 1 When *Apollo* mission astronaut-geologists were on the Moon, NASA developed a cordless tool for them to be able to drill core samples from beneath the lunar surface.

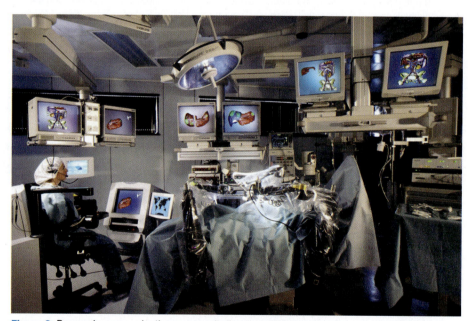

Figure 2 Researchers are adapting space robotic arm technology developed in Canada to create the Neuroarm. This will be a robot that uses miniaturized tools to conduct highly accurate surgeries.

Global Positioning Systems

People commonly use maps and compasses to determine exactly where they are on Earth and to guide them to where they want to go. In 1978, space exploration led to the development of the first satellite-based navigation system, called a global positioning system (GPS). Originally intended for military use, GPS technology was made available to the general public in the late 1980s. Today, a network of 24 GPS satellites orbit Earth twice daily (Figure 3). The system is used to determine latitude and longitude of receivers located anywhere on Earth, in any weather conditions, any time of day.

Figure 3 The wing-like solar panels of a GPS satellite absorb radiation from the Sun, which is then stored in batteries.

Three or more GPS satellites send a signal to a receiver on Earth, which uses triangulation to calculate its location (Figure 4). GPS handheld receivers (Figure 5) and those mounted in vehicles can also help people locate directions to a particular destination. All GPS satellites transmit continuously; a GPS receiver has to detect signals from at least three satellites to obtain an accurate positioning. GPS satellites transmit a code in their signals that indicates the precise time the signal was emitted. This is compared with the clock in the GPS receiver and allows it to reset its own clock so that it matches the time on the satellite's highly accurate atomic clock. The GPS receiver can then make accurate measurements of the distance between it and the GPS satellites.

DID YOU KNOW?

GPS Accuracy

How accurate are today's GPS receivers? Most receivers can locate an object's position anywhere on Earth to within a few metres. Certain high-end engineering and military units can locate objects as small as a few centimetres in diameter, or the size of a dollar coin. These accuracies depend on having a clear overhead sky with no obstructions.

Figure 4 A GPS receiver's position on Earth can be accurately determined by using triangulation, which is the measure of the distance between the receiver and three orbiting GPS satellites.

Figure 5 A handheld GPS receiver

TRY THIS: HOW DOES GPS WORK?

SKILLS: Observing, Evaluating

SKILLS HANDBOOK
3.B.

GPS technology has revolutionized navigation on Earth in the twenty-first century. Anyone with a GPS receiver can now accurately find his or her latitude and longitude anywhere on the planet as long as three GPS satellites can communicate with the receiver. In this activity, you will perform a simple exercise that shows how satellites are able to pinpoint the exact location of a person on the surface of Earth. You will also investigate the practical applications of this space technology.

Equipment and Materials: map of Ontario from your teacher; geometry compass

1. Look at the map of Ontario your teacher has given you. Imagine you are travelling in Ontario and have become lost.
2. The first satellite tells you that you are 340 km from Toronto. Looking at the scale on the map, determine how far that is on the map.
3. With a compass, draw a circle around Toronto that represents a radius of 340 km. The information from the first satellite tells you that you are somewhere along this circle.
4. A second satellite gives you the information that you are 800 km from Thunder Bay.
5. Draw a circle around Thunder Bay that represents a radius of 800 km. There are two points where the circles intersect.
6. A third satellite tells you that your location is 300 km from Timmins.
7. Draw a circle around Timmins that represents a radius of 300 km.
8. Determine your location.
A. Explain why you need a minimum of three satellites to determine your location using GPS technology. T/I
B. What kind of signal does the GPS satellite beam down to the receiver, and what information does it contain? K/U T/I
C. What determines the level of accuracy of locating an object on the ground? K/U T/I
D. If you had a GPS receiver on a cellphone, what might you use it for? A

RADARSAT

RADARSAT is Canada's first series of remote sensing Earth observation satellites. RADARSAT-1 has recently released the most detailed map of Antarctica ever created (Figure 6). From the RADARSAT map, scientists have been able to better study this mysterious continent, including information about how ancient ice-shelves are crumbling. RADARSAT-2 is a commercial radar satellite that resulted from a unique collaboration between government and industry. The satellite is used globally for marine surveillance, ice monitoring, disaster management, environmental monitoring, and resource management and mapping.

To learn more about RADARSAT,
GO TO NELSON SCIENCE

Figure 6 A high-resolution RADARSAT image of Antarctica. Earth's southernmost continent is so cold and inhospitable that much of it remains unexplored. From space, though, it is possible to map this entire region by radar by systematically noting how long it takes for radio waves to reflect off the terrain.

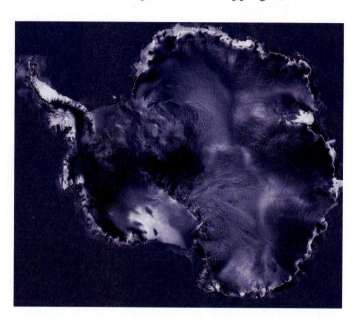

> **DID YOU KNOW?**
>
> **UNISPACE**
>
> UNISPACE is the United Nations Conference on the Exploration and Peaceful Uses of Outer Space. UNISPACE is committed to using space technology to assist any country in the world during a major disaster. For example, during the flooding of the Red River in southern Manitoba in 2009, scientists used RADARSAT images to track the flooded area and collect data for future flood management. RADARSAT also monitored the flooding of the Yangtze River in China in 1998 and the rising waters in and around New Orleans caused by Hurricane Katrina in 2005.

SCISAT

Another Canadian satellite used to monitor Earth's environment is SCISAT. It was launched in 2003 to monitor the state of Earth's atmosphere over the Arctic. Canadian and international scientists are concerned about the ozone layer, which provides Earth with some protection from the Sun's harmful ultraviolet rays. Scientists are learning more about changes in our atmosphere as SCISAT measures how much ozone is in Earth's atmosphere and how much it is being reduced due to human-produced pollution.

Health Benefits

Some of the investments in space research and exploration by governments, companies, and universities result in technologies that can be used to monitor human health. For example, home blood pressure testing devices provide us with an immediate measurement of our blood pressure (Figure 7). This everyday technology is a spinoff of the device created by NASA to measure the blood pressure of astronauts during launches, when they are exposed to powerful forces and vibrations.

In sports medicine today, an athlete's internal body temperature can be monitored easily with an electronic thermometer pill that he or she swallows. The pill helps physicians know when there is danger of exhaustion or heatstroke while playing sports. This tiny device was first developed in the

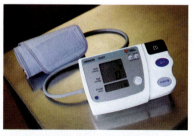

Figure 7 The portable device used to monitor blood pressure is called a sphygmomanometer.

early 1980s by NASA to check the body temperatures of astronauts during spacewalks so they would not overexert themselves.

These are just two examples of the many inventions created for space exploration that provide effective ways to monitor people's health.

The Environment

The microgravity environment on the ISS is used to investigate various factors related to the environment (Figure 8). One such investigation examines how the energy industry can achieve more environmentally sound oil exploration. Researchers have sent samples of crude oil into space to figure out how oil behaves below Earth's surface. The results provide geologists with a better idea of the size of underground oil reserves. This information helps oil companies drill into them more accurately while minimizing the disturbance of the surrounding environment.

> **READING TIP**
>
> **Explaining Relationships**
> While reading a text you might recall something you read or viewed in another text. For example, you may have viewed a documentary about Colony Collapse Disorder, in which massive numbers of honeybee colonies are dying off around the world. You can combine this fact with the information in the text and conclude that beeswax may not be available in the future in large enough quantities to use for oil spills in the oceans.

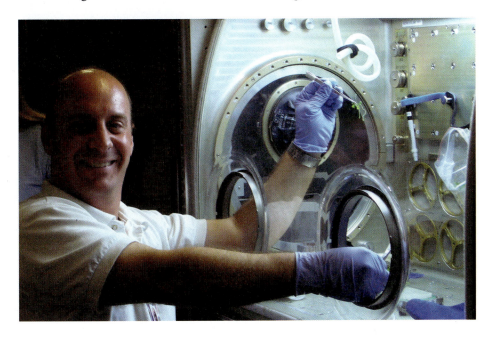

Figure 8 Astronaut Garrett Reisman examines the effects of microgravity on plant cell walls.

Our oceans and lakes are better protected from toxic oil spills thanks to a special technology developed through research carried out in space. In this technology, millions of tiny balls of beeswax are dumped into water polluted by an oil spill. The wax attracts the oil, which becomes trapped inside the balls. Instead of the oil destroying fragile ecosystems in and around the ocean or lake, it is contained and easier to clean up.

Improved Consumer Goods

Some technologies, such as personal computers, high-tech running shoes, dehydrated foods, and water purifiers, were developed for space exploration but are available to all today. Other space technologies have been adapted for use in consumer goods that have wider applications. Some examples are listed below:

- Aerodynamic technology has led to golfballs that can soar through the air faster and with greater accuracy.
- Fire-resistant fabrics were developed for spacesuits but have been adapted for use in suits for racecar drivers and hazardous material handlers (Figure 9).

Figure 9 NASA technology used in spacesuits worn by astronauts during spacewalks is being used to design fire-retardant suits worn by racecar drivers.

- Space technology created shock-absorbing helmets that have been adapted for use in sports or certain work environments (Figure 10).

Figure 10 Foam that NASA developed for aircraft seats is used to make protective helmets.

UNIT TASK Bookmark

You can apply what you have learned about everyday applications of space research to your work on the Unit Task described on page 446.

IN SUMMARY

- Technologies originally designed for space exploration have made their way into everyday use.
- GPS technology, common in many cellphones and vehicles, helps people navigate and determine their location.
- Canadian technologies such as RADARSAT and the Canadarm have led to the development of sophisticated applications in everyday life.
- Benefits to health and the environment have come about due to space technology.

CHECK YOUR LEARNING

1. What is the spinoff of space technology shown in Figure 11?
2. Write a paragraph as if you were posting information on a blog, explaining the relationship between the Canadarm and the new field of robotic surgery.
3. How does the global positioning system (GPS) work?
4. Why is GPS a valuable technology?
5. What are five geographical features of Earth that are monitored by RADARSAT? Why are they being monitored?
6. Which layer of Earth's atmosphere do scientists study with the SCISAT satellite? What purpose does it serve?

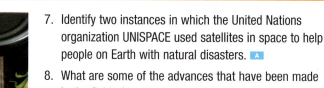

Figure 11

7. Identify two instances in which the United Nations organization UNISPACE used satellites in space to help people on Earth with natural disasters.
8. What are some of the advances that have been made in the field of health sciences due to space science technology?
9. Identify two ways in which technologies invented for space exploration have benefited the study of Earth's fragile environment.
10. Describe three objects people might find in their homes that use technology originally developed for the space program. Add one sentence of description for each object, explaining its origin.
11. Describe two ways in which experiments in space have led to more environmentally friendly means of extracting oil and cleaning up oil spills.

SCIENCE WORKS ✓ OSSLT

Canadian Weather Station on Mars

In May 2008, after a 10 month, 700 million km journey, the robotic spacecraft named *Phoenix* finally landed on Mars. This latest mission to the Red Planet was the first to explore the high Arctic region of Mars (Figure 1) and the first time in history that Canadian-built science instruments touched down on another planet.

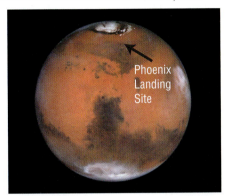

Figure 1 By studying the atmosphere on Mars, scientists hope to learn more about our own atmosphere and the way it affects climate.

Landing during springtime, *Phoenix* explored a frigid, open plain located about 70 degrees latitude on the edge of the permanently frozen North Polar Cap. On Earth, the equivalent latitude runs through the northernmost parts of the Yukon, the Northwest Territories, and Nunavut.

Powered solely by solar arrays and batteries, *Phoenix* lasted 5 months before the dark, cold winter set in, ending the mission. Although the little spacecraft's lifetime was short, it put Canadian scientists and engineers at the forefront of discovery as they led the first study of Mars's polar climate.

One of the main goals of *Phoenix* was to search for water and understand daily weather and seasonal climate patterns on Mars. Scientists believe that any new finding made by this robotic probe will also directly benefit our ability to recognize how Earth's climate works.

Scientists think that millions of years ago, Mars may have been much more Earth-like, with a thicker atmosphere and warmer weather than it has today, as well as large bodies of liquid water on its surface. *Phoenix* gathered data that may explain what caused the Martian environment to change to a cold, dry world. The climate data gathered by *Phoenix* may help meteorologists create computer models to predict Earth's climate change.

A team made up of scientists from Canadian universities, companies, and the CSA developed and built a $37 million weather centre that sits on the tabletop deck of *Phoenix*. It is called the meteorological station (MET). Throughout the mission, MET recorded daily weather at the landing site, using temperature, wind, and pressure sensors, as well as a unique laser instrument called a light detection and ranging instrument, or LIDAR.

The LIDAR laser is a bright green laser beam that is used by a world-wide network of observatories to detect light and measure its distance. In September, 2008, NASA's Phoenix Lander used LIDAR to detect snow in the atmosphere of Mars. By sending pulses of laser light into Mars' atmosphere, LIDAR was able to determine the composition, movement, and size of particles in the atmosphere. The snow never reaches the ground because it sublimates as it falls though the atmosphere.

The LIDAR does not rely on visible light to analyze the environment, and it makes more accurate and higher resolution measurements than optical instruments. As a result, LIDAR technology is already being used to study Earth (Figure 2). LIDAR provides meteorologists with a better understanding of storm clouds and their moisture content and gives forest conservationists a clearer picture of the health of tropical rain forests around the world. Researchers are hoping that Canadian LIDAR technology will be used in other planetary missions. Space agencies are now looking at using this laser technology as part of a guidance and navigation system for future spacecraft.

Figure 2 A LIDAR instrument at Davis station in Antarctica is used to study the middle atmosphere. The aurora australis is also visible in the sky.

It will take years for scientists to understand the many gigabytes of data from *Phoenix*. However, the information it has provided will improve our understanding of how the Martian climate works and help us plan upcoming robotic and possible human exploration missions. Understanding why Mars is the way it is will also help atmospheric scientists studying our own planet. This knowledge will benefit everyone on Earth.

10.4 EXPLORE AN ISSUE CRITICALLY

Should Canadians Be Paying for Space Research?

Is there life on Mars? Scientists have been researching this question for decades. In 2008, Canadian researchers contributed to a successful mission to Mars called *Mars Phoenix*. This space probe was designed to hunt for signs of water and the presence of microbial life.

Canadian scientists were responsible for designing and building *Phoenix's* meteorological station (MET), dedicated to recording the daily weather on Mars (Figure 1). This information helps us understand how the surface of Mars has changed over time. Scientists at York University in Toronto oversaw the development of the instruments, designed to measure the wind, temperature, pressure, and atmospheric conditions on Mars. The CSA contributed funds for the project, and private Canadian companies, such as Optech, and MacDonald, Dettwiler and Associates, built the MET.

Since the CSA is a government organization, it paid for the mission with money collected from Canadian taxpayers. In total, the CSA spent $37 million on the MET.

Mars Phoenix was launched on August 4, 2007, and landed near the Martian North Pole on May 25, 2008. The probe collected data on the surface of Mars for over five months before powering down. This was much longer than mission scientists predicted. The Canadian-built MET was successful in creating the first profile of an extraterrestrial atmosphere. *Mars Phoenix* confirmed that Mars has ice just below the surface, a sign that life may have existed in the past.

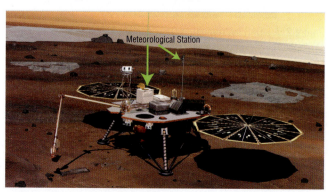

Figure 1 The meteorological station on the *Mars Phoenix* lander was paid for and created by the CSA. Was it worth the money?

The Issue

Because of our strong space program and a long history of Canadian discoveries in astronomy, Canada is known to scientists as one of the best countries in which to perform research in astronomy and space science. However, to remain a world leader, the Canadian government must continue to spend money on research and development. Some people believe that it would be better to spend taxpayer money on other things, such as health care and education.

As a consultant hired by the Canadian government, you must research the benefits and drawbacks of spending money on space exploration projects such as the *Mars Phoenix* MET. You will report back with a decision on whether to continue investing money in such programs.

Goal

To create a pamphlet, website, written report, or electronic presentation designed to summarize your findings on the benefits and drawbacks of spending money on space science research in Canada.

Gather Information

SKILLS HANDBOOK
4.A., 4.B.

You will need to research the ways the CSA has partnered with Canadian companies, government agencies, and universities for different space projects. You can focus on the recent success of the CSA in exploring Mars or identify other projects Canadian scientists are involved in, such as the following:

- RADARSAT-1 or -2
- MOST
- DEXTRE
- SCISAT-1 (Figure 2 on the next page)
- Canadarm
- International Space Station

 GO TO NELSON SCIENCE

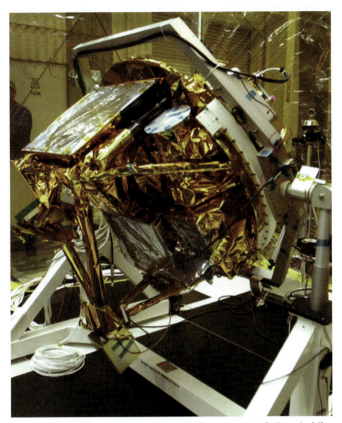

Figure 2 SCISAT measures and studies the processes that control the distribution of ozone in Earth's atmosphere.

Before you decide on a recommendation for the Canadian government, you need to research more about the other financial responsibilities the Canadian government has.

- Where else could the government spend the money they would normally spend on space science research?
- What other areas do people think the government should spend money on?

In addition, you might want to conduct a survey to learn what the public's opinion is on the issue of funding for space research. You may consult your classmates, teachers, or family members.

Identify a Solution

For the project you have chosen, answer the following questions:

- How has the project been a benefit to the field of space exploration?
- How has the project been a benefit to society in general?
- What negative effects has the project had on society, and could there be any negative effects in the future?
- How much has been spent by the CSA on this type of research?
- In what ways does the government stand to make money from its investment in Canadian space science research?
- What kinds of costs might there be for investing in future research projects like this?
- What benefits might there be for investing in future research projects like this?

Make a Decision

Decide whether you will recommend to the Canadian government to continue funding the CSA at current levels or to increase or decrease the amount of money spent on space research.

Communicate

Create your presentation and share it with your classmates in small groups. Some of your classmates might have different opinions concerning your recommendation to the government. Consider their points of view and listen to their arguments carefully. Your teacher might set up a formal debate so you can listen to all the issues.

When you present your recommendation, be sure to support your position using facts. If you are presenting the results of a survey, make sure you present your findings in an appropriate format, such as a graph. T/I C A

10.5 The Future of Space Exploration

Space exploration is relatively new. Before the 1950s, information about space was obtained only through telescopes. However, in the past 50 years, humans have visited the Moon, sent robots to distant planets, invented the first space plane, and lived and worked in orbit aboard space stations.

What do you think will be the next steps for space exploration? The next two decades may see human expeditions going out into the Solar System. Can you imagine humans establishing a colony on the Moon or landing on Mars (Figure 1)? Will we have faster, cheaper ways to travel in space? Will people vacation on other planets? These are only a few of the possible innovations in space exploration.

Figure 1 Mars will be a prime destination for future astronauts.

After the Space Shuttle

The space shuttle is the main vehicle used by astronauts to go into space. Flying since 1981, all three remaining shuttles are scheduled for retirement in 2010 after they haul the last component up to the International Space Station to finish its construction.

To replace the space shuttle system, NASA is in the early stages of developing a pair of new reusable launch vehicles. NASA's *Ares* rockets, named for the Greek god associated with Mars, will return humans to the moon and later take them to Mars and other destinations. The crew launch vehicle will be a smaller rocket called *Ares I* (Figure 2(a)). The cargo launch vehicle will be a large rocket booster called *Ares V* (Figure 2(b)). *Ares V* will carry many tonnes of cargo into low Earth orbit using the old space shuttle launcher design: fuel tank and solid rocket boosters. Both new launch rockets are going to be based on parts of the current space shuttle launch vehicle. Once the boosters have lifted their cargo into orbit, they will re-enter the atmosphere and use parachutes to land in the ocean for collection. Sitting atop the *Ares I* booster will be a crew capsule called *Orion* (Figure 2(c)). The first human mission aboard the *Orion–Ares* launch system is scheduled for 2014, in preparation for the first mission to the Moon since the *Apollo* program in the late 1960s.

> **LEARNING TIP**
>
> **Ares Series**
>
> *Ares* is named after the Greek god of war, the god associated with the planet Mars. The *Ares* spacecraft is designed to take humans beyond Earth orbit, hopefully all the way to Mars and back.

(a)

(b)

(c)

Figure 2 (a) The *Ares I* crew launcher is designed to carry humans and cargo into space. (b) The *Ares V* will lift heavy cargo into low Earth orbit. (c) The *Orion* crew exploration vehicle is capable of carrying as many as six astronauts.

The Moon: A Testing Ground for Mars Exploration

Some scientists are planning to send a human-occupied spacecraft to Mars to explore its surface. They also intend to set up a base station there, where astronauts can live for extended periods of time. To reach this goal, astronauts will first travel to the Moon, where they will live and train for their eventual mission to Mars. Current plans call for humans to return to the surface of the Moon by 2020. Crews of three to six will stay as long as two weeks, conducting experiments and testing technologies in preparation for the longer and more dangerous journey to Mars (Figure 3).

> **READING TIP**
>
> **Using Graphs and Illustrations**
> Graphs, tables, and images can help you to visualize the main idea of a text. For example, the images of the *Ares* boosters and *Orion* spacecraft can help you synthesize what you are reading with what you have already learned about the space shuttle, *Apollo*, and the *Saturn V* rocket. This synthesis will help you understand how future space missions will differ from past missions.

Figure 3 Long-term plans call for building permanently occupied international science bases on the Moon, as is done in Antarctica.

The Moon has been chosen as a testing ground for future Mars exploration because it is closer to Earth and has characteristics similar to those of Mars. For example, both the Moon and Mars are much colder than Earth and are blanketed by a fine layer of silt-like dust particles. They also both have less gravity than Earth does. However, the Moon can be reached by spacecraft in less than four days, whereas travelling to Mars takes six to ten months. Sending equipment and people to Mars is far more time consuming and risky than a trip to the Moon.

Challenges and Dangers of Exploring the Moon and Mars

The fine layer of dust blanketing the Moon and Mars may be 10 m deep in places. This presents challenges for exploration. Because the dust is so fine, it can filter into the living spaces of the astronauts and irritate their skin, eyes, and lungs. It can also damage scientific equipment and spacesuits. Landing rockets will stir up the dust, creating clouds of dust on the surface of the planet or the Moon, and surface vehicles will need special tires that do not sink into the deep layers of dust.

Other challenges for humans colonizing the Moon and Mars include the difficulties in obtaining a constant supply of water, oxygen, and food.

Figure 4 U.S. billionaire Dennis Tito received technical and safety training as an astronaut for six months to prepare for his ride into space.

Protection from the cold and the Sun's ultraviolet and X-ray radiation will also be necessary. Heated houses might melt the surface on which they are built, causing the ground to collapse. These types of potential dangers will be checked first by robotic explorers. Plans are already underway to send robots on reconnaissance missions to both the Moon and Mars over the course of the next decade to lay the foundation for the safest possible human missions.

Space Tourism

Travelling beyond Earth's gravity and into space has captured the imagination of generations of people. However, until recently, only a select few—specialized military or highly trained astronauts—have had a chance to travel into space. In 2001, U.S. businessman Dennis Tito became the world's first space tourist when he blasted into orbit aboard a Russian rocket (Figure 4). He stayed aboard the ISS for one week. As of spring 2009, six individuals who are not professional astronauts have paid to go into space. The 2009 price tag for a week-long trip into space was around $30 million.

Although space tourism opportunities today are limited and expensive, companies in Canada, the United States, and England are working on building and flying low-cost rockets to take tourists on a more affordable, 2.5 hour suborbital trip (Figure 5). Passengers will be able to experience free fall for a few minutes when they fly to an altitude of 100 km—the official boundary between Earth's atmosphere and outer space. The space flight ends with the vehicle gliding back down to Earth.

Figure 5 British company Virgin Galactic has already flown a prototype of its suborbital space plane and plans to fly its first space tourists in 2010.

The Space Elevator

Scientists in Canada and around the world are competing to develop the first elevator to space. Elevators in buildings on Earth consist of

- a compartment or car for people or cargo
- a steel cable or tether anchored to the bottom and top of an elevator shaft
- a counterweight

An ordinary elevator car is powered by electricity. It is raised and lowered by traction. The counterweight, which weighs about the same as the car, makes it possible to raise and lower the elevator using less energy.

Similarly, the design of a space elevator consists of a very long cable or tether anchored from a location on Earth's surface and extending 36 000 km into space (Figure 6).

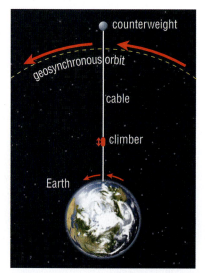

Figure 6 Concepts call for the elevator to rise vertically along a cable to Earth orbit.

436   Chapter 10 • Space Research and Exploration

This tether will be anchored by an ocean ship and attached to a counterweight in space (Figure 7(a)). Earth's rotation and the counterweight in space will keep the cable taut. (The counterweight is always pulling upward on the ocean ship, much like a pail of water at the end of a rope pulls against your hand if you whirl it around in a circle.) The space elevator's car is called a climber. Powered by laser or light energy, the climber is used to move cargo—satellites or people—to Earth orbit. Once the cargo reaches orbit, it can be launched beyond Earth (Figure 7(b)).

The benefits of space elevators are many. The propulsion technology used today to launch spacecraft is expensive. The estimated cost to build a space elevator is $10 billion. This is much less than the cost of the ISS and space shuttle systems. The space elevator is operated from Earth and not from space, which may also lower costs.

However, the technology to make a fully functional space elevator does not yet exist. Currently, no material is strong enough and light enough to make the cable stretch from Earth into space. Scientists have also not figured out how to provide the climber with enough energy to climb up the cable against Earth's gravity all the way to space.

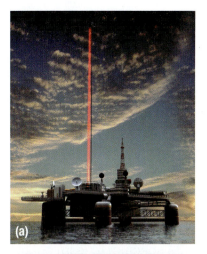

Figure 7 (a) Current designs call for a ship to anchor the space elevator tether to Earth and house the laser to power it. (b) The elevator rises vertically to Earth orbit with its cargo and people.

UNIT TASK Bookmark

You can apply what you have learned about space tourism and the challenges and dangers of space exploration to the Unit Task described on page 446.

IN SUMMARY

- New vehicles and methods are being designed to shuttle people and materials into space.
- Space travel is dangerous and very expensive.
- Tourists have begun to travel into space for pleasure and might continue to do so in the future.

CHECK YOUR LEARNING

1. Compare and contrast the *Ares V* spacecraft with the *Ares I* spacecraft. K/U
2. Why is the Moon a suitable testing ground for Mars exploration? T/I
3. Describe NASA's plans to replace the space shuttle. K/U
4. There are many challenges to space exploration. Rank the following criteria from most important (1) to least important (7). Explain your reasoning in two to three sentences. T/I
   - constant supply of water
   - protection from dust storms on Mars
   - constant supply of oxygen
   - constant supply of food
   - dealing with feelings of isolation and claustrophobia
   - protection from the cold
   - protection from the Sun's radiation
5. How long does it take to get to the Moon compared with the time it takes to get to Mars? K/U
6. Why does the dust on the Moon and Mars make colonization and exploration more difficult? K/U A
7. Who was the first person to travel into space as a tourist, and how much did he pay for this privilege? K/U
8. Summarize the future plans to create journeys into space for tourists. K/U
9. What are the components of the proposed space elevator? Explain the purpose of the elevator with the use of a diagram. K/U C
10. What are the benefits of space elevators? What are the drawbacks of space elevators? K/U A

CHAPTER 10

LOOKING BACK

KEY CONCEPTS SUMMARY

Humans use telescopes that gather energy from the entire EM spectrum to learn about objects in the Universe.

- Galileo first used a refracting telescope to gather and focus light 400 years ago. (10.1)
- Astronomers use radio telescopes to collect radio waves emitted from distant stars and galaxies. (10.1)
- High-energy EM radiation, such as gamma rays and X-rays, are being collected by the Chandra, INTEGRAL, and SWIFT space telescopes. (10.1)

Humans use space probes to learn more about the Solar System.

- Rovers have been exploring the surface of Mars and sending back information on Martian climate and geology since 2004. (10.1)
- The *Cassini* spacecraft reached Saturn in 2004. (10.1)
- Space probes have travelled to all of the planets in the Solar System, giving us valuable information about their composition. (10.1)

Canada is a world leader in developing technology for space exploration.

- Canada is partnering with many countries on projects such as the *Mars Phoenix* meteorological station, the ISS, the Square Kilometre Array telescope, and others. (10.1, 10.4)
- The Canadarm2 and DEXTRE are used on the space shuttle and the ISS to move things around in space. (10.1)
- RADARSAT-1 and RADARSAT-2 satellites track features and climate change on Earth. (10.3)

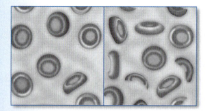

The environment in space is hostile to living things.

- Astronauts in the ISS exercise to keep their bones and muscles strong. (10.2)
- Scientists are always working on ways to protect astronauts from radiation, cold temperatures, and hostile environments in space. (10.2, 10.5)
- Humans need oxygen, food, and water to survive in space, so they must either take them with them or manufacture them on the way. (10.5)
- When astronauts live and work in space, they are in continuous free fall around Earth. (10.2)

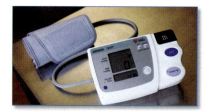

New technologies and information from the space program have benefited our everyday lives.

- Space technologies have led to advances in medical science. (10.3)
- Consumer products have been improved with the aid of space program technologies. (10.3)
- GPS satellites are used in conjunction with cellphones, computers, GPS receivers in vehicles, and many other applications. (10.3)
- Weather satellites are continuously monitoring Earth's environment. (10.3)

Space exploration is technically challenging and expensive.

- NASA, the CSA, and the European Space Agency have spent billions of dollars on space exploration. (10.1, 10.4, 10.5)
- Spacecraft designers must address many challenges to safely take humans to their celestial destinations and back. (10.2)
- Any new spacecraft launches must be calculated to avoid the thousands of pieces of space debris orbiting Earth. (10.2)

WHAT DO YOU THINK NOW?

You thought about the following statements at the beginning of the chapter. You may have encountered these ideas in school, at home, or in the world around you. Consider them again and decide whether you agree or disagree with each one.

Vocabulary

refracting telescope (p. 410)
reflecting telescope (p. 410)
spacecraft (p. 411)
space probe (p. 415)
microgravity environment (p. 421)
space junk (p. 424)
spinoff (p. 426)

1 Sending humans into space is important for research and discovery.
Agree/disagree?

4 Spending billions of dollars on robots that explore the surface of Mars is worthwhile.
Agree/disagree?

2 All celestial objects can be viewed from Earth using a telescope.
Agree/disagree?

5 Technology developed for space exploration has many practical uses here on Earth.
Agree/disagree?

3 Space exploration encourages international cooperation and understanding.
Agree/disagree?

6 Astronauts "float" in space because there is no gravity.
Agree/disagree?

How have your answers changed since then? What new understanding do you have?

BIG Ideas

- Different types of celestial objects in the Solar System and Universe have distinct properties that can be investigated and quantified.
- People use observational evidence of the properties of the Solar System and the Universe to develop theories to explain their formation and evolution.
- ✓ Space exploration has generated valuable knowledge but at enormous cost.

CHAPTER 10 REVIEW

The following icons indicate the Achievement Chart category addressed by each question.

K/U Knowledge/Understanding   **T/I** Thinking/Investigation
C Communication   **A** Application

What Do You Remember?

1. Who was the first person to look up into the night sky with a telescope, and in what year did he do this? (10.1) **K/U**

2. What happened to the two space shuttles *Columbia* and *Challenger*? (10.1) **K/U**

3. Name four planets that scientists have sent space probes to. (10.1) **K/U**

4. What is "space junk," and why is it problematic for astronauts? (10.2) **K/U**

5. Describe the theory of a space elevator and identify two obstacles to building it. (10.5) **K/U**

6. What do the letters GPS stand for, and what does this technology do? (10.3) **K/U**

7. In what ways is the Canadian satellite RADARSAT useful to people on Earth? (10.3) **A**

8. In the 1960s, what was the goal of the *Apollo* space program? Was it successful in meeting that goal? (10.1) **K/U**

9. What are the names of the two spacecraft designed by NASA to replace the space shuttle fleet in 2010? (10.5) **K/U**

What Do You Understand?

10. One of Canada's greatest contributions to research in outer space is studying the disease osteoporosis. What does this disease affect in the human body, and why is outer space a good place to study it? (10.2, 10.3) **K/U** **A**

11. Give an example of something in your home that was created by research in outer space. (10.3) **A**

12. Sketch a diagram of a satellite in orbit around Earth in free fall. Label it with the following words: force of gravity, speed, orbital path, satellite, Earth. (10.1) **K/U** **C**

13. Explain the difference between the *Saturn V* rockets that took the *Apollo* astronauts to the Moon and the space shuttle currently in use. (10.1) **K/U**

14. Why do astronauts' muscles and bones weaken when they live aboard the ISS? (10.2) **K/U** **A**

15. Explain the advantages and disadvantages of using nuclear power to produce electricity to power the instruments on spacecraft. (10.2) **A**

16. Gamma-ray telescopes can collect electromagnetic radiation that human beings cannot see. Give two examples of gamma-ray telescopes and explain what they are used for. (10.1) **K/U**

17. Compare the results of viewing a distant celestial object with a visual telescope with those of an X-ray telescope. (10.1) **K/U**

18. In what ways have explorations of outer space helped tackle Earth's environmental problems? (10.3) **A**

19. Match the list of astronauts on the left with their contribution to space exploration on the right. (10.1) **K/U**

    (a) Julie Payette   (i) was the first Canadian in space
    (b) Chris Hadfield   (ii) was the first Canadian to operate Canadarm
    (c) Marc Garneau   (iii) stayed on the ISS with six other astronauts
    (d) Robert Thirsk   (iv) helped build the ISS

Solve a Problem

20. Imagine you are hired to clean up a toxic waste spill in northern Ontario. Explain why you would want some of the engineers who designed the Canadarm to design a device to help with the cleanup effort. (10.1, 10.3) **A**

21. What is the difference between "space tourism" and "space exploration"? How are the motivations behind each different? How are they similar? (10.1, 10.5) **A**

22. After a particularly heavy year of rain, the river in your city is overflowing. How could space technology be used to track the damage to the surrounding area? Why would this knowledge be useful? (10.3) **A**

23. NASA is preparing for its next mission to Mars and needs to find a place on Earth to train astronauts who are going to be working on the Martian surface. Suggest a place that might be similar to Mars and explain your reasoning. (10.1, 10.5) T/I A

24. Imagine you are a scientist planning a future mission to the planet Mars. Suggest reasons why you would want to use the Moon as a base for launching your rocket to Mars. (10.5) T/I

25. Suppose there is a young protostar hidden in a certain gas cloud in our galaxy. Suggest a way that we could gather data from the star without relying on visible light. (10.1) T/I

Create and Evaluate

26. Write a letter to a family member as if you were a space tourist visiting the Moon for a couple of days. Describe your surroundings and how being on the Moon makes you feel. (10.5) K/U C

27. Write an e-mail to a friend as if you were a scientist living for the past six months in the first settlement on Mars (Figure 1). Describe the settlement on Mars and what the planet is like. (10.1, 10.5) K/U C

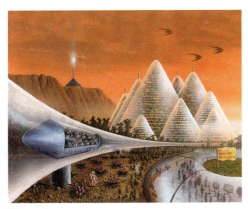

Figure 1

28. Sending human beings into space is much more expensive than sending robots into space. Debate the pros and cons of human space travel with your classmates. Be sure to back up your opinion with facts. (10.1, 10.2, 10.3, 10.4 10.5) C A

29. Write a letter to one of the astronauts mentioned in this section asking some specific questions about what his or her space experience was like. (10.1) C

30. UNISPACE is devoted to using space technology for peaceful means for the benefit of all humanity. Brainstorm ways in which space technology could be used to help people and encourage cooperation between people on planet Earth. (10.3) A

31. Suggest an experiment to be performed on the next space shuttle mission into space. Be sure to explain why you think this experiment is important and what you hope to achieve. (10.1, 10.3) C A

Reflect On Your Learning

32. What information from this chapter did you already know before you studied the content? How has your knowledge of this topic changed in the last couple of weeks? C

33. Name something that you learned in this chapter that changed the way you think about space or space exploration. Explain how it changed your thinking or understanding. C

34. Which topic in this chapter did you find particularly difficult to understand? Come up with two strategies to help you understand this topic before the next test or exam. T/I

Web Connections

35. The Mars Society is an organization that supports the idea of human beings living and working on Mars. Their mandate is "to further the goal of the exploration and settlement of the Red Planet." Do you think humans should live on Mars? Support your decision with research. T/I A

36. Research Galileo's contributions to developing telescope technology. How is the work that Galileo performed a product of the country and time period in which he lived? (10.1) T/I A

37. Space suits are a marvel of modern technology and are very carefully and intricately designed. Research the characteristics and design of space suits and how much they cost to develop. Summarize your findings in a brochure or poster. (10.2) T/I A C

CHAPTER 10 SELF-QUIZ

The following icons indicate the Achievement Chart category addressed by each question. | **K/U** Knowledge/Understanding | **T/I** Thinking/Investigation | **C** Communication | **A** Application

For each question, select the best answer from the four alternatives.

1. Which statement accurately describes the relationship between a spacecraft and a space probe? (10.1) K/U
   (a) A spacecraft is a reusable space probe.
   (b) A spacecraft carries cargo on a space probe.
   (c) A space probe is a type of robotic spacecraft.
   (d) A space probe is a human-occupied spacecraft.

2. The Canadarm, Canadarm2, and DEXTRE are
   (a) robotic tools used on space shuttle missions and the ISS
   (b) space probes used to explore the Solar System
   (c) satellite-based navigation systems
   (d) Canadian-built space shuttles (10.1) K/U

3. During which of the following missions did the first humans walk on the moon? (10.1) K/U
   (a) *Mariner 4*
   (b) *Saturn V*
   (c) *Apollo 11*
   (d) *Apollo 13*

Indicate whether each of the statements is TRUE or FALSE. If you think the statement is false, rewrite it to make it true.

4. Galileo Galilei invented the first telescope. (10.1) K/U

5. Many technologies originally designed for space have made their way into everyday use. (10.3) K/U

6. In 1997, the NASA *Pathfinder* was used to explore Saturn. (10.1) K/U

Copy each of the following statements into your notebook. Fill in the blanks with a word or phrase that correctly completes the sentence.

7. In a _____ environment, objects behave as though there is very little gravity affecting them. (10.2) K/U

8. A telescope that uses mirrors to gather and focus light is a _____ telescope. (10.1) K/U

9. _____ are human-occupied or robotic vehicles used to explore space or celestial objects. (10.1) K/U

Match each term on the left with the appropriate definition on the right.

10. (a) refracting telescope
    (b) radio telescope
    (c) X-ray telescope
    (d) gamma ray telescope

    (i) forms an image from the visible light spectrum
    (ii) orbits Earth and detects rays that are not detectable from Earth's surface
    (iii) detects the highest-energy EM waves
    (iv) collects waves from space and transmits them to a receiver, which amplifies the signal (10.1) K/U

Write a short answer to each of these questions.

11. Why is the moon being used as a testing ground for Mars exploration? (10.1) K/U

12. List three areas of scientific research that contribute to or benefit from space research and exploration. (10.1, 10.2, 10.3, 10.5) K/U

13. Name one Canadian astronaut and describe his or her contribution to space exploration. (10.1, 10.2) K/U

14. Why do astronauts aboard the International Space Station appear to float? (10.1, 10.2) K/U

15. Describe how the study of space differs from other areas of scientific research in terms of observation methods. (10.1) T/I

16. Many countries compete with each other to develop new space technologies. However, space exploration also provides opportunities for countries to work together. Provide evidence to show that space exploration can encourage cooperation among nations. (10.1, 10.2) T/I

17. A cost-benefit analysis is a technique used to determine if the benefits that could arise from completing a task outweigh the costs of the task. (10.1, 10.2, 10.3, 10.4) K/U A
    (a) Name two of the economic benefits of space exploration.
    (b) Name two of the costs associated with space exploration.
    (c) Do the economic benefits of space exploration outweigh the costs? Justify your answer.

18. Which of the following is the most important possible outcome of space exploration? Write a paragraph for your school newspaper justifying your answer. (10.1, 10.2, 10.4, 10.5) C
    (a) developing a space tourism industry
    (b) finding other forms of life
    (c) establishing a human civilization on another planet

19. (a) List three problems scientists face when living in space. (10.1, 10.2) K/U T/I
    (b) Choose one problem that you have listed and describe the solution scientists have developed for the problem.

20. Explain how satellite technology can help people in each of the following situations. (10.3) A
    (a) A family wants to get directions to a vacation destination.
    (b) An ecologist wants to collect data on various Canadian coastlines.
    (c) A scientist wants to monitor changes in the ozone layer.

21. Imagine you are an astronaut living on the International Space Station. Write a brief journal entry describing your daily activities. (10.1, 10.2) C

22. In the next 15 years, NASA plans to build an outpost on the Moon where scientists will conduct research. Some people see this as the first steps toward lunar colonization. Name three challenges that need to be addressed before humans can build colonies on the Moon. (10.2, 10.5) A

23. Find evidence to support the statement "Space travel has generated valuable knowledge, but at an enormous cost." (10.1, 10.2, 10.4, 10.5) T/I

24. Canada is a world leader in developing technology for space exploration and research from space. In a paragraph, describe Canada's contributions to space exploration and research. (10.1, 10.2, 10.3, 10.4, 10.5) T/I

25. You are a doctor with a patient suffering from osteoporosis. Based on what you know about astronauts' routines in space, suggest one thing your patient could do to help fight weakening bones. (10.2) T/I

UNIT D

LOOKING BACK

UNIT D: The Study of the Universe

CHAPTER 8: Our Place in Space

KEY CONCEPTS

 Careful observation of the night sky can offer clues about the motion of celestial objects.

 Celestial objects in the Solar System have unique properties.

 Some celestial objects can be seen with the unaided eye and can be identified by their motion.

 The Sun emits light and other forms of radiant energy that are necessary for life to exist on Earth.

 Satellites have useful applications for technologies on Earth.

 The study of the night sky has influenced the culture and lifestyles of many civilizations.

CHAPTER 9: Beyond the Solar System

KEY CONCEPTS

 Stars differ in colour, size, and temperature.

 Stars change their physical properties over time.

 There is strong evidence that the Universe had its origin around 13.7 billion years ago.

 Stars are clustered into galaxies, which we can observe using telescopes.

 Black holes, neutron stars, and supernovas are part of the life cycle of massive stars.

 Evidence collected by satellites and Earth-based observations support theories of the origin, evolution, and large-scale structure of the Universe.

CHAPTER 10: Space Research and Exploration

KEY CONCEPTS

 Humans use telescopes that gather energy from the entire EM spectrum to learn about objects in the Universe.

 Humans use space probes to learn more about the Solar System.

 Canada is a world leader in developing technology for space exploration.

 The environment in space is hostile to living things.

 New technologies and information from the space program have benefited our everyday lives.

 Space exploration is technically challenging and expensive.

MAKE A SUMMARY

In recent years, websites such as Wikipedia have grown in popularity. Wikipedia is a powerful tool because the entries are written by many people who combine their knowledge to create thorough definitions of different concepts. It is often true that the collective knowledge of many people is more thorough than any one individual, even if that person is an expert (Figure 1). But be careful! Wikipedia can be edited by anyone, so you should cross-reference any information you find there to make sure it is accurate. In this activity, you will use the Wikipedia model to develop definitions of concepts learned in this unit that rely on contributions from all of your classmates.

1. Your class will be divided into groups of three people each. Each group will be given a coloured marker.

2. Your teacher has placed chart paper around the room. One of the following topics is written on each of the pieces of paper: "Stars," "Galaxies," "Planets," "Big Bang," "Satellites," "The Moon," and "The Sun."

3. Begin at one of the pieces of paper with your group members. Write any information you remember about that particular topic. You can add
   - key words
   - drawings
   - definitions
   - graphs
   - equations
   - websites

4. After a few minutes, your teacher will tell you to move on to the next piece of paper. Take your coloured marker with you so you can recognize your contributions.

5. Read the information the previous group(s) wrote. Add any new information you can think of. You can even edit their content if you wish.

6. Continue around the room contributing to all of the topics.

7. When you have written on all of the topics, post the papers around the room. With your classmates, you have now summarized a unit's worth of material in useful categories.

Figure 1

Questions

1. Look at the summaries around the room. Which of the topics did you find the most interesting?
2. Which of the topics did you find the most difficult?
3. Make a list of the top 10 word definitions you want to memorize for the Unit Test.
4. Which of the topics are best understood with pictures?

CAREER LINKS

List the careers mentioned in this unit. Choose two of the careers that interest you or choose two other careers that relate to the study of the Universe. For each of these careers, research the following information:

- educational requirements (secondary and post-secondary)
- skill/personality/aptitude requirements
- potential employers
- salary
- duties/responsibilities

Assemble the information you have discovered into a brochure. Your brochure should compare your two chosen careers and explain how they connect to The Study of the Universe.

UNIT TASK

Celestial Travel Agency

Our knowledge of space and astronomy has greatly increased since the first satellite was sent into orbit 50 years ago. Today, astronauts live and work on the International Space Station for six months at a time, and space tourists blast off to experience a microgravity environment. Many scientists think that in the near future, travel into space for vacations will become commonplace.

However, outer space is not hospitable for human habitation, and space travel remains dangerous and expensive. In order for space travel to become more common, scientists will have to figure out a way to make it safer, cheaper, and faster.

SKILLS MENU
- Questioning
- Hypothesizing
- Predicting
- Planning
- Controlling Variables
- Performing
- Observing
- Analyzing
- Evaluating
- Communicating

Purpose

In this activity, you will be working for a space travel company in the year 2309. You must design an attractive package to advertise travel to a destination in space for space tourists.

Equipment and Materials

- star map
- compass
- pencil
- poster paper
- markers
- Internet access

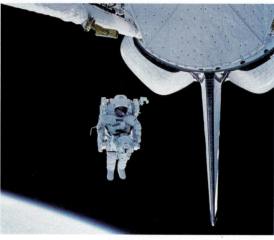

Figure 1

Procedure

Part A: What will your company offer?

1. With a partner, decide on a name for your company and choose one of the three space travel options your company offers to the public, based on research described in the next few steps:
   - Short journey: travelling into orbit for a vacation in microgravity on a space station orbiting Earth like the International Space Station (Figure 1)
   - Medium journey: travelling to a planet in the Solar System that is visible in the night sky
   - Long journey: travelling into deep space to a distant star system that is visible in the night sky

2. Conduct some market research to determine which of the three space journey choices would be most successful. Ask your classmates, friends, and family whether they would select the short journey, medium journey, or long journey option, if offered the opportunity to vacation in space.

3. Create a table to gather responses to this question.

4. Create a bar graph showing how popular each of the journeys into space is.

5. Select the space travel option your company will offer.

Part B: How will your company advertise?

6. Research the destination of the travel option you selected. Find information about the following:
   - What are the physical properties of the celestial object you are visiting?
   - How far away is it?
   - How long will the journey take compared with current journeys currently underway?
   - What are the dangers involved in the journey?

446 Unit D • The Study of the Universe

- What will be some of the physical challenges facing travellers on such a journey?
- How much will the journey cost compared with similar journeys currently underway?
- How have Canadian scientists contributed to research on the planet or star system being visited?
- What is the history of the planet or star system? How was it created? How old is it? How long have we known about it? What methods have humans used to study it?

7. Design a logo for your company and include accurate images of your destination to use in your presentation. Provide references for your image sources.

8. Decide on a way to advertise the travel option to the public. You might choose to present the information about your chosen destination by designing a poster, a pamphlet, a PowerPoint presentation, an interactive whiteboard presentation, or creating a television commercial or song. Talk to your teacher if you have any other ideas.

9. Using your star map, determine where in the night sky the public can look to see their destination from Earth. You will need to go out at night and draw a horizon diagram to help people locate the object from a particular location.

10. Identify the altitude and azimuth of your object from a particular location at a particular time of night. Include this information in your advertisement.

Analyze and Evaluate

(a) Present your advertising campaign to the class. What are the differences or similarities among the various advertising campaigns?

(b) After the presentations, conduct a poll of your classmates on which advertisement they feel is most realistic. Which one is most attractive? Which destination would your classmates most like to visit? Have their answers changed since before the presentations?

(c) In the future, do you think people will take vacations into space like this? What are some of the complications people might encounter in planning space vacations?

Apply and Extend

(d) Now that you have seen other campaigns, if you had additional resources, what would you add to your advertising campaign to make it more appealing to the public?

(e) Identify three ways in which your knowledge of the objects in the night sky has changed by completing this project.

ASSESSMENT CHECKLIST

Your completed Performance Task will be evaluated according to the following criteria:

Knowledge/Understanding
- ☑ Display a collection of data on the space travel destination.
- ☑ Show an understanding of the physical properties and history of the destination star system or planet.

Thinking/Investigation
- ☑ Plan an advertising campaign and divide the workload accordingly with your partner.
- ☑ Show an understanding of how to find information on the Internet and evaluate the quality of information found.
- ☑ Show sources and references and display them in the advertisement.
- ☑ Create a horizon diagram.
- ☑ Measure altitude and azimuth.

Communication
- ☑ Use correct vocabulary and terminology regarding astronomy and space science.
- ☑ Present graphical information in a clear way in a bar graph and a table.
- ☑ Develop an advertisement that communicates information effectively to the public.

Application
- ☑ Include the costs and dangers of space travel.
- ☑ Include information on Canadian contributions to space travel and astronomy research.

UNIT D REVIEW

What Do You Remember?

For each question, select the best answer from the four alternatives.

1. An example of a Canadian contribution to space science research is the
   (a) Canadarm2
   (b) meteorological station on *Mars Phoenix*
   (c) RADARSAT-2
   (d) all of the above (10.1, 10.3) K/U

2. Constellations in the night sky appear to move throughout the night because
   (a) Earth is rotating on its axis
   (b) the stars actually move together through space
   (c) the Moon pulls on the stars with its gravity
   (d) Earth is changing its distance from the Sun (8.5, 8.9) K/U

3. A pulsar is
   (a) a star just about to be born
   (b) a neutron star that emits beams of energy at regular intervals
   (c) a type of black hole
   (d) a supernova (9.4) K/U

4. A lenticular galaxy is
   (a) a galaxy with a spiral shape, but the spiral arms are absent
   (b) a galaxy with no regular shape
   (c) a galaxy with an elliptical or oval shape
   (d) a galaxy with a spiral shape and spiral arms (9.6) K/U

5. What are the uses of the *Ares V* and *Ares I* spacecraft? (10.5) K/U
   (a) to study environmental changes and natural resources in Canada
   (b) to send astronauts and equipment into orbit around Earth or to the Moon
   (c) to send astronauts to Mars
   (d) to analyze the atmosphere of distant planets, such as Saturn

6. Which of the following is *not* one of the reasons ancient cultures studied the movement of stars in the sky? (8.6) K/U
   (a) Farmers planted their crops according to the positions of the stars.
   (b) Ancient mariners charted their journey across the ocean using the stars as a guide.
   (c) Structures were built to coincide with the positions of celestial objects.
   (d) People accurately predicted future occurrences in people's lives by mapping the constellations.

7. Which statement accurately describes a light year? (9.1) K/U
   (a) The distance light travels in one year.
   (b) The distance light travels between Earth and the Sun.
   (c) The time it takes for light to travel 10 trillion km.
   (d) The time it takes for light to travel from Earth to the Sun.

8. Which statement correctly describes Earth's location in our Solar System? (8.1) K/U
   (a) Earth is the second planet from the Sun.
   (b) Earth is the third planet from the Sun.
   (c) Earth is the fourth planet from the Sun.
   (d) Earth is the fifth planet from the Sun.

9. Which of the following is a condition astronauts experience when spending long periods of time in a microgravity environment? (10.2) K/U
   (a) bone loss
   (b) weight gain
   (c) poor eyesight
   (d) difficulty breathing

10. Which of the following best describes the shape of a spiral galaxy? (9.6) K/U
    (a) a flattened disc with no spiral arms
    (b) a flattened disc with two to four spiral arms
    (c) a large group of stars with no definite shape
    (d) a large group of stars with an elliptical or oval shape

Indicate whether each of the statements is TRUE or FALSE. If you think the statement is false, rewrite it to make it true.

11. A star with an absolute magnitude of 4.5 gives off more light than a star with an absolute magnitude of –1.5. (9.2) K/U

12. The greater the volume of an object, the stronger its gravitational force. (8.5) K/U

13. Astronomy is the study of only the objects that are in the Solar System. (8.1) K/U

14. The Universe was born in an expansion around 13.7 billion years ago known as the Big Bang. (9.7) K/U

15. Earth's wobble or precession causes Earth's axis to trace a circle every 12 000 years. (8.5) K/U

16. Hot stars appear red, and cool stars appear blue. (9.2, 9.4) K/U

17. Nebulas are the birthplace of stars. (9.5) K/U

18. Astronauts encounter health risks when they are in space, such as bone loss and changes in blood pressure. (10.2) K/U

19. Astronomers often use the light year as unit of time. (9.1) K/U

20. Satellites in geostationary orbit travel around Earth up to 14 times a day and provide continuous viewing of the same area. (8.11) K/U

21. The Sun creates its energy through a process known as nuclear fusion. (8.2) K/U

22. Parallax is a method astronomers use to measure the temperatures of stars. (9.1) K/U

23. Pluto is no longer considered a planet because it does not orbit the Sun. (8.1, 8.3) K/U

Copy each of the following statements into your notebook. Fill in the blanks with a word or phrase that correctly completes the sentence.

24. A _____ telescope uses a series of mirrors to collect and focus light from objects in space. (10.1) K/U

25. During the _____ phase of the Moon, the Moon is not visible because the side facing Earth is not illuminated. (8.5) K/U

26. _____ is the only star visible in the northern hemisphere that does not move throughout the night. (8.5) K/U

27. _____ keeps Earth and all the planets in orbit around the Sun. (8.5) K/U

28. During a solar eclipse, the _____ is directly between Earth and the _____. (8.5) K/U

29. During a lunar eclipse, the shadow of _____ covers the Moon. (8.5) K/U

30. _____ are relatively cool regions on the Sun's surface. (8.2) K/U

31. An astronomical unit is the average distance between _____ and _____. (8.3) K/U

32. Astronauts working in the ISS need to exercise daily to fight the loss of bone mass, a disease known as _____. (10.2, 10.3) K/U

33. The famous astronomer _____ used a refracting telescope to observe spots on the Sun. (8.2, 10.1) K/U

34. A _____ is an enormous collection of billions of stars. (9.6) K/U

35. _____ is a measure of how bright stars appear to be from the surface of Earth. (9.2) K/U

Match the term on the left with the appropriate definition on the right.

36. (a) white dwarf
    (b) black hole
    (c) protostar
    (d) red giant
    (e) quasar

    (i) an extremely dense quantity of matter in space from which no light or matter can escape
    (ii) a small, dim, hot star that forms from the remains of a red giant
    (iii) a dense region of gas and dust that develops into a star
    (iv) a distant, young galaxy that emits large amounts of energy
    (v) a star near the end of its life cycle with a mass equal to or smaller than the Sun's (9.4) K/U

What Do You Understand?

Write a short answer to each of these questions.

37. Name two ways space scientists on Earth can gather information about objects in space. (10.1) K/U

38. The mass of the Sun is 2×10^{30} kg. Another star has a mass twice that of the Sun. Can you conclude that this star is also twice as large as the Sun? Justify your answer. (9.4) T/I

39. Complete Table 1 in your notebook. The first entry is completed for you. (8.1) K/U

    Table 1

Celestial body	Description	Example
star	a giant ball of hot gases that produce and give off energy	the Sun
planet		
dwarf planet		
natural satellite		
comet		

40. (a) Explain the difference between Earth's rotation and its revolution.
    (b) Which movement is responsible for our day and night?
    (c) Which movement is responsible for our year? (8.5) K/U

41. In the early days of the International Space Station, astronauts living on it could not recycle the water they used. Describe two problems caused by not having a water-purifying system on board the ISS. (10.1) A

42. Name three different ways Earth moves in space. (8.5) K/U

43. What characteristics do the inner and outer planets have in common? How are they different? (8.3) K/U

44. Name three types of stars and list two physical properties of each. (9.4) K/U

45. Complete Table 2 in your notebook, contrasting the heliocentric and geocentric models of the Universe. (8.3, 8.5, 8.9) T/I

    Table 2

	Heliocentric	Geocentric
What do the planets orbit?		
When was this theory proposed?		
What evidence led people to believe this?		
Where are the stars located?		

46. Explain the difference between absolute and apparent magnitude. (9.2) K/U

47. Describe how various pieces of evidence provide support for the Big Bang theory of the origin of the Universe. (9.7) K/U

48. In what ways do Canadian scientists gather information about various structures in the Universe? (8.11, 9.7, 10.1, 10.3) T/I

49. Define the terms "altitude" and "azimuth." (8.9) K/U

50. Describe three products currently on the market that have been developed as spinoffs of space program technology. (10.3) A

51. Sketch the configurations of the Sun, Earth, and the Moon that cause
    (a) a lunar eclipse
    (b) a solar eclipse (8.5) K/U C

52. Define the term "zodiac" and explain why it has been important to cultures all over the world. (8.6) K/U A

53. Define the following terms: meteor, meteorite, meteoroid, comet, asteroid. (8.3) K/U

54. Table 3 lists data gathered from a collection of stars. Organize them on an H–R diagram. Create a set of axes with temperature on the *x*-axis (horizontal) and absolute magnitude on the *y*-axis (vertical), similar to the H–R diagram in Section 9.4. On the axes, plot the stars from the table below: (9.4) T/I C

Table 3

Star	Temperature (°C)	Absolute magnitude
Canopus	7350	−5.53
Alpha Centauri A	5790	4.38
Arcturus	4300	−0.29
Vega	9600	0.58
Capella	5700	0.20
Rigel	11 000	−6.7
Sirius A	9940	1.42
Betelgeuse	3500	−5.14
Achernar	14 500	−2.77
Aldebaran	4100	−0.63
Spica	22 400	−3.55
Procyon B	7740	2.65

(a) Does the distribution of these stars match the pattern of the stars in the H–R diagram in Section 9.4?
(b) Which stars are not on the main sequence?
(c) What is the relationship between the temperature of a star and its absolute magnitude?

Solve a Problem

55. A space probe has travelled approximately 2 250 000 000 km from Earth. (9.1) T/I
    (a) What is this distance, expressed in scientific notation?
    (b) What is this distance in astronomical units (AU)? (1 AU = 150 000 000 km)

56. The Orion Nebula is approximately 1600 light years (ly) from Earth (Figure 1). How far is this nebula from Earth in kilometres? (1 ly = 9.46×10^{12} km) (9.1) T/I

Figure 1

57. You have seen photos of small objects floating in the air inside the International Space Station. Some people say these objects are weightless. Does this mean the objects have no weight? Explain your answer. (10.2) K/U

58. You are driving through central Canada. From your car window, you see mostly flat fields of grain. At one point, you see two trees in the distance. (9.1) A
    (a) Explain how you could use parallax to tell which tree was farther away from you.
    (b) You decide to stop and get out of the car to better observe the trees. In which direction, relative to the trees, would you need to move in order to increase the parallax?

59. An astronaut working with the CSA has been in space for six months on the International Space Station. List two things about the trip that might cause dangers for his health. What is the cause of these dangers? (10.2) T/I

60. A satellite in orbit around Earth observes the same spot on Earth's surface continuously. (8.11) K/U T/I
    (a) What kind of orbit is the object in: low, mid, or geostationary?
    (b) If scientists wanted the satellite to orbit Earth three times a day, how would they adjust the satellite's altitude?

61. An astronomer has discovered a distant galaxy and wishes to collect data on its properties (Figure 2). (9.1, 9.4, 9.6, 10.1) K/U T/I A

Figure 2

(a) Describe three regions of the electromagnetic spectrum in which she could observe the galaxy.
(b) Figure 1 is pictured in visible light. What classification should she give the galaxy?
(c) The astronomer notices some stars in the galaxy that she classifies as red giants. Using the H–R diagram from Section 9.4, suggest a range of temperatures for these stars. Are these stars at the end of their life or the beginning?
(d) The astronomer makes careful measurements of the galaxy's distance and comes up with a value of 12.2 million ly. How many metres away is this? Express your answer in scientific notation.

62. Identify the following features of the Sun shown in Figure 3: sunspot, solar flare, the chromosphere, solar prominence. (8.2) K/U

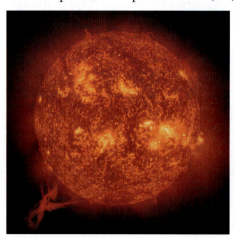

Figure 3

63. (a) On days when a solar flare is particularly large, what recommendations would you make to companies with communications satellites and to people flying in airplanes that week? (8.2) A

(b) How might you observe the effects of this radiation from the Sun at the North and South Poles? (8.2, 8.9) T/I

64. Through careful observation over many nights, you observe a new object in the Solar System that is moving slowly from night to night against the background stars from west to east. (8.1, 8.3, 8.5, 8.9) T/I C

(a) After months of observing, you notice the object appears to move westward for a few nights before resuming its eastward motion. What can you conclude about the object's orbital radius compared to Earth?
(b) You observe the light coming from it and realize it is not a spherical object. Classify it as either a dwarf planet, an asteroid, or a planet.
(c) You wish to announce to the press that you have found a new object in the Solar System. Prepare a statement for the press with the following words: ecliptic, orbiting, celestial body, revolving, reflecting light.

Create and Evaluate

65. You have probably heard the expression, "The Sun rises in the east and sets in the west." How could you demonstrate that this statement is not technically true? (8.5) T/I

66. In this unit, you read that some companies want to start taking paying passengers into space. Briefly explain your opinion of space tourism. (10.5) C

67. Create a poster titled, "The Life Cycle of the Sun." Draw and label pictures or diagrams to represent all the predicted stages in the Sun's life cycle. (9.4) K/U C

68. Astronomers believe that distant galaxies are moving away from the Milky Way at a faster rate than galaxies closer to the Milky Way. Design a presentation involving a balloon, a magic marker, and a ruler to demonstrate this phenomenon. (9.7) T/I C

69. You are designing a spacecraft to explore Mars. What considerations must you make to ensure the safety of the astronauts on their trip to Mars and back? (10.2) T/I

70. Imagine you had to explain the Big Bang theory of the beginning of the Universe to elementary schoolchildren. How would you do it? Come up with a plan for a lesson, including at least two hands-on activities that will help them understand the concepts. (9.7) C A

71. Describe how you could create a scale model of the Solar System using common fruits and vegetables. Try to include the nearest star in your model. (8.3, 8.4) T/I C

72. What evidence supports the statement, "astronomy is the oldest science"? (8.6) T/I

Reflect On Your Learning

73. Create a visual representation that illustrates the main concepts discussed in each chapter. K/U C

74. What unanswered questions do you still have about astronomy?

75. Think back to a movie or television program you saw about a fictional account of space or space travel.
    (a) Describe at least two things you saw that were accurate about space or space travel.
    (b) Describe at least two things you saw that were not accurate about space or space travel.

76. Identify one topic about space that you would like to investigate further. Explain how you could find out more about this topic.

Web Connections

77. Go to the Nelson Science website and download "Google Earth." Explore the Google Universe options, including the stars, planets, and satellites. How realistic is this representation? Where does Google get its data? T/I

78. Canada's amateur astronomy association is called the Royal Astronomical Society of Canada. What role do amateur astronomers play in helping professional institutions in their astronomy research? A

79. The world's largest scale model of the Solar System is the Sweden Solar System. It is modelled on a scale of 1:20 000 000. It uses the Ericsson Globe (Figure 4) to represent the Sun. Research the Sweden Solar System to find out how the planets and other celestial bodies are represented in the city of Stockholm, Sweden. T/I

Figure 4

For all Nelson Web Connections, **GO TO NELSON SCIENCE**

Unit D Review 453

UNIT D

SELF-QUIZ

The following icons indicate the Achievement Chart category addressed by each question.

K/U Knowledge/Understanding   **T/I** Thinking/Investigation
C Communication   **A** Application

For each question, select the best answer from the four alternatives.

1. Which of these devices was originally developed as a result of space exploration programs? (10.3) **K/U**
   (a) hand-held calculators
   (b) nuclear fission reactors
   (c) GPS navigation devices
   (d) jet engine propulsion systems

2. What is the minimum number of satellites needed to fix a position on Earth using GPS navigation? (10.3) **K/U**
   (a) 1
   (b) 2
   (c) 3
   (d) 4

3. Among the following people, who is credited with pioneering the study of objects in the sky with a telescope? (10.1) **K/U**
   (a) Isaac Newton
   (b) Galileo Galilei
   (c) Neil Armstrong
   (d) Wernher von Braun

4. To which of the following planets have scientists sent space probes? (10.1) **K/U**
   (a) only Mars
   (b) all of the planets
   (c) all of the planets except Neptune
   (d) only the inner planets

5. Which of the following is caused by the rotation of Earth on its axis? (8.9) **K/U**
   (a) red shift
   (b) changing ocean tides
   (c) star constellations moving across the sky
   (d) phases of the Moon repeating every 28 days

6. Which type of galaxy is the Milky Way? (9.6) **K/U**
   (a) elliptical
   (b) irregular
   (c) lenticular
   (d) spiral

Indicate whether each of the statements is TRUE or FALSE. If you think the statement is false, rewrite it to make it true.

7. Proxima Centauri, Earth's second-nearest star, is about ten times farther away from Earth than the Sun is. (9.1) **K/U**

8. Red stars are the hottest stars. (9.2) **K/U**

9. The energy radiated from the Sun is the result of hydrogen combustion. (8.2) **K/U**

Copy each of the following statements into your notebook. Fill in the blanks with a word or phrase that correctly completes the sentence.

10. The unit of measurement that is based on the distance from Earth to the Sun is called the _____. (9.1) **K/U**

11. The currently accepted view of the Solar System is the heliocentric model, which replaced the old _____ model. (8.5) **K/U**

12. The small, rocky objects orbiting the Sun between the orbits of Mars and Jupiter are called _____ . (8.3) **K/U**

Match the term on the left with the most appropriate description on the right.

13. (a) pulsar   (i) a cloud of dust and gas where stars are born
    (b) supernova (ii) the remains of a star that has collapsed under its own gravity
    (c) nebula   (iii) a neutron star that emits bursts of energy at regular intervals
    (d) black hole (iv) an exploding star (9.4) **K/U**

Write a short answer to each of these questions.

14. What does the astronomical measurement "light-year" represent? (9.1) K/U

15. Suppose you read in the newspaper that one of the planets will be very bright in the night sky because it will be closer to Earth than usual. The newspaper indicates that the planet will be above the horizon to the East at 10:00 P.M. Explain how you could distinguish the planet from surrounding stars. Describe any diagrams, charts, or photographs you might use to prove which point of light was the planet. Justify your method of proof. (8.2, 8.6, 8.7, 8.9) T/I

16. The planet Jupiter is 4.2 AU from Earth (remember that 1 AU = 150 000 000 km). Light travels at a speed of about 300 000 km/s. If an event occurred on Jupiter that could be seen from Earth, how long after the event would we see it? (9.1) T/I

17. Explain how you could use a toy spinning top to demonstrate each of the following: (8.5) T/I
    a) Earth's rotation on its axis
    b) the tilt of Earth's axis
    c) the precession of Earth's axis

18. Identify four devices or types of technology that you have used in the last year that were spinoffs from technologies developed for space exploration. (10.3) A

19. Sketch a diagram of the Sun and include the following labels: photosphere, chromosphere, corona, solar prominence. (8.2) C

20. You want to study the relationship between phases of the Moon and heights of ocean tides in your area. (8.5) A
    (a) Describe how you could carry out this study. Be sure to identify where you would get your information and the period of time for which you would gather data.
    (b) Describe one relationship you would expect your data to demonstrate.

21. (a) Describe the positions of the Earth, Sun, and Moon at the time of a lunar eclipse.
    (b) Explain why eclipses of the Moon are always at the time of a full moon. (8.5) T/I

22. Draw up a plan for a model of the Earth–Moon–Sun system. Describe the objects you would use. Be sure your model demonstrates how light is emitted and reflected, and how each of the three bodies would look from outside the system, as from a satellite. (8.5) T/I A

23. The Doppler effect is the basis for perceived changes in sound waves as well as the red shift of electromagnetic radiation. Explain the principle that underlies both of these conditions and describe an example of each in the real world. K/U A

UNIT E: The Characteristics of Electricity

OVERALL Expectations

- assess some of the costs and benefits associated with the production of electrical energy from renewable and non-renewable sources, and analyze how electrical efficiencies and savings can be achieved, through both the design of technological devices and practices in the home
- investigate, through inquiry, various aspects of electricity, including the properties of static and current electricity, and the quantitative relationships between potential difference, current, and resistance in electrical circuits
- demonstrate an understanding of the principles of static and current electricity

BIG Ideas

- Electricity is a form of energy produced from a variety of non-renewable and renewable sources.
- The production and consumption of electrical energy has social, economic, and environmental implications.
- Static and current electricity have distinct properties that determine how they are used.

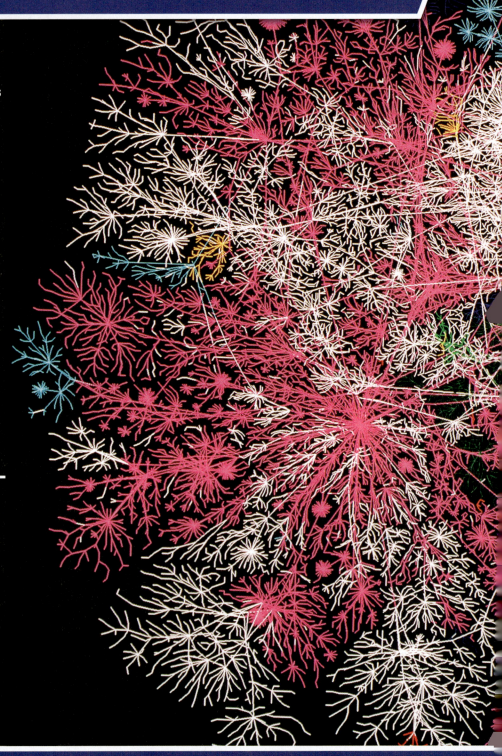

Focus on STSE

POWERING THE WORLD WIDE WEB

People all over the world use the World Wide Web (also called "the Web") to communicate. How did this technology begin? Where does it get its electricity from?

The Internet consists of a worldwide collection of computers and sub-networks that exchange data. In the early 1990s, the European Organization for Nuclear Research (CERN) discovered a way to link all of the resources on the Internet (documents, images, etc.) using a "web" of hypertext links and URLs. Web pages containing this information are collected in "websites," which make up what is called the World Wide Web. By November 2008, there were over 185 million websites. That number continues to grow. Today, most cellphones have Web browsers and many electronics, such as Blu-ray disc players, use the Internet.

The global network of computers that run the Internet consumes a large amount of electricity. Not only is your computer using electricity as it sends and receives information, the servers that control the transfer of information need electricity to run and to keep them cool. Even if your household uses only non-polluting sources of electricity, such as electricity produced from a wind turbine, using the Web involves many computers around the world, all of which use electricity. Most computers use electricity generated by burning fossil fuels, which produces air pollutants and greenhouse gases. Scientists and governments are now researching more efficient ways to generate electricity.

Using the World Wide Web

1. How much time do you spend accessing the Web daily? Share your answer with a partner. How do your activities on the Web compare with your partner's activities? C

2. Using your answer from question 1, estimate how much time Canadians use the Web annually. Does this estimate surprise you? Explain. T/I

3. In what ways could you reduce your personal Web and electricity usage? A

4. What would you suggest citizens, computer manufacturers, and governments do to reduce Web and electricity use? Discuss your ideas with a partner. A C

UNIT E: LOOKING AHEAD

UNIT E
The Characteristics of Electricity

CHAPTER 11
Static Electricity

Static electricity results from an imbalance of electric charge on the surface of an object.

CHAPTER 12
Electrical Energy Production

The production of electrical energy has social, economic, and environmental implications.

CHAPTER 13
Electrical Quantities in Circuits

Circuits are a part of all electrical technology. Circuits have loads that are connected in series or in parallel.

UNIT TASK Preview

Communicate with Electricity

In this unit, you will investigate the properties of static and current electricity, and learn about the production of electrical energy from various sources. You will also learn how to safely handle circuits and how to determine which electrical quantities are affected in circuits that operate properly and in circuits that do not operate properly.

In this Unit Task, you and a partner will play the role of engineers that have been asked to design, test, and build an electric communication device. Your device must run on electricity and include an electric circuit with a number of loads that work together to produce a communication output.

The company that has hired you is also concerned about how the manufacture, use, and eventual disposal of your product may impact the environment. Your device must live up to the company's commitment to environmentally responsible design.

UNIT TASK Bookmark

The Unit Task is described in detail on page 586. As you work through the unit, look for this bookmark and see how the section relates to the Unit Task.

ASSESSMENT

You will be assessed on how well you

- understand the characteristics of electricity in circuits
- design and carry out procedures for testing circuits
- communicate a societal, economic, and environmental issue to an audience

What Do You Already Know?

PREREQUISITES

Concepts
- Demonstrate an understanding of the principles of electrical energy and its transformation into and from other forms of energy
- Electrical energy plays a significant role in society, and its production has an impact on the environment
- Recognize relationships between variables

Skills
- Manipulate a mathematical equation to solve for an unknown
- Analyze data tables and create graphs to show the relationship between variables
- Evaluate the impact of the use of electricity on both the way we live and the environment
- Investigate the characteristics of static and current electricity, and construct simple circuits
- Communicate scientific ideas using a variety of forms

1. Figure 1 shows two electrical generating stations. Which one has an obvious environmental impact? Which one seems harmless? Is it harmless? Explain. **A**

Figure 1 (a) A nuclear generating station and (b) a wind farm both produce electrical energy.

2. There are often warnings on the back of electrical components, similar to that shown in Figure 2. What causes the dangers that they warn against? **A**

Figure 2

3. Does the design of a building affect how much energy it uses (Figure 3)? Explain. **A**

Figure 3

4. What is unsafe about the actions of the person shown in Figure 4? **A**

Figure 4

5. Make a statement about the relationship between x and y for the graphs shown in Figures 5 and 6. For example, as x increases, y _____ . **T/I**

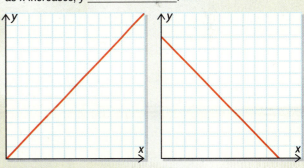

Figure 5   **Figure 6**

6. Plot the data in Table 1 in a scatter plot and determine if the relationship between money and time is linear. Plot money on the dependent axis (y) and time on the independent axis (x). **T/I** **C**

Table 1 Money and Time Data

Money ($)	Time (h)
10	2
35	7
55	11
70	14
90	18

Looking Ahead   459

CHAPTER 11

Static Electricity

KEY QUESTION: How do electrons interact with each other and with objects around them?

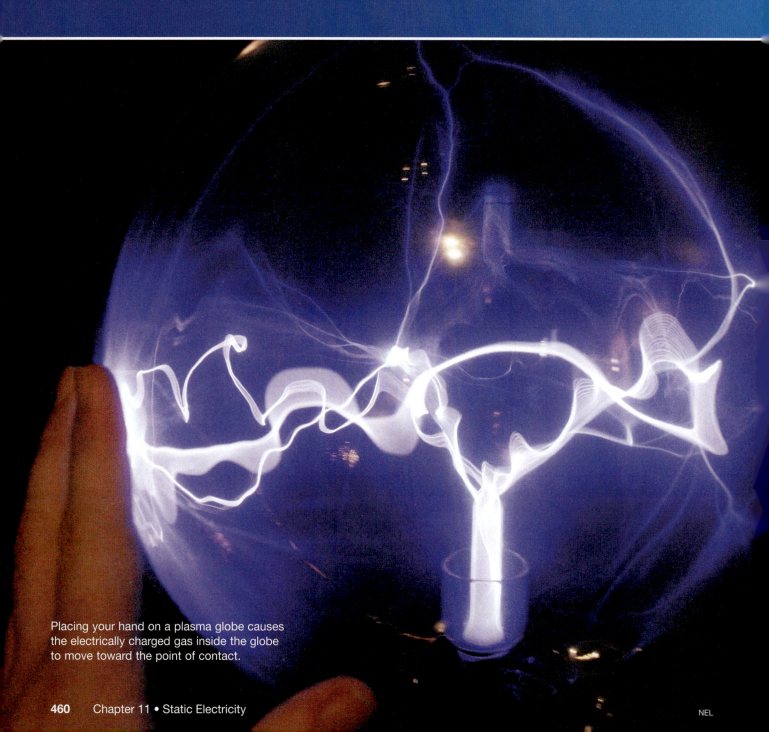

Placing your hand on a plasma globe causes the electrically charged gas inside the globe to move toward the point of contact.

UNIT E
The Characteristics of Electricity

- **CHAPTER 11** — Static Electricity
- **CHAPTER 12** — Electrical Energy Production
- **CHAPTER 13** — Electrical Quantities in Circuits

KEY CONCEPTS

Static electricity results from an imbalance of electric charge on the surface of an object.

There are many useful applications of static electricity.

Objects can be charged by contact or by induction.

Electric discharge can be harmful and must be treated with caution.

Materials can be conductors or insulators.

Simple procedures can be used to test the ability of materials to hold or transfer electric charges.

ENGAGE IN SCIENCE

PRINTER PROBLEM

Wow! Look at all of these printers. I had no idea that there are so many different kinds. We need help!

I had just finished my essay and was thinking to myself, "What a relief to have that project finished!" Suddenly, my printer made a horrible grinding noise. I turned it off, but it was too late. The paper was mangled inside. When I opened the printer to try and clear the jam, I could see that the inside of the machine was covered in ink. What a mess!

Uh-oh. I think this printer is really broken.

"Mom, the printer just broke, and I don't think it can be fixed. Can we please buy a new one?" I asked.

We went to the local electronics store the next day. The printer aisle had an entire wall of different choices at different prices. We could get black and white or colour printers. We could also get all-in-one machines with scanners and fax machines included. We could choose between laser and inkjet printers. There were so many options!

A salesperson recommended a laser printer. It was more expensive than the inkjet printer. Was it really better? We decided to go home and think about it.

I asked myself, "Why should we buy the more expensive laser printer? How does a laser printer even work?" I checked the Internet to research how laser printers work. The websites mentioned drums, toner, fusers, and the laser. They also talked about static electricity. "Science knowledge might help me understand this technology," I thought.

The websites explained that printer drums are made from a special material called a photoconductive insulator. "What does that mean?" I thought. "*Photo* must have something to do with light. I think *conductive* could have something to do with conducting electricity."

The website explained that the laser shines the image onto the drum. Then the drum rolls over powdered toner, which is a kind of printer ink. The toner is attracted to the drum because of the charge of electrons. Finally, the drum rolls over the paper. Because of the electric charges, the toner is attracted to the paper. The whole process ends when the toner is melted onto the paper by the fuser.

Photoconductive insulator? Toner? Fuser? Electrically charged? What do all these terms mean?

My science background helped me understand what I read on the Internet. Learning about how a laser printer works helped me understand why the salesperson recommended this type of printer. When I went back to the store, I was able to ask better questions. In the end, I purchased the right printer for my needs.

1. This reading discussed the use of electrical charges in a laser printer. What other practical applications do you think electrical charges might have?
2. With a partner, discuss ways in which your previous science learning has helped you in making everyday decisions.

WHAT DO YOU THINK?

Many of the ideas you will explore in this chapter are ideas that you have already encountered. You may have encountered these ideas in school, at home, or in the world around you. Not all of the following statements are true. Consider each statement and decide whether you agree or disagree with it.

1 Clothes always get static cling when dried in a clothes dryer.
Agree/disagree?

4 Dust sticks to all charged objects.
Agree/disagree?

2 Static electricity always stays in one place and never moves from object to object.
Agree/disagree?

5 Going down a plastic slide can make your hair stand up.
Agree/disagree?

3 Static electricity can damage electronic devices.
Agree/disagree?

6 Static electricity is a useful form of electricity.
Agree/disagree?

FOCUS ON WRITING

Writing a Persuasive Text

By using reasoning and effective organization, you can persuade an audience to accept your opinion about a topic or issue. The following is an example of a persuasive text written about conductors and insulators in Section 11.4. Beside it are the strategies used to write the persuasive text effectively.

WRITING TIP

As you work through the chapter, look for tips like this. They will help you develop literacy strategies.

Replace Old Wiring

Most new homes today are wired with a non-metallic sheathed cable (NMC). This type of cable consists of insulated copper black (power) and white (neutral) conductors and a bare conductor (ground) covered in a flame resistant and water repellent thermoplastic insulation sheath. Unfortunately, many older homes still use outdated wiring called "knob and tube." Homeowners with this type of wiring should have it inspected and possibly replaced.

Knob and tube wiring consists of two separate copper conductors (one hot and one neutral) secured to studs and joists by porcelain knobs and tubes. There is no ground wire. In today's homes, the ungrounded knob and tube wiring is unsafe in kitchens, bathrooms, and laundry rooms that have heavy electrical loads.

Knob and tube wiring was used originally with 60 amp services, not with the 100 or 200 amp services that are common in homes today. More electrical appliances and devices are used today than were used between 1880 and 1930. Power bars and extension cords are commonly used to expand electrical outlets. Consequently, knob and tube wiring must bear increased loads in modern homes.

The most important reason to replace knob and tube wiring is possible fires. The cloth and soft rubber insulation (loom) on knob and tube wiring can degrade over time. As well, the inline splices are not protected by junction boxes. Squirrels or raccoons can chew off the insulation. Thus, a spark caused by arcing to exposed conductors could cause a serious fire.

Most insurance companies today do not provide coverage to homes with knob and tube wiring for safety reasons. Homeowners with knob and tube wiring behind walls or above ceilings are strongly advised to have it inspected or replaced by a licensed electrician.

- First paragraph introduces the topic.
- Connecting word shows the relationship between ideas.
- Clear opinion is stated concisely.
- First reason to accept the opinion is stated clearly.
- Details are used to support the key point of this paragraph.
- Second reason to accept the opinion is stated clearly.
- Third reason to accept the opinion is stated clearly.
- Connecting words show the relationship between ideas.
- Concluding paragraph explains how the ideas are connected and reinforces the opinion.

What Is Static Electricity? 11.1

Have you ever had your hair stand up after putting on, or taking off, a sweater (Figure 1(a))? Or have you perhaps noticed that a balloon placed near your head will attract your hair (Figure 1(b))? Why does this happen?

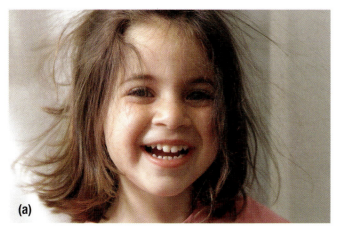

Figure 1 (a) Have you ever removed a sweater and noticed that your hair stands straight up? (b) What causes this girl's hair to become attracted to a rubber balloon?

Atomic Structure and Electric Charge

All matter is made up of atoms. Atoms contain smaller particles: protons, neutrons, and electrons. Some of these particles have an electric charge. Protons have a positive charge (+), electrons have a negative charge (−), and neutrons have no charge (0).

According to the Bohr-Rutherford model of the atom, protons and neutrons are located in the nucleus, or centre, of the atom and are held in place by very strong forces. Under normal circumstances, they cannot be removed from an atom. Electrons, however, can move in the space surrounding the nucleus and can be added to or removed from atoms (Figure 2). Table 1 summarizes the properties of the particles that make up an atom.

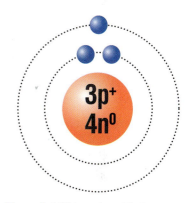

Figure 2 A lithium atom. Electrons (blue) move in the space surrounding the nucleus of an atom and can be transferred from atom to atom. Electrons and protons carry the electric charges within an atom.

Table 1 Properties of the Particles of an Atom

Particle	Electric charge	Location	Particle symbol
proton	positive	nucleus	p^+
neutron	no charge	nucleus	n^0
electron	negative	outside nucleus	e^-

If an atom has the same number of protons and electrons, the positive and negative charges balance and the atom has no overall charge—it is neutral. If an atom does not have an equal number of protons and electrons, it has an **electric charge**. Recall from Chapter 7 that an atom that has an electric charge is called an ion. A negative ion is an atom that has picked up one or more electrons. A positive ion is an atom that has lost one or more electrons.

electric charge a form of charge, either positive or negative, that exerts an electric force

Positive, Negative, and Neutral Objects

Everyday objects such as combs, rulers, clothing, airplanes, and clouds are made up of billions and billions of atoms—each containing a number of positive and negative charges. For this reason, it is impossible to show individual atoms or their charged particles in diagrams of large objects.

In this unit, we will use charge symbols to represent charges. Each charge symbol represents a very large number of protons (+) or electrons (−). A "+" symbol will be used to represent a large number of protons (carrying positive charges) and a "−" symbol will be used to represent an equally large number of electrons (carrying negative charges). These symbols will be used to show the relative abundance and distribution of charges on an object (Figure 3). The overall electric charge of an object can be determined simply by comparing the number of positive and negative symbols drawn on the object.

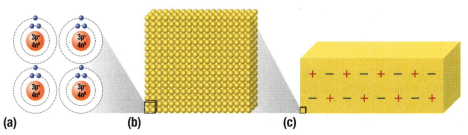

Figure 3 (a) A Bohr–Rutherford diagram can be used to show the electrons and protons in an individual lithium atom. (b) Objects contain billions of individual atoms. (c) The "+" and "−" symbols on the object represent the relative numbers of protons and electrons and their distribution.

neutral object an object that has equal numbers of protons and electrons

negatively charged object an object that has more electrons than protons

positively charged object an object that has fewer electrons than protons

A **neutral object** is an object that has an equal number of protons and electrons (Figure 4(a)). A **negatively charged object** is an object that has more electrons than protons (Figure 4(b)). A **positively charged object** is an object that has fewer electrons than protons (Figure 4(c)).

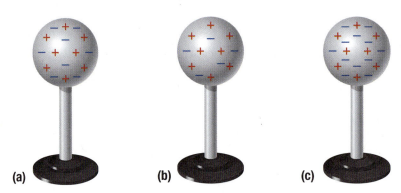

Figure 4 (a) Neutral objects have equal numbers of protons and electrons. (b) Negatively charged objects have more electrons than protons. (c) Positively charged objects have fewer electrons than protons. (Note: Each charge symbol represents a very large number of protons (+) or electrons (−).)

Objects may also become charged when electrons are transferred to or from another object. For example, a neutral or positively charged object that gains electrons will become negatively charged if its total number of electrons is more than its total number of protons. Similarly, a neutral or negatively charged object that loses electrons will become positively charged if its total number of electrons is less than its total number of protons. Most of the objects and materials we interact with daily are electrically neutral.

When two neutral objects made of different materials come in contact, such as the hair and the rubber balloon in Figure 1, electrons can be transferred from one object to the other. Both objects become charged. In Figure 1, the balloon gains electrons and becomes negatively charged, and the hair loses electrons and becomes positively charged. Each hair on the girl's head becomes positively charged, which creates static electricity. **Static electricity** is an imbalance of electric charge on the surface of an object. The positive charges on the individual hairs react in such a way as to cause the hairs to move as far away from each other as they can. Static electricity produces what we call "static charges" because the charges are at rest on the surface of the object.

static electricity an imbalance of electric charge on the surface of an object

Detecting Static Electric Charges

Scientists can detect the presence of electric charges using an instrument called an electroscope.

One type of electroscope is a pith ball electroscope. A pith ball electroscope consists of a ball of pith (plant material) suspended from a stand by a thread (Figure 5). It can be used to test for the presence and type of electric charge on an object. This is done by bringing an object near the neutral pith ball. If the object is charged, the pith ball will be attracted to it. You will learn about another type of electroscope, the metal leaf electroscope, later on in this section.

Figure 5 A pith ball electroscope is a simple device that can be used to detect the presence of electric charges.

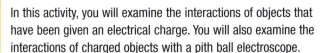

TRY THIS: POSITIVE AND NEGATIVE CHARGES

SKILLS: Hypothesizing, Observing

SKILLS HANDBOOK
3.B.3., 3.B.6.

In this activity, you will examine the interactions of objects that have been given an electrical charge. You will also examine the interactions of charged objects with a pith ball electroscope.

Equipment and Materials: acetate strips (2); ring clamp; retort stand; pith ball electroscope; thread; tape; vinyl strips (2); paper towels.

1. Use a piece of thread and tape to suspend and balance an acetate strip from a ring clamp (Figure 6). Rub the acetate strip at both ends with a paper towel. You can assume that the acetate has become positively charged.

2. Rub a second acetate strip with a paper towel and bring the acetate strip toward the hanging acetate strip. Observe and record what happens.

3. Repeat steps 1 and 2 using two vinyl strips. Assume that the charge placed on the vinyl strip is negative.

4. Repeat steps 1 and 2, this time using one vinyl strip and one acetate strip.

5. Obtain a pith ball electroscope. Slowly bring a charged acetate strip up to the pith ball. Observe what happens before and after the pith ball contacts the acetate strip.

6. Gently roll the pith ball between your fingers to remove any charge from the pith ball. Then repeat step 5 with the electroscope and a charged vinyl strip.

A. Based on your observations, how do objects with similar charges behave? How do objects with different charges behave? T/I

B. How did the pith ball react to the positively and negatively charged strips? T/I C

C. Make a hypothesis to explain what happened to the pith ball after it made contact with a charged strip. How could you test your hypothesis? T/I C

Figure 6

In the Try This activity, you saw that two positively charged acetate strips repelled each other but were attracted to a charged vinyl strip. This means that the vinyl strip had a different type of electrical charge than the acetate strip—the vinyl strip was negatively charged. The activity shows us that two different types of electrical charge exist (positive and negative).

The Law of Electric Charges

A charged object exerts an **electric force**, which can be either an attractive force (pulling together) or a repulsive force (pushing apart). This is summarized in the Law of Electric Charges, which states that

- objects that have like charges repel each other (Figure 7(a))
- objects that have opposite charges attract each other (Figure 7(b))

electric force the force exerted by an object with an electric charge; can be a force of attraction or a force of repulsion

LEARNING TIP
Opposites Attract
To remember the Law of Electric Charges, remember that "opposites attract." Opposite charges attract and like charges repel. You can observe this in magnets: opposite poles of two magnets will attract and like poles of two magnets will repel.

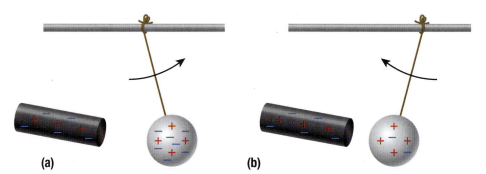

Figure 7 (a) Both objects are negatively charged and repel each other. This causes the pith ball to move away from the metal rod. (b) When the pith ball and the metal rod are oppositely charged, they are attracted to each other and the pith ball moves toward the rod.

The strength of the electric force is related to both the amount of charge on each object and the distance between the charged objects. Electric force increases with increasing electrical charges and decreases with increasing distance.

Attraction of Neutral Objects to Charged Objects

A neutral object has an equal number of positive and negative electric charges, which means that when two neutral objects are brought together, they are neither attracted nor repelled from one another. What happens when a charged object is brought toward a neutral object? When a charged object is brought near a neutral object, it causes (induces) the electrons to shift in position. The induced movement of electrons in a neutral object by a nearby charged object is called an **induced charge separation**. The movement of electrons occurs according to the Law of Electric Charges. If the charged object is positively charged, it will induce electrons in the neutral object to move toward it. If the charged object is negatively charged, if will induce electrons in the neutral object to move away from it (Figure 8). After the electrons shift position, the side of the neutral object closest to the charged object will be attracted to the charged object. This force of attraction can cause a neutral object to move toward a charged object. Although there is a shift in the positions of the electrons in the neutral object, it does not gain or lose any electrons. Once the charged object is moved away from the neutral object, the electrons return to their original positions.

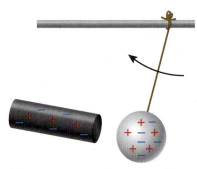

Figure 8 The side of the pith ball facing the negatively charged metal rod now has a local positive charge and is attracted to the metal rod. The force of attraction is greater than the force of repulsion because of the differences in distances between the charges.

induced charge separation a shift in the position of electrons in a neutral object that occurs when a charged object is brought near it

TRY THIS | CHARGING OBJECTS

SKILLS: Hypothesizing, Observing

SKILLS HANDBOOK
3.B.3., 3.B.6.

In this activity, you will examine the interactions of charged objects and neutral objects and use neutral objects to detect an electrical charge.

Equipment and Materials: wool; wooden stick; Petri dish; acetate strip; copper pipe; 10 paper circles (from a hole punch); balloon; paper towel.

1. Use a piece of wool to rub and charge a wooden stick. Then bring the wooden stick close to the paper circles (Figure 9). Record your observations.

Figure 9

2. Blow up the balloon and rub it against your hair. Bring the rubbed surface of the balloon close to the paper circles. Record your observations.

3. Repeat step 1 using paper towel and the acetate strip, and wool and the copper pipe.

A. Based on your observations, what general conclusion can you make regarding neutral objects and charged objects? T/I

B. Did the paper circles behave in an identical manner with all the charged objects? Suggest an explanation for your answer. T/I C

C. Did the behaviour of the paper circles allow you to tell what kind of charge (positive or negative) was on the object? T/I

D. Suggest a method you might be able to use to measure the amount of charge on an object. T/I C

Using a Metal Leaf Electroscope to Detect Electric Charge

A metal leaf electroscope is more sensitive to electric charge than a pith ball electroscope and is more commonly used to detect electric charge. It consists of a vertical rod, which has two thin pieces of metal foil (often gold) known as "the leaves". The charge to be tested is applied to a metal terminal attached to the top of the vertical rod (Figure 10(a)). When the metal terminal is touched with a charged object, electrons are transferred to (or from) the leaves of the electroscope. The leaves receive the same charge, which causes them to repel each other and spread apart.

A metal leaf electroscope can also be charged without touching it to a charged object. For example, if a negatively charged object is brought near the electroscope terminal, the electrons in the terminal are repelled by the electrons in the negatively charged object and move down onto the leaves. This causes a separation of charge in the electroscope—the terminal becomes positively charged, and the leaves become negatively charged. The negatively charged leaves repel each other and spread apart (Figure 10(b)). When the charged object is moved away from the electroscope, the electrons and the leaves return to their original positions (Figure 10(c)).

> **WRITING TIP**
>
> **Connecting to the Main Idea**
> Conclude your persuasive text by connecting the main idea and key points. For example, for a persuasive text on a metal leaf electroscope, you might conclude by saying a metal leaf electroscope is a valuable tool to tell if a charge on an object is positive or negative because of the way the "leaves" move.

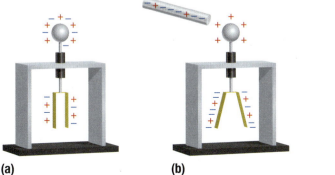 

(a)   (b)   (c)

Figure 10 A metal leaf electroscope (a) is used to detect electric charge. When a negatively charged object is brought near the electroscope electrons are transferred into the leaves, which causes them to repel and spread apart (b) until the charged object is removed (c).

11.1 What Is Static Electricity?

Using Static Charges

Scientists and engineers can use the properties of static charges in many useful ways in a branch of science called electrostatics. Electrostatics is the branch of science that deals with static charges and static electricity.

Electrostatic Paint Sprayers

If you have ever used a can of spray paint, you know that some of the paint can miss the target and spray into the air or onto surrounding objects. This can be messy and wasteful. In addition, solvent-based paints are damaging to the environment and to human health. Many industries use electrostatic paint sprayers to reduce the amount of wasted paint. Electrostatic paint sprayers use the properties of static charges to more efficiently paint objects. The paint is given a charge as it leaves the nozzle of the sprayer, and the object to be painted is given the opposite charge (Figure 11). The charged paint particles are attracted to the object, which minimizes the amount of wasted paint and ensures that the object receives an even coat of paint. Electrostatic paint sprayers are especially useful for painting curved objects (Figure 12).

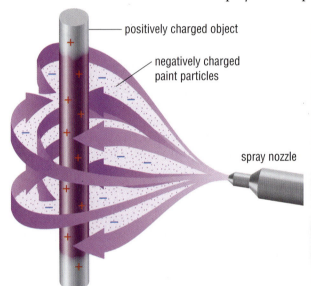

Figure 11 The charged paint particles are attracted to the oppositely charged object.

Figure 12 Curved objects and objects with open spaces (such as fencing) are often painted using an electrostatic paint sprayer. The attraction of the paint particles to the oppositely charged object ensures that paint will wrap around the curved portions of the object, rather than fall through the open spaces.

RESEARCH THIS: POWDER COATING

SKILLS: Researching, Analyzing the Issue, Communicating, Evaluating

SKILLS HANDBOOK 4.A.

Manufacturers have developed dry paint, which is paint that comes in a powder form. Painting objects with dry paint is called powder coating. Powder paint does not use toxic solvents like liquid paint does. How are these dry paints applied? What benefits does dry paint have compared with liquid paint?

1. Research the process of powder coating using electric charges.
2. Research some applications of powder paint.
3. Research the negative effects of paint solvents on human health and ecosystems.
4. Research the environmental benefits of using powdered paint compared with solvent-based paint.

 **GO TO NELSON SCIENCE**

A. Draw a diagram showing the charges to explain how dry paint works. Include electric charges in your diagram and explain their significance. T/I C

B. How has the use of solvents been affected by the use of powder-coating technologies? How has the use of powder–coating technologies impacted the environment? T/I C A

UNIT TASK Bookmark

How can you apply what you have learned about static electricity to the Unit Task described on page 586?

IN SUMMARY

- Static electricity is an imbalance of electric charge on the surface of an object.
- Electric charge is a property of the particles of atoms. Protons have a positive charge, electrons have a negative charge, and neutrons have no charge.
- Electrons can be added to, or removed from, atoms.
- When two different materials come in contact, electrons can be transferred between them, causing an imbalance in charge.
- Neutral objects have equal numbers of protons and electrons. Negatively charged objects have more electrons than protons, and positively charged objects have fewer electrons than protons.
- Objects can become positively charged by losing electrons so that there are fewer electrons than protons.
- Objects can become negatively charged by gaining electrons so that there are more electrons than protons.
- The Law of Electric Charges states that like charges repel while unlike charges attract.
- A charged object can be used to induce a charge separation in a neutral object
- A metal leaf electroscope can be used to detect the presence of an electric charge.
- Static electricity has practical applications.

CHECK YOUR LEARNING

1. Describe a concept from this section that was already familiar to you. How did your previous knowledge of this concept help with your understanding of other concepts in this section?

2. In your own words, define static electricity.

3. (a) Which particle(s) are difficult to add to or remove from an atom?
   (b) Which particle(s) are easier to add to or remove from an atom?
   (c) How do your answers to (a) and (b) explain the formation of positively and negatively charged objects?

4. Describe the total charge on each of the following objects as either neutral, positive, or negative. Explain your reasoning.

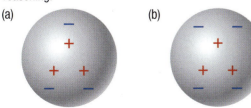

Figure 13   **Figure 14**

5. What would you do to the object in Figure 15 to make it neutral?

Figure 15

6. What would you do to the object in Figure 16 to make it positively charged? What would you do to the object in Figure 16 to make it negatively charged?

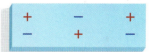

Figure 16

7. Would the following objects repel or attract? Explain your answer using the Law of Electric Charges.
   (a) A positively charged object is placed beside a negatively charged object.
   (b) A negatively charged object is placed beside a negatively charged object.

8. (a) In your own words, summarize how an electrostatic paint sprayer works.
   (b) Are electrostatic paint sprayers beneficial? Why or why not?

9. Use diagrams to illustrate how a positively charged object can be used to induce a charge separation and attract a neutral object.

11.2 Charging by Contact

Over 2500 years ago, Thales of Miletus, a Greek philosopher, noticed something unusual when he rubbed a piece of amber with a piece of fur. He noticed that after contact with the fur, the amber attracted objects such as feathers and pieces of straw (Figure 1). Since then, scientists have learned that neutral objects can become charged through direct contact in several different ways. Two common methods of charging by contact are charging by friction and charging by conduction.

Figure 1 Amber is fossilized tree sap. This photo shows a piece of charged amber attracting feathers.

charging by friction the transfer of electrons between two neutral objects (made from different materials) that occurs when they are rubbed together or come in contact (touch)

Charging Objects by Friction

Charging by friction occurs when two different neutral materials are rubbed together or come in contact (touch) and electric charges are transferred from one object to the other. One material is more likely to attract extra electrons and become negatively charged, while the other material is more likely to give up electrons and become positively charged. This is because some kinds of atoms are more strongly attracted to electrons than others. During contact, or when being rubbed together, each material is charged. For example, in Figure 2(a), the hair and the comb are both neutral. When they are rubbed together, the atoms in the comb gain electrons and the atoms in the hair lose electrons (Figure 2(b)).

LEARNING TIP

Word Origins
The words "electron" and "electricity" come from the Greek word for amber, *elektron*.

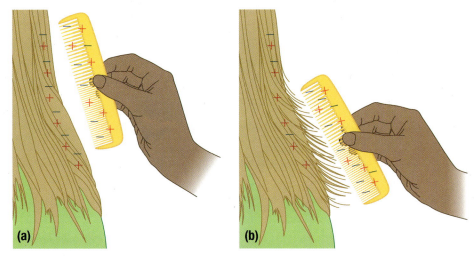

Figure 2 (a) The comb and the hair are both neutral. (b) After being rubbed together, the comb is negatively charged and the hair is positively charged.

You may have noticed that objects become charged more easily in the winter when the air is dry than in the summer when the air is humid. When the air is humid, it contains more water vapour than usual. Water molecules in the water vapour can collide with a nearby charged object, transferring electrons from the charged object to the water molecules. Thus, a charged object will lose its charge quickly in humid weather. In dry weather, the air has fewer water molecules. Therefore, a charged object has less chance of losing its charge due to collisions with water molecules in the air.

Rubbing objects together is effective but not always necessary to develop a charge imbalance. Sometimes you can charge materials by simply touching the different materials together and then separating them.

The Electrostatic Series

The **electrostatic series**, or triboelectric series, is a list of materials in order of increasing tendency to gain electrons (Table 1). As you move further down the list, the materials increase in their tendency to gain extra electrons.

We can use the electrostatic series to predict the charge that will be gained by two objects or materials that are rubbed together. For example, if two materials from the series are rubbed together, the material that is higher on the series will lose electrons and become positively charged and the material that is lower on the series will gain electrons and become negatively charged.

electrostatic series a list of materials arranged in order of their tendency to gain electrons

Table 1 Electrostatic Series

Material	Charge tendency
human skin	+ (weaker tendency to gain electrons)
rabbit fur	
acetate	
glass	
human hair	
nylon	
wool	
cat fur	
silk	
paper	
cotton	
wood	
amber	
rubber balloon	(stronger tendency to gain electrons)
vinyl	
polyester	
ebonite	−

SAMPLE PROBLEM 1 Using the Electrostatic Series

You rub a piece of wool against your skin. What charge does each material now have?

Step 1 Compare the positions of the materials on the electrostatic series.
Wool is lower on the list than human skin.

Step 2 Compare the attraction for electrons of the two materials.
Since wool is lower on the list than human skin, it has a greater attraction for electrons than human skin.

The wool will gain electrons from your skin and become negatively charged. Your skin will lose electrons to the wool and become positively charged.

Practice

You grab a rubber balloon with a wool glove on your hand. What charge does each material now have?

TRY THIS CHARGING BY FRICTION

SKILLS: Performing, Observing, Analyzing, Communicating

SKILLS HANDBOOK
3.B.5., 3.B.6.

Materials do not always need to be rubbed together to create a charge imbalance. In this activity, you will explore charging by friction through simple contact.

Equipment and Materials: roll of clear, plastic adhesive tape

1. Pull two 8–10 cm pieces of tape from the roll.
2. Hold one piece of tape in each hand and bring the two shiny (non-sticky) sides of the tape close together without letting them touch (Figure 3). Record your observations.

Figure 3

3. Exhale onto both sides of each piece of tape several times (over its entire length). Bring the two shiny sides of the tape close together again without letting them touch. Observe what happens.

4. Using the same two pieces of tape, allow each piece of tape to stick to the top of a clean desk without rubbing. Then quickly pull the pieces off the desk. Bring the shiny, non-sticky side of one of the pieces close to the edge of the desk without letting it touch the desk. Observe what happens.

5. Quickly bring the shiny, non-sticky sides of both pieces of tape close to each other without letting them touch. Observe what happens.

A. What do your observations in step 2 indicate about the electric charge on the pieces of tape when they were first pulled off the roll? Explain. T/I C

B. What do your observations in step 3 indicate about the electric charge on the pieces of tape? Explain. T/I

C. Why is there a difference in the electric charge on the pieces of tape between steps 2 and 3? T/I

D. Write a question you have about the observations you made in this activity. Exchange questions with a classmate and decide how you may find answers to your questions. Then design and carry out simple experiments to answer your questions. T/I

charging by conduction charging an object by contact with a charged object

Charging Objects by Conduction

In addition to charging by friction, objects can also be charged by conduction. **Charging by conduction** occurs when two objects with different amounts of electric charge come in contact and electrons move from one object to the other. This is illustrated in Figure 4, which shows a negatively charged bar touching a neutral sphere. As a result of the contact, some of the excess electrons on the charged bar which are repelling each other, move onto the sphere. As a result, the sphere becomes negatively charged.

Figure 4 A neutral metal sphere (a) becomes charged by conduction when brought in contact with a negatively charged bar. Some of the electrons are transferred from the bar to the sphere, resulting in an overall negative charge on the sphere (b).

Charging by conduction, however, does not always involve a charged object and a neutral object. Two charged objects may come in contact, and electrons may move from one object to the other. Electrons always move from the object with a larger negative charge (less positive) to the object with the smaller negative charge (more positive). This produces a more even distribution of electric charge between the two objects. For example, if a positively charged piece of metal comes in contact with another identically sized piece of metal that has less positive charge (Figure 5(a)), electrons will travel from the less positively charged piece to the more positively charged piece until both pieces have an even distribution of charge (Figure 5(b)). After contact, the two pieces of metal each still have a positive charge (Figure 5(c)). The electric charge is not neutralized in the two pieces of metal after they come in contact, but they both now have the same type and amount of charge.

WRITING TIP

Identify the Topic
Use the first paragraph to identify the topic and state your main idea/opinion concisely. For example, "Can an object be charged by conduction?" Yes—when two charged objects come in contact electrons may move from one object to the other. After contact, the two objects will have the same type and amount of charge.

metal rod X (overall +3 charge)   metal rod Y (overall +1 charge)
(a)

metal rod X (gaining an electron)   metal rod Y (losing an electron)
(b)

metal rod X (overall +2 charge)   metal rod Y (overall +2 charge)
(c)

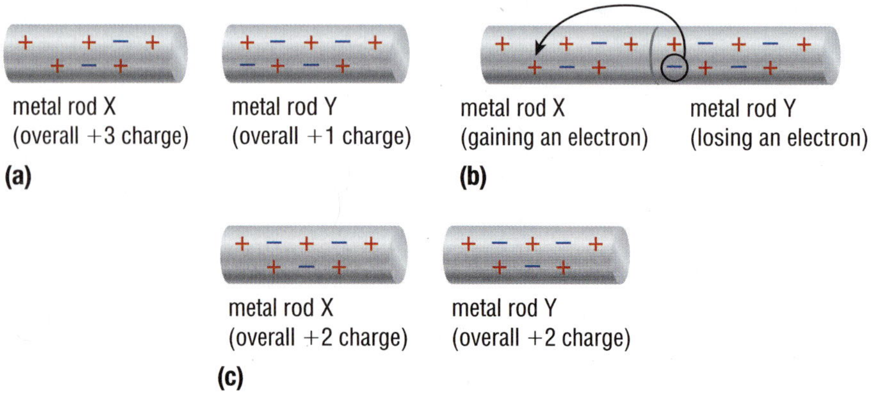

Figure 5 Two metal rods being charged by conduction

Grounding

Objects with an excess electric charge—either positive or negative—can have the excess charge removed by a process called **grounding**. Grounding an object involves removing the excess charge by transferring electrons between the object and a large neutral object such as Earth (the ground). Any object that serves as a seemingly infinite reservoir of electrons can be used as a "ground" for electric charges. For example, because Earth is so large, any excess charge is spread over a huge area and is effectively neutralized.

When a positively charged object is grounded, electrons from the ground travel up to the positively charged object until the object is neutral. When a negatively charged object is grounded, electrons travel from the object into the ground until the object is neutral. Figure 6 shows the symbol for grounding.

What happens during the grounding of a charged object or person? Figure 7 shows what happens when a student carrying a negative charge on his hand comes in contact with a neutral faucet that is grounded. If the student reaches for the faucet with his hand, he will receive a shock as the excess electrons are transferred from his hand to the faucet. After grounding, the hand loses the negative charge and is neutral. The charge imbalance between the hand and the faucet is gone.

grounding connecting an object to a large body, like Earth, that is capable of effectively removing an electric charge that the object might have

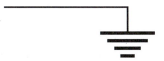

Figure 6 The symbol for grounding

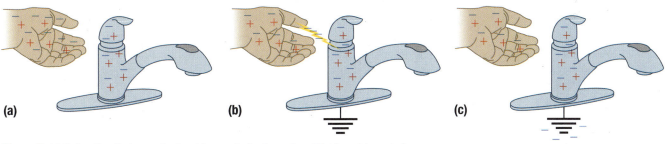

Figure 7 (a) Before the discharge, the hand is negatively charged and the faucet is neutral. (b) During the discharge, excess electrons are transferred from the hand into the ground. (c) After the discharge, the negative charge on the hand has been removed.

Putting Charged Objects to Work

Advances in electrostatics have led to the development of many useful products and technologies. Several of these involve charging objects by contact (friction or conduction).

Electrostatic Dusters

Electrostatic dusters depend on charging by friction to attract dust. When you use an electrostatic duster, you gently sweep it across an object, causing a buildup of charge on the duster. The dust is attracted to the electrostatic duster and "jumps" off the dusty surface onto the duster. Natural electrostatic dusters have been used since the 1800s. Ostrich feathers, like human hair, have a natural tendency to become charged when rubbed against a surface (Figure 8). This causes dust to be attracted to the feathers.

Figure 8 Charged ostrich feather dusters keep dust trapped on the feathers.

RESEARCH THIS | FABRIC SOFTENER SHEETS

SKILLS: Analyzing the Issue, Communicating, Evaluating

SKILLS HANDBOOK
4.A.

Many people use fabric softener dryer sheets to control static charge buildup on clothes. As clothes made of different materials tumble inside a clothes dryer, they rub together and become charged by friction. Fabric softener sheets prevent the buildup of static charges.

1. Research how fabric softener sheets prevent the buildup of static charges.
2. Research what chemicals are used in fabric softener sheets and their effects on people and the environment.

A. Analyze how this technology works to hinder the effect of charging by friction. Draw a diagram to support your analysis. T/I C

B. List some benefits and drawbacks of using fabric softener sheets. Suggest alternatives to using fabric softener sheets. A

C. Based on your research, decide whether fabric softener sheets are necessary. Support your opinion. T/I A

GO TO NELSON SCIENCE

WRITING TIP

Use Evidence

To support the main idea or opinion, use facts, statistics, examples, or reasons. For example, in a persuasive essay about electrostatic precipitators (ESPs) for household use, you might give statistics to show that ESP filters are the best and cheapest way of cleaning the air from forced-air furnaces.

Electrostatic Precipitators

Almost half of the world's electrical energy is generated by burning coal, oil, and natural gas. These processes release small particles of soot, dust, and other substances that can pose health risks and cause environmental damage. These particles are expelled from large chimneystacks into the air.

An electrostatic precipitator is a device that uses the properties of electrostatic charge to remove particles from the air. Electrostatic precipitators are used in large power plants, manufacturing plants, and incinerators. Figure 9 shows how an electrostatic precipitator can be used to filter particles from smokestack emissions. When smoke passes through the negatively charged plates, the particles in the smoke become negatively charged by conduction. The particles then pass between positively charged plates, which they stick to, due to the attraction of opposite charges. The particles fall onto the collection plate, allowing them to be safely removed. Electrostatic precipitators use very little electricity and are very good filtration devices. About 99 % of the particles in smokestack emissions can be removed by an electrostatic precipitator. Many household air purifiers also use electrostatic precipitators to clean the air.

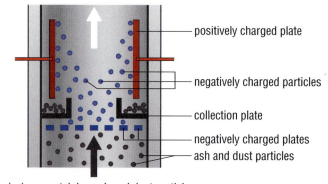

Figure 9 Electrostatic precipitators use charged plates to filter the particles out of flowing gases such as exhaust and household air.

IN SUMMARY

- Charging by friction occurs when two neutral objects made of different materials rub against or touch each other and electrons are transferred between them.
- When objects are charged by friction, one material is more likely to accept electrons, while the other is more likely to give up electrons. This is because some kinds of atoms are more strongly attracted to electrons than others.
- The electrostatic series is a list that ranks the tendency of different materials to gain electrons. It can be used to predict the charge that will be gained by two objects (made from different materials) when they come in contact.
- Charging by conduction occurs when two objects with different amounts of electric charge come in contact and electrons are transferred from one object to the other.
- A neutral object is charged by conduction when a charged object touches it. The neutral object becomes charged with the same charge as the object that touched it.
- When two charged objects with different amounts of electric charge come in contact, electrons are transferred between them.
- When we ground an object, we transfer electrons between the object and a large neutral object such as Earth (the ground).
- Charging objects by friction or by conduction has practical applications.

CHECK YOUR LEARNING

1. Consider the following pairs of materials. Using the electrostatic series, determine the charge that each material will gain when the two are rubbed together. K/U
   (a) glass and silk
   (b) ebonite and fur
   (c) human hair and a rubber balloon
   (d) amber and cotton

2. Why do objects made from different materials develop an electric charge when rubbed together? What is this method of charging called? Use a diagram to illustrate your answer. K/U C

3. In your own words, explain charging by conduction. Include diagrams showing how a positively charged object can be used to charge a neutral object. K/U C

4. Use a graphic organizer to compare charging by conduction to charging by friction. K/U C

5. A rod, "X," has a positive charge of 8. An otherwise identical rod, "Y," has a negative charge of 4. The rods are touched together and then separated.
   (a) When they touched, what particles moved between them? K/U
   (b) Did the particles move from "X" to "Y" or from "Y" to "X"? K/U

6. Describe how electrons travel when a positively charged object is grounded. K/U

7. Why do you think that some factories might not use electrostatic precipitators? A

8. Describe a situation in which static electricity was a nuisance for you. How did the objects involved become charged? A

11.3 CONDUCT AN INVESTIGATION

Predicting Charges

Earlier, you learned how objects made from different materials can become charged by friction when rubbed, or touched, together. This causes electrons to be transferred from one object to the other, producing a charge imbalance in the objects. For example, experiments show that a vinyl strip becomes negatively charged when rubbed with wool, while the wool becomes positively charged. This means that electrons are transferred from the wool to the vinyl.

As you learned in Section 11.1, a pith ball electroscope (Figure 1) can be used to detect a charge imbalance. When a charged object is brought near the pith ball, the pith ball moves in response to the charge on the object. It is important to note that, unless they touch, both the object and the pith ball keep their respective charges because no electrons transfer between them. If the pith ball and the object come in contact, electrons will be transferred between the pith ball and the object until both have the same type of charge.

In this investigation, you will use an electrically charged pith ball to explore charge interactions. You will then use the Law of Electric Charges to determine the overall charge on several test objects.

SKILLS MENU
- Questioning
- Hypothesizing
- Predicting
- Planning
- Controlling Variables
- Performing
- Observing
- Analyzing
- Evaluating
- Communicating

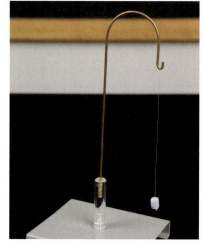

Figure 1 A pith ball electroscope has a small pith ball hanging from a thread.

Testable Question

How can the Law of Electric Charges be used to determine the charge on an object?

Hypothesis/Prediction

Read the Experimental Design and Procedure, and then formulate a hypothesis based on the Testable Question. Your hypothesis should include a prediction and reasons for your prediction.

Experimental Design

Working with a partner, you will observe charge interactions. You will begin by charging a test object by rubbing it with an object made from a different material. You will then use a pith ball electroscope to observe charge interactions and to determine the overall charge on several other test objects charged by friction with a variety of materials.

Equipment and Materials

- pith ball electroscope or pith ball on a thread
- ring stand (optional)
- test materials (vinyl strip, acetate strip, glass, wood, ebonite)
- rubbing materials (wool, paper, polyester)

Procedure

1. Copy Table 1 into your notebook. Add any additional objects to test provided by your teacher.

Table 1 Charges Placed on Objects

		Rubbing materials		
		wool	paper	polyester
Objects to be charged	vinyl			
	acetate			
	glass			
	wood			
	ebonite			

2. Set up a pith ball electroscope on the lab bench. Otherwise, suspend a pith ball from a ring stand with a string. Ground (remove the charge from) the pith ball by touching it with your hand.

3. Rub a vinyl strip with wool. The strip is now negatively charged. Record the charge of the vinyl strip in Table 1.

4. Bring the charged vinyl strip close to the pith ball. Do not allow the strip to touch the pith ball. Record your observations in your notebook.

5. Ground the pith ball with your hand to ensure it is not charged.

6. Rub the vinyl strip again with wool. Charge the pith ball by touching it with the vinyl strip. Be careful not to accidentally ground the pith ball.

7. Make a prediction about what would happen if the charged vinyl strip were brought near the charged pith ball without making contact. Try it. Record your observations in your notebook. Be careful not to accidentally ground the pith ball.

8. Next, rub an acetate strip with wool. Bring the strip close to the charged pith ball. (Recharge the pith ball with the charged vinyl strip if necessary.) Record your observations in your notebook. Use your observations to determine the charge on the acetate strip. Record the charge in Table 1.

9. Test each pair of materials given in Table 1. Record the charge of each object in your table.

Analyze and Evaluate

(a) Review the Testable Question and your hypothesis. Explain whether your results support your hypothesis. Remember to state your evidence.

(b) Explain your observations in step 4.

(c) Explain your observations in step 7.

(d) Summarize how the charged pith ball electroscope is used to determine the charge on an object.

Apply and Extend

(e) What would happen if a charged strip is placed between two electrically neutral pith balls hanging side by side as shown in Figure 2?

(f) Predict what would happen if the following combinations of charged strips are used to simultaneously touch the pith balls in Figure 2 and then are taken away. Justify your prediction in each case.

  (i) two negatively charged strips
  (ii) two positively charged strips
  (iii) one negatively charged strip and one positively charged strip

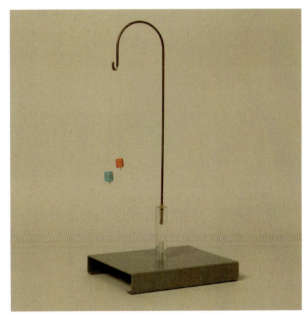

Figure 2 A double pith ball electroscope consists of two neutral pith balls hanging side by side.

11.4 Conductors and Insulators

conductor a material that lets electrons move easily through it

insulator a material that does not easily allow the movement of electrons through it

Scientists have categorized most materials into two categories: conductors and insulators (Figure 1). **Conductors** are materials that allow the movement of electrons. **Insulators** are materials that inhibit or prevent the movement of electrons. Both conductors and insulators have practical applications.

DID YOU KNOW?

Superconductors
As electrons move in a conductor, some of their energy is lost as thermal energy. Scientists have discovered that some conductors, when cooled below a critical temperature, conduct electricity without losing energy. These conductors are called superconductors and are essential in high-speed computers. Modern superconductors work best at temperatures of about −150 °C. However, chilling these devices is costly. So, Canadian researchers, like Michael Thewalt at Simon Fraser University in British Columbia, are developing a room temperature superconductor.

(a)

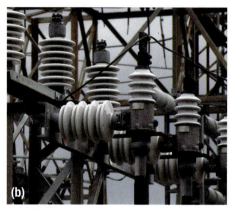
(b)

Figure 1 (a) Copper wire is a conductor. (b) Ceramic is an insulator.

Conductors

Good conductors allow electrons to move through them with ease. Fair conductors allow electrons to move through them with a small amount of difficulty. The most familiar conductors are metals such as copper and aluminum. These metals can be used in the wiring that appears in electrical cords in appliances, such as lamps and televisions, and in the walls of your home. Some non-metals such as graphite (a form of carbon) are also reasonably good conductors of electricity. Graphite and silicon are examples of materials called semiconductors because they allow electrons to move through them, although not as easily as in a good conductor. Table 1 lists some common conductors.

Could you place a charge on a conductor if you were holding it in your hand? No—any charge would immediately pass through the conductor into your hand. If you think back to the Try This Activity in Section 11.1, this is why the copper pipe did not pick up any paper pieces. Any excess charge travelled to your hand.

You have probably heard or read warning labels stating that you should never use an electrical appliance near water. Why? Consider the conductive property of water. Table 1 lists salt water as a fair conductor. This is because it contains ions (recall that ions are charged atoms). Ions pass through salt water easily. Pure water, which has no ions, is non-conductive. The water in a typical home contains minerals and ions and is a fair conductor of water. Touching an electrical appliance when you are near water, or damp from a bath or shower, puts you at greater risk of injury from the movement of a large amount of electric charge through your body.

Table 1 Common Conductors

Good conductors	Fair conductors
silver	graphite (carbon)
copper	nichrome
gold	the human body
aluminum	damp skin
magnesium	acid solutions
tungsten	salt water
nickel	Earth
mercury	water vapour in air
platinum	silicon
iron	germanium

Insulators

Insulators are the answer to using electricity safely. In an insulator, the electrons are tightly bound to the atoms that make up the material, so they are not free to move to neighbouring atoms. Insulators include non-metals such as plastic, ebonite, wood, and glass (Table 2).

From about 1880 to the 1930s, copper conductors in the walls of homes were passed through protective porcelain insulating tubes, and supported along their length on nailed-down porcelain knob insulators. This type of electrical wiring is known as "knob and tube" wiring (Figure 2). Today, electrical wires in homes are coated with an insulating material, such as plastic. Electricity distribution also uses insulators. For example, electricity transmission wires are connected to poles using insulating materials (Figure 3). Insulators protect us from the danger of having large amounts of electric charge move through our bodies if we come in contact with a conductor. Placing a plastic insulator around a copper wire prevents electrons from escaping the conductor. In this way, the electrons stay in the conductor and discharges are avoided.

Table 2 Common Insulators

Good insulators	
oil	plastic
fur	wood
silk	paper
wool	wax
rubber	ebonite
porcelain, glass	pure water

Figure 2 Knob and tube wiring was common in homes up to about 1930.

Figure 3 These workers are installing new wires and insulators on this hydro pole.

Using Conductors and Insulators

Different types of technology use conducting and insulating materials to work properly. For example, technologies such as lightning rods and automobile frames and bodies are made of conducting materials. Glass has been used as an insulator in telegraph wiring since the 1800s. The telecommunications industry uses teflon and silicon dioxide as electrical insulators to protect against exposed wiring.

A laser printer is a common household electrical device that requires the properties of both insulators and conductors.

Laser Printers

Laser printers use conductors and insulators, as well as the Law of Electric Charges, to work (Figure 4). A laser printer consists of a drum made of a positively charged photoconductor. A photoconductor is a special class of conductor that conducts electrons only when a light shines on it. If no light shines on it, it remains an insulator.

Figure 4 Laser printers use conductors and insulators to produce images.

11.4 Conductors and Insulators

WRITING TIP

State Key Points
When writing persuasive text, use the body of the text to explain the key points of your argument. State your point clearly, and then support your point with detailed evidence. Use connecting words to show the relationship between ideas.

In this case, the light is a laser. The laser light quickly "draws" the image to be printed across a positively charged selenium drum, causing these areas to become negatively charged (Figure 5(a)). The drum is then rolled across positively charged toner particles that are attracted to the negatively charged areas on the drum and are repelled by the positive areas (Figure 5(b)). When the drum is rolled over paper that has been given a larger negative charge than the drum, the positive toner particles become attracted to the negative paper (Figure 5(c)). The paper then passes through a fuser that melts the toner particles, which are made from plastic, onto the paper using temperatures of over 200 °C. The paper does not catch fire because of how quickly this process happens. There is also a fan that keeps the area around the fuser cool.

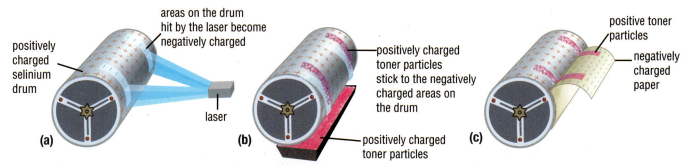

Figure 5 (a) The laser "draws" the image on the drum, making these areas negatively charged. (b) The drum rolls across the positively charged toner, which sticks to the negatively charged laser "drawing." (c) The drum rolls across paper with a higher negative charge and the toner particles "stick" to the paper.

UNIT TASK Bookmark

How can you apply what you have learned about conductors and insulators to the Unit Task described on page 586?

IN SUMMARY

- Conductors allow electrons to pass through them easily.
- Insulators inhibit or prevent the movement of electrons.
- Semiconductors have special properties that make them fair conductors, not good conductors.
- Technologies such as laser printing use conductors and insulators.

CHECK YOUR LEARNING

1. Provide a brief explanation of conductors and insulators, giving at least two examples of each. K/U

2. Compare the conductivity of pure water and salt water. Relate this information to the conductivity of water in and around your home. T/I A

3. If you use an insulator to touch a conductor that is carrying electricity, will you receive a shock? Explain. K/U

4. Two new materials have been discovered. One is shiny and has a metallic look, while the other is dull and has a non-metallic look. Although you think that one is a conductor and the other an insulator, you want to be certain. Describe a test you could do to test the conductivity of these two materials. T/I

5. Create a pamphlet to explain how laser printers operate using electric charges. Include diagrams in your explanation. K/U C

6. Electricians often use screwdrivers that have thick rubber handles. Explain why. K/U

7. A golfer and her caddy see lightning nearby. The golfer is about to take a shot with a metal club, while her caddy is holding a plastic-handled umbrella. Which person is at greater risk? Explain your answer. K/U

AWESOME SCIENCE ✓ OSSLT

A Hair-Raising Experience

"Whatever you do don't let go." The Ontario Science Center technician hits the switch on the Van de Graaff generator. Soon you feel a bizarre tingling sensation as all your hair is standing on end (Figure 1). Have a look in the mirror—nice hairstyle!

Figure 1 Known as "the ball that makes your hair stand up", a Van de Graaff generator is a common static electricity display at science centres. The Ontario Science Centre, for example, has one on display.

How does it work?

A Van de Graaff generator consists of a large hollow metal sphere resting on cylinder made from an insulating material. Inside the cylinder, a belt is stretched tightly across two pulleys. Friction between the belt and the pulleys either transfers electrons onto the belt or removes them. The belt acts like an electron "escalator"— carrying electrons either to the sphere or taking them away from the sphere. This escalator runs in only one direction. The direction of electron movement determines the charge of the sphere. As electrons build up on the sphere, the sphere becomes negatively charged. The amount of charge that collects depends on the size and shape of the sphere. Soon, there so many charges that the sphere can no longer hold onto them. When this happens, charges "leak" onto nearby air molecules (Figure 2).

Figure 2 When electrons are picked up by air molecules the energy is released in a flash of light, or a "spark".

The Generator

The Van de Graaff generator was invented by American physicist Robert Jemison Van de Graaff in 1931. It was designed to create large amounts of static electricity for use in experimentation. The largest Van de Graaff generator in the world, built by Dr. Van de Graaff himself, is on display at Boston's Museum of Science (Figure 3). The 4.5 m-high structure can produce up to 2 million volts of electricity. Today, Van de Graaf generators can produce up to 20 million volts of electricity!

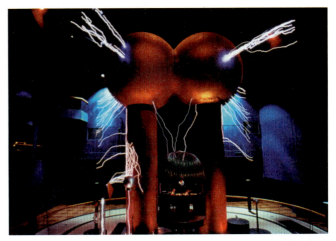

Figure 3 The Boston Museum of Science houses the largest Van de Graaff generator in the world.

Van de Graaff generators are also used to supply the high energy needed for particle accelerators, or "atom smashers." These devices accelerate sub-atomic particles to very high speeds and then "smash" them into target atoms. The collisions that result create new sub-atomic particles and high-energy radiation such as X-rays. This process is the founding principle of a branch of science known as particle and nuclear physics.

Spiked hair

As long as your hand stays on the sphere, the surface of your body has the same charge as the sphere. If the sphere is negatively charged, electrons from the sphere quickly cover the surface of your body, including every strand of hair. Your new "energized" hairstyle is a result of the negatively charged strands of hair repelling one another.

11.5 PERFORM AN ACTIVITY

Testing for Conductors and Insulators

Conductors and insulators can be tested by using a metal leaf electroscope (Figure 1). When a metal leaf electroscope is charged, its leaves (carrying the same charge) will repel each other. If a conductor comes in contact with the sphere of the electroscope, the charged metal leaf electroscope will discharge, the leaves will fall, and its charge will be lost as electrons pass through the conductor to or from your body. If the material being tested is an insulator, the leaf electroscope will retain its charge and the leaves will remain raised. In this activity, you will explore why this happens.

SKILLS MENU
- Questioning
- Hypothesizing
- Predicting
- Planning
- Controlling Variables
- Performing
- Observing
- Analyzing
- Evaluating
- Communicating

Figure 1

In this activity, you will induce a negative charge on an electroscope by touching the top of it with a charged ebonite rod. You will know the electroscope is charged because the leaves of the electroscope will rise. You will then test each material to determine whether it is a conductor or an insulator.

Purpose

To identify what types of materials are conductors and what types are insulators.

Equipment and Materials

- metal leaf electroscope
- ebonite rod
- piece of fur (or wool)
- testing materials, such as a copper penny, chalk, an iron rod, a plastic drinking straw, a glass rod, pencil "lead" (graphite)

Procedure

1. Obtain the objects that you want to test and place them at your work area.
2. Copy Table 1 into your notebook.

Table 1 Test Results

Object	Position of electroscope leaves after touching the object (raised or lowered)	Electroscope remains charged? (yes or no)	Conductor or insulator?
copper penny			
chalk			
iron rod			
plastic drinking straw			
glass rod			
pencil "lead"			

484 Chapter 11 • Static Electricity

3. Charge the ebonite rod by rubbing it with a piece of fur (or wool). Touch the ebonite rod to the electroscope terminal in order to charge the electroscope (Figure 2). The leaves should be in the raised position now. If not, repeat the charging process. Do not touch the electroscope terminal with your hands at any point during the testing process.

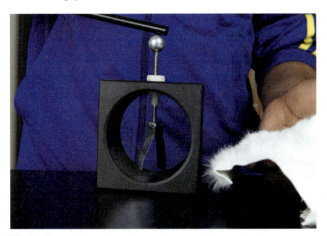

Figure 2

4. Touch the copper penny to the electroscope terminal. Record your observations in the third column of your table.

5. If the electroscope leaves have returned to the neutral position (hanging down), you must recharge the electroscope as you did in step 3. If the leaves remain raised, the metal leaf electroscope is still charged and you do not need to recharge it.

6. Repeat steps 4 and 5 with the remaining materials and record your observations in your table.

Analyze and Evaluate

(a) Use your observations to complete the last two columns in your table. **T/I**

(b) Explain how your observations helped you to make your decisions in (a). **T/I** **C**

(c) Why do you think you were cautioned not to touch the electroscope terminal with your hands during the testing? What do you think would have happened if you did? Discuss your answer with a classmate. **T/I** **C**

(d) Create a t-chart to sort the materials that were conductors from the materials that were insulators. Do the conductors have something in common? How about the insulators? **T/I** **C**

Apply and Extend

(e) Would the metal leaf electroscope work if the terminal were made from an insulating material? Explain your answer. **T/I** **C**

(f) The leaves of the electroscope repelled each other when given a negative charge. Do you think it would be possible to get the leaves of the electroscope to attract each other? Explain. **T/I**

(g) Does this activity have any way of indicating how good an electrical conductor or insulator the material is? If yes, indicate how. If no, what would be required? **T/I**

(h) How might the results of this experiment have been influenced by differences in humidity on the day it was performed?

11.6 Charging by Induction

You have learned that objects can be charged by conduction when they come in contact with a charged object. However, the same charged object can also be used to charge a neutral object without contact. This process is called **charging by induction**. Objects can be temporarily or permanently charged by induction.

charging by induction charging a neutral object by bringing another charged object close to, but not touching, the neutral object

Charging Objects Temporarily by Induction

Recall from Section 11.1 that when a charged object is brought near a neutral object it causes (induces) the electrons to shift in position, resulting in an uneven distribution of charges. This will only be temporary as the electrons will move back to their original positions once the charged object is taken away.

Figure 1 shows a negatively charged balloon that is brought near a neutral wall. The electrons in the balloon repel the electrons in the wall, causing an induced charge separation in the wall (the electrons in the wall move away from the balloon). This creates a positive charge on the surface of the wall, which the negatively charged balloon is attracted to. The result is that the balloon moves toward the wall. The wall remains neutral because it still contains the same number of positive charges as negative charges.

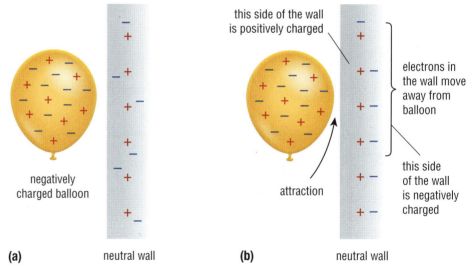

Figure 1 A negatively charged balloon (a) is brought near a neutral wall, causing an induced charge separation in the wall. (b) The wall becomes temporarily charged by induction.

An everyday example of charging by induction occurs with the buildup of dust on the screen of a television or computer monitor (Figure 2). When a computer monitor or television screen is turned on it begins to build up a charge. When a neutral dust particle comes near the screen, the charge on the screen induces an opposite charge on the near side of the dust particle and a charge, similar to that on the screen, on the far side. The result is that the dust is attracted to the screen.

Figure 2 After being charged by induction, the dust is attracted to the computer screen.

TRY THIS | BENDING WATER

SKILLS: Predicting, Observing, Communicating

SKILLS HANDBOOK
3.B.3., 3.B.6., 3.B.9.

You have read about charge interactions between solid objects. In this activity, you will observe charge interactions between a solid and a liquid.

Equipment and Materials: faucet; balloon

1. Blow up the balloon.
2. Run a gentle stream of water from a faucet. Place the balloon beside, but not touching, the stream of water. Record your observations in a diagram.
3. Rub the balloon against your hair to charge it.
4. Predict what you think will happen when you bring the charged balloon near, but not touching, the stream of water. Test your predictions and use a diagram to record your observations.
5. Try moving the charged balloon to the other side of the stream of water. Does the same thing happen?

A. Does your observation from step 2 prove that both the balloon and the water are neutral? Explain. T/I
B. Use your knowledge of electrons to explain your observations in step 4. T/I
C. Predict whether it is possible to bend water away from a charged object. Explain your reasoning. T/I

Charging Objects Permanently by Induction

An object can be permanently charged by induction by grounding the object. For example, consider a negatively charged ebonite rod and a neutral pith ball. When the rod is brought near but not touching the pith ball, the electrons in the pith ball are repelled by the electrons in the rod. As a result, the side of the pith ball closest to the rod becomes temporarily positively charged, while the side farthest from the rod becomes temporarily negatively charged (Figure 3(a)). If you then ground the negatively charged side with your hand, some of the electrons travel from the pith ball into your hand, and the pith ball is left with a positive charge. You could also remove the electrons by connecting a conducting wire to the ground (Figure 3(b)). When the conducting wire is disconnected from the pith ball, the pith ball is left with a permanent positive charge (Figure 3(c)). For the charge to be permanent, the ground must be disconnected or removed before the charged object is removed.

To learn more about charging by induction,

GO TO NELSON SCIENCE

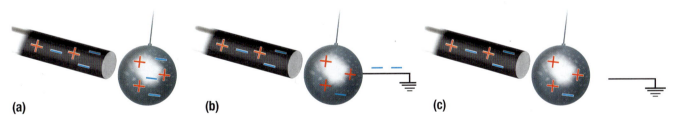

Figure 3 (a) When a negatively charged ebonite rod is brought near a neutral pith ball, the electrons in the pith ball are repelled and it becomes temporarily negatively charged on its right side. (b) Attaching a ground wire to the pith ball conducts the repelled electrons on the right side into the ground. (c) After removing the ground wire, the pith ball remains permanently positively charged.

Charging by induction always results in two objects with opposite charges. The object that induces the charge keeps its original charge, while the object whose charge was induced receives the opposite charge.

Technological Applications of Charging by Induction

Charging by induction has many useful applications, including forensics and air-cleaning technologies.

Electrostatic Lifting Apparatus

Footprints are often left behind at crime scenes. Investigators can use this important evidence to help determine who was present at the time of the crime. But how can you make a copy of a footprint if it is very difficult to see? Investigators use an electrostatic lifting apparatus (ESLA) (Figure 4). Special film or foil is placed over the footprint. The black side of the film is placed over the footprint. The film is then electrostatically charged. The dust and dirt particles from the footprint are attracted to the black side of the film. The dust particles "jump" off the floor onto the black film, revealing the details of the footprint. Now investigators have a copy of the footprint on the film that they can take to a laboratory to analyze.

Figure 4 An electrostatic lifting apparatus is used in forensics to create a copy of a footprint from a crime scene.

To learn more about being a forensic investigator,

GO TO NELSON SCIENCE

Electrostatic Speakers

You have probably already heard the results of one application of charging by induction—that of electrostatic loudspeakers (Figure 5). These speakers are constructed of three thin layers. The outer two layers, called stators, are fixed in place and are made of a porous material. The inner layer is a flexible film called the diaphragm. In order to produce sound, the diaphragm must vibrate. This is accomplished using the principles of induction and the Law of Electric Charges. First, the inner surface is given a permanent electrical charge. Then an audio transformer is used to induce opposite charges in the two outer plates. This causes the diaphragm to move—as it is simultaneously attracted to one outer plate and repelled by the other (Figure 6(a)). The audio transformer then rapidly induces the static charges in the outer plates to reverse themselves. This causes the diaphragm to now move toward the opposite outer plate (Figure 6(b)). This reversing of charges on the outer plates happens repeatedly and at variable frequencies causing the diaphragm to rapidly vibrate back and forth between the plates, producing sound waves in the air.

Electrostatic speakers have the advantage of being extremely thin and light weight. A disadvantage is their poor bass response.

Figure 5 Electrostatic loudspeakers operate on the principles of induction.

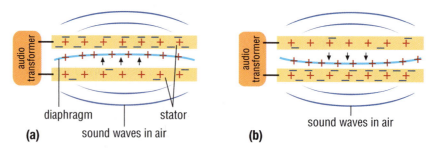

Figure 6 (a) In an electrostatic speaker, an audio transformer induces opposite electrical charges on two outer plates (stators) and then (b) reverses them. This process is repeated rapidly causing a flexible and charged inner membrane (diaphragm) to vibrate back and forth between the plates, producing sound waves in the air.

IN SUMMARY

- A temporary charge imbalance can be induced in a neutral object by bringing a charged object near it.
- Objects can be charged permanently by induction by bringing a charged object near a neutral object and then grounding the neutral object.
- Charging by induction produces a separation of charge in the object that is charged.
- Charging by induction has many applications, including electrostatic loud speakers and lifting footprints.

CHECK YOUR LEARNING

1. Identify a concept that you found particularly difficult or confusing. Have a classmate explain the concept to you. Write a brief description of the concept with your new understanding of it. Then help your classmate understand a concept he or she found challenging. K/U C

2. Copy and complete Table 1 in your notebook. Assume that the object getting charged is neutral just prior to using the charging method. K/U

   Table 1 Charges on Objects

Charging method	Object doing the charging	Object getting charged	Explanation of the movement of charge
charging by induction (temporary)	positive		
charging by induction (temporary)	negative		
charging by induction (permanent)	positive		
charging by induction (permanent)	negative		

3. Use diagrams to show how you would
   (a) induce a positive, temporary charge on the right side of a metal ball
   (b) induce a permanent negative charge on a metal ball K/U C

4. Identify and explain one kind of technology that uses charging by induction. A

5. The leaves on a metal leaf electroscope repel each other even though no other objects are near it. With the use of diagrams, explain what must have happened if no charged object ever touched the electroscope. T/I C

6. (a) What would happen if you charged a balloon by rubbing it against your hair and then brought it near another balloon that is neutral? Explain. K/U
   (b) What would happen if the charged balloon were allowed to touch the neutral balloon? Explain. K/U

11.7 PERFORM AN ACTIVITY

Charging Objects by Induction

You may have noticed that the screens on some electronic devices, such as TVs, DVD players, and computer monitors, seem to attract dust. The neutral dust particles have been attracted to the screen due to a charge imbalance.

In this activity, you will use a metal leaf electroscope to observe charging by induction temporarily and permanently.

SKILLS MENU
- Questioning
- Hypothesizing
- Predicting
- Planning
- Controlling Variables
- Performing
- Observing
- Analyzing
- Evaluating
- Communicating

Purpose

To observe charging by induction using a metal leaf electroscope.

Equipment and Materials

- metal leaf electroscope
- ebonite rod
- fur (or wool)

Procedure

SKILLS HANDBOOK 3.B.5.

Part A: Temporarily Charging by Induction

1. In your notebook, create a table to record your observations.
2. Be sure that the metal leaf electroscope is neutral by touching (grounding) it with your fingers.
3. Rub the ebonite rod with the fur to make the ebonite rod negatively charged.
4. Bring the charged ebonite rod close to, but not touching, the terminal of the electroscope. Observe the leaves of the electroscope. Record your observations in your table.
5. Move the ebonite rod closer to (but not too close) and farther away from the electroscope terminal and observe what happens to the leaves. Record your observations.

Part B: Permanently Charging by Induction

6. In your notebook, create a table to record your observations.
7. Repeat steps 2 and 3 from Part A.
8. Bring the charged ebonite rod close to, but not touching, the electroscope terminal. Without moving the ebonite rod, touch your finger to the top of the electroscope terminal opposite where the ebonite rod is. Move your finger away and observe the electroscope leaves. Record your observations in your table.
9. Move the ebonite rod away from the electroscope terminal. Observe the leaves and record your observations.
10. Charge the ebonite rod again with the fur. Bring the rod close to, but not touching, the electroscope terminal and observe the leaves. Record your observations.
11. Try moving the ebonite rod closer to (but not too close) and farther away from the electroscope and observe the leaves. Record your observations.

Analyze and Evaluate

(a) For step 4 in Part A, draw a series of diagrams showing how the charges in the electroscope were affected by the negatively charged ebonite rod. K/U C

(b) Refer to your observations table. What did you notice happening in step 5? What do you think was happening to the charges in the leaves of the electroscope to cause what you observed? Discuss your answer with a classmate. T/I

(c) For step 8 in Part B, draw a series of diagrams showing how the charges in the electroscope were affected by the negatively charged ebonite rod and the presence of your finger. T/I C

(d) For step 9 in Part B, draw a diagram showing how the charges are now distributed in the electroscope leaves.

(e) Compare your observations from steps 5 and 11. Suggest a reason for the similarities and differences.

Apply and Extend

(f) If you had to repeat this experiment starting with a positively charged ebonite rod, what do you think would happen to your results? Use diagrams to support your answer.

(g) If you were wearing rubber gloves when you were performing step 8 of the procedure in Part B, would you expect similar or different results? Explain your reasoning.

(h) The same activity could be performed with a pith ball electroscope. Write a short report explaining the results of the activity and any challenges you would experience when performing the activity.

(i) Suggest at least one advantage of using a metal leaf electroscope over a pith ball electroscope.

(j) Use Figure 1 to answer the following questions.

Figure 1

(i) Is the electroscope charged or neutral? Explain your reasoning.

(ii) Copy Figure 1 into your notebook and add symbols to show the charges on each object. Assume that the object is positively charged.

11.8 Electric Discharge

In many hospitals, medical staff members are required to wear shoes with special soles to prevent the buildup of a static charge as they walk. The sudden transfer of a static charge could damage sensitive medical equipment (Figure 1). Damage to electronic devices can occur when a negatively or positively charged object touches the device. What is really happening when static charges are transferred? Are all transfers of static charges dangerous and destructive? Can they be useful?

DID YOU KNOW?
Memory Loss
Electric discharges can damage some forms of computer memory and can also cause damage to computer software. In particular, this can happen when the components of the computer are exposed.

For this reason, computer technicians wear antistatic wrist straps when building computer hardware. This prevents any electric charges that have built up on the body from being discharged onto the delicate parts of the computer.

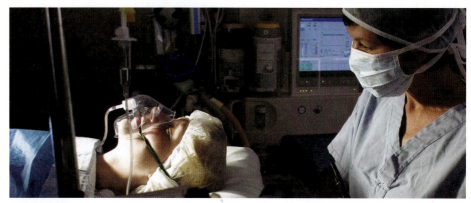

Figure 1 Electronic equipment in hospitals could be damaged by an electric discharge, potentially harming the patient.

When two objects that have a charge imbalance are brought close together or come in contact, electrons are transferred from one object to the other object. We call this rapid transfer of excess charge an **electric discharge**. Electrons leave one object and pass into another object. Electrons always move from the object with the more negative charge (less positive) to the object with the less negative charge (more positive). Discharges can sometimes be seen as sparks. The greater the charge imbalance, the larger and more noticeable the discharge will be. Small discharges are not dangerous or painful, such as when you remove your sweater and hear a crackling sound. However, a larger discharge can sometimes hurt. When an electric discharge occurs, the air temperature around the discharge increases. If the discharge is large enough, this increase can cause a burn on the skin. Discharges can also damage circuits in electronic equipment, such as medical equipment or computer hardware.

electric discharge the rapid transfer of electric charge from one object to another

To learn more about being a computer technician,

GO TO NELSON SCIENCE

RESEARCH THIS | ELECTRIC DISCHARGE

SKILLS: Researching, Communicating

SKILLS HANDBOOK
4.A.

It is important to prevent electric discharges when repairing electronics. Electric discharge can damage some of the components. This is of particular importance to computers because their sensitive electronics can be ruined by a static discharge.

1. Research how you can prevent an electric discharge.
2. Antistatic devices are items that reduce the buildup of static electricity on objects. Research two different antistatic devices.
3. Research antistatic devices that are specifically used to reduce static electricity in electronic devices.

A. Use a diagram to explain electric discharge. K/U C
B. Explain how each of the antistatic devices you researched in step 2 works. What are the practical applications of each? K/U
C. Which of the antistatic devices in step 3 do you think is most useful for preventing static discharge in electronic devices? Why? T/I

 GO TO NELSON SCIENCE

Have you ever received an electric shock from touching a car door handle? Suppose a car door handle becomes negatively charged. When you reach out to touch the car door handle, you receive a shock. Figure 2 explains why this happens.

Your hand is neutral and the car door handle is negatively charged (Figure 2(a)). When your hand comes close to the car door handle, electrons are transferred from the car door handle to your hand, causing an electric discharge (Figure 2(b)). After the discharge, the car door handle and your hand have the same charge (Figure 2(c)). It is important to remember that the small numbers of symbols in the diagrams represent much larger numbers of charges.

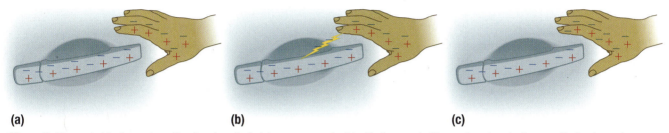

(a) (b) (c)

Figure 2 Charged objects, such as the door handle in (a), can cause electric discharges. In (b), coming close to the negatively charged handle enables the excess electrons to be transferred to the hand, producing an electric discharge. (c) After the discharge, the imbalance of electrons between the hand and the handle is gone.

Lightning

Lightning is a very dramatic electric discharge. Although the process is not fully understood by scientists, it occurs because of a charge imbalance between clouds, or between clouds and the ground. One theory suggests that when water droplets in clouds move past one another, they become charged. Electrons are transferred from rising water molecules to falling water droplets. As a result, the negatively charged water molecules collect at the bottom of the cloud (Figure 3).

To learn more about lightning,
GO TO NELSON SCIENCE

Figure 3 Lightning is an electric discharge that occurs between clouds or between clouds and the ground.

The excess negative charge at the bottom of the cloud repels the electrons at Earth's surface. Electrons move away from the area on Earth's surface near the cloud, causing it to become positively charged. The overall result is a charge imbalance between the bottom of the cloud and Earth's surface. If the charge imbalance becomes great enough, the excess electrons may be rapidly transferred from the cloud to the ground in the form of lightning. The resulting large transfer of electrons causes the surrounding air to become superheated. This produces both the flash of light and the rumbling sound of thunder. Lightning strikes can occur from cloud to cloud, from cloud to Earth, and from Earth to cloud (Figure 4).

Figure 4 Notice that the lightning is travelling between clouds as well as to the ground.

Lightning Rods

Lightning is dangerous to people, structures, and technology. To minimize these dangers, lightning rods are often placed on top of buildings to provide a safe path for lightning to follow to the ground (Figure 5). Although tall objects such as trees and towers get hit by lightning more often than shorter objects, lightning can also sometimes strike the ground beside tall objects. When lightning strikes, it usually travels along the path that most easily transfers electrons to or from the ground.

A lightning rod is usually made of metal, such as iron or copper. There is also a wire that goes from the rod into the ground.

Figure 5 The lightning rod on this home directs any lightning strikes safely into the ground.

Figure 6 shows the typical placement of a lightning rod on a house. When lightning strikes, the house is protected. The lightning safely passes through the rod and down the wire into the ground, instead of through the building.

To learn more about lightning rods,
GO TO NELSON SCIENCE

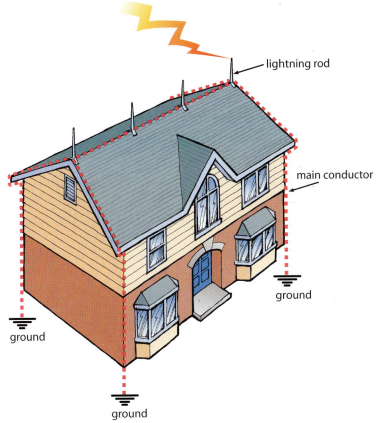

Figure 6 Lightning rods are common on homes and barns in the country, where large open areas make these buildings more vulnerable to lightning strikes.

UNIT TASK Bookmark

How can you apply what you have learned about grounding and releasing charges to the Unit Task described on page 586?

IN SUMMARY

- An electric discharge is the rapid transfer of electrons from one object to another.
- Electric discharges can produce visible sparks.
- Lightning is a visible and dramatic electric discharge.
- Lightning rods are used to safely direct lightning into the ground.

CHECK YOUR LEARNING

1. What precautions should you take before working with electronic equipment? Explain why. K/U

2. Suppose that you took off your sweater and noticed that your hair was standing up. Using six electrons as an example, and assuming that the electrons are transferred from the sweater to your hair, draw a series of diagrams showing how the excess electrons in your hair could be discharged
   (a) to your neutral hand
   (b) to the neutral ground K/U C

3. Recall a personal experience that involved an electric discharge. Explain how the information in this section has helped you understand your experience. C

4. What role does lightning play in nature? How do you think this might affect the economy? K/U A

5. Lightning sometimes happens between clouds. Suggest a possible way that a discharge can occur between clouds. K/U

11.8 Electric Discharge

CHAPTER 11 LOOKING BACK

KEY CONCEPTS SUMMARY

Static electricity results from an imbalance of electric charge on the surface of an object.

- Electrons can move from object to object. (11.1)
- The number of electrons and protons determines whether an object is negatively charged, positively charged, or neutral. (11.1)
- The Law of Electric Charges states that like charges repel, opposite charges attract. (11.1)
- Neutral objects can be attracted to charged objects. (11.1)

There are many useful applications of static electricity.

- Electrostatic paint sprayers use static electricity to attract the paint to the object to be painted. (11.1)
- Examples of technologies that use electrostatic charges include electrostatic dusters and electrostatic precipitators. (11.1, 11.2)
- Laser printers use both insulators and conductors to function. (11.6)

Objects can be charged by contact or by induction.

- Two different neutral objects become oppositely charged when they come in contact or rub against each other. (11.2)
- A neutral object may also become charged if it comes in contact with an already charged object. (11.1)
- When two charged objects come in contact, electrons may move from one object to the other. (11.2, 11.3)
- Objects and parts of objects can be temporarily or permanently charged by induction. (11.6)

Electric discharge can be harmful and must be treated with caution.

- An electric discharge occurs when a large charge imbalance is released. (11.8)
- Electric discharges often produce a visible spark. (11.8)
- Grounding is the safe discharging of excess charge using the ground. (11.8)
- Lightning is a very dramatic electric discharge that occurs because of a huge charge imbalance between clouds or between clouds and the ground. (11.8)

Materials can be conductors or insulators.

- Conductors such as copper wiring allow electrons to move through them easily. Semiconductors, such as carbon and silicon, allow electrons to move through them with some difficulty. (11.4)
- Insulators such as plastic, ebonite, wood, and glass reduce the movement of electrons. (11.4)

Simple procedures can be used to test the ability of materials to hold or transfer electric charges.

- A pith ball electroscope can be used to detect the presence of an electric charge. A charged object will cause the pith ball to move. (11.1)
- A metal leaf electroscope can also be used to determine the presence of an electric charge, either by contact, or by bringing a charged object near the electroscope terminal. (11.1, 11.7)

WHAT DO YOU THINK NOW?

You thought about the following statements at the beginning of the chapter. You may have encountered these ideas in school, at home, or in the world around you. Consider them again and decide whether you agree or disagree with each one.

Vocabulary

electric charge (p.465)
neutral object (p.466)
negatively charged object (p.466)
positively charged object (p.466)
static electricity (p.467)
electric force (p.468)
induced charge separation (p.468)
charging by friction (p.472)
electrostatic series (p.473)
charging by conduction (p.474)
grounding (p.475)
conductor (p.480)
insulator (p.480)
charging by induction (p.486)
electric discharge (p.492)

1 Clothes always get static cling when dried in a clothes dryer. **Agree/disagree?**

4 Dust sticks to all charged objects. **Agree/disagree?**

2 Static electricity always stays in one place and never moves from object to object. **Agree/disagree?**

5 Going down a plastic slide can make your hair stand up. **Agree/disagree?**

3 Static electricity can damage electronic devices. **Agree/disagree?**

6 Static electricity is a useful form of electricity. **Agree/disagree?**

How have your answers changed since then? What new understanding do you have?

BIG Ideas

- Electricity is a form of energy produced from a variety of non-renewable and renewable sources.
- The production and consumption of electrical energy has social, economic, and environmental implications.
- ✓ Static and current electricity have distinct properties that determine how they are used.

CHAPTER 11 REVIEW

The following icons indicate the Achievement Chart category addressed by each question.

- **K/U** Knowledge/Understanding
- **T/I** Thinking/Investigation
- **C** Communication
- **A** Application

What Do You Remember?

1. Describe the following terms in your own words: (11.1, 11.2, 11.4, 11.6) **K/U**
   (a) the Law of Electric Charges
   (b) charging by friction
   (c) charging by conduction
   (d) conductor
   (e) insulator
   (f) charging by induction

2. Describe the difference between charging by friction and temporarily charging by induction. (11.2, 11.6) **K/U**

3. Explain how the following devices work. Use diagrams in your explanations. (11.1) **K/U** **C**
   (a) a pith ball electroscope
   (b) a metal leaf electroscope

4. Describe how two different materials are charged by friction. (11.2) **K/U**

5. Draw a diagram using the appropriate charges showing (11.1) **K/U** **C**
   (a) an attraction between charges
   (b) a repulsion between charges
   (c) an attraction of a charged object to a neutral object

6. Determine the type of charge on the following objects. An object having (11.1) **K/U**
   (a) 4 positive charges and 3 negative charges
   (b) 7 positive charges and 9 negative charges

7. List two applications of electrostatics. (11.1, 11.2, 11.6) **A**

8. Explain the operation of a laser printer. (11.6) **K/U** **C**

9. Copy and complete the following graphic organizer (11.1, 11.2, 11.8) **K/U** **C**

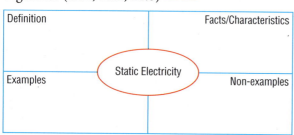

What Do You Understand?

10. Does static electricity mean that the charges never move? Explain. (11.1) **K/U** **C**

11. Explain, using a diagram, why electrons travel easily between atoms, while protons do not. (11.1) **K/U** **C**

12. There are plastic decals that you can attach to a window without using glue (Figure 1). Explain how these decals stay on the window. (11.1, 11.2, 11.6) **C** **A**

Figure 1

13. Can charging by friction occur only in solids? Explain using an example. (11.2) **T/I** **C**

14. During electrostatic experiments, you charged handheld insulators. Explain why you did not use conductors. (11.3, 11.4, 11.5) **K/U**

15. Draw diagrams showing how an object can be charged (11.1, 11.2, 11.6) **K/U** **C**
    (a) positively by temporary induction
    (b) negatively by conduction
    (c) neutral by grounding

16. Consult the electrostatic series on page 473 and determine the charge that each of these materials would receive: (11.3) **K/U**
    (a) polyester rubbed with nylon
    (b) wool rubbed with acetate
    (c) silk rubbed with glass
    (d) cotton rubbed with cotton

17. Two solids are both positively charged. Solid A is more positively charged than solid B. Using diagrams, show what would happen if the two materials came in contact and you observed a discharge. (11.8) **K/U** **C**

18. Neutral objects are attracted to charged objects. Explain why. (11.1) **K/U** **C**

498 Chapter 11 • Static Electricity

19. A neutral particle is travelling past an object that has a negative charge. What might happen to the neutral particle? Explain using the correct terms. (11.2, 11.6) K/U C

20. Explain the difference between how a metal leaf electroscope and a pith ball electroscope detect the presence of a charge imbalance. (11.1) K/U

Solve a Problem

21. A movie plot involves using a fictitious material called Element X. Describe how the scientist in the movie might use an electroscope to test whether Element X was a conductor or an insulator. (11.1, 11.2) T/I

22. A cat rubs against a student's rubber boot. The boot then touches the neutral pith ball of an electroscope. A piece of ebonite is brought near the pith ball electroscope, and the pith ball is observed to repel. What can you conclude about the type of charge on the ebonite? Explain your reasoning. (11.1, 11.2 11.3) T/I C

23. An experiment involves three charged objects: A, B, and C. Object A repels object B and attracts object C. Object C is repelled by ebonite charged with fur. What is the charge on each object? Explain your reasoning. (11.1, 11.3) T/I C

24. Carpets are usually made from different materials than socks. Explain what could be done to reduce the effect of static discharge when you walk across a carpet wearing socks. (11.1, 11.2) K/U

25. Suppose that you were sent to an area where flammable chemicals were being stored. The area has a cold, dry climate. What types of clothes would you wear to minimize the chance of an electrostatic discharge? Be specific. (11.2) K/U C

26. Why would a factory install an electrostatic precipitator on its chimney when there are financial costs associated with both the installation and the operation of the device? Discuss your answer with a classmate. (11.2) A

Create and Evaluate

27. Design your own lightning rod. Provide a labelled diagram. (11.8) T/I C

28. Use diagrams to explain how an electric air cleaner works. Your answer should include a discussion on the Law of Electric Charges. (11.2) K/U C

29. Design a procedure that would allow you to determine whether an unknown material is charged negatively or positively. You may use any equipment discussed in this chapter. (11.1, 11.2, 11.3) T/I C

30. Many different types of products remove dust from your furniture at home. Many of the products use the principle of static charges causing attraction of the dust to the product. Evaluate the effectiveness of different kinds of electrostatic dusters based on your own criteria. Suggest an environmentally responsible choice. (11.2) T/I C A

Reflect On Your Learning

31. In this chapter, you have learned that objects can be charged by induction—either permanently or temporarily. (11.1, 11.6)
    (a) How are temporary and permanent charging similar? How do they differ? K/U
    (b) Do you find these techniques easy to understand and explain?
    (c) How can using diagrams enhance your ability to communicate your understanding?

Web Connections

32. In 2003, NASA sent two rovers to Mars to collect data about the planet. Research how scientists may have had to deal with an imbalance of electrons. Write a short summary reporting on what you learned. (11.8) T/I C

33. Researchers are designing less expensive colour laser printers so that laser printers are as affordable as inkjet printers. Research some of the latest developments. Write a short summary of your findings. (11.4) T/I C

34. Compare the operation of a laser printer with the operation of a photocopier. List any differences or similarities. (11.4) C A

CHAPTER 11 SELF-QUIZ

The following icons indicate the Achievement Chart category addressed by each question.
K/U Knowledge/Understanding **T/I** Thinking/Investigation **C** Communication **A** Application

For each question, select the best answer from the four alternatives.

1. Rabbit fur is above rubber in the electrostatic series. What would happen if rabbit fur were rubbed with a rubber balloon? (11.2) **K/U**
   (a) The rabbit fur and the balloon would become positively charged.
   (b) The rabbit fur and the balloon would become negatively charged.
   (c) The rabbit fur would take on a positive charge, and the balloon would take on a negative charge.
   (d) The rabbit fur would take on a negative charge, and the balloon would take on a positive charge.

2. Which of these materials is the best insulator? (11.4) **K/U**
   (a) aluminum   (c) salt water
   (b) steel wool   (d) glass

3. A balloon is rubbed on carpet and assumes a charge. The balloon will stick to all of these except (11.6, 11.8) **K/U**
   (a) a metal door   (c) a glass window
   (b) a plaster wall (d) a wooden cabinet

4. Object A has a negative charge. Object A is repelled by object B. Object B is attracted to object C and repelled by object D. Object C is attracted to object D. What are the charges on objects B, C, and D? (11.1, 11.2) **K/U**
   (a) B, C, and D are all negative.
   (b) B and D are negative, and C is positive.
   (c) B is negative, and C and D are positive.
   (d) B and D are positive, and C is negative.

5. Metal ball A has a charge of +8, and metal ball B has a charge of +2. What will the charges on the balls be after they come in contact while remaining insulated from their surroundings? (11.1) **T/I**
   (a) A +2, B +8
   (b) A +2, B −4
   (c) A +5, B +5
   (d) A +8, B −6

Indicate whether each of the statements is TRUE or FALSE. If you think the statement is false, rewrite it to make it true.

6. Oppositely charged objects are attracted to each other. (11.1) **K/U**

7. In the electrostatic series, materials are arranged in order of their electrical conductivity. (11.2) **K/U**

8. Electricians use tools with metal handles for safety. (11.4) **K/U**

Copy each of the following statements into your notebook. Fill in the blanks with a word or phrase that correctly completes the sentence.

9. The charge on an object changes when atoms within the object gain or lose _____ . (11.1) **K/U**

10. When a charged object loses its charge by making contact with a large neutral object such as Earth, the process is called _____ . (11.8) **K/U**

11. Water that contains impurities conducts electricity because _____ carry the charge. (11.4) **K/U**

Match each term on the left with the most appropriate description on the right.

12. (a) electron   (i)   a term describing charges on objects that are fixed, rather than flowing as a current
    (b) electroscope   (ii) a material that prevents the flow of electrical current
    (c) electrostatic   (iii) a particle with a negative electrical charge
    (d) induction   (iv) a device used to demonstrate static charges
    (e) insulator   (v)   the process of charging an object from a distance
                              (11.1, 11.2, 11.4, 11.6) **K/U**

Write a short answer to each of these questions.

13. Explain what, on an atomic level, causes the buildup of an electrostatic charge. (11.1) K/U

14. Why are electrical wires coated with rubber or plastic? (11.4) K/U

15. Imagine you are installing lightning rods on a farmer's barn. (11.8)
    (a) Describe where you would position the rods and why.
    (b) Describe additional steps you would take to protect the barn from lightning strikes. T/I C

16. Suppose you are employed by an electrical company to repair damage to power lines. It is raining and you receive a call from the company to go out and repair a power line that fell down when a tree landed on it. Describe what you would wear on your hands and feet as you repaired the fallen power line. Justify your answer. (11.4) T/I C

17. You are performing an experiment with an electroscope (Figure 1).

Figure 1

   (a) Describe how you would charge the electroscope, and explain how you would know if you were successful.
   (b) Describe how you would discharge the electroscope, and explain how you would know if you were successful. (11.1, 11.2, 11.6, 11.8) T/I C

18. Figure 2 shows three balls hanging from strings. Redraw the diagram to show how the balls would hang if the centre ball and the left ball were positively charged and the ball on the right were negatively charged. Use the symbols "+" and "−" to mark the charge of each ball. (11.1, 11.3) T/I C

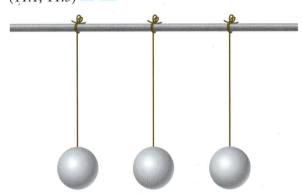

Figure 2

19. Using a diagram, explain how lightning occurs. (11.8) T/I C

20. When plastic adhesive tape that is stuck to glass is quickly pulled away, it becomes negatively charged. (11.1, 11.2)
    (a) Explain what happened to the tape in terms of electrons and protons. K/U
    (b) Explain what happened to the glass in terms of electrons and protons. C A

21. Explain why clothes cling to each other after tumbling around in a clothes dryer. (11.2) C A

22. Vinyl is near the bottom of the electrostatic series, while glass is near the top of the series. Design an experiment using glass and vinyl to find out where cat fur and hard rubber would be placed in the electrostatic series. (11.2) T/I A

23. Explain why trucks carrying flammable materials have what appears to be a metal chain attached to the underside that drags along the ground. (11.8) A

CHAPTER 12

Electrical Energy Production

KEY QUESTION: How do we meet our electrical energy demands?

UNIT E
The Characteristics of Electricity

CHAPTER 11 — Static Electricity

CHAPTER 12 — Electrical Energy Production

CHAPTER 13 — Electrical Quantities in Circuits

KEY CONCEPTS

The motion of electrons can be controlled.

An electric circuit requires a source of electrical energy, conductors, a control device, and a load.

Electrical energy is produced by energy transformations.

Electrical energy is produced from renewable and non-renewable resources.

Electrical power is the rate at which electrical energy is produced or consumed in a given time.

Electrical energy consumption should be reduced.

Looking Ahead

ENGAGE IN SCIENCE

Midnight Sun

I am an engineering student at the University of Waterloo. For the past two years, I have been working with a team of students to develop a solar-powered vehicle, *Midnight Sun*. Every two years, we enter the North American Solar Challenge. This competition encourages engineering students from colleges and universities across North America to build, design, test, and race solar-powered vehicles in a long distance road rally event.

The event also raises awareness of energy efficiency. We start in Dallas, Texas and finish in Calgary, Alberta 10 days later. That is a total of 3862.43 km. What a trip!

The North American Solar Challenge Route Map

Midnight Sun Team Blog
A race to win the North American Solar Challenge...

Day 10 We finished! We completed the last leg of the race in Calgary, Alberta, crossing the finish line at about 4:30 p.m. as a crowd of people cheered at the finish line.

Despite our flat tire yesterday, *Midnight Sun* performed well, reaching a maximum speed of 130 km/h! Unfortunately, it was not fast enough to win this year. When we return to Waterloo, we will refine our technology and see if we can get our racer to go faster.

Day 8 We have crossed the border into Manitoba! After heavy downpours and intense thunderstorms last night, we were worried that we had not harnessed enough energy from the Sun. Luckily, the sky cleared and allowed for a great charging session. On cloudy days, the car is powered by batteries. If the clouds stayed for a long time, we would not have enough stored electrical energy in the batteries to continue the race.

Day 6 We made it to Sioux Falls, South Dakota! *Midnight Sun* is currently in seventh place and is running extremely efficiently. Let me quickly explain how this works! Solar energy is collected from panels on the roof of the car and converted to electrical energy. This energy is then either stored in our batteries or used in our motor to power the car. The stored electrical energy is saved for cloudy days when the solar panels do not collect much power from the Sun. When it is sunny, the energy collected by the solar panels goes directly to the motor, which, in turn, moves the car—and we are off!

Day 1 This is the starting line for the North American Solar Challenge. We are anxious to get our solar-powered ride out on the road!

1. What do you think are the advantages and disadvantages of solar-powered vehicles?
2. Work with a partner to research events that are held in your community to raise awareness of energy consumption and efficiency. Share your findings with the class and generate a class bulletin board on the topic of energy efficiency. As you work through this chapter, revisit your ideas.

WHAT DO YOU THINK?

Many of the ideas you will explore in this chapter are ideas that you have already encountered. You may have encountered these ideas in school, at home, or in the world around you. Not all of the following statements are true. Consider each statement and decide whether you agree or disagree with it.

1 Using electricity creates no pollution.
Agree/disagree?

4 An incandescent light bulb is an efficient way to produce light.
Agree/disagree?

2 Batteries are a waste of money.
Agree/disagree?

5 Computers use very little energy.
Agree/disagree?

3 The EnerGuide label on an appliance indicates how efficient the appliance is.
Agree/disagree?

6 We can never run out of electricity.
Agree/disagree?

FOCUS ON READING

Evaluating

When you evaluate a text, you make judgments about its ideas, information, and credibility by applying your prior knowledge and analyzing the content. Support your judgments with strong reasons and specific details or quotations from the text. Ask the following questions when evaluating texts:

- Do I agree with the main idea? Does it seem reasonable?
- Is the information accurate and reliable?
- Do I agree with the author's point of view on the subject?
- Is the text biased? What values does it promote? What clues can I find?

READING TIP
As you work through the chapter, look for tips like this. They will help you develop literacy strategies.

Timely Use of Electrical Energy

The cost of electricity during the peak times can be three times more expensive. This means that if you use electrical energy outside of the peak times, such as running a dishwasher at night, you can save money and reduce pollution.

By 2010, all homes and businesses in Ontario will have smart meters (Figure 1). The smart meter is an electrical energy meter connected to your home that measures the amount of electrical energy being used as well as the time at which it is being used.

Figure 1 Many communities in Ontario have already started implementing smart meters.

Evaluating *in Action*

Evaluating means you question the information in a text in order to form an opinion or make a judgment. Here's how one student used the strategies to evaluate the selection about solar panels.

Questions I Asked of the Text	Prior Knowledge / Clues in the Text	My Final Opinion / Judgment
Do I agree with the main idea?	Using electricity during a peak period costs more and produces more pollution.	Monitoring both how much and when electricity is used is important.
Is the information accurate?	Actually, there are three periods when the cost of electricity can be more expensive: off-peak, mid-peak, and peak.	Increased demand during off-peak periods will make them mid-peak and peak periods.
Do I agree with the author's point of view?	The author appears to support smart meters as a way of conserving energy.	Smart meters just shift demand times, they do not conserve energy.
Is the text biased?	The text gives only the benefits, not the costs, of smart meters.	The costs of implementing smart meters will be recovered in the long run.

506 Chapter 12 • Electrical Energy Production

12.1 Introducing Current Electricity

Electricity is an important aspect of our daily lives (Figure 1). Most appliances use electricity that flows through conducting wires in the walls of our homes. What kind of electricity is this, and how do we produce it?

The Flow of Electrons

In Chapter 11, you learned that electric charges can build up on the surface of an object until they are discharged. An example of electric discharge can be observed when lightning moves from cloud to cloud, or from a cloud to Earth. Unlike static electricity, **current electricity** refers to electric charges (electrons) that flow through a conductor in a controlled way.

Electrons are always moving. The difference between static and current electricity is that in static electricity the electrons gather in one place (the surface of an object) and move randomly in all directions (Figure 2(a)), whereas in current electricity there is a steady flow of electric charge (electrons) through a conductor (Figure 2(b)).

Figure 1 With the flick of a switch, a room is filled with light, water in a kettle begins to boil, or a computer powers up. All this is possible because of electricity—the most widely used form of energy in the world.

current electricity the controlled flow of electrons through a conductor

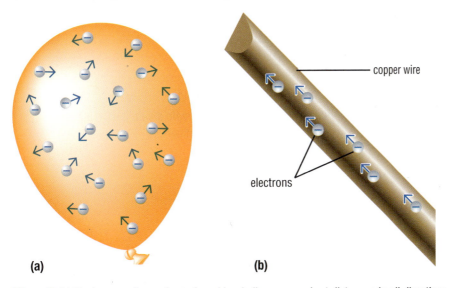

Figure 2 (a) Electrons on the surface of a rubber balloon move short distances in all directions. (b) Electrons flow through a copper wire in a controlled way.

LEARNING TIP

Word Orgins
The word "static" means something that does not move.

Although electric charges flow between two objects during an electrostatic discharge, this flow follows an unpredictable path, and occurs only for a very short period of time. The steady flow of electrons in current electricity, however, means that charges are moving for much longer. This steady flow can be directed and used to power devices, which you will learn more about later in this chapter.

As you learned in Chapter 11, electrons move easily through conductors, such as copper and aluminum. Since human skin is a fair conductor, it is very dangerous to touch a conducting wire that has electricity moving through it. Too many electrons entering your body too quickly could be fatal. For safety, conducting wires are wrapped with an insulator, such as plastic, which prevents the flow of electrons from entering your body when you handle the cables (Figure 3).

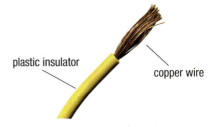

Figure 3 Copper wire, often used in electrical cables, is insulated with plastic.

Making Electrons Move

What makes electrons flow in a conductor? Consider an MP3 player and a television. A fully charged battery is needed to operate the MP3 player. If the battery is charged, electrons can flow through the MP3 player and it works. If the battery is dead, the electrons do not flow and the MP3 player does not work. Not all devices require a battery to create the flow of electrons. A television, for example, is usually plugged into a wall outlet. In this case, the flow of electrons is produced at an electric generating station. The electrons eventually flow into the wires in your home and then into your TV. If there is a blackout, electrons do not flow and the TV does not work. No matter what the electrical device is, to make it operate you need a source of electrical energy.

TRY THIS: MODEL ELECTRON FLOW

SKILLS: Observing, Communicating

SKILLS HANDBOOK
3.B.6., 3.B.9.

In this activity, you will model the flow of electrons in a conductor using marbles to represent electrons.

Equipment and Materials: 2 rulers; 10 marbles; tape

1. Make a track for the marbles by taping the two rulers to a desk. The rulers must be parallel to each other and marble-width apart.
2. Place the 10 marbles at one end of the track, touching each other.
3. Push the first marble in the row. Keep pushing this marble until it has travelled the full length of the track. Record your observations. This trial represents a fully charged battery.
4. Repeat step 3 but this time push the first marble with less force. Record your observations. This trial represents a partly charged battery.

A. What determines the speed of the electrons? Use your observations to explain your answer. T/I C
B. Do you think the results would have been different if you used 1000 marbles instead of 10? Explain. T/I
C. Would it be possible to create a path that would allow the marbles to travel continuously? Explain. T/I

IN SUMMARY

- Static electricity involves the movement of electrons in an uncontrolled way.
- Current electricity involves the controlled flow of electrons through a conductor.
- Current electricity moves easily through a conductor and poorly through an insulator.
- Current electricity requires a source of electrical energy to create a flow of electrons.

CHECK YOUR LEARNING

1. What information in this section helped you understand the difference between static and current electricity? Discuss your answer with a classmate. K/U C
2. What are some of the useful properties of current electricity? Explain why. K/U
3. Explain why conductors and insulators are both required to construct the electrical wiring in your home. K/U
4. What is required for electricity to flow? Explain. K/U
5. What is the difference between static and current electricity? Give an example of each. K/U C
6. Why would it be challenging to use static electricity in electrical devices, such as your television or stereo? Explain. A

Electric Circuits

12.2

What do MP3 players, televisions, and flashlights have in common? They all require a flow of electrons in order to operate. When you turn on an MP3 player, electrons flow from the battery through conductors and other electrical components of the player. The electrons eventually return to the battery. The flow continues for as long as the MP3 player is on, providing that the battery does not become dead. This continuous path for electron flow is an **electric circuit**.

To understand how electrons flow through an electric circuit, you can think of how blood flows in the circulatory system of your body. In the circulatory system, your blood vessels are filled with blood. For example, when the heart pumps, some blood moves through blood vessels from the heart to the tips of your toes and then back to the heart. A looped set of pipes (blood vessels) creates the continuous path needed for blood to flow in the circulatory system. In the same way, current electricity must have a continuous path in order for electrons to flow. A simple electric circuit includes an energy source, a load, conducting wires, and, sometimes, a switch (Figure 1).

The energy source can be as small and portable as a battery, or as large as a generating station. The **load** is a device that transforms electrical energy into other usable forms of energy. Loads can be heaters, lamps, fans, computer hard drives, or microchips. A light bulb is an example of a load that converts electrical energy into light energy, as well as some thermal energy.

Conducting wires (also called "connecting wires") join all the parts of an electric circuit together. They provide a pathway for electrons to flow from one component of the circuit to another. Connecting wires are usually made of insulated copper or aluminum wires.

A **switch** controls current flow in an electric circuit. When the switch is "on," the electric circuit is closed, providing a complete path for electron flow. When the switch is "off," the electric circuit is open and the path is incomplete. Hence, there is no electric current. "On/Off" switches are used to control household devices like the flashlight in Figure 2. Switches are sometimes hidden inside an appliance, such as the switch that shuts off the hard drive of a computer when it is idle for a period of time.

electric circuit a continuous path in which electrons can flow

LEARNING TIP
Word Origins
The word "circuit" comes from the Latin word for "circuitus," which means "going around." If a circuit is open, the electrons cannot flow.

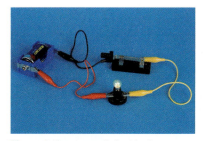

Figure 1 A source of electrical energy, conducting wires, a control device, and a load are required for a functioning electric circuit.

load the part of an electric circuit that converts electrical energy into other forms of energy

switch a device in an electric circuit that controls the flow of electrons by opening (or closing) the circuit

(a)

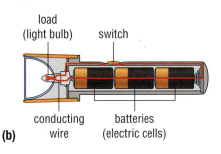
(b)

Figure 2 (a) A flashlight contains an electric circuit. (b) The batteries are the energy source, the bulb is the load, the switch controls the flow of electrons, and the connecting wires provide a closed path that joins all of the parts of the circuit together. The red line represents the continuous flow of electrons in the circuit.

TRY THIS: BUILDING CIRCUITS

SKILLS: Observing, Communicating

SKILLS HANDBOOK
2.D., 3.B.6, 3.B.9.

In this activity, you will practise building your own circuits.

Equipment and Materials: 4 conducting wires with alligator clips; battery; switch; various loads (for example, miniature lamp, light-emitting diode [LED], small motor, portable electric fan)

Figure 3

1. Using the alligator clips, connect the battery to one of the electrical devices in such a way that the device works. Create a diagram to show how your circuit is connected. Label all parts.
2. Repeat step 1 using a different load.
3. Connect more than one load in your circuit and try to get the device to operate. Draw a diagram showing how you connected the circuit. Label the parts.

A. Use your diagrams to explain what components were required for your circuits to work. T/I C

B. Sort the components you used into four categories: sources of electrical energy, loads, control devices, and conductors. K/U T/I

C. What do you think would happen if you connected more than one battery together in your circuits? Try it and see. Does this affect how the loads operate? T/I

UNIT TASK Bookmark

How can you apply what you have learned about electrical circuits to the Unit Task on page 586?

IN SUMMARY

- An electric circuit is a continuous path that allows electrons to flow. It is made up of an energy source, conducting wires, a load, and a switch.
- Energy sources include batteries and generating stations.
- The load in a circuit can be any electrical device that converts electrical energy into other usable forms of energy, such as a light bulb.
- A switch is used on electric circuits to enable you to turn the circuit on or off.

CHECK YOUR LEARNING

1. Did you already know something about circuits before you read this section? How has the reading changed what you know about circuits? C
2. Copy and complete Table 1 in your notebook. K/U

   Table 1 Parts of a Circuit

Part	Function
switch	
	provides a path for the electrons to flow
	transforms electrical energy into other types of energy
source of electrical energy	

3. In your own words, describe an electrical circuit. K/U C
4. A student connects a light bulb directly to an electric cell using connecting wires. She notices that the light bulb lights up. Is anything missing? Why would you consider the missing part necessary? T/I
5. A student builds a circuit that has a source of electrical energy, connecting wires, a switch, and a load. However, the circuit does not work. Suggest three possible reasons why the circuit does not work. T/I
6. Which of the following parts of a circuit would be considered a load? Justify your answer by explaining your choice(s). T/I C
   - light
   - switch
   - motor
   - battery

12.3

Electrical Energy

Every day, people all over the world use electrical devices to make their lives easier (Figure 1). Electric stoves cook our food, electric lights brighten our homes at night, and electric heaters keep us warm in the winter. What makes it possible for all these electrical devices to function?

Figure 1 Electricity plays an important role in our daily lives. Can you imagine life without it?

In Section 12.1, you learned that to produce current electricity, you need a source of energy. This energy source enables electrons to flow through the electric circuits in these devices. This process involves energy transformations.

Electrical energy is the energy that is provided by the flow of electrons in an electric circuit. As with all other types of energy, electrical energy is measured in joules (J). Electrical devices use electrical energy to do work. Recall from Grade 8 that work occurs when one form of energy is converted into another form of energy. Electric lights, for example, convert electrical energy into light energy. Work is done when the lights are on. In a motor, work is done when electrical energy is converted into motion.

Although electrical energy exists naturally in the form of lightning, we cannot yet harness the energy in lightning as a source of electricity. Fortunately, there are many other forms of energy that can be transformed into electrical energy.

electrical energy the energy provided by the flow of electrons in an electric circuit

DID YOU **KNOW?**

Alessandro Volta
The first electric battery was invented by Alessandro Volta in 1799. This was the first practical method of generating electricity.

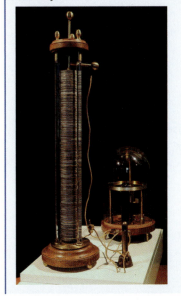

Sources of Electrical Energy

You may not realize it, but you use electrical energy every day. When you charge the batteries in your cellphone or plug your computer into a wall outlet, you are using electrical energy. Two main sources of electrical energy are batteries and electric generating stations. Batteries supply small amounts of electrical energy. Electric generating stations deliver large amounts of electrical energy along thick metal wires called transmission lines to your home. (You will learn more about this later in the chapter.)

Electric Cells

Portable devices, such as cameras, cellphones, and flashlights, require a portable source of electrical energy. An **electric cell** is a portable device that converts chemical energy into electrical energy. An electric cell consists of two electrodes in a conducting solution, called an electrolyte. The electrodes are conductors. One electrode can be easily positively charged and the other can be easily negatively charged. When the ends of the electrodes are joined to the conducting wires of a circuit, electrons begin to flow from one electrode to the other (Figure 2(a)). The electrons in the electrolyte are repelled by the negative electrode and attracted to the positive electrode.

In everyday language, we use the word battery to refer to an electric cell. However, in science, the term "battery" refers to two or more electric cells in combination (Figure 2(b)).

electric cell a device that converts chemical energy into electrical energy

LEARNING TIP
Forces of Attraction
Recall from Chapter 11 that like charges repel and opposite charges attract.

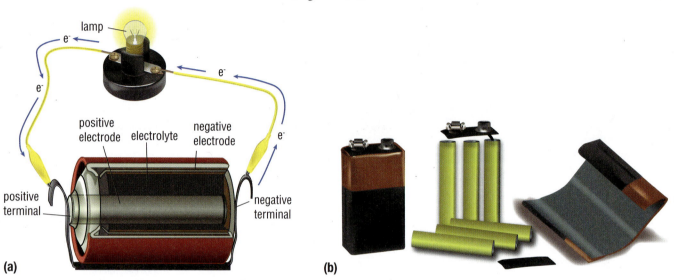

Figure 2 (a) An electric cell contains an electrolyte that provides a source of electrons. When the two electrodes of the cell are joined to the conducting wires of a circuit, electrons flow from the negative electrode through the circuit to the positive electrode. (b) A 9 V battery consists of six 1.5 V electric cells. The types of electrodes used and the chemicals in the electrolyte affect how strong the flow of charge in an electric cell will be and how long that flow will last.

Electric cells come in many different shapes and sizes. Some electric cells can be recharged and used over and over, while others can only be used once. **Primary cells** are electric cells that cannot be recharged because the chemical reactions that produce the flow of electrons are not reversible. Primary cells must be replaced. Examples of primary cells include zinc chloride, alkaline, and lithium cells. If they are not in use, primary cells have a shelf life of up to five years.

Secondary cells are electric cells that can be recharged and reused many times before they are recycled. Electrical energy from a wall outlet can be used to reverse the chemical reactions that take place in a secondary cell. Lead-acid batteries (in cars) and nickel metal hydride (NiMH) batteries are examples of secondary cells. They have a shelf life of six months. Lithium secondary cells are the most expensive type of electrical cell, but they have a shelf life of up to one year. All electric cells contain chemicals that are toxic to the environment. They should always be properly recycled.

primary cell an electric cell that may only be used once

secondary cell an electric cell that can be recharged

TRY THIS: THE PICKLE CELL

SKILLS: Observing, Analyzing, Evaluating

SKILLS HANDBOOK
2.D., 3.B.

In this activity, you will make an electric cell, using a copper penny and a metal paper clip as the two electrodes. The juice inside a pickle will act as the conducting solution. The battery tester shows whether electrons are flowing through the pickle cell.

Equipment and Materials: copper penny; metal paper clip; battery tester with connecting wires; pickle

1. Insert the copper penny into one side of the pickle.
2. Next, insert the metal paper clip into the other side of the pickle.
3. Connect the copper penny and then the paper clip to the battery tester using the connecting wires (Figure 3). Observe what readings the battery tester displays.

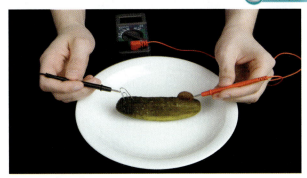

Figure 3 Step 3

A. What do you think would happen if you used two pennies or two paper clips? Try it. T/I

B. Do you think the pickle cell would work if you used a plastic paper clip? Explain your reasoning. Try it. T/I

C. Would any other type of fruit or vegetable work? Explain your reasoning. T/I

Fuel Cells

A fuel cell is a special kind of electric cell through which a continuous supply of chemicals is pumped as the cell operates. During this process, any waste products are removed. As a result, a fuel cell can operate for far longer than a conventional (or typical) electric cell. The hydrogen fuel cell is an example of a fuel cell. It produces electrical energy by converting hydrogen and oxygen into water. The fuel (hydrogen and oxygen) react to produce electrical energy.

Fuel Cell Vehicles (FCVs) are vehicles that are propelled by electric motors that obtain energy from fuel cells, rather than burning gas. The fuel cells extract electricity from hydrogen fuel quietly and efficiently. These vehicles are becoming increasingly popular because they produce no tailpipe emissions.

The Canadian government is developing a fuel cell vehicle program in Vancouver, British Columbia.

The waste product of the reaction in a hydrogen fuel cell, water, has no direct negative impact on the environment. However, a closer look at the process reveals a different story. Although hydrogen is the most common element in the universe, it does not occur naturally on Earth in its elemental form. The hydrogen used in most fuel cells must be removed from compounds that are rich in hydrogen, namely fossil fuels. This process uses lots of energy: energy to extract the fossil fuel and energy to remove the hydrogen atoms. This process creates pollution and contributes to climate change. For this reason, a fuel cell is not as "green" as you may think.

Fuel cells can be compact and lightweight, which makes them useful as power sources in remote locations. The U.S. space program, for example, uses fuel cells to supply electricity to spacecraft.

> **DID YOU KNOW?**
>
> **The Invention of the Fuel Cell**
> Sir William Grove invented the first fuel cell in 1839. Grove knew that water could be broken down into hydrogen and oxygen by sending an electric current through it (a process called electrolysis). He hypothesized that by reversing the procedure you could produce electricity and water. He created a primitive fuel cell that supported his hypothesis. He called his model a gas voltaic battery, which later became known as a fuel cell.

To learn more about fuel cells,

To learn more about becoming an astronaut,

IN SUMMARY

- Electrical energy is the energy provided by the flow of electrons in a circuit. Electrical energy is measured in joules (J).
- Electrical cells produce small quantities of electrical energy and are useful for portable electrical devices. Electric generating stations provide large amounts of electrical energy.
- Primary cells cannot be recharged. Disposable batteries are an example of primary cells.
- Secondary cells can be recharged and reused many times before being recycled.
- Fuel cells require a continuous supply of fuel, such as hydrogen and oxygen, to provide electrical energy.

CHECK YOUR LEARNING

1. A battery (electric cell) is a portable source of electrical energy. Describe how this device relates to your life outside school. A C

2. In your opinion, what are the five most important uses of electrical energy? Compare your list to a classmate's. A

3. Identify ten devices that use cells or batteries. What types of cells or batteries are used in each device? Are the cells or batteries used the best choice? Explain. A

4. What type of electric cell would be best used in a flashlight or smoke detector? Explain each choice. A

5. Fuel cells are used on the space shuttle to provide the shuttle with all of its electrical energy. Explain why. A

6. Describe any differences and similarities between the following pairs of terms. K/U
   (a) joule and watt
   (b) a lithium battery and an alkaline battery
   (c) a fuel cell and a primary cell

7. Describe what parts are required to make an electric cell. K/U

8. A student tries to make an electric cell using a lemon and two identical nails poked into the lemon. If a battery tester were connected to it, would you expect it to register any current electricity? Justify your answer. T/I

Forms of Current Electricity

12.4

You may have noticed that some loads, such as an MP3 player, require a battery in order to work, while others, such as a television, must be plugged into a wall outlet. These two ways to access electrical energy are related to how electricity is produced. Current electricity can be produced directly (for example, from a battery) or it can be generated in a generating station (for example, the electricity you obtain from your wall outlet). Each of these methods produces a different type of current electricity.

There are two forms of current electricity: direct current and alternating current. In **direct current (DC)**, electrons flow in one direction only (Figure 1). Direct current is produced by an electric cell, such as a battery, to power portable electrical devices.

direct current (DC) a flow of electrons in one direction through an electric circuit

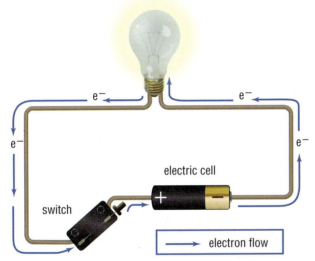

Figure 1 In a DC circuit, the electrons move in one direction only. The electrons move from the negative end of the energy source, through the circuit, to the positive end of the energy source.

In **alternating current (AC)**, electrons move back and forth, alternating their direction (Figure 2). Alternating current is produced by generators at electric generating stations. Alternating current is used in electric generating stations because it is a more efficient method of distributing electrical energy over long distances than DC. Wall outlets provide alternating current. Many devices, such as lights, ovens, and clothes dryers, use alternating current.

alternating current (AC) a flow of electrons that alternates in direction in an electric circuit

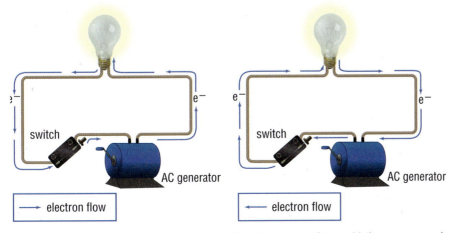

Figure 2 In an AC circuit, the direction of electron flow changes as often as 60 times per second.

> **DID YOU KNOW?**
>
> **AC Circuits**
> In an AC circuit, as the electrons move back and forth, there is an instant at which the electrons stop to change direction. During this instant of time, there is no electric current. Although it happens too quickly for you to see, the lights in your home turn on and off 60 times per second.

12.4 Forms of Current Electricity 515

Generating Electricity

Electricity does not exist as a primary form of energy. It is produced when one type of energy is converted into electrical energy. Electric generating stations convert mechanical energy into electrical energy. An external energy source is used to push on the blades of a fan-like device called a turbine, causing it to turn. The turbine is connected to the movable parts of another device called a generator. Inside the generator, a coil of wire is held between the two poles of a magnet. The wire is made of copper because copper is a good conductor of electricity. When the turbine rotates, the coil of wire in the generator also rotates. Since the moving coil is held near a magnet, electrons begin to flow in the wire. These electrons eventually move in transmission lines to the electrical outlets in your home. The process of generating electrical energy in a generating station is shown in Figure 3.

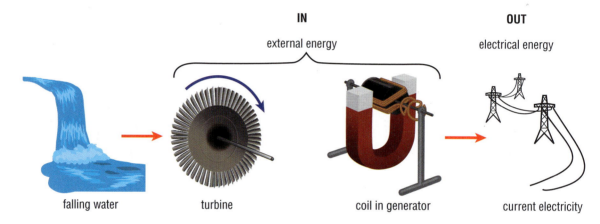

Figure 3 An electrical generator obtains energy from a non-electrical source of energy and converts it into electrical energy. The external energy source spins a turbine consisting of a set of fan-like blades. The turbine spins a coil of wire inside a magnet, producing electrical energy.

A generator makes it possible to produce a constant flow of electrons whenever it is needed. Every time you plug a load into a wall outlet, you are accessing electrical energy produced using a generator. This phenomenon was first observed by Michael Faraday (Figure 4).

Electrical energy is special because, unlike other forms of energy, its production can be controlled and it can travel long distances to where it will eventually be used. Since the flow of electrons is controlled, engineers can design devices that convert the electrical energy more easily into other useful forms of energy, such as motion, thermal energy, or light. An electric blender, for example, is a device that converts electrical energy into motion. A toaster, for example, converts electrical energy into thermal energy.

Faraday's discovery was so important that it revolutionized the way humans use electricity. Before this discovery, electrical energy could not be produced for long enough periods of time. The only available electrical energy at the time came from early versions of batteries.

Figure 4 Sir Michael Faraday

IN SUMMARY

- Current electricity can be produced directly (direct current) or it can be generated in a generating plant (alternating current).
- In direct current (DC), electrons flow in one direction only. Electric cells produce direct current.
- In alternating current (AC), electrons flow back and forth. Alternating current is produced in generating plants.
- Electrical generating plants convert mechanical energy into electrical energy.
- An electrical device that uses DC electricity requires a DC energy source, such as a battery. A device that uses AC electricity must be plugged into a wall outlet.

CHECK YOUR LEARNING

1. What did you learn about the different types of current that surprised you? Explain. K/U C
2. What is the difference between alternating current and direct current? K/U
3. Draw a flow chart showing how a generator works. K/U C
4. Each of the images shown in Figure 5 shows electrical devices or electricity in some form. Identify the type of current electricity represented in each image as being either direct current (DC) or alternating (AC) current. K/U A

(a)

(b)

(c)

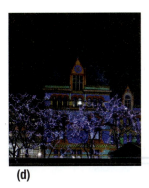
(d)

Figure 5

12.5 Generating Current Electricity

Today, most of the energy provided by a wall outlet is produced in electric generating stations. Recall that a generator is attached to a turbine. The turbine requires an external energy source to push the blades and rotate the generator. The external energy may come from moving water, steam, or wind.

Light can also be used to produce electrical energy directly without the use of a generator, which you will learn more about later on in this section.

Generating Electrical Energy Using Moving Water: Hydro-Electric Generation

Humans have used the energy of falling water since ancient times. For thousands of years, people built wheels near small waterfalls. The wheels were most often attached to grain mills. When the falling water pushed the wheel, grain was ground into a powder called flour.

The energy of falling water was eventually used to move the large blades of water turbines. In 1882, the first hydro-electric generating stations was built at Chaudière Falls on the Ottawa River. Hydro-electric generation was one of the first methods used for generating electricity and continues to be an important method to this day.

Hydro-electric generating stations use the fast-moving water of a waterfall or river, or the water stored in a reservoir behind a dam (Figure 1). Today, there are 180 hydro-electric generating stations in Ontario, 58 of which are connected to the electrical distribution grid (Figure 2). The grid is a network of transmission towers and power lines that provide electrical energy to homes and businesses in a geographic area.

To learn more about the history of electrical energy generation in Ontario,
GO TO NELSON SCIENCE

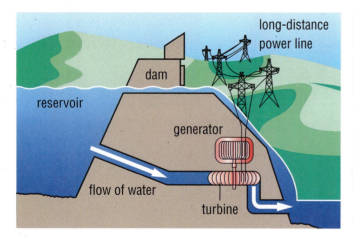

Figure 1 Hydro-electric dams divert and control the flow of water so that it flows past a turbine connected to a generator.

Figure 2 The Adam Beck 2 generating station in Niagara Falls is the largest hydro-electric generating station in Ontario.

Waterfalls and rivers seem to have an endless supply of water. This is because they are part of the water cycle. The energy in sunlight evaporates water in lakes and rivers. Water vapour in clouds condenses to form rain and snow that fall to the ground. When human activities do not alter the landscape, this moisture replenishes the rivers at the tops of waterfalls.

Since the water in a waterfall can be used over and over again to generate electrical energy, hydro-electric generation is considered to be a renewable energy source. **Renewable resources** are natural resources that are continually replenished and will never run out.

renewable resource natural energy resource that is unlimited (for example, energy from the Sun or wind) or can be replenished by natural processes in a relatively short period of time (for example, biomass)

The Pros and Cons of Hydro-Electric Generation

Electricity generated using falling water does not pollute air or water. However, large hydro-electric generating stations artificially create large volumes of falling water by using dams. Dams change the way water flows in a particular area. This results in changes in the ecology of a watershed. Rivers no longer flow the same way because water is diverted toward the dam. The migration of fish is often disrupted and large areas of land may become flooded (Figure 3). To reduce the impact on fish, some dams have fish ladders, which allow fish to swim around the dam (Figure 4).

Figure 3 In the 1980's, the James Bay project flooded over 11 000 km^2 of forest, displacing all the animals that once lived there. As a result of the flooding, the Cree lost a significant amount of their hunting grounds.

Figure 4 This fish ladder allows salmon to swim around the dam so that they can reach their spawning location.

Dams have many other "invisible" costs associated with them. The building of dams uses vast amounts of resources such as concrete, steel, specialized machinery, and fossil fuels. All these factors make large-scale hydro-electric generating plants very expensive for taxpayers and the environment.

A better way to use falling water to generate electricity involves using small-scale hydro-electric generating plants. These operations have far fewer environmental impacts in an area because dams are not needed. The Ontario government is now encouraging the use and development of small-scale hydro-electric projects.

Another drawback to using falling water to generate electricity is that there are not always suitable locations available. Hydro-electric generating stations can only be built in areas near fast-flowing water. Ontario's natural landscape, however, has many lakes and rivers. So we are fortunate to have many possible locations for hydro-electric generating plants.

> **DID YOU KNOW?**
> **Pushing and Pulling**
> Tides are the periodic rising and falling of sea levels each day. Many factors account for the formation of tides. The biggest one is the difference between the Moon's gravity and the Sun's gravity pulling on Earth's oceans.

Generating Electrical Energy Using Moving Water: Tidal Generation

A newer, alternative method for generating electrical energy using moving water is tidal energy. Tidal energy is the energy of moving ocean water. Just as a water fall or river can be used to generate electricity, the movement of the tides can also be used to rotate a water turbine.

Tides occur twice daily at specific times, in a cycle. This means that a tidal generating station can only operate as the tide comes in or goes out. The station is on for 5 h and off for 7 h. The generation cycle repeats when the tides change from low tide to high tide. The tide height in Ontario lakes does not change enough to effectively use tidal energy. However, there are many other areas in Canada where tidal generation is possible (Figure 5).

Figure 5 The Bay of Fundy has the highest tides in the world, up to 17 m. The tidal generating station located in Annapolis, Nova Scotia, uses a tiny fraction of the tidal energy available in the Bay of Fundy.

Since tides are cyclical, the movement of the tides can be used over and over again to generate electricity. For this reason, tidal generation is considered to be a renewable energy source.

The Pros and Cons of Tidal Generation

Using tidal sources to generate electrical energy produces no pollution. However, a major drawback of tidal generation is that tidal generating stations can only be built near coastlines that experience significant changes in tides.

Generating Electrical Energy Using Steam: Thermal Generation

Thermal generation involves heating water to produce steam. The steam is then used to rotate a turbine to generate electricity. The common ways to produce steam involve the burning of fossil fuels or using radioactive materials to heat water. There are also three alternative ways to produce steam. One way involves burning plant or animal materials. Another way involves using thermal energy deep in Earth's crust, and the third way involves using sunlight to heat water.

Using Fossil Fuels in Thermal Generation

Coal, oil, and natural gas are all fossil fuels. They were formed from plants, animals, and micro-organisms that lived millions of years ago. When these organisms died, the energy in their cells remained "locked up" in the form of carbon and hydrogen. These fuels can be burned to heat water.

When fossil fuels are burned, they produce significant amounts of energy. For this reason, fossil fuels have historically been an important resource for energy production. However, fossil fuels take millions of years to form and reserves are being depleted much faster than new ones are being formed. A resource that cannot be replenished as quickly as it is being consumed is called a **non-renewable energy resource**. Fossil fuels are nonrenewable resources.

non-renewable resource a resource that cannot be replaced as quickly as it is consumed

Using Radioactive Materials in Thermal Generation

After World War II, generating electricity using uranium became popular. In this process, long tubes called feeder rods are filled with the radioactive form of uranium. These rods are then placed inside a reactor. When a neutron collides with the radioactive uranium, a very high energy reaction called nuclear fission takes place. The nuclei of the uranium atoms break apart and release huge amounts of energy in the process. This energy is then used to heat water, producing steam (Figure 6). Thermal generating stations that use the nuclear reactions to produce energy are called nuclear generating stations (or power plants). Uranium, just like fossil fuels, is a non-renewable energy source because there is a limited amount of uranium on Earth.

To learn more about nuclear generation,

DID YOU KNOW?
More Efficient Energy?
It is a common misconception that uranium is a more efficient energy source than fossil fuels. It is true that one gram of radioactive uranium provides as much electricity as 15 tonnes of coal. However, only 0.1 % of uranium ore contains radioactive uranium. That means that you would need to mine 100 tonnes of uranium ore and purify it to get 0.1 tonne of radioactive uranium.

Figure 6 The Pickering Nuclear Generating Station is one of the largest nuclear facilities in the world.

Using Biomass in Thermal Generation

For centuries, people have used biomass as an energy source for heating and cooking. **Biomass** is any form of plant or animal matter, including wood, straw, manure, plant-based oils, and decaying natural materials. Biomass may also include biodegradable wastes that can be burned as fuel (such as sewage gas and landfill gas).

biomass any biological material (including plants and animals)

12.5 Generating Current Electricity

Biomass, also known as biofuel, is renewable because trees and crops are continuously being planted and because plant and animal waste is continuously being produced. During the process of photosynthesis, plants convert sunlight into sugar and absorb carbon dioxide from the atmosphere. When biomass is burned, carbon dioxide is produced. However, the amount of carbon dioxide produced is offset by the amount of carbon dioxide the plants absorbed during their lifespan. For this reason, fossil fuels (organic materials that have been transformed over very long periods of time by geological processes) are *not* considered to be biomass.

More recently, biomass has been used as an alternative fuel for generating electrical energy. The biomass used for electricity generation can be produced in two ways: by using "leftover" biomass or by growing biofuel crops. Leftover biomass is dead or decaying plant or animal material, or post-consumer biomass, such as used cooking oils. In this process, thermal energy is transferred to water in a central tower. The water in the tower becomes steam, which is then used to move a turbine.

There are several ways that biomass can be used as an energy source. Some of these methods are listed below:

- Biomass can be burned.
- Methane gas is released when plant and animal matter decays on farms, in forests, in sewage treatment plants, and in most landfill sites. The methane gas can be collected and burned as a fuel.

> **DID YOU KNOW?**
> **Co-Generation**
> Manufacturing plants and generating stations often produce steam as a waste product. Co-generation involves using this waste product to either generate electricity or heat buildings. With co-generation, small turbines are used to generate electricity. This method of energy capture is being used very successfully in many European countries like Germany to reduce the use of non-renewable energy sources.

Using Geothermal Energy in Thermal Generation

Geothermal generation uses the thermal energy from deep in Earth's crust, in the form of hot springs and geysers. Liquid is circulated through pipes that are deep in the ground. Thermal energy is transferred from the ground to the liquid. The liquid is then pumped to a central tower containing water. The water is heated to produce steam, which then rotates a turbine and generator to produce electricity. Currently, there are no geothermal generating stations in Ontario, although there are plans to build some in British Columbia. Geothermal energy is renewable because thermal energy is always present deep in Earth's crust.

To learn more about geothermal projects in Ontario and worldwide,
GO TO NELSON SCIENCE

The Pros and Cons of Thermal Generation

Thermal generation produces a large amount of energy. However, the production and use of fossil fuels or radioactive materials raises environmental concerns.

When fossil fuels burn, carbon dioxide is released in the atmosphere. There, the carbon dioxide traps energy radiated from Earth's surface and re-radiates that energy in all directions. About half of the radiation gets sent back toward Earth's surface. This increases Earth's temperature. This energy-trapping process is called the greenhouse effect, and gases that contribute to this effect are called greenhouse gases. Natural processes can only absorb about half of the carbon dioxide produced from the burning of fossil fuels each year. This results in a net increase of atmospheric carbon dioxide, which is a greenhouse gas. Consequently, the burning of fossil fuels is contributing very significantly to climate change.

Fossil fuels are found below ground, sometimes even below the ocean floor. These non-renewable energy sources must be extracted and, except for coal, must also be refined. The machinery used to extract fossil fuels is itself often powered by fossil fuels. The extraction and refining of fossil fuels creates air pollution and destroys natural habitats. It can take decades for a used mining site to be restored to a natural state. Very few, if any, large-scale abandoned mines have been successfully restored (Figure 7).

Figure 7 This aerial photo shows an abandoned open cut mine in North Queensland, Australia.

COAL

Coal is relatively easy and inexpensive to mine. For these reasons, many countries use coal-fired generating stations to produce electrical energy. However, burning coal produces airborne particles such as ash, smoke, sulfur dioxide, and nitrogen dioxide. The result is air pollution, smog, and acid precipitation (Figure 8). Coal ash irritates eyes and skin, and the pollution from coal-fired generating station contributes to breathing problems for people living in the areas where the pollution settles. In addition, the burning of coal releases mercury, which contaminates bodies of water.

OIL AND NATURAL GAS

Burning oil and natural gas also produces pollutants that contribute to acid precipitation, smog, and climate change. Oil is usually transported using large ships called supertankers. Supertankers can crash and cause an oil spill, which can have devastating effects on ecosystems.

Some people think that burning natural gas produces less air pollution than coal or oil because it burns more cleanly. This is true; however, there is a limited supply of natural gas. Natural gas reserves are declining worldwide. In fact, half of Earth's available oil reserves have been used up. The continued use of oil is not only an environmental problem, it is also going to become increasingly more expensive since our demand for oil is increasing even though oil is becoming a scarce resource.

Figure 8 The Nanticoke generating station uses coal as an energy source, and is Canada's largest contributor to air pollution. Currently, there are four coal-fired generating stations in Ontario.

To learn more about the environmental impact of oil spills,

GO TO NELSON SCIENCE

RADIOACTIVE MATERIALS

Like fossil fuels, uranium ore is found below ground and must be mined, which results in pollution and habitat loss. Even though generating electricity in a nuclear reactor does not directly produce air pollution or contribute to acid precipitation, thermal pollution is produced. Nuclear generating stations produce steam using water from a lake. After the reaction process, the water is returned to the lake at a higher temperature. Less oxygen is able to dissolve in water at higher temperatures, so aquatic life and lake ecosystems are affected.

In addition, electricity generated using uranium produces large amounts of radioactive waste (Figure 9), which has serious environmental and health consequences. Radioactive waste releases radiation for several thousands of years. Exposure to nuclear radiation at high levels is known to cause cancer or organ failure.

> **DID YOU KNOW?**
>
> **Nuclear Waste and Nuclear Weapons**
> Nuclear waste from nuclear reactors is often used in the manufacture of nuclear weapons. Not only are nuclear weapons an environmental problem, they are a security risk. Responsible choices for generating electricity are very much tied in with responsible government policies that minimize security risks.

Figure 9 Radioactive waste from a nuclear reactor is stored underwater at the reactor site for over 10 years, after which it is stored in dry storage behind 1 m of concrete. After another 60 years, the canisters can be moved to long-term storage, but the waste is still unsafe.

Lastly, nuclear generating stations are very expensive to build and maintain compared with fossil-fuel generating stations. There are safety concerns associated with nuclear reactors. There is a risk of accidents caused by human error, equipment failure, and factors such as earthquakes. When accidents have occurred, they had damaging consequences to the environment and involved expensive repair bills for taxpayers.

BIOMASS GENERATION

By using renewable energy sources to turn steam turbines, many countries have actively reduced greenhouse gas emissions and their dependency on fossil fuels. Turning leftover biomass into fuel is an excellent form of recycling, but it may not be enough to supply our ever-increasing demand for energy. When we grow crops to be used as fuel, we are reducing the amount of land available to grow food. The issue of biofuels is very controversial. That is why energy conservation must be considered as being even more important than using renewable energy sources.

GEOTHERMAL GENERATION

Using geothermal energy sources to generate electrical energy produces no pollution and does not contribute to climate change. However, because geothermal energy is generated deep within Earth's core, it can be challenging to obtain. Although geothermal energy can be tapped for heating in almost any location, geothermal generating stations must be built in places where the temperature deep in the ground is enough to produce steam.

Generating Electrical Energy Using Wind

As with water, humans have used the energy of wind throughout history. For thousands of years, wind moved sailboats and was used to pump water. With the invention of the wind turbine, wind could be used to generate electricity. A wind turbine consists of large blades mounted on a tall tower. As the wind blows, it turns the blades of the turbine, which is connected to a generator. Today, a single wind turbine can generate enough electricity for about 250 Canadian homes. When many wind generators are connected to the grid, a significant amount of electricity can be generated (Figure 10).

> **DID YOU KNOW?**
>
> **Geothermal Energy for Heating and Cooling**
> Geothermal energy can be used to heat and cool buildings. Since it is not necessary to produce steam, this type of renewable energy source can be used almost anywhere. A pump circulates a liquid through piping in the ground. The thermal energy of the liquid then gets transferred to the flooring of the building, creating warmth in winter and cooling in summer. Currently, Sweden has the largest number of buildings using this technology.

To learn more about a career in wind generation,
GO TO NELSON SCIENCE

Figure 10 The effect that wind turbines have on bird populations is taken into account when choosing a site for a wind farm. At Erie Shores wind farm, the turbines are widely spaced and more than 41 m above ground. Studies there have shown no change in bird populations in the two years since the wind farm started operation.

The Pros and Cons of Generating Electrical Energy Using Wind

The installation and maintenance costs for a wind turbine are significantly cheaper than those for generating stations that use fossil fuels or uranium; wind is free. Wind turbines can be added to or removed from the grid more easily, without any major disruptions in the electricity supply. Also, wind turbines can be located much closer to homes and businesses than generating plants that use fossil fuels or uranium. This means that transmission lines can be much shorter, minimizing losses in electricity.

Using wind to generate electrical energy does not produce any air or water pollution. Unlike water-driven turbines, wind turbines can be located almost anywhere, provided the wind speed is at least 5 m/s. Since wind speeds change continually, it is sometimes difficult to generate a steady supply of electricity at a particular location. For this reason, the installation of many wind turbines across a large area is required. These turbines form wind farms, and these farms can generate a significant amount of electricity when they are connected to the grid. The best location for a wind farm is offshore in a lake or in the ocean, where wind speeds are greater and the wind blows more continuously. An offshore wind farm in Lake Ontario, consisting of 142 turbines, has been proposed.

Harnessing wind power is also available on a smaller scale. Home and business owners can buy small-scale wind turbines to generate electricity for their own use (Figure 11). Small wind turbines are becoming quite common and can be purchased at the local hardware store for a few thousand dollars.

Wind farms are often criticized as being noisy, and dangerous to birds. However, many of today's wind turbines have been engineered to be quiet.

Figure 11 Small-scale wind turbines, like this one, can supply a household with about 50 % of its electricity needs, depending on the loads being used.

RESEARCH THIS: WIND POWER ON LAKE ONTARIO

SKILLS: Researching, Analyzing the Issue, Communicating

SKILLS HANDBOOK
4.C.2., 4.C.4., 4.C.6.

Wind farm projects on Lake Ontario were stopped until the Ministry of Natural Resources could study the effects on bats, butterflies, migratory birds, and aquatic species. The studies are now close to completion, allowing a number of wind farm projects to seek approval once again.

1. Research the proposed wind farms that are being developed on Lake Ontario.
2. Research the various stakeholders and their viewpoints. Stakeholders may include, but are not limited to, residents of the area, government, and electric generation companies.

A. List the wind farms being developed. How much electricity will they generate? How many turbines are involved, and where will they be located? Write a summary of your findings. T/I C

B. What are some expenses for developing an offshore wind energy project? A

C. What role does the government play in the development of projects like these? A

D. List the concerns of each stakeholder that you researched. A

GO TO NELSON SCIENCE

526 Chapter 12 • Electrical Energy Production

Producing Electrical Energy from Light

Electrical energy can be produced directly without the use of a turbine or generator. A **photovoltaic cell** is a device that converts light directly into electrical energy. One type of photovoltaic cell is the solar cell, which captures energy from the Sun and converts it into electrical energy. In a solar cell, sunlight shining on the cell creates a flow of electrons because the materials in a solar cell can convert sunlight into electrical energy. Since the Sun radiates light every day, sunlight is a renewable energy source.

photovoltaic cell a device that converts light energy from any light source directly into electrical energy

The Pros and Cons of Producing Electrical Energy from Sunlight

Solar photovoltaic systems can be installed at any sunny location and do not produce air or water pollution. However, some pollution is produced in the manufacture of solar panels and in the disposal of broken solar panels. The advantage of solar photovoltaic systems is that they can operate independently of the power grid or be connected to it. These systems are also maintenance-free when they are properly installed, and can last more than 40 years.

There are several limitations to using solar panels to produce electricity. Solar panels are only about 30% efficient at converting sunlight into electricity because of the limitations of the materials used to make the panels. The cost of solar panels can be expensive. Also, solar panels do not produce large amounts of electricity immediately. Instead, each solar cell needs several hours of sunlight in order to produce a significant amount of electricity. This limitation, though, can be overcome by having very large numbers of solar panels connected to the electrical distribution grid. Large-scale solar farms with many solar panels linked together can produce electrical energy for entire communities (Figure 12).

Figure 12 The solar panels on this solar farm produce a significant amount of electricity. Each solar panel consists of many solar cells connected together.

To learn more about solar power projects,

GO TO NELSON SCIENCE

The Future of Energy Production

In Ontario, about 26 % of our electrical energy comes from burning coal and natural gas. Over 50 % of our electrical energy comes from nuclear reactions using uranium (Figure 13). This means that more than 76 % of Ontario's electrical energy is produced using non-renewable energy sources.

When you compare generators with direct ways to produce electricity, it is important to consider environmental issues, technological limitations, and socio-economic factors. All sources of energy have environmental consequences that we, as global citizens, must face. Some sources are in limited supply. As our electrical requirements increase, governments and citizens need to make choices on how to best meet that demand. These choices involve developing better and more efficient ways to generate electrical energy while also considering the source and effects of its use on the environment. Another important consideration is to develop ways to use less electrical energy.

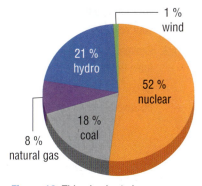

Figure 13 This pie chart shows how Ontario's electrical energy was generated in 2007.

UNIT TASK Bookmark

How can you apply what you have learned about the environmental impacts of electrical energy production to the Unit Task on page 586?

IN SUMMARY

- Turbines are connected to generators to produce electrical energy.
- Conventional ways to generate electrical energy include hydro-electric, thermal, and nuclear generation.
- Alternative ways to generate electrical energy include tidal, wind, biomass, geothermal, and solar energy generation.
- Resources used to generate electrical energy can be renewable or non-renewable.
- Thermal generation commonly uses coal, oil, natural gas, and uranium, which are non-renewable resources.
- Fast-moving water, tides, biomass, geothermal energy, wind, and sunlight can be used to rotate a turbine and are renewable energy resources.
- Fossil fuels and radioactive materials are non-renewable resources used in thermal generation.
- Solar cells produce electricity directly, by converting light energy into electrical energy.
- Producing energy using either renewable or non-renewable resources has advantages and disadvantages for society and for the environment.
- Societies choose which method of electric generation, or combination of methods, to use based on many factors.

CHECK YOUR LEARNING

1. Write a journal entry describing your thoughts and concerns after reading about the environmental issues associated with generating electricity. C

2. Did it surprise you to find that most of the world's electrical energy production happens because of simple materials such as wire and magnets? How does this idea change your perspective on electrical energy production? Discuss your responses with a classmate. C

3. Explain three disadvantages associated with burning fossil fuels to generate electricity. K/U

4. Create a table in your notebook, and list the advantages and disadvantages of each renewable resource. K/U C

5. Explain why renewable resources are becoming more popular. K/U

6. Why do we still use non-renewable resources to generate electrical energy? Discuss your views with a classmate. A

7. Approximately 18 % of electricity generated in Ontario comes from coal-fired generating stations. One of the main goals of the government is to replace coal-fired generating stations in the shortest possible time. Why do you think this is such a high priority? Explain. A C

8. List some technologies that you could use at home to reduce your dependence on fossil-fuelled electrical energy. Discuss your ideas with a classmate. A C

EXPLORE AN ISSUE CRITICALLY 12.6

Not in My Backyard!

Recently, Canadians have become more aware of where their electrical energy comes from and the costs—economic, environmental, and social—associated with generating that energy.

SKILLS MENU
- Defining the Issue
- Researching
- Identifying Alternatives
- Analyzing the Issue
- Defending a Decision
- Communicating
- Evaluating

The Issue

A subdivision of new homes is being built in your neighbourhood (Figure 1). Your municipal government has proposed building either a new nuclear reactor or a solar farm to meet the increasing demands of the neighbourhood. Municipal government representatives have called on members of the community to make recommendations on the type of generating station that will be built.

Figure 1 New homes require a source of electrical energy

Goal

To make a recommendation for the type of generating station that should be developed in your neighbourhood. You must present your recommendation to representatives of your municipal government. As a concerned member of the community group, you will compare the advantages and disadvantages associated with each of the proposed types of generating stations.

Gather Information

Work in pairs or small groups to learn about the issues associated with type of electricity generation. Consider these questions as you gather your information:

- What are the benefits and limitations to each type of generating station?
- How will the type of generating station selected affect the land and local ecosystems?
- What hazardous substances are used to manufacture the parts that run in each type of generating station?
- What types of emissions, if any, are produced by each type of generating station? Do these emissions affect human health?
- What are the long-term financial costs associated with maintaining each type of generating station?

Identify Solutions

Take notes or create a graphic organizer to help you organize your thoughts. Compare and contrast the two electrical generating stations carefully. Using the comparison, identify which of the two options is the better choice for the community.

Make a Decision

Make a clear recommendation to your municipal government representatives about which type of electrical generating station you believe should be built in your local community. Clearly state the criteria that you used to arrive at your recommendation.

Communicate

Prepare a campaign to encourage your community to choose the better option for a new generating station. Your campaign must identify all of the concerns and issues related to each type of generating plant and should indicate the factors that you used to analyze the issue. Summarize all of the issues you came across in your research, but focus on what criteria you used to make your recommendation.

Your campaign could be presented in the form of a report, newspaper article, dramatization, blog, etc. Your campaign should end with a clear recommendation and how you arrived at this decision.

12.7 Electrical Power and Efficiency

You may have noticed that electrical devices are labelled with a power rating. For example, a compact fluorescent light bulb (CFLs) may be labelled 15 W, while a hair dryer might be labelled 1200 W. What is an electrical power rating, and what does it mean?

Electrical Power

Power is the rate at which energy is transformed or the rate at which work is done. **Electrical power** is the rate at which electrical energy is produced or consumed in a given time. The unit of measurement for electrical power is the watt (W). One watt is the equivalent of one joule per second (J/s). The higher the power rating value, or "wattage," the more electrical energy a device produces (or uses to operate).

Consider a 60 W incandescent light bulb and a 15 W compact fluorescent bulb (CFL). The incandescent bulb uses more electrical energy than the CFL to produce light. However, each produces about the same amount of light. So where does the extra energy used by the incandescent bulb go? It is converted into thermal energy, instead of light.

Measuring Electrical Energy Usage

Electric generating stations have an electrical power rating of megawatts (MW, or millions of watts) or gigawatts (GW, or billions of watts). The joule is a relatively small unit of electrical energy, so we often measure larger amounts of electrical energy in watt·hours (W·h), kilowatt·hours (kW·h), or gigawatt·hours (GW·h). A watt·hour is 3600 times greater than a joule. The **kilowatt·hour** is the SI unit used to measure energy usage. A kilowatt·hour is 1000 times greater than the watt·hour, while a gigawatt·hour is 1 000 000 times greater than a kilowatt·hour. Electricity meters keep track of how much electrical energy is used in homes, schools, and businesses in units of kW·h.

We often think of how we can generate more electrical energy rather than think of ways that we can conserve it. An average Canadian family consumes over 16 000 kW·h of electrical energy in one year. That is a staggering amount of energy!

Efficient Devices

Not all electrical devices use electrical energy efficiently. **Efficiency** is a measure of how much useful energy an electrical device produces compared with the amount of energy that was supplied to the device. For example, an older clothes dryer might use 800 kW·h of electrical energy in one year, while a new model might use 300 kW·h in one year (Figure 1). Both clothes dryers perform the same task, but the newer model is more efficient because it uses less electrical energy than the older model. The difference in energy use is more than 60 %.

> **MATH TIP**
>
> **Orders of Magnitude**
> The prefix kilo means a magnitude of 1000. So a kilowatt (kW) is 1000 watts. Mega means a magnitude of one million. So a megawatt (MW) is 10^6 watts. Giga means a magnitude of one billion. So a gigawatt (GW) is 10^9 watts.

electrical power the rate at which electrical energy is produced or used

kilowatt·hour (kW·h) the SI unit for measuring electrical energy usage; the use of one kilowatt of power for one hour

efficiency comparison of the energy output of a device with the energy supplied

For more information on how to conserve energy,

GO TO NELSON SCIENCE

Figure 1 Newer appliances are more efficient than older models.

Evaluating Efficiency in Devices

Many people use computers. A computer can use up to 600 kW·h of electrical energy in one year if it is left on when not in use. If the computer is put in sleep mode, the amount is reduced to 20 kW·h of electrical energy per year. If you use a computer 1 h for every 4 h that it is on, you would conserve electrical energy by putting your computer into sleep mode when you are not using it or by turning the computer off.

Researching the type of computer that you need may also help save energy. Notebook computers use far less energy than desktop computers. Notebook computers are designed to use energy efficiently, since they often run on rechargeable batteries, which are a limited energy source. Newer LCD screens use less electrical energy than conventional computer monitors and televisions.

We use a lot of electrical energy to provide light. Not all light bulbs use electrical energy efficiently. The incandescent light bulb uses electrical energy to heat a wire, called a filament, inside a glass bulb. The heated filament produces bright light but also produces lots of thermal energy. A typical incandescent light bulb converts about 90 % of the electrical energy into thermal energy, whereas only 10 % is converted into light. The incandescent light bulb is actually better at producing thermal energy than light!

A 100 W incandescent light bulb uses about 40 kW·h of electrical energy in 400 h. A comparable compact fluorescent light bulb (CFL), which emits the same amount of light, would be rated at 25 W but would only use 10 kW·h if it were on for the same length of time (Figure 2). The Ontario government plans to ban the incandescent light bulb by 2012.

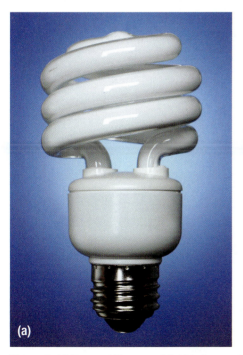
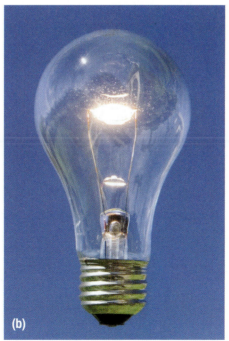

Figure 2 (a) Compact fluorescent light bulbs are 20 % to 30 % more efficient at producing light than (b) incandescent light bulbs.

Figure 3 Although LED bulbs are still expensive to buy, their very low energy use over their lifetime makes them a reasonable alternative.

Although efficient, CFLs (and regular fluorescent tubes) contain small amounts of mercury. Mercury is toxic and can harm living things. CFLs must be disposed of properly. Some stores that sell CFLs provide a place for consumers to return their used bulbs for proper recycling, which recovers virtually all the mercury. Interestingly, switching to CFLs can actually reduce mercury pollution by reducing the need for electrical energy, a third of which is produced by burning coal (recall that mercury is emitted by burning coal). The burning of coal is a major source of mercury pollution in Canada.

Recently, a new way of producing light has been developed. Light-emitting diodes (LEDs) need less electrical energy to produce light than any other type of light bulb (Figure 3). They last a very long time and produce almost no heat, making them very efficient. LEDs are used in many applications, such as in Christmas tree lights, computers, traffic lights, billboards, and cars—even your home gaming system may use LEDs.

EnerGuide and Energy Star Labels

When buying any electrical device, it is important to consider both the price of the device and the cost of operating the device over time. A less expensive device might be a tempting purchase, but it may use much more electrical energy than a more expensive device. All household appliances are sold with an EnerGuide label (Figure 4) to help consumers make informed choices. This label provides an estimate of how much electrical energy (measured in kilowatt·hours) the appliance will use in one year. Consumers can compare EnerGuide labels to help make energy-wise choices when buying new appliances.

Some energy-efficient appliances are labelled with an Energy Star® symbol as well as an EnerGuide label (Figure 5). The Energy Star® symbol is used to identify products that meet a minimum level of efficiency.

To learn more about EnerGuide and Energy Star, **GO TO NELSON SCIENCE**

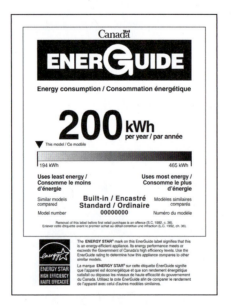

Figure 4 The EnerGuide labelling system by Natural Resources Canada helps consumers make informed choices when buying a new appliance.

Figure 5 The Energy Star® program was created in 1972 by the U.S. Environmental Protection Agency and the U.S. Department of Energy.

Calculating the Efficiency of a Device

The higher the percentage, the more efficient the device is. You can calculate the percent efficiency of a device using the equation:

$$\text{percent efficiency} = \frac{\text{energy out}}{\text{energy in}} \times 100\ \%$$

$$\%\ \text{efficiency} = \frac{E_{out}}{E_{in}} \times 100\ \%$$

Energy out is a measure of how much useful energy the device puts out to do its task. For example, a light bulb might produce 35 J of light energy. *Energy in* is a measure of how much energy the device requires. For example, the same light bulb might require 100 J of electrical energy.

> **SAMPLE PROBLEM 1** Calculating the Efficiency of a Light Bulb
>
> A light bulb uses 100 J of electrical energy and produces 35 J of light energy. Calculate the percent efficiency of the light bulb.
>
> **Given:** $E_{out} = 35$ J
> $E_{in} = 100$ J
>
> **Required:** percent efficiency (% efficiency)
>
> **Analysis:** $\%\ \text{efficiency} = \frac{E_{out}}{E_{in}} \times 100\ \%$
>
> $\%\ \text{efficiency} = \frac{35\ \text{J}}{100\ \text{J}} \times 100\ \%$
>
> **Solution:** $\%\ \text{efficiency} = 0.35 \times 100\ \%$
> $\%\ \text{efficiency} = 35\ \%$
>
> **Statement:** The efficiency of the light bulb is 35 %.
>
> **Practice**
> A toaster oven uses 1200 J of energy to produce 850 J of thermal energy. Calculate the percent efficiency of the toaster oven.

Timely Use of Electrical Energy

The price of electrical energy changes throughout the day. The most expensive time for electrical energy production is generally weekdays from 9 a.m. to 5 p.m., when the demand for electricity is the greatest. During peak times, fossil fuel plants are run in addition to nuclear and hydro-electric generating stations to supply the extra energy. The cost of electricity during the peak times can be three times more expensive. This means that if you use electrical energy outside of the peak times, such as running a dishwasher at night, you can save money and reduce pollution.

By 2010, all homes and businesses in Ontario will have smart meters (Figure 6). The smart meter is an electrical energy meter connected to your home that measures the amount of electrical energy being used as well as the time at which it is being used.

Efficient appliances and smart meters are two ways to conserve electrical energy. The ultimate power-saving feature is your common sense. When you are not using a device, it makes sense to turn it off, and when you buy a new device, consider all the costs and choose wisely.

Figure 6 Many communities in Ontario have already started implementing smart meters.

Cost of Electricity

In Ontario, the current cost of electricity for homeowners is regulated at 5.6 ¢/kW·h for the first 1000 kW·h during winter and the first 600 kW·h during summer. After 1000 kW·h, the cost increases to 6.5 ¢/kW·h. How much would it cost to use a laptop computer for a year? To figure that out, you would need to know the power rating of the laptop in kilowatts and the length of time that it is operating in hours. You would use this equation:

$$\text{cost to operate} = \text{power used} \times \text{time} \times \text{cost of electricity}$$

SAMPLE PROBLEM 2 Calculating the Cost of Operating a Laptop Computer

A laptop computer uses a 75 W adapter when it is plugged in. Electricity costs 5.6 ¢/kW·h. Calculate how much it would cost to operate the laptop for 1 year for 24 hours per day.

Given: power = 75 W (converted to kW = $75 \text{ W} \times \frac{1 \text{ kW}}{1000 \text{ W}} = 0.075 \text{ kW}$)

time = 24 hours per day for 365 days = 8760 hours

cost of electricity = 5.6 ¢/kW·h

Required: cost to operate

Analysis: cost to operate = power used × time × cost of electricity

Solution: cost to operate = $0.075 \text{ kW} \times 8760 \text{ h} \times \frac{5.6 \text{ ¢}}{\text{kW} \cdot \text{h}} = 3679$ ¢

Statement: It would cost 3679 ¢ (or $36.79) to operate a laptop computer for 24 hours per day for 365 days.

Practice

Calculate the cost of operating a 1500 W hair dryer to dry your hair for 6 minutes per day for 3 days. The cost of electricity is 5.6 ¢/kW·h.

From Sample Problem 2, you can see that operating one electrical device for a year is fairly inexpensive. However, we use a lot of technology in our homes. Even when a device is off, it may use a few watts in standby mode, waiting to be used. When you add up all the watts used, they could add up to hundreds of dollars per year.

There are other costs associated with electrical energy. The utility companies that provide the electrical energy also charge you for the distribution and transmission of the electrical energy. This can add a significant amount to your family's electricity bill. In fact, the rate at least doubles when you include these charges.

Not only is there a financial cost to electrical energy, but there is also an environmental cost as well. Remember that many of the ways we generate electrical energy produce some form of pollution. This pollution has a social cost because it can affect people's health and the environment in a negative way.

CITIZEN ACTION

Is Your School Conserving Electrical Energy?

Find a group of people who are interested in reducing the amount of electrical energy used by your school. The group might already exist in the form of an environmental club. Meet with this group or ask to attend an environmental club meeting. Brainstorm ideas about how you could find information on how electrical energy is used in your school. You may wish to develop a survey that could be passed along to students, teachers, administrators, and support staff.

Analyze the information that you collect for common areas where electrical energy seems to be wasted or used inefficiently. Again, brainstorm to suggest ways to reduce electrical energy consumption. Consider replacing inefficient devices with energy-efficient ones. You may also suggest some alternative energy production systems to lower the dependence of your school on the province's electrical grid.

Once you have completed putting together your ideas, prioritize them into categories. These categories may include, but not be limited to, ideas that cost the least, ideas that involve the most energy-use reduction, and ideas that are the most ambitious. Once you have sorted out the ideas, draft a proposal that you could take to your teacher to demonstrate that it could be implemented to help the school conserve energy, save money, and reduce environmental pollution. Present your plan in a report to your teacher.

As an extension to this activity, you could vote on the best proposal as a class. This proposal could be presented to the school's principal. If the principal approves of the plan, the plan could be presented at a teacher's staff meeting and actually be implemented. If successful, it could be a model that can be shared with other schools in your district.

IN SUMMARY

- Electrical power is the rate at which electrical energy produced or used.
- Electrical energy should be conserved and used wisely.
- Purchasing electrical energy–efficient devices saves money and is better for the environment.
- Percent efficiency can be calculated using the equation:

$$\% \text{ efficiency} = \frac{E_{out}}{E_{in}} \times 100\ \%$$

- When purchasing an electrical device, you should also consider the cost of operating it.
- The time of day affects the costs of producing and purchasing the electrical energy that you use.
- The cost of operating an electrical device can be calculated using the equation:

$$\text{cost to operate} = \text{power used} \times \text{time} \times \text{cost of electricity}$$

CHECK YOUR LEARNING

1. Describe an idea that you read about in this section and found to be particularly important. Why do you think this idea is important, and how will you act on it? C

2. Identify three ways that you can conserve electrical energy and produce fewer greenhouse gases. K/U

3. Explain how EnerGuide labels are useful to consumers. K/U

4. When purchasing an electrical device, what are the two financial costs you need to consider? Are there any environmental considerations you would make? Explain your reasoning. A

5. Calculate the efficiency of a compact fluorescent light bulb if it produces 30 J of light energy, while using 95 J of electrical energy. T/I

6. Calculate the cost of operating the following devices. The cost of electricity is 12 ¢/kW·h. T/I

    (a) a 100 W incandescent light bulb for 1000 hours
    (b) a 13 W CFL for 1000 hours
    (c) a 400 W computer for 600 hours
    (d) a refrigerator operating at its peak power of 750 W for one year

7. Calculate the difference between the operating cost of a 60 W incandescent light bulb and a 13 W CFL, each operating for 100 hours. The cost of electricity is 11 ¢/kW·h. T/I

12.8 PERFORM AN ACTIVITY

Examining Electrical Energy Production

In this activity, you will analyze how much electrical energy is generated and how that energy production affects the environment.

SKILLS MENU
- Questioning
- Hypothesizing
- Predicting
- Planning
- Controlling Variables
- Performing
- Observing
- Analyzing
- Evaluating
- Communicating

Purpose
To analyze electrical energy production in Canada.

Equipment and Materials
- calculator
- graph paper

Procedure
1. Examine Figure 1 and Tables 1 and 2. Recall that 1 GW·h is equal to 1 000 000 kW·h.

Analyze and Evaluate

SKILLS HANDBOOK 3.B.7, 3.B.8.

Answer (a) to (d) by referring to Figure 1.

(a) Which provinces use mainly hydro-electric power to produce electrical energy? T/I

(b) Which province uses mainly coal to produce electrical energy? T/I

(c) Which provinces generate the most electrical energy? T/I

(d) Which provinces/territories use little to no nuclear energy to produce electrical energy? T/I

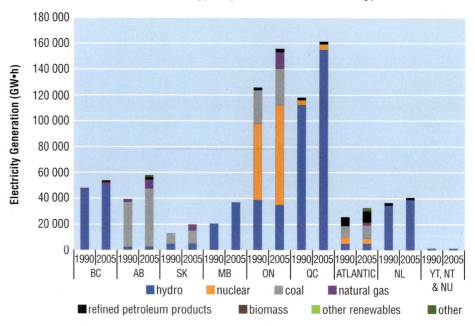

Figure 1 Electrical energy production by province for the years 1990 and 2005. Note that "Refined Petroleum Products" refers to oil, while "Other Renewables" refers to wind and tidal energy combined.

Table 1 Canadian Electrical Energy Generation in GW·h

Source	1995	1996	1997	1998	1999	2000	2001	2002	2003	2004	2005
coal	83 351	84 548	92 555	99 236	99 591	107 680	107 779	106 913	100 391	94 872	96 620
oil	8 418	6 282	10 109	14 537	11 749	10 807	13 252	10 793	12 561	12 799	11 887
natural gas	14 189	10 777	15 099	18 904	19 688	25 881	27 280	26 389	26 244	25 340	27 947
nuclear	92 306	87 510	77 857	67 467	69 331	68 675	72 353	71 251	70 653	85 240	86 830
hydro	299 738	321 414	315 959	299 123	309 334	323 468	299 604	314 555	302 437	303 591	327 171
biomass	–	–	–	1 703	1 743	1 911	2 116	2 182	2 137	1 995	1 794
wind and tidal	33	32	82	84	278	264	366	435	704	970	1 578

Table 2 Canadian Greenhouse Gas Emissions by Source and Year in kt CO_2 Equivalent*

Source	1995	1996	1997	1998	1999	2000	2001	2002	2003	2004	2005
coal	82 998	84 835	91 375	97 044	96 681	104 770	103 365	101 949	101 260	92 862	96 022
oil	6 514	5 126	7 720	11 486	9 111	8 345	10 156	8 123	9 867	9 708	9 514
natural gas	7 001	5 575	7 444	9 641	9 735	12 906	13 863	12 698	13 575	12 576	14 058
nuclear	–	–	–	–	–	–	–	–	–	–	–
hydro-electric[1]	–	–	–	–	–	–	–	–	–	–	–
biomass[2]	–	–	–	–	–	–	–	–	–	–	–
wind and tidal	–	–	–	–	–	–	–	–	–	–	–

*Note that the units kt CO_2 equivalent refers to a kilotonne of gases equal to CO_2
[1] Greenhouse gas emissions not included from flooded lands or hydro dams
[2] Greenhouse gas emissions not included

Answer (e) to (k) by referring to Table 1.

(e) Which source of electrical energy is used most in Canada? Suggest a reason for this.

(f) What was the trend for coal electrical energy production from 1995 to 2001? What was the trend after 2001? Suggest reasons for this.

(g) What is the trend for overall electricity generation in Canada from 1995 to 2001? Calculate this by adding the amount of energy for each source each year. Suggest reasons for this.

(h) Which fossil fuel has become more used over the time period represented in Table 1? Suggest a reason for this increased use in the fossil fuel. You may need to refer to Section 12.5.

(i) Has the amount of electrical energy produced by nuclear sources changed significantly between 1995 and 2005? Explain your answer.

(j) What is the general trend in using alternative energy sources for electrical energy production? Suggest a reason for this.

(k) Across Canada, how does the use of alternative energy sources compare with the use of conventional energy sources? Suggest a reason for this.

Answer (l) to (o) by referring to Table 2.

(l) Plot a graph of greenhouse gas emissions versus year. You can use a bar graph or a line graph. Plot all three sources on the same graph.

(m) Describe the trend in greenhouse gas emissions related to burning coal. Does this trend connect with any other trend you have found?

(n) Does natural gas show similar trends to coal? Explain.

(o) Why are the values for nuclear, hydro-electric, wind, and tidal left blank?

Apply and Extend

(p) Why do you think that the provinces you listed for part (j) generate the most electrical energy?

(q) Suggest a reason for the biggest difference in electricity generation technologies used in Ontario between 1990 and 2005.

(r) Why do you think the provinces/territories you listed for part (l) do not use nuclear energy?

(s) Write a paragraph summarizing what you have discovered about electrical energy production and greenhouse gas emissions in Canada.

(t) Suppose that the government authorized the development of a new electricity generation plant very close to where you live. Which type of electricity generation technology would you prefer to be developed? Justify your answer by explaining how it may affect your neighbourhood. Include how it may affect the financial situation of family members, your community, and the local environment.

TECH CONNECT

Energy for Spacecraft

Here on Earth, we can easily switch on lights, computers, and heating and cooling systems because we have access to electrical energy—whether in the form of wall outlets or batteries. But what happens in space?

Each space mission relies on electrical energy to run all the equipment. In space, electrical energy is needed not only to keep the lights and computers working but also to provide clean air and water for astronauts. On board the space shuttle, fuel cells provide the electrical energy for the length of the mission. A large array of solar panels provides the electrical energy on the International Space Station (Figure 1). If there is a problem between the solar panels and the connection to the space station, nothing will work. An electrical failure in space can be very dangerous.

Figure 1 An array of solar panels on the International Space Station provides electrical energy.

Scientists in Italy, working with NASA, set out to see if they could generate electrical energy in space. They developed a system that consisted of a tether attached to a satellite. The tether was a 20 km long cable made out of a conducting material. During one of the space shuttle flights, this satellite–tether system would be deployed while the shuttle was in orbit.

The intent was to use planet Earth itself to generate the electrical energy. You may not know that Earth is a giant magnet. Scientists hypothesized that if the satellite dragged the tether through Earth's magnetic field, a flow of electrons could be produced in the tether. Recall from section 12.4 that when a conductor moves between the poles of a magnet, electrons begin to flow in the conductor. This is how all generators work.

To complete the circuit, the current from the tether would flow through the shuttle into Earth's ionosphere, the part of the atmosphere that contains charged particles. The ionosphere conducts electricity, but not as well as the conductor in the tether.

The mission in February 1996 started out as planned. As the tether was being unrolled, scientists on Earth were thrilled to see that the electric current produced exceeded their expectations. However, just before deployment of the satellite–tether system was complete, the tether suddenly broke (Figure 2). The system began drifting away into space. For a while, the satellite could still be tracked by radio, but eventually it moved beyond the range of radio contact.

Figure 2 The satellite, which was tethered to the space shuttle Columbia, floated off into space after the tether broke.

When the shuttle returned to Earth, scientists carefully examined the broken end of the tether. They determined that the innermost part of the tether, an insulator, was the culprit. During its manufacture, many air bubbles were trapped within the material. When the insulator entered the ionosphere, the trapped air escaped through pinholes and became a better electrical conductor than the conducting wires in the tether itself. The current through the insulation was enough to melt the tether.

Although the tether experiment had a disappointing end, it was not a total loss. Scientists learned a great deal about generating electrical energy during the mission. In future space missions, this technology may be used to generate electrical energy during emergencies or to supplement the solar panels when the electrical energy supply is low.

PERFORM AN ACTIVITY 12.9

Performing an Electrical Energy Audit

Have you ever wondered how much electrical energy you actually use? In this activity, you will determine the amount of electrical energy you use on a weekly basis. You will then determine how much electrical energy you used for each device and its cost.

SKILLS MENU
- Questioning
- Hypothesizing
- Predicting
- Planning
- Controlling Variables
- Performing
- Observing
- Analyzing
- Evaluating
- Communicating

Purpose
To determine the amount of electrical energy you use in one week on a selection of devices.

Equipment and Materials
- watch
- watt meter (optional)
- logbook

Procedure
1. Prepare a table similar to Table 1 in your logbook.

Table 1 Electrical Energy Used in One Week

Date	Device	Power rating of device (kW)	Time used (h)	Kilowatt·hours of energy	Operating cost

2. List five electrical devices you use every day. Keep track of the amount of time that you use the device. Record the information in the logbook.

3. Record the power rating (wattage) of each device. This information can be found directly on the device. Convert the watts to kW (1000 W = 1 kW). Verify the amount of power used by your device with a watt meter.

4. Calculate the kilowatt·hours of electrical energy used by each device by multiplying the power rating of the device in kW by the time used in hours.

5. At the end of each day, for 7 days, total the electrical energy used in kW·h in your logbook.

Analyze and Evaluate SKILLS HANDBOOK 3.B.7., 3.B.8.

(a) Calculate the total electrical energy used each day. What do you notice about your daily use of electrical energy?

(b) Calculate the total electrical energy that you used for the week.

(c) Calculate the total cost of operating all the appliances that you used. Use a recent electricity bill to determine the cost of electricity per kilowatt·hour for your family.

(d) What is the total electrical energy consumed by your class? Calculate a class average. Do you use more or less electrical energy than your classmates?

Apply and Extend

(e) In 2006, Canadians used, approximately 5.37×10^{11} kW·h of electricity. Assuming the population in Canada was approximately 33 million people, what was the average use of electrical energy per person? Assume that your energy consumption for the week continued for 51 more weeks. How does your consumption compare with the average Canadian? Is this a fair comparison?

(f) Use a watt meter to measure the actual electrical energy a device in your home uses. You can borrow a watt meter from your local public library. Identify the devices that use the most electrical energy. You may also wish to investigate how much electrical energy these devices use when in "standby" mode. Many devices that have timers in them, such as VCRs, DVD recorders, clock radios, and TVs, use electrical energy when they are off. This is known as a "phantom load" and can amount to 7 % of a household's total energy usage. Use the watt meter to determine exactly how much electrical energy your devices use as phantom loads. How could you eliminate phantom loads in your household?

Chapter 12 LOOKING BACK

KEY CONCEPTS SUMMARY

The motion of electrons can be controlled.

- Current electricity is the flow of electrons through a conductor. (12.1)
- Electron flow through conductors can be regulated and controlled. (12.1)
- A flow of electrons requires an energy source. (12.2)
- In direct current (DC) electrons flow in one direction only.
- In alternating current (AC) electrons flow back and forth, alternating their direction.

An electric circuit requires a source of electrical energy, conductors, a control device, and a load.

- A circuit is a closed path through which electrons flow. (12.2)
- A control device can regulate the flow of electrons through a circuit or direct the electron flow through connecting wires. (12.2)
- A load transforms electrical energy into other forms of energy. (12.2)

Electrical energy is produced by energy transformations.

- Electrical energy does not exist as a primary form of energy. (12.3)
- Many forms of energy can be transformed into electrical energy. (12.3, 12.5)
- Generators, solar cells, and batteries all involve energy transformations. (12.5)
- An electric cell converts the energy stored in chemicals into electrical energy. (12.3)

Electrical energy is produced from renewable and non-renewable resources.

- Electrical energy is produced by conventional and alternative methods. (12.5)
- Renewable resources include falling water, wind, sunlight, biomass, geothermal, and tidal. (12.5)
- Non-renewable resources include fossil fuels such as coal, oil, natural gas, and uranium. (12.5)

Electrical power is the rate at which electrical energy is produced or consumed in a given time.

- The unit of energy, including electrical energy, is the joule (J). (12.3)
- The unit of electrical power is the watt (W). A watt is the equivalent of a joule per second (J/s). (12.7)
- A common unit for electrical energy is the kilowatt hour (kW·h). (12.7)

Electrical energy consumption should be reduced.

- Government, business, and individuals all have roles in reducing electrical energy consumption. (12.7)
- Efficient devices use less energy to do the same tasks. (12.7)
- Percent efficiency of an electrical device is calculated using the equation

  % efficiency = $\dfrac{E_{out}}{E_{in}} \times 100\,\%$ (12.7)
- The cost of operating an electrical device is calculated using the equation:

  cost to operate = power used × time × cost of electricity (12.7)

WHAT DO YOU THINK NOW?

You thought about the following statements at the beginning of the chapter. You may have encountered these ideas in school, at home, or in the world around you. Consider them again and decide whether you agree or disagree with each one.

1 Using electricity creates no pollution.
Agree/disagree?

4 An incandescent light bulb is an efficient way to produce light.
Agree/disagree?

2 Batteries are a waste of money.
Agree/disagree?

5 Computers use very little energy.
Agree/disagree?

3 The EnerGuide label on an appliance indicates how efficient the appliance is.
Agree/disagree?

6 We can never run out of electricity.
Agree/disagree?

How have your answers changed since then?
What new understanding do you have?

Vocabulary

current electricity (p. 507)
electric circuit (p. 509)
load (p. 509)
switch (p. 509)
electrical energy (p. 511)
electric cell (p. 512)
primary cell (p. 512)
secondary cell (p. 512)
direct current (DC) (p. 515)
alternating current (AC) (p. 515)
renewable resource (p. 519)
non-renewable resource (p. 521)
biomass (p. 521)
photovoltaic cell (p. 527)
electrical power (p. 530)
kilowatt·hour (kW·h) (p. 530)
efficiency (p. 530)

BIG Ideas

- ✓ Electricity is a form of energy produced from a variety of non-renewable and renewable sources.
- ✓ The production and consumption of electrical energy has social, economic, and environmental implications.
- • Static and current electricity have distinct properties that determine how they are used.

Looking Back

CHAPTER 12 REVIEW

The following icons indicate the Achievement Chart category addressed by each question.

K/U Knowledge/Understanding **T/I** Thinking/Investigation **C** Communication **A** Application

What Do You Remember?

1. Can electrical energy be created? Explain. (12.3, 12.5) K/U

2. Identify two main advantages that are common to all renewable energy sources used to generate electrical energy. (12.5) K/U

3. Identify how you could generate electrical energy for each of the following locations or devices. Justify your answer. (12.5) K/U
   (a) two cities near a river
   (b) an MP3 player
   (c) your home
   (d) a cottage
   (e) a spacecraft
   (f) an electric car
   (g) a flashlight that is used infrequently

4. Ontario has the capacity to produce 27 000 MW of electrical power. The demand for electrical power on a typical summer day is 24 000 MW. Ontario produced 175 000 GW·h. Name what electrical quantities these values represent. (12.7) K/U

What Do You Understand?

5. A light bulb is labelled 60 W. What does this mean? (12.7) K/U

6. Explain why fossil fuel generating stations are so common throughout the world. (12.5) K/U C

7. Compare one non-renewable energy source with one renewable energy source. Consider pollution, cost, and how electrical energy is produced. (12.5) C A

8. You have just been elected mayor of Futuretown. The suburban town has 100 000 residents and a small ceramics factory. There is a wide, slow-moving river nearby. Your residents need a source of electrical energy. In two or three paragraphs, describe what sources you would use and why. (12.5) A C

9. Draft a plan on how you could reduce your personal electrical energy consumption. (12.7) A C

10. You are presented with two EnerGuide labels (Figures 1 and 2). What factors will you consider before you purchase one of these devices? (12.7) T/I

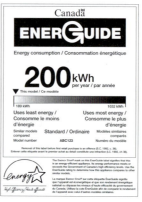

Figure 1

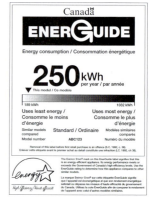

Figure 2

Solve a Problem

11. Calculate the percent efficiency of a motor that produces 4500 J of mechanical energy, while using 6500 J electrical energy. (12.7) T/I

12. (a) Calculate the total annual cost of operating the electrical appliances listed below. The cost of electricity is 11 ¢/ kW·h. (12.7) T/I

Table 1 Electrical Appliance Data

Electrical appliance	Power (W)	Time per year (h)
stove	12 000	300
microwave	1 000	12
refrigerator	400	2 500
coffee maker	120	100
TV	300	700
DVD player	50	350
computer	250	700
washer	1 300	60
dryer	5 600	70

(b) Do the data in Table 1 relate to your household? Is the table missing any electrical appliances? Are the total hours used per year reasonable? Explain your answers. T/I

13. You have been hired as the energy officer for a company that produces steel. Making steel is very energy intensive and requires a large amount of electrical energy. Your manager has requested that you estimate the cost of electrical energy that the company will need so that they can develop a budget. What information should you collect in order to answer your manager's request? Suggest a sample calculation of how you would determine the answer. (12.7)

14. The Ministry of the Environment would like to know how much less greenhouse gas would be emitted by switching from coal-powered plants to natural gas-powered plants to generate electricity. Show how you would calculate this value. The ministry has the following information. Burning coal to generate 1 kW·h of electrical energy produces 0.95 kg of CO_2. Burning natural gas to generate 1 kW·h of electrical energy produces 0.59 kg of CO_2. Assume that Ontario generated 150 000 000 kW·h of electrical energy and that 18 % of it is produced by burning coal. Once you have finished your calculation, justify whether or not you think the decision to switch to natural gas is worthwhile. (12.5, 12.7)

15. Oil is being transported across the Atlantic Ocean to reach the shores of Canada. This oil is going to be used to generate electricity in the northern part of the country. Other sources are not as easily available, so oil is an important commodity. Unfortunately, the tanker that was transporting the oil has crashed into the shore during a storm, and oil has now spilled into the ocean. Brainstorm how this accident has impacted society near the site of the accident as well as in the North, where the oil was going to be used. How will it be possible for the communities in the North to generate their much-needed electricity? Write a proposal to the government on how to help the community with its needs. (12.5)

Create and Evaluate

16. Would it be better to generate more electrical energy from renewable energy sources or conserve it? Discuss your opinion in a paragraph with supporting information. (12.5, 12.7)

17. You are spending a few weeks at a friend's cottage, which is very isolated. Since it is not possible to connect to the power grid for electrical energy, the family generates their own. They have been using a gasoline-powered generator, but it is starting to fail. The family is considering replacing it with a new one and have asked for your opinion. What would you say to them? In a paragraph or two, outline some of the things you would mention to the family. (12.5)

Reflect On Your Learning

18. Describe an idea about the generation of electrical energy that was new to you. Will you change anything in your everyday life because of this idea? Explain what you will change.

Web Connections

19. Thomas Edison sought a patent for a practical working source of electric light in 1879. This was the incandescent light bulb. This design has remained essentially unchanged since then. Ontario is banning this type of light bulb because it is very inefficient. Find out how this ban will affect how much electrical energy the average home can save. Consider the environmental impact of using less electrical energy. Write a paragraph about what you find. (12.5, 12.7)

20. Sir Adam Beck was an important Canadian. Two hydro-electric generating stations are named after him. Research to determine why the stations are named in his honour. Write a paragraph about what you find. (12.5)

21. Compact fluorescent light bulbs (CFL) contain mercury. Mercury is an element that is poisonous to organisms. Research the environmental impact and health concerns of mercury in CFL bulbs. Write a paragraph to summarize your findings.

CHAPTER 12 SELF-QUIZ

The following icons indicate the Achievement Chart category addressed by each question.
- K/U Knowledge/Understanding
- T/I Thinking/Investigation
- C Communication
- A Application

For each question, select the best answer from the four alternatives.

1. Which phrase correctly describes a primary cell? (12.3) K/U
   (a) It can be recharged many times.
   (b) It is used in disposable batteries.
   (c) It has a shelf life of six months to one year.
   (d) It uses chemical reactions that are reversible.

2. Which unit is used to measure electrical power? (12.7) K/U
   (a) watt
   (b) joule
   (c) kilowatt·hour
   (d) % efficiency

3. Which of the following is a drawback to solar generation? (12.5) K/U
   (a) Solar power is a non-renewable energy source.
   (b) Solar cells are about 30 % efficient.
   (c) Solar farms release hot water into nearby lakes.
   (d) The production of electrical energy on solar farms is noisy.

4. Which method of generating electrical energy involves neutrons colliding with a radioactive material? (12.5) K/U
   (a) tidal
   (b) nuclear
   (c) geothermal
   (d) hydro-electric

Indicate whether each of the statements is TRUE or FALSE. If you think the statement is false, rewrite it to make it true.

5. Nuclear generation does not contribute to air pollution. (12.5) K/U

6. Electricity is most costly to produce at night. (12.7) K/U

Copy each statement into your notebook. Fill in the blanks with a word or phrase that correctly completes the sentence.

7. A fuel cell uses _____ energy to produce _____ energy. (12.3) K/U

8. Energy resources, such as coal, oil, and natural gas, that were formed from the remains of organisms that lived millions of years ago are called _____. (12.5) K/U

Match each term on the left with the most appropriate description on the right.

9. (a) switch
   (b) photovoltaic cell
   (c) battery
   (d) load
   (e) CFL

   (i) a part of an electric circuit that converts electrical energy into another form of energy
   (ii) a device that controls the flow of electrons by opening or closing a circuit
   (iii) a combination of two or more electric cells
   (iv) a type of light bulb that is more efficient than an incandescent light bulb
   (v) a device that converts light energy from a light source directly into electrical energy (12.2, 12.3, 12.5, 12.7) K/U

10. How many watts are in each of the following units? (12.7) K/U
    (a) 1 kilowatt
    (b) 1 megawatt
    (c) 1 gigawatt

Write a short answer to each of these questions.

11. You notice that the insulation on the cord attached to your friend's electric toaster has cracked and the bare copper wire is exposed. What should you tell your friend to do about the toaster? Why? (12.1) A C

12. In your own words, explain how hydro-electric generation, thermal generation, and wind generation are similar in the way they produce electricity. (12.5) K/U C

13. (a) Your microwave oven is 64 % efficient. How much electrical energy must it use to produce 900 J of radiant energy? (12.7) **T/I**

    (b) Your neighbour's microwave oven uses 1500 J of electrical energy to produce 900 J of radiant energy. Is it more or less efficient than your microwave oven? **T/I**

14. An electric clothes dryer used 4400 W to dry two loads of laundry. Electricity costs 5.6 ¢/kW·h. If it cost 75¢ to use the dryer, for approximately how many hours did the dryer operate? (12.7) **T/I**

15. A conservation organization has purchased time on a local radio station. Write a paragraph of text for their brief commercial, explaining why people should consider buying energy-efficient appliances, even if they are initially more expensive than less efficient models. (12.7) **C** **A**

16. Make a list of at least five things you have done during the past week to conserve electrical energy. (12.7) **A**

17. An electronics store advertises a stereo speaker that produces 30 W of power. Does the speaker use 30 W of power to generate this output? Explain your answer. (12.7) **T/I**

18. Copy and complete Table 1 in your notebook. (12.1) **K/U**

    Table 1

	Example	How the electrons move	Length of electron flow
static electricity	lightning		
current electricity		electrons flow steadily through a conductor	

19. You tell your younger sister that electricity flows from a wall outlet through a plug and a cord and into your radio. She now thinks that when you unplug the radio, the electricity will spill out of the outlet and onto the floor. Explain to her why this does not happen. (12.1, 12.2, 12.4) **K/U** **C**

20. Someone who opposes the construction of something near his or her home is sometimes referred to as a NIMBY, which stands for Not In My Back Yard. Imagine that you are a NIMBY who lives in a windy community but does not want a wind farm built near your home.

    (a) What reasons could you give to justify your opposition to the wind farm? (12.5) **A**

    (b) What reasons might someone give to contradict your opinion and support the wind farm? (12.5) **A**

21. It is important to consider energy efficiency when purchasing a new appliance, such as a washer, dryer, or refrigerator. It is also important to consider how to dispose of old appliances. (12.7) **A**

    (a) Explain what a consumer could do with an old appliance that would help the community or the environment.

    (b) Explain what manufacturers could do with old appliances that would help the community or the environment.

22. Choose one alternative method for generating electrical energy, and list several reasons why this method would be a good choice for your community. (12.5) **A**

23. (a) Explain how growing biofuel crops can affect the world's food supply. (12.5) **A**

    (b) Do you think the change you described in part (a) would also affect the prices of crops and of land? Explain your answer. (12.5) **A**

24. Some floor lamps have caused fires when placed too close to curtains or drapes. Use what you know about the efficiency of light bulbs to explain why this can happen. (12.7) **A**

CHAPTER 13

Electrical Quantities in Circuits

KEY QUESTION: How are electrical circuits used?

The circuits on this circuit board are stamped on, but function the same as circuits made with wires and leads.

UNIT E: The Characteristics of Electricity

- **CHAPTER 11** — Static Electricity
- **CHAPTER 12** — Electrical Energy Production
- **CHAPTER 13** — Electrical Quantities in Circuits

KEY CONCEPTS

Circuits are a part of electrical technology. They can be connected in series and in parallel.

Current is the rate of electron flow.

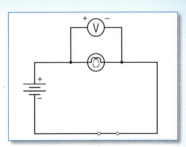

Potential difference is a measure of the difference in electric potential energy per unit of charge between two points.

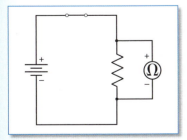

Resistance is a measure of the opposition to the flow of electrons.

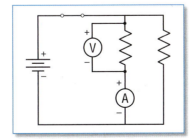

Connecting loads in series or parallel affects the current, potential difference, and total resistance.

Current, voltage, and resistance measurements can be used to test circuits.

Looking Ahead

ENGAGE IN SCIENCE

Suspected Electrical Fire Destroys Homes

A three-alarm fire raged through an apartment complex in the early hours of the morning on December 2nd. Local residents say that they saw smoke and flames from many blocks away. City firefighters fought the blaze for several hours before finally getting the fire under control.

The fire marshall, who was interviewed at the scene, said, "Fortunately, all of the tenants were able to escape safely because they all had working smoke alarms. At this time, we suspect faulty electrical wiring may have been the cause of the fire. Many of the residents had been complaining that their lights were dimming when they turned on an appliance. That is a sign of an overloaded circuit."

Fire Marshall Investigates Electrical Fire

An overloaded circuit has been confirmed as the cause of an apartment complex fire on December 2nd.

"December is often the worst month for electrical fires," said the fire marshall. "The weather is cold, and families are inside and preparing for the holidays. Plugging too many things into one electrical outlet can overload electric circuits and cause fires."

According to insurance statistics, misused electrical extension cords, overloaded circuits, or poorly maintained electric cords cause almost 30 % of electrical fires in homes. Frayed or loose electrical cords or outlets can spark and cause fires.

Fire authorities have provided the following prevention tips to protect your home from electrical fires:

- Never overload extension cords or electrical outlets.
- Never force a three-pronged (grounded) electrical cord into a two-slot outlet or extension cord.
- Regularly inspect electrical cords for wear. Discard any cords with frayed, cracked, or worn insulation.

Signs that you have an overloaded electric circuit include flickering or dimming lights, sparks, warm electrical cords, and frequent blown fuses or tripped circuit breakers. If you experience any of these warning signs, unplug the faulty device immediately or contact a certified electrician for professional help.

> Think about some of the electrical circuits that play a role in your daily life. How are these electrical circuits used?

WHAT DO YOU THINK?

Many of the ideas you will explore in this chapter are ideas that you have already encountered. You may have encountered these ideas in school, at home, or in the world around you. Not all of the following statements are true. Consider each statement and decide whether you agree or disagree with it.

1 There is one wire that connects all the electrical devices in my home.
Agree/disagree?

4 Electric current flows just like water flows in a river.
Agree/disagree?

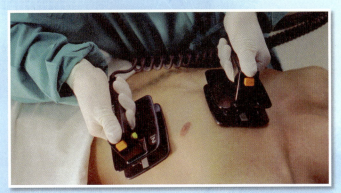

2 An electric shock is always dangerous.
Agree/disagree?

5 It is always safe to plug as many devices into a power bar as there are outlets.
Agree/disagree?

3 All electrical devices get warm when they are operating.
Agree/disagree?

6 An electric stove requires more energy to operate than an electric fan.
Agree/disagree?

FOCUS ON WRITING

Writing a Critical Analysis

WRITING TIP

As you work through the chapter, look for tips like this. They will help you develop literacy strategies.

When you write a critical analysis, you examine an issue in depth by raising questions about the accuracy and consistency of the information and the reliability of the sources. The following is an example of a critical analysis written about two passages in this chapter. Use the strategies listed next to the text to improve your critical writing.

Resistance Is Not Futile

After reading the Engage in Science about the electrical fire caused by an overloaded circuit, and then the information in my textbook about resistance and safety, I wondered why the fuses or circuit breakers did not prevent the fire. "Aren't they supposed to interrupt the power to a circuit?" I asked myself.

First paragraph establishes context of critical analysis.

A question is asked about missing information.

My initial research revealed that an overloaded circuit might be caused by too many devices plugged into the outlets for the circuit, a short circuit, a ground fault, or a loose connection. Still, this information did not answer my question.

Conduct initial research to collect information.

Question is still unanswered.

Further research revealed the missing information. There are 20 000 house fires annually in Canada. Approximately one third of them are caused by arc faults. An arc fault produces an "unintentional electrical discharge." This can occur between the line and neutral conductors (parallel fault), or from a broken wire, faulty switch, or loose connection (series fault). Arc faults can be caused when insulation around a wire becomes damaged and exposes the bare copper conductor. Arc faults create heat and sparks that can make flammable materials ignite quickly.

Conduct further research to uncover important information that is omitted or is inconsistent.

Details are given to explain the information.

The Arc Fault Circuit Interrupter (AFCI) has been developed to detect arc faults and trip the circuit to protect against fires. Additionally, it will trip in the event of an overloaded or short circuit. In Canada AFCIs must be installed in bedroom circuits of new homes, but can also be used for extra protection in older homes.

Identify other people's perspectives on the issue.

The news report in the Engage in Science initially stated that "faulty electrical wiring" was thought to be the cause of the fire. If this were true, then AFCIs might have prevented the fire. Or if an overloaded circuit actually did cause the fire, AFCIs might have interrupted the circuit and prevented the fire.

Final paragraph gives possible solutions to the problem.

13.1 Circuits and Circuit Diagrams

In Chapter 12, you learned that a circuit is a continuous loop consisting of an energy source, conducting wires, one or more loads, and a switch. Before you can learn the properties of simple circuits, it is important to learn how to draw circuit diagrams. A **circuit diagram** is a standard way of drawing an electrical circuit.

In a circuit diagram, symbols are used to represent the different types of components that are linked together in a circuit. By understanding what these symbols mean, you can interpret how the circuit is supposed to function (Figure 1). Table 1 lists some common circuit symbols. You can find a complete list in the Skills Handbook, Section 2.D.

circuit diagram a way of drawing an electric circuit using standard symbols

To learn more about being an electrical engineer,

GO TO NELSON SCIENCE

Table 1 Common Circuit Diagram Symbols

	Part of circuit and symbol		
sources of electrical energy	electric cell ⊢⊣	three-cell battery ⊢⊢⊢⊣	variable DC power supply
electrical conductor	connecting wire		
control device	open switch	closed switch	
loads	lamp	electric motor —(M)—	

Figure 1 To fix an electrical fault in a house, an electrician must know how to interpret the wiring diagram on the fuse panel.

When you are asked to draw a circuit diagram, you are usually given a written description of the circuit. From the description, identify what devices are in the circuit. Find the symbols for those devices and then draw them, arranged as is required by the written description. Finally, connect all the devices with connecting wire lines.

Notice that positive and negative signs are used to identify the two terminals of an energy source. For an electric cell or battery, the longer line (one electrode) is positive, while the shorter line (the other electrode) is negative.

Most circuits used in everyday life have more than one load. These loads may be connected in two ways: in series or in parallel.

WRITING TIP

Writing a Critical Analysis
If writing a critical analysis of a circuit diagram, you might refer to the information about common circuit diagram symbols when describing the context for your analysis.

Series Circuits

In a **series circuit**, the flow of electrons follows only one path (Figure 2). This is achieved by connecting the loads in a chain, one after another in one continuous loop. For example, when you plug in a toaster and an electric kettle into the same outlet, the toaster and the kettle are two loads on the same circuit. The toaster and kettle are two loads in series. If the fuse for that outlet blows, neither device will work.

series circuit a circuit in which the loads are connected end to end so that there is only one path for electrons to flow

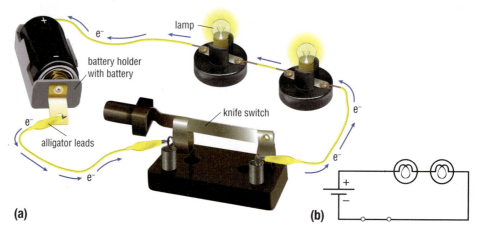

Figure 2 (a) There is only one path for the electrons to follow in a series circuit with two loads. The blue arrows show the direction of electron flow. (b) Circuit diagram for the circuit shown in (a).

Parallel Circuits

In a **parallel circuit**, electrons can flow more than one way (Figure 3). In this type of circuit, the loads are on at least two different branches of wires that connect to an energy source. When the electrons reach a branch, they separate just like a river splits into two tributaries. Some electrons will follow one branch, while the others will follow the other branch. These separate paths then merge before returning to the energy source.

Consider the electrical panel in your home. Each fuse in the panel regulates a series circuit, but all these series circuits are connected in parallel to each other. For example, when the fuse that regulates your stove blows, your refrigerator still works. The reason is that the refrigerator is regulated by a different fuse. The stove and refrigerator are two loads in parallel.

parallel circuit a circuit in which the loads are connected by branches so that there are two or more paths for electrons to flow

LEARNING TIP

Is It Series or Parallel?
You might not easily see if loads are in series or in parallel when you look at a circuit. Use your finger to trace the path from the energy source. If you can make it all the way back to the energy source without coming to a branch in the path, it is a series circuit.

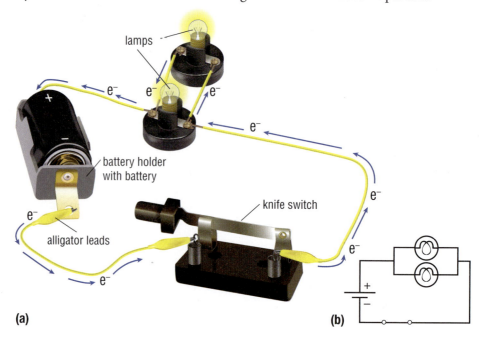

Figure 3 (a) In this parallel circuit, there are two separate paths for the electrons to follow with a single load in each path. (b) Circuit diagram for the circuit shown in (a).

The following sample problems show how to draw circuit diagrams of series and parallel circuits.

SAMPLE PROBLEM 1 Drawing Circuit Diagrams with Loads Connected in Series

Draw a circuit diagram showing an electric cell, a switch, and three lamps connected in series.

Step 1 Determine the circuit diagram symbols for each device (Figure 4).

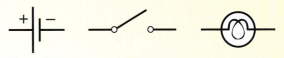

Figure 4

Step 2 Draw the loads (the lamps) in series (Figure 5).

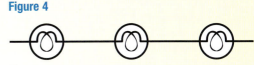

Figure 5

Step 3 Add the switch and electric cell. Remember to close the loop of the circuit (Figure 6).

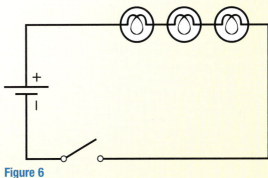

Figure 6

Practice

Draw a circuit diagram with a three-cell battery in series with a switch and two lamps.

SAMPLE PROBLEM 2 Drawing Circuit Diagrams for Loads Connected in Parallel

Draw a circuit diagram showing a two-cell battery with three lamps connected in parallel. Include a switch for controlling each lamp.

Step 1 Determine the circuit diagram symbols for each device (Figure 7).

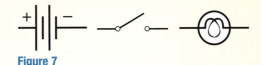

Figure 7

Step 2 Draw the lamps in parallel (Figure 8).

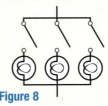

Figure 8

Step 3 Join the symbols together to close the loop of the circuit (Figure 9).

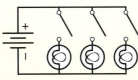

Figure 9

Practice

Draw a circuit diagram showing a two-cell battery, two lamps, and one motor. The motor and lamps must all be connected in parallel.

UNIT TASK Bookmark

You can apply what you have learned about series and parallel circuits to the Unit Task described on page 586.

IN SUMMARY

- Circuits can be represented using diagrams and circuit symbols.
- Loads in a circuit may be connected in series or in parallel.
- There is only one path for the electrons to follow in a series circuit, while in a parallel circuit, there is more than one path.

CHECK YOUR LEARNING

1. How many complete pathways are there for electrons to follow in each circuit shown in Figure 10? K/U

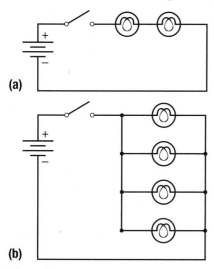

Figure 10

2. Identify whether each circuit in Figure 11 is series or parallel. K/U

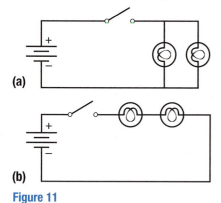

Figure 11

3. Draw circuit symbols for the following: K/U C
   (a) a switch
   (b) a four-cell battery with the cells connected in series with each other
   (c) a motor
   (d) a lamp
   (e) a connecting wire
   (f) a two-cell battery with the cells connected in parallel with each other

4. Draw a circuit diagram of
   (a) a circuit that contains a two-cell battery, two lamps, and a switch all connected in series
   (b) a circuit that contains a two-cell battery, three lamps in parallel, and a switch that controls the whole circuit
   (c) a circuit that contains a two-cell battery, three lamps in parallel, and enough switches to control each lamp independently
   (d) a circuit that contains a two-cell battery, five lamps in series, and a switch to control the circuit K/U C

5. Why are the outlets in homes never wired in series? What problems might this present? K/U A

6. A parallel connection of several lamps in a home allows you to control each lamp independently with switches. How is this possible? K/U T/I

7. A switch is closed, and a lamp turns on. The switch is opened, and the lamp turns off. Explain what is meant by these two statements. K/U

PERFORM AN ACTIVITY 13.2

Connecting Multiple Loads

If you dried your hair with a hairdryer or used a computer, you were using circuits that contained more than one load. In this activity, you will design two circuits to light two lamps. Your circuits will consist of a power supply, connecting wires, and lamps.

SKILLS MENU
- Questioning
- Hypothesizing
- Predicting
- Planning
- Controlling Variables
- Performing
- Observing
- Analyzing
- Evaluating
- Communicating

Purpose
To construct and operate series and parallel circuits.

Equipment and Materials
- a battery with an external switch (or a power supply with a built-in switch)
- 6 to 10 connecting wires
- 2 identical lamps

 Always disconnect the energy source before connecting a circuit. Do not adjust the power supply unless instructed to do so by your teacher. If your circuits do not work, ask your teacher for assistance.

Procedure

SKILLS HANDBOOK 2.D.

1. Set the power supply as instructed by your teacher *while it is turned off*. Do not change the setting during this activity.
2. Connect one lamp to the power supply. Turn on the power supply using the switch to check that the circuit is working. Note the brightness of the lamp. Turn off the power supply.
3. Connect two lamps to the power supply using the connecting wires. There should be only one path for the current to follow.
4. Turn on the power supply. Note the brightness of the lamps. Draw a circuit diagram of the circuit.
5. Remove one lamp from the socket and observe what happens. Turn off the power supply.
6. Connect two lamps so that each lamp has its own path to the power supply. Turn on the power supply. Note the brightness of the lamps. Draw a circuit diagram of the circuit.
7. Remove one of the lamps from the socket and observe what happens. Turn the power supply off.

Analyze and Evaluate

(a) How did the brightness of the two lamps in step 4 compare with the brightness of the single lamp in step 2? T/I

(b) What happened when you removed one lamp in step 5? T/I

(c) How did the brightness of the lamps in step 6 compare with the brightness of the lamps in step 4? T/I

(d) What happened when you removed one lamp in step 7? T/I

(e) Which circuit was a series circuit? Explain. T/I

(f) Which circuit was a parallel circuit? Explain. T/I

(g) Which circuit with two lamps provided the brightest light? T/I

(h) Which circuit with two lamps allowed you to operate each lamp on its own? T/I

Apply and Extend

(i) Suppose you wanted to operate each lamp independently with a switch for each. Draw a diagram of this circuit. T/I C

(j) Think about how electricity is distributed to the various rooms in your home. Do you think the circuits in your home are in series or in parallel? Explain. A C

(k) Think about how a light switch controls an overhead light. Is this a series connection or a parallel connection? Explain. A

UNIT TASK Bookmark

You can apply what you have learned about connecting multiple loads to the Unit Task described on page 586.

13.3 Electric Current

For any electrical device to operate, there must be a flow of electrons. When designing or troubleshooting a circuit, you need to know how much electric current is flowing through the different loads of the circuit. **Electric current** is the rate of electron flow past a specific point in a circuit. To better understand the concept of electric current, consider a waterfall. Imagine that you are standing near the top of the cliff. If you counted the number of water molecules flowing past the top of the cliff during a given time period, you would get the rate at which the water is flowing past that point. In the same way, electric current is a measure of the rate at which a large number of electrons are flowing past a specific point in a circuit. French physicist André-Marie Ampère (1775–1836) devised a way to measure electric current. The unit for electric current is called the ampere (A), named in honour of Ampère. The symbol for current is I.

electric current (I) a measure of the rate of electron flow past a given point in a circuit; measured in amperes (A)

To learn more about being a physicist,
GO TO NELSON SCIENCE

Measuring Current

When an electrical circuit does not work, an electrician, technician, or engineer must troubleshoot the circuit to find the problem. To do this, they must measure the current flowing through the different loads of the circuit. An **ammeter** is the device designed for this purpose (Figure 1(a)). The circuit diagram symbol for an ammeter is shown in Figure 1(b).

An ammeter must be connected in series with a load to measure the current flowing through the load. For example, to measure the current through the lamp in Figure 2, you must connect the ammeter in series with the lamp. This ensures that all of the electrons that flow through the lamp will also flow through the ammeter.

ammeter a device used to measure electric current

(a)

(b)

Figure 1 (a) An ammeter (b) The circuit diagram symbol for an ammeter

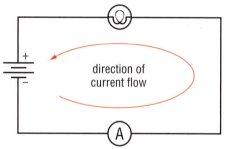

Figure 2 An ammeter must be connected in series with a load to measure the current through the load. Note that the negative side of the battery is connected to the negative side of the ammeter.

Typical electric currents involve the flow of huge numbers of electrons. Consider, for example, when an ammeter indicates that there is a current of 1 A through a circuit. This means that approximately 6.2×10^{18} electrons (over 6 billion billion electrons) are flowing through the ammeter and the rest of the circuit each second!

Safety with Electric Currents

Very large currents can damage electrical devices and cause an electrical fire. That is why every home has a distribution panel with circuit breakers or, in older homes, fuses. The circuit breakers or fuses are connected in series with the circuits leading to the appliances or wall outlets in your home.

If there is too much current through the circuit breaker or fuse, it trips or blows and behaves like an open switch. No current flows through it. This protects electrical devices, such as a washing machine or computer, from becoming damaged by electric currents.

There are two very important safety tips to follow when measuring current:

- Always set the ammeter to the highest current setting. Too low a setting can damage the meter.
- To prevent an electric shock, never touch the tips of the ammeter leads when they are connected to a circuit.

The Human Body and Electrical Shock

The brain coordinates the action of your muscles through electrical signals sent through the nervous system. For example, electrical signals from your brain are stimulating the muscles of your eyes, making them move to scan this page for information. Electrical signals are sent through the nervous system using charged particles called ions rather than electrons.

Even small electrical shocks can be dangerous. An electric current of about 0.001 A passing through the body may give you a tingling sensation. A current of 0.050–0.150 A can cause muscles to contract or convulse out of control. This amount of current is sometimes called the "let-go" threshold because beyond this value you can no longer let go of the object that is shocking you. A current of 1.0–4.3 A passing through the chest will stop your heart. A wall outlet that powers a computer can deliver 15 A. This is why you should never touch a circuit with a current going through it. Remember, current kills!

> **WRITING TIP**
>
> **Asking Questions for Critical Analysis**
> If writing a critical analysis of an article about common household electrical hazards, you might ask questions such as "How might these hazards be prevented?" By conducting research, you might discover that the use of electrical outlet covers can prevent young children from sticking their fingers or a fork into an uncovered electrical socket and receiving a severe electrical shock.

To learn more about safety tips for preventing electric shocks,

IN SUMMARY

- Current is a measure of the rate of electron flow past a given point in a circuit.
- Current is measured in amperes (A) by using a device called an ammeter.
- Ammeters must be connected in series with a load when measuring current.
- Circuit breakers and fuses protect electrical devices from excess current.
- A small amount of electric current can be dangerous to the human body.

CHECK YOUR LEARNING

1. Copy and complete Table 1 in your notebook.

   Table 1 Electric Current

Electrical quantity	Electrical quantity symbol	Unit of measurement	Unit of measurement symbol
electric current			

2. List two important things to remember when using an ammeter.

3. A student connected an ammeter as shown in Figure 3. Did the student connect the ammeter correctly? Explain.

   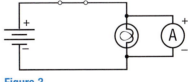

   Figure 3

4. Describe why electric currents can be dangerous.

13.4 PERFORM AN ACTIVITY

Comparing the Conductivity of Conductors

SKILLS MENU
- Questioning
- Hypothesizing
- Predicting
- Planning
- Controlling Variables
- Performing
- Observing
- Analyzing
- Evaluating
- Communicating

Electric currents do not flow through all materials with the same effectiveness. Electrical conductivity describes how well a material allows a current to flow through it. In this activity, you will design a circuit to measure the current flowing through a conductor to compare the electrical conductivity of different materials.

Purpose
To design a circuit to measure conductivity and compare the conductivity of different materials.

Equipment and Materials
- ammeter (or multimeter)
- variable DC power supply
- switch
- 4 to 6 connecting wires
- various materials of similar dimensions (i.e., strips of copper, aluminum, zinc, graphite)

 Always disconnect the energy source before connecting or disconnecting any meters or parts to a circuit.

Procedure

 SKILLS HANDBOOK 2.D., 2.E.

Part A: Designing an Electrical Conductivity Apparatus

1. On a piece of paper, draw a circuit diagram that contains a variable DC power supply, a switch, connecting wires, and an ammeter (or multimeter). Remember to add a place to insert the material you are testing.
2. Design a procedure to get the most consistent results when measuring the current through the test conductor.
3. Before proceeding, have your teacher approve your circuit diagram and design.
4. Once your design is approved, obtain the necessary materials and connect the circuit.

Part B: Measuring Conductivity

5. Copy Table 1 into your notebook.

Table 1 Observations

Material being tested for conductivity	Current (A)

6. Test each material using your conductivity apparatus. Be sure to turn off the circuit before switching to the next material.

Analyze and Evaluate

(a) Rate the materials from best to worst in terms of electrical conductivity. T/I
(b) Why is electric current a good measure of the conductivity of a material? T/I
(c) If you repeated this experiment with insulating materials, do you think you would get any results? Justify your answer. T/I A

Apply and Extend

(d) Do you think that electrical conductivity is the only factor to consider when choosing electrical conductors? Explain your thoughts. C A
(e) If you did this activity with materials that were different dimensions, do you think you would get the same results? Explain your thoughts. If time permits and the materials are present, test your predictions by using materials of different diameters. T/I

TECH CONNECT ✓ OSSLT

Heart Technologies

How is your body able to move? How are you able to lift things? Your body is able to carry out many of its functions because of its muscles. Muscles are controlled by your nervous system through small electrical impulses.

A very important organ in your body is made of muscle—your heart. The heart pumps blood throughout the body. Heart muscles have a natural rhythm that is controlled by a specialized group of heart cells called the sinus node. When you are at rest, your heart beats slowly. However, when you start to do physical work or get anxious, your heartbeat increases. The sinus node, the heart's natural pacemaker, sends electrical signals to the heart's muscles to increase its pumping action (Figure 1).

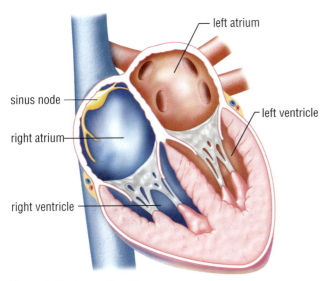

Figure 1 The human heart

When you are healthy, your heart speeds up and slows down when it needs to. If there is something wrong with your sinus node, then the rhythm of your heart does not match your physical activity. It may beat quickly when you are relaxing or beat slowly during physical activity. A slow heartbeat during physical activity can cause you to lose your breath. You might even lose consciousness.

An artificial pacemaker is necessary for people with an abnormally slow heartbeat. This is an electrical device that is implanted under the skin near the heart. It regulates the heartbeat so that the heart beats properly.

A pacemaker contains a battery, a small sensor (to determine if the beating is too slow), a small computer chip (to control the pacemaker), and leads (small electrical wires connected directly to the heart) (Figure 2). A pacemaker can last many years without needing replacement.

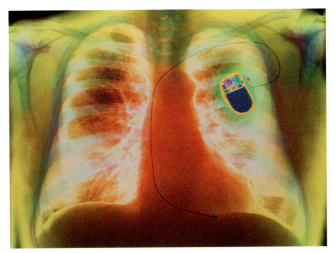

Figure 2 This coloured X-ray shows an implanted pacemaker in a person's chest.

A heart that is beating too quickly and irregularly can cause fibrillation. Fibrillation is uncoordinated, quick contractions of the heart muscles instead of a steady, regular pumping action. If this happens, a person's heart can suddenly stop. If a person's heart is fibrillating, a device called a defibrillator can be used to shock the heart back into a normal rhythm (Figure 3). These lifesaving devices are available in many public places, such as public swimming pools, government buildings, and large arenas.

Figure 3 Portable defibrillators, like this one, are now available in many public buildings.

In some cases, people have severe heart disease. Their hearts have slow rhythms and can also beat abnormally fast. For these people, scientists have developed a special kind of pacemaker that also has a defibrillator. This pacemaker is called an implantable cardiac defibrillator. When the device senses fibrillation, it will shock the heart to correct its rhythm.

The scientific understanding of the electrical nature of our muscles, particularly the heart, has led to these inventions that have prolonged and saved the lives of many people.

13.5 Potential Difference

In Section 13.3, you learned that electricians, technicians, and engineers measure current when designing and troubleshooting circuits. They also measure another quantity called potential difference. **Potential difference**, or **voltage**, is defined as the difference in electric potential energy per unit of charge measured at two points. The electric potential energy per unit charge is often abbreviated to just "electric potential." The unit for voltage is the volt (V).

potential difference (voltage) (*V*) the difference in electric potential energy per unit charge measured at two different points; measured in volts (V)

A Model for Electric Potential Energy

Figure 1 shows a simple circuit in which a motor is connected to an electric cell. Stored electrical energy in the cell causes the shaft of the motor to spin. Let's use an analogy to help visualize what's happening inside this circuit.

For centuries people have used the energy of falling water to push waterwheels. This is possible because water above the wheel has more gravitational potential energy than it does below the wheel. As water falls, some of this energy is used to spin the waterwheel. To keep the wheel spinning you need a steady supply of falling water such as a fast-flowing stream. If this is not available, a pump can be used to push water up to its original position. As water is pumped to its original position, its gravitational potential energy also increases to its original amount (Figure 2(a)).

Similarly, there is a potential difference between the two terminals of an electric cell. Electrons leave the negative terminal with electric potential energy that can be used to operate a motor. As a result, the electrons return to the positive terminal of the cell with less electric potential energy than they started with, because some of their energy was used to run the motor (Figure 2(b)). Once inside the cell, chemical reactions "re-energize" the electrons and send them out the negative terminal again. In this way, the electric cell acts like the pump in Figure 2(a).

> **LEARNING TIP**
>
> **Volts and Voltage**
> In science, we use an italic *V* to represent potential difference, or voltage, to avoid confusing it with the symbol for volt (V).

Figure 1 A battery connected to a motor

> **DID YOU KNOW?**
>
> **Measuring the Voltage of Batteries**
> When you check if a battery is still good, you need to check the voltage drop across the two terminals. To measure the voltage, connect the positive lead of a voltmeter to the positive electrode of the battery. The positive electrode is usually marked with a + sign. Then connect the negative lead of the voltmeter to the negative electrode of the battery.
> If you measure a voltage less than the rated voltage for the battery, it means that the battery is weak. If you measure 0 V, the battery is dead. Recall that a voltage of zero means that no current will flow.

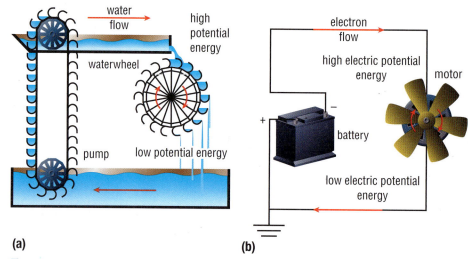

Figure 2 (a) A pump provides potential energy to the water, which can then turn a waterwheel. (b) A potential difference is necessary for current to flow in a circuit. Within the battery, the electrons flow from the negative electrode (higher electric potential energy) to the positive electrode (lower electric potential energy).

Measuring Potential Difference

When an electrician, technician, or engineer troubleshoots a circuit, the voltage as well as the current at different parts of the circuit must be measured. A **voltmeter** is the device designed to measure potential difference (Figure 3(a)). The circuit diagram symbol for a voltmeter is shown in Figure 3(b).

voltmeter a device used to measure potential difference (voltage)

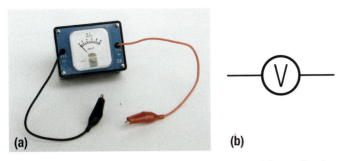

Figure 3 (a) A voltmeter (b) The circuit diagram symbol for a voltmeter

Unlike an ammeter, a voltmeter must be connected in parallel with a load or an energy source. The reason for this is that voltage is relative to two points. There is always a drop in voltage across a load or energy source.

For example, to measure the voltage across the lamp in Figure 4, connect the voltmeter in parallel with the lamp. Since the lamp is the only load in the circuit, the voltage displayed on the voltmeter will be the same as the voltage across the two terminals of the battery.

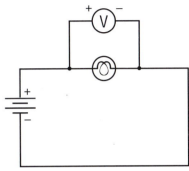

Figure 4 A voltmeter measures the potential difference across the lamp. Note that the negative side of the battery is connected to the negative side of the voltmeter.

IN SUMMARY

- Potential difference, or voltage, is the difference in electric potential energy per unit charge measured at two points.
- Electrons flow from a point of higher electric potential energy to a point of lower electric potential energy.
- Potential difference is measured in volts (V) using a device called a voltmeter.
- Voltmeters must be connected in parallel with either a load or an energy source when measuring potential difference.

CHECK YOUR LEARNING

1. Describe one idea about measuring potential difference that you think you will need to spend more time on to learn. How do you plan to study this idea? K/U C
2. In science, we use symbols to represent quantities. We assign these quantities specific measurement units. Copy and complete Table 1 in your notebook. K/U

   Table 1 Potential Difference

Electrical quantity	Electrical quantity symbol	Unit of measurement	Unit of measurement symbol
potential difference			

3. Give an example of how the potential difference would be measured in a circuit. K/U
4. A student connected a voltmeter into a circuit as shown in Figure 5. Is this the correct way to connect a voltmeter? Explain your answer. K/U

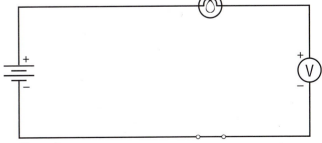

Figure 5

13.6 PERFORM AN ACTIVITY

Measuring Voltage and Current in Circuits

Expensive devices with electric circuits, such as cars, furnaces, and refrigerators, are too costly to throw away when they suddenly stop working. Instead, we can hire technicians to test the circuits and make repairs.

In this activity, you will test a DC circuit. You will measure the current and the potential difference using an ammeter and a voltmeter, or a multimeter. Meters allow you to determine which part of a circuit is causing the problem.

SKILLS MENU
- Questioning
- Hypothesizing
- Predicting
- Planning
- Controlling Variables
- Performing
- Observing
- Analyzing
- Evaluating
- Communicating

Purpose

To measure electric currents and potential differences in DC circuits using meters.

Equipment and Materials

- variable DC power supply
- 2 lamps
- switch
- 6 to 8 connecting wires
- ammeter (or multimeter)
- voltmeter (or multimeter)

 Always disconnect the energy source before connecting or disconnecting any meters or parts to a circuit. Once you have connected a circuit and any of the meters, turn the power supply on. If your circuit does not work, ask your teacher for assistance.

Procedure

 SKILLS HANDBOOK 2.D., 2.E.

Part A: Measuring Quantities in a Series Circuit

1. Obtain the required materials from your teacher. Design a table to record your observations.
2. Turn off the power supply. Connect the circuit shown in Figure 1.

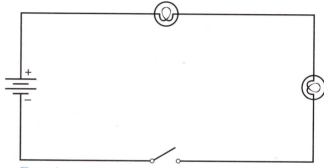

Figure 1

3. Make sure the power supply is at its lowest setting as instructed by your teacher. Close the switch.
4. Slowly increase the power until the lamps turn on. Once they are lit, leave the power supply at this setting.
5. Use an ammeter to measure the current going into the first lamp (Figure 2). Record the measurement in your table. Repeat this step for the second lamp.

Figure 2

6. Use a voltmeter to measure the potential difference across the first lamp. Record the measurement in your table. Repeat this step for the second lamp and for the power supply.

7. Slightly increase the setting on the power supply. Connect the voltmeter so that you can measure the potential difference across the second lamp. Note the brightness of the lamps. Record the potential difference in your table.

8. Turn off the power supply and disconnect your circuit.

Part B: Measuring Quantities in a Parallel Circuit

9. Construct a table to record your observations. Turn off the power supply. Connect the circuit shown in Figure 3.

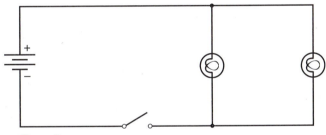

Figure 3

10. Make sure the power supply is at its lowest setting. Close the switch.

11. Slowly turn up the power until the lamps turn on. Once they are lit, leave the power supply at this setting.

12. Remove one of the lamps from the socket. Use an ammeter to measure the current going through the lamp that is still lit. Record the measurement in your table.

13. Use an ammeter to measure the current going through the unlit lamp. Record the measurement in your table.

14. Use a voltmeter to measure the potential difference across the lamp that is still lit (Figure 4). Record the measurement in your table. Repeat this for the unlit lamp.

Figure 4

15. Turn off the power supply and disconnect your circuit.

Analyze and Evaluate

(a) Write a concluding statement about the electric current in a series circuit. T/I C

(b) Write a concluding statement about the electric current in a parallel circuit. T/I C

(c) How does the potential difference affect the brightness of a light bulb? T/I

(d) In step 13, you were asked to measure the current going through the unlit lamp. Refer to your table for this data. Did your results for this measurement surprise you? Explain. T/I

Apply and Extend

(e) It is easy to test whether or not a lamp is working. However, there are some devices in electric circuits that do not move, make sounds, or produce light. Each of these devices does perform a specific task. How would you apply the procedure in Part B to a malfunctioning circuit that did not make sounds or produce light? T/I

UNIT TASK Bookmark

How can you apply what you have learned about measuring voltage and current in circuits to the Unit Task described on page 586?

13.6 Perform an Activity

13.7 Resistance in Circuits

Have you ever noticed that when you recharge your cellphone, MP3 player, or laptop computer, the adaptor gets warm? The warmth is caused by the resistance experienced by the electric current flowing through the adapter.

Electrical resistance is the opposition to the movement of electrons as they flow through a circuit. The symbol for resistance is R. The unit for resistance is the ohm (Ω). To better understand the concept of resistance, imagine that you are kicking a soccer ball. If the ball is on a smooth, hard surface like pavement, the ball will roll easily. If the ball is on a rough surface like tall grass, you would have to kick the ball much harder just to make it roll.

In the same way, when electrons flow through a material that has many "bumps" along its path, there will be more resistance than if the material is "smooth." For example, insulators tend to minimize the amount of electron flow, so the internal resistance of an insulator is quite high. Conversely, a conductor like copper has a very low internal resistance. This is why electrons flow so easily in copper wire.

electrical resistance (R) the ability of a material to oppose the flow of electric current; measured in ohms (Ω)

DID YOU KNOW?

Internal Resistance in Everyday Materials
Many devices that you use every day use materials with high internal resistance. For example, a toaster consists of nichrome wires, which have a high internal resistance. The electrical energy through the wires gets converted into light (the red glow) and thermal energy. It is the thermal energy that toasts bread.

GO TO NELSON SCIENCE

Factors that Affect Resistance

All materials have some internal resistance. The greater the resistance, the lower the current, and the warmer the material becomes when current flows through it. This happens because, as the electrons move through the material, they bump into the atoms that make up the material. In the process, electrical energy is converted into thermal energy.

Internal resistance depends on many factors. We will look at four of these factors: type of material, cross-sectional area, length, and temperature.

Type of Material

The ability of a material to conduct electricity is determined by how freely electrons can move within the material. Copper is used in circuits because it is an excellent conductor. It has a low electrical resistance. Silver is a better conductor because its resistance is even lower. Silver, however, is an expensive material and is not suitable for low-cost electrical devices.

Cross-Sectional Area

When you cut through a wire, you can see its cross-section more easily. The diameter of the cross-section gives you a sense of how thick the wire is. Thicker wires have less internal resistance than thinner ones. Electrons flowing through a thicker wire have more room to move freely (Figure 1). This is similar to what happens with water in a pipe. The greater the diameter of a pipe, the greater the water flow will be.

Figure 1 The wire at the left has a smaller diameter cross-section than the wire at the right. The thinner wire will have more internal resistance than the thicker wire.

Length

As you increase the length of a wire, its internal resistance increases. This happens because electrons have to travel through more material. You can see how this applies if you examine extension cords.

The longer the extension cord, the more resistance it has, and the warmer it will get while being used. This can be dangerous because the cord can overheat and potentially cause a fire. Manufacturers of extension cords can avoid this hazard by using a larger-diameter wire, which lowers the electrical resistance.

Temperature

As you have learned, resistance increases when electrons bump into atoms as they move through a material. When a wire gets warmer, the atoms that make up the wire gain energy and vibrate faster. The increased vibration results in more collisions between the atoms and the free-flowing electrons in the current. Since greater vibrations cause more collisions, resistance increases with temperature.

Measuring Resistance

Just as current and voltage are useful quantities to measure when troubleshooting a circuit, so is resistance. An **ohmmeter** is the device designed to measure resistance. The circuit diagram symbol for an ohmmeter is shown in Figure 2(a).

As with a voltmeter, an ohmmeter must be connected in parallel with a load (Figure 2(b)). However, you do not power up the circuit to measure resistance. The reason is that the ohmmeter is powered and provides an electric current through the load.

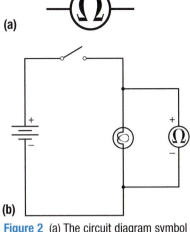

Figure 2 (a) The circuit diagram symbol for an ohmmeter (b) An ohmmeter measures the electrical resistance across a load. It is placed in parallel with the load.

ohmmeter a device used to measure resistance

Resistors in Circuits

You may have a lamp at home with three brightness settings. Or you may have a dimmer switch on the wall that allows you to control the brightness of the light. This is possible because these devices contain resistors. A **resistor** is an electrical device that reduces the current in a circuit. There are many different types of resistors (Figure 3). For example, lightweight carbon resistors are commonly used in electronics; heavier ceramic resistors are used in larger circuits. Dimmer switches and the volume controls on a stereo amplifier are another type of resistor called variable resistors. A variable resistor allows you to change the resistance in a circuit. For example, by turning a dimmer switch, you can dim the light in a room. Figure 4 shows the circuit diagram symbol for a resistor.

resistor a device that reduces the flow of electric current

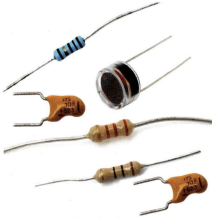

Figure 3 A variety of resistors

Figure 4 The circuit diagram symbol for a resistor

TRY THIS: MEASURING RESISTANCE WITH AN OHMMETER

SKILLS: Performing, Observing, Communicating

SKILLS HANDBOOK
2.D.

You can test the resistance of a resistor with an ohmmeter. In this activity, you will use an ohmmeter to test resistance in a variety of resistors.

Equipment and Materials: resistors of different types and resistances; ohmmeter with leads

1. Design a table in which to record your observations.
2. Record the resistance of one of the resistors in your table. The resistance is usually printed on the resistor. If not, your teacher will provide you with the value.
3. Connect the ohmmeter with one lead on one side of the resistor and the other lead on the opposite side. Record the resistance in ohms.
4. Repeat steps 2 and 3 using another resistor.

A. How did the rated resistance compare with the measured value? K/U T/I

B. Predict what would happen to the resistance if you reversed the position of the leads. Test your prediction. T/I

IN SUMMARY

- Electrical resistance is the opposition to the flow of electrons.
- The internal resistance of a wire increases by decreasing its cross-sectional area, lengthening the wire, and increasing its temperature.
- All materials have some internal resistance. Materials that have less resistance are usually used as conductors.
- Resistance causes electrical devices to warm up when a circuit is functioning.
- Resistance is measured in units of ohms (Ω) using a meter called an ohmmeter.
- Resistors are electrical devices that affect the electric current in a circuit.
- Ohmmeters are connected in parallel with a load when measuring resistance. The circuit must be turned off to measure the resistance.

CHECK YOUR LEARNING

1. Have you had an experience that relates to the reading about resistors? How did the reading help you understand your experience? C

2. In science, we measure quantities and use symbols to represent quantities. We also assign measurement units. Copy and complete Table 1 in your notebook. K/U

    Table 1 Internal Resistance

Electrical quantity	Electrical quantity symbol	Unit of measurement	Unit of measurement symbol
resistance			

3. Which material would you expect to have greater resistance, plastic or silver? Explain your choice. K/U

4. Draw a circuit diagram that shows a two-cell battery in series with a switch and two lamps in parallel. Include an ohmmeter correctly connected to one of the lamps. K/U C

5. Identify a situation in which you would want to have a high resistance and a situation in which you would want to have a low resistance. A

6. What effect would the following changes have on a conductor's resistance? In each situation, explain why the change occurs. T/I

    (a) decreasing the diameter of a conductor
    (b) placing an extension cord outside in the winter
    (c) plugging two identical extension cords together to make it longer
    (d) changing from a copper conductor to a silver conductor

7. An extension cord that you would use for a lamp is much thinner than an extension cord recommended for use with large appliances. If you plugged in a refrigerator using a lamp extension cord, the plastic coating of the cord could melt and perhaps start a fire. Why does this occur? How can this hazard be prevented? K/U A

PERFORM AN ACTIVITY **13.8**

Determining the Relationship Between Current and Potential Difference

SKILLS MENU
- Questioning
- Hypothesizing
- Predicting
- Planning
- Controlling Variables
- Performing
- Observing
- Analyzing
- Evaluating
- Communicating

Some electrical devices, such as clothes dryers, require more potential difference to operate than cellphone chargers. What effect does a higher potential difference have on the current through the wires of each device?

In this activity, you will use a resistor to determine the relationship between the current and the potential difference as the values of I and V are changed.

Purpose

To determine the relationship between the current and the potential difference.

Equipment and Materials

- variable DC power supply
- switch
- 6 connecting wires
- voltmeter (or multimeter)
- ammeter (or multimeter)
- resistor
- graph paper

Procedure

1. Construct a table to record your observations.
2. Make sure that the power supply is turned off. Record the resistance of the resistor (in ohms) in your table. The resistance value is printed on the resistor, or your teacher will provide you with the rating.
3. Connect the circuit shown in Figure 1.

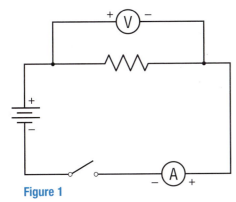

Figure 1

4. Turn the power supply on, and set it to a low setting as instructed by your teacher. Close the switch. Record the potential difference (in volts) and the current (in amperes) in your table. Then open the switch.
5. Increase the setting on the power supply and close the switch again. Record the potential difference and the current in your table. Then open the switch.
6. Repeat step 5 until you have five sets of data.

Analyze and Evaluate

SKILLS HANDBOOK
6., 8.B.

(a) Plot a graph of potential difference (on the y-axis) versus current (on the x-axis). T/I
(b) Draw a line of best fit on your graph. You do not necessarily need to place the line through zero. T/I
(c) Calculate the slope of the line of best fit. T/I
(d) Compare the value of the slope with the resistance value of the resistor. What do you notice? T/I
(e) What is the relationship between current and potential difference? T/I
(f) Compare the relationships between current, potential difference, and resistance. Use a graphic organizer to help you. T/I C

Apply and Extend

(g) Suppose you performed the activity with another resistor placed in series in the circuit. Would you expect to record a graph with a slope that is higher or lower than with one resistor? Why? Design and carry out an activity to test your hypothesis. T/I A

13.9 Relating Current, Voltage, and Resistance

In Activity 13.8, you completed an activity very similar to one done by German physicist Georg Ohm (1787–1854). Ohm discovered a mathematical relationship between potential difference and current. Instead of varying the potential difference and measuring the current as you did, Ohm kept those variables constant. Instead, he varied the length of the conductor. He noticed that as the length of the wire increased, the current decreased. Recall from Section 13.7 that increasing the length of a wire increases its resistance.

Ohm's findings apply only to certain types of materials. For these materials, when you plot a graph of the voltage versus the current, you get a straight-line relationship (Figure 1). The slope of the straight line represents the resistance of the material. You confirmed this relationship in Activity 13.8 when you identified the slope of the line of best fit as the resistance of the load. The steeper the slope of the straight line, the greater the resistance.

The relationship among voltage, current, and resistance can be written mathematically as an equation:

$$R = \frac{V}{I}$$

where R is the resistance in ohms (Ω), V is the voltage or potential difference in volts (V), and I is the current in amperes (A). This relationship is now called **Ohm's law**. Ohm's law states that as the potential difference across a load increases, so does the current. Ohm's law greatly increases our understanding of currents and potential differences in circuits.

Sample Problem 1 shows how to use Ohm's relationship to calculate the resistance of a load in a circuit.

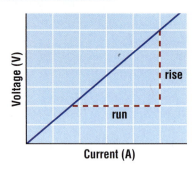

$$\text{slope} = \frac{\text{rise}}{\text{run}}$$
$$\text{slope} = \frac{\text{voltage }(V)}{\text{current }(I)}$$
$$\text{slope} = \text{resistance }(R)$$

Figure 1 A graphical representation of Ohm's findings

Ohm's law the straight-line relationship between voltage and current; $R = \frac{V}{I}$

SAMPLE PROBLEM 1 Calculating the Resistance of a Load

A load has 1.2 A of current flowing through it. The voltage across the load is 6.0 V. Calculate the resistance of the load.

Given: $I = 1.2$ A
$V = 6.0$ V

Required: resistance (R)

Analysis: $R = \frac{V}{I}$

Solution: $R = \frac{6.0 \text{ V}}{1.2 \text{ A}}$

$R = 5.0 \ \Omega$

Statement: The resistance of the load is 5.0 Ω.

Practice

A hair dryer is plugged into a wall outlet that has a voltage of 115 V. When the hair dryer is on, a current of 4.0 A flows through it. Calculate the resistance of the hair dryer.

Ohm's equation can be rearranged to solve for current:

$$I = \frac{V}{R}$$

Sample Problem 2 shows how to calculate the current through a resistor.

SAMPLE PROBLEM 2 Calculating the Current through a Resistor

A 110 Ω resistor is connected to a power supply set at 12 V. Calculate the current going through the resistor.

Given: $R = 110\ \Omega$
$V = 12\ V$

Required: electric current (I)

Analysis: $I = \dfrac{V}{R}$

Solution: $I = \dfrac{12\ V}{110\ \Omega}$
$I = 0.11\ A$

Statement: The current going through the resistor is approximately 0.11 A.

Practice
A device creates a potential difference of 1500 V across two points at the surface of human skin. Dry human skin has a resistance of about 125 000 Ω. Calculate the current moving through the skin.

The voltage across a load can be calculated by rearranging Ohm's equation:

$$V = IR$$

Sample Problem 3 shows how to calculate the voltage across a load.

SAMPLE PROBLEM 3 Calculating the Potential Difference across a Resistor

A toaster oven has a 24.0 Ω resistor that has 5.00 A of current going through it when the toaster is on. Calculate the potential difference across the resistor.

Given: $R = 24.0\ \Omega$
$I = 5.00\ A$

Required: potential difference (V)

Analysis: $V = IR$

Solution: $V = (5.00\ A)(24.0\ \Omega)$
$V = 120\ V$

Statement: The potential difference across the resistor is 120 V.

Practice
A compact fluorescent light bulb has a resistance of 1100 Ω. The current going through the bulb is 0.11 A. Calculate the potential difference across the light bulb.

IN SUMMARY

- The current through a load depends on its resistance and the voltage drop across it.
- As the potential difference across a load is increased, so is the current going through the load.
- Electrical resistance can be represented graphically or using Ohm's equation.

CHECK YOUR LEARNING

1. Ohm's law was stated in words, shown as a graph, and written mathematically. Which form did you find the easiest to understand? Explain.

2. Which of the following graphs shows a load with a greater resistance (Figure 2)? Explain your answer.

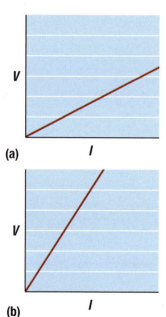

Figure 2

3. A student is investigating a resistor. She has collected the data shown in Table 1. Plot the data on a graph. Then calculate the resistance.

Table 1 Potential Difference and Current Data

Potential difference (V)	Current (A)
2.5	0.002
6.0	0.005
8.7	0.007
11.6	0.009
14.5	0.012

4. Copy Table 2 into your notebook and use Ohm's equation to fill in the blanks.

Table 2 Data for Question 4

Resistance (Ω)	Potential difference (V)	Current (A)
	120	0.25
270		1.25
110	25	
1 000 001	120	
370		6.0

5. A resistor is connected to a 36 V power supply. An ammeter measures a current of 2.0 A going through it. Determine the resistance in ohms.

6. A laptop computer adapter has a voltage of 19 V. It has a resistance of 4.0 Ω. The adapter gets warm when operating. Determine the current going through the adapter.

7. Typical household circuits can carry a maximum current of 15 A. If a wire has a resistance of 8.0 Ω, determine the voltage across the energy source.

8. When you turn the key to start a car, it completes a circuit. The starter motor in the car is part of the circuit and has a voltage of 12 V. The starter motor requires a very large current of 505 A. This current flows only while the car is starting. Calculate the resistance of the starter motor.

9. Use Ohm's law to explain why an electric shock applied on wet skin is more dangerous than the same electric shock on dry skin.

13.10 How Series and Parallel Circuits Differ

In Activity 13.2, you observed that when the two lamps were connected in series, the brightness of the lamps was less than when the lamps were connected in parallel. By using knowledge of voltage, current, and resistance, it is possible to explain these results. Let's start by examining loads in series.

Loads in Series

When you connect an energy source to a series circuit, a voltage or potential difference is created and electrons begin to flow. No matter how many loads are connected in series, there is only one path that the current can follow.

Current through Loads in Series

If you have a circuit with one load (Figure 1(a)), the total resistance of the circuit will be different than if you have two or more of those loads connected in series (Figure 1(b)). The electrons have only one path to follow and with two or more loads, they have more "bumps" to deal with along the way. Because of this, the current flowing through the circuit in Figure 1(b) will be less than the current flowing through the circuit in Figure 1(a). The next sample problem illustrates this result.

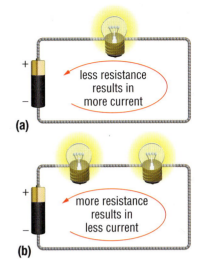

Figure 1 From Ohm's relationship, the current in a circuit depends on the resistance in that circuit.

SAMPLE PROBLEM 1 Comparing the Current in Two Series Circuits

The same type of lamp is used in two series circuits. The first circuit has two identical lamps (Figure 2). The second circuit has three identical lamps (Figure 3). The potential difference across the battery is 10 V. The circuit on the left has a total resistance of 10 Ω. The circuit on the right has a total resistance of 15 Ω. Use the total resistance given to calculate the current through each circuit.

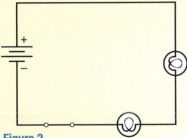

Figure 2

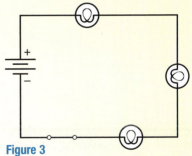

Figure 3

Given:	$R_T = 10\ \Omega$ $V = 10$ V		Given:	$R_T = 15\ \Omega$ $V = 10$ V
Required:	electric current (I)		Required:	electric current (I)
Analysis:	$I = \dfrac{V}{R_T}$		Analysis:	$I = \dfrac{V}{R_T}$
Solution:	$I = \dfrac{10\ \text{V}}{10\ \Omega}$ $I = 1$ A		Solution:	$I = \dfrac{10\ \text{V}}{15\ \Omega}$ $I = 0.7$ A

Statement: The current through each lamp in Figure 2 is 1 A, while in Figure 3 the current is about 0.7 A.

Practice

A series circuit has four identical lamps. The potential difference of the energy source is 60 V. The total resistance of the lamps is 20 Ω. Calculate the current through each lamp.

> **WRITING TIP**
>
> **Supporting the Main Idea**
> A main idea, expressed clearly and directly, is a powerful communication tool. But often a main idea needs supporting details such as examples, reasons, or data to help readers understand what it means. For instance, when writing a critical analysis about voltage across loads in series, use an example like several lamps connected in series, or even a diagram, to help explain the main idea.

> **DID YOU KNOW?**
>
> **Uneven Loads**
> When the loads in a series circuit are not identical, the load with the greatest resistance causes the greatest decrease in electric potential energy for an electron passing through it. The reason is that the electron has to convert more electric potential energy into kinetic energy just to get through all the "bumps" of the load.

Voltage across Loads in Series

In Activity 13.2, you observed that the brightness of each lamp in a series circuit decreased as you connected more lamps. To understand why this happens, let's examine the electric potential energy of the circuit. The battery converts chemical energy into electric potential energy. This potential energy gets transferred to all the electrons that leave the battery and creates the current flow in the circuit. As the electrons move through the circuit, the electric potential energy gets converted into kinetic energy.

Suppose the circuit has only one load—a lamp. When an electron passes through the load, all the electric potential energy that the electron received from the battery gets converted into light and heat. After the electron completes one loop, it returns to the battery, only to be given more electric potential energy. This allows the electron to continue moving through the loop of the circuit over and over again. Since voltage is related to electric potential energy, the voltage drop across the one load will be the same as the voltage drop across the battery.

When two identical loads such as two lamps are connected in series, an electron leaving the battery will have the same amount of electric potential energy as if there were only one load. The reason is that the battery can only supply so much potential energy. For the electron to go through each of the lamps, only half of the electric potential energy gets converted into light and heat. Since only half of the potential energy gets converted through each lamp, the voltage drop across each lamp is half the voltage drop across the battery. More identical lamps connected in series means less of an electron's electric potential energy gets converted into heat and light. The voltage drop across each load decreases. This result can be written mathematically:

$$V_{load} = \frac{V_{source}}{\# \text{ of loads}}$$

where V_{load} is the voltage drop across each identical load (in volts), V_{source} is the voltage drop across the energy source (in volts), and # of loads is the total number of identical loads. Sample Problem 2 shows how to apply this relationship.

SAMPLE PROBLEM 2   Calculating the Voltage in a Series Circuit

A series circuit contains three identical lamps (Figure 4). The potential difference of the battery is 30 V. Calculate the potential difference across each lamp.

Given: $V_{source} = 30$ V

Required: voltage drop across each lamp (V_{load})

Analysis: $V_{load} = \dfrac{V_{source}}{\# \text{ of loads}}$

Solution: $V_{load} = \dfrac{30 \text{ V}}{3}$

$V_{load} = 10$ V

Statement: The potential difference across each lamp is 10 V.

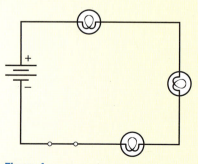

Figure 4

Practice

A series circuit contains four identical lamps. The voltage across the energy source is 96 V. Calculate the voltage across each lamp.

Loads in Parallel

When an energy source is connected to a parallel circuit, a potential difference is created across the two terminals of the energy source. The potential difference causes electrons to flow. Unlike a series circuit, the electrons have different paths to follow. The number of paths depends on the number of loads connected in parallel.

Current through Loads in Parallel

In Figure 5(a), two identical loads are connected in parallel. Electrons leaving the battery have two possible paths to follow. Since each of those paths has the same load, the current splits in two.

In Figure 5(b), three identical loads are connected in parallel. This time, electrons leaving the battery have three possible paths. Each of those paths has the same load, so the current splits in three. The splitting of the current is exactly what happens when a river flows into two or more tributaries. The quantity of water that flows in each tributary is less than the quantity in the original river. This result can be written mathematically:

$$I_{load} = \frac{I_{source}}{\text{\# of loads}}$$

where I_{load} is the current through each identical load (in amperes), I_{source} is the current coming out of the energy source, and # of loads is the total number of identical loads. Sample Problem 3 shows how to apply this relationship.

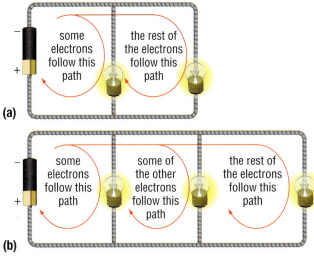

Figure 5 The more loads that are connected in parallel, the more paths electrons have to follow.

SAMPLE PROBLEM 3 Calculating the Current in a Parallel Circuit

The total resistance in the circuit in Figure 5(b) is 2 Ω. The potential difference of the battery is 18 V. Calculate the current through each lamp.

Given: $R_T = 2\ \Omega$
$V = 18\ V$

Required: current through each lamp (I_{load})

Analysis: $I_{source} = \frac{V}{R_T}$ (to calculate the current coming out of the energy source) and

$I_{load} = \frac{I_{source}}{\text{\# of loads}}$ (to calculate the current through each identical load)

Solution: $I_{source} = \frac{18\ V}{2\ \Omega}$

$I_{source} = 9\ A$

$I_{load} = \frac{9\ A}{3}$

$I_{load} = 3\ A$

Statement: The current through each lamp is 3 A.

Practice

A parallel circuit contains four identical lamps. The potential difference across the energy source is 48 V. The total resistance of the lamps is 12 Ω. Calculate the current through each lamp.

> **WRITING TIP**
>
> **Conducting Research for Critical Analysis**
>
> In a critical analysis of voltage across loads in parallel circuits, you might ask yourself how many appliances you can safely plug into an electrical outlet. You might conduct research to find out that you can divide the number of watts of power on a circuit by the voltage in the house to determine the current used. You might conclude that if the current used is greater than the circuit can handle, the fuse will blow or the circuit breaker will trip.

Voltage across Loads in Parallel

An interesting difference between series and parallel circuits involves total resistance. In both parallel circuits shown in Figure 5 on the previous page, the electrons that flow through any one branch go through only one load. The resistance that the electrons in that branch experience will be less than if the loads were connected in series. This means that when you connect loads in parallel, the total resistance will be less than if the loads were connected in series.

Another difference between series and parallel circuits involves the voltage drop across loads. As stated above, an electron that flows through any one branch of the parallel circuits in Figure 5 only passes through one load. All the electric potential energy that the electron receives from the battery gets converted into light and heat.

Since voltage is related to electric potential energy, the voltage drop across each parallel load will be the same as the voltage drop across the battery. In other words, for any parallel circuit, the voltage drop across each parallel branch will be the same as the voltage drop across the energy source. That is why, in Activity 13.2, the brightness of the lamps did not change when you connected more lamps in parallel. Sample Problem 4 shows how to apply this result to a parallel circuit.

SAMPLE PROBLEM 4 Calculating the Voltage in a Parallel Circuit

A parallel circuit contains three identical lamps (Figure 6). The current coming out of the energy source is 2.5 A. The total resistance of the circuit is 6.0 Ω. Calculate the voltage across the energy source and across each lamp.

Given: $I_{source} = 2.5$ A
$R_T = 6.0$ Ω

Required: voltage drop across the energy source (V_{source}) and across each lamp (V_{load})

Analysis: $V_{source} = V_{load}$ (loads are in parallel to energy source)

$V_{source} = (I_{source})(R_T)$

Solution: $V_{source} = (2.5$ A$)(6.0$ Ω$)$

$V_{source} = 15$ V

$V_{load} = 15$ V

Statement: The voltage across the energy source is 15 V, and the voltage across each lamp is also 15 V.

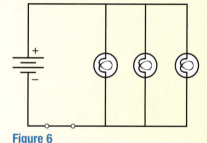

Figure 6

Practice

A parallel circuit contains 10 identical lamps. The current through the energy source is 3.0 A. The total resistance of the circuit is 15 Ω. Calculate the voltage across the energy source and across each lamp.

Resistance, Current, and Voltage in Circuits

Table 1 summarizes how voltage, current, and total resistance differ in series and parallel circuits.

Table 1 The Relationships of Loads in Series and Parallel Circuits

Quantity	Series circuits	Parallel circuits
total resistance of circuit (R_T)	increases	decreases
current through loads (I_{load})	I_{source} decreases as more loads are added	I_{source} splits among loads based on the number of branches in parallel
voltage across loads (V_{load})	V_{source} splits based on the number of loads	voltage of each parallel branch is the same as V_{source}

574 Chapter 13 • Electrical Quantities in Circuits

UNIT TASK Bookmark

You can apply what you have learned about series and parallel circuits to the Unit Task described on page 586.

IN SUMMARY

- In a parallel circuit, the total resistance is less than if the loads were connected in series.
- The more loads that are connected in series, the lower the current will be in the circuit.
- The more loads that are connected in parallel, the lower the current will be through each branch.
- In a series circuit with identical loads, the voltage drop across the energy source is split equally among the loads.
- In a parallel circuit, the voltage drop across each parallel branch will be the same as the voltage drop across the energy source.

CHECK YOUR LEARNING

1. Which did you find easier to understand in a parallel circuit, voltage or current? What would you ask your teacher to help you understand the more challenging concept? C

2. Compare the total resistance of loads connected in series with those loads connected in parallel. K/U

3. Why is it a bad idea to connect too many devices in parallel to an energy source, such as a wall outlet? K/U

4. (a) What would happen to the voltage drop across each lamp if you kept adding lamps to a series circuit?
   (b) What do you think you would observe in terms of the brightness of the lamps? K/U

5. The total resistance of the circuit in Figure 7 is 25 Ω. The voltage drop across the battery is 6.0 V. T/I
   (a) Calculate the current in the circuit.
   (b) Calculate the voltage drop across each lamp.

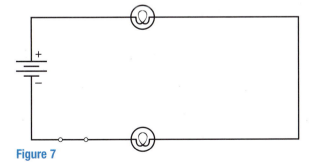

Figure 7

6. A house has a lamp in every room. The circuit for the lamps is shown in Figure 8. The voltage drop across the energy source is 120 V. The total resistance is 10 Ω. T/I
   (a) Calculate the current through each lamp.
   (b) Calculate the voltage drop across each lamp.

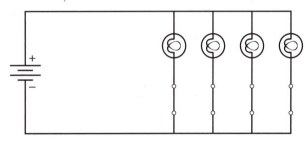

Figure 8

7. A battery-powered set of five patio lanterns is connected in series. An ammeter measures the current through the battery as 0.75 A. The total resistance of the circuit is 52 Ω. T/I
   (a) Calculate the voltage drop across the battery.
   (b) Calculate the voltage drop across each load.

13.11 CONDUCT AN INVESTIGATION

The Effect of Increasing the Number of Loads in a Circuit

There are two main ways to connect circuits. In this investigation, you will connect loads in series and in parallel arrangements to observe any differences in current, voltage, and total resistance in a circuit. You will use an appropriate meter to measure current, voltage, and resistance.

SKILLS MENU
- Questioning
- Hypothesizing
- Predicting
- Planning
- Controlling Variables
- Performing
- Observing
- Analyzing
- Evaluating
- Communicating

Testable Question
How do the total resistance, current, and voltage compare in series and parallel circuits?

Hypothesis/Prediction

Make and record a hypothesis for the Testable Question. Your hypothesis should include a prediction and a reason for your prediction.

Experimental Design
You will connect one load and then two loads in series. Then you will measure the voltage drop across each resistor, the current through each resistor, and the resistance of all the resistors in the circuit. You will repeat the process for a circuit of two loads in parallel. The voltage drop across the power supply will remain constant throughout the investigation.

Equipment and Materials
- variable DC power supply
- voltmeter (or multimeter)
- ammeter (or multimeter)
- ohmmeter (or multimeter)
- connecting leads
- switch
- 2 identical resistors

Procedure

Part A: Resistors in Series
1. Obtain the equipment and place it at your workbench.
2. Copy Tables 1 and 2 into your notebook.

Table 1 One Resistor

One resistor	voltage (V)	current (A)
resistor 1		

Table 2 Two Resistors in Series

Two resistors	voltage (V)	current (A)	total resistance (Ω)
resistor 1			
resistor 2			

3. Connect the circuit shown in Figure 1.

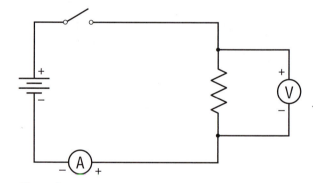

Figure 1

4. Close the switch. Set the power supply to 2 V or as specified by your teacher. Record this number as the voltage drop across the energy source.
5. Record the voltage drop across and the current through the resistor in your table.
6. Turn off the power supply. Then open the switch.
7. Add another resistor in series to the circuit. Repeat steps 4 to 6 for each resistor.

8. Remove the voltmeter and ammeter. Measure the total resistance across both resistors by placing an ohmmeter as shown in Figure 2.

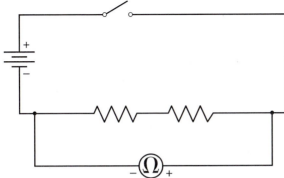

Figure 2

Part B: Resistors in Parallel

9. Copy Table 3 into your notebook.

Table 3 Resistors in Parallel

	Voltage (V)	Current (A)	Total resistance (Ω)
resistor 1			
resistor 2			

10. Connect the circuit shown in Figure 3.

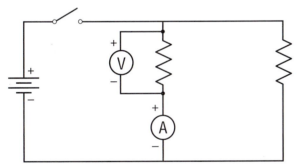

Figure 3

11. Close the switch. Set the power supply to 2 V or as specified by your teacher. Record this number as the voltage drop across the energy source.
12. Measure and record the voltage drop across and the current through resistor 1.
13. Move the meters to measure the values for the second resistor.
14. Turn off the power supply. Then open the switch.

15. Remove the voltmeter and ammeter. Measure the total resistance across both resistors by placing an ohmmeter as shown in Figure 4. Record the total resistance.

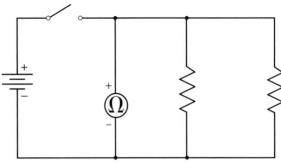

Figure 4

Analyze and Evaluate

SKILLS HANDBOOK 3.B.7.

(a) Answer the Testable Question. T/I
(b) Did evidence support your hypothesis? Explain. T/I
(c) Make a statement about the voltage drop across resistors. Compare resistors in series with resistors in parallel. Also make a statement comparing the voltage drop across the energy source with the voltage drop across the resistor. T/I
(d) Make a statement about the currents through resistors. Compare resistors in series with resistors in parallel. T/I
(e) Which circuit, series or parallel, would allow more current to flow if you added one more resistor? Explain your answer. T/I
(f) Which arrangement of resistors has a greater total resistance, series or parallel? Explain your choice. Was your hypothesis correct? T/I

Apply and Extend

(g) Fire departments warn against plugging too many devices into one outlet. Using what you have learned in this investigation, explain why fire departments issue this warning. A
(h) If you have hot wires, then electrical energy is being used to heat the wires. Is this an efficient use of electricity? Explain. A
(i) What would be the advantage of using parallel circuits instead of series circuits in your home? A

CHAPTER 13 LOOKING BACK

KEY CONCEPTS SUMMARY

Circuits are a part of electrical technology. They can be connected in series and in parallel.

- Series circuits allow electrons to flow along one path only. (13.1, 13.10)
- Parallel circuits allow electrons to flow along more than one path. (13.1, 13.10)
- When a load fails in a series circuit, all the loads will stop working. (13.1, 13.10)
- When a load fails in a parallel circuit, the rest of the loads will continue to work. (13.1, 13.10)

Current is the rate of electron flow.

- Electric current (I) is the amount of electrons passing a point in a circuit in a given time period. (13.3)
- Electric current is measured in amperes (A) using an ammeter connected in series. (13.3)

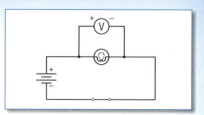

Potential difference is a measure of the difference in electric potential energy per unit of charge between two points.

- The potential difference between two points in a circuit determines whether or not charges will flow. (13.5)
- The potential difference across a load or an energy source is measured using a voltmeter connected in parallel. (13.5)

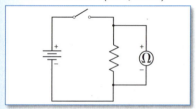

Resistance is a measure of the opposition to the flow of electrons.

- Resistance (R) is the amount of opposition to the flow of current. Resistors are electrical devices that oppose the flow of electric current. (13.7)
- The resistance of a conductor is affected by the cross-sectional area, the type of material used, the temperature, and the length of the conductor. (13.7)
- For conductors, the graph of voltage versus current yields a straight line. The slope of the graph represents the resistance of the load, $R = \dfrac{V}{I}$. (13.9)

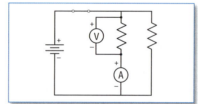

Connecting loads in series or parallel affects the current, potential difference, and total resistance.

- Connecting loads in series increases the total resistance and decreases the electric current coming from the energy source. (13.10)
- The voltage drop across the energy source is split between the loads connected in series, while the current is constant throughout the circuit. (13.10)
- Connecting loads in parallel decreases the total resistance of the circuit. (13.10)
- The current is split among loads connected in parallel, while the voltage drop across each parallel branch remains the same. (13.10)

Current, voltage, and resistance measurements can be used to test circuits.

- Hypotheses can be written to answer a Testable Question to compare total resistance, current, and potential difference across resistors. (13.11)
- Hypotheses and predictions are tested using voltmeters, ammeters, ohmmeters, or multimeters to measure current, potential difference, and resistance. (13.11)

WHAT DO YOU THINK NOW?

You thought about the following statements at the beginning of the chapter. You may have encountered these ideas in school, at home, or in the world around you. Consider them again and decide whether you agree or disagree with each one.

Vocabulary

circuit diagram (p. 551)
series circuit (p. 552)
parallel circuit (p. 552)
electric current (I) (p. 556)
ammeter (p. 556)
potential difference (voltage) (V) (p. 560)
voltmeter (p. 561)
electrical resistance (R) (p. 564)
ohmmeter (p. 565)
resistor (p. 565)
Ohm's law (p. 568)

1 There is one wire that connects all the electrical devices in my home.
Agree/disagree?

4 Electric current flows just like water flows in a river.
Agree/disagree?

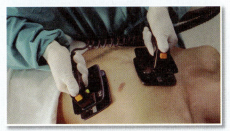

2 An electric shock is always dangerous.
Agree/disagree?

5 It is always safe to plug as many devices into a power bar as there are outlets.
Agree/disagree?

3 All electrical devices get warm when they are operating.
Agree/disagree?

6 An electric stove requires more energy to operate than an electric fan.
Agree/disagree?

How have your answers changed since then?
What new understanding do you have?

BIG Ideas

- Electricity is a form of energy produced from a variety of non-renewable and renewable sources.
- The production and consumption of electrical energy has social, economic, and environmental implications.
- ✓ Static and current electricity have distinct properties that determine how they are used.

CHAPTER 13 REVIEW

What Do You Remember?

1. Draw a circuit diagram for each of the following: (13.1)
   (a) a two-cell battery with three light bulbs in series and a switch
   (b) a two-cell battery with three light bulbs in parallel with a switch to control all the bulbs
   (c) a two-cell battery with three light bulbs in parallel with a switch to control the first two lights only

2. Identify the following circuits as series or parallel. (13.1)

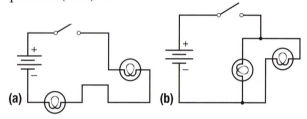

3. Draw a parallel circuit with a correctly placed ammeter and voltmeter. (13.3, 13.5)

4. Sketch a voltage–current graph of a resistor that has the following: (13.9)
   (a) a high resistance
   (b) a low resistance

5. Describe resistance. (13.7)

6. Describe the function of fuses and circuit breakers. (13.3)

7. What may happen when too much current flows through a conductor? (13.3)

8. Explain how an ohmmeter should be connected in a circuit. Should the circuit be turned on or off? (13.7)

9. In the two circuits shown, all of the resistors are identical. Which circuit would have a greater total resistance? Explain your choice. (13.10)

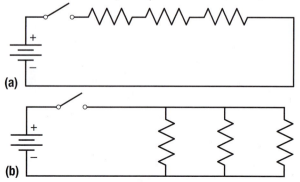

What Do You Understand?

10. Electric current is described as the rate of flow of electrons. Is there anything wrong with saying that the electrons flow from the energy source to the load? Explain. (13.3)

11. The voltage drop across a battery is typically 1.5 V. What does this mean? (13.5)

12. Why would touching a circuit with a current going through it be dangerous? Explain. (13.3)

13. Is a load a resistor? Explain. (13.7)

14. Suppose that you wanted to design an extension cord that is going to be used for long lengths. What properties would you choose to make it safe and effective? (13.7)

15. Wires used for high-voltage power lines are made from aluminum and have several small-diameter wires twisted together, effectively becoming a large-diameter wire. Why do you think the wire is designed this way? (13.7)

16. All branches in a parallel circuit have the same voltage drop as the energy source even if the branches have different types of loads. Does this mean the loads have the same amount of current going through them? Explain. (13.10)

17. Your skin has a much higher resistance when it is dry than when it is wet. If you touched a wire with current going through it, would you rather have wet skin or dry skin? Explain. (13.7)

18. Some people bypass fuses by using a conductor in the place of a fuse. Why is this an unsafe practice? (13.3)

19. You are an apprentice to an electrician. You have been given the task of rewiring a living room. The electrician asks you whether the room should be wired in series or parallel. Which would you choose? Justify your answer. (13.1, 13.10)

20. You go to a friend's house to play on her gaming console. You noticed that she has her TV, surround sound receiver, DVD player, cable box, subwoofer, and gaming console all hooked up to one outlet. This setup does not trip the circuit breaker. Why not? (13.1, 13.10)

21. Analyze the following tables and figures. Identify any values that do not make sense and explain. (13.9, 13.10) T/I

    (a) **Table 1** Data Table

Location	Current (A)	Voltage (V)	Total resistance (Ω)
Total	3.0 (at source)	6.0 (across source)	2.0
R1	1.0	6.0	4.0
R2	1.0	6.0	6.0
R3	1.0	6.0	8.0

    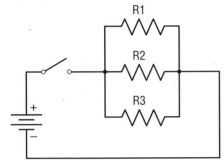

    (b) **Table 2** Data Table

Location	Current (A)	Voltage (V)	Total resistance (Ω)
Total	3.0 (at source)	6.0 (across source)	5.0
R1	3.0	2.0	5.0
R2	3.0	2.0	5.0
R3	3.0	2.0	5.0

    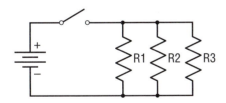

Solve a Problem

22. Your MP3 player has stopped working. Describe some possible tests you could do to determine what the problem is. Include the use of a voltmeter. (13.11) T/I

23. A microwave oven has a current of 5.0 A going through it. It is plugged into a wall outlet with a voltage of 120 V. Calculate the resistance of the oven. (13.9) T/I

24. A 25 Ω resistor has a voltage drop of 12 V across it. Calculate the current flowing through the resistor. (13.9) T/I

25. A 35 mA current is flowing through a lamp with a resistance of 120 Ω. Calculate the voltage drop across the lamp. (13.9) T/I

26. Figure 1 shows a graph of potential difference versus current for a resistor. Calculate the resistance. (13.9) T/I

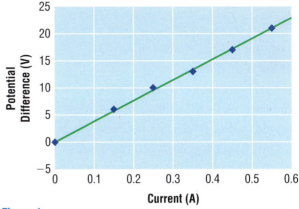

Figure 1

Create and Evaluate

27. Design a circuit that operates three different motors independently. When each motor is operating, a lamp should light, indicating that the motor is operational. There should also be a main switch to operate the whole circuit. (13.11) K/U

Reflect On Your Learning

28. Electric potential difference is a concept that is difficult for some people to understand. What part did you find challenging? Write out a list of ideas that helped you understand electric potential difference. K/U C

Web Connections

29. In the Tech Connect: Heart Technologies, you learned that the sinus node in the human body sends electrical signals to the heart's muscles to regulate its pumping action. Research another control mechanism in the human body that also uses electrical signals to regulate a particular function. How does that mechanism work? Are there any artificial devices that can be used when that mechanism does not work properly? Write a paragraph of your findings. (13.3) T/I C A

To do an online self-quiz or for all other Nelson Web Connections,
GO TO NELSON SCIENCE

CHAPTER 13 SELF-QUIZ

The following icons indicate the Achievement Chart category addressed by each question.

K/U Knowledge/Understanding **T/I** Thinking/Investigation **C** Communication **A** Application

For each question, select the best answer from the four alternatives.

1. The potential difference of a circuit is a measure of the
   (a) rate of flow of electrons
   (b) opposition to the flow of electrons
   (c) difference in electric potential energy per unit of charge between two points
   (d) number of electrons flowing from the negative electrode to the positive electrode (13.5) K/U

2. A wire with which of the following attributes would tend to have the most electrical resistance? (13.7) K/U
   (a) long
   (b) thick
   (c) cold
   (d) copper

3. Electrical resistance is calculated by
   (a) subtracting current from potential difference
   (b) dividing potential difference by current
   (c) multiplying potential difference and current
   (d) adding current and potential difference (13.9) K/U

Indicate whether each of the statements is TRUE or FALSE. If you think the statement is false, rewrite it to make it true.

4. An ammeter measures the number of electrons flowing in a circuit. (13.3) K/U

5. When a load fails in a series circuit, all loads will stop working. (13.1) K/U

6. The human body can withstand electric currents of up to 15 A. (13.3) K/U

Copy each of the following statements into your notebook. Fill in the blanks with a word or phrase that correctly completes the sentence.

7. An ohmmeter is a device used to measure _____. (13.9) K/U

8. According to Ohm's law potential difference varies directly with _____. (13.9) K/U

9. A _____ is a method of drawing an electric circuit using standard symbols. (13.1) K/U

Match the term on the left with the appropriate example on the right.

10. (a) energy source   (i) switch
    (b) electrical conductor (ii) copper wire
    (c) load   (iii) lamp
    (d) control device   (iv) battery (13.1) K/U

Write a short answer to each of these questions.

11. Why must an ammeter be connected in series with a load to measure the current flowing through the load? (13.3) K/U

12. Explain why a burnt-out filament in a bulb has the same effect as an open switch. (13.3) K/U

13. How are electrical systems useful in body functions? (13.3, Tech Connect) K/U

14. Describe the relationship among voltage, volt, and potential difference. (13.5) K/U

15. Two identical resistors are connected in series to a 12.0 V battery. The total resistance of the circuit is 16 Ω. (13.10) T/I
    (a) Calculate the voltage drop across each resistor.
    (b) Calculate the current through each resistor.

582 Chapter 13 • Electrical Quantities in Circuits

16. A blender has a 35.0 Ω resistor that has 5.0 A of current going through it. What is the potential difference across the resistor? (13.9) T/I

17. How is the flow of electrons in a circuit similar to the flow of water down a river? (13.3) K/U T/I

18. Explain why the words "across" and "through" cannot be interchanged in the statement: "As the potential difference across a load increases, so does the current going through the load." (13.3, 13.5) K/U

19. (a) In your own words, explain the difference between a parallel circuit and a series circuit.
    (b) Draw an example of each type of circuit using the standard symbols of a circuit diagram. (13.1, 13.10) K/U C

20. Suppose you are going to interview an electrician about the safety hazards of working with electricity. What are four questions you would include in your interview? (13.3) C

21. (a) Decorator lights are often connected in a series circuit. What happens when one light goes out? Why?
    (b) What is the advantage of connecting decorator lights in a series circuit? (13.1, 13.10) A

22. Imagine that your bedroom has three lamps connected in series. Describe what would happen to the brightness and the voltage drop across each lamp if you added two more lamps in the series. Explain your answer. (13.10) T/I

23. An engineer is testing a cable that connects a television to an electrical outlet. She notices that the cable tends to overheat after being plugged in for a few hours. Describe two ways she could modify the cable to solve this problem. (13.7) T/I

24. (a) You have noticed that turning on a hairdryer often causes the lights, the computer, and several other appliances in your home to turn off. Explain why this happens.
    (b) What is the risk of not having a fuse panel or circuit breaker panel in your home? (13.3, 13.10) A

UNIT E LOOKING BACK

UNIT E: The Characteristics of Electricity

CHAPTER 11: Static Electricity

KEY CONCEPTS

 Static electricity results from an imbalance of electric charge on the surface of an object.

 There are many useful applications of static electricity.

 Objects can be charged by contact or by induction.

 Electric discharge can be harmful and must be treated with caution.

 Materials can be conductors or insulators.

 Simple procedures can be used to test the ability of materials to hold or transfer electric charges.

CHAPTER 12: Electrical Energy Production

KEY CONCEPTS

 The motion of electrons can be controlled.

 An electric circuit requires a source of electrical energy, conductors, a control device, and a load.

 Electrical energy is produced by energy transformations.

 Electrical energy is produced from renewable and non-renewable resources.

 Electrical power is the rate at which electrical energy is produced or consumed in a given time.

 Electrical energy consumption should be reduced.

CHAPTER 13: Electrical Quantities in Circuits

KEY CONCEPTS

 Circuits are a part of electrical technology. They can be connected in series and in parallel.

 Current is the rate of electron flow.

 Potential difference is a measure of the difference in electric potential energy per unit of charge between two points.

 Resistance is a measure of the opposition to the flow of electrons.

 Connecting loads in series or parallel affects the current, potential difference, and total resistance.

 Current, voltage, and resistance measurements can be used to test circuits.

MAKE A SUMMARY

Imagine that you are working for a firm called Ontario Electricity Production that is hiring students for summer jobs. Your task is to help find a student who understands both static and current electricity as they relate to circuits used in electrical energy production.

An interviewer will ask each candidate a series of eight questions. Part of your job is to add one or two questions to each main question that will help distinguish between people who have the basic skills and deep knowledge of electricity from those who do not. The main questions are listed below.

1. What is your interest in working for Ontario Electricity Production?
2. What type of circuits are you familiar with?
3. How does static electricity pose a danger for circuits?
4. At Ontario Electricity Production, we generate electrical energy using both renewable and non-renewable sources. Which of these interests you more? Explain your answer.
5. Ontario Electricity Production prides itself on being respectful of the environment. How would you support this idea?
6. You have been sent with a work crew to carry out repairs on a circuit. What safety precautions should you take before carrying out the work?
7. What things would you like our company to know about your understanding of electricity?
8. Is there anything that you would like to learn from a job with our company?

Once you have finished adding your questions, prepare an answer to each question.

As an extension, work with a partner and alternate role-playing. One student can act as the interviewer, while the other acts as the potential student employee. Once you have finished carrying out the interview with one set of questions, you can switch roles and repeat the interview role-play with the other set of questions. You may want to take on the role of a potential student employee or the interviewer. Consider what your responses to each of these questions will be and be sure to record them.

CAREER LINKS

List the careers mentioned in this unit. Choose two of the careers that interest you or choose two other careers that relate to The Characteristics of Electricity. For each of these careers, research the following information:

- educational requirements (secondary and post-secondary)
- skill/personality/aptitude requirements
- potential employers
- salary
- duties/responsibilities

Assemble the information you have discovered into a poster. Your poster should compare your two chosen careers and explain how they connect to The Characteristics of Electricity.

GO TO NELSON SCIENCE

UNIT E — UNIT TASK

Communicate With Electricity

Communication is a big part of everyday life. People communicate with each other all the time and for different reasons. We may communicate to share new ideas, tell stories, or warn each other of dangers. The simplest way to communicate is by speaking to each other. However, in some cases, person to person communication is not possible or practical. For example, it would be difficult and dangerous for people to control traffic in a big city by standing at intersections and shouting or using hand signals all day and night. Therefore, we use electric traffic lights instead (Figure 1). Ambulances and fire trucks need to move quickly through busy streets to reach the scene of accidents and fires. They use flashing electric lights and sirens to do this more safely. Electric billboards use many small light bulbs or LEDs to communicate information about products or upcoming events. Electricity has revolutionized the way people communicate.

SKILLS MENU
- Questioning
- Hypothesizing
- Predicting
- Planning
- Controlling Variables
- Performing
- Observing
- Analyzing
- Evaluating
- Communicating

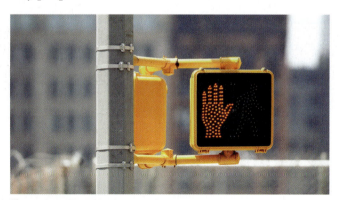

Figure 1 Traffic lights help us communicate important signals safely and effectively.

In this activity, you and a partner are part of a team of engineers and technicians who are responsible for designing a new electric communication system. The company you work for produces all sorts of electric communication appliances, and designs and builds them according to their customers' needs and desires. The company is also concerned about how the manufacture, use, and eventual disposal of their products may impact the environment. The company always makes an effort to use as little material as possible in the design of their products, and to use recyclable components as much as possible. They also use designs that reduce energy consumption and, where possible, rely on renewable energy supplies.

You and your partner will identify a situation that may benefit from a particular type of electric communication device. You will then design, build, and test the device. Your device will try to live up to the company's commitment to environmentally responsible design. The device must run on electricity, and must include an electric circuit in which a number of loads (such as, lamps, buzzers, LEDs) work to produce the sort of communication you planned to accomplish. Switches may be used to control various parts of the circuit.

After you have designed, built, and tested your device, you will present it to your classmates with an explanation of the purpose of your device (the type of communication desired), the circuit used to accomplish the task, and the environmentally friendly features of the device. You will also describe the device's shortcomings and possible changes that may improve its function.

Purpose

To design, build, and test a simple electric communication device.

Equipment and Materials

- electrical energy source
- connecting wires
- various loads (lamps, LEDs, motors, resistors)
- switches
- voltmeter, ammeter, or multimeter
- watt meter
- other necessary equipment and materials

 Always disconnect the energy source before constructing your circuit. If your circuits do not work, ask your teacher for assistance.

Procedure

1. Identify a situation that could benefit from an electric communication device.
2. Design an electric circuit that may accomplish the task identified in step 1.
3. Draw a circuit diagram of your design for approval by your teacher.
4. Build the circuit that you have designed. Be sure that the energy source is disconnected or the power supply is off. Once you have completed the circuit, ask for your teacher's approval to turn it on.
5. Test your device to determine if it accomplishes the task identified in step 1.
6. If a part of your circuit is not working, try to identify the problem and possible solutions. You may ask your teacher for assistance.
7. Use the voltmeter and ammeter to determine the total voltage drop and total current of the circuit when it is in use.
8. Use a watt meter to determine the energy consumption of your device.
9. For your presentation, choose how you will share information with your classmates.

    Select one of these options:
    - brochure
    - radio broadcast
    - classroom presentation
    - newspaper article and ad for the local newspaper
    - informational poster

Analyze and Evaluate

(a) Analyze the circuit in your device. Did you use series connections, parallel connections, or both? Explain why.
(b) What type of energy source did you use? Why?
(c) How much electrical power does your communication device consume?
(d) Describe some of the successes and some of the drawbacks of your device in terms of
   (i) electrical function
   (ii) environmental friendliness
   (iii) ability to communicate according to goals identified in step 1

Apply and Extend

(e) Identify an electric communication device in real life that may accomplish the task you identified in step 1. Describe some similarities and differences between your device and the other device.
(f) The disposal of the components of your circuit can have environmental impacts. Research some alternatives to landfilling the components of your circuit. Can any of the components be recycled or reused? Incorporate this information into your presentation.
(g) Some highway signs are now powered with solar panels. What are some advantages of this design?
(h) Research additional advances in communication technology. Report on any innovations designed to reduce the environmental impact of their use, such as improvements in energy efficiency or reduction of the use of hazardous materials.

ASSESSMENT CHECKLIST

Your completed Performance Task will be assessed according to the following criteria:

Knowledge/Understanding
- ☑ Understand how to connect loads in series and in parallel.
- ☑ Know how to use electric meters safely.
- ☑ Construct a functioning electrical circuit.

Thinking/Investigation
- ☑ Design a working circuit for a specific purpose.
- ☑ Measure the electrical power used by the circuit.
- ☑ Research and gather information about environmental effects and disposal options for the components of a circuit.

Communication
- ☑ Express ideas using an appropriate format in oral, written, and/or visual form.

Application
- ☑ Apply knowledge to an unfamiliar context.
- ☑ Make connections between society and the use of technology and its impact on the environment.
- ☑ Propose ways to improve upon a circuit design to better its functionality or minimize its impact on the environment.

UNIT E REVIEW

The following icons indicate the Achievement Chart category addressed by each question.

K/U Knowledge/Understanding
T/I Thinking/Investigation
C Communication
A Application

What Do You Remember?

For each question, select the best answer from the four alternatives.

1. The law of electric charges describes how
   (a) like charges attract each other
   (b) unlike charges repel each other
   (c) like charges repel each other
   (d) some materials are conductors, while others are insulators (11.1) **K/U**

2. Electrostatic painting works because
   (a) the paint and the object to be painted have opposite charges
   (b) the paint is difficult to charge, while the object to be painted is not
   (c) the paint is easily charged, while the object to be painted is not
   (d) none of the above (11.1) **K/U**

3. Conductors of electricity
   (a) allow protons to move through them easily
   (b) allow electrons to move through them easily
   (c) allow positive charges to move through them easily
   (d) do not allow charges of any kind to move through them easily (11.4) **K/U**

4. Charging by induction best explains why
   (a) rubbing two different objects together produces oppositely charged objects
   (b) lightning discharges into the ground
   (c) your hair stands up if you go down a plastic tube slide
   (d) neutral objects are attracted to charged objects (11.6) **K/U**

5. Power is
   (a) the electrical energy stored in electric charges
   (b) the electrical energy used by electrical devices over time
   (c) the electrical energy produced by electrical generators over time
   (d) both (b) and (c) (12.3) **K/U**

6. Non-renewable energy sources that are used to generate electrical energy are
   (a) coal, wind, uranium, and tidal
   (b) coal, natural gas, oil, and uranium
   (c) coal, sunlight, oil, and uranium
   (d) coal, oil, hydro-electric, and geothermal (12.5) **K/U**

7. Renewable energy sources that are used to generate electrical energy are
   (a) wind, tidal, sunlight, uranium, and hydro-electric
   (b) wind, tidal, natural gas, uranium, and sunlight
   (c) wind, tidal, geothermal, biomass, and hydro-electric
   (d) water, uranium, sunlight, and wind (12.5) **K/U**

8. Which of the following energy sources produces the least amount of air pollution when used to generate electrical energy? (12.5) **K/U**
   (a) coal
   (b) hydro-electric
   (c) oil
   (d) natural gas

9. Which of the following energy sources is not used to heat water during the generation of electrical energy? (12.5) **K/U**
   (a) hydro-electric
   (b) uranium
   (c) coal
   (d) oil

10. Which of the following would you consider to be a load in a circuit? (12.2) **K/U**
    (a) switch
    (b) battery
    (c) voltmeter
    (d) lamp

11. A series circuit
    (a) has a power source, connecting wires, a control device, and several paths for the electric current to follow
    (b) has a power source, connecting wires, a control device, and one path for the electric current to follow
    (c) has a power source, connecting wires, a control device, a load, and several paths for the electric current to follow
    (d) has a power source, connecting wires, a control device, a load, and one path for the electric current to follow (13.1) K/U

12. Which of the following correctly relates to Ohm's law? (13.9) K/U
    (a) Electrical resistance is equal to the ratio of voltage to current.
    (b) The slope of a line from a voltage versus a current graph gives the electrical resistance.
    (c) As the electric potential difference is increased, the current increases.
    (d) all of the above

13. Loads connected in series
    (a) increase the total resistance of the circuit
    (b) decrease the total resistance of the circuit
    (c) do not affect the total resistance of the circuit
    (d) none of the above (13.10) K/U

14. The electric potential difference of the power source in a series circuit is
    (a) greater than the voltage drop across each load added together
    (b) equal to the voltage drop across each load added together
    (c) less than the voltage drop across each load added together
    (d) not related to the voltage drop across each load added together (13.10) K/U

15. Which of the following sets of materials consists of good electrical conductors? (11.4) K/U
    (a) copper, aluminum, nickel
    (b) platinum, rubber, ebonite
    (c) plastic, wood, paper
    (d) porcelain, wax, carbon

16. Which of the following devices uses the properties of electrostatic charge to remove particles from the air? (11.2) K/U
    (a) electromagnet
    (b) electroscope
    (c) electrostatic precipitator
    (d) electrostatic lifting apparatus

17. Which of the following best describes a fuel cell? (12.3) K/U
    (a) an electric cell that consumes reactants from an outside source
    (b) an electric cell that converts water into hydrogen and oxygen
    (c) an electric cell that produces fossil fuels through combustion
    (d) an electric cell that relies on heat instead of chemicals

18. Which of the following best describes how length and temperature affect the resistance of a wire? (13.7) K/U
    (a) As length and temperature increase, the internal resistance decreases.
    (b) As length and temperature increase, the internal resistance increases.
    (c) As length increases and temperature decreases, the internal resistance decreases.
    (d) As length increases and temperature decreases, the internal resistance increases.

Indicate whether each of the statements is TRUE or FALSE. If you think the statement is false, rewrite it to make it true.

19. In a parallel circuit, electrons can flow along more than one path. (13.1) K/U

20. Nuclear power relies on renewable resources to generate electricity. (12.5) K/U

21. Electrostatic dusters use the process of charging by conduction to attract dust. (11.2) K/U

22. A generator does not require an external energy source to generate electrical energy. (12.4) K/U

23. AC is a flow of electrons in one direction. (12.4) K/U

24. DC is produced by devices that do not have moving parts. (12.4) K/U

25. To make an object positively charged, you should add electrons to it. (11.1)

26. When the number of positive charges exceeds the number of negative charges, the object is considered negative. (11.1)

27. A discharge occurs when two neutral objects are brought together. (11.8)

28. The electric current going through a wire is composed of moving electrons. (12.1, 12.4, 13.3)

29. Potential difference is the difference in electric potential energy per unit charge measured at two different points in a circuit. (13.5)

30. The use of coal to produce electrical energy is considered to have a small environmental impact. (12.5)

31. Using uranium to generate electrical energy does not contribute to climate change. (12.5)

32. Solar cells use energy from the Sun to turn a generator. (12.5)

33. Using biomass energy to generate electrical energy releases greenhouse gases into the atmosphere. (12.5)

34. Ammeters measure electric current and are connected in parallel to a circuit. (13.3)

35. Electrical resistance is the opposition to electric current. (13.7)

36. A load produces electrical energy to do something useful. (12.4)

37. Dividing the current by the potential difference of a load will give you the resistance. (13.9)

38. The total resistance of a circuit increases by adding more loads in parallel. (13.10)

Copy each of the following statements into your notebook. Fill in the blanks with a word or phrase that correctly completes the sentence.

39. The rapid transfer of electrons from one object to another produces a(n) _____. (11.8)

40. Electricians use a(n) _____ to measure the current flowing through a circuit. (13.3)

41. _____ is a type of current in which electrons flow in only one direction. (12.4)

42. An object can be made negative by _____ electrons. (11.1)

43. An insulator _____ allow electrons to travel through it easily. (11.4)

44. Conducting excess charge into the _____ is called _____. (11.2)

45. When a charged object touches another object, it is called charging by _____. (11.2)

46. Nuclear generating plants use the potential energy stored in _____ to generate electrical energy. (12.5)

47. Solar cells convert the energy from the _____ into electrical energy. (12.5)

48. _____ circuits have several paths for the electric current to follow, while _____ circuits have one path for the electric current to follow. (13.1, 13.10)

49. Voltmeters measure _____ and are connected in _____ with a load or power source. (13.5)

50. The _____ of a line of best fit on a voltage–current graph gives the electrical resistance. (13.9)

51. The voltage drop across each branch in a parallel circuit is _____. (13.10)

52. The electric current going through each load in a series circuit is _____. (13.10)

53. One disadvantage of fossil fuel electrical energy production is the _____ of greenhouse gases into the atmosphere. (12.5)

Match the term on the left with the appropriate definition on the right.

54. (a) static electricity
    (b) current electricity
    (c) electrical resistance
    (d) electrical circuit
    (e) electrical conductor

    (i) the imbalance of electric charge that builds up on the surface of objects
    (ii) a continuous path of electron flow that includes an energy source, switch, load, and wire
    (iii) the controlled flow of electrons through a conductor
    (iv) a material that allows the easy movement of electrons
    (v) a material's ability to oppose the movement of electrons (11.1, 11.4, 13.3, 13.7) K/U

What Do You Understand?

Write a short answer to each of these questions.

55. Summarize the properties of the three subatomic particles. (11.1) K/U
    (a) neutrons
    (b) protons
    (c) electrons

56. Describe how you could make an object
    (a) neutral
    (b) positively charged
    (c) negatively charged (11.1) K/U

57. Restate the Law of Electric Charges. (11.1) K/U C

58. Draw a series of diagrams showing how you would permanently induce a positive charge on a pith ball electroscope. (11.6) K/U C

59. Briefly explain why lightning rods are usually made of metal. (11.8) K/U

60. A certain light bulb uses 100 J of electrical energy to produce 40 J of light energy. What happens to the rest of the energy in the system? (12.7) K/U

61. Describe the relationship between the voltage drop and the number of loads in a series circuit. (13.10) K/U

62. A negatively charged metal rod is brought near a neutral aluminum can. Draw a diagram to show the position of electrons on the can's surface before and after the induced charge separation. (11.1) C

63. Use Ohm's Law to find the missing information for each of the following circuits. (13.9) T/I K/U

	Resistance (Ω)	Voltage (V)	Current (A)
circuit A		90	2.5
circuit B	350		0.5
circuit C	150	120	

64. You want to warm up a slice of leftover pizza for lunch. Explain why a toaster oven would be more efficient than a regular oven for this task. (12.7) A

Solve a Problem

65. An electroscope is positively charged. Describe a method that could be used to make it negatively charged. (11.6) K/U T/I

66. You have been given some rabbit fur, an ebonite rod, a glass rod, and a piece of polyester. Your teacher asks you to charge the rods negatively by friction. What combinations will you use? Explain why this will work. (11.2) K/U T/I

67. An object is brought near the following electroscopes. What charge, if any, is on the object? (11.6) K/U T/I

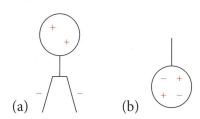

(a)   (b)

68. Determine the voltage drop across and the current through each load in the circuit. V_{source} = 6.0 V and I_{source} = 1.0 A for both circuits. Assume the loads are identical. (13.10) T/I

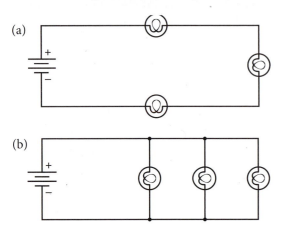

(a)

(b)

69. Calculate the resistances of each load and the total resistance of each circuit in Question 48. (13.10) T/I

70. Calculate the resistance of the load given the following graph. (13.9) T/I

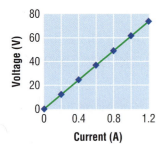

71. An electric toaster has a current of 12.0 A going through it. It is plugged into a wall outlet. The voltage drop across the toaster is 115 V. Calculate the resistance of the toaster. (13.9) T/I

72. An MP3 phone has a 3.6 V battery. This means that the phone has a potential difference of 3.6 V to operate. The resistance of the phone is 50.0 Ω. Calculate the current going through this circuit. (13.9) T/I

73. A current of 20 mA goes through the circuit of a digital watch. The resistance of the watch is 75 Ω. Calculate the voltage drop across the watch's battery. (13.9) T/I

74. Electronics manufacturers often ship their products in anti-static bags. Give a likely reason that they have adopted this policy. (11.8) T/I

75. A student decides to carry out a study of her energy usage. She creates a log that shows how much power certain electrical appliances consume, and how much she uses the appliances each month. (12.8) T/I

	Power (W)	Time per month (h)
television	210	20
computer	150	32
video Game System	180	8

(a) Calculate the cost of operating each of the appliances per month if the cost of electricity in her area is $0.12 per kW·h.
(b) Although the student personally uses the computer only 32 hours per month, she leaves it plugged in and on for 24 hours per day. Calculate how many kilowatt hours the computer actually consumes each month (assume 30 days).

76. The Three Gorges Dam in China is one of the world's largest hydroelectric power projects. Though it promises to generate enormous amounts of energy, the dam is controversial because of its expected environmental consequences. Explain how the dam could negatively affect the surrounding environment. (12.5) A

Create and Evaluate

77. Devise a plan that manufacturers of electrical consumer goods could use to reduce electrical energy consumption. In your plan, include ways to reduce consumption during both the production and the use of the electrical consumer good. (12.5, 12.7, 12.8) K/U T/I C

78. Suppose that you could design an automated house. What things would you include so that it would minimize electrical energy consumption? (12.7, 12.8) K/U A C

79. In a paragraph, justify the continued use of non-renewable resources for electrical energy production. (12.5)

80. Discuss some of the difficulties with implementing the use of renewable energy sources to generate electrical energy. (12.5)

81. Design a lint brush that can remove human and pet hair from clothes. Provide a labelled diagram and justify the materials in your design. (11.1, 11.2)

82. An engineer wants to design a new type of space heater for use in schools that do not have central heating systems. Will his design most likely rely on static electricity or current electricity? Explain your answer. (11.1, 12.1)

83. You want to create a circuit from the following materials: a battery, copper wire, a switch, and three light bulbs. Design your circuit so that the light bulbs shine as brightly as possible. Draw a circuit diagram to reflect your design. (13.10, 13.11)

84. Design an apparatus made from a battery, a light bulb, and copper wire that could be used to test the conductivity of various materials. Describe your design and explain how you would interpret the results of the tests. (11.6)

Reflect On Your Learning

85. Do you support electrical energy production through the use of non-renewable sources? Do you support the idea of electrical energy conservation? Discuss your thoughts about these two issues with your classmate and describe similarities and differences in your thoughts.

86. (a) What occupations require an understanding of electricity? Why?
    (b) Consider your own career goals. Will they require you to understand principles of electricity? If so, how?

Web Connections

87. Suppose that you were elected the premier of Ontario. As part of your job, you will need to appoint ministers to become experts in different aspects of running the province. What qualifications would you want to have in your minister of energy, minister of transportation, and minister of natural resources? Research these ministries so that you know what their responsibilities are. What responsibilities should the minister have? Write a paragraph about each ministry and how it relates to what was learned in this unit of study. (12.5, 12.7, 12.8)

UNIT E SELF-QUIZ

The following icons indicate the Achievement Chart category addressed by each question.

- K/U Knowledge/Understanding
- T/I Thinking/Investigation
- C Communication
- A Application

For each question, select the best answer from the four alternatives.

1. Which of the following atoms has an electric charge? (11.1) K/U
   (a) a sodium atom with 11 protons, 11 electrons, and 12 neutrons
   (b) a calcium atom with 20 protons, 18 electrons, and 20 neutrons
   (c) a helium atom with 2 protons, 2 electrons, and 2 neutrons
   (d) an iron atom with 26 protons, 26 electrons, and 30 neutrons

2. Which of the following best describes the process of charging by conduction? (11.2) K/U
   (a) A highly charged object is connected to a large neutral object that removes the excess charge.
   (b) A highly charged object is connected to another highly charged object so that the charges neutralize each other.
   (c) Two neutral objects of different materials transfer electrons when they are rubbed against each other.
   (d) Two objects with different amounts of electric charge transfer electrons when they contact each other.

3. Which part of an electric circuit transforms electrical energy into other usable forms of energy? (12.2) K/U
   (a) load
   (b) switch
   (c) power source
   (d) connecting wire

4. Which of the following statements best summarizes Ohm's Law? (13.9) K/U
   (a) In an electric circuit, the resistance equals the voltage divided by the current.
   (b) In an electric circuit, the current equals the voltage multiplied by the resistance.
   (c) In an electric circuit, the resistance is independent of voltage and current.
   (d) In an electric circuit, the voltage is twice the product of the current and resistance.

Indicate whether each of the statements is TRUE or FALSE. If you think the statement is false, rewrite it to make it true.

5. A charged object can cause electrons in a neutral object to shift position. (11.1) K/U

6. One disadvantage of tidal generation is that it contributes to water pollution. (12.5) K/U

7. An electric charge is a positive or negative charge that exerts an electric force. (11.1) K/U

Copy each of the following statements into your notebook. Fill in the blanks with a word or phrase that correctly completes the sentence.

8. According to the Law of Electric Charges, objects that have opposite charges _____ each other. (11.1) K/U

9. _____ are materials that prevent the movement of electrons. (11.4) K/U

10. _____ are electric cells that can be recharged, while _____ are electric cells that can be used only once. (12.3) K/U

Match each term on the left with the appropriate statements on the right.

11. (a) coal   (i)   relies on a renewable resource; can cause flooding
    (b) nuclear   (ii) causes little or no air pollution; produces hazardous waste
    (c) hydro-electric (iii) recycles waste; affects the world's food supply
    (d) wind   (iv) is easy to mine; contributes to smog
    (e) biomass   (v)   produces no pollution; can be noisy and harm birds (12.5) K/U

Write a short answer to each of these questions.

12. Give three examples of good conductors and three examples of good insulators. (11.4) K/U

13. Why is the electrostatic series an important resource for engineers? (11.2) K/U

14. Name four factors that affect the internal resistance of a material. (13.7) K/U

15. In your own words, explain the difference between static electricity and current electricity. (11.1, 12.1) K/U C

16. People often complain about static cling, which occurs when clothes stick to people's legs or arms. (11.1, 11.2) T/I
    (a) Explain why static cling occurs using your knowledge of static electricity.
    (b) Which of the following fabrics would most likely cause static cling: polyester, silk, or wool? Explain your answer.

17. You want to buy a Bunsen burner to take on an upcoming camping trip. You are considering three different models. Calculate the percent efficiency of each of the models in the table below, in order to find the model that is the most energy-efficient. (12.7) T/I

	Energy input	Useful energy output
model A	850 J	350 J
model B	600 J	175 J
model C	750 J	250 J

18. Strong electrical discharges can sometimes cause electrical fires. Explain why it could be dangerous to throw water on an electrical fire in your home. (11.8) T/I

19. A scientist is investigating two resistors. She has collected the data in the following table. Which resistor has the greater resistance? (13.9) T/I

Resistor A		Resistor B	
Voltage (V)	Current (A)	Voltage (V)	Current (A)
45	0.55	30	0.27
65	0.80	55	0.50
85	1.05	70	0.65

20. As you rest your hand on a metal railing outside your home, you receive a shock. Draw a diagram to illustrate the process of electric discharge that has just occurred. (11.8) C

21. Name a situation in your daily life that demonstrates each of the following processes. For each situation, describe the movement of electrons. (11.2, 11.6) A
    (a) charging by friction
    (b) charging by induction
    (c) grounding

22. Suppose you are an engineer who needs to develop a simple circuit to test a new light bulb. You have the following materials available to create the wire for this circuit: copper, silicon, rubber, and wax. Design a wire for the circuit that would be both safe and effective. Explain your response. (11.4, 12.1, 13.7) T/I

23. Suppose you work for a wind power company. Your company has recently proposed the construction of a new wind farm in a rural Ontario community. Before construction begins, you want to survey residents about their attitudes toward the proposed wind farm. (12.5) C
    (a) What are three questions you would ask residents to get their opinions about the new wind farm and about wind power in general?
    (b) How could you explain the benefits of wind energy to residents who seem opposed to the new wind farm?

24. Name one object in your home that uses a circuit. Describe the structure of this circuit, identifying the load and energy source. (12.2, 13.1) A

APPENDIX A — Skills Handbook

CONTENTS

1. **Safe Science**
   - **1.A.** Having a Safe Attitude 598
   - **1.B.** Specific Safety Hazards 599
   - **1.C.** Accidents Can Happen 600
   - **1.D.** Safety Conventions and Symbols 601

2. **Scientific Tools and Equipment**
   - **2.A.** Using the Microscope 602
   - **2.B.** Testing for Electrical Conductivity 604
   - **2.C.** Using Star Maps 604
   - **2.D.** Drawing and Constructing Circuits 606
   - **2.E.** Using the Voltmeter, Ammeter, and Digital Multimeter 607
   - **2.F.** Using Other Scientific Equipment 607

3. **Scientific Inquiry Skills**
   - **3.A.** Thinking as a Scientist 609
   - **3.B.** Scientific Inquiry 609

4. **Research Skills**
   - **4.A.** General Research Skills 617
   - **4.B.** Using the Internet 618
   - **4.C.** Exploring an Issue Critically 619

5. **Using Mathematics in Science**
   - **5.A.** SI Units 623
   - **5.B.** Solving Numerical Problems Using the GRASS Method . . 625
   - **5.C.** Scientific Notation 625
   - **5.D.** Uncertainty in Measurement . .627
   - **5.E.** Using the Calculator 630

6. **Data Tables and Graphs**
   - **6.A.** Graphing Data 632
   - **6.B.** Using Computers for Graphing 634
   - **6.C.** Interpreting Graphs 634

7. **Study Skills**
   - **7.A.** Working Together 635
   - **7.B.** Setting Goals and Monitoring Progress 636
   - **7.C.** Good Study Habits 638

8. **Literacy**
   - **8.A.** Reading Strategies 639
   - **8.B.** Graphic Organizers 640

9. **Latin and Greek Root Words** 643

10. **Periodic Table** 644

1 SAFE SCIENCE

1.A. Having a Safe Attitude

Science investigations can be a lot of fun, but certain safety hazards exist in any laboratory. You should know about them and about the precautions you must take to reduce the risk of an accident.

Why is safety so important? Think about the safety measures you already take in your daily life. Your school laboratory, like your kitchen, need not be a dangerous place. In any situation, you can avoid accidents when you understand how to use equipment and materials and follow proper procedures.

For example, you can take a hot pizza out of the oven safely if you take a few common-sense safety precautions. Similarly, corrosive acids can be used safely if appropriate safety precautions are taken.

Safety in the laboratory combines common sense with the foresight to consider the worst-case scenario. The activities and investigations in this textbook are safe, as long as you follow proper safety precautions. General laboratory safety rules are outlined below (Table 1). Your teacher may provide you with additional safety rules for specific tasks.

Table 1 Practise Safe Science in the Classroom

Be science ready	Follow instructions	Act responsibly
• Come prepared with your textbook, notebook, pencil, and anything else you need.	• Do not enter a laboratory unless a teacher is present or you have permission to do so.	• Pay attention to your own safety and the safety of others.
• Tell your teacher about any allergies or medical problems.	• Listen to your teacher's directions. Read written instructions. Follow them carefully.	• Know the location of MSDS (Material Safety Data Sheet) information, exits, and all safety equipment, such as the first-aid kit, fire blanket, fire extinguisher, and eyewash station.
• Keep yourself and your work area tidy and clean. Keep aisles clear.	• Ask your teacher for directions if you are not sure what to do.	• Alert your teacher immediately if you see a safety hazard, such as broken glass, a spill, or unsafe behaviour.
• Keep your clothing and hair out of the way. Roll up your sleeves, tuck in loose clothing, and tie back loose hair. Remove any loose jewellery.	• Wear eye protection or other safety equipment when instructed by your teacher.	• Stand while handling equipment and materials.
• Wear closed shoes (not sandals).	• Never change anything, or start an activity or investigation on your own, without your teacher's approval.	• Avoid sudden or rapid motion in the laboratory, especially near chemicals or sharp instruments.
• Do not wear contact lenses while doing investigations.	• Get your teacher's approval before you start an investigation that you have designed yourself.	• Never eat, drink, or chew gum in the laboratory.
• Read all written instructions carefully before you start an activity or investigation.		• Do not taste, touch, or smell any substance in the laboratory unless your teacher asks you to do so.
		• Clean up and put away any equipment after you are finished.
		• Wash your hands with soap and water at the end of each activity or investigation.

1.B. Specific Safety Hazards

Follow these instructions to use materials and equipment safely in the science classroom.

1.B.1. Chemicals

Some of the chemicals recommended for use in this course are dangerous if used incorrectly. Be sure to follow these rules to avoid accidents.

- Assume that any unknown chemicals are hazardous.
- Reduce exposure to chemicals to the absolute minimum. Avoid direct skin contact, if possible.
- When taking a chemical from a container, first check the label to be sure you are taking the correct substance. Replace the lid securely when you have taken what you need.
- Never use the contents of a container that has no label or has an illegible label. Give any such containers to your teacher.
- Place test tubes in a rack before pouring liquids into them. If you must hold a test tube, tilt it away from you, and others, before pouring in a liquid.
- Pour liquid chemicals carefully (down the side of the receiving container or down a stirring rod) to avoid splashing. Always pour from the side opposite the label so that drips will occur only on the side away from your hand.
- When you are instructed to smell a chemical by your teacher, first take a deep breath to fill your lungs with just air, then waft or fan the vapours toward your nose.
- Do not return surplus chemicals to stock bottles and do not pour them down the drain. Dispose of excess chemicals as directed by your teacher.
- If any part of your body comes in contact with a chemical, wash the area immediately and thoroughly with cool water. Rinse affected eyes for at least 15 minutes. Alert your teacher.

1.B.2. Heat Sources

Heat sources, such as hot plates, light bulbs, and Bunsen burners, can cause painful burns. Use caution when there are hot objects around.

- Secure your Bunsen burner to a retort stand with a clamp before lighting it.
- Your teacher will show you the proper method of lighting and adjusting the Bunsen burner. Always follow this method.
- Never leave a lighted burner unattended as a clean blue flame is almost invisible.
- Never heat a flammable material over a Bunsen burner. Make sure there are no flammable materials nearby.
- Never look down the barrel of a burner.
- When heating liquids in glass containers, use only heat-resistant glassware. If a liquid is to be heated to boiling, use boiling chips to prevent "bumping." Keep the open end of the container away from yourself and others. Never allow a container to boil dry.
- When heating a test tube over a Bunsen burner, use a test-tube holder and a spurt cap. Hold the test tube at an angle, with the opening facing away from yourself and others. Heat the upper half of the liquid first and then move it gently into the flame to distribute the heat evenly.
- Always turn off the gas at the valve, not using the gas-adjustment screw of the Bunsen burner.
- If you burn yourself, immediately apply cool water to the affected area and inform your teacher.

1.B.3. Glass and Sharp Objects

Handle glass carefully: it can break, leaving sharp edges and splinters.

- Never use glassware that is broken, cracked, or chipped.
- Never pick up broken glass with your fingers. Use gloves as well as a broom and a dustpan to remove glass from the area.
- Dispose of glass fragments in special containers marked "Broken Glass."
- If you cut yourself, inform your teacher immediately. Embedded glass or continued bleeding requires medical attention.
- Select the appropriate instrument for the task. Never use a knife when scissors would work better.
- Never carry a scalpel in the laboratory with an exposed blade; transport it in a dissection case or box or on a dissection tray.

- Make sure your cutting instruments are sharp. Dull cutting instruments require more pressure than sharp instruments and are, therefore, much more likely to slip. If your scalpel blade needs to be changed, ask your teacher to change it for you.
- Always cut away from yourself and others. Cut downward, on a tray, cutting board, or paper towel.

1.B.4. Electricity and Light

- Never touch an electrical device, electrical cord, or outlet with wet hands.
- Keep water away from electrical equipment.
- Do not use the equipment if wires or plugs are damaged or if the ground pin has been removed.
- If using a light source, check that the wires of the light fixture are not frayed and that the bulb socket is in good shape and is well secured to a stand.
- Make sure that electrical cords are not a tripping hazard.
- When unplugging equipment, hold the plug to remove it from the socket. Do not pull on the cord.
- When using variable power supplies, start at low voltage and increase slowly.
- Do not look directly at any bright source of light. You cannot rely on the sensation of pain to protect your eyes.
- Never point a laser beam (directly or after reflection) at anybody's eyes.

1.B.5. Living Things

- Treat all living things with care and respect. Never treat an animal in a way that would cause it injury or harm.
- Animals that live in the classroom should be kept in a clean, healthy environment.
- Wear gloves and wash your hands before and after feeding or handling an animal, touching materials from the animal's cage or tank, or handling bacterial cultures.
- Human blood, urine, or saliva samples should not be tested or used in investigations due to the risk of contracting a disease from these fluids.

1.B.6. Observing Space

- Never look directly at a solar eclipse or at the Sun. Specially designed filters must be used to protect the eyes when viewing the Sun directly.
- When going out at night to observe stars, always be accompanied by an adult. Let someone else know where you are going and when you expect to return.

1.C. Accidents Can Happen

The following guidelines apply if an injury occurs, such as a burn, a cut, a chemical spill, ingestion, inhalation, an electrical accident, or a splash in the eyes.

- If an injury occurs, inform your teacher immediately.
- If the injury is from contact with a chemical, wash the affected area with a continuous flow of cool water for at least 15 minutes. Remove contaminated clothing. Consult the Material Safety Data Sheet (MSDS) for the chemical; this sheet provides information about the first-aid requirements. If the chemical is splashed into your eyes, have another student assist you in getting to the eyewash station immediately. Rinse with your eyes open for at least 15 minutes.
- If you have ingested or inhaled a hazardous substance, inform your teacher immediately. Consult the MSDS for the first-aid requirements for the substance.
- If the injury is from a burn, immediately immerse the affected area in cold water. This will reduce the temperature and prevent further tissue damage.
- In the event of electrical shock, do not touch the affected person or the equipment the person was using. Break contact by switching off the source of electricity or by removing the plug.

1.D. Safety Conventions and Symbols

The activities and investigations in *Nelson Science Perspectives 9* are safe to perform provided precautions are taken. This is why general safety hazards are identified with caution symbols (Figure 1).

Figure 1 Safety hazards are identified with caution symbols.

More specific hazards, related to dangerous chemicals, are indicated with the appropriate WHMIS symbol. Make sure that you read the information in red type carefully. You must understand and follow these instructions to perform the activity or investigation safely. Check with your teacher if you are unsure.

1.D.1. Workplace Hazardous Materials Information System (WHMIS) Symbols

The Workplace Hazardous Materials Information System (WHMIS) provides information about hazardous products. Clear and standardized labels must be present on the product's container and must be added to other containers if the product is transferred. If the material is hazardous, the label will include one or more of the WHMIS symbols (Figure 2).

Figure 2 WHMIS symbols identify dangerous materials that are used in all workplaces, including schools.

1.D.2. Hazardous Household Product Symbols (HHPS)

The *Canadian Hazardous Products Act* requires manufacturers of consumer products to include a symbol that specifies both the nature and the degree of any hazard. The Hazardous Household Products Symbols (HHPS) were designed to do this. The symbol is made up of a picture and a frame. The picture tells you the type of danger. The frame tells you whether it is the contents or the container that poses the hazard (Figure 3).

Symbol	Danger
	Explosive This container can explode if it is heated or punctured.
	Corrosive This product will burn skin or eyes on contact, or throat and stomach if swallowed.
	Flammable This product, or its fumes, will catch fire easily if exposed to heat, flames, or sparks.
	Poisonous Licking, eating, drinking, or sometimes smelling, this product is likely to cause illness or death.

Figure 3 Household Hazardous Products Symbols (HHPS) appear on many products that are used in the home. A triangular frame indicates that the container is potentially dangerous. An octagonal frame indicates that the product inside the container poses a hazard.

2 SCIENTIFIC TOOLS AND EQUIPMENT

Selecting the correct tools and equipment and using them properly are essential for your safety and the safety of your classmates.

2.A. Using the Microscope

To view cells and other small objects closely, you will use a microscope (Figure 1). You will most likely use a compound light microscope that uses more than one lens and a light source to make an object appear larger. The object is magnified first by the lens just above the object: the objective lens. Then that image is magnified by the eyepiece: the ocular lens. The comparison of the actual size of the object with the size of its image is called the magnification. The parts of a compound light microscope and their functions are listed in Table 1.

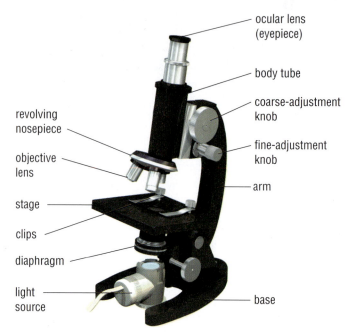

Figure 1 A compound light microscope

Table 1 Parts of a Light Microscope

Part	Function
stage	• supports the microscope slide • has a central opening that allows light to pass through the slide
clips	• hold the slide in position on the stage
diaphragm	• controls the amount of light that reaches the object being viewed
objective lenses	• magnify the object • have three possible magnifications: low power (4×), medium power (10×), and high power (40×)
revolving nosepiece	• holds the objective lenses • rotates, allowing the objective lenses to be changed
body tube	• contains the eyepiece (ocular lens) • supports the objective lenses
eyepiece (ocular lens)	• is the part you look through to view the object • magnifies the image of the object, usually by 10×
coarse-adjustment knob	• moves the body tube up or down to get the object into focus • is used with the low-power objective lens only
fine-adjustment knob	• moves the tube to get the object into sharp focus • is used with medium-power and high-power magnification • is used only after the object has been located, centred, and focused under lower power magnification using the coarse-adjustment knob
light source	• may be an electric light bulb or a mirror that can be angled to direct light through the object being viewed

2.A.1. Microscope Skills

The basic microscope skills are presented as instructions. This allows you to practise these skills before you need to use them in the activities in *Nelson Science Perspectives 9*.

VIEWING OBJECTS UNDER THE MICROSCOPE

1. Make sure that the low-power objective lens is positioned over the diaphragm. Either raise the objective lens or lower the microscope stage as far as possible. Place your dry-mount prepared slide in the centre of the stage. Use the stage clips to hold the slide in position. Turn on the light source (Figure 2).

Figure 2

2. View the microscope stage from the side. Then, using the coarse-adjustment knob, lower the low-power objective lens until it is close to the object. (Some microscopes have moveable stages, rather than moveable lenses.) Do not allow the lens to touch the cover slip (Figure 3). Make sure that you know which way to turn the knob to raise the objective lens.

Figure 3

3. Look through the eyepiece. Slowly raise the objective lens using the coarse-adjustment knob until the image is in focus. Note that the object is facing the "wrong" way and is upside down. The area you can see is called the *field of view*.

4. Draw a circle in your notebook to represent the field of view. Look through the microscope and draw what you see. Make the object fill the same proportion of area in your drawing as it does in the microscope.

5. While you look through the microscope, slowly move the slide horizontally away from you. Note that the object appears to move toward you. Now move the slide to the left. Notice that the object appears to move to the right.

6. Rotate the nosepiece to the medium-power objective lens. Use the fine-adjustment knob to bring the object into focus. Notice that the object appears larger. Always use the fine adjustment when the medium- or high-power objectives are in place; the coarse adjustment may damage the slide or lenses.

7. Adjust the position of the object so that it is directly in the centre of the field of view. Rotate the high-power objective lens into place. Again, use the fine-adjustment knob to focus the image. Notice that you see less of the object than you did under medium-power magnification. Also notice that the object appears larger.

PUTTING AWAY THE MICROSCOPE

After you have completed an activity using a microscope, follow these steps:

1. Rotate the nosepiece to the low-power objective lens.
2. Raise the lenses (or lower the stage) as far as possible.
3. Remove the slide and the cover slip (if applicable).
4. Clean the slide and the cover slip and return them to their appropriate location.
5. Return the microscope, using two hands, to the storage area.

2.B. Testing for Electrical Conductivity

Before you start any conductivity testing, ask your teacher for specific operating instructions on your school's equipment (Figure 4). Devices used to test electrical conductivity vary considerably. Two metal electrodes are inserted into the sample to be tested.

Figure 4 You should use only low-voltage (battery-powered) conductivity testers.

In many cases, if the sample conducts electricity, the conductivity tester gives a positive result (e.g., a light bulb turns on). Some conductivity testers give variable results (Table 2).

Table 2 Results of Conductivity Testing

Observation	Sample
bright glow	good conductor
faint glow	poor conductor
no glow	non-conductor (insulator)

2.C. Using Star Maps

2.C.1. What Is a Star Map?

A star map is a map of the night sky that shows the relative positions of the stars in a particular part of the sky. Each star map is designed for a specific range of latitudes, such as latitudes 45° north of the equator. Thus, a star map designed for southern Canada cannot be used in Australia.

2.C.2. Maps for All Seasons

The visible parts of the sky vary according to time and date. Different star maps have been designed for different seasons or months (Figure 5 on the next page). Planispheres, which combine a star map with a frame, can be adjusted to display only the stars visible at a particular time and date. Planispheres are helpful tools that stargazers can use to recognize stars and constellations.

2.C.3. Stargazing Trips

When you want to observe the night sky, consider these tips:
- Plan your stargazing trip in advance, taking into consideration the weather forecast, safety, transportation, location, what to wear, and what to bring.
- Choose a location that is far away, or at least screened away, from bright lights.
- Be prepared to record your observations.
- Before viewing, allow your eyes at least 10 minutes to adapt to the dark.
- Use a flashlight covered with red cellophane to view your star map.

2.C.4. Using a Star Map

To use a seasonal star map, follow these steps:
- Select the star map for the appropriate month or season.
- Hold the map, facing downward, above your head.
- Rotate the map so the top part (marked "Facing North") is facing north.
- Compare what you see in the sky with what is on the map.
- Notice any planets or other objects, besides stars.

2.C.5. Keeping Records

Use a table to record your observations. Possible titles for the columns are shown in Table 3. Be careful when recording dates because, for example, December 15 becomes December 16 after midnight.

Table 3

Date	Object seen	Description (including a diagram)	Location (including angles)	Questions I want answered

2.C.6. Star Map

When to use this map:

January
early 10 pm
late 9 pm

February
early 8 pm
late dusk

also

October
early 5 am
late 4 am

November
early 2 am
late 1 am

December
early 12 am

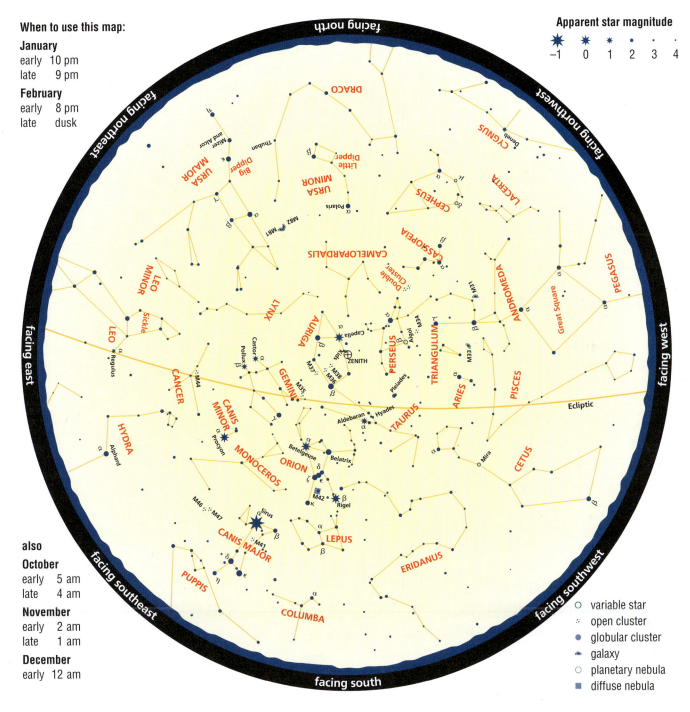

Figure 5 A star map drawn specifically for 45° north of the equator

2.D. Drawing and Constructing Circuits

2.D.1. Sources of Electrical Energy

To provide the electrical energy in most of the circuits you create, you will be using a combination of dry cells or a special device called a power supply.

The source in the circuits you construct and test will be a direct current, or DC, source of electrical energy. In a DC circuit, the current flows in only one direction.

Wall outlets provide a different kind of electrical energy known as an alternating current, or AC, source. In an AC circuit, the electric current reverses its direction 60 times a second.

2.D.2. Drawing Circuit Diagrams

There are some conventions to follow when drawing a circuit diagram: connecting wires are generally shown as straight lines or 90° angles, and symbols (Table 4) are used to represent all the components.

2.D.3. Safety Considerations

It is important to follow safety procedures, especially when you see a WHMIS safety symbol.

- Always ensure that your hands are dry and that you are standing on a dry surface.
- Do not use faulty dry cells or batteries, and do not connect different makes of dry cells in the same battery. Avoid connecting partially discharged dry cells to fully charged cells.
- Do not use frayed or damaged connectors.
- Only operate a circuit after it has been approved by your teacher.

2.D.4. Constructing a Circuit

When working with circuits, always follow the instructions. If you are unsure of the instructions, ask for help. Check that all the components are working properly.

- Check the connections carefully when linking connecting dry cells in series or in parallel. Incorrect connections could cause shorted circuits or explosions. Ask your teacher for clarification if you are unsure.
- When attaching connecting wires to a meter, connect a red wire to the positive terminal and a black wire to the negative terminal of the meter.
- Sometimes the ends of connecting wires do not have the correct attachments to connect to a device or meter. Use approved connectors, such as alligator clips, but be careful to position the connectors so that they cannot touch one another.
- Open the switch before altering a meter connection or adding new wiring or components.
- If a circuit does not operate correctly, open the switch and check the wiring and all the connections to the terminals. If you cannot find the problem, ask your teacher to inspect the circuit again.

Table 4 Circuit Diagram Symbols

	Part of circuit	Symbol
sources of electrical energy	electric cell	
	three-cell battery	
	AC power supply	
	variable DC power supply	
electrical conductor	connecting wire	
control device	switch (open)	switch (open)
	switch (closed)	switch (closed)
loads	lamp	
	electric motor	
	resistor	
meters	ammeter	
	voltmeter	
	ohmmeter	

2.E. Using the Voltmeter, Ammeter, and Digital Multimeter

We rely on instruments that can detect and measure electricity. Two common instruments are the voltmeter and the ammeter. These instruments may be digital (providing a digital readout) or analog (indicating voltage or current by the movement of a needle across a scale). You may also use a digital multimeter, which measures voltage, current, and resistance.

2.E.1. The Voltmeter

A voltmeter measures the voltage difference between two different points in a circuit. A voltmeter can be connected across the terminals of a cell to measure the voltage output of the cell, or across another component of a circuit to measure the voltage drop across this component. In other words, a voltmeter is always connected in parallel with the component you want to investigate.

2.E.2. Reading an Analog Voltmeter

The needle on a voltmeter usually moves from left to right, with the zero voltage on the left side of the scale and the maximum voltage on the right. If the voltmeter scale has only one set of numbers, it is relatively easy to measure the voltage. If the voltmeter has several sets of numbers, identify which set of numbers matches the voltage range selected by the switch on the voltmeter.

The negative terminal of the voltmeter must be connected to the more negative part of the circuit. If the leads are attached incorrectly, the needle will be unable to move and give a reading.

2.E.3. The Ammeter

An ammeter measures the amount of electric current that is flowing in a circuit. To measure the electric current, connect the ammeter directly into the circuit. Disconnect a wire from the part of the circuit you wish to measure and connect the ammeter, in series, to complete the circuit. Reading an analog ammeter is very similar to reading an analog voltmeter. The unit of current is the ampere (A) or milliampere (mA).

2.E.4. The Digital Multimeter

A digital multimeter can measure the voltage, current, or resistance in a circuit (Figure 6). Using the selector knob, select the electrical value and range that you wish to measure. Then connect the wire leads properly to the appropriate place in the circuit.

Remember to connect the meter in parallel for measuring voltage or in series for measuring current. The multimeter will provide a digital readout of the measurement of the electrical value.

Figure 6 A digital multimeter

2.F. Using Other Scientific Equipment

For some labs in this textbook, a list of required equipment and materials is provided. In others, you have to decide what equipment and materials are necessary. Always keep safety in mind as you make your selections. Be sure to include appropriate safety equipment, such as eye protection, gloves, or lab aprons, in your planning. For more information on safety, refer to the Safe Science section (page 598).

Figure 7 shows some of the equipment required for the labs in this book.

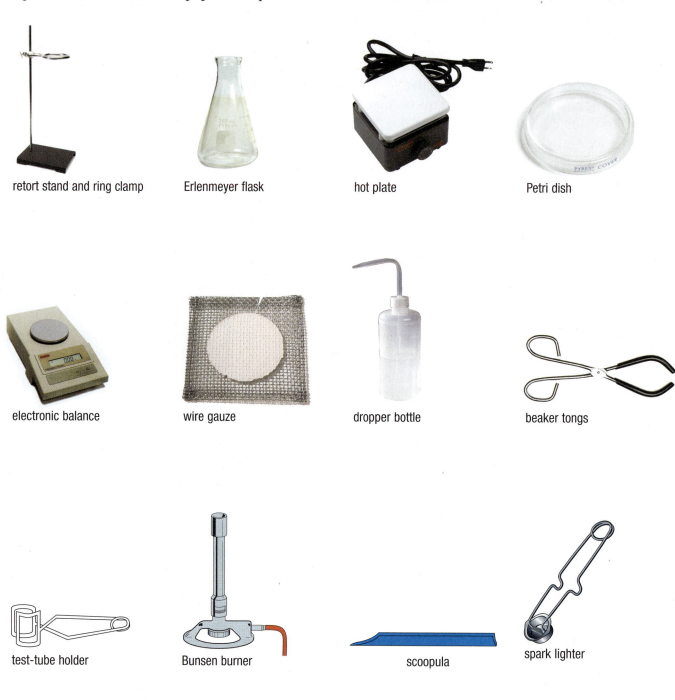

spectroscope

Figure 7 Some typical laboratory equipment

SCIENTIFIC INQUIRY SKILLS 3

3.A. Thinking as a Scientist

Imagine that you are planning to buy a new personal electronic device. First, you write a list of questions. Then you check print and Internet sources, visit stores, and talk to your friends to find out which is the best purchase. When you try to solve problems in this way, you are conducting an investigation and thinking like a scientist.

- Scientists investigate the natural world to describe it. For example, climatologists study the growth rings of trees to make inferences about what the climate was like in the past (Figure 1).

Figure 1 Looking at and measuring tree rings tells scientists how climate affects the growth of trees.

- Scientists investigate objects to classify them. For example, chemists classify substances according to their properties (Figure 2).

Figure 2 These compounds are similar because they are insoluble in water.

- Scientists investigate the natural world to test their ideas about it. For example, biologists ask cause-and-effect questions about the impact of climate change on water in the Arctic (Figure 3). They also propose hypotheses to answer their questions. Then they design investigations to test their hypotheses. This process leads scientists to come up with new ideas to be tested and more questions that need answers.

Figure 3 Water samples can be tested for pH, dissolved substances, and other impurities.

3.B. Scientific Inquiry

You need to use a variety of skills when you do scientific inquiry and design or carry out an experiment. Refer to this section when you have questions about any of the following skills and processes:

- questioning
- hypothesizing
- predicting
- planning
- controlling variables
- performing (Figure 4)
- observing
- analyzing
- evaluating
- communicating

Figure 4 In each unit you will have an opportunity to develop new skills of scientific inquiry.

3.B.1. Questioning

Scientific investigations result from our curiosity about the natural world. Our observations lead us to wonder "Why does that happen?" or "What would happen if I ...?" Before you can conduct an investigation, you must have a clear idea of what you want to know. This will help you develop a question that will lead you to the information you want. Try to avoid questions that lead simply to "yes" and "no" answers. Instead, develop questions that are testable and lead to an investigation.

Sometimes, an investigation starts with a special type of question called a "cause-and-effect" question. A cause-and-effect question asks whether something is causing something else to happen. It might start in one of the following ways: What causes ...? How does ... affect ...?

3.B.2. Controlling Variables

Considering the variables involved is an important step in designing an effective investigation. Variables are any factors that could affect the outcome of an investigation.

There are three kinds of variables in an investigation:

1. The variable that is changed is called the independent variable, or cause variable.
2. The variable that is affected by a change is called the dependent variable, or effect variable. This is the variable you measure to see how it was affected by the independent variable.
3. All the other conditions that remain unchanged and did not affect the outcome of the investigation are called the controlled variables.

For example, consider the variables involved in an investigation to determine the effects of the mass of dissolved salt on the boiling point of water (Figure 5). In this investigation,
- the mass of dissolved salt is the independent variable
- the boiling point is the dependent variable
- the amount of water in the beaker and the type of salt used are two controlled variables

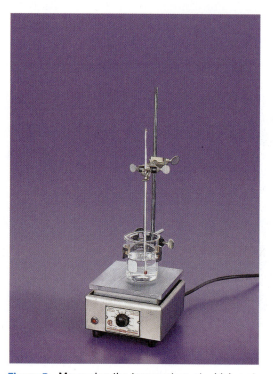

Figure 5 Measuring the temperature at which water, containing a known mass of salt, boils.

When scientists conduct controlled experiments (described in Section 1.1), they make sure that they change only one independent variable at a time. This way, they may assume that their results are caused by the variable they changed and not by any of the other variables they identified.

610 Skills Handbook

3.B.3. Predicting and Hypothesizing

A prediction states what is likely to happen as the result of a controlled experiment. Scientists base predictions on their observations and knowledge. They look for patterns in the data they gather to help them understand what might happen next or in a similar situation (Figure 6). A prediction may be written as an "if … then …" statement. For this investigation, your prediction might be, "If the amount of dissolved salt is increased, then the boiling point will also increase."

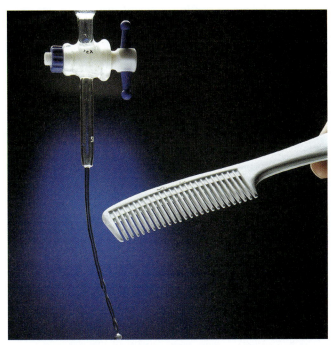

Figure 6 A positively charged piece of plastic causes a stream of tap water to bend. What would you predict might happen if a negatively charged object is brought near the stream of water?

In summary, a prediction states what you think will happen. Remember, however, that predictions are not guesses. They are suggestions based on prior knowledge and logical reasoning.

A prediction can be used to generate a hypothesis. A hypothesis is a tentative answer about the outcome of a controlled experiment along with an explanation for the outcome. A hypothesis may be written in the form of an "if … then … because …" statement. *If* the cause variable is changed in a particular way, *then* the effect variable will change in a particular way, and this change occurs *because* of certain reasons.

For example, "If the amount of salt is increased, then the boiling point will also increase because salt forms attractions with water molecules and prevents them from changing into gas." If your observations confirm your prediction, then they support your hypothesis. You can create more than one hypothesis from the same question or prediction. Another student might test the hypothesis, "If the amount of salt is increased, then the boiling point will not change because the attractions that salt forms with water are very weak." Of course, both of you cannot be correct. When you conduct an investigation, your observations do not always confirm your prediction. Sometimes, they show that your hypothesis is incorrect. An investigation that does not support your hypothesis is not a bad investigation or a waste of time. It has contributed to your scientific knowledge. You can re-evaluate your hypothesis and design a new investigation.

3.B.4. Planning

You have been asked to design and carry out the boiling point investigation described in 3.B.3. First, you create an experimental design. To conduct a controlled experiment that tests your hypothesis, you decide to change only one variable—the amount of dissolved salt. This is your independent variable. You dissolve 5.0 g, 10.0 g, 15.0 g, and 20.0 g samples of salt in identical beakers containing exactly 100 mL of water. You also prepare a beaker containing water with no salt. Each solution is then heated using the same hot plate until it boils. A thermometer that is suspended in each solution measures the temperature at which boiling occurs—the dependent variable.

Now consider the equipment and materials you need. Be sure to include any safety equipment, such as an apron or eye protection, in your equipment and materials list. How will you secure the beaker so that it doesn't accidentally fall off the hot plate as it boils? Will you use tap water or distilled water?

You need to write a procedure—a step-by-step description of how you will perform your investigation. A procedure should be written as a series of numbered steps, with only one instruction per step. For example:

1. Put on safety goggles, a protective apron, and heat-resistant gloves.
2. Set up the equipment as shown in the diagram. (Include a clearly labelled diagram.)
3. Add 100 mL of distilled water to the beaker.
4. Add one boiling chip to the beaker to ensure that the water boils gently and does not bump.
5. Dissolve 5.0 g of sodium chloride in the water.
6. Set the heating control on the hot plate to 50 %.
7. Wait for the mixture to boil.
8. Turn off the heat and allow the beaker to cool.
9. Repeat steps 3–8 using 10.0 g, 15.0 g, and 20.0 g of sodium chloride.
10. Allow the apparatus to cool before dismantling it.
11. Once cool, return all equipment to the storage cabinet.

Your procedure must be clear enough for someone else to follow, and it must explain how you will deal with each of the variables in your investigation. The first step in a procedure usually refers to any safety precautions that need to be addressed, and the last step relates to any cleanup that needs to be done. Your teacher must approve your procedure and list of equipment and materials before you begin.

3.B.5. Performing

As you carry out an investigation, be sure to follow the steps in the procedure carefully and thoroughly. Check with your teacher if you find that significant alterations to your procedure are required. Use all equipment and materials safely, accurately, and with precision. Be sure to take detailed, careful notes and to record all of your observations. Record numerical data in a table.

3.B.6. Observing

When you observe something, you use your senses to learn about it. You can also use tools, such as a balance or a microscope. Observations of measured quantities such as temperature, volume, and mass are called quantitative observations. Numerical data from quantitative observations are usually recorded in data tables or graphs.

Other observations describe characteristics that cannot be expressed in numbers. These are called qualitative observations (Figure 7). Colour, smell, clarity, and state of matter are common examples of qualitative observations. Qualitative observations can be recorded using words, pictures, or labelled diagrams.

Figure 7 Qualitative observations such as a colour change, bubbles, an irritating odour, or a sizzling sound suggest that a chemical change is occurring when certain chemicals are mixed.

As you work through an investigation, be sure to record all of your observations, both qualitative and quantitative, clearly and carefully. If a data table is appropriate for your investigation, use it to organize your observations and measurements. (See Data Tables and Graphs, page 632.) Include all observations and measurements in your final lab report or presentation. It is important to remain impartial when recording observations. Record exactly what you observe. Observations from an experiment may not always be what you expect them to be.

SCIENTIFIC DRAWINGS

Scientific drawings are done to record observations as accurately as possible. They are also used to communicate, which means that they must be clear, well labelled, and easy to understand. Below are some tips that will help you produce useful scientific drawings.

Getting Started

The following materials and ideas will help you get started:

- Use plain, blank paper. Lines might obscure your drawing or make your labels confusing.
- Use a sharp pencil rather than a pen or marker as you will probably need to erase parts of your drawing and do them over again (Figure 8).
- Observe and study your specimen or equipment carefully, noting details and proportions, before you begin the drawing.

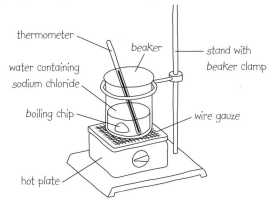

Figure 8 Your drawing of an experimental setup should show how the equipment was assembled.

- Create drawings large enough to show details. For example, a third of a page might be appropriate for a diagram of a single cell or a unicellular organism. When drawing lab equipment, do not include unnecessary details.
- Label your diagram clearly. Use a ruler to draw the label lines.

Scale Ratio

You may want to indicate the actual size of your object on your drawing. To do this, use a ratio called the scale ratio.

- If your diagram is 10 times larger than the real object (e.g., a tiny organism), your scale ratio is 10×.
- In general, scale ratio = $\dfrac{\text{size of drawing}}{\text{actual size of object}}$

You can also show the actual size of the object on your diagram (Figure 9).

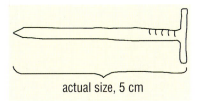

Figure 9 This is an example of a scientific drawing that indicates the actual size of the object.

Checklist for Scientific Drawing

✔ Use plain, blank (unlined) paper and a sharp, hard pencil.

✔ Draw as large as necessary to show details clearly.

✔ Do not shade or colour.

✔ Draw label lines that are straight and run outside your drawing. Use a ruler.

✔ Include labels, a title or caption, and, if appropriate, the magnification of the microscope you are using.

HORIZON DIAGRAMS

Horizon diagrams are a way to record your observations of celestial objects relative to the landmarks on the horizon (Figure 10). When you go out to observe the night sky, make sure you are accompanied by an adult. Refer to the section on Using Star Maps on page 604 for tips on planning a stargazing trip.

The following steps will help you create a horizon diagram:

- On a blank sheet of paper, draw a line at the bottom to represent the horizon.
- Sketch any landmarks, such as trees or buildings, you observe along the horizon.
- Sketch all the objects you observe in the night sky, positioning them relative to the landmarks.
- Record the date, exact time, and cloud conditions of your observations. Include a detailed description of the location where you made your observations.

Figure 10

3.B.7. Analyzing

You analyze data from an investigation to make sense of it. You examine and compare the measurements you have made. You look for patterns and relationships that will help you explain your results and give you new information about the question you are investigating.

Once you have analyzed your data, you can tell whether your prediction or hypothesis is correct. You can also write a conclusion that indicates whether or not the data supports your hypothesis (Figure 11). You may even come up with a new hypothesis that can be tested in a new investigation.

Figure 11 Students analyze data and make notes about their observations in order to find any patterns.

3.B.8. Evaluating

How useful is the evidence from an investigation? You need quality evidence before you can evaluate your prediction or hypothesis. If the evidence is poor or unreliable, you can identify areas of improvement for when the investigation is repeated.

Below are some things to consider when evaluating the results of an investigation:

- *Plan:* Were there any problems with the way you planned your experiment or your procedure? Did you control for all the variables, except the independent variable?
- *Equipment and Materials:* Could better or more accurate equipment have been used? Was something used incorrectly? Did you have difficulty with a piece of equipment?
- *Observations:* Did you record all the observations that you could have? Or did you ignore or overlook some observations that might have been important?
- *Skills:* Did you have the appropriate skills for the investigation? Did you have to use a skill that you were just beginning to learn?

Once you have identified areas in which errors could have been made, you can judge the quality of your evidence.

3.B.9. Communicating

When you plan and carry out your own investigation, it is important to share both your process and your findings. Other people may want to repeat your investigation, or they may want to use or apply your findings in another situation. Your write-up or report should reflect the process of scientific inquiry that you used in your investigation.

SAMPLE LAB REPORT

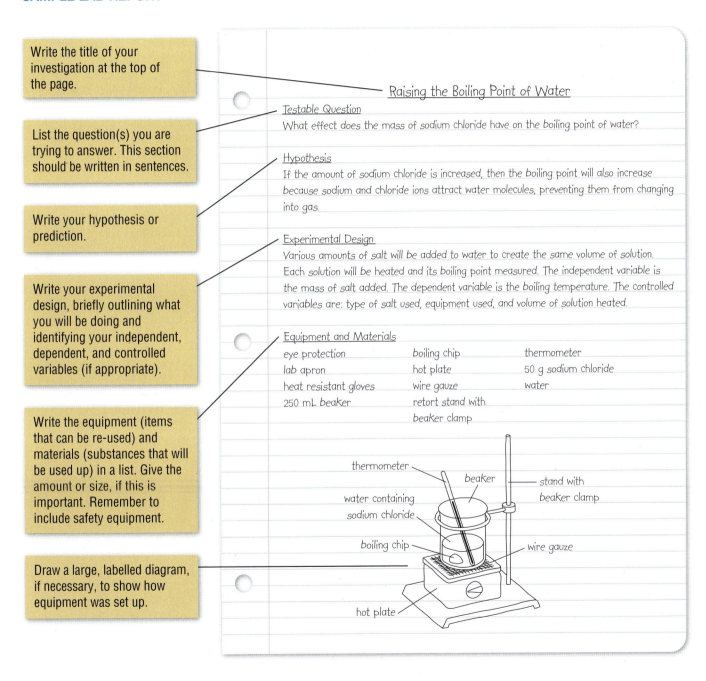

Write the title of your investigation at the top of the page.

List the question(s) you are trying to answer. This section should be written in sentences.

Write your hypothesis or prediction.

Write your experimental design, briefly outlining what you will be doing and identifying your independent, dependent, and controlled variables (if appropriate).

Write the equipment (items that can be re-used) and materials (substances that will be used up) in a list. Give the amount or size, if this is important. Remember to include safety equipment.

Draw a large, labelled diagram, if necessary, to show how equipment was set up.

SAMPLE LAB REPORT (CONTINUED)

Procedure
1. Safety goggles, a protective apron, and heat resistant gloves were obtained.
2. The equipment was set up as shown in the diagram.
3. 100 mL distilled water was added to the beaker.
4. One boiling chip was added to the beaker to prevent the solution from bumping.
5. 5.0 g sodium chloride was dissolved in the water.
6. The hot plate heater control was set to 50 %.
7. The temperature at which the water boiled was recorded.
8. The heat was turned off and the beaker was allowed to cool.
9. Steps 3–8 were repeated using 10.0 g, 15.0 g, and 20.0 g of sodium chloride.
10. The apparatus was allowed to cool before it was returned.

Observations
Table of Observations

Mass of sodium chloride (g)	Boiling point (°C)
5	100.4
10	100.8
15	101.4
20	101.8

In each case, the water boiled gently. Small bubbles formed throughout the solution as it boiled.

Analysis and Evaluation
The boiling point of the solution increased as more salt was dissolved in it. Water condensing on the thermometer made some of the temperatures difficult to read. This problem could be overcome by using a temperature probe instead of a thermometer. The data clearly support the hypothesis that the boiling point of water increases as the mass of dissolved sodium chloride increases.

Applications and Extensions
Adding table salt (sodium chloride) to boiling water does not increase the boiling point of the water a great deal. Therefore, adding a little salt to boiling water when cooking will not speed up the cooking process.

Describe the procedure using numbered steps. Each step should start on a new line. Write the steps as they occurred, using past tense and passive voice. Make sure that your steps are clear so that someone else could repeat your investigation. Include any safety precautions.

Present your observations in a form that is easily understood. Quantitative observations should be recorded in one or more tables, with units included. Qualitative observations can be recorded in words or drawings.

Analyze your results and evaluate your procedure. If you have created graphs, refer to them here. If necessary, include them on a separate piece of graph paper. Write a conclusion indicating whether or not your results support your hypothesis or prediction. Answer any Analyze and Evaluate questions here.

Describe how the knowledge you gained from your investigation relates to real-life situations. How can this knowledge be used? Answer any Apply and Extend questions here.

RESEARCH SKILLS 4

In modern society, you are constantly bombarded with information. Some of this information is reliable, and some is not. Trying to find the "right" information to conduct scientific research may seem overwhelming. However, the task is less daunting when you learn how to search efficiently for the information you need. Then you must know how to assess its credibility. Here are some tips that will help you in your research.

4.A. General Research Skills

4.A.1. Identify the Information You Need

- Identify your research topic.
- Identify the purpose of your research.
- Identify what you already know about the topic.
- Identify what you do not yet know.
- Develop a list of key questions that you need to answer.
- Identify categories based on your key questions.
- Use these categories to identify key search words.

4.A.2. Identify Sources of Information

Identify places where you could look for information about your topic. These places might include programs on television, people in your community, print sources (Figure 1), and electronic sources (such as CD-ROMs and Internet sites).

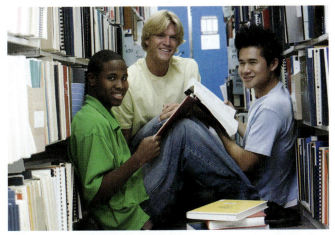

Figure 1 Your school library and local public library are both excellent sources of information.

Refer to "Using the Internet" on page 618. Remember, gathering information from a variety of sources will improve the quality of your research.

4.A.3. Evaluate the Sources of Information

Read through your sources of information and decide whether they are useful and reliable. Here are five things to consider:

- *Authority:* Who wrote or developed the information or who sponsors the website? What are the qualifications of this person or group?
- *Accuracy:* Are there any obvious errors or inconsistencies in the information? Does the information agree with that of other reliable sources?
- *Currency:* Is the information up to date? Has recent scientific information been included?
- *Suitability:* Does the information make sense to someone with your experience or of your age? Do you understand it? Is the information well organized?
- *Bias:* Are facts reported fairly? Are there reasons why your sources might express some bias? Are facts deliberately left out?

4.A.4. Record and Organize the Information

After you have gathered and evaluated your sources of information, you can start organizing your research. Identify categories or headings for note taking. In your notebook, use point-form notes to record information in your own words under each heading. You must be careful not to copy information directly from your sources. If you quote a source, use quotation marks. Record the title, author, publisher, page number, and date for each of your sources. For websites, record the URL (website address). All of these details are necessary to help you keep track of your sources of information so that you can go back to them to clarify any points in the future. You will also need this information to create a bibliography. If necessary, add to your list of questions as you find new information.

To help you organize the information further, you may want to use pictures, graphic organizers, and diagrams. (See the Literacy section on page 639.)

4.A.5. Make a Conclusion

Look at your original research question. What did you learn from the information you gathered? Can you state and explain a conclusion based on that information? Do you need further information? If so, where would you look for that information? Do you have an informed opinion on the research topic that you did not have before you started? If not, what additional information do you need to reach such an opinion?

4.A.6. Evaluate Your Research

Now that your research is complete, reflect on how you gathered and organized the information (Figure 2). Can you think of ways to improve the research process for next time? How valuable were the sources of information you selected?

Figure 2 Keep track of your sources so you do not forget where you found your information.

4.A.7. Communicate Your Conclusions

Choose a format for communication that suits your audience, your purpose, and the information you gathered. Are labelled diagrams, graphs, or charts appropriate?

4.B. Using the Internet

The Internet is a vast and constantly growing network of information. You can use search engines to help you find what you need, but keep in mind that not everything you find will be useful, reliable, or true.

4.B.1. Search Results

Once you have entered your search word or phrase, a list of web page "matches" will appear. If your keywords are general, you are likely to get a high number of matches. Therefore, you need to refine your search. Most search engines provide online help and search tips. Look at these to find ideas for better searching.

Every web page has a URL (universal resource locator). The URL may tell you the name of the organization hosting the web page, or it may indicate that you are looking at a personal page (often indicated by the ~ character in the URL). The URL also includes a domain name, which provides clues about the organization hosting the web page (Table 1). For example, a URL that includes "ec.gc.ca" indicates that the content is hosted by Environment Canada—a reliable source.

Table 1 Common URL Codes and Organizations

Code	Organization
ca	Canada
com or co	commercial
edu or ac	educational
org	nonprofit
net	networking provider
mil	military
gov	government
int	international organization

4.B.2. Evaluating Internet Resources

Anyone can post information on the Internet without verifying its accuracy. Therefore, you must learn to evaluate the information that you find on the Internet as coming from dependable and legitimate sources.

Use the following questions to help determine the quality of an Internet source. The greater the number of questions answered "yes," the more likely it is that the source is of high quality.

- Is it clear who is sponsoring the page? Does the site seem to be permanent or sponsored by a reputable organization?
- Is there information about the sponsoring organization? For example, is a telephone number or address given to contact for more information?
- Is it clear who developed and wrote the information? Are the author's qualifications provided?
- Are the sources for factual information given so that they can be checked?
- Are there dates to indicate when the page was written, placed online, or last revised?
- Is the page presented as a public service? Does it present balanced points of view?

4.B.3. Using School Library Resources

Many schools and school boards have access to online encyclopedias with science sections in them. Find out if your school or board has a website where you can access these resources. You may need a password.

4.B.4. Using the Nelson Website

When you see the Nelson Science icon in your textbook, you can go to the Nelson website and find links to useful sources of information.

 GO TO NELSON SCIENCE

4.C. Exploring an Issue Critically

An issue is a situation in which several points of view need to be considered in order to make a decision. It is often difficult to come to a decision that everyone agrees with. When a decision affects many people or the environment, it is important to explore the issue critically. Think about all the possible solutions and try to understand all the different perspectives—not just your own point of view. Consider the risks and benefits of each possible solution. Put yourself in the place of several of the stakeholders, to try to understand their positions.

Exploring an issue critically also means researching and investigating your ideas and communicating with others. Figure 3 shows all the steps in the process.

Figure 3 You may perform some or all of these steps as you explore an issue critically.

4.C.1. Defining the Issue

To explore an issue, first identify what the issue is. An issue has more than one solution, and there are different points of view about which solution is the best. Rephrase the issue as a question: "What should …?" The issue can also include information about the *role* a person takes when thinking about an issue. For instance, you may think about the issue from someone else's point of view—you may take the role of a landowner, a government worker, or a tour guide. The issue can also include a description of who your audience will be—will it be other students, a meeting of government officials, or your parents? Be sure to take into account your role and audience when defining your issue.

4.C.2. Researching

Ensure that the decision you reach is based on a good understanding of the issue. You must be in a position to choose the most appropriate solution. To do this, you need to gather factual information that represents all the different points of view. Develop good questions and a plan for your research. Your research may include talking to people, reading about the topic in newspapers and magazines, and doing Internet research.

As you collect information, make sure that it is reliable, accurate, and current. Avoid biased information that favours only one side of the issue. It is important to ensure that the information you have gathered represents all aspects of an issue. Are the sources valuable? Could you find better information elsewhere?

4.C.3. Identifying Alternatives

Consider possible solutions to the issue. Different stakeholders may have different ideas on this. Consider all reasonable options. Be creative about combining the suggestions. For example, suppose that your municipal council is trying to decide how to use some vacant land next to your school. You and other students have asked the council to use the land as a nature park. Another group is proposing that the land be used to build a seniors' home because there is a shortage of this kind of housing. The school board would like to use the land to build a track for sporting events.

After defining the issue and researching, you can now generate a list of possible solutions. You might, for example, come up with the following choices for the land-use issue:

- Turn the land into a nature park for the community and the school.
- Use the land as a playing field and track for the community and the school.
- Create a combination of a nature park and a playing field.
- Use the land to build a seniors' home with a nature park.

4.C.4. Analyzing the Issue

Develop criteria to evaluate each possibility. For example, should the solution be the one that has the most community support, or should it be the one that best protects the environment? Should it be the least costly, financially, or the one that creates the most jobs? You need to decide which criteria you will use to evaluate the alternatives so that you can decide which solution is the best.

4.C.5. Defending a Decision

This is the stage where everyone gets a chance to share ideas and information gathered about the issue. Then the group needs to evaluate all the possible alternatives and decide on one solution based on the criteria.

COST–BENEFIT ANALYSIS

A cost–benefit analysis can help you determine the best solution to a complex problem when a number of solutions are possible. First, research possible costs and benefits associated with a proposed solution. Costs are not always financial. You may be comparing advantages and disadvantages. Then, based on your research, try to decide the level of importance of each cost and benefit. This is often a matter of opinion. However, your opinions should be informed by researched facts.

Once you have completed your research and identified costs and benefits, you may conduct the cost–benefit analysis as follows:

1. Create a table similar to Table 2.
2. List costs and benefits.
3. Rate each cost and benefit on a scale from 1 to 5, where 1 represents the least important cost or benefit and 5 represents the most important cost or benefit.
4. After rating each cost and benefit, add up the results to obtain totals. If the total benefits outweigh the total costs, you may decide to go ahead with the proposed solution.

Table 2 Cost–Benefit Analysis of Using Land to Build a Seniors' Home with a Nature Park

Costs		Benefits	
Possible result	Cost of result (rate 1 to 5)	Possible result	Benefit of result (rate 1 to 5)
land cannot be used for sports	2	seniors' home provides necessary housing	5
expensive to maintain	4	park preserves some habitat for plants and animals	4
nature park will be very small	3	park increases value of seniors home	3
Total cost value	9	**Total benefit value**	12

4.C.6. Communicating

You might be told how you will communicate your decision. For example, your class might hold a formal debate. Alternatively, you might be free to choose your own method of communication.

You could choose one of the following methods of communicating your decision:
- Write a report.
- Give an oral presentation.
- Make a poster.
- Prepare a slide show.
- Create a video (Figure 4).
- Organize a town hall or panel discussion.
- Create a blog or webcast.
- Write a newspaper article.

Figure 4 Creating a video is an effective way to communicate information about an issue.

Choose a type of presentation that will share your decision or recommendation in a way that is suitable for your audience. For example, if your audience is small, it might be easiest to present your decision in person. An oral presentation is a good way to present your decision to many people at one time. If your presentation includes visuals, ensure that they are large and clear enough for your audience to see. Creating a poster or a blog allows people to read your recommendation on their own, but you must find a way to let others know where to find this information.

Whatever means you use, however, you should

- state your position clearly, considering your audience;
- support it with objective data if possible, and with a persuasive argument; and
- be prepared to defend your position against opposition (Figure 5).

Figure 5 Share your information in a way that makes it easy for your audience to understand.

4.C.7. Evaluating

The final step of the decision-making process includes evaluating the decision itself and the process used to reach the decision. After you have made a decision, carefully examine the thinking that led to this decision. Some questions to guide your evaluation include the following:

- What was my initial perspective on the issue? How has my perspective changed since I first began to explore the issue?
- How did I gather information about the issue? What criteria did I use to evaluate the information? How satisfied am I with the quality of my information?
- What information did I consider to be the most important when making my decision?
- How did I make my decision? What process did I use? What steps did I follow?
- To what extent were my arguments factually accurate and persuasively made? (Figure 6)
- In what ways does my decision resolve the issue?
- What are the likely short-term and long-term effects of my decision?
- How might my decision affect the various stakeholders?
- To what extent am I satisfied with my decision?
- If I had to make this decision again, what would I do differently?

Figure 6 Were the arguments clearly presented and backed up by evidence?

USING MATHEMATICS IN SCIENCE

Effective communication of experimental data is an important part of science. To avoid confusion when reporting measurements or using measurements in calculations, there are a few accepted conventions and practices that should be followed.

5.A. SI Units

The scientific communities of many countries, including Canada, have agreed on a system of measurement called SI (Système international d'unités). This system consists of the seven fundamental SI units, called base units, shown in Table 1.

Table 1 The Seven SI Base Units

Quantity name	Unit name	Unit symbol
length	metre	m
mass	kilogram	kg*
time	second	s
electric current	ampere	A
temperature	kelvin	K**
amount of substance	mole	mol
light intensity	candela	cd

*The kilogram is the only base unit that contains a prefix.
**Although the base unit for temperature is a kelvin (K), the common unit for temperature is a degree Celsius (°C).

All other physical quantities can be expressed as a combination of these seven SI base units. For example, the speed of an object is determined by the distance it travels in a specified time period. Therefore, the unit for speed is metres (distance) per second (time) or m/s. Units that are formed using two or more base units are called derived units. Some derived units have special names and symbols. For example, the unit of force that causes a mass of 1 kg to accelerate at a rate of 1 m/s² (metre per second per second) is known as a newton (N). In base units, the newton is m·kg/s². The dot between m and kg means "multiplied by," but m·kg is simply read as "metre kilogram." The slash means "divided by" and is read "per." The whole unit is read "metre kilogram per second squared." You can see why a special name and symbol are given to some derived units.

Some common quantities and their units are listed in Table 2. Note that the symbols representing the quantities are italicized, whereas the unit symbols are not.

Table 2 Common Quantities and Units

Quantity name	Quantity symbol	Unit name	Unit symbol
distance	d	metre	m
area	A	square metre	m²
volume	V	cubic metre	m³
		litre	L
speed	v	metre per second	m/s
acceleration	a	metre per second per second	m/s²
concentration	c	gram per litre	g/L
temperature	t	degree Celsius	°C
pressure	p	pascal	Pa
energy	E	joule	J
work	W	joule	J
power	P	watt	W
electric potential	V	volt	V
electrical resistance	R	ohm	Ω
current	I	ampere	A

5.A.1 Converting Units

An important feature of SI is the use of prefixes to express small or large sizes of any quantity conveniently. SI prefixes act as multipliers to increase or decrease the value of a number in multiples of 10 (Table 3). The most common prefixes change the size in multiples of 1000 (10^3 or 10^{-3}), except for *centi* (10^{-2}), as in centimetre.

SI prefixes are also used to create conversion factors (ratios) to convert between larger or smaller values of a unit. For example,

1 km = 1000 m

Therefore, $\dfrac{1 \text{ kg}}{1000 \text{ g}} = \dfrac{1000 \text{ g}}{1 \text{ kg}} = 1$

Multiplying by a conversion factor is like multiplying by 1: it does not change the quantity, only the unit in which it is expressed. Let's see how to convert from one unit to another.

Table 3 Common SI Prefixes

Prefix	Symbol	Factor by which unit is multiplied	Example
giga	G	1 000 000 000	1 000 000 000 m = 1 Gm
mega	M	1 000 000	1 000 000 m = 1 Mm
kilo	k	1 000	1 000 m = 1 km
hecto	h	100	100 m = 1 hm
deca	da	10	10 m = 1 dam
		1	
deci	d	0.1	0.1 m = 1 dm
centi	c	0.01	0.01 m = 1 cm
milli	m	0.001	0.001 m = 1 mm
micro	µ	0.000 001	0.000 001 m = 1 µm
nano	n	0.000 000 001	0.000 000 001 m = 1 nm

SAMPLE PROBLEM 1 Using Conversion Factors

A block of cheese at a grocery store has a mass of 1 256 g. Its price is $15.00/kg. What is the price of the block of cheese?

First, convert the mass from grams to kilograms.
There are two possible conversion factors between g and kg, as shown above.
Always choose the conversion factor that cancels the original unit. In this case, the original unit is g, so the correct conversion factor is $\dfrac{1 \text{ kg}}{1000 \text{ g}}$

$1256 \text{ g} \times \dfrac{1 \text{ kg}}{1000 \text{ g}} = 1.256 \text{ kg}$

The original units, g, cancel (divide to give 1), leaving kg as the new unit.
Now that you have determined the mass in kg, multiply the price per kg by the mass in kg.

$15.00/\text{kg} \times 1.256 \text{ kg} = \$18.84$

The price of the block of cheese is $18.84.

PRACTICE

Conversions

Make the following conversions. Refer to Table 3 if necessary.

(a) Write 3.5 s in ms.
(b) Change 5.2 A to mA.
(c) Convert 7.5 µg to ng.

Convenient conversion factors to convert between millimetres and metres are

$\dfrac{1000 \text{ mm}}{1 \text{ m}}$ and $\dfrac{1 \text{ m}}{1000 \text{ mm}}$

Conversion factors can be used for any unit equality, such as 1 h = 60 min and 1 min = 60 s.

5.B. Solving Numerical Problems Using the GRASS Method

In science and technology, you sometimes have problems that involve quantities (numbers), units, and mathematical equations. An effective method for solving these problems is the GRASS method. This always involves five steps: Given, Required, Analysis, Solution, and Statement.

Given: Read the problem carefully and list all of the values that are given. Remember to include units.

Required: Read the problem again and identify the value that the question is asking you to find.

Analysis: Read the problem again and think about the relationship between the given values and the required value. There may be a mathematical equation you could use to calculate the required value using the given values. If so, write the equation down in this step. Sometimes it helps to sketch a diagram of the problem.

Solution: Use the equation you identified in the "Analysis" step to solve the problem. Usually, you substitute the given values into the equation and calculate the required value. Do not forget to include units and to round off your answer to an appropriate number of digits. (See Sections 5.C. and 5.D. on significant digits and scientific notation for help.)

Statement: Write a sentence that describes your answer to the question you identified in the "Required" step.

5.C. Scientific Notation

Scientists often work with very large or very small numbers. Such numbers are difficult to work with when they are written in common decimal notation. For example, the speed of light is about 300 000 000 m/s. There are many zeros to keep track of if you have to multiply or divide this number by another number.

Sometimes it is possible to change a very large or very small number, so that the number falls between 0.1 and 1000, by changing the SI prefix. For example, 237 000 000 mm can be converted to 237 km, and 0.000 895 kg can be expressed as 895 mg.

Alternatively, very large or very small numbers can be written using scientific notation. Scientific notation expresses a number by writing it in the form $a \times 10^n$, where the letter a, referred to as the coefficient, is a value that is at least 1 and less than 10. The number 10 is the base, and n represents the exponent. The base and the exponent are read as "10 to the power of n." Powers of 10 and their decimal equivalents are shown in Table 4 on page 626.

SAMPLE PROBLEM 2 Determining the Number of Neutrons

Determine the number of neutrons in the most common isotope of aluminum.

Given:	atomic mass of Al = 26.98 u
	atomic number = 13
Required:	number of neutrons
Analysis:	Round the atomic mass of the element to the nearest whole number to get the mass number of the most common isotope.
	mass number of Al = 27 u (rounded up)
	mass number − atomic number = number of neutrons
Solution:	27 − 13 = 14
Statement:	The most common isotope of aluminum contains 14 neutrons.

Table 4 Powers of 10 and Decimal Equivalents

Power of 10	Decimal equivalent
10^9	1 000 000 000
10^8	100 000 000
10^7	10 000 000
10^6	1 000 000
10^5	100 000
10^4	10 000
10^3	1 000
10^2	100
10^1	10
10^0	1
10^{-1}	0.1
10^{-2}	0.01
10^{-3}	0.001
10^{-4}	0.000 1
10^{-5}	0.000 01
10^{-6}	0.000 001
10^{-7}	0.000 000 1
10^{-8}	0.000 000 01
10^{-9}	0.000 000 001

To write a large number in scientific notation, follow these steps:

1. To determine the exponent, count the number of places you have to move the decimal point to the left, to give a number between 1 and 10. For example, when writing the speed of light (300 000 000 m/s) in scientific notation, you have to move the decimal point eight places to the left. The exponent is therefore 8.

2. To form the coefficient, place the decimal point after the first digit. Now drop all the trailing zeros *unless* all the numbers after the decimal are zeros, in which case, keep one zero. In our example, the coefficient is 3.0.

3. Combine the coefficient with the base, 10, and the exponent. For example, the speed of light is 3.0×10^8 m/s.

Very small numbers (less than 1) can also be expressed in scientific notation. To find the exponent, count the number of places you move the decimal point to the right, to give a coefficient between 1 and 10. For very small numbers, the base (10) must be given a negative exponent.

For example, a millionth of a second, 0.000 001 s, can be written in scientific notation as 1×10^{-6} s. Table 5 shows several examples of large and small numbers expressed in scientific notation.

To multiply numbers in scientific notation, multiply the coefficients and add the exponents. For example,

$$(3 \times 10^3)(5 \times 10^4) = 15 \times 10^{3+4}$$
$$= 15 \times 10^7$$
$$= 1.5 \times 10^8$$

Note that when writing a number in scientific notation, the coefficient should be at least 1 and less than 10.

When dividing numbers in scientific notation, divide the coefficients and subtract the exponents. For example,

$$\frac{8 \times 10^6}{2 \times 10^4} = 4 \times 10^{6-4}$$
$$= 4 \times 10^2$$

Table 5 Numbers Expressed in Scientific Notation

Large or small number	Common decimal notation	Scientific notation
124.5 million km	124 500 000 km	1.245×10^8 km
154 thousand nm	154 000 nm	1.54×10^5 nm
753 trillionths of a kg	0.000 000 000 753 kg	7.53×10^{-10} kg
315 billionths of a m	0.000 000 315 m	3.15×10^{-7} m

5.D. Uncertainty in Measurement

There are two types of quantities used in science: exact values and measurements. Exact values include defined quantities: those obtained from SI prefix definitions (such as 1 km = 1000 m) and those obtained from other definitions (such as 1 h = 60 min).

Exact values also include counted values, such as 5 beakers or 10 cells. All exact values are considered completely certain. In other words, 1 km is exactly 1000 m, not 999.9 m or 1000.2 m. Similarly, 5 beakers could not be 4.9 or 5.1 beakers; 5 beakers are exactly 5 beakers.

Measurements, however, have some uncertainty. The uncertainty depends on the limitations of the measuring instrument and the skill of the measurer.

5.D.1. Significant Digits

The certainty of any measurement is communicated by the number of significant digits in the measurement. In a measured or calculated value, significant digits are the digits that are certain, plus one estimated (uncertain) digit. Significant digits include all the digits that are correctly reported from a measurement.

For example, 10 different people independently reading the water volume in the graduated cylinder in Figure 1 would all agree that the volume is at least 50 mL. In other words, the "5" is certain. However, there is some uncertainty in the next digit. The observers might report the volume as being 56 mL, 57 mL, or 58 mL. The only way to know for sure is to measure with a more precise measuring device. Therefore, we say that a measurement such as 57 mL has two significant digits: one that is certain (5) and the other that is uncertain (7). The last digit in a measurement is always the uncertain digit. For example, 115.6 g contains three certain digits (115) and one uncertain digit (6).

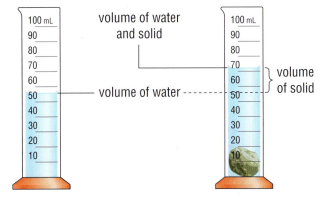

Figure 1 Measuring volume

Table 6 provides the guidelines for determining the number of significant digits, along with examples to illustrate each guideline.

Table 6 Guidelines for Determining Significant Digits

Guideline	Example Number	Number of significant digits
Count from left to right, beginning with the first non-zero digit.	345	3
	457.35	5
Zeros at the beginning of a number are never significant.	0.235	3
	0.003	1
All non-zero digits in a number are significant.	1.1223	4
	76.2	2
Zeros between digits are significant.	107.05	5
	0.02094	4
Zeros at the end of a number with a decimal point are significant.	10.0	3
	303.0	4
Zeros at the end of a number without a decimal point are ambiguous.	5400	at least 2
	200 000	at least 1
All digits in the coefficient of a number written in scientific notation are significant.	5.4×10^3	2
	5.400×10^3	4

ROUNDING

Use these rules when rounding answers to the correct number of significant digits:

1. When the first digit discarded is less than 5, the last digit kept (i.e., the one before the discarded digit) stays the same.

    Example:

    3.141 326 rounded to four digits is 3.141.

2. When the first digit discarded is greater than 5, or when it is 5 followed by at least one digit other than zero, the last digit kept increases by one unit.

    Examples:

    2.221 372 rounded to five digits is 2.2214.
    4.168 501 rounded to four digits is 4.169.

3. When the first digit discarded is 5 followed by only zeros, the last digit kept is increased by one unit if it is odd but not changed if it is even. Note that when this rule is followed, the last digit in the final number is always even.

    Examples:

    2.35 rounded to two digits is 2.4
    2.45 rounded to two digits is 2.4
    6.75 rounded to two digits is 6.8

4. When adding or subtracting measured quantities, look for the quantity with the fewest number of digits to the right of the decimal point. The answer can have no more digits to the right of the decimal point than this quantity has. In other words, the answer cannot be more precise than the least precise value.

    Example:

    ```
      12.52 g
    + 349.0 g
    +   8.24 g
    ─────────
     369.76 g
    ```

    Because 349.0 g is the quantity with the fewest digits to the right of the decimal point, the answer must be rounded to 369.8 g.

    Example:

    ```
      157.85 mL
    −  32.4 mL
    ──────────
      125.45 mL
    ```

    Because 32.4 mL has the fewest decimal places, the answer must be rounded to 125.4 mL. Note that Rule 3 applies.

5. When multiplying or dividing, the answer must contain no more significant digits than the quantity with the fewest number of significant digits.

    Examples:

    $$m = \frac{1.15 \text{ g}}{\text{cm}^3} \times 16 \text{ cm}^3 = 18 \text{ g}$$

    $$\Delta t = 1.25 \text{ h} \times \frac{60 \text{ min}}{1 \text{ h}} = 75.0 \text{ min}$$

    In other words, the answer cannot be more certain than the least certain value.

    Note, in the second example, 1.25 h is a measurement and, as a result, contains uncertainty. However, 60 min/h is an exact quantity. Since it has no uncertainty, the number of significant digits in the final answer is based only on the measured value of 1.25 h.

    Rule 5 also applies when multiplying or dividing measurements expressed in scientific notation. For example,

    $$(3.5 \times 10^3 \text{ km})(7.4 \times 10^2 \text{ km}) = 25.9 \times 10^5 \text{ km}^2$$
    $$= 2.59 \times 10^6 \text{ km}^2$$
    $$= 2.6 \times 10^6 \text{ km}^2$$

    The coefficient should be rounded to the same number of significant digits as the measurement with the least number of significant digits (the least certain value). In this example, both measurements have only two significant digits, so the coefficient 2.59 should be rounded to 2.6 to give a final answer of $2.6 \times 10^6 \text{ km}^2$.

    Similarly,

    $$\frac{3.9 \times 10^6 \text{ m}}{5.3 \times 10^3 \text{ s}} = 0.7377 \times 10^3 \text{ m/s}$$
    $$= 7.377 \times 10^2 \text{ m/s}$$
    $$= 7.4 \times 10^2 \text{ m/s}$$

5.D.2. Measurement Errors

There are two types of error that can occur when measurements are taken: random and systematic. Random error results when an estimate is made to obtain the last significant digit for a measurement. The size of the random error is determined by the precision of the measuring instrument. For example, when measuring length, it is necessary to estimate between the marks on the measuring tape. If these marks are 1 cm apart, the random error is greater and the precision is less than if the marks were 1 mm apart. Systematic error is caused by a problem with the measuring system itself, such as equipment not set up correctly. For example, if a balance is not tared (re-set to zero) at the beginning, all the measurements taken with the balance will have a systematic error.

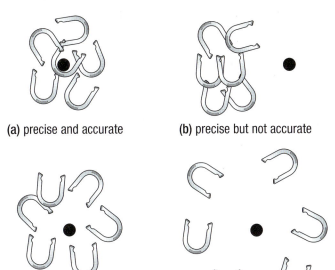

(a) precise and accurate (b) precise but not accurate

(c) accurate but not precise (d) neither accurate nor precise

Figure 2 The patterns of the horseshoes illustrate the comparison between accuracy and precision.

The precision of measurements depends on the markings (gradations) of the measuring device. Precision is the place value of the last measurable digit. For example, a measurement of 12.74 cm is more precise than a measurement of 127.4 cm because 12.74 was measured to hundredths of a centimetre, whereas 127.4 was measured to tenths of a centimetre.

When adding or subtracting measurements with different precisions, round the answer to the same precision as the least precise measurement. Consider the following:

```
   11.7   cm
    3.29  cm
+   0.542 cm
   15.532 cm
```

The first measurement, 11.7 cm, is measured to one decimal place and is the least precise. The answer must be rounded to one decimal place, or 15.5 cm.

No matter how precise a measurement is, it still may not be accurate. Accuracy refers to how close a value is to its accepted value. Figure 2 uses the results of a horseshoe game to explain precision and accuracy.

How certain you are about a measurement depends on two factors: the precision of the instrument and the size of the measured quantity. Instruments that are more precise give more certain values. For example, a measurement of 13 g is less precise than one of 12.76 g because the second measurement has more decimal places than the first. Certainty also depends on the size of the measurement. For example, consider the measurements 0.4 cm and 15.9 cm. Both have the same precision (number of decimal places): both are measured to the nearest tenth of a centimetre. Imagine that the measuring instrument is precise to ± 0.1 cm, however. An error of 0.1 cm is much more significant for the 0.4 cm measurement than it is for the 15.9 cm measurement because the second measurement is much larger than the first. For both factors—the precision of the instrument used and the value of the measured quantity—the more digits there are in a measurement, the more certain you are about the measurement.

5.E. Using the Calculator

A calculator is a very useful device that makes calculations easier, faster, and probably more accurate. However, like any other electronic instrument, you need to learn how to use it. These guidelines apply to a basic scientific calculator. If your calculator is different, such as a graphing calculator, some of the instructions and operations may use different keys or sequences, so always check the manual.

GENERAL POINTS

- Most calculators follow the usual mathematical rules for order of operations—multiplication/division before addition/subtraction. For example, if you are calculating y using $y = mx + b$, you can enter the values of m times x plus b in one sequence. The calculator will "know" that m and x must be multiplied first before b is added.
- Calculators do not keep track of significant digits. For example, 12.0 is the same as 12 for a calculator.
- Some calculator keys such as $\boxed{\tfrac{1}{x}}$ $\boxed{+/-}$, $\boxed{x^2}$, (and its second function, $\boxed{\sqrt{x}}$) apply the operation only to the value in the display regardless of other operations in progress. This means you can quickly change the sign of the number, convert to the reciprocal, square the number, or determine the square root while inputting a sequence of calculations.
- Do not clear all numbers from the calculator until you are completely finished with a question: the result of one calculation can be reused to start the next one.
- All scientific calculators have at least one memory location where you can store a number (M+ and STO are common keys) and recall it later (usually with MR or RCL). Use it to avoid re-entering many digits.

MULTIPLICATION AND DIVISION

- Division is the inverse of multiplication. This means that dividing by a number is the equivalent to multiplying by the inverse of the same number. For example,

$\dfrac{12 \text{ km}}{0.75 \text{ h}}$ is the same as $12 \text{ km} \times \dfrac{1}{0.75 \text{ h}}$

and equals $16 \dfrac{\text{km}}{\text{h}}$.

This is particularly useful when you want to divide by a number that is currently in the display of your calculator. For example, you have just finished converting 45 min to 0.75 h in your calculator and now you want to calculate the speed.

Display: 0.75

Press:

New display: 16

- Brackets, ( ), are useful to force the calculator to perform the operation(s) inside the brackets first, before continuing with the calculation. The calculation of the slope of a line is a good example.

$$\text{slope} = \dfrac{\Delta d}{\Delta t}$$
$$= \dfrac{(15.2 - 4.1) \text{ m}}{(6.5 - 3.6) \text{ s}}$$
$$= 3.8 \dfrac{\text{m}}{\text{s}}$$

If you do not use the brackets on your calculator, you will have to calculate the numerator and denominator separately and then divide.

SCIENTIFIC NOTATION

On many calculators, scientific notation is entered using a special key, labelled EXP or EE. This key includes "×10" from the scientific notation, and you need to enter only the exponent. For example, to enter

7.5×10^4 press [7] [.] [5] [EXP] [4]

3.6×10^{-3} press [3] [.] [6] [EXP] [+/−] [3]

CALCULATING A MEAN

There are many statistical methods of analyzing experimental evidence. One of the most common and important is calculating an arithmetic mean, or simply a mean. The mean of a set of values is the sum of all reasonable values divided by the total number of values. (This is also commonly known as the average of a set of values, but this term is not recommended because it is too vague and open to different interpretations.)

Suppose you measure the root growth of five seedlings. Your root measurements after three days are 1.7 mm, 1.6 mm, 1.8 mm, 0.4 mm, and 1.6 mm. What is the mean root growth? Inspection of the measurements shows that the 0.4 mm measurement clearly does not fit with the rest. Perhaps this seedling was infected with a fungus, or some other problem occurred. You should not include this result in your mean but leave it in your evidence table so everyone can see the decision that you made. Using only reasonable values,

mean root growth = $\frac{1.7 \text{ mm} + 1.6 \text{ mm} + 1.8 \text{ mm} + 1.6 \text{ mm}}{4}$
= 1.7 mm

Means are important in all areas of science because multiple measurements or trials are widely used to increase the reliability of the results.

EQUATIONS

Algebra is a set of rules and procedures for working with mathematical equations. In general, your equations will contain one unknown quantity. Whatever mathematical operation is performed to one side of an equation must be performed to the other side. To solve for an unknown value, you need to isolate it on one side of the equal sign. To accomplish this, you should follow three rules:

1. The same quantity can be added or subtracted from both sides of the equation without changing the equality.

   The following examples illustrate this rule:

100 m	= 100 m	$x + b = y$
100 m − 5 m	= 100 m − 5 m	$x + b - b = y - b$
95 m	= 95 m	$x = y - b$

   The example on the left shows the application of this rule using quantities with numbers and units. The rule works equally well with quantity symbols. The example on the right shows how to isolate x.

2. The same quantity can be multiplied or divided on both sides of the equation without changing the equality.

   The following examples illustrate this rule. The example on the left shows the use of this rule with known quantities. Use the same rule with an equation containing quantity symbols to isolate one of the quantities. To solve for d in the example on the right, multiply both sides by t. Notice that t divided by t equals 1. Multiplying or dividing any quantity by 1 does not change the quantity; therefore, $d \times 1 = d$.

   $$120 \text{ m} = 120 \text{ m} \qquad v = \frac{d}{t}$$
   $$\frac{120 \text{ m}}{8.0 \text{ s}} = \frac{120 \text{ m}}{8.0 \text{ s}} \qquad v \times t = \frac{d}{\cancel{t}} \times \cancel{t}$$
   $$15 \text{ m/s} = 15 \text{ m/s} \qquad vt = d \text{ or } d = vt$$

3. The same power (such as square or square root) can be applied to both sides of the equation without changing the equality.

   The following examples illustrate this rule:

   $$25 \text{ s}^2 = 25 \text{ s}^2 \qquad b^2 = A$$
   $$\sqrt{25 \text{ s}^2} = \sqrt{25 \text{ s}^2} \qquad \sqrt{b^2} = \sqrt{A}$$
   $$5.0 \text{ s} = 5.0 \text{ s} \qquad b = \sqrt{A}$$

   If several of the rules listed above are required to isolate an unknown quantity, then you should apply Rule 1 first, whenever possible, and then apply Rule 2. In general, Rule 3 should be used last.

6 DATA TABLES AND GRAPHS

Data tables are an effective means of recording both qualitative and quantitative observations. Making a data table should be one of your first steps as you prepare to conduct an investigation. It is particularly useful for recording the values of the independent variable (the cause) and the dependent variables (the effects), as shown in Table 1.

Table 1 A Running White-Tailed Deer

Time (s)	Distance (m)
0	0
1.0	13
2.0	25
3.0	40
4.0	51
5.0	66
6.0	78

Follow these guidelines to make a data table:
- Use a ruler to make your table.
- Write a title that precisely describes your data.
- Include the units of measurement for each variable, when appropriate.
- List the values of the independent variable in the left-hand column.
- List the values of the dependent variable(s) in the column(s) to the right.

6.A. Graphing Data

A graph is a visual representation of quantitative data. Graphing data often makes it easier to identify a trend or pattern in the data, which indicates a relationship between the variables. There are many types of graphs that can be used to organize data. Three of the most useful kinds of graphs are bar graphs, circle graphs, and point-and-line graphs. Each kind of graph has its own special uses. You need to identify which type of graph is most appropriate for the data you have collected. Then you can construct the graph.

BAR GRAPHS

A bar graph helps you make comparisons when one variable is in numbers (e.g., rainfall) and the other variable is not (e.g., month of the year). Figure 1 shows a bar graph of the distribution of rainfall over a 12-month period in Ottawa. Each bar stands for a different month. This graph clearly shows that rainfall is higher in the summer months and is lower in the winter months.

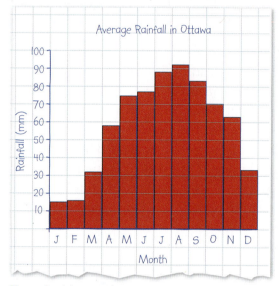

Figure 1 A bar graph

CIRCLE GRAPHS

Circle graphs (pie charts) are useful to show how the whole of something is divided into many parts. For example, the circle graph in Figure 2 shows that transportation is the largest source of greenhouse gas emissions by individuals.

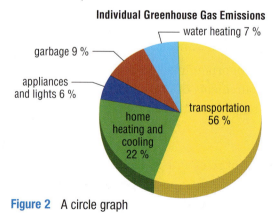

Figure 2 A circle graph

POINT-AND-LINE GRAPHS

When both variables are quantitative, use a point-and-line graph. This format shows all the data points and a line of best fit indicating any relationship

632 Skills Handbook

between the variables. For example, we can use the following guidelines and the data in Table 1 to construct the line graph in Figure 3.

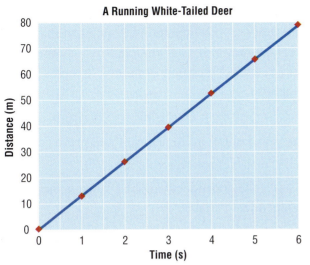

Figure 3 A point-and-line graph of the data from Table 1

Making Point-and-Line Graphs

1. Use graph paper. Draw a horizontal line close to the bottom of the paper as the *x*-axis and a vertical line close to the left edge as the *y*-axis.
2. You will generally plot the independent variable along the *x*-axis and the dependent variable along the *y*-axis. The exception is when one variable is time: always plot time on the *x*-axis. The slope of the graph then always represents a rate. Label each axis, including the units.
3. Give your graph a title: a short, accurate description of the data represented by the graph.
4. Determine the range of values for each variable. The range is the difference between the greatest and least values. Graphs often include a little extra length on each axis.
5. Choose a scale for each axis. The scale will depend on how much space you have and the range of values for each axis. Spaces on the grid usually represent equal increments, such as 1, 2, 5, 10, or 100.
6. Plot the points. In Figure 3, the first set of points is 0 on the *x*-axis and 0 on the *y*-axis: the origin.
7. After all the points are plotted, try to visualize a line through the points to show the relationship between the variables. Not all points may lie exactly on a line. Draw the line of best fit—a straight or smoothly curving line through the points so that there is approximately the same number of points on each side of the line. The line's purpose is to show the overall pattern of the data.
8. If you are plotting more than one set of data on the same graph, use different colours or symbols for each. Provide a legend.

CALCULATING SLOPES

If the line of best fit on a graph is a straight line, there is a simple relationship between the two variables. You can represent this linear (straight-line) relationship with this mathematical equation:

$$y = mx + b$$

where *y* is the dependent variable (on the *y*-axis), *x* is the independent variable (on the *x*-axis), *m* is the slope of the line, and *b* is the *y*-intercept (the point where the line touches the *y*-axis). To determine the slope, choose two points on the line, (x_1, y_1) and (x_2, y_2). (Note that these are not necessarily data points: just two points on the line.) The slope is equal to

$$m = \frac{rise}{run} = \frac{y_2 - y_1}{x_2 - x_1}$$

For the graph in Figure 4, suppose you choose the two points (1.5, 20) and (5.5, 72). You can use them to calculate the slope.

$$m = \frac{y_2 - y_1}{x_2 - x_1}$$
$$= \frac{72 \text{ m} - 20 \text{ m}}{5.5 \text{ s} - 1.5 \text{ s}}$$
$$= \frac{52 \text{ m}}{4.0 \text{ s}}$$
$$= 13 \text{ m/s}$$

The slope of the line is 13 m/s. The positive number indicates a direct relationship.

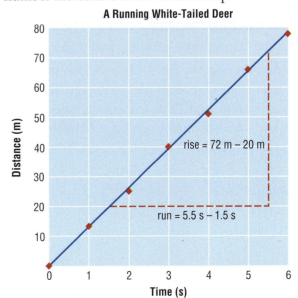

Figure 4 Calculating the slope of a line of best fit

6. Data Tables and Graphs 633

6.B. Using Computers for Graphing

You can use spreadsheet or graphing programs on your computer to construct bar, circle, and point-and-line graphs. In addition, such programs can use statistical analysis to compute the line of best fit. The following instructions guide you to produce best-fit line graph values for two variables on a spreadsheet.

STEPS FOR CREATING A GRAPH IN A SPREADSHEET PROGRAM

1. Start the spreadsheet program. Enter the data with the values for x (the first column in your data table) that start in cell A2. Now enter the values for y from the other column that start in B2.

2. Highlight the two columns containing your data (A2 … B6). From the toolbar, select the button for creating graphs. The cursor may change to a symbol, which allows you to "click and drag" to choose the size of your graph.

3. A series of choices will now be presented to you, so you can specify what kind of graph you want to create. A scatter graph is most appropriate for our data.
   - Click in the "Title" box and type the title of the graph.
   - Click in the "Value (X)" box and type the label and units for the x-axis.
   - Click the "Finish" button.

4. To add a line of best fit, point your cursor to one of the highlighted data points and right click once. Select the "Trendline" option. Make sure that the "Linear" box is highlighted under Type.

5. To find the slope or y-intercept, click on the "Options" tab and select the box beside "Display Equation on Chart." This will put the values for $y = mx + b$ on the graph, giving you the slope and y-intercept value.

6. Save your spreadsheet before you close the program.

6.C. Interpreting Graphs

When data from an investigation are plotted on an appropriate graph, patterns and relationships become easier to see and interpret. You can more easily tell if the data support your hypothesis. Looking at the data on the graph may also lead you to a new hypothesis. Or you can extract meaning from other people's investigations.

WHAT TO LOOK FOR WHEN READING GRAPHS

Here are some questions to help you interpret a graph:
- What variables are represented?
- What is the dependent variable? What is the independent variable?
- Are the variables quantitative or qualitative?
- If the data are quantitative, what are the units of measurement?
- What do the highest and lowest values represent on the graph?
- What is the range between the highest and the lowest values on each axis?
- Are the axes continuous? Do they start at zero?
- What patterns or trends exist between the variables?
- If there is a linear relationship, what might the slope of the line tell you?

USING GRAPHS FOR PREDICTING

If a graph shows a regular pattern, you can use it to make predictions. For example, you could use the graph in Figure 3 on page 633 to predict the distance travelled by the deer in 8.0 s. To do this, you extrapolate the graph (extended beyond the measured points), assuming that the observed trend would continue. You should be careful when predicting values outside of your measured range. The farther you are from the known values, the less reliable your prediction will be.

STUDY SKILLS 7

7.A. Working Together

Teamwork is just as important in science as it is on the playing field or in the gym. Scientific investigations are usually carried out by teams of people working together. Ideas are shared, experiments are designed, data are analyzed, and results are evaluated and shared with other investigators. Group work is necessary and is usually more productive than working alone.

Several times throughout the year, you may be asked to work with one or more of your classmates. Whatever the task that your group is assigned, you need to follow a few guidelines to ensure a productive and successful experience.

7.A.1. General Guidelines for Effective Teamwork

- Keep an open mind. Everyone's ideas deserve consideration.
- Divide the task among all the group members. Choose a role best suited to your particular strengths.
- Work together and take turns. Encourage, listen, clarify, help, and trust one another.
- Remember that the success of the team is everyone's responsibility. Every member needs to be able to demonstrate what the team has learned and support the team's final decision.

7.A.2. Exchanging Information Orally

You will be involved in a number of different activities during your science course. These activities help you express your ideas and learn about new ones. Here are three of the more successful discussion formats that your group may use to share ideas:

- In a **think-pair-share activity**, you and your partner are given a problem. Each of you develops a response (usually within a time limit). Then share your ideas with each other to resolve the problem. You may also be asked to share your results with a larger group or the class.

- In a **jigsaw activity**, you are an active member in two teams: your home team and your expert team. Each member of the home team chooses or is assigned to a particular area of research. Each home team member then meets with an expert team in which everyone is working on the same area of research. In your expert team, you may work together to come up with answers to questions in your area of research. Once you have accomplished your task as a team, you return to your home team, and it is your responsibility to teach what you have learned to the members of your home team. Each person on the home team will do the same thing. The guidelines of teamwork are important with the jigsaw, so try to keep them in mind as you work with your expert and home teams.

- A **round table activity** can be used to give your group an opportunity to review what they know. Your group is given a pen, paper, and a question or questions. Pass the pen and paper around and take turns writing one line of the solution. You can pass on your turn if you wish. Keep working until the solution is complete. Check to ensure that everyone understands the solution. Finally, working as a team, review the steps to the solution.

7.A.3. Investigations and Activities

This kind of work is most effective when completed by small groups. Here are some suggestions for effective group performance during investigations and activities:

- Make sure that each group member understands and agrees to the role assigned to him or her.
- Take turns doing the work during similar and repeated activities.
- Safety must come first. Be aware of where other group members are and what they are doing. Ask yourself, what potential hazards are involved in this activity? What are the safety procedures?

- Take responsibility for your own learning by making notes and contributing to any discussions about the activities. Make your own observations and compare them with the observations of other group members (Figure 1).

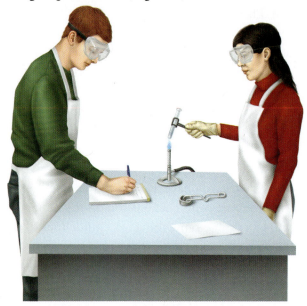

Figure 1 When you are working with one student or several, you can share and compare your results.

7.A.4. Explore an Issue

Follow these guidelines when you are conducting research with a group:

- Divide the topic into several topics and assign one topic to each group member.
- Keep records of the sources used by each group member.
- Decide on a format for exchanging information (such as photocopies of notes, oral discussion, electronically).
- When the time comes to make a decision and take a position on an issue, allow for the contributions of each group member. Make decisions by compromise and consensus.
- Communicating your position should also be a group effort.

7.A.5. Evaluating Teamwork

After you have completed a task with your group, evaluate your team's effectiveness using these criteria: strengths and weaknesses, opportunities, and challenges.

Reflect on your experience by asking yourself the following questions:

- What were the strengths of your teamwork?
- What were the weaknesses of your teamwork?
- What opportunities were provided by working with your group?
- What challenges did you encounter as a member of a group?

7.B. Setting Goals and Monitoring Progress

Think back to your last school year. What classes did you do well in? Why do you think you were successful? In which classes did you have difficulty? Why do you think you had difficulty? What could you do differently this year to improve your performance? Use your answers to these questions to reflect on your past experiences to make new and positive changes. Things that you want to accomplish today, this week, and this year are all called goals. Learning to set goals and to make a plan to achieve them takes skill, patience, and practice.

7.B.1. Setting Goals

ASSESS YOUR STRENGTHS AND WEAKNESSES

The process of setting goals starts with honest reflection. Maybe you have noticed that you do better on projects than you do on tests and exams. You may perform better when you are not pressured by time. Inattention in class and poor study habits may be weaknesses that result in poor performance.

REALISTIC GOALS THAT YOU CAN MEASURE

Do not set yourself up to fail by setting goals that you cannot possibly achieve. Saying, "I will have the best mark in the class at the end of the semester" may not be realistic. Setting a goal to increase your test marks by 10 % this semester may be achievable, however. You will find it easier to reach your goals if you can tell whether you are getting closer to them. A goal to increase your test marks by 10 % this semester is easy to measure. When you are thinking of setting goals, remember the acronym SMART: Specific, Measurable, Attainable, Realistic, and Time-limited.

SHARE YOUR GOALS

People whom you respect can often help you set and clarify goals. Someone who knows your strengths and weaknesses may be able to think of possibilities you may not have considered. Sharing your goals with a trusted friend or adult will often provide needed support to help you reach your goals.

PLANNING TO MEET YOUR GOALS

Once you have made a list of realistic goals, create a plan to achieve them. A successful plan usually consists of two parts: an action plan and target dates.

THE ACTION PLAN

First, make a list of the actions or behaviours that might help you reach your goals (Figure 2).

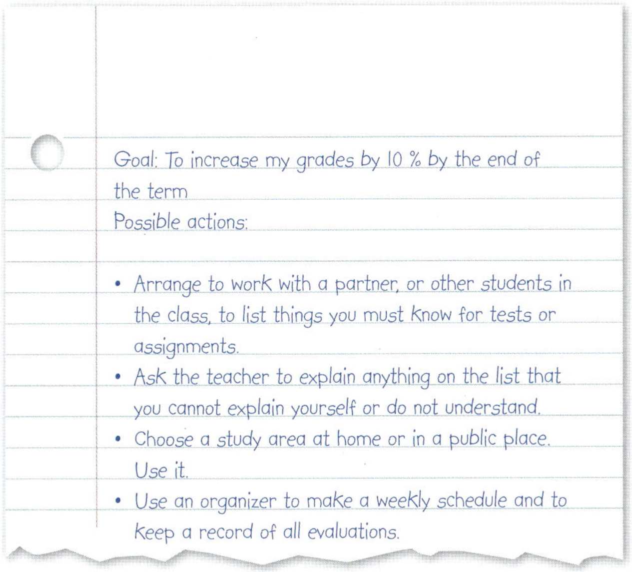

Figure 2 One student's goal and action plan to improve test results

If you have made an honest assessment of your strengths and weaknesses, then you know what you have to do to improve. If you want to improve your test marks, you could try to work with others to prepare for tests. You could also use a weekly planner or reorganize your study area at home.

Identify what is preventing you from achieving your goal. Think of ways to overcome these obstacles. Ask friends for tips on the different ways that they maintain good study habits and improve test results.

SETTING TARGET DATES

Suppose you want to improve test results by 10 % by the end of the semester. How much time do you have? How many tests are scheduled between now and then? Work back from your target date at the end of the semester. Determine the dates of all the tests between now and the end of the semester. These dates will give you short-term targets that, if you hit them, will make it easier for you to reach your overall target. Figure 3 shows an example of a working schedule.

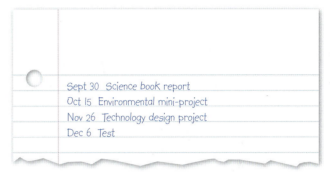

Figure 3 Test dates

Once you complete your schedule in your planner, transfer it to a calendar in your study area. Refer to either your planner or your calendar every day.

7.B.2. Monitoring Progress

Remember to measure your progress along the way. It is always important to look at and monitor the results of your tests and activities during the school year rather than just at the end of it. You might decide, for example, to check your progress after the first test. Did these results meet your short-term target to improve test results by 10 % by the end of the semester? If you do not seem to be on track to meet your goal, then you may need to change your plan. For example, perhaps you need to study by yourself or with one friend rather than with a group of friends.

It is always possible to change your plan or even adjust your goal. The most important thing is to keep moving forward and to remain committed to improvement.

7.C. Good Study Habits

Studying takes many forms. Developing good study skills can help you study and learn more successfully. Below are some tips to help you achieve better study habits.

7.C.1. Your Study Space

- *Organize your work area.* The place where you study should be tidy and organized. Place all papers, books, magazines, and pictures in appropriate areas of your study area (e.g., keep books in a bookcase or crate and magazines in a stack on the floor). This will make it easier for you to focus on your school work.
- *Maintain a quiet work area.* Where possible, make sure that your work area is free from distractions—telephone, music, television, and other family members. If there are too many distractions at home, you can usually find an appropriate space at the school or public library. Any quiet space, free from interruptions, can be a productive work area.
- *Make sure that you are comfortable in your work area.* If possible, personalize your work area. For example, make sure that the light is right for your needs. Decide what works best for you and create a study area that provides a productive and positive environment in which you feel comfortable.
- *Be prepared—bring everything you need.* It is important to have all the necessary materials and equipment that you will need when you begin to study. You can easily increase your productivity by gathering materials such as pens, pencils, notebooks, or textbooks in one place near your computer or on your desk. Continually getting up to find something you need decreases your ability to stay on task and be productive.

7.C.2. Study Habits

- *Take notes.* Take notes during class. Outside class, review the appropriate section of the textbook. Read or view additional material on the topic from other sources, such as newspapers, magazines, the Internet, and television. Ask a friend to share notes with you.
- *Use graphic organizers.* You can use a variety of graphic organizers to help you summarize a concept or unit (page 640). They also help you more easily connect different concepts.
- *Schedule your study time.* Use a daily planner and take it with you to class. Write any homework assignments, tests, or projects in it. Use it to create a daily "to do" list. This will help you complete work to hand in and avoid last-minute panic. Also, jot down in your planner when, where, and with whom you plan to study for certain topics.
- *Take study breaks.* It is important to schedule breaks into your study time. For example, you could decide to take a study break after completing one or two items on your "to do" list. Taking breaks allows you to relax and to recharge your brain so that you can keep focused when you return to your assignments.

8.A. Reading Strategies

The skills and strategies that you use to help you read depend on the type of material you are reading. Reading a science book is different from reading a novel. When you are reading a science book, you are reading for information. Here are some strategies to help you read for information.

8.A.1. Before Reading

Skim the section you are going to read. Look at the illustrations, headings, and subheadings.

- *Preview.* What is this section about? How is it organized?
- *Make connections.* What do I already know about the topic? How is it connected to other topics I already know about?
- *Predict.* What information will I find in this section? Which parts provide the most information?
- *Set a purpose.* What questions do I have about the topic?

8.A.2. During Reading

Pause and think as you read. Spend time on the photographs, illustrations, tables, and graphs, as well as on the words.

- *Check your understanding.* What are the main ideas in this section? How would I state them in my own words? What questions do I still have? Should I reread? Do I need to read more slowly, or can I read more quickly?
- *Determine the meanings of key science terms.* Can I figure out the meanings of terms from context clues in the words or illustrations? Do I understand the definitions of terms in bold type? Is there something about the structure of a new term that will help me remember its meaning? Which terms should I look up in the glossary?
- *Make inferences.* What conclusions can I make from what I am reading? Can I make any conclusions by "reading between the lines"?
- *Visualize.* What mental pictures can I make to help me understand and remember what I am reading? Should I make a sketch?
- *Make connections.* How is the information in this section like information I already know?
- *Interpret visuals and graphics.* What additional information can I get from the photographs, illustrations, tables, or graphs?

8.A.3. After Reading

Many of the strategies you use during reading can also be used after reading. For example, this textbook provides summaries and questions at the ends of sections. These questions will help you check your understanding and make connections to information that you have just read or to other parts in the textbook.

At the end of each chapter are a Key Concepts Summary and a Vocabulary list, followed by a Chapter Review and Chapter Self-Quiz.

- *Locate needed information.* Where can I find the information I need to answer the questions? Under what heading might I find the information? What terms in bold type should I look for? What details do I need to include in my answers?
- *Synthesize.* How can I organize the information? What graphic organizer could I use? What headings or categories could I use?
- *React.* What are my opinions about this information? How does it, or might it, affect my life or my community? Do other students agree with my reactions? Why or why not?
- *Evaluate information.* What do I know now that I did not know before? Have any of my ideas changed because of what I have read? What questions do I still have?

8.B. Graphic Organizers

Diagrams that are used to organize and display ideas visually are called graphic organizers. Graphic organizers are especially useful in science and technology studies when you are trying to connect together different concepts, ideas, and data. Different graphic organizers have different purposes. They can be used to

- show processes
- organize ideas and thinking
- compare and contrast
- show properties or characteristics
- review words and terms
- collaborate and share ideas

TO SHOW PROCESSES

Graphic organizers can show the stages in a process (Figure 1).

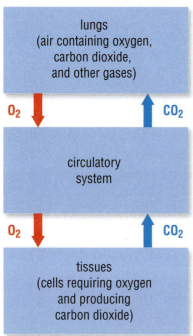

Figure 1 This organizer shows that oxygen and carbon dioxide are transported around the body.

TO ORGANIZE IDEAS AND THINKING

A **concept map** is a diagram showing the relationships between ideas (Figure 2). Words or pictures representing the ideas are connected by arrows and words or expressions that explain the connections. You can use a concept map to brainstorm what you already know, to map your thinking, or to summarize what you have learned.

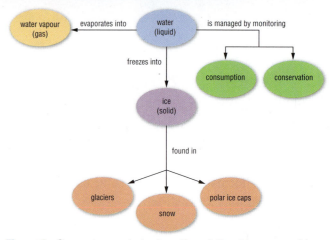

Figure 2 Concept maps help show the relationships among ideas.

Mind maps are similar to concept maps, but they do not have explanations for the connections between ideas.

You can use a **tree diagram** to show concepts that can be broken down into smaller categories (Figure 3).

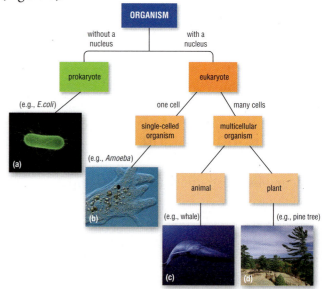

Figure 3 Tree diagrams are very useful for classification.

You can use a **fishbone diagram** to organize the important ideas under the major concepts of a topic you are studying (Figure 4).

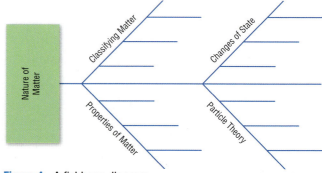

Figure 4 A fishbone diagram

640 Skills Handbook

What do we Know?	What do we Want to find out?	What did we Learn?
Carbon dioxide is a greenhouse gas.	Are there any other important greenhouse gases? If so, where do they come from?	Methane, water, and nitrous oxide are other common greenhouse gases. Methane is released from decaying organic matter and from the digestive tracts of grazing animals. Nitrous oxide is released in automobile emissions.
The greenhouse effect traps solar energy in the atmosphere.	If the energy of the Sun can get through the atmosphere and warm the Earth's surface, how is the energy trapped by greenhouse gases?	The atmosphere is transparent to light, allowing light rays from the Sun to strike Earth's surface. This energy is absorbed and then released as infrared waves. Because the atmosphere is not transparent to infrared waves, energy trapped in the atmosphere warms Earth.

Figure 5 A K-W-L chart

You can use a **K-W-L** chart to write down what you know (K), what you want (W) to find out, and, afterwards, what you have learned (L) (Figure 5).

TO COMPARE AND CONTRAST

You can use a **comparison matrix** (a type of table) to compare related concepts (Table 1).

Table 1 Subatomic Particles

	Proton	Neutron	Electron
electrical charge	positive	neutral	negative
symbol	p$^+$	n^0	e$^-$
location	nucleus	nucleus	orbit around the nucleus

You can use a **Venn diagram** to show similarities and differences (Figure 6).

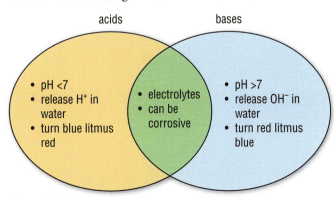

Figure 6 A Venn diagram

You can use a **compare-and-contrast chart** to show similarities and differences between two substances, actions, ideas, and so on (Figure 7).

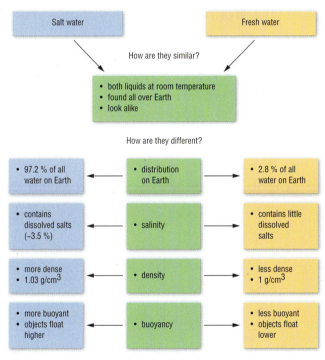

Figure 7 A compare-and-contrast chart

8. Literacy 641

TO SHOW PROPERTIES OR CHARACTERISTICS

You can use a **bubble map** to show properties or characteristics (Figure 8).

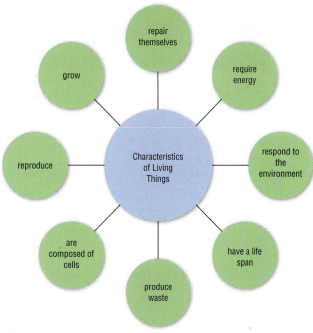

Figure 8 A bubble map

TO REVIEW WORDS AND TERMS

You can use a **word wall** to list, in no particular order, the key words and concepts for a topic (Figure 9).

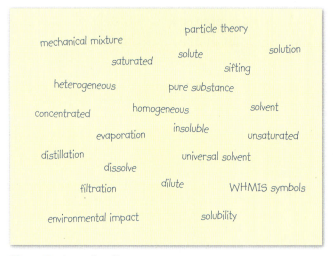

Figure 9 A word wall

TO COLLABORATE AND SHARE IDEAS

A **placemat organizer** gives students in a small group a space to write down what they know about a certain topic. Then group members discuss their answers and write in the middle section what they have in common (Figure 10).

Before:

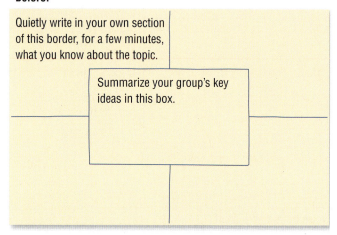

After:

Figure 10 A placemat organizer

LATIN AND GREEK ROOT WORDS

Prefix	Latin or Greek	Meaning	Example
ant-	Greek	opposing	antacid; Antarctic
anthrop-	Greek	related to people	anthropogenic
aqu-	Latin	water	aqueous solution; aquatic ecosystem
bio-	Greek	life	biology; biosphere
cardio-	Greek	related to the heart	cardiology
co-	Latin	with or together	covalent bond; converging lens
di-	Greek	two	diatomic molecule
epi-	Greek	upon or over	epidermis
hal-	Greek	sea salt	halogen; halophile
hemo- (haemo-)	Greek	related to blood	hemoglobin
hydro-	Greek	related to water	hydrosphere; hydroelectric
hyper-	Greek	more or excessively	hyperopia; hyperactive
infra-	Latin	below or lower	infrared radiation
inter-	Latin	between	interphase; interglacial period
litho-	Greek	rock	lithosphere
lum-	Latin	related to light	luminous
meta-	Greek	beyond	metamorphosis
micro-	Greek	small	micro-organism
mono-	Greek	one	carbon monoxide
peri-	Greek	around	peripheral nervous system; periderm
poly-	Greek	many	polyatomic ion
pro-	Greek and Latin	before	prophase; propagate
pseudo-	Greek	false	pseudoscience
retro-	Latin	turned back	retro-reflector; retrograde
sal-	Latin	related to salt	saline solution
therm-	Greek	heat	thermometer; thermocline
trans-	Latin	through or across	transparent; transistor
ultra-	Latin	above or beyond	ultraviolet radiation
xeno-	Greek	foreign or other	xenotransplantation
Suffix	**Latin or Greek**	**Meaning**	**Example**
-gen	Greek	to make or generate	carcinogen; halogen
-meter	Greek	measure	conductivity meter
-ology	Greek	study of	geology
-stasis	Greek	location	metastasis

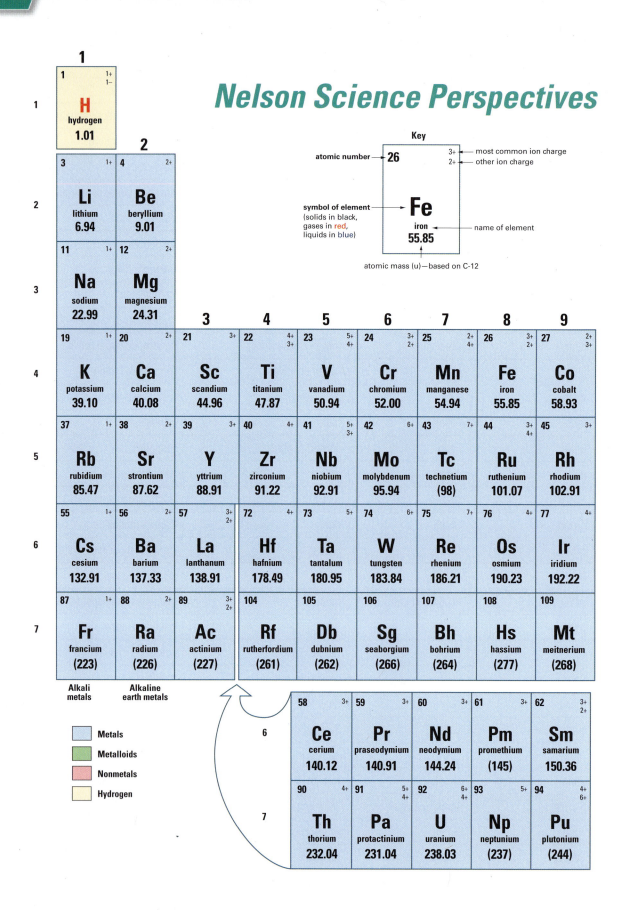

Periodic Table of the Elements

Measured values are subject to change as experimental techniques improve. Atomic molar mass values in this table are based on IUPAC Web site values (2005).

13	14	15	16	17	18
					2 — **He** helium 4.00
5 — **B** boron 10.81	6 — **C** carbon 12.01	7 3− **N** nitrogen 14.01	8 2− **O** oxygen 16.00	9 1− **F** fluorine 19.00	10 — **Ne** neon 20.18
13 3+ **Al** aluminium 26.98	14 — **Si** silicon 28.09	15 3− **P** phosphorus 30.97	16 2− **S** sulfur 32.07	17 1− **Cl** chlorine 35.45	18 — **Ar** argon 39.95

10	11	12						
28 2+/3+ **Ni** nickel 58.69	29 2+/1+ **Cu** copper 63.55	30 2+ **Zn** zinc 65.41	31 3+ **Ga** gallium 69.72	32 4+ **Ge** germanium 72.64	33 3− **As** arsenic 74.92	34 2− **Se** selenium 78.96	35 1− **Br** bromine 79.90	36 — **Kr** krypton 83.80
46 2+/3+ **Pd** palladium 106.42	47 1+ **Ag** silver 107.87	48 2+ **Cd** cadmium 112.41	49 3+ **In** indium 114.82	50 4+/2+ **Sn** tin 118.71	51 3+/5+ **Sb** antimony 121.76	52 2− **Te** tellurium 127.60	53 1− **I** iodine 126.90	54 — **Xe** xenon 131.29
78 4+/2+ **Pt** platinum 195.08	79 3+/1+ **Au** gold 196.97	80 2+/1+ **Hg** mercury 200.59	81 1+/3+ **Tl** thallium 204.38	82 2+/4+ **Pb** lead 207.2	83 3+/5+ **Bi** bismuth 208.98	84 2+/4+ **Po** polonium (209)	85 1− **At** astatine (210)	86 — **Rn** radon (222)
110 **Ds** darmstadtium (281)	111 **Rg** roentgenium (272)	112 **Uub** ununbium (285)	113 **Uut** ununtrium (284)	114 **Uuq** ununquadium (289)	115 **Uup** ununpentium (288)	116 **Uuh** ununhexium (291)	117 **Uus** ununseptium	118 **Uuo** ununoctium (294)

Halogens | Noble gases

63 3+/2+ **Eu** europium 151.96	64 3+ **Gd** gadolinium 157.25	65 3+ **Tb** terbium 158.93	66 3+ **Dy** dysprosium 162.50	67 3+ **Ho** holmium 164.93	68 3+ **Er** erbium 167.26	69 3+ **Tm** thulium 168.93	70 3+/2+ **Yb** ytterbium 173.04	71 2+ **Lu** lutetium 174.97
95 3+/4+ **Am** americium (243)	96 3+ **Cm** curium (247)	97 3+/4+ **Bk** berkelium (247)	98 3+ **Cf** californium (251)	99 3+ **Es** einsteinium (252)	100 3+ **Fm** fermium (257)	101 2+/3+ **Md** mendelevium (258)	102 2+/3+ **No** nobelium (259)	103 3+ **Lr** lawrencium (262)

APPENDIX B: What Is Science?

THE NATURE OF SCIENCE

What Is Science?

Science is often thought of as a collection of facts, laws, and theories. While this is partially true, science is much more. Science, or more specifically, **scientific inquiry**, is a way of learning about the natural world by observing things, asking questions, proposing answers, and testing those answers. The main goal of science is to understand the natural world, and the main product is knowledge in the form of facts, laws, and theories. For example, we know that Earth revolves around the Sun. This is accepted as a scientific fact. It came from repeated observations and analysis of the Sun and the night sky.

Science is also the processes that are used to gather and organize this knowledge about the natural world. Science, then, is both our present understanding of the natural world and the processes that led to this understanding.

Scientists make some basic assumptions about science and the natural world. Scientists believe the following:

- We live in a physical world, not a mental construction (Figure 1).
- The natural world is mostly structured and understandable; however, there is some randomness in the natural world. Therefore, the natural world is not always entirely predictable.
- Human intelligence enables us to understand the natural world, and scientific research is the best way to advance this understanding.
- Our knowledge is constructed, tentative, and incomplete. It will likely change or be modified, and grow through continued research.

Figure 1 The world around us is a natural, physical place.

The Characteristics of Science

Science has some characteristics and methods that are similar to other areas of study. Science is also unique in some ways.

Science Starts from Observations That Lead to Questions

Scientific investigations always start from observations that lead to an appropriate question. The question may arise from curiosity or from a serious problem that faces society. For example, you may ask the question "What kind of seeds do the birds at my bird feeder prefer?" A scientist may ask, "What causes cancer?" or "How can we cure cancer?" Scientists gather many observations or set up experiments to provide evidence that may help them answer their questions.

Scientists use what they observe to describe the natural world. **Observations** are evidence gained using the five senses—touch, smell, taste, vision, and hearing. Scientists often use special tools and equipment (such as microscopes, telescopes, meters, radar, and sensors) to expand or extend the capabilities of their senses, to quantify their observations, and to make observations in locations where they are not able to go themselves.

Scientists cannot always directly observe the subject of their investigation (Figure 2). However, scientists have been able to devise clever ways of figuring out the characteristics of things that cannot be directly observed. For example, scientists cannot see the particles of an atom, yet scientists have used indirect evidence to infer the existence and characteristics of protons, neutrons, and electrons. An **inference** is a tentative conclusion based on logic or reasoning.

Figure 2 (a) A radio antenna and (b) a multimeter are examples of tools scientists use to detect and measure phenomena that cannot be observed directly.

Scientific Knowledge Comes from Observations

Scientific knowledge is acquired by careful observation and experimentation. Knowledge that is obtained in this way is generally referred to as **empirical knowledge**. Thus, empirical knowledge is knowledge gained through experiences.

People often think that empirical knowledge is produced only by science done in laboratories. There are other very important sources of empirical knowledge, however. Traditional or indigenous knowledge has been around much longer than modern science and provides a wealth of empirical knowledge that could not possibly be obtained through laboratory studies. Indigenous knowledge can help us learn better ways to live in harmony with our world.

Aboriginal peoples lived in their traditional territories long before the first explorers and immigrants arrived in North America. Indigenous peoples, and later the Métis people, lived very closely with nature. From their experiences they developed very detailed knowledge about their environment, including knowledge about plants, animals, weather, and landforms. For this reason, indigenous knowledge is also known as **Traditional Ecological Knowledge and Wisdom (TEKW)**.

Indigenous peoples have an oral tradition: their empirical knowledge has been carefully told by one generation to the next so that it will not be lost. The Elders are the people in each community who have this important knowledge. They teach it to the young people so that future generations can live carefully and successfully in the environment.

Traditional Ecological Knowledge and Wisdom has benefited humanity in numerous ways. For example, many modern drugs have been developed based on the knowledge of indigenous peoples. Lidocaine, one of the most common local anesthetics (pain relievers) used by doctors and dentists, is derived from the coca shrub. This shrub is native to the Andes Mountains in South America, and indigenous peoples were well aware of its anesthetic properties. There is evidence that the leaves of this shrub were chewed for pain relief as early as 300 BCE. Quinine, made from the bark of the cinchona tree, is used to treat malaria. Curare, derived from the bark of the strychnos plant, was used as arrow poison by indigenous South American tribes. Today, curare is used during surgery as a muscle relaxant. The discovery of many of these drugs has been claimed by modern science even though the real discoverers were people who experimented with these natural remedies long before modern medical science learned of their existence.

As they have done for centuries, indigenous peoples around the world continue to observe, describe, explain, predict, and work with the natural world. Indigenous knowledge is increasingly valued as people around the world are becoming aware of, and concerned about, environmental issues.

In Ontario, as in other places around the world, more and more scientists are working with Aboriginal communities and tapping into indigenous knowledge. Indigenous knowledge has made many contributions to modern society and continues to provide important biological and ecological insights.

Scientific Knowledge Is Tentative But Reliable

Scientists analyze their observations by looking for patterns. If a pattern is discovered, a law may be formulated. A **law** is a general statement, based on extensive empirical data, about *what* has happened; it does not explain *why* this happened.

A law can be used to make predictions. The law of gravity is a good example. Through extensive and repeated observations, Sir Isaac Newton concluded that all objects in the Universe exert a gravitational force on one another, and that the strength of the force depends on two factors: the masses of the objects and the distance between the objects. According to Newton's universal law of gravitation, the attractive force between two objects increases as mass increases, and also increases as the objects get closer together. The universal law of gravity does not explain gravity, but it describes the effects of gravity. It can also be expressed mathematically.

A law cannot be proven to be true because it is impossible to test every possible situation in which the law might apply. A law can be proven to be false, however, if evidence is provided that contradicts or disagrees with it. For example, imagine that a location is found in a remote part of our solar system where objects "fall" upward. This contradicts the law of gravity as we know it, and the law would have to be rejected or modified to account for the new observations. Even though scientific laws cannot be proven true, you should be confident in their validity because of the vast number of observations that support them.

Unlike a law, which is determined by careful analysis of observations, a theory is a product of a scientist's creativity and inventiveness. It is important to not confuse the everyday use of the word "theory" with the scientific definition of the word. In everyday use, a theory is simply a hunch or a guess. For example, you might say, "I have a theory about why Aaron did not go to the school dance." In this case, a theory is a personal opinion that attempts to explain a simple, single event.

In science, a **theory** is an explanation of observations (or of a law). In attempting to develop a theory, a scientist first suggests a tentative answer or an untested explanation called a **hypothesis**. The most important characteristic of a hypothesis is its testability, or the ability to obtain evidence that will test the explanation. Such a test can never prove a hypothesis to be true. A test can either support a hypothesis or refute it—that is, prove it to be unacceptable.

For an explanation (hypothesis) to become a theory, it must be widely accepted in the scientific community as the best possible explanation after being tested extensively through repeated observations. In some cases, theories can be tested by experimentation under controlled conditions. In other cases, scientists must rely on observations of natural phenomena instead (Figure 3).

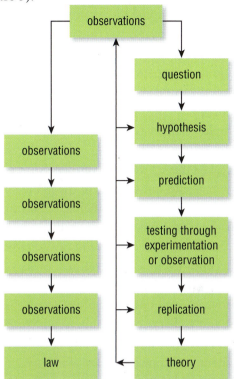

Figure 3 A scientific law is a description. A scientific theory is an explanation.

For example, when formulating the theory of plate tectonics (the large-scale movement of Earth's land masses), scientists could not control conditions. The movements of tectonic plates take place very slowly, over long periods of time, and they are impossible to control. Scientists therefore relied on direct observations and indirect evidence rather than experiments to support their hypotheses. This is frequently the case in sciences such as geology and astronomy.

Science Is Progressive

As already mentioned, a scientific theory is a current but tentative explanation—it is not a fact. Theories are never final and are always being examined and questioned. This means that a theory can change. Even though scientific knowledge is very reliable, it is also tentative, meaning that it can change with new evidence. A theory is valid only as long as every new piece of evidence supports it. This is one characteristic that distinguishes scientific knowledge from other types of knowledge.

A theory is accepted by the scientific community when all available empirical knowledge supports it. A theory, however, can change when new scientific evidence suggests that a change is justified. This does not make the original theory any less valuable. Just because a theory is not 100 % correct does not automatically make it 100 % wrong. Even a theory that is not quite right provides a basis for further scientific investigations, which often results in an improved theory.

For example, in the early 1600s, Galileo believed that the force of gravity depended only on the mass of an object. His theory did not completely explain the arrangement and movements of the planets in the Solar System. Newton and other scientists used Galileo's work as the basis for further investigations, which eventually led to modification of Galileo's theory to account for new observations. Even today, scientists continue to investigate gravity to try to improve current understanding. Science is progressive and can advance because scientists build on existing knowledge and are willing to change their thinking when new knowledge is available.

Science Is Repeatable, Self-Correcting, and Not Based on Authority

There are many other characteristics of science that make it the best way of understanding the natural world. For example, in order to be authentic and valid, scientific investigations must be repeatable. Under the same conditions, scientists should be able to conduct the experiments carried out by other scientists and obtain the same results. This characteristic ensures that science is self-correcting. If a scientist's explanations are incorrect or the evidence used to support the hypothesis is not valid, other scientists will test the explanation and reveal that it is incorrect. This self-correction may take a long time; the modern atomic theory that explains the nature of matter is still being modified after more than 200 years.

Even though we often look to experts in science as sources of accurate information, science does not rely on authority. Even the most famous and highly regarded scientists cannot decide or proclaim the correct explanation of a natural phenomenon. No scientist or authority, and no theory, regardless of how widely accepted, can escape the scrutiny of other scientists. A scientific theory is not judged on the credentials or reputation of the scientist who creates it but rather on its ability to answer questions and explain some aspect of the world around us.

The Scientific Method

Although we often hear of the scientific method, there is no single, correct way to do science that is followed by all scientists. It is not a recipe to be followed step by step, exactly the same in each investigation. Scientific inquiry is generally described as a series of steps:

1. A scientist first asks a testable question and develops a hypothesis or possible answer.
2. Then the scientist designs and performs an investigation, makes observations, and analyzes them.
3. Finally, the scientist draws a conclusion based on the evidence, and compares the conclusion with the hypothesis to determine whether the evidence supports the hypothesis.

Variations of this method are common in science, because scientists can approach problems differently. Furthermore, the various scientific fields require different approaches, and scientists use skills and technologies specific to their field of study. If you asked 10 scientists to describe their method, you would likely get 10 unique descriptions. There would be similarities, however. There is no doubt that science is methodical. It follows procedures that generally lead to a logical conclusion, which addresses the goal of the investigation and hopefully answers the question. Nevertheless, there is no single scientific method that all scientists follow step by step. The term **scientific method** refers to the general types of mental and physical activities that scientists use to create, refine, extend, and apply knowledge.

Classification of Science

There are many ways of classifying the activities that we refer to as science. Science tries to understand nature, which really means the whole universe. Science is usually divided into three main branches—life science, physical science, and Earth and space science.

Life science is the study of living things and is often referred to as biology. It has many branches, including botany, the study of plants; zoology, the study of animals; and ecology, the study of the natural environment. Medicine and agricultural science are also branches of life science because they deal with the study of living things.

Physical science has two main branches—chemistry and physics. Chemistry is the study of matter and its changes, and physics is the study of forces and energy.

The branches of Earth and space science include geology, the study of the physical nature and history of Earth; meteorology, the study of the atmosphere and weather; astronomy, the study of celestial objects (such as galaxies, stars, planets, and comets); and oceanography, the study of oceans and ocean life.

This classification of the branches of science appears simple and neat, like the drawers in which nuts and bolts are arranged in a hardware store (Figure 4). In reality, science is not quite so simple. As science progresses over time, the branches of science expand and cross over so that they no longer fit into the neat categories. For example, some chemists study chemicals that make up living things (such as DNA). This is the science of biochemistry, the study of the matter of living things. It is both a life science and a physical science. In the same way, geophysics, the study of forces that affect Earth, is both an Earth science and a physical science.

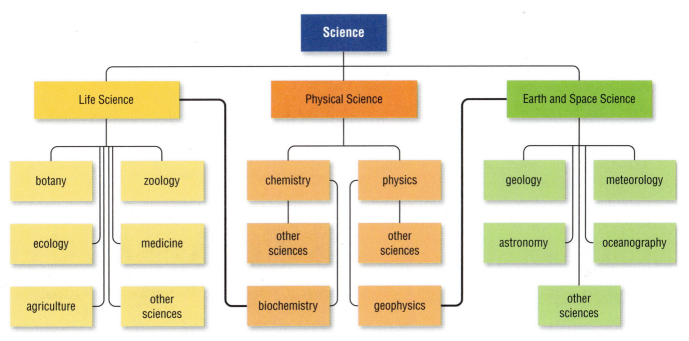

Figure 4 There are many separate branches of science, but there are also many interactions among the various branches.

RESEARCH THIS   IDENTIFYING SCIENCES

SKILLS: Researching, Communicating

The branches of science and their interactions are too numerous to identify in this text. However, you can get an idea of the nature of a branch of science from its name.

1. Examine the following list of science branches and classify them as a life, physical, and/or Earth and space science.
2. For each of the sciences, analyze the name and write a one-sentence description that you think matches the science.
A. Compare your definitions with those of your classmates, and then, using a dictionary or the Internet, find a formal definition for each of the sciences. Compare your definition with the formal definition.

Some branches can fit into more than one category.
- aerobiology
- biophysics
- genetics
- microbiology
- astrobiology
- environmental science
- marine biology
- optics
- analytical chemistry
- electrostatics
- geochemistry
- mineralogy
- astrochemistry
- fluid dynamics
- materials science
- sedimentology

 GO TO NELSON SCIENCE

The Nature of Science

SCIENCE, TECHNOLOGY, SOCIETY, AND THE ENVIRONMENT (STSE)

Science and technology are very different activities, but we often hear about them together. This is because they are closely related and often go hand in hand. Scientists rely on technologies to further their research, and technologists rely on scientists to understand the scientific basis of technological developments.

A Partnership: Science and Technology

Scientific research produces knowledge, or understanding of natural phenomena. Technologists and engineers look for ways to apply this knowledge in the development of practical products and processes. For example, scientists want to know how the force of gravity changes as you travel farther from Earth (Figure 5). Technologists and engineers use that knowledge to calculate how much power will be needed to put a spacecraft in orbit around Earth or how much fuel will be required to send a spacecraft to the Moon or Mars. Scientists use technologies when doing their research. Engineers and technologists may use scientific knowledge and principles when designing and developing new technologies.

In some cases, technological inventions and innovations occur before the scientific principles are known. Native peoples understood and managed natural resources in a sustainable way long before the science of ecology came into existence. They also designed effective tools that used scientific principles, even though they may not have understood these principles. For example, Australian indigenous peoples designed the boomerang long before the principles of air pressure and flight were understood and described.

In other cases, scientific discoveries are made because of technological inventions. For example, the invention of glass lenses led to the development of telescopes, which allowed astronomers to observe our solar system and the Universe (Figure 6). The telescope, in turn, led to more accurate astronomical observations and measurements, which contributed to the change from an Earth-centred scientific model of the Universe to a Sun-centred model. Many other technologies, including the thermometer and the computer, have been greatly helpful.

Figure 5 The amount of fuel used to launch a shuttle into space must be carefully calculated to ensure there is not too much or too little.

Figure 6 The development of the telescope followed the understanding of the structure of glass lenses and the behaviour of light as it passed through these lenses.

Sometimes, technological inventions follow scientific discoveries (Table 1). For example, the television was invented after scientists had created theories to explain the structure of the atom and understood electrons, current electricity, and electromagnetism. The relationship between science and technology is mutually beneficial; scientific discoveries lead to technological advances, which lead to further scientific discoveries, and so on.

> **DID YOU KNOW?**
> **The First Telescope**
> Contrary to popular belief, the first telescope was not invented by Galileo. Hans Lippershey, a Dutch eyeglass maker, invented the telescope when he put a convex and a concave lens together to achieve 3X magnification. His patent application was denied because it was felt that the device could not be kept secret. In 1609, Galileo learned of the invention and immediately saw its potential. He experimented with different lens curvatures and arrangements, and achieved a magnification of 9X—an invaluable asset to scientific and military endeavours.
>
> **GO TO NELSON SCIENCE**

Table 1 Examples of the Science–Technology Relationship

Science	Technology	Example	
Scientists learn that atoms contain protons and electrons. Electrons contain negative charges and these charges can be transferred from one atom to another.	Technologists use the knowledge about electrical charges to develop ways to store energy and to transfer energy from one place to another.	Lithium ion cell (battery)	
Chemists learn about the physical properties of substances, such as their density, melting point, and boiling point.	Technologists use substances with specific physical properties for specific purposes.	Solutions with low freezing point used as antifreeze	
Biologists learn about the nutrients that plants require in order to grow.	Technologists develop environmentally friendly soil supplements to increase crop yields.	New carbon-negative soil supplements—biochar or agrichar	
Physicists discover how light behaves when it passes through or reflects off different materials.	Technologists design lenses and mirrors for telescopes, microscopes, and other optical devices.	Optical telescope	

Science, Technology, Society, and the Environment (STSE)

Science, Technology, and Society

You do not have to look far to find evidence of the impact of science and technology on society. Many important discoveries and inventions have occurred within the last century—vaccines, antibiotics, organ transplants, reproductive technologies, genetic engineering, pesticides, atomic weapons, computers, lasers, plastics, televisions, communication satellites, and the Internet. The homes we live in, the food we eat, the vehicles we drive, and the electronic gadgets we use are all products of scientific and technological achievement (Figure 7).

Figure 7 Examples of science and technology in our daily lives: (a) microwave (b) vehicle GPS.

There are obvious influences of science and technology on society. But science and technology are influenced by society as well. The values and priorities of society at a particular time can influence the direction and progress of developments in both science and technology. For example, our desire to reduce greenhouse gas emissions is encouraging the rapid development of vehicles and machinery that use alternative fuels.

Scientific and technological research is very expensive. Research facilities employ highly paid and highly skilled professionals. The facilities consume large amounts of energy and require expensive and sophisticated tools and equipment. Funding for research comes from both private and public sources. It may be a long time between the beginning of research and development and the release of a new product or process.

Basic Research

Basic research helps people learn more about how the natural world works and is essentially the same as scientific investigation. Basic research—in areas such as biochemistry, particle physics, astronomy, and geology—usually receives funding from government grant agencies. This type of research often produces knowledge that engineers and technologists can use to develop practical solutions to everyday problems. The priorities of government, representing the priorities of the public, determine which areas of research are funded.

Applied Research

Applied research is primarily focused on developing new and better solutions to practical problems. Applied research can be equated with the development of technology. Research into the development of technology—such as new cosmetics, sports equipment, telephones, automobiles, computer software and hardware, medical equipment, and pharmaceuticals—is usually carried out by privately owned companies. The marketplace, or the public demand for new products, will obviously influence which areas of research private companies fund. If market analysis shows that there is a demand for better cellphones or more fuel-efficient cars, research and development in these areas will be supported.

There are risks and benefits associated with many scientific and technological developments. Although most technologies are developed with the intention of solving problems, there are often unintended consequences associated with their use. Social networking sites and other Internet technologies can greatly facilitate communication; however, the same technology is also used to commit crimes.

It is difficult, if not impossible, to foresee all of the consequences of technologies. Often, new applications of scientific and technological knowledge are found long after the knowledge was first acquired. An important question remains up for debate. Who is responsible for the negative impacts of science and technology on society: scientists, science and technology, or human use of science and technology?

Science, Technology, and the Environment

Since the beginning of the Industrial Revolution in the late 1700s, the industrialized world has used natural resources at an increasing rate. With the use of resources, the development of technology, and the human population all increasing, we are producing waste faster than ever.

Many of the by-products of human industry and technology become pollutants. For example, there are hundreds of millions of discarded cellphones in North America, most of which end up in landfills (Figure 8).

Figure 8 Cellphones contain hazardous substances such as lead, mercury, and arsenic. These metals can leach into groundwater if they are dumped into landfills.

The impact of science and technology on the environment is not always negative. There are many positive effects of science and technology. New knowledge improves our understanding of the natural world, and new technologies allow us to live with the environment in a more sustainable way.

Figure 9 illustrates how technology can transform the energy of the Sun into electrical energy. This technology is generally more environmentally friendly than the use of fossil fuels. Researchers in the auto industry are busy developing vehicles that use alternative sources of energy and are more fuel efficient. The intention is that these vehicles will have less of an impact on the environment.

Figure 9 (a) Small solar panels can charge your cellphone. (b) Large panels can heat water or power your home.

The relationships among science, technology, society, and the environment are complex. To make wise personal decisions and to act as responsible citizens we must think carefully about the influences of each of these elements on the others. We each have a responsibility to ourselves, to society, and to future generations to become scientifically and technologically literate.

TRY THIS: IS IT SCIENCE OR TECHNOLOGY?

SKILLS: Analyzing, Evaluating, Communicating

Some achievements are clearly in the realm of science (discoveries) and others are clearly examples of technology (inventions). Others are more difficult to classify.

1. In a small group, examine each of the following descriptions and decide whether it is an example of science, technology, or both. In each case, give reasons for your choice.

    (a) A laboratory at the University of Toronto has identified the properties of a new chemical element on the periodic table.

    (b) A material was developed that conducts electricity with very little resistance.

    (c) The wires used in the International Space Station are made from a material that conducts electricity with very little resistance.

    (d) A space probe will be sent to Mars to study its surface.

    (e) The silk in a spider's web is, in equivalent sizes, six times stronger than steel.

    (f) After studying millions of fingerprints, it is generally accepted that every fingerprint is unique.

    (g) People can communicate by sending radio waves from one location to be received at a different location.

    (h) It has been determined that the average global temperature has risen during the last century.

    (i) When electricity flows through a copper wire, the temperature of the wire rises.

    (j) Some plants absorb metals from the soil and are used to clean up polluted areas.

    A. Which examples of science and technology were the most difficult to identify? Why?

Numerical and Short Answers

This section includes numerical and short answers to questions in Check Your Learning, Chapter Review, Chapter Self-Quiz, Unit Review, and Unit Self-Quiz.

Unit B

Section 2.2, p. 35
2. yes
3. backyard pond, tree, schoolyard, potted plant

Section 2.4, p. 41
1. 0.023 %
2. sugars
4. plant growth and maintenance

Section 2.5, p. 47
5. first trophic level
6. third and fourth trophic levels
8. (b) rabbit, mouse, squirrel
   (c) grasses, berries, tree seeds

Section 2.6, p. 51
3. burning fossil fuels, deforestation
7. burning fewer fossil fuels, planting more trees
10. decomposition

Section 2.7, p. 55
8. (a) yes

Section 2.8, p. 59
3. deciduous forest
4. climate

Section 2.9, p. 62
2. salt concentrations, nutrient levels, temperature, light

Chapter 2 Review, pp. 68–69
1. (a) lithosphere
   (b) cellular respiration; photosynthesis
   (c) hydrosphere
   (d) sustainable
   (e) carrying capacity
   (f) marine
   (g) ecosystem
   (h) hydrosphere; lithosphere; thermal
   (i) photosynthesis

3. (c) photosynthesis and cellular respiration
   (d) cellular respiration
   (e) only for photosynthesis
   (f) only during cellular respiration
   (g) only for photosynthesis
   (h) only for photosynthesis
6. (a) (iii)
   (b) (iv)
   (c) (i)
   (d) (v)
   (e) (ii)
11. the Sun
13. the Sun
16. dolphin, seal, small fish

Chapter 2 Self-Quiz, pp. 70–71
1. (b)
2. (a)
3. (b)
4. (c)
5. (c)
6. (a)
7. F
8. F
9. T
10. photosynthesis
11. oxygen
12. species
13. (a) (ii)
    (b) (iii)
    (c) (iv)
    (d) (i)
14. oxygen
17. (a) increase
    (b) increase
25. (a) vegetation, animals
    (b) temperature, amount of rainfall
26. (a) e.g., bees and flowering plants
    (b) e.g., nesting birds and trees

Section 3.2, p. 82
3. secondary

Section 3.3, p. 86
4. (a) habitat destruction, overexploitation
6. extirpated, endangered, threatened, special concern

Section 3.4, p. 90
4. Africa, Latin America, the Caribbean

Section 3.5, p. 94
1. no
7. chemical, mechanical, and biological control

Section 3.6, p. 101
1. human population increase, modern industry
2. sulfur dioxide and nitrogen oxide
4. limestone
6. skimming, burning, detergents, bioremediation

Section 3.7, p. 105
1. wood, wildlife
2. clear-cutting, shelterwood cutting, selective cutting
8. clear-cutting

Chapter 3 Review, pp. 110–111
1. (a) services
   (b) secondary
   (c) rainforests
   (d) risk
   (e) habitat fragmentation
   (f) habitat
   (g) invasive
2. (a) (iii)
   (b) (iv)
   (c) (ii)
   (d) (v)
   (e) (i)
3. (a) tropical rainforest
   (b) near the equator
5. skimming, burning, detergents, bioremediation

6. (a) cultural service
   (b) product
   (c) other service
   (d) product
   (e) cultural service
   (f) other service
   (g) product
7. (a) (v)
   (b) (iv)
   (c) (i)
   (d) (ii)
   (e) (iii)
12. biological, mechanical, and chemical control

Chapter 3 Self-Quiz, pp. 112–113
1. (b)
2. (a)
3. (c)
4. (b)
5. primary
6. acid precipitation
7. equilibrium
8. F
9. F
10. F
11. (a) (iv)
    (b) (i)
    (c) (v)
    (d) (ii)
    (e) (iii)
19. yes

Section 4.1, p. 122
5. no
7. no

Section 4.2, p. 128
8. crop rotation, crop selection, no-till farming

Section 4.4, p. 134
1. lack of biodiversity
2. a species that reduces crop yield
3. herbicides, insecticides, rodenticides, fungicides, molluscicides

Section 4.5, p. 140
1. less crop damage, greater crop yield, insect population control

Chapter 4 Review, pp. 150–151
1. (a) pest
   (b) natural
   (c) pesticides
   (d) bioaccumulate
   (e) narrow-spectrum
   (f) integrated pest management
   (g) monoculture
   (h) engineered
2. (a) natural
   (b) monoculture
   (c) monoculture
   (d) monoculture
   (e) monoculture
   (f) monoculture
4. (a) (v)
   (b) (iv)
   (c) (iii)
   (d) (i)
   (e) (ii)
7. tools, concentrated energy sources
8. engineered
15. to reduce soil compaction and erosion

Chapter 4 Self-Quiz, pp. 152–153
1. (c)
2. (a)
3. (b)
4. (d)
5. F
6. T
7. engineered
8. monoculture
9. leaching
10. organic farming
11. (a) (iii)
    (b) (i)
    (c) (ii)
    (d) (iv)
18. natural forest
27. nitrogen, phosphorus, potassium
30. (a) allow nutrients, water, and oxygen to reach roots

Unit B Review, pp. 158–163
1. (b)
2. (d)
3. (b)
4. (d)
5. (d)
6. (c)
7. (c)
8. (c)
9. (b)
10. (d)
11. (c)
12. (d)
13. (c)
14. (b)
15. (d)
16. (b)
17. T
18. T
19. F
20. T
21. F
22. T
23. F
24. F
25. F
26. T
27. F
28. T
29. tolerance range
30. carrying capacity
31. marine
32. clearing and burning
33. estuary
34. biological control
35. sulfur dioxide; nitrogen
36. selective cutting
37. thermal
38. nitrogen; phosphorus; potassium
39. tundra
40. ecological footprint
41. (a) (iv)
    (b) (ii)
    (c) (i)
    (d) (v)
    (e) (iii)
42. the Sun
57. (a) increase
64. (b) yes
66. (a) just after the red crab migration
67. (a) an inverse relationship
    (b) after
68. (b) grasses

Unit B Self-Quiz, pp. 164–165
1. (b)
2. (c)
3. (b)
4. (d)
5. F
6. F
7. T
8. leaching
9. ecosystem
10. (a) (v)
    (b) (ii)
    (c) (iv)
    (d) (vi)
    (e) (iii)
    (f) (i)
17. (a) soil compaction
18. (a) temperate deciduous forest
    (b) tundra

Unit C
Section 5.1, p. 178
8. (a) solution
   (b) mechanical mixture
   (c) solution
   (d) mechanical mixture
   (d) mechanical mixture
10. lead poisoning

Section 5.2, p. 182
2. (a) qualitative
   (b) quantitative
   (c) qualitative
   (d) quantitative

Section 5.3, p. 186
3. (a) physical
   (b) physical
   (c) chemical
   (d) chemical
4. (a) physical
   (b) physical
   (c) chemical
   (d) physical
   (e) chemical
5. emission of light

Section 5.6, p. 198
2. liquid at room temperature, conducts electricity
3. sink
4. 0.48 cm^3
5. 8.96 g/cm^3, copper
6. 33.6 g
7. 2.70 g/cm^3, aluminum
8. 0.63 cm^3
9. 11.36 g/cm^3, lead
10. 14 252.77 g

Chapter 5 Review, pp. 202–203
3. boiling point, melting point, density
5. Workplace Hazardous Materials Information System
7. (a) qualitative
   (b) quantitative
   (c) qualitative
   (d) quantitative
   (e) qualitative
10. (a) physical
    (b) physical
    (c) physical
    (d) chemical
    (e) physical
    (f) chemical
    (g) chemical
11. chemical
12. physical
16. 69 g
17. 6.86 kg/L or 6.86 g/cm^3
18. 0.75 g/mL or 0.75 g/cm^3; yes
19. 187 cm^3

Chapter 5 Self-Quiz, pp. 204–205
1. (b)
2. (d)
3. (a)
4. (c)
5. F
6. T
7. faster
8. physical
9. (a) (iv)
   (b) (i)
   (c) (ii)
   (d) (iii)
   (e) (v)
10. colour and odour change
12. 5 g of liquid gold
15. 2.70 g/cm^3
16. 2.52 g
18. no
24. float
25. (a) physical
    (b) chemical
    (c) chemical
    (d) physical
    (e) physical

Section 6.1, p. 215
1. (b) tin
   (c) chromium
   (f) arsenic
   (g) nickel
3. no
5. metallic
6. high lustre, good conductivity, malleable, ductile
7. low lustre, poor conductivity, brittle
8. (a) metals
   (b) non-metals
   (c) metals
   (d) non-metals
   (e) non-metals
   (f) non-metals
   (g) metals
   (h) metals
9. (a) good conductors of thermal energy
   (b) malleable, lustrous, do not corrode
   (c) poor conductor of thermal energy
10. (b) thermal and electrical conductivity

Section 6.4, p. 225
1. (a) incorrect
   (b) incorrect
   (c) correct
   (d) correct
2. (a) halogens
   (b) alkaline earth metals
   (c) alkali metals
   (d) noble gases
3. (a) all but one of the names end in "-ium"
   (b) hydrogen
4. highly reactive, too soft
6. Group 17, halogens

Section 6.6, p. 233

2. (b) electrons
   (c) embedded in the atom
   (d) atoms are neutral
   (e) surrounding the electrons
3. 3
5. (c) protons and electrons
   (d) protons in nucleus, electrons surround nucleus

Section 6.7, p. 240

1. yes
2. no
3. (a) correct
   (b) incorrect
   (c) correct
   (d) incorrect
   (e) correct
   (f) correct
4. 30
5. 2, 8, 8
8. (a) F
   (b) T
   (c) T
   (d) T

Section 6.8, p. 244

6. (a) graphite
   (b) diamond
   (c) charcoal

Chapter 6 Review, pp. 248–249

1. (a) halogens
   (b) alkaline earth metals
   (c) alkali metals
   (d) noble gases
2. divides metals from non-metals
3. Dmitri Mendeleev
14. (a) 2, 8, 8
    (b) 18
19. (a) C; B
    (b) B; C

Chapter 6 Self-Quiz, pp. 250–251

1. (a)
2. (d)
3. (b)
4. (c)
5. F
6. T
7. metals, non-metals
8. halogens
9. isotopes
10. (a) (v)
    (b) (i)
    (c) (ii)
    (d) (iii)
    (e) (iv)
12. (a) negative
    (b) no
13. fluorine
14. (a) 78
    (b) platinum
16. 3
19. (a) they would conduct thermal energy
    (b) non-conducting material, e.g. plastic
22. 0
23. less reactive, more stable

Section 7.1, p. 261

1. (a) 4
   (b) 1 Na atom, 1 H atom, 1 C atom, 3 O atoms
   (c) yes
2. (a) H_2, S_8, Ne
   (b) CO_2, C_3H_8
   (c) Ne
   (d) H_2, CO_2, S_8, C_3H_8
4. H_2, N_2, O_2, F_2, Cl_2, Br_2, I_2
5. they contain only hydrogen and carbon
7. (a) NH_3
   (b) CO_2
   (c) CO
   (d) H_2O
   (e) C_3H_8O
8. ion with +2 electrical charge
9. Ca^{2+}
11. (a) molecular element
    (b) ionic compound
    (c) molecular compound
    (d) molecular compound

Section 7.3, p. 266

1. (a) 58 %
   (b) no
3. (a) lye or caustic soda
   (b) ozone
   (c) table salt
   (d) as a solid, dry ice; as a gas, carbon dioxide
   (e) baking soda
   (f) as a mineral, limestone or marble
4. (a) hydrochloric acid
   (b) acetic acid
   (c) potassium carbonate
   (d) calcium oxide
   (e) magnesium hydroxide
   (f) methane
5. (c) yes
10. non-metallic atoms

Section 7.6, p. 273

3. bleaches colour
4. oxygen produced kills bacteria
5. to prevent breakdown into water and oxygen

Chapter 7 Review, pp. 280–281

2. (b) H_2, N_2, O_2, F_2, Cl_2, Br_2, I_2
3. (a) ionic
   (b) ionic
   (c) molecular
4. (a) CO_2
   (b) H_2
   (c) O_2
7. non-metallic elements
8. metallic elements and non-metallic elements
9. FeO and Fe_2O_3
10. compound; production of gas
15. 2 H_2O, 1 CO_2

Chapter 7 Self-Quiz, pp. 282–283

1. (d)
2. (a)
3. (c)
4. (a)
5. (b)
6. F
7. F
8. T
9. hydrocarbons
10. covalent
11. cation
12. (a) (v)
    (b) (i)
    (c) (ii)
    (d) (iii)
    (e) (iv)
13. full outer electron orbits

15. (a) one electron transferred from K to Cl
    (b) 18
    (c) 18
17. (a) burning splint test
18. (a) both react with oxygen

Unit C Review, pp. 288–293
1. (c)
2. (a)
3. (b)
4. (c)
5. (c)
6. (c)
7. (b)
8. (b)
9. (d)
10. (d)
11. (c)
12. (d)
13. T
14. F
15. F
16. F
17. T
18. F
19. T
20. F
21. F
22. F
23. T
24. T
25. T
26. mass
27. density
28. Workplace Hazardous Materials Information System
29. wafting
30. physical
31. chemical
32. oxygen
33. vinegar
34. table salt
35. J.J. Thomson
36. sulfur
37. decomposition
38. hydrogen
39. limewater
40. (a) (iv)
    (b) (vii)
    (c) (i)
    (d) (viii)
    (e) (ii)
    (f) (iii)
    (g) (v)
    (h) (vi)
41. (a) (iii)
    (b) (iv)
    (c) (i)
    (d) (ii)
42. (a) (ii)
    (b) (iii)
    (c) (iv)
    (d) (v)
    (e) (i)
44. (a) physical
    (b) physical
    (c) chemical
    (d) physical
    (e) chemical
    (f) physical
    (g) physical
54. mass number
59. metals
65. low or no reactivity
66. (a) the same
72. (a) ductility
74. (a) B
    (b) A
75. (a) burning splint test
    (b) there are no carbon atoms
76. eagles

Unit C Self-Quiz, pp. 294–295
1. (c)
2. (b)
3. (a)
4. (b)
5. F
6. T
7. F
8. cations
9. qualitative; quantitative
10. (a) (iii)
    (b) (ii)
    (c) (i)
    (d) (iv)
11. (a) physical
    (b) chemical
    (c) chemical
    (d) physical
    (e) physical
13. (a) agree
    (b) agree
    (c) disagree
16. (a) $CaCl_2$
    (b) N_2O_4
    (c) Na_2S_2
20. physical
23. (a) d = 19.33 g/cm³, gold
    (b) d = 5.02 g/cm³, pyrite

Unit D
Section 8.1, p. 308
1. celestial objects
3. it reflects sunlight
8. Milky Way galaxy
9. Universe, galaxy, star, planet, moon

Section 8.2, p. 312
4. 2 000 000–15 000 000 °C
5. approximately 25 days
6. Galileo Galilei
8. North and South poles

Section 8.3, p. 317
3. (b) 150 000 000 km
4. dwarf planet
5. (a) meteor
   (b) meteorite
   (c) meteoroid
8. (a) 2061
   (b) 4377

Section 8.5, p. 328
1. gas giants
5. (a) gravity
   (b) objects fall to the ground
   (c) Sir Isaac Newton
6. above the North Pole; Polaris
9. (a) around June 21 and December 21
10. (a) summer
    (b) fall
    (c) winter
    (d) spring

11. (a) (iv)
    (b) (iii)
    (c) (i)
    (d) (ii)
12. 8
13. solar, lunar
14. alignment of the Sun, the Moon, and Earth; no

Section 8.6, p. 333
2. yes; the shape will change
4. North Star or Polaris

Section 8.9, p. 343
1. the Moon, Polaris, comets
3. ecliptic
6. retrograde motion
7. they have a slower orbit than Earth
9. 180°
10. 90°
12. 4°
13. 3

Section 8.11, p. 351
3. (a) 1957; Soviet Union
   (b) 1962; Alouette 1
9. Global Positioning System

Chapter 8 Review, pp. 356–357
2. (a) chromosphere
   (b) photosphere
   (c) core
   (d) corona
   (e) convective zone
   (f) radiative zone
   (g) solar prominence
   (h) solar flare
3. from east to west, along ecliptic
8. small, rocky celestial object; in orbit between Mars and Jupiter
9. large chunk of ice, rock, and dust; beyond orbit of Neptune
10. apparent backward motion of a planet
11. gravitational force
23. 35 790 km
24. altitude, azimuth; date, time, location
25. Polaris
30. equator

Chapter 8 Self-Quiz, pp. 358–359
1. (d)
2. (d)
3. (a)
4. (b)
5. T
6. F
7. Sun
8. summer
9. luminous
10. (a) (v)
    (b) (i)
    (c) (ii)
    (d) (iv)
    (e) (iii)
12. (a) 30.0 AU
    (b) 108 000 000 km
19. yes
23. no

Section 9.1, p. 369
2. 5.7×10^{15} km
3. 430 ly
5. star positions, parallaxes, and motions
6. 1.3×10^{26} m
7. 61.2 AU
10. 30 cm
11. (a) AU
    (b) m or km
    (c) ly or AU
    (d) ly
    (e) m or km

Section 9.2, p. 373
2. the total amount of energy produced per second
4. Hipparchus; 2100 years ago
6. 33 ly from Earth; yes
7. 3100 °C, red; 4800 °C, orange; 8000 °C, yellow; 10 200 °C, blue

Section 9.4, p. 382
1. 15 million °C (1.5×10^7 °C)
4. (b) no
7. (a) solar mass

Section 9.5, p. 384
5. hydrogen, helium
6. a shockwave from a nearby supernova; contraction

Section 9.6, p. 391
4. (a) 1935, 2005
8. Local Group, 35 galaxies
9. Virgo Supercluster, thousands of galaxies

Section 9.7, p. 397
4. the Universe's rate of expansion is increasing
5. 13.6–13.8 billion years ago; Georges Lemaître
6. no
8. COBE and WMAP; temperature variations

Chapter 9 Review, pp. 400–401
1. a massive cloud of interstellar gases and dust
3. elliptical, spiral, lenticular, and irregular
4. distance from Earth, size
7. above and below the disc of the Milky Way
8. a huge, energy-rich galaxy with a black hole in its centre
12. nebula, protostar, nuclear fusion, supernova, neutron star
14. 5.9×10^{15} km
21. red giant, orange main sequence star, yellow main sequence star, white dwarf, blue supergiant
22. 1.1×10^{23} km
23. radiation
24. 25 000 °C
25. +5
26. Virgo Supercluster, spiral galaxy, nebula, red supergiant, white dwarf, neutron star
27. spiral
32. it is moving toward Earth

Chapter 9 Self-Quiz, pp. 402–403

1. (d)
2. (b)
3. (a)
4. (d)
5. F
6. T
7. T
8. parallax
9. luminosity
10. neutron stars
11. (a) (iv)
    (b) (ii)
    (c) (i)
    (d) (iii)
13. apparent magnitude scale
16. 3.78×10^{15} km
18. black hole or neutron star
20. (a) red shift
21. 10 Lacertra, Canopus, Aldebaran, Antares
23. (a) light years
    (b) AU
    (c) kilometres
24. (b) galaxy is moving toward Earth

Section 10.1, p. 418

2. radio waves, X-rays, gamma rays, ultraviolet and infrared rays
7. over 16

Section 10.2, p. 425

2. small robotic probes

Section 10.3, p. 430

1. barcode scanning
5. coastlines, glaciers, oceans, land, freshwater resources
6. ozone layer

Section 10.5, p. 437

2. it has similar characteristics
5. 4 days vs. 6–10 months
7. Dennis Tito, $30 million
9. counterweight, tether, climber

Chapter 10 Review, pp. 440–441

1. Galileo Galilei, 1609
6. Global Positioning System
9. *Ares V* and *Ares I*
10. bone density
14. microgravity environment
19. (a) (iv)
    (b) (ii)
    (c) (i)
    (d) (iii)

Chapter 10 Self-Quiz, pp. 442–443

1. (c)
2. (a)
3. (c)
4. F
5. T
6. F
7. microgravity
8. reflecting
9. spacecraft
10. (a) (i)
    (b) (iv)
    (c) (ii)
    (d) (iii)
11. it has similar characteristics
14. they are in free fall
25. using resistance devices, running on treadmill

Unit D Review, pp. 448–453

1. (d)
2. (a)
3. (b)
4. (a)
5. (b)
6. (d)
7. (a)
8. (b)
9. (a)
10. (b)
11. F
12. F
13. F
14. T
15. F
16. F
17. T
18. T
19. F
20. F
21. T
22. F
23. F
24. reflecting
25. new moon
26. Polaris
27. gravity
28. Moon; Sun
29. Earth
30. sunspots
31. the Sun; Earth
32. osteoporosis
33. Galileo
34. galaxy
35. apparent magnitude
36. (a) (ii)
    (b) (i)
    (c) (iii)
    (d) (v)
    (e) (iv)
38. no
40. (b) rotation
    (c) revolution
55. (a) 2.25×10^9 km
    (b) 15 AU
56. 1.51×10^{16} km
57. no
58. (b) toward the trees
60. (a) geostationary
    (b) decrease its altitude
61. (b) irregular
    (c) 3000–4000 °C; end of their life
    (d) 1.2×10^{23} m
63. (b) auroras would be visible
64. (a) it is larger than Earth's
    (b) asteroid

Unit D Self-Quiz, pp. 454–455

1. (c)
2. (c)
3. (b)
4. (b)
5. (c)
6. (d)
7. F
8. F
9. F

10. astronomical unit or AU
11. geocentric
12. asteroids
13. (a) (iii)
    (b) (iv)
    (c) (i)
    (d) (ii)
14. distance light travels in a vacuum in 1 year
16. 35 min
21. (a) Earth is between the Sun and the Moon

Unit E

Section 11.1, p. 471
3. (a) protons, neutrons
   (b) electrons
4. (a) neutral
   (b) negative
5. add 2 electrons
6. remove 1 or more electrons; add 1 or more electrons
7. (a) attract
   (b) repel

Section 11.2, p. 477
5. (a) electrons
   (b) from "Y" to "X"

Section 11.4, p. 482
3. no
6. rubber is an insulator
7. golfer

Section 11.6, p. 489
6. (a) attract

Chapter 11 Review, pp. 498–499
6. (a) overall positive charge
   (b) overall negative charge
10. no
13. no
18. induced charge separation
22. it is negatively charged

Chapter 11 Self-Quiz, pp. 500–501
1. (c)
2. (d)
3. (a)
4. (b)
5. (c)
6. T
7. F
8. F
9. electrons
10. grounding
11. ions
12. (a) (iii)
    (b) (iv)
    (c) (i)
    (d) (v)
    (e) (ii)
14. they are insulators
15. (a) highest part of barn
17. (a) the leaves will spread apart
    (b) the leaves will fall

Section 12.1, p. 508
4. source of electrical energy

Section 12.2, p. 510
4. switch
6. light, motor

Section 12.3, p. 514
4. primary cell
8. no

Section 12.4, p. 517
4. (a) AC
   (b) DC
   (c) DC
   (d) AC

Section 12.7, p. 535
4. price, cost of operating
5. 32 %
6. (a) $12.00
   (b) $1.56
   (c) $28.80
   (d) $788.40
7. 51.7 ¢

Chapter 12 Review, pp. 542–543
1. no
2. renewable, non-polluting
5. it uses 60 J/s
11. 69 %
12. (a) $604.62

Chapter 12 Self-Quiz, pp. 544–545
1. (b)
2. (a)
3. (b)
4. (b)
5. T
6. F
7. chemical; electrical
8. fossil fuels
9. (a) (ii)
   (b) (v)
   (c) (iii)
   (d) (i)
   (e) (iv)
10. (a) 1000 or 10^3
    (b) 1 000 000 or 10^6
    (c) 1 000 000 000 or 10^9
13. (a) 1406 J
    (b) less
14. 3 hours
17. no

Section 13.1, p. 554
1. (a) 1
   (b) 4
2. (a) parallel
   (b) series

Section 13.3, p. 557
3. no

Section 13.5, p. 561
4. no

Section 13.7, p. 566
3. plastic
6. (a) increase
   (b) decrease
   (c) increase
   (d) decrease

Section 13.9, p. 570
2. (b)
3. approximately 1200 Ω
5. 18 Ω
6. 4.75 A
7. 120 V
8. 0.024 Ω

Section 13.10, p. 575

4. (a) decrease
   (b) decrease
5. (a) 0.24 A
   (b) 3 V
6. (a) 3 A
   (b) 120 V
7. (a) 69.3 V
   (b) 13.86 V

Chapter 13 Review, pp. 580–581

2. (a) series
   (b) parallel
7. its temperature will increase
8. in parallel; off
9. (a)
13. yes
16. no
17. dry
19. parallel
23. 24 Ω
24. 0.48 A
25. 0.42 V
26. 40 Ω

Chapter 13 Self-Quiz, pp. 582–583

1. (c)
2. (a)
3. (b)
4. F
5. T
6. F
7. resistance
8. current
9. circuit diagram
10. (a) (iv)
    (b) (ii)
    (c) (iii)
    (d) (i)
15. (a) 6 V
    (b) 0.75 A
16. 175 V
21. (a) all lights would go out
22. both would decrease

Unit E Review, pp. 588–593

1. (c)
2. (a)
3. (b)
4. (d)
5. (d)
6. (b)
7. (c)
8. (b)
9. (a)
10. (d)
11. (d)
12. (d)
13. (a)
14. (b)
15. (a)
16. (c)
17. (a)
18. (b)
19. T
20. F
21. F
22. F
23. F
24. T
25. F
26. F
27. F
28. T
29. F
30. F
31. T
32. F
33. T
34. F
35. T
36. F
37. F
38. F
39. electric discharge
40. ammeter
41. direct current
42. gaining
43. does not
44. ground; grounding
45. contact
46. atoms
47. Sun
48. parallel; series
49. potential difference; parallel
50. slope
51. equal
52. equal
53. release or emission
54. (a) (i)
    (b) (iii)
    (c) (v)
    (d) (ii)
    (e) (iv)
67. (a) negative
    (b) neutral
68. (a) V_{load} = 2.0 V, I_{load} = 1.0 A
    (b) V_{load} = 6.0 V, I_{load} = 0.33 A
69. (a) R_{load} = 2.0 Ω, R_{total} = 6.0 Ω
    (b) R_{load} = 18.0 Ω, R_{total} = 6.0 Ω
70. approximately 65 Ω
71. 9.58 Ω
72. 0.072 A or 72 mA
73. 1.5 V
75. (b) 108 kW·h/month
82. current electricity

Unit E Self-Quiz, pp. 594–595

1. (b)
2. (d)
3. (a)
4. (a)
5. T
6. F
7. T
8. attract
9. insulators
10. secondary cells; primary cells
11. (a) (iv)
    (b) (ii)
    (c) (i)
    (d) (v)
    (e) (iii)
16. (b) polyester
19. B

Appendix A

Section 5.A, p. 624

(a) 3500 ms
(b) 5200 mA
(c) 7500 ng

Glossary

A

abiotic factors [AY-bye-aw-tik FAK-turz] the non-living physical and chemical components of an ecosystem (p. 32)

absolute magnitude the brightness of stars as if they were all located 33 ly from Earth; for example, Polaris's absolute magnitude is −3.63 (p. 371)

acid precipitation [A-sid pruh-si-puh-TAY-shuhn] precipitation that has been made more acidic than usual by the combination of certain chemicals in the air with water vapour (p. 96)

agroecosystem [AG-ro-EE-ko-sis-tuhm] an agricultural ecosystem (p. 119)

alkali metal [AL-kuh-lee MET-uhl] an element in Group 1 of the periodic table (p. 220)

alkaline earth metal [AL-kuh-lyen URTH met-uhl] an element in Group 2 of the periodic table (p. 221)

alloy [AL-oy] a solid solution of two or more metals (p. 176)

alternating current (AC) a flow of electrons that alternates in direction in an electric circuit (p. 515)

altitude the angular height a celestial object appears to be above the horizon; measured vertically from the horizon (p. 341)

ammeter [AM-eet-ur] a device used to measure electric current (p. 556)

anion [AN-eye-awn] a negatively charged ion (p. 260)

apparent magnitude the brightness of stars in the night sky as they appear from Earth; for example, Polaris's apparent magnitude is 1.97 (p. 371)

astronomical unit [AS-truh-NAW-mik-uhl YOO-nit] approximately 150 million kilometres; the average distance from Earth to the Sun (p. 313)

astronomy the scientific study of what is beyond Earth (p. 305)

atmosphere the layer of gases surrounding Earth (p. 29)

atom the smallest unit of an element (p. 228)

atomic mass [uh-TAW-mik MAS] the mass of an atom in atomic mass units (u) (p. 235)

atomic number the number of protons in an atom's nucleus (p. 234)

aurora borealis [uh-ROH-ruh bo-ree-AL-is] a display of shifting colours in the northern sky caused by solar particles colliding with matter in Earth's upper atmosphere (p. 311)

azimuth [AH-zi-muhth] the horizontal angular distance from north measured eastward along the horizon to a point directly below a celestial body (p. 341)

B

Big Bang theory the theory that the Universe began in an incredibly hot, dense expansion approximately 13.7 billion years ago (p. 395)

bioaccumulation [BYE-o-uh-kyoo-myoo-LAY-shuhn] the concentration of a substance, such as a pesticide, in the body of an organism (p. 136)

bioamplification [BYE-o-am-pluh-fuh-KAY-shuhn] the increase in concentration of a substance, such as a pesticide, as it moves higher up the food web (p. 136)

biodiversity [BYE-o-duh-VUR-suh-tee] the variety of life in a particular ecosystem; also known as biological diversity (p. 83)

biogeochemical cycle [BYE-o-jee-o-KEM-uh-kuhl SYE-kuhl] the movement of matter through the biotic and abiotic environment (p. 48)

biomass [BYE-o-mas] the mass of living organisms in a given area (p. 46); any biological material (including plants and animals) (p. 521)

biome [BYE-om] a large geographical region defined by climate (precipitation and temperature) with a specific set of biotic and abiotic features (p. 56)

bioremediation [BYE-o-ruh-mee-dee-AY-shuhn] the use of micro-organisms to consume and break down environmental pollutants (p. 99)

biosphere [BYE-os-feer] the zone around Earth where life can exist (p. 30)

biotic factors [bye-AW-tik FAK-turz] living things, their remains, and features, such as nests, associated with their activities (p. 32)

black hole an extremely dense quantity of matter in space from which no light or matter can escape (p. 380)

Bohr–Rutherford [BOR RUH-thuhr-furd] **diagram** a simple drawing that shows the numbers and locations of protons, neutrons, and electrons in an atom (p. 236)

boiling point the temperature at which a substance changes state rapidly from a liquid to a gas (p. 194)

broad-spectrum pesticide a pesticide that is effective against many types of pest (p. 133)

bubble map a graphic organizer that shows properties or characteristics (p. 642)

C

carbon cycle the biogeochemical cycle in which carbon is cycled through the lithosphere, atmosphere, hydrosphere, and biosphere (p. 49)

carrying capacity the maximum population size of a particular species that a given ecosystem can sustain (p. 55)

catalyst [KAT-uh-list]   a substance that speeds up a chemical reaction but is not used up in the reaction (p. 273)

cation [KAT-eye-awn]   a positively charged ion (p. 260)

celestial navigation [suh-LES-tee-uhl nav-uh-GAY-shuhn]   the use of positions of stars to determine location and direction when travelling (p. 331)

celestial object   any object that exists in space (p. 305)

celestial sphere   the imaginary sphere that rotates around Earth, onto which all celestial objects are projected (p. 331)

cellular respiration   the process by which sugar is converted into carbon dioxide, water, and energy (p. 40)

characteristic physical property   a physical property that is unique to a substance and that can be used to identify the substance (p. 192)

charging by conduction   charging an object by contact with a charged object (p. 474)

charging by friction   the transfer of electrons between two neutral objects (made from different materials) that occurs when they are rubbed together or come in contact (touch) (p. 472)

charging by induction   charging a neutral object by bringing another charged object close to, but not touching, the neutral object (p. 486)

chemical change   a change in the starting substance or substances and the production of one or more new substances (p. 184)

chemical family   a column of elements with similar properties on the periodic table (p. 220)

chemical formula   notation that indicates the type and number of atoms in a pure substance (p. 257)

chemical property   a characteristic of a substance that is determined when the composition of the substance is changed and one or more new substances are produced (p. 183)

circuit [SUR-kuht] **diagram**   a way of drawing an electric circuit using standard symbols (p. 551)

comet [KAW-met]   a chunk of ice and dust that travels in a very long orbit around the Sun (p. 316)

compare-and-contrast chart   a graphic organizer that shows similarities and differences (p. 641)

comparison matrix   a graphic organizer used to record and compare observations or results (p. 641)

compound   a pure substance composed of two or more different elements that are chemically joined (p. 211)

concept map [KAWN-sept map]   a graphic organizer that shows relationships between ideas; the ideas are connected by arrows and words or expressions that explain the connections (p. 640)

conductor   a material that lets electrons move easily through it (p. 480)

constellation [kawn-stell-AY-shun]   a grouping of stars, as observed from Earth (p. 329)

consumer   an organism that obtains its energy from consuming other organisms (p. 41)

controlled experiment   an experiment in which the independent variable is purposely changed to find out what change, if any, occurs in the dependent variable (p. 8)

covalent [ko-VAY-lent] **bond**   a bond formed when two non-metal atoms share electrons (p. 265)

current electricity   the controlled flow of electrons through a conductor (p. 507)

D

density   a measure of how much mass is contained in a given unit volume of a substance; calculated by dividing the mass of a sample by its volume (p. 192)

dependent variable   a variable that changes in response to the change in the independent variable (p. 8)

direct current (DC)   a flow of electrons in one direction through an electric circuit (p. 515)

dwarf planet   a celestial object that orbits the Sun and has a spherical shape but does not dominate its orbit (p. 314)

E

eclipse [ee-KLIPS]   a darkening of a celestial object due to the position of another celestial object (p. 325)

ecliptic [ee-KLIP-tik]   the path across the sky that the Sun, the Moon, the planets, and the zodiac constellations appear to follow over the course of a year (p. 339)

ecological niche [EE-kuh-LAW-ji-kuhl NEESH]   the function a species serves in its ecosystem, including what it eats, what eats it, and how it behaves (p. 42)

ecological pyramid   a representation of energy, numbers, or biomass relationships in ecosystems (p. 45)

ecosystem   all the living organisms and their physical and chemical environment (p. 32)

efficiency [ih-FISH-en-see]   comparison of the energy output of a device with the energy supplied (p. 530)

electric cell   a device that converts chemical energy into electrical energy (p. 512)

electric charge   a form of charge, either positive or negative, that exerts an electric force (p. 465)

electric circuit [e-LEK-trik SUR-kuht]   a continuous path in which electrons can flow (p. 509)

electric current (I)   a measure of the rate of electron flow past a given point in a circuit; measured in amperes (A) (p. 556)

electric discharge   the rapid transfer of electric charge from one object to another (p. 492)

electric force the force exerted by an object with an electric charge; can be a force of attraction or a force of repulsion (p. 468)

electrical energy the energy provided by the flow of electrons in an electric circuit (p. 511)

electrical power the rate at which electrical energy is produced or used (p. 530)

electrical resistance (R) the ability of a material to oppose the flow of electric current; measured in ohms (Ω) (p. 564)

electromagnetic (EM) radiation energy emitted from matter, consisting of electromagnetic waves that travel at the speed of light (p. 309)

electromagnetic (EM) spectrum the range of wavelengths of electromagnetic radiation, extending from radio waves to gamma rays, and including visible light (p. 309)

electron a negatively charged particle in an atom (p. 229)

electrostatic series a list of materials arranged in order of their tendency to gain electrons (p. 473)

element a pure substance that cannot be broken down into a simpler chemical substance by any physical or chemical means (p. 211)

element symbol an abbreviation for a chemical element (p. 211)

elliptical galaxy [ee-LIP-tuh-kuhl GA-luhk-see] a large group of stars that together make an elliptical or oval shape (p. 386)

empirical knowledge scientific knowledge acquired by careful observation and experimentation (p. 648)

endangered a species facing imminent extirpation or extinction (p. 85)

equilibrium describes the state of an ecosystem with relatively constant conditions over a period of time (p. 80)

equinox [EK-wuh-nawks] the time of year when the hours of daylight equal the hours of darkness (p. 323)

eutrophic [yoo-TRO-fik] a body of water that is rich in nutrients (p. 60)

experimental design a brief description of the procedure by which a hypothesis is tested (p. 11)

extinct refers to a species that has died out and no longer occurs on Earth (p. 83)

extirpated [EKS-tur-pay-ted] a species that no longer exists in a specific area (p. 85)

F

fishbone diagram a graphic organizer that shows the important ideas under the major concepts of a topic (p. 640)

food chain a sequence of organisms, each feeding on the next, showing how energy is transferred from one organism to another (p. 43)

food web a representation of the feeding relationships within a community (p. 43)

freezing point the temperature at which a substance changes state from a liquid to a solid; melting point and freezing point are the same temperature for a substance (p. 194)

G

galaxy [GA-luhk-see] a huge, rotating collection of gas, dust, and billions of stars, planets, and other celestial objects (p. 307)

geostationary orbit [jee-o-STAY-shuh-neh-ree OR-bit] an orbital path directly over Earth's equator with a period equal to the period of Earth's rotation (p. 350)

global positioning system (GPS) a group of satellites that work together to determine the positions of given objects on the surface of Earth (p. 349)

gravitational force the force of attraction between all masses in the Universe (p. 321)

grounding connecting an object to a large body, like Earth, that is capable of effectively removing an electric charge that the object might have (p. 475)

H

halogen [HA-luh-jen] an element in Group 17 of the periodic table (p. 222)

hydrosphere [HYE-druhs-feer] all of Earth's water in solid, liquid, and gas form (p. 30)

hypothesis [hye-PAW-thuh-sis] a possible answer or untested explanation that relates to the initial question in an experiment (p. 10, p. 649)

I

independent variable a variable that is changed by the investigator (p. 8)

induced charge separation a shift in the position of electrons in a neutral object that occurs when a charged object is brought near it (p. 468)

inference [IN-fuh-rens] a tentative conclusion based on logic or reasoning (p. 648)

insulator a material that does not easily allow the movement of electrons through it (p. 480)

integrated pest management a strategy to control pests that uses a combination of physical, chemical, and biological controls (p. 140)

invasive species a non-native species whose intentional or accidental introduction negatively impacts the natural environment (p. 91)

ion [EYE-awn] a particle that has either a positive or a negative charge (p. 260)

ionic compound [eye-AW-nik KAWM-pownd] a compound that consists of positively and negatively charged ions (p. 261)

irregular galaxy a large group of stars that together make an irregular shape (p. 386)

isotope [EYE-suh-top] an atom with the same number of protons but a different number of neutrons (p. 235)

J

jigsaw activity an activity where each member of the team goes to a different group to do an activity and then returns to the team to share what was learned (p. 635)

K

kilowatt·hour (kW·h) [KIL-uh-waht OW-ur] the SI unit for measuring electrical energy usage; the use of one kilowatt of power for one hour (p. 533)

K-W-L chart a graphic organizer that shows what you know (K), what you want (W) to find out, and what you have learned (L) (p. 641)

L

law a general statement that describes a commonly occurring natural event; a law does not explain why or how a natural event occurs; it just describes what happens in detail (p. 648)

leaching the process by which nutrients are removed from the soil as water passes through it (p. 125)

lenticular galaxy [len-TIK-yoo-lur GA-luhk-see] a large group of stars that together make a shape that has a central bulge but no spiral arms (p. 386)

light energy visible forms of radiant energy (p. 38)

light year the distance that light travels in a vacuum in 1 year (365.25 days); equal to 9.46×10^{12} km (p. 365)

limiting factor any factor that restricts the size of a population (p. 52)

lithosphere [LI-thuhs-feer] Earth's solid outer layer (p. 30)

load the part of an electric circuit that converts electrical energy into other forms of energy (p. 509)

luminosity the total amount of energy produced by a star per second (p. 370)

luminous producing and giving off light; shining (p. 305)

lunar cycle all of the phases of the Moon (p. 324)

M

main sequence [mayn SEE-kwens] the stars (including the Sun) that form a narrow band across the H–R diagram from the upper left to the lower right (p. 377)

mass number the number of protons and neutrons in an atom's nucleus (p. 235)

mechanical mixture a mixture in which you can distinguish between different types of matter (p. 176)

melting point the temperature at which a substance changes state from a solid to a liquid (p. 194)

metal an element that is lustrous, malleable, and ductile, and conducts heat and electricity (p. 212)

metalloid an element that has properties of both metals and non-metals (p. 213)

microgravity environment an environment in which objects behave as though there is very little gravity affecting them (p. 421)

mind map a graphic organizer that shows relationships between ideas; words or pictures representing ideas are connected by arrows; similar to a concept map, but does not include explanations for connections (p. 640)

mixture a substance that is made up of at least two different types of particles (p. 176)

molecular compound a molecule that consists of two or more different elements (p. 258)

molecular element a molecule consisting of atoms of the same element (p. 257)

molecule two or more atoms of the same or different elements that are chemically joined together in a unit (p. 257)

monoculture [MAWN-o-kul-chur] the cultivation of a single crop in an area (p. 121)

N

narrow-spectrum pesticide a pesticide that is effective against only a few types of pest (p. 133)

natural fertilizer plant nutrients that have been obtained from natural sources and have not been chemically altered by humans (p. 124)

nebula [NEB-yoo-luh] a massive cloud of interstellar gas and dust; the beginning of a star (p. 375)

negatively charged object an object that has more electrons than protons (p. 466)

neutral object an object that has equal numbers of protons and electrons (p. 466)

neutralize [NOO-truh-lyez] counteract the chemical properties of an acid (p. 96)

neutron [NYOO-trawn] a neutral particle in the atom's nucleus (p. 231)

neutron star an extremely dense star made up of tightly packed neutrons; results when a star over 10 solar masses collapses (p. 380)

nitrogen cycle the series of processes in which nitrogen compounds are moved through the biotic and abiotic environment (p. 50)

noble gas an element in Group 18 of the periodic table (p. 221)

non-metal an element, usually a gas or a dull powdery solid, that does not conduct heat or electricity (p. 213)

non-renewable resource a resource that cannot be replaced as quickly as it is consumed (p. 521)

O

observational study the careful watching and recording of a subject or phenomenon to gather scientific information to answer a question (p. 9)

observations evidence gained by using the five senses and special tools (p. 647)

Ohm's law [O-uhmz LAW] the straight line relationship between voltage and current; $R = V/I$ (p. 568)

ohmmeter [O-uhm-mee-tur] a device used to measure resistance (p. 565)

oligotrophic [AW-li-go-TRO-fik] a body of water that is low in nutrients (p. 60)

orbit the closed path of a celestial object or satellite as it travels around another celestial object (p. 306)

orbital radius the average distance between an object in the Solar System and the Sun (p. 320)

organic farming the system of agriculture that relies on non-synthetic pesticides and fertilizers (p. 139)

P

particle theory of matter a theory that describes the composition and behaviour of matter (p. 175)

parallax the apparent change in position of an object as viewed from two different locations that are not on a line with the object (p. 367)

parallel circuit a circuit in which the loads are connected in branches so that there are two or more paths for electrons to flow (p. 552)

period a row on the periodic table (p. 222)

pest any plant, animal, or other organism that causes illness, harm, or annoyance to humans (p. 121)

pesticide a substance used to kill a pest (p. 133)

photosynthesis [fo-to-SIN-thuh-sis] the process in which the Sun's energy is converted into chemical energy (p. 38)

photovoltaic cell [fo-to-vol-TAY-ik sel] a device that converts light energy from any light source directly into electrical energy (p. 527)

physical change a change in which the composition of the substance remains unaltered and no new substances are produced (p. 181)

physical property a characteristic of a substance that can be determined without changing the composition of that substance (p. 179)

placemat organizer a graphic organizer that gives each student in a group space to write down what he or she knows about a topic; the group then discusses the answers and writes what they have in common in the middle section (p. 642)

planet a large, round celestial object that travels around a star (p. 306)

pollution harmful contaminants released into the environment (p. 96)

positively charged object an object that has fewer electrons than protons (p. 466)

potential difference (voltage) (V) the difference in electrical potential energy per unit charge measured at two different points; measured in volts (V) (p. 560)

precession [pruh-SESH-uhn] the changing direction of Earth's axis (p. 324)

precipitate [pruh-SIP-uh-tayt] a solid that separates from a solution (p. 184)

prediction a statement that predicts the outcome of a controlled experiment (p. 10)

primary cell an electric cell that may only be used once (p. 512)

primary succession succession on newly exposed ground, such as following a volcanic eruption (p. 80)

producer an organism that makes its own energy-rich food compounds using the Sun's energy (p. 38)

proton a positively charged particle in the atom's nucleus (p. 230)

protostar a massive concentration of gas and dust thought to eventually develop into a star after the nebula collapses (p. 375)

pure substance a substance that is made up of only one type of particle (p. 175)

Q

qualitative observation a non-numerical observation that describes the qualities of objects or events (p. 12)

qualitative property a property of a substance that is not measured and does not have a numerical value, such as colour, odour, and texture (p. 179)

quantitative observation a numerical observation based on measurements or counting (p. 12)

quantitative property a property of a substance that is measured and has a numerical value, such as temperature, height, and mass (p. 179)

quasar [KWAY-zahr] a distant, young galaxy that emits large amounts of energy produced by a supermassive black hole at its centre (p. 389)

R

radiant energy energy that travels through empty space (p. 38)

red giant a star near the end of its life cycle with a mass that is equal to or smaller than that of the Sun; becomes larger and redder as it runs out of hydrogen fuel (p. 377)

red shift the phenomenon of light from galaxies shifting toward the red end of the visible spectrum, indicating that the galaxies are moving away from Earth (p. 394)

red supergiant a star near the end of its life cycle with a mass that is 10 times (or more) larger than that of the Sun; becomes larger and redder as it runs out of hydrogen fuel (p. 378)

reflecting telescope [ri-FLEK-ting TEL-uh-skop] an optical telescope that uses mirrors to gather and focus light (p. 410)

refracting telescope [ri-FRAK-ting TEL-uh-skop] an optical telescope that uses glass lenses to gather and focus light (p. 410)

renewable resource a natural resource that is unlimited (for example, energy from the Sun or wind) or can be replenished by natural processes in a relatively short period of time (for example, biomass) (p. 519)

resistor a device that reduces the flow of electric current (p. 565)

retrograde motion [RET-ro-grayd MO-shuhn] the apparent motion of an object in the sky, usually a planet, from east to west, rather than in its normal motion from west to east (p. 340)

round table activity an activity that allows a group to review what they know, with each member sharing an idea until a solution is found (p. 635)

S

satellite a celestial object that travels around a planet or dwarf planet (p. 306)

scientific inquiry [sye-en-TIF-ik IN-kwuh-ree] the process of exploring the world, asking questions, and searching for answers that increase our understanding (p. 647)

scientific method the general types of mental and physical activities that scientists use to create, refine, extend, and apply knowledge (p. 650)

secondary cell an electric cell that can be recharged (p. 512)

secondary succession succession in a partially disturbed ecosystem, such as following a forest fire (p. 80)

series circuit a circuit in which the loads are connected end to end so that there is only one path for the electrons to flow (p. 552)

solar flare gases and charged particles expelled above an active sunspot (p. 310)

solar mass a value used to describe the masses of galaxies and stars other than our Sun; equal to the mass of the Sun (2×10^{30} kg) (p. 373)

solar prominence [SO-lur PRAW-muh-nens] low-energy gas eruptions from the Sun's surface that extend thousands of kilometres into space (p. 310)

Solar System the Sun and all the objects that travel around it (p. 306)

solstice [SOL-stis] an astronomical event that occurs two times each year, when the tilt of Earth's axis is most inclined toward or away from the Sun, causing the position of the Sun in the sky to appear to reach its northernmost or southernmost extreme (p. 323)

solution a uniform mixture of two or more substances (p. 176)

space junk debris from artificial objects orbiting Earth (p. 424)

space probe a robotic spacecraft sent into space to explore celestial objects such as planets, moons, asteroids, and comets (p. 415)

spacecraft a human-occupied or robotic vehicle used to explore space or celestial objects (p. 411)

special concern a species that may become threatened or endangered because of a combination of factors (p. 85)

species richness the number of species in an area (p. 83)

spinoff a technology originally designed for a particular purpose, such as space technology, that has made its way into everyday use (p. 426)

spiral galaxy [SPYE-ruhl GA-luhk-see] a large group of stars that together make a spiral shape, such as the Milky Way (p. 386)

star a massive collection of gases, held together by its own gravity and emitting huge amounts of energy (p. 305)

star cluster a group of stars held together by gravity (p. 385)

static electricity an imbalance of electric charge on the surface of an object (p. 467)

stewardship [STOO-uhrd-ship] taking responsibility for managing and protecting the environment (p. 105)

succession [suk-SESH-uhn] the gradual and usually predictable changes in the composition of a community and the abiotic conditions following a disturbance (p. 80)

sunspots dark spots appearing on the Sun's surface that are cooler than the area surrounding them (p. 310)

supernova a stellar explosion that occurs at the end of a massive star's life (p. 379)

sustainability [suhs-TAY-nuh-BI-luh-tee] the ability to maintain an ecological balance (p. 35)

sustainable ecosystem an ecosystem that is maintained through natural processes (p. 34)

switch a device in an electric circuit that controls the flow of electrons by opening (or closing) the circuit (p. 509)

synthetic fertilizer [sin-THET-ik FUR-tuh-lye-zur] fertilizers that are manufactured using chemical processes (p. 124)

T

theory an explanation of an observation (p. 649)

thermal energy the form of energy transferred during heating or cooling (p. 38)

think-pair-share activity an activity where students think about a topic, pair with a partner to discuss it, and share their conclusions with the class (p. 635)

threatened a species that is likely to become endangered if factors reducing its survival are not changed (p. 85)

tide the alternate rising and falling of the surface of large bodies of water; caused by the interaction between Earth, the Moon, and the Sun (p. 327)

tolerance range [TAWL-uh-rens raynj] the abiotic conditions within which a species can survive (p. 52)

Traditional Ecological Knowledge and Wisdom (TEKW) the detailed knowledge of Aboriginal peoples about plants, animals, weather, landforms, and other things in the environment (p. 648)

tree diagram a graphic organizer showing concepts that can be broken down into smaller categories (p. 640)

trophic level [TRO-fik le-vuhl] the level of an organism in an ecosystem depending on its feeding position along a food chain (p. 43)

U

Universe everything that exists, including all energy, matter, and space (p. 305)

V

variable any condition that changes or varies the outcome of a scientific inquiry (p. 8)

Venn diagram a graphic organizer that shows similarities and differences (p. 641)

viscosity [vis-KAW-suh-tee] the degree to which a fluid resists flow (p. 180)

voltmeter [VOLT-mee-tur] a device used to measure potential difference (voltage) (p. 561)

W

water cycle the series of processes that cycle water through the environment (p. 48)

watershed the land area drained by a particular river; also called a drainage basin (p. 60)

white dwarf a small, hot, dim star created by the remaining material that is left when a red giant dies (p. 378)

word wall a graphic organizer that lists the key words and concepts for a topic (p. 642)

Index

A

Abiotic factors, 32, 35, 52–53, 56–57
Aboriginal peoples
    and aurora borealis, 311
    and environment, 96
    and Gulf of St. Lawrence, 61
    and stars, 332–333
    traditional knowledge of, 648
    and wildlife, 104–105
Absolute magnitude, 371
Accidents, 600
Acid precipitation, 96–98, 130–131, 523, 524
Agriculture
    and biodiversity, 121, 132
    black gold, 129
    engineered biosystems and, 119–122
    modern, 119–122
    no-tillage, 128
    pests in, 132–134
    and rainforest, 89
    and water, 125, 126–127
    and wetlands, 90
Agroecosystems, 119, 121–122, 132
Air
    as mixture of gases, 270
    pollution, 142, 523, 524
Algae, 125
Alkali metals, 220–221, 226–227, 238
Alkaline earth metals, 221, 226–227
Alloys, 176–177, 263
Alternating current (AC), 515
Altitude, 341
Amazon, 88–89
Ammeter, 556, 561, 562–563, 607
Ampere (A), 556
Analyses
    cost-benefit, 620–621
    critical, writing of, 550
    of data, 13–14, 614
    of issues, 620
Andromeda galaxy, 390, 393
Anions, 260, 263
Antibiotics, 133
Antifreeze, 191
Apparent magnitude, 371
Aquatic ecosystems, 36–37, 39, 60–62
    abiotic factors in, 53
    acid precipitation and, 96
    and ecological succession, 81
    fertilizers and, 125
    loss of, 90
Arctic ecosystems, biomagnification and, 138
Aristotle, 228
Arnott, Shelley, 95
Artificial satellites, 311–312, 346–351
Asteroids, 315
Astronauts, health of, 420, 421–423, 424
Astronomical unit (AU), 313, 365, 366
Astronomy, 305, 409
Atmosphere, 29
Atom
    theories of, 228–233
Atomic mass, 235
Atomic number, 234–235
Atoms, 228, 465
Aurora australis, 311
Aurora borealis, 311
Azimuth, 341

B

Bacteria
    denitrifying, 50
    nitrogen-fixing, 128
Baker, Scott, 63
Batteries, 512, 560
Big Bang theory, 395–396
Bioaccumulation, 136, 137
Bioamplification, 136–138, 277
Biochar, 129
Biodiesel, 286
Biodiversity, 83–86
    agroecosystems and, 132
    food production and, 132
    of forests, 103
    in modern agriculture, 121
Biofuels, 524
Biogeochemical cycles, 48–51, 124, 126, 128
Biomass, 46, 521–522, 524
Biomes, 56–59
Bioremediation, 99
Biosphere, 30, 155
Biosystems, engineered. *See* Engineered biosystems
Biotic factors, 32, 35, 54–55, 56–57
Black holes, 362, 380–381, 389
Blood cells, 422
Blue supergiants, 377
Bohr, Niels, 232
Bohr–Rutherford diagrams, 236–238
Bohr–Rutherford model, 232–233, 465
Boiling point, 194
Bolton, Charles Thomas, 362
Bondar, Roberta, 421
Bones, of astronauts, 420, 422
Boreal forest, 56–57, 58
Broad-spectrum pesticides, 133–134
Bubble maps, 642

C

Calculators, 630–631
Canadarm, 414–415, 426
Carbohydrates, 39
Carbon, 214
    and charcoal, 129
    cycle, 49–50
    deposits, 49
    in nuclear fusion, 379
    sinks, 49
Carbon dioxide
    and biomass energy generation, 522
    properties of, 270
Carbonic acid, 98
Carrying capacity, 55
Catalysts, 273
Cations, 260, 263
Cause variables. *See* Independent variables
Celestial navigation, 331
Celestial objects, 305, 313, 338–343, 366
Celestial sphere, 331
Cellphones, 214
Cells, replacement of, 48
Cellular respiration, 40–41
Chadwick, James, 231
Characteristic physical properties, 192–198
Charcoal, 129, 213, 241
Charging by conduction, 474
Charging by friction, 472–473
Charging by induction, 486–491
Cheko, Lake, 302, 315
Chemical changes, 184–185
Chemical energy, 38, 39
Chemical families, 220–222, 226–227
Chemical formulas, 257, 268
Chemical properties, 183
    of gases, 270–271
Chemicals, safety with, 599
Chlorine, 222, 260
Chlorophyll, 39
Chromosphere, 309
Circuit diagrams, 551–554
Cities, 142–145
Climate change, and abiotic factors, 49
Coal, 523, 532
Coma, of comet, 316
Comets, 316–317, 353
Commensalism, 54
Communication
    of decisions on issues, 621–622
    of research conclusions, 618
    of results of investigations, 14–15, 615–616
Communications
    electricity in, 586
    Sun and, 311–312
Compare-and-contrast charts, 641
Comparison matrices, 641
Composting, 185
Compounds, 211
Computers, 531
Concept maps, 640
Conclusions, in research, 618
Conduction, charging by, 474

Conductors, 480, 481–482, 484–485, 512, 558
Constellations, 329–330
Consumer products
   electrostatically charged, 475–476
   "greening" of, 286
   space technology spinoffs, 429–430
Consumers, 41
Consumption, and ecological footprint, 141
Controlled experiments, 8, 9, 10, 610
Cootes Paradise, 93
Copernicus, Nicholas, 322
Copper, 213, 234, 254
Coral reefs, 61, 85
Corona, 309, 310
Corrosion, 254
Cost-benefit analyses, 620–621
Covalent bonds, 265
Crops
   and food webs, 132
   rotation of, 128
Current electricity, 507–508
   forms of, 515–517
   generating, 518–528

D

Dalton, John, 229, 234
Dams and reservoirs, 127, 518–519
Dark energy, 395, 396
Dark matter, 394–395, 396
Data
   analysis of, 13–14, 614
   tables, 632–634
Davies, Peter, 191
DDT (dichlorodiphenyltrichloroethane), 133, 135–136, 137, 276–277
Decision making, 619–622
Decomposers, 185
Defibrillators, 559
Deforestation, 50
Democritus, 228, 229
Denim, 172, 183
Density, 192–193
Dependent variables, 8, 610
Dewailly, Eric, 276
DEXTRE, 415
Diamonds, 241, 242–244
Diapers, disposable, 187
Diaphragm, 488
Direct current (DC), 515
DNA fingerprinting technology, 63
Drawings, scientific, 613
Ductility, 177, 181
Dunes, succession of, 81
Dwarf planets, 314

E

Earth
   distance of stars from, 366
   distance of Sun from, 365
   formation of, 384
   life on, 29–31
   magnetic field, 311
   as planet, 306, 313
   revolution of, 320–321, 339–340
   rotation of, 320, 339–340
   spheres of, 29–30
   Sun and, 310–312
   tilt of, 322–324
   viewing celestial objects from, 338–343
Eclipses, 325–326
Ecliptic, 339
Ecological footprint, 141
Ecological niches, 42, 85
Ecological pyramids, 45–47
Ecosystems, 32–35
   abiotic factors and, 52–53
   agroecosystems compared to, 121–122
   aquatic, 36–37. See Aquatic ecosystems
   artificial, 35
   biotic factors and, 54–55
   cycling of matter in, 48–51
   energy flow in, 38–41
   features of, 33–34
   human activities and, 35
   marine, 74
   products, 78
   protective function of, 78
   services from, 77–79
   sustainability of, 34–35
   terrestrial, 56–59, 87–89, 97
   urban, 142–145
Ecotourism, 77
Edison, Thomas, 195
Effect variables. See Dependent variables
Efficiency, of electrical energy, 521, 530–533
Einstein, Albert, 393
Electric cells, 512, 560
Electric charges, 465
   by conduction, 474
   by contact, 472–477
   detection of, 467, 469
   by friction, 472–473
   by induction, 486–491
   law of, 468, 488
   predicting, 478–479
Electric circuits, 508–509, 515, 551–554. See also Circuit diagrams
   breakers, 556–557
   connecting multiple loads, 555
   drawing and constructing, 606
   increasing number of loads in, 576–577
   measuring current in, 562–563
   resistance in, 564–566
Electric current, 556–557, 562–563, 567, 568–570
   through loads in parallel, 573
   through loads in series, 571
Electric discharges, 492–495, 507
Electric forces, 468
Electric potential. See Potential difference
Electrical conductivity, 558, 604

Electrical energy, 511–514
   audits, 539
   conservation in schools, 535
   efficiency of, 521, 530–533
   examining production of, 536–537
   sources of, 508, 511–514
   and spacecraft, 538
Electrical fires, 548
Electrical power, 530–535
Electrical propulsion systems, 420
Electrical resistance, 564–566, 568–570
Electrical shocks, 493, 557, 600
Electricity
   in communications, 586
   cost of, 533–534
   current. See Current electricity
   generating, 516–517
   safety with, 600
Electrolytes, 512
Electromagnetic (EM) radiation, 309, 409, 411
Electromagnetic (EM) spectrum, 309
Electron flow
   in circuits, 552
   in current electricity, 507–508, 509, 515
   in electric currents, 556
   in electrical resistance, 564–566
   in generators, 516
Electrons, 229, 231, 465
   and charging by conduction, 474
   and charging by contact, 472
   in charging by induction, 486–487
   and conductors, 480
   and electric charges, 466–467, 468, 469, 472
   in electric discharges, 492–495
   and electrostatic series, 473
   and grounding, 475
   in lightning, 493–494
   orbits of, 232–233, 238
Electroscopes, 467, 469, 478–479
Electrostatic lifting apparatus (ESLA), 488
Electrostatic precipitators, 476
Electrostatic series, 473
Electrostatic speakers, 488
Electrostatics, 470
Element symbols, 211
Elements, 211
   periodic table of, 211–215, 220–225, 234–239
Elliptical galaxies, 386
Empirical knowledge, 648
Endangered species, 85
Energy flow
   in ecological pyramids, 46
   in ecosystems, 38–41
Engineered biosystems, 119–120
Environment
   Aboriginal peoples and, 96
   batteries and, 512
   cities and, 144

diamond mining and, 243–244
energy production and, 527
fertilizers and, 125–126
human activities and, 120
lifestyles and, 21, 116
plastics and, 167
relationship of humans with, 96
space technology spinoffs and, 429
thermal energy generation and, 522–523
Equations, 631
Equilibrium, 80
Equinoxes, 323, 332
Estuaries, 61
Eutrophic bodies of water, 60
Evaluation
of decisions, 622
of evidence, 13–14, 614
of Internet resources, 619
of research, 618
of sources of information, 617
of teamwork, 636
of texts, 506
Experimental design, 11
Extinction, 83–84
Extirpated species, 85

F

Faraday, Michael, 516
Fertilizers, 124–126, 125–126, 167
Fires
electrical, 548
forest, 103
Fireworks, 183
Fishbone diagrams, 640
Fluorescent light bulbs, 530, 531–532
Food
modern agriculture and, 120–122
sustainable production, 122
Food chains, 43–44
bioamplification and, 136–137
toxins and, 276–277
Food webs, 43–44
agriculture and, 132
Forensic chemistry, 190
Forests and forestry, 102–104, 106–107
acid precipitation and, 97
pests in, 133
Fossil fuels, 49, 50, 120
in thermal energy generation, 521, 522–523
Free fall, 421
Freezing point, 194
Friction, charging by, 472–473
Fuel cells, 513–514
Fundy, Bay of, 62, 520
Fuses, 552, 556–557

G

Gaia hypothesis, 30
Galaxies, 307, 385–391, 393–395
Galileo Galilei, 310, 322, 409, 650, 653

Gamma ray telescopes, 411
Garneau, Marc, 412, 414
Gases
air as mixture of, 270
densities of, 193
properties of, 270–271
as state of matter, 175
Generators, 516
Geocentric model, of Solar System, 322
Geostationary orbit satellites, 350
Geothermal energy, 522, 525
Gigawatts (GW), 530
Glass, safety with, 599–600
Global positioning systems (GPS), 349, 426–427
Goals, setting of, 636–637
Gold, 193, 213, 234
Granules, 310
Graphic organizers, 640–642
Graphite, 241
Graphs, 632–634
GRASS method, 625
Grassland, 56–57, 58–59
Gravitational forces. See also Microgravity environment
of black holes, 380
and dark energy, 396
and health issues in space exploration, 422
Moon and, 29
and orbits, 321
and orbits of satellites, 347
and space travel, 419
and stars, 375, 376, 378, 380
and tides, 327
and weightlessness, 420–421
Gravitational potential energy, 560
Greenhouse effect, 522
Greenhouse gases, 522
Grounding, 475

H

Habitats, 29, 87–90. See also Wildlife
Hadfield, Chris, 414, 415
Halogens, 222
Hazardous Household Product Symbols (HHPS), 601
Health
of astronauts, 420, 421–423, 424
spinoffs from space technology, 428–429
Heart, 559
Heat, safety with, 599
Heavy metals, 193
Heliocentric model, of Solar System, 322
Helium, 221
and main sequence, 377
in nuclear fusion, 376, 378, 379
Herbicides, 133
Hertzsprung, Ejnar, 376
Hertzsprung–Russell diagram (H–R diagram), 376–377

Hipparchus, 371
Homes
in cities, 144
"green," 144
properties of chemicals in, 218
Horizon diagrams, 614
Hubble, Edwin, 386, 393, 394, 411
Hubble Space Telescope (HST), 411
Hubble's Law, 394
Human activities
and carbon cycle, 49–50
and ecosystems, 35
and environment, 120
and extinction, 84
and sustainability, 35
Hydro-electric energy generation, 518–519
Hydrogen
fuel cells, 513–514
and main sequence, 377
in nuclear fusion, 376, 378, 379
properties of, 270
Hydrogen peroxide, 272–273
Hydrosphere, 30
Hypotheses, 10–11, 14, 611, 649

I

Ice, salt and, 195–196
Incandescent light bulbs, 194–195, 530, 531
Independent variables, 8, 610
Indigenous peoples. See also Aboriginal peoples
and agriculture, 129
Induced charge separation, 468
Induction, charging by, 486–491
Inferences, 210, 648
Insulators, 480, 481–482, 484–485
Integrated pest management (IPM), 140
International Space Station (ISS), 346, 348, 412–413, 437, 538
Internet, 618–619
Intertidal zones, 62
Inuit, 138, 276–277
Invasive species, 91–95, 95
Iodine, 222, 260
Ionic compounds, 260–261, 263–264
names of, 264
Ions, 260, 263, 465
and electrical conduction, 480
Iqaluit, 144, 244
Iron, 254
Iron oxide, 262
Irregular galaxies, 386
Irrigation, 126–127
Isotopes, 235
Issues, exploration of, 619–622, 636

J

Jigsaw activities, 635
Joules (J), 511, 530
Jupiter, 313, 340, 416

674 Index

K

K-W-L charts, 641
Kevlar, 245
Kilowatt-hours (kW-h), 530
Knowledge, scientific, 648–650
Kuiper Belt, 314, 353

L

Laser light, 431, 482, 600
Laser printers, 462, 481–482
Latin and Greek root words, 643
Law of Electric Charges, 468, 488
Laws, 648–649
Leaching, 125
Lead, 176–177
Lemaître, Georges, 395
Lenticular galaxies, 386
Libraries, 619
LIDAR technology, 431
Light
    bulbs, 194–195, 509, 530, 531–532
    laser, 431, 482, 600
    safety with, 600
    as source of electrical energy, 527
    visible, 306
    white, 232, 372
Light-emitting diodes (LEDs), 532
Light energy, 38, 39
    and composition of stars, 372
Light minutes (lm), 366
Light seconds (ls), 366
Light years, 365–366
Lightning, 493–495, 511
Limestone, 49, 96
Limiting factors, 52
Lippershey, Hans, 409, 653
Liquids, 175
    densities of, 193
Lithium, 221, 235, 238, 512
Lithosphere, 30
Living things, in laboratory, 600
Loads, 509, 515
    in circuits, 551, 552, 571–572, 573–574
    increasing number of, 576–577
    multiple, 555
    voltage and, 561, 574
Local Group, 390
Loggerhead shrike, 89
Lovelock, James, 30
Luminosity, 370–371, 378
Luminous, defined, 305
Lunar cycle, 324–325, 327
Lunar eclipse, 326

M

Magnetic fields
    of Earth, 311
    of Sun, 310
Main idea, finding, 304
Main sequence, 377
Mangroves, 61
Manicouagan Crater, 315
Marine ecosystems, 61–62, 74
Mars, 313, 340, 384
    colonization of, 435–436
    exploration of, 416–417, 423–424, 431, 432, 434–436
    rovers, 416–417, 420
Mass number, 235
Material Safety Data Sheet (MSDS), 600
Mathematics, 623–631
Mather, John, 396
Matter
    Aristotle's theory of, 228
    particle theory of, 175
Measurement, 627–629
Mechanical mixtures, 176
Megawatts (MW), 530
Melting point, 194
Mendeleev, Dmitri, 223, 226
Mercury
    as element, 195, 208, 532
    as planet, 313, 384
Metal leaf electroscopes, 469
Metalloids, 213–214
Metals, 212–214
    alloys, 263
    as conductors, 480
    densities of, 193
    detecting, 216–217
    melting points of, 195
Meteor showers, 316
Meteorites, 315
Meteoroids, 315–316
Microgravity environment, 421–422, 429
Microscopes, 602–603
Milky Way, 307, 385, 388–389, 393
Mind maps, 640
Mixtures, 176
Molecular compounds, 258–259
Molecular elements, 257
Molecular models, 268–269
Molecules, 257
Monocultures, 121, 122, 132–133, 142
Moon, 29, 306
    colonization of, 435–436
    eclipse of, 326
    missions to, 406, 412, 434–436
    phases of, 324–325
    revolution of, 321
    rotation of, 321
    size of, 306
    and tides, 327
Moons, 306
Mountain forest, 56–57
Multimeter, 562–563, 607
Muscles, of astronauts, 422
Mutualism, 54

N

Nanotechnology, 267
Narrow-spectrum pesticides, 133–134
Natural fertilizers, 124–126
Natural gas, 523
NDD (nature deficit disorder), 145
Nebulas, 375, 378, 383, 393
Negatively charged objects, 466–467, 468, 469, 472, 474, 475, 482
Neptune, 313, 340, 416
Neutral objects, 466–467, 468, 474, 475
Neutralization, 96
Neutron stars, 380
Neutrons, 231, 465
Newton, Sir Isaac, 410, 649, 650
Night sky, 614
    Aboriginal peoples and, 332–333
    changing views of, 339–340
    components of, 305–308
    finding objects in, 344–345
    light pollution and, 338
    modelling motion in, 336–337
    navigating, 341
    patterns in, 329–333
Nitrogen, 50–51
    fixation, 50
    in soil, 128
    in water, 125
Nitrogen cycle, 50–51
Noble gases, 221
Non-fiction text, reading, 7
Non-luminous, defined, 306
Non-metals, 213–214, 265
    as conductors, 480
Non-renewable energy sources, 521, 523
Nuclear energy, 420
Nuclear fission, 521
Nuclear fusion, 309, 376, 379
Nuclear generating stations, 521, 524
Nuclear reaction, 235
Nucleus, of atom, 230, 232–233

O

Oak Ridges Moraine, 143
Observational studies, 9
Observations, 11–12, 76, 612–614, 647–648
Oceans
    carbonic acid in, 98
    plastics in, 100–101
Ohm, Georg, 568
Ohmmeter, 565
Ohm's law, 568–570
Oil
    exploration, 429
    spills, 99–100, 429
    in thermal energy generation, 523
Oligotrophic bodies of water, 60
Orbital radius, 320

Orbits, 306
- of planets, 320–321
- polar, 349
- of satellites, 347–350

Organic farming, 139–140
Osteoporosis, 422
Oxygen. See also Corrosion; Rust
- in hydrogen peroxide, 272–273
- properties of, 270

Ozone, 258, 428

P

Pacemakers, 559
Paint sprayers, electrostatic, 470
Parallax, 367–368
Parallel circuits, 552–554, 561, 573–574, 576–577
Parasitism, 54
Particle theory of matter, 175
Payette, Julie, 412
Peebles, Jim, 395
Peer review, 14–15
Penzias, Arno, 395–396
Periodic table of elements, 211–215, 220–225, 234–239, 644–645
Periods, 222
Permafrost, 58, 243
Persuasive texts, writing, 464
Pesticides, 93, 128, 133–134, 167
- and bioaccumulation, 136, 137
- and bioamplification, 136–138
- reducing dependence on, 139–140
- resistance to, 139
- and target vs. non-target species, 135–136

Pests, 121, 122, 132–133
Photoconductors, 481
Photosphere, 309, 310, 376
Photosynthesis, 38–39, 40–41, 522
Photovoltaic cells, 527
Physical changes, 181–182
Physical properties, 179–182, 192–198
Pith ball electroscopes, 467, 478–479
Placemat organizers, 642
Planets, 306, 313
- dwarf, 314
- formation of, 384
- orbits of, 320–321
- view from Earth, 338

Planispheres, 330
Plants, and ecological succession, 81
Plaskett, John S., 388
Plastics, 100–101, 167
Pluto, 314, 353, 416
Polar orbits, 349
Polaris, 324, 371
Pollution, 96–101
- air, 142, 523, 524
- thermal, 524
- water, 99–101

Polychlorinated biphenyls (PCBs), 276–277

Positively charged objects, 466–467, 468, 474, 475, 482
Potential difference, 560–561, 567, 568–570
Precession, 324
Precipitates, 184
Predation, 54
Predictions, 9, 10–11, 14, 611
- graphs for, 634

Primary cells, 512
Primary succession, 80–82
Printers, 462
- laser, 462, 481–482

Prisms, 232, 372
Protons, 230, 231, 465
Protostars, 375, 376, 383, 384
Proxima Centauri, 365
Ptolemy, Claudius, 321–322
Pulsars, 380
Pulses, 380
Pure substances, 175

Q

Qualitative observations, 12, 612
Qualitative properties, 179
Quantitative observations, 12, 612
Quantitative properties, 179
Quasars, 389–390
Questioning, 610
Questions, 647
- in reading texts, 28
- in scientific inquiry and investigations, 9, 10, 14

R

Radiant energy, 38, 306
Radio telescopes, 410
Radio waves, 410
Radioactive materials, 521, 524
Rainforests, 83, 88–89
Reading, 639
- and asking questions, 28
- non-fiction text, 7

Recording
- of observations, 12, 76
- research information, 617–618

Red giants, 377–378
Red shift, 394
Red supergiants, 378
Reflecting telescopes, 410
Refracting telescopes, 410
RENEW (REcovery of Nationally Endangered Wildlife), 85
Renewable energy sources, 519, 522, 524
Reports, science, 256
Research
- applied, 654
- basic, 654
- in groups, 636
- of issues, 620
- skills, 617–622

Resistance, 571–575. See Electrical resistance
Resistors, 565
Respiration, cellular, 40–41
Retrograde motion, 340
Revolution
- of Earth, 320–321, 339–340
- of Moon, 321

Robotic technology, 415–418, 426
Rockets, 419, 434
Root words, Latin and Greek, 643
Rotation, 310
- of Earth, 320, 339–340
- of Moon, 321

Rotting, 185, 274–275
Round table activities, 635
Russell, Henry Norris, 376
Rust, 254, 262
Rutherford, Ernest, 230

S

Safety
- with electric currents, 556–557
- science, 11, 188–189, 598–602

Salt
- and ice, 195–196
- on roads, 199
- table, 260, 261

Satellites, 306, 346
- Cosmic Background Explorer (COBE), 396
- geostationary orbit, 350
- global positioning system (GPS), 350, 426–427
- HIPPARCOS, 368
- orbits of, 347–350
- RADARSAT, 349, 352, 428
- SCISAT, 428
- security, 352
- Wilkinson Microwave Anisotropy Probe (WMAP), 396

Saturn, 297, 313, 340, 416, 417–418
Science, 647
- classification of, 650–651
- equipment and tools, 602–608
- literacy in, 16–18
- and technology, 652–653

Science, technology, and society, 654
Science, technology, and the environment, 654–655
Science, technology, society, and the environment (STSE), 652–655
Scientific inquiry, 8–9, 609–616, 647
Scientific investigations, 9–15, 610–616, 635–636, 647–648
Scientific method, 650
Scientific notation, 625–626, 631
Scientists, thinking as, 609
Seasons, tilt of Earth and, 322–323
Secondary cells, 512
Secondary succession, 80–82

Series circuits, 552–554, 571–572, 576–577
Sharks, 74
Sharp objects, 599–600
Shelton, Ian, 379
Shooting stars, 315
SI units (Système international d'unités), 623–624
Significant digits, 627–628, 630
Sirius, 370
Smart meters, 533
Smoot, George, 396
Sodium, 221, 260
Sodium chloride, 261, 263
Soil(s), 123
    acid precipitation and, 97, 130–131
    air spaces in, 127
    charcoal in, 129
    compaction of, 127
    as mixture, 130–131
    nutrients, 124–126
    water and, 126–127
Solar eclipse, 325–326
Solar energy
    for spacecraft, 420
    for vehicles, 504
Solar flares, 310
Solar mass, 373
Solar Nebula theory, 383–384, 395
Solar photovoltaic systems, 527
Solar prominences, 310
Solar propulsion systems, 420
Solar radiation, 309, 312, 316, 423
Solar storms, 310, 312
Solar System, 306, 313–317
    formation of, 383–384
    geocentric model, 322
    heliocentric model, 322
    motion in, 321–322
    scale model of, 318–319
Solar winds, 306, 310, 311, 316
Solder, 176–177
Solids
    densities of, 193
    as state of matter, 175
Solstices, 323, 332
Solutions, 176
Space, safety in observing, 600
Space elevators, 436–437
Space exploration, 409–418, 434–437
Space junk, 424
Space probes, 415–418
Space propulsion technologies, 419–420, 437
Space research, 432–433
Space shuttles, 346, 412, 434, 437, 538
Space technology spinoffs, 426–430
Space tools, 414–415
Space tourism, 436
Space travel, challenges of, 413, 419–425

Spacecraft, 346, 348, 411–412, 538
    fuel for, 419–420
Speakers, electrostatic, 488
Species, 29
    of concern, 22, 156–157
    of grassland, 59
    imported, 120–121
    invasive, 91–94
    non-native, 91–95
    richness, 83–86
    at risk, 85
    of special concern, 85
    of tundra, 58
Spectrographs, 372
Spinoffs, 426–430
Spiral galaxies, 386
Star clusters, 385
Stars, 305–306. See also Constellations
    in Andromeda galaxy, 390
    brightness of, 370–371, 374
    characteristics of, 370–374
    colour of, 372
    composition of, 372
    distance between, 367–368
    distance from Earth to, 366
    life cycle of, 375–382
    luminosity of, 378
    maps, 330, 334–335, 604–605
    mass of, 373, 378, 379, 380, 384
    of Milky Way, 388
    temperature of, 372, 376
    view from Earth, 338
Static electricity, 467
    and printers, 462
Stators, 488
Stewardship, 105
Study skills, 635–639
Succession, ecological, 80–82
Sudbury Basin, 315
Sulfur dioxide, 97, 98
Summarizing, 364
Sun, 305–306, 309–312, 313. See also Headings beginning Solar
    absolute magnitude of, 371
    apparent magnitude of, 371
    communications disruptions caused by, 311–312
    distance from Earth, 365
    and Earth, 310–312
    eclipse of, 325–326
    formation of, 384
    luminosity of, 370
    magnetic field, 310
    mass of, 373, 378, 379
    and Milky Way, 388
    as red giant, 378
    in Solar Nebula theory, 384
Sunspots, 310
Supernovas, 379, 380, 383

Sustainability
    of cities, 144
    of ecosystems, 34–35
    in food production, 122
    of forests, 103
    fragmented ecosystems and, 87–89
    human activities and, 35
    and humans–environment relationship, 120
    of soil resources, 123
    succession and, 82
    of wildlife, 104–105
Switches, 509, 565
Synthesizing, 408
Synthetic fertilizers, 124–126

T

Tantalum, 214
Teamwork, 635, 636
Telescopes, 409–411, 653
    and galaxies, 386
    gamma ray, 411
    Hubble Space Telescope (HST), 392, 411
    MOST, 348, 392
    and parallax, 368
    radio, 410
    Spitzer Space Telescope, 409
    X-ray, 411
    XMM-Newton Observatory, 409
Temperate deciduous forest, 56–57, 59
Temperate rainforest, 56–57
Theories, 228, 649–650
Thermal electrical energy, 520–525
Thermal energy, 38, 46, 214
Think-pair-share activities, 635
Thirsk, Robert, 413
Thomson, J.J., 229, 230
Threatened species, 85
Thunder, 494
Tides
    and electrical energy generation, 520
    Moon and, 327
Titan (moon of Saturn), 297
Tito, Dennis, 436
Tolerance ranges, 52
Traditional Ecological Knowledge and Wisdom (TEKW), 648
Transpiration, 48–49
Tree diagrams, 640
Trees. See also Forests
    of boreal forest, 58
    of temperate deciduous forest, 59
Triangulation, 368, 427
Triboelectric series, 473
Trophic levels, 43–44, 45–47
Tundra, 56–58
Tunguska event, 302, 315
Turbines, 516, 525–526

U

Universe, 305
    evolution of, 395–396
    expansion of, 393–395, 396
    future of, 396
Uranium, 521, 524
Uranus, 313, 340, 416
URLs (universal resource locators), 618–619

V

Van de Graaff generator, 483
Variables, 8, 9, 610
Vega, 366
Venus, 313, 384
Virgo Supercluster, 390
Viscosity, 180
Visible light, 306
Voltage, 560–561, 562–563, 568–570
    across loads in parallel, 574
    across loads in series, 572
Voltmeter, 561, 562–563, 607

W

Waste
    management, 146–147
    radioactive, 524
Water
    agriculture and, 125, 126–127
    as compound, 258, 265
    conductivity of, 480
    eutrophic bodies of, 60
    and generation of electrical energy, 518–519
    nitrogen in, 125
    oligotrophic bodies of, 60
    pollution of, 99–101, 142
    properties of, 196–197
    in space exploration, 413
    vapour, 48–49, 413, 472
Water cycle, 48–49, 128
Watersheds, 60–61
Waterwheels, 560
Watts (W), 530
Weightlessness, 420–421
Wetlands, 60, 90
    agriculture and, 127
Whales, 63
White dwarfs, 377, 378
Wildlife. *See also* Habitats
    agriculture and, 127
    management of, 104–105
Wilson, Robert, 395–396
Wind power, 6, 525–526
Word walls, 642
Workplace Hazardous Materials Information System (WHMIS), 601, 606
Writing
    critical analyses, 550
    of observations, 76
    persuasive texts, 464
    science reports, 256

X

X-ray radiation, 380
X-ray telescopes, 411

Credits

This page constitutes an extension of the copyright page. We have made every effort to trace the ownership of all copyrighted material and to secure permission from copyright holders. In the event of any question arising as to the use of any material, we will be pleased to make the necessary corrections in future printings. Thanks are due to the following authors, publishers, and agents for permission to use the material indicated.

Table of Contents. iv ©Arco Images GmbH/Alamy; vi NASA; viii NASA; x CAIDA/Photo Researchers, Inc.

Unit A

2–3 Science Source/Photo Researchers, Inc.; 3 top to bottom ©Arco Images GmbH/Alamy, NASA, NASA, CAIDA/Photo Researchers, Inc.

Chapter 1. 4 Martin Sheilds/Photo Researchers, Inc.; 5 top left to right ©Rubens Abboud/Alamy, ©LatinStock Collection/Alamy, ©Bloomimage/Corbis, bottom left to right FLIP NICKLIN/MINDEN PHOTOGRAPHY/National Geographic Stock, ©Michael Blann/Alamy, Getty Images; 6 background rahulred/Shutterstock, bottom left ©Rubens Abboud/Alamy, bottom right ©Derek Croucher; 7 background rahulred/Shutterstock; 8 ©Roger Ressmeyer/Corbis; 10 Keren Su/Getty Images; 11 top ©John Madere/Corbis, bottom ©LatinStock Collection/Alamy; 12 top FLIP NICKLIN/MINDEN PHOTOGRAPHY/National Geographic Stock, bottom ©Bloomimage/Corbis; 14 Michael Blann/Getty Images; 16 Diego Cervo/Shutterstock; 17 top left to right Laurence Gough/Shutterstock, ©RIA Novosti/Alamy, National Geographic/Getty Images, bottom Getty Images; 18 ©Bernhard Classen/Alamy.

Unit A Looking Back. 19 top left to right ©Rubens Abboud/Alamy, ©LatinStock Collection/Alamy, ©Bloomimage/Corbis, bottom left to right FLIP NICKLIN/MINDEN PHOTOGRAPHY/National Geographic Stock, Michael Blann/Getty Images, Getty Images.

Unit B

20 ©Arco Images GmbH/Alamy; 21 top to bottom ©Bill Brooks/Alamy, ©Bloomimages/Corbis, Grant Faint/Getty Images, ©ACE STOCK LIMITED/Alamy; 22 top left to right ©Danny Lehaman/Corbis, ©Ralph A. Clevenger/CORBIS, ©Radius Images/Alamy, bottom Nobert Rosing/Getty Images; 23 top left Vast Photography & Productions/First Light, top right Gary Pearl/Getty Images.

Chapter 2. 24 ©Danny Lehman/CORBIS; 25 top left to right ©NASA/Courtesy of Apercu/CP Images, ©Nick Hawkes;Ecoscene/CORBIS, ©Digital Vision/Alamy, bottom left to right Larry Macdougal/Getty Images, iStock, Aurora/Getty Images; 26 top ANT Photo Library/Photo Researchers, Inc., bottom left Fred Bruemmer/Peter Arnold Inc., bottom right Jeff Sherman/Getty Images; 27 left top to bottom Ragnarock/Shutterstock, WS Productions/Getty Images, James Steidl/Shutterstock, right top to bottom wave/Firstlight, Oxana Zuboff/Shutterstock, Kristin Smith/Shutterstock; 28 background, rahulred/Shutterstock, foreground Erkki & Hanna; 29 top left Erkki & Hanna/Shutterstock, top right Paul B. Moore/Shutterstock, bottom ©NASA/Courtesty Apercu/CP Images; 32 left ©David L. Moore-Washington/Alamy, right Greg Stott/Masterfile; 34 left Courtesy of Verena Tunicliffe, right top to bottom ©Gunter Marx/Alamy, ©Brandon Cole Marine Photography/Alamy, ©imac/Alamy, Michael P. Gadomski/Photo Researchers, Inc.; 39 top ©Nick Hawkes;Ecoscene/CORBIS, bottom Gail Jankus/Photo Researchers, Inc.; 41 icyimage/Shutterstock; 42 top and bottom ©Bill Brooks/Alamy; 47 ©blickwinkel/Alamy; 48 left Morgan Lane Photography/Shutterstock, right ©Digital Vision/Alamy; 49 Larry Mcdougal/Getty Images; 53 top left to right iStockphotos, ©Don Johnston/Alamy, Don Johnston/A.G.E. Fotostock/First Light; 54 left Gilbert S. Grant/Photo Researchers, Inc., right Michael Poliza/Getty Images; 58 top Aurora/Getty Images, bottom Garry Black/Masterfile; 59 left ©John E Marriott/Alamy, right ©Radius Images/Alamy; 60 top ©Clint Farlinger/Alamy, bottom left ©Skye Hohmann/Alamy, bottom right ©Bill Brooks/Alamy; 61 top ©All Canada Photos/Alamy, bottom ©Visuals&Written SL/Alamy; 62 ©Brandon Cole Marine Photography/Alamy; 63 background dwphotos/Shutterstock, left ©Stephen Frink Collection/Alamy, top right EarthTrust Campaign Archives/Honolulu, bottom right Paul Chesley/Getty Images; 64 ©Stephen Frink Collection/iStockphotos; 66 top left to right ©NASA/Courtesty Apercu/CP Images, ©Nick Hawkes;Ecoscene/CORBIS, ©Digital Vision/Alamy, bottom left to right Larry Mcdougal/Getty Images, iStockphotos, Aurora/Getty Images; 67 left top to bottom Ragnarock/Shutterstock, WS Productions/Getty Images, James Steidl/Shutterstock, right top to bottom wave/Firstlight, Oxana Zuboff/Shutterstock, Kristin Smith/Shutterstock.

Chapter 3. 72 ©Ralph A. Clevenger/CORBIS; 73 top left to right Doug Hamilton/Getty Images, Altrendo Travel/Getty Images, Richard Ellis/Photo Researchers, bottom left to right ©Michael Freeman/CORBIS, ASSOCIATED PRESS, ©Alt-6/Alamy; 74 top background Henrik Anderson/Shutterstock, bottom background Specta/Shutterstock, top Michael Patrick O'Neill/Photo Researchers, Inc., bottom Jeff Hunter/Getty Images; 75 top left to right Dan70/Shutterstock, Mike Neale/Shutterstock, David Leach/Getty Images, middle left Larry Goldstein/Getty Images, middle right IntraClique LLC/Shutterstock, bottom left to right ©Mark Conlin/Alamy, Yuri Arcurs/Shutterstock, ©Galen Rowell/CORBIS; 76 background rahulred/Shutterstock ; 77 left Richard Smith, right Doug Hamilton/Getty Images; 78 left to right ©Tom Craig/Alamy, ©Mike Grandmaison/CORBIS, ©Dean Conger/CORBIS; 79 Pavel Cheiko/Shutterstock; 80 top ©Hulton-Deutsch Collection/CORBIS, bottom Altrendo Travel/Getty Images; 82 Courtesy Dufferin Aggregates; 83 Thomas Kitchin & Victoria Hurst/Getty Images; 84 left After George Edward Lodge/Getty Images, right Richard Ellis/Photo Researchers, Inc.; 86 ©Jim West/Alamy; 87 top ©Michael Freeman/CORBIS, bottom left and right Re-drawn from The Big Picture 2002, based on analysis by the Nature Conservancy of Canada and the Natural Heritage Information Centre; 88 left and right WWF 2006, Global Environment Outlook 4, United Nations Environment Programme; 89 ©Patrick Lynch/Alamy; 90 top ©John Bradbury/Alamy, bottom Doug Fraser; 91 top left ©blickwinkel/Alamy, top right ©NaturePics/Purcell/CORBIS, bottom ©Carl & Ann/CORBIS; 92 top to bottom John Mitchell/Photo Researchers, Inc., ©BRUCE COLEMAN/Alamy, AFP/Getty Images, ©Frank Blackburn/Alamy; 93 Patrick Hirlehey; 94 top ©john t. fowler/Alamy, bottom Doug Fraser; 95 background Terrance Emerson/Shutterstock, top left NOOA Research, top right Stephen Wild, bottom ©Pat Conova/Alamy; 96 left Thomas Kitchin & Victoria Hurst/Getty Images, middle BROOK/Science Photo Library, right Donald Neusbaum/Getty Images; 97 left ©orchidpoet/Alamy, right ©Mark Leach/Alamy; 99 ASSOCIATED PRESS; 100 NOOA Research; 101 John Cancalosi/Peter Arnold Inc.; 103 top left Karen Huntt/Getty Images, top right Alt-6/Alamy, bottom Courtesy Canadian Standards Association; 104 National Geographic/Getty Images; 108 top left to right Doug Hamilton/Getty Images, Altrendo Travel/Getty Images, Richard Ellis/Photo Researchers, bottom left to right ©Michael Freeman/CORBIS, ASSOCIATED PRESS, ©Alt-6/Alamy; 109 left to right Dan70/Shutterstock, Mike Neale/Shutterstock, David Leach/Getty Images, middle left Larry Goldstein/Getty Images, middle right IntraClique LLC/Shutterstock, bottom left to right ©Mark Conlin/Alamy, Yuri Arcurs/Shutterstock, ©Galen Rowell/CORBIS.

Chapter 4. 114 ©Radius Images/Alamy; 115 top left to right ©Ron Watts/CORBIS, ©Nigel Cattlin/Alamy, Sally Scott/Shutterstock, bottom left to right Paul Grebliunas/Getty Images, ©BRUCE COLEMAN INC./Alamy, ©Oleksiy Maksymenko/Alamy; 116 background Nicemonkey/Shutterstock, top altrendo images/Getty Images, middle ©Gloria H. Chomica/Masterfile, bottom left to right Daley Mikalson, ©Arctic-Images/CORBIS, ©Phillippe Renault/Hemis/CORBIS; 117 left top to bottom Laurent Renault/Shutterstock, ©Andrew Robtsov/Alamy, James H. Robinson, top right ©CanStock Images/Alamy, middle right ©Brownstock Inc./Alamy; 118 background rahulred/Shutterstock, foreground ©AGStockUSA, Inc./Alamy; 119 top to bottom ©Greg Vaughn/Alamy, ©Keith Douglas/Alamy, ©Ron Watts/Corbis; 120 iStockphotos; 122 szarzynski/Shutterstock; 123 ©L. Clarke/CORBIS; 125 top left and right ©Nigel Cattlin/Alamy, bottom Lance Rider/Shutterstock; 126 left Sally Scott/Shutterstock, right BONNIE WATTON/Shutterstock; 127 Francois Gohier/Photo Researchers, Inc.; 128 top and bottom ©AGStockUSA, Inc./Alamy; 129 background Svetlana Privezentseva/Shutterstock, left Dr. Bruna Glaser, right Courtesy BEST Energies, Inc.; 132 left ©Marvin Dembinsky Photo Associates/Alamy, right ©Don Johnston/Alamy; 133 top E.R. Degginger/Photo Researchers, Inc., bottom Paul Grebliunas/Getty Images; 135 top left ©Scott Camazine/Alamy, top right EYE OF SCIENCE/Science Photo Library, bottom Doug Fraser; 137 ©BRUCE COLEMAN INC./Alamy; 138 © louise murray/Alamy; 139 Australian National Botanic Gardens; 140 iStockphotos; 142 top ©Bill Brooks/Alamy, bottom left ©orchidpoet/Alamy, bottom right ©Oleksiy Maksymenko/Alamy; 144 Big Sky Aerial Photography/Benson Steel Ltd.; 145 Image Source/Getty Images; 146 top photodisc/First Light, bottom left ©PBWPIX/Alamy, bottom right Courtesy Taylors Recycled Plastic Products Inc.; 147 CP PHOTO/Toronto Sun/Bill Sanford; 148 left to right ©Ron Watts/CORBIS, ©Nigel Cattlin/Alamy, Sally Scott/Shutterstock, bottom left to right Paul Grebliunas/Getty Images, ©BRUCE COLEMAN INC./Alamy, ©Oleksiy Maksymenko/Alamy; 149 left top to bottom Laurent Renault/Shutterstock, ©Andrew Robtsov/Alamy, James H. Robinson, top right ©CanStock Images/Alamy, middle right ©Brownstock Inc./Alamy; 150 left ©Steve Allen Travel Photography/Alamy, middle and right Doug Fraser; 151 ©Dennis MacDonald/Alamy.

Unit B Looking Back. 154 left top to bottom ©NASA/Courtesy of Apercu/CP Images, ©Nick Hawkes;Ecoscene/CORBIS, ©Digital Vision/Alamy, Larry Macdougal/Getty Images, iStockphotos, Aurora/Getty Images, middle top to bottom Doug Hamilton/Getty Images, Altrendo Travel/Getty Images, Richard Ellis/Photo Researchers, ©Michael Freeman/CORBIS, ASSOCIATED PRESS, ©Alt-6/Alamy, right top to bottom ©Ron Watts/CORBIS, ©Nigel Cattlin/Alamy, Sally Scott/Shutterstock, Paul Grebliunas/Getty Images, ©BRUCE COLEMAN INC./Alamy, ©Oleksiy Maksymenko/Alamy; 155 ASSOCIATED PRESS; 156 U.S Fish and Wildlife Services; 157 top PARTSCH/Science Photo Library, bottom stockbyte/First Light; 161 iStockphotos; 162 Douglas D. Seifert/Getty Images.

Unit C

166 NASA; 167 top Image Source/Getty Images, middle Health Protection Agency, bottom Getty Images; 168 top left to right Orla/Shutterstock, ©PHOTOTAKE INC./Alamy, ©Ashley Cooper/Alamy, bottom ©JUPITERIMAGES/Creatas/Alamy; 169 left ©Ed Bock/CORBIS, right Johner Royalty-Free/Getty Images.

Chapter 5. 170 Orla/Shutterstock; 171 top left to right Pritmova Svetlana/Shutterstock, ©Images Etc Ltd/Alamy, David Troud/Getty Images, bottom left Medford Taylor/Getty Images, bottom right ©Robert Sciarrino/Star Ledger/CORBIS; 172 left iStockphotos, right ©Marion Bull/Alamy; 173 left top to bottom picamaniac/Shutterstock, Sebastian Duda/Shutterstock, Scientifica/Getty Images, right top to bottom, ©vario images BmbH & Co. KG/Alamy, Michael Rosenfeld/Getty Images, ©Mark Fairey/Alamy ; 174 background rahulred/Shutterstock; 176 top ©Broad Spektrum/Alamy, bottom left ©Da Costa/photocuisine/CORBIS, bottom right ©Edd Westmacott/Alamy; 177 top left ©Dennis MacDonald/Alamy; 179 top ©Alex Serge/Alamy, bottom left Andrew Lambert Photography/Science Photo Library, bottom right ©D. Hurst/Alamy; 180 Dole/Shutterstock; 181 top left to right terekhov igor/Shutterstock, M L Harris/Getty Images, Jan Bruggeman/Getty Images, iStockphotos, bottom left to right Joel Arem/Science Photo Library, ©Indigo Photo Agency/Alamy, Charles D. Winters/Photo Researchers/First Light, iStockphotos; 182 Leonard Lessin/Peter Arnold Inc; 183 ©Images Etc Ltd/Alamy; 185 top left to right letty17/Shutterstock, ©mediacolor's/Alamy, ©Phil Degginger/Alamy, bottom Mark Boulton/Photo Researchers, Inc.; 186 iStockphotos; 187 background Terrance Emerson/Shutterstock, bottom right Chris Sattlberger/Photo Researchers, Inc; 191 background Svetlana Privezentseva/Shutterstock, top T-Pool/Getty Images, middle Image provided by Dr. Peter L. Davies, Queen's University, Kingston ON-Produced by Dr. Michael Kuiper, Molecular Modelling Scientist, Melbourne, Australia, using VMD(Visual Molecular Dynamics), bottom ©blickwinkel/Alamy; 192 David Trood/Getty Images; 194 top ©Carl & Ann Purcell/CORBIS, middle Photos.com, bottom ©Leslie Garland Picture Library/Alamy; 195 top ©Dennis MacDonald/Alamy, bottom Charles D. Winters/Photo Researchers, Inc.; 196 Medford Taylor/Getty Images; 197 bottom left and right ©Phil Degginger/Alamy; 199 ©Robert Sciarrino/Star Ledger/CORBIS; 200 left to right Pritmova Svetlana/Shutterstock, ©Images Etc Ltd/Alamy, David Troud/Getty Images, bottom left Medford Taylor/Getty Images, bottom right ©Robert Sciarrino/Star Ledger/CORBIS; 201 left top to bottom picamaniac/Shutterstock, Sebastian Duda/Shutterstock, Scientifica/Getty Images, right top to bottom, ©vario images BmbH & Co. KG/Alamy, Michael Rosenfeld/Getty Images, ©Mark Fairey/Alamy; 202 Scientifica/Getty Images.

Chapter 6. 206 ©PHOTOTAKE INC./Alamy; 207 top left Andrew Lambert Photography/Science Photo Library, top middle Richard Treptow/Visuals Unlimited, bottom left ©Simon Reddy/Alamy, bottom right ©Reuters/CORBIS; 208 ©Pictures Contact/Alamy; 209 left top to bottom ©Stock Connection Blue/Alamy, Charles D. Winters/Photo Researchers, Inc., Scott Rothstein/Shutterstock, right top to bottom iStock, ©Charles Phillip Cangialosi/CORBIS, Sheila Terry/Science Photo Library; 210 background rahulred/Shutterstock, ©Reuters/CORBIS; 211 bottom ©Lester V. Bergman/CORBIS; 212 bottom Andrew Lambert Photography/Science Photo Library; 213 top Richard Treptow/Visuals Unlimited, Inc., bottom left Franck Bichon/Getty Images, bottom right Mark Schneider/Visuals Unlimited, Inc.; 214 left Theodore Gray/Visuals Unlimited, Inc., right iStockphotos; 218 ©Aardvark/Alamy; 221 top E.R. Degginger/Photo Researchers, Inc., middle left Andrew Lambert Photography/Science Photo Library, middle right ©PHOTOTAKE INC./Alamy, bottom ©B.A.E. Inc/Alamy; 222 left ©Glenn Harper/Alamy, right Andrew Lambert Photography/Science Photo Library; 223 Ria Novosti/Science Photo Library; 229 top iStockphotos, bottom left ©sciencephotos/Alamy, bottom right ©Simon Reddy/Alamy; 232 left ©Photodisc/Alamy, right ©Phil Degginger/Alamy; 234 left to right Charles D. Winters/Photo Researchers, Inc., E.R. Degginger/Photo Researchers, Inc., Charles D. Winters/Photo Researchers, Inc.; 236 The Alchymist, 1771 (oil on canvas), Wright of Derby, Joseph (1734-97) / Derby Museum and Art Gallery, UK / The Bridgeman Art Library; 238 Getty Images; 241 left to right ©Hugh Threlfall/Alamy, robophobic/Shutterstock, South 12th Photography/Shutterstock; 243 CPPHOTO/Adrian Wyld; 244 ©Reuters/CORBIS; 245 background dwphotos/Shutterstock, top marckuliasz/Shutterstock, bottom left to right fotoadamzyk/Shutterstock, Sinclair Stammers/Science Photo Library, Christophe Launay/Getty Images; 246 top left Andrew Lambert Photography/Science Photo Library, top middle Richard Treptow/Visuals Unlimited, Inc., bottom left ©Simon Reddy/Alamy, bottom right ©Reuters/CORBIS; 247 left top to bottom ©Stock Connection Blue/Alamy, Charles D. Winters/Photo Researchers, Inc., Scott Rothstein/Shutterstock, right top to bottom iStockphotos, ©Charles Phillip Cangialosi/CORBIS, Sheila Terry/Science Photo Library.

Chapter 7. 252 ©Ashley Cooper/Alamy; 253 top middle Herman Eisenbeiss/Photo Researchers, Inc., top right ©Rob Howard/CORBIS, bottom left ©BroadSpektrum/Alamy; 254 background iStockphotos, top to bottom Anson Hung/Shutterstock, iStockphotos, ©Bill Ross/CORBIS; 255 left top to bottom Peter Gudella/Shutterstock, ©Al Francekevich/Alamy, Martyn F. Chillmaid, right top to bottom ©Mark Boulton/Alamy, ©Phil Degginger/Alamy, Purestock/Getty Images; 256 background rahulred/Shutterstock; 258 George Bailey/Shutterstock; 263 top iStockphotos, bottom Herman Eisenbeiss/Photo Researchers, Inc.; top ©Image Source Pink/Alamy, bottom ©BroadSpektrum/Alamy; 267 background Terrance Emerson/Shutterstock, top ©Simone Brandt/Alamy, middle Equinox Graphics/Science Photo Library, bottom Air Force/Science Photo Library; 272 top ©Image Source Pink, bottom ©BroadSpektrum/Alamy; 275 ©Jack Sullivan/Alamy; 276 ©Rob Howard/CORBIS; 278 top middle Herman Eisenbeiss/Photo Researchers, Inc., top right ©Rob Howard/CORBIS, ©BroadSpektrum/Alamy ; 279 top to bottom Peter Gudella/Shutterstock, ©Al Francekevich/CORBIS, Martyn F. Chillmaid, right to bottom ©Mark Boulton/Alamy, ©Phil Degginger/Alamy, Purestock/Getty Images.

Unit C Looking Back. 284 left top to bottom Pritmova Svetlana/Shutterstock, ©Images Etc Ltd/Alamy, David Troud/Getty Images, Medford Taylor/Getty Images, ©Robert Sciarrino/Star Ledger/CORBIS, middle top to bottom Lambert Photography/Science Photo Library, Richard Treptow/Visuals Unlimited, Inc., ©Simon Reddy/Alamy, ©Reuters/CORBIS, right top to bottom Herman Eisenbeiss/Photo Researchers, Inc., ©Rob Howard/CORBIS, ©BroadSpektrum/Alamy; 286 ©Mark Boulton; 287 ©Tony Hertz/Alamy; 292 ©Fabrizio Troinai/Alamy; 293 ©John Boud/Alamy; 295 Reika/Shutterstock.

Unit D

296-297 NASA; 297 top to bottom ©NASA/JPL/ESA/CNP/CORBIS, ASSOCIATED PRESS, NASA/JPL/Space Science Institute/Science Photo Library, ©E.S.A./CORBIS SYGMA; 298 top left to right Dr. Juerg Alean/Photo Researchers, Inc., ©Stocktrek Images/CORBIS, Science Source/Photo Researchers, Inc., bottom left Scaled Composites/Photo Researchers, Inc., bottom right ©Steve Bloom Images/Alamy; 299 top left to right ©pronature/Alamy, ©Laura Coelho/Alamy, Royal Astronomical Society/Photo Researchers, Inc., middle Science Source/Photo Researchers, Inc., bottom left ©Tim Kiuslass/CORBIS, bottom right Dic Liew/Shutterstock.

Chapter 8. 300 Dr. Juerg Alean/Photo Researchers, Inc.; 301 top left to right KARIM JAAFAR/AFP/Getty Images, NASA, Tunc Tezel, bottom left to right ©Troy and Mary Parlee/Alamy, Erik Simonsen/Getty Images, ©Andrew parker/Alamy; 302 background Ivan Cholakov Gostock-dot-net, top Ria Novosti/Science Photo Library, bottom right Tunguska/University of Bologna-Department of Physics; 303 left top to bottom J. Bell(Cornell University)/M. Wolff/Hubble Heritage Team STScI/AURA/NASA/ESA/Science Photo Library, NASA, Science Source/Photo Researchers, Inc., right top to bottom Eckhard Slawick/Photo Researchers, Inc., ©Rob Krist/CORBIS, epa/CORBIS; 304 background rahulred/Shutterstock; 305 top PeKKa Parviainen/Science Photo Library, bottom NASA/National Geographic Stock; 306 ©Goodshoot/CORBIS; 307 The Solar System ©Corbis Premium RF/Alamy, Milky Way galaxy Mark Garlick/Science Photo Library, the Universe top left Tony Hallas, top right NASA/Science Photo Library, bottom Mark Slawick/Science Photo Library; 308 Allan Morton/Dennis Milon/Science Photo Library; 309 top ©Troy and Mary Parlee/Alamy; 310 John Chumack/Photo Researchers, Inc.; 311 top right ©Daniel J. Cox/CORBIS, bottom ©David H. Wells/CORBIS; 315 top NASA, bottom left ASSOCIATED PRESS, bottom right PLANETOBSERVER/Science Photo Library; 316 top Tony & Daphne Hallas/Science Photo Library, bottom left and middle David Jewitt, Rachel Stevenson, Pedro Lacenda and Jan Kleyan, University of Hawaii, bottom right ©Roger Ressmeyer/Science Faction/CORBIS; 317 Mount Stromlo and Siding Springs Observatories, ANU/Photo Researchers, Inc.; 324 ©Louie Psihoyos/Science Faction/CORBIS; 325 Eckhard Slawick/Science Photo Library; 326 left Dr. Fred Espenak/Science Photo Library, top right Royal Astronomical Society/Science Photo Library, bottom right KARIM JAAFAR/AFP/Getty Images; 328 Mark Garlick/Photo Researchers; 329 top John R. Foster/Photo Researchers, Inc., bottom left and right Eckhard Slawick/Photo Researchers, Inc.; 331 bottom Adam Hart-Davis/Photo Researchers, Inc.; 332 left ELIZABETH RUIZ/AFP/Getty Images, right ©andrew parker/Alamy; 333 Larry Landolfi/Photo Researchers; 338 bottom Gerard Lodriguss/Photo Researchers; 340 top Tunc Tezel; 341-342 starry background ©Steve Bloom Images/Alamy; 343 top Frank Zullo/Photo Researchers, bottom Tunc Tezel; 346 left Erik Simonsen/Getty Images; 350 left NOOA/Science Photo Library, right ©Michel Setboun/CORBIS; 351 NASA/Science Photo Library; 353 background Svetlana Privezentseva/Shutterstock, top NASA/John Hopkins University Applied Physics Laboratory/Southwest Research Institute, middle Mark Garlick/Science Photo Library, bottom Robert McNaught/Science Photo Library; 354 top left to right KARIM JAAFAR/AFP/Getty Images, NASA, Tunc Tezel, bottom left to right ©Troy and Mary Parlee/Alamy, Erik Simonsen/Getty Images, ©Andrew parker/Alamy; 355 left to right J. Bell(Cornell University)/M. Wolff/Hubble Heritage Team STScI/AURA/NASA/ESA/Science Photo Library, NASA, Science Source/Photo Researchers, Inc., right to bottom Eckhard Slawick/Photo Researchers, Inc., ©Rob Krist/CORBIS, epa/CORBIS; 358 ©fotoshoot/Alamy.

Chapter 9. 360 ©Stocktrek Images/CORBIS; 361 top middle European Southern Observatory/Photo Researchers, Inc., top right Science Source/Photo Researchers, Inc., bottom left to right ©STScI/NASA/Photo Researchers, Inc., Science Source/Photo Researchers, Inc., NASA/ESA/M. Robberto(STScI/ESA) and The Hubble Space Telescope Orion Treasury Project Team; 362 ESA/Hubble; 363 left top to bottom ©UVimages/amanaimages/CORBIS, Atlas Photo Bank/Photo Researchers, Inc., ©Bettman/CORBIS, right top to bottom Science Source/Photo Researchers, Inc., ©NASA/JPL-Caltech/CORBIS, NASA; 364 background rahulred/Shutterstock; 366 John Sanford/Science Photo Library; 367 left and right background ©Radius

Images/Alamy; 368 bottom European Space Agency/Science Photo Library; 370 Eckhard Slawik/Photo Researchers, Inc.; 371 Getty Images; 372 top John Chumack/Photo Researchers, Inc., bottom left and right Physics Dept., Imperial College/Science Photo Library; 375 top left NASA/ESA/M. Robberto(STScI/ESA) and The Hubble Space Telescope Orion Treasury Project Team, top right NASA/JPL-Caltech/University of Colorado, bottom European Southern Observatory/Photo Researchers, Inc.; 377 Science Source/Photo Researchers, Inc.; 378 NASA/ESA/HEIC/The Hubble Heritage Team(STScI/AURA); 379 Science Source/Photo Researchers, Inc.; 380 top left Science Source/Photo Researchers, Inc., bottom left NASA/CXC/M. Weiss; 383 left ©PCN Chrome/Alamy, right D. Golimowski, S. Durrance & M. Clampin/Science Photo Library; 384 Mark Garlick/Photo Researchers, Inc.; 385 left Davide De Martin/Photo Researchers, Inc., right Celestial Image Co./Photo Researchers, Inc.; 386 top Emilio Segrè Visual Archives/American Institute of Physics/Photo Researchers, Inc., middle left to right Science Source/Photo Researchers, Inc., John Chumack/Photo Researchers, Inc., ©STScI/NASA/CORBIS, bottom left W. Keel/U. Alabama/NASA/ESA/STScI/Photo Researchers, Inc., bottom right Science Source/Photo Researchers, Inc.; 388 Jerry Schad/Photo Researchers, Inc.; 389 top left left NASA/JPL-Caltech, top right Chris Butler/Photo Researchers, Inc., bottom left Mark Garlick/Science Photo Library, bottom right NASA/ESA/STScI/J. Bahcall, Princeton IAS/Science Photo Library; 391 Royal Observatory, Edinburgh/Anglo-Australian Telescope Board/Photo Researchers, Inc.; 392 background dwphotos/Shutterstock, top Jaymie Matthews/MOST Science Team, bottom Science Source/Photo Researchers, Inc.; 393 Robert Gendler; 394 Volker Steger/Science Photo Library; 395 ©NASA/Courtesy of Apercu/The Canadian Press; 396 left and right Science Source/Photo Researchers, Inc.; 398 top middle European Southern Observatory/Photo Researchers, Inc., top right Science Source/Photo Researchers, Inc., bottom left to right ©STScI/NASA/Photo Researchers, Inc., Science Source/Photo Researchers, Inc., NASA/ESA/M. Robberto(STScI/ESA) and The Hubble Space Telescope Orion Treasury Project Team; 399 background dwphotos/Shutterstock, left top to bottom ©UVimages/amanaimages/CORBIS, Atlas Photo Bank/Photo Researchers, Inc., ©Bettman/CORBIS, right top to bottom Science Source/Photo Researchers, Inc., ©NASA/JPL-Caltech/CORBIS, NASA; 400 National Optical Astronomy Observatories/Coloured by Science Photo Library.

Chapter 10. 404 Science Source/Photo Researchers, Inc.; 405 top left to right David Ducros/Photo Researchers, Inc., Science Source/Photo Researchers, Inc., NASA, bottom left to right National Institute of Diabetes & Kidney Diseases, National Institutes of Health, GUSTOIMAGES/Science Photo Library, Science Source/Photo Researchers; 406 background William Attard McCarthy/Alamy, top NASA/Science Photo Library, bottom ©Bettman/CORBIS; 407 left to bottom NASA/Science Photo Library, ©Roger Ressmeyer/CORBIS, AFP/Getty Images, right top to bottom ©NASA/Handout/Reuters/CORBIS, ©Rick Gayle/CORBIS, ©NASA/Roger Ressmeyer/CORBIS; 408 background rahulred/Shutterstock; 409 top ©NASA/Handout/CNP/CORBIS, bottom left ©NASA-CAL/Handout/Reuters/CORBIS, bottom right David Ducros/Photo Researchers, Inc.; 410 Dr. Seth Shostak/Photo Researchers, Inc.; 411 top NASA/Goddard Space Flight Center Scientific Visualization Studio, middle NASA/YPOP, bottom NASA/STScI; 412 top ©JUSTIN DERNIER/epa/CORBIS, bottom NASA; 413 ©SERGEI REMEZOV/Reuters/CORBIS; 414 NASA; 415 top NASA, bottom ©Stocktrek Images, Inc./Alamy; 416 bottom Science Source/Photo Researchers, Inc.; 417 top ©Reuters/CORBIS, bottom Science Source/Photo Researchers, Inc.; 418 NASA/JPL/GSFC; 419 top Ramon Santos/Photo Researchers, Inc., bottom Science Source/Photo Researchers, Inc.; 420 NASA/JPL-Caltech; 421 NASA/Science Photo Library; 422 middle left and right National Institute of Diabetes & Kidney Diseases, National Institutes of Health, bottom NASA; 423 NASA; 424 Chris Butler/Photo Researchers, Inc.; 426 top Science Source/Photo Researchers, Inc., middle Patrick Landmann/Photo Researchers, Inc., bottom Detlev van Ravenswaay/Photo Researchers, Inc.; 427 right ©Hugh Threlfall/Alamy; 428 top NASA/Goddard Space Flight Center Scientific Visualization Studio, bottom GUSTOIMAGES/Science Photo Library; 429 left Science Source/Photo Researchers, Inc., right ©John Buxton/Alamy; 430 top Dmitry Yashkin/Shutterstock, bottom ©Phil Degginger/Alamy; 431 background Terrance Emerson/Shutterstock, top David Crisp and the WFPC2 Science Team(JPL/Caltech/NASA), bottom ©Andrew Dowdy Australian Antarctic Division/Handout/Reuters/CORBIS; 432 ©Stocktrek Images, Inc./Alamy; 433 NASA; 434 top NASA/Science Photo Library, bottom left and right LOCKHEED MARTIN CORPORATION/NASA/Science Photo Library, bottom middle NASA/MSFC/Science Photo Library; 435 Doug Allan/Getty Images; 436 top ©Reuters/CORBIS, bottom Getty Images; 437 top and bottom The Spaceward Foundations; 438 top left to right David Ducros/Photo Researchers, Inc., Science Source/Photo Researchers, Inc., NASA, bottom left to right National Institute of Diabetes & Kidney Diseases, National Institutes of Health, GUSTOIMAGES/Science Photo Library, Science Source/Photo Researchers; 439 left top to bottom NASA/Science Photo Library, ©Roger Ressmeyer/CORBIS, AFP/Getty Images, right top to bottom ©NASA/Handout/Reuters/CORBIS, ©Rick Gayle/CORBIS, ©NASA/Roger Ressmeyer/CORBIS; 441 Richard Bizley; 444 left top to bottom KARIM JAAFAR/AFP/Getty Images, NASA, Tunc Tezel, ©Troy and Mary Parlee/Alamy, Erik Simonsen/Getty Images, ©Andrew parker/Alamy, middle top to bottom European Southern Observatory/Photo Researchers, Inc., Science Source/Photo Researchers, Inc., ©STScI/NASA/Photo Researchers, Inc., Science Source/Photo Researchers, Inc., NASA/ESA/M. Robberto(STScI/ESA) and The Hubble Space Telescope Orion Treasury Project Team, right top to bottom David Ducros/Photo Researchers, Inc., Science Source/Photo Researchers, Inc., NASA, National Institute of Diabetes & Kidney Diseases, National Institutes of Health, GUSTOIMAGES/Science Photo Library, Science Source/Photo Researchers.

Unit D Looking Back. 445 ©John Henley/CORBIS; 446 StockTrek/Getty Images; 451 Jerry Lodriguss/Photo Researchers, Inc.; 452 top National Optical Astronomy Observatories/Science Photo Library, bottom Science Source/Photo Researchers, Inc.; 453 ©Rolf Adlercreutz/Alamy.

Unit E
456-457 CAIDA/Photo Researchers, Inc.; 457 top to bottom Johann Helgason/Shutterstock, ©David J. Green/Alamy, ©James Cheadle/LOOP IMAGES/CORBIS, ©Ian Shaw/Alamy; 458 left ©David Wall/Alamy, middle NASA/Goddard Space Flight Center Scientific Visualization Studio, right Fedor Selevanov/Shutterstock; 459 top left ©Grant Heilman Photography/Alamy, top middle ©K-PHOTOS/Alamy, bottom ©Joe Belanger/Alamy.

Chapter 11. 460 ©David Wall/Alamy; 461 top left ©Lourens Smak/Alamy, top middle ©Horizon International Images Limited/Alamy, bottom left mediacolor's/Alamy, bottom middle Fotosearch, LLC; 463 top left Michael Newman/PhotoEdit Inc., bottom left Uwe Krejci/Getty Images, middle right ©Sherry Moore/Alamy; 464 background rahulred/Shutterstock; 465 left ©Frank Siteman-Doctor Stock/Science Faction/CORBIS, right ©Lourens Smak/Alamy; 470 ©Horizon International Images Limited/Alamy; 472 Clive Streeter ©Dorling Kindersley; 480 left ©Phillip Duff/Alamy, right Fotosearch, LLC; 481 top right CP PHOTO/Dimitri Papadopoulos, bottom Robert Milek/Shutterstock; 483 background Svetlana Privezentseva/Shutterstock, top left ©Paul A. Souders/CORBIS, top right ©sciencephotos/Alamy, bottom Peter Menzel/Science Photo Library; 488 top Jim Varney/Photo Researchers, Inc., bottom ©Nick Schroedl/Alamy; 492 ©Jim Craigmyle/CORBIS; 494 top RENE JOHNSTON/Toronto Star, bottom ©mediacolor's/Alamy; 496 top ©Lourens Smak/Alamy, top middle ©Horizon International Images Limited/Alamy, bottom left mediacolor's/Alamy, bottom middle Fotosearch, LLC; 497 top left Michael Newman/PhotoEdit Inc., bottom left Uwe Krejci/Getty Images, middle right ©Sherry Moore/Alamy; 498 Digital Vision/Getty Images.

Chapter 12. 502 NASA/Goddard Space Flight Center Scientific Visualization Studio; 503 top left Shutterstock, bottom left to right Chantal Cameron/Niagara Falls (Ontario) Public Library, William Caram/Alamy, ©Richard Levine/Alamy; 504 Courtesy Midnight Sun Solar Race Team-University of Waterloo; 505 left top to bottom ©José Fuste Ragu/Zeta/CORBIS, ©Nick Turner/Alamy, Norm Betts, right top to bottom ©TOM MARESCHAL/Alamy, ©Keren Su/China Span/Alamy, ©NASA/CORBIS; 506 background rahulred/Shutterstock, foreground Alexandra McLeod; 507 top Dennis O'Clair/Getty Images, bottom Shutterstock; 509 left ©Judith Collins/Alamy; 511 top ©Tibor Bognar/Alamy, bottom ©The Art Archives/CORBIS; 518 Chantal Cameron/Niagara Falls (Ontario) Public Library; 516 ©Michael Nicholson/CORBIS; 517 left to right ©Chesh/Alamy, ©Hugh Threlfall/Alamy, ©Caro/Alamy, ©Ling Xia/Alamy; 519 left ©Karen Tweedy-Holmes/CORBIS, right ©TOM MARESCHAL/Alamy; 520 CP PHOTO/Mike Dembeck; 521 CP PHOTO/Kevin Frayer; 523 right CP PHOTO/Dave Chidley, left ©aeropix/Alamy; 524 Francois Mori/AP Photo/CP Photo; 525 ©Frank Vetere/Alamy; 526 ©Clynt Garnham/Alamy; 527 FSG/A.G.E.Fotostock/First Light; 529©Bill Brooks/Alamy; 530 Derek Capitaine; 531 bottom left Jonathan Vasata/Shutterstock, bottom right ©FloridaStock/Shutterstock; 532 top iStockphotos, bottom left Bill Aron/Photo Edit, bottom right Michael Newman/Photo Edit; 533 Alexandra McLeod; 538 background dwphotos/Shutterstock, left ©Stocktrek, Inc/Alamy, right ASSOCIATED PRESS; 540 top left Shutterstock, bottom left to right Chantal Cameron/Niagara Falls (Ontario) Public Library, William Caram/Alamy, ©Richard Levine/Alamy; 541left top to bottom ©José Fuste Ragu/Zeta/CORBIS, ©Nick Turner/Alamy, Norm Betts, right top to bottom ©TOM MARESCHAL/Alamy, ©Keren Su/China Span/Alamy, ©NASA/CORBIS; 542 Reproduced with permission of the Minister of Natural Resources, Canada 2009.

Chapter 13. 546 Fedor Selivanov/Shutterstock; 547 top left ©CORBIS; 548 top CP PHOTO/Sun-Dale MacMillan, bottom ©Rick Gayle Studio/CORBIS; 549 left top to bottom ©Keith van Loen/Alamy, ©JUPITERIMAGES/PIXLAND/Alamy, ©Jeff Greenberg, right top to bottom Elena Elisseeva/Shutterstock, ©Joson/Zeta/CORBIS; 550 background rahulred/Shutterstock; 551 ©CORBIS; 559 background dwphotos/Shutterstock, top Salisbury District Hospital/Photo Researchers, Inc., bottom ©Ivo Roospold/Alamy; 560 ©Dorling Kindersley; 564 ©David Zimmerman/CORBIS; 565 ©Helene Rogers/Alamy; 578 top left ©CORBIS; 579 left top to bottom ©Keith van Loen/Alamy, ©JUPITERIMAGES/PIXLAND/Alamy, ©Jeff Greenberg, right top to bottom Elena Elisseeva/Shutterstock, ©Joson/Zeta/CORBIS.

Unit E Looking Back. 584 left from top to bottom ©Lourens Smak/Alamy, ©Horizon International Images Limited/Alamy, mediacolor's/Alamy, Fotosearch, LLC, middle from top to bottom Shutterstock, Chantal Cameron/Niagra Falls (Ontario) Public Library, William Caram/Alamy, ©Richard Levine/Alamy; right top ©CORBIS, 586 Thomas Northcut.

Appendix A

596 ©Monkey Business Images/Dreamstime.com; 597 top to bottom ©sciencephotos/Alamy, ©Steve Allen/Brand X/CORBIS, ©Biodisc/Visuals Unlimited/Alamy, ©Corbis Premium RF/Alamy; 604 tomek/Shutterstock; 608 bottom Corpid/Shutterstock; 609 top left Laurence Gough/Shutterstock, top right ©Patrick Robert/CORBIS; 611 ©Denkou Images/Alamy; 612 Martyn Chillmaid/Oxford Scientific/Photolibrary; 614 Laurence Gough/Shutterstock; 617 Monkey Business Images/Shutterstock; 618 Sofos Design/Shutterstock; 621 ©Zeffs/Dreamstine; 622 left ©iStockphotos/Chris Schmidt, right ©Denkou Images/Alamy; 640 left to right ©Eraxion/Dreamstime, Wim van Egmond/Visuals Unlimited/Getty Images, ©2009 David B Fleetham/Oxford Scientific/Jupiterimages Corporation, ©All Canada Photos.

Appendix B

646 Noel Powell, Schaumburg/Shutterstock; 647 ©Wave Royalty Free/Alamy; 648 left ©Gunter Marx Photography, ©Wayne Higgins/Alamy; 652 left AFP/Getty Images, Carlo Antonio Buttieri/Getty Images; 653 top to bottom ©fine art/Alamy, ©Clark Brennan/Alamy, Getty Images, ©Roger Ressmeyer/CORBIS; 654 left ©photostock1/Alamy, right ©Micheal Ventura/Alamy; 655 left ©Ianni Dimitrov/Alamy, middle ©David Burton/Beateworks/CORBIS, right Tom Uhlenberg/Shutterstock.

Text Credits

Chapter 3. 106 Table 1: P.A. Quinby, F. McGuiness and R. Hall. 1996. Forest Landscape Baseline No.13. Brief Progress and Summary Reports; Table 2: Swan, D., B. Freedman, and T. Dilworth. 1984. Effects of various hardwood forest management practices on small mammals in central Nova Scotia. Can. Field-Nat., 98:362–364; 107; Table 3: Freedman, B., R. Morash, and A. J. Hanson. 1981. Biomass and nutrient removals by conventional and whole-tree clear-cutting of a red spruce-balsam fir stand in central Nova Scotia. Canadian J. For. Res., 11:249–257.

Chapter 12. 536 Figure 1 ©Her Majesty The Queen in Right of Canada, Environment Canada, 2007. Reproduced with the permission of the Minister of Public Works and Government Services Canada.

Studio Photographer: Dave Starrett

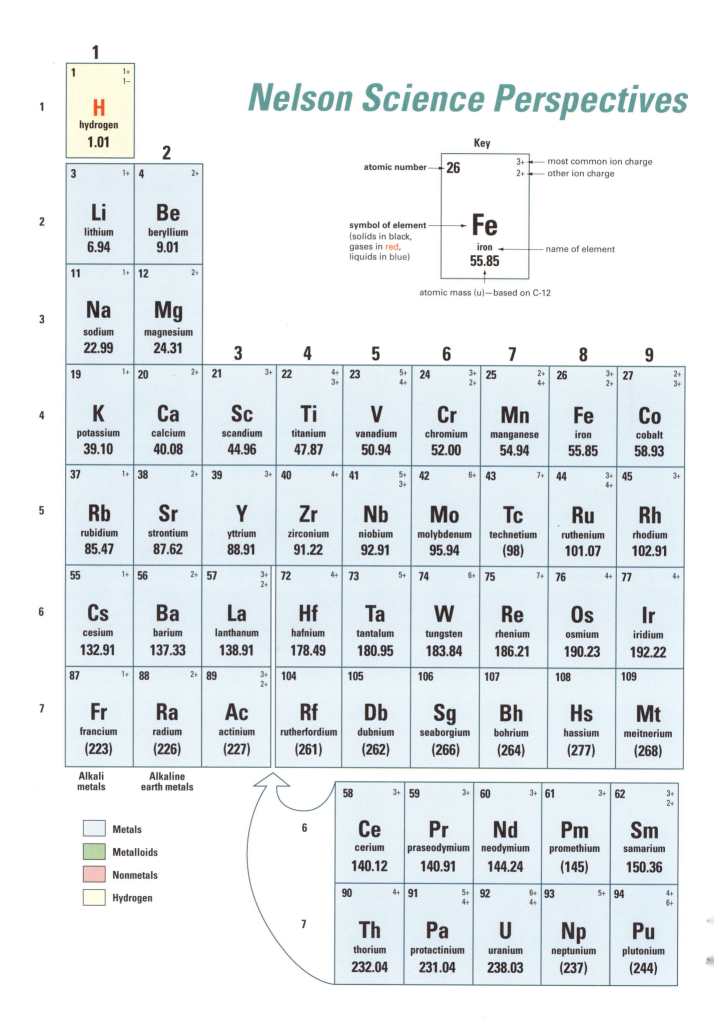